Gmelin Handbook of Inorganic Chemistry

8th Edition

The following Gmelin Formula Index volumes have been published up to now:

Formula Index

Volume 1	Ac–Au
Volume 2	B–Br_2
Volume 3	Br_3–C_3
Volume 4	C_4–C_7
Volume 5	C_8–C_{12}
Volume 6	C_{13}–C_{23}
Volume 7	C_{24}–Ca
Volume 8	Cb–Cl
Volume 9	Cm–Fr
Volume 10	Ga–I
Volume 11	In–Ns
Volume 12	O–Zr
	Elements 104–132

Formula Index 1st Supplement

Volume 1	Ac–Au
Volume 2	B–$B_{1.9}$
Volume 3	B_2–B_{100}
Volume 4	Ba–C_7
Volume 5	C_8–C_{17} (present volume)

Gmelin Handbook of Inorganic Chemistry

8th Edition

Gmelin Handbuch der Anorganischen Chemie

Achte, völlig neu bearbeitete Auflage

Prepared and issued by

Gmelin-Institut für Anorganische Chemie
der Max-Planck-Gesellschaft
zur Förderung der Wissenschaften
Director: Ekkehard Fluck

Founded by — Leopold Gmelin

8th Edition — 8th Edition begun under the auspices of the Deutsche Chemische Gesellschaft by R. J. Meyer

Continued by — E.H.E. Pietsch and A. Kotowski, and by Margot Becke-Goehring

Springer-Verlag Berlin · Heidelberg · New York · Tokyo 1985

Gmelin Handbook of Inorganic Chemistry

8th Edition

INDEX

Formula Index

1st Supplement Volume 5

C_8–C_{17}

AUTHORS Marie-Louise Gerwien, Helga Hartwig, Uwe Nohl, Hans-Jürgen Richter-Ditten, Paul Velić, Rudolf Warncke

CHIEF EDITOR Rudolf Warncke

Springer Verlag Berlin · Heidelberg · New York · Tokyo 1985

The volumes of the Gmelin Handbook are evaluated from 1974 up to the end of 1979

Library of Congress Catalog Card Number: Agr 25-1383

ISBN 3-540-93485-5 Springer-Verlag, Berlin · Heidelberg · New York · Tokyo
ISBN 0-387-93485-5 Springer-Verlag, New York · Heidelberg · Berlin · Tokyo

Typesetting, printing and bookbinding: Universitätsdruckerei H. Stürtz AG, Würzburg

Foreword

The Gmelin Formula Index published between 1975 and 1980 covered all volumes of the Eighth Edition of the Gmelin Handbook that had appeared up to the end of 1974 in the case of Main Volumes and up to the end of 1973 in the case of Supplement Volumes. The Gmelin Formula Index, First Supplement, continues from there and covers the handbook volumes published up to the end of 1979.

This First Supplement will consist of eight volumes, which will appear at intervals of four to six months. The basic structure of the Formula Index has been fully retained in the First Supplement: The index lists all elements, compounds, ions, and systems having definite composition that are described in the handbook text. The first column gives the empirical formula, while the second gives the conventional formula. The third column lists the pertinent pages. (The details are available in "Instructions for the Formula Index", on the next pages.)

This First Supplement was prepared and printed with extensive use of computers. In the future this will allow publication of cumulative indexes. The procedures were worked out together with the Technical Section of the *Gesellschaft für Information und Dokumentation mbH (GID)*, Frankfurt, and I take this occasion to thank them for their generous help. I would also like to thank our printers, *Universitätsdruckerei H. Stürtz AG*, Würzburg, for their advice and cooperation.

Frankfurt am Main
September 1983

Rudolf Warncke

Instructions for the Formula Index

The formula index consists of three columns. The first gives the empirical formula, the second gives the conventional formula, as well as any supplementary information or subdivisions, and the third gives the pertinent volume and page numbers.

First Column (Empirical Formula)

In the empirical formula the symbols of the elements are arranged alphabetically; C and H are not placed first. The list of empirical formulas is arranged alphabetically and by the magnitude of the subscripts. Any indefinite subscripts are placed last. Ions are always placed after the neutral species, positive ions preceding the negative.

The unsubscripted symbol is used for the element unless a specific diatomic or polyatomic species is meant (e.g., Br_2, Br_3). Transuranium elements that do not yet have an internationally recognized symbol are listed under their atomic number and placed at the end of the index. Special superscripted symbols are not used for isotopes.

H_2O is included in the empirical formula only if it is an integral part of a complex as written in the second column. Polymers of the type $(AB)_n$ are listed under AB. Multicomponent systems (mixed crystals, melts, etc.) are found under the empirical formulas of their components. However, solutions are found only under the solute.

Second Column (Conventional Formula)

The formula is written as it appears in the handbook text. However, in many cases another form is shown if there is adequate space and if it presents additional structural details. If this is not possible for isomers, then they are numbered consecutively. For elements the name is given in the second column.

Entries having the same empirical formula are arranged as follows: compounds, isotopic species, polymers, hydrates, multicomponent systems. Elements are treated in the same way.

For multicomponent systems the components are arranged in the sequence "inorganic components – organic components – water." The inorganic components are arranged alphabetically; the organic components are arranged by number of carbon atoms. If an element is a component, it is always represented by its unsubscripted symbol. Isotopic species are listed immediately after the normal species.

The concept *system* is used in its restricted sense in this index: it represents equilibrium mixtures described in phase diagrams. Ionic systems are included under their parent compounds.

The location of solubility data for compounds mentioned only briefly in the text is included under the main empirical-formula entry.

Entries for elements and compounds treated extensively in the handbook are subdivided by topics, e.g., geochemistry, preparation, or toxicity.

The concepts *solubility, solutions,* and *systems* partially overlap, and in these cases the user should always look at all three places. That is also true for the concepts *diffusion* and *systems* and for the concepts *sorption* and *system*.

In referrals to another entry in the index both the empirical and conventional formula are given. For example, "see $Al_2Na_2O_4$...$Na_2O \cdot Al_2O_3$." For referrals within the topics of a particular compound, then only the topic is given. For example "see Deposits."

Third Column (Volume and Page Numbers)

The first symbol is that of the element to which the volume belongs. Next is the abbreviated form of the type of volume followed usually by the Part or Section. The page numbers are given after a hyphen. The following abbreviations are used for type of volume:

MVol.	Main Volume (Hauptband)
SVol.	Supplement Volume (Ergänzungsband)
Org.Comp.	Organic Compounds
Org.Verb.	Organische Verbindungen
PerFHalOrg.	Perfluorohalogenoorganic Compounds of Main Group Elements
SVol.GD	Gmelin-Durrer, Metallurgy of Iron
TrU.	Transuranium Elements
Water Desalt.	Water Desalting

For example, the entry "Ag: MVol.B7-237/9" indicates that the information is to be found on pages 237 to 239 of the Silver Main Volume B 7. The entry "Fe: Org.Comp.C3-89" indicates that the desired information is to be found on page 89 of "Organic Compounds C 3" for the element iron (Fe).

Comments on System Numbers and Element Symbols

In the Formula Index itself the volume and page number citation was based on the traditional System Number (Main Volume Series) or the New Supplement Series Volume number. Today the volumes are usually arranged by the symbols of the elements. However, the symbols can be deduced easily from the system numbers. Most citations have the element symbol immediately after the system number. For example, 61 (Ag) refers to the silver volumes. The exceptions are

1 (EG)	is now	He
8 (J)	is now	I
39 (SE)	is now	Sc
69 (Ma)	is now	Tc

The old abbreviation for MVol was Hb (Hauptband), and the old abbreviation for SVol was Eb (Ergänzungsband). The volumes of the New Supplement Series are associated with the symbols of the elements as follows:

Erg.W. 1	He
Erg.W. 2	V
Erg.W. 3	Cr
Erg.W. 4, 7a, 7b, 8	Np
Erg.W. 5, 6	Co
Erg.W. 9, 12	F
Erg.W. 10	Zr
Erg.W. 11	Hf

The New Supplement Series Volumes 2 and 3 and the Volumes 10 and 11 are bound together as double volumes.

C_8 — C_{17}

Formula	Compound	Reference
$C_8CaFe_2H_2O_8$	$Ca[HFe(CO)_4]_2$	Fe: Org.Verb.B2-18, 28/9
$C_8CdFe_2K_2O_8$	$K_2[(CO)_4FeCdFe(CO)_4]$	Fe: Org.Verb.C1-58/9
$C_8CdFe_2O_8{}^{2-}$	$[(CO)_4FeCdFe(CO)_4]^{2-}$	Fe: Org.Verb.C1-58/9
$C_8CeCl_6H_{24}N_2$	$[(CH_3)_4N]_2CeCl_6$	Sc: MVol.C5-223
$C_8CeCl_7H_{32}N_4$	$CeCl_3 \cdot 4\,[(CH_3)_2NH_2]Cl$	Sc: MVol.C5-213/5, 219/20
–	$CeCl_3 \cdot 4\,[(CH_3)_2NH_2]Cl \cdot 2\,H_2O$	Sc: MVol.C5-213/5
$C_8CfH_8O_{12}{}^{-}$	$Cf(C_4H_4O_6)_2{}^{-}$	Np: TrU.D1-159
$C_8ClFH_{10}Sn$	$(CH_3)_2(FC_6H_4)SnCl$	Sn: Org.Verb.5-194
$C_8ClFH_{18}Sn$	$(C_4H_9)_2SnFCl$	Sn: Org.Verb.6-293
–	$(t\text{-}C_4H_9)_2SnFCl$	Sn: Org.Verb.6-293/4
$C_8ClFH_{24}Ni_2P_2$	$(CH_3Ni(P(CH_3)_3)F)(CH_3Ni(P(CH_3)_3)Cl)$	Ni: Org.Verb.2-253/4
$C_8ClF_2H_5O_2$	$ClC_6H_4COOCF_2H$	F: PerFHalOrg.5-18
$C_8ClF_2H_{15}Si$	$CF_2CClSi(C_2H_5)_3$	F: PerFHalOrg.4-19
$C_8ClF_3HNO_2$	$C_6F_3ClC(O)NHC(O)$	F: PerFHalOrg.6-124
$C_8ClF_3H_5NOS$	$C_2H_5C(O)SC_5F_3ClN$	F: PerFHalOrg.5-216
$C_8ClF_3H_9N_5$	$(CF_3)(NH_2)((CH_3)_2CNNH)C_4ClN_2$	F: PerFHalOrg.6-71
$C_8ClF_4H_5N_2$	$ClCF_2CF_2NNC_6H_5$	F: PerFHalOrg.7-192
$C_8ClF_4H_7Si$	$C_6F_4ClSi(CH_3)_2H$	F: PerFHalOrg.4-77
$C_8ClF_4H_{10}NO_2S$	$FSO_2NC_5H_4Cl(CH_3)_2CF_3$	S: S-N-Verb.1-117
C_8ClF_5	$ClCCC_6F_5$	F: PerFHalOrg.4-17
$C_8ClF_5H_6Si$	$C_6F_5Si(CH_3)_2Cl$	F: PerFHalOrg.4-19
$C_8ClF_5H_6Sn$	$(C_6F_5)(CH_3)_2SnCl$	F: PerFHalOrg.4-30 Sn: Org.Verb.5-194, 198
$C_8ClF_5H_7N_5$	$C_6F_5NHC(NH)NHC(NH)NH_2 \cdot HCl$	F: PerFHalOrg.7-3/4
$C_8ClF_5N_2$	$C_6F_3ClCFCFNN$	F: PerFHalOrg.6-134, 147
–	$C_6F_3ClNCFCFN$	F: PerFHalOrg.6-133/4, 146
–	$NC_5F_3CFNCFCCl$	F: PerFHalOrg.6-135, 148
C_8ClF_7	$FClCCFC_6F_5$	F: PerFHalOrg.4-16
$C_8ClF_7H_2O$	$CClF_2CH(OH)C_6F_5$	F: PerFHalOrg.4-101
$C_8ClF_7H_3NO$	$(CH_3O)(CF_3CFCl)C_5F_3N$	F: PerFHalOrg.5-177
$C_8ClF_{11}H_2N_2$	$Cl(CF_2)_5C(NH_2)CFCN$	F: PerFHalOrg.7-3
$C_8ClF_{12}H_9N_2OSi$	$(CH_3)_2SiClCH(ON(CF_3)_2)CH_2N(CF_3)_2$	F: PerFHalOrg.7-143
$C_8ClF_{12}H_9N_2O_2Si$	$(CH_3)_2SiClCH[ON(CF_3)_2]CH_2ON(CF_3)_2$	F: PerFHalOrg.7-127/8
$C_8ClF_{14}H_5OSi$	$(C_3F_7)_2Si(OC_2H_5)Cl$	F: PerFHalOrg.4-18
$C_8ClF_{15}N_2O_2$	$(CF_3)_2CNOCF_2CFClONC(CF_3)_2$	F: PerFHalOrg.7-96/7, 114
$C_8ClFeH_5O_4$	$ClCHCHCHCH_2Fe(CO)_4$	Fe: Org.Verb.B4-238, 251
$C_8ClFeH_5O_5$	$ClCHCHCOCH_3Fe(CO)_4$	Fe: Org.Verb.B4-267/8, 271, 288/9
$C_8ClFeH_6KO_6$	$K[ClCH_2CH(OH)CH_2C(O)Fe(CO)_4]$	Fe: Org.Verb.B4-154, 162

Formula	Compound		Reference
$C_8ClFeH_6O_6^-$	$[ClCH_2CH(OH)CH_2C(O)Fe(CO)_4]^-$	Fe:	Org.Verb.B4-154, 162
$C_8ClFeH_7O_4$	$[CH_3C_3H_4Fe(CO)_4]Cl$	Fe:	Org.Verb.B5-100, 102
$C_8ClFeH_8NO_7$	$[C_3H_5Fe(CO)_3]ClO_4 \cdot CH_3CN$	Fe:	Org.Verb.B5-85
$C_8ClFeH_9O_3$	$(CH_3)_2C_3H_3Fe(CO)_3Cl$	Fe:	Org.Verb.B5-54, 64
$C_8ClFeH_9O_4$	$C_2H_5OC_3H_4Fe(CO)_3Cl$	Fe:	Org.Verb.B5-53
$C_8ClFeH_9O_7$	$[(CH_3)_2C_3H_3Fe(CO)_3]ClO_4$	Fe:	Org.Verb.B5-82/6
$C_8ClFeH_{13}NO_6P$	$ClC_3H_4C(O)Fe(CO)(NO)P(OCH_3)_3$	Fe:	Org.Verb.B5-108/11
$C_8ClH_6S_3^-$	$[S_2CSCH_2C_6H_4Cl]^-$	C:	MVol.D4-268
$C_8ClH_7S_3$	$SC(SCH_2C_6H_4Cl)(SH)$	C:	MVol.D4-264/6
$C_8ClH_8NO_3S$	$CH_3OC_6H_4CHNSO_2Cl$	S:	S-N-Verb.1-126
$C_8ClH_8NO_4S$	$ClSO_2NHCOOCH_2C_6H_5$	S:	S-N-Verb.1-124/5
$C_8ClH_9N_2O_5S_2$	$ClSO_2NHCONCH_3SO_2C_6H_5$	S:	S-N-Verb.1-125
–	$ClSO_2NHCONHSO_2C_6H_4CH_3$	S:	S-N-Verb.1-125
$C_8ClH_{10}NiO_5P$	$(CO)_3NiP[(OCH_2)_2C(CH_3)_2]Cl$	Ni:	Org.Verb.1-241
$C_8ClH_{11}Sn$	$(CH_3)_2(C_6H_5)SnCl$	Sn:	Org.Verb.5-194, 198
–	$(CH_3)_2(ClC_6H_4)SnH$	Sn:	Org.Verb.4-90
$C_8ClH_{12}Ni$	$(C_8H_{12}NiCl)_x$	Ni:	Org.Verb.2-399
$C_8ClH_{14}NSn$	$(CH_3)_3SnCl \cdot C_5H_5N$	Sn:	Org.Verb.5-87
$C_8ClH_{14}NiO_3P$	$C_5H_5Ni(P(OCH_3)_3)Cl$	Ni:	Org.Verb.2-117/9
$C_8ClH_{15}Sn$	$(CH_2CH)_2(C_4H_9)SnCl$	Sn:	Org.Verb.5-222, 224
–	$(C_2H_5)_3SnCCCl$	Sn:	Org.Verb.2-221, 234
$C_8ClH_{16}NiO_2^+$	$[(CH_3)_4C_4Ni(H_2O)_2Cl]^+$	Ni:	Org.Verb.2-325
$C_8ClH_{17}IrN_5O_6$	$H[Ir(NO_2)(CH_3C(NOH)C(NOH)CH_3)_2Cl]$	Ir:	SVol.2-62
$C_8ClH_{17}N_2Sn$	$(CH_3)_3SnCCl(CN)CH_2N(CH_3)_2$	Sn:	Org.Verb.2-38
$C_8ClH_{17}Sn$	$(CH_3)_2(C_6H_{11})SnCl$	Sn:	Org.Verb.5-194
–	$(CH_3)_3SnCHCHCH_2CH_2CH_2Cl$	Sn:	Org.Verb.2-79
–	$(C_2H_5)(CH_2CH)(C_4H_9)SnCl$	Sn:	Org.Verb.5-226
$C_8ClH_{18}ISn$	$(C_4H_9)_2SnClI$	Sn:	Org.Verb.6-88
$C_8ClH_{18}NO_2S$	$ClSO_2N(C_4H_9)_2$	S:	S-N-Verb.1-124
–	$ClSO_2NHC_8H_{17}$	S:	S-N-Verb.1-123
$C_8ClH_{18}Sn^+$	$(C_4H_9)_2SnCl^+$	Sn:	Org.Verb.5-124, 138
		Sn:	Org.Verb.6-82
$C_8ClH_{18}Ti$	$(C_4H_9)_2TiCl$	Ti:	Org.Verb.1-70
–	$(i\text{-}C_4H_9)_2TiCl$	Ti:	Org.Verb.1-70
$C_8ClH_{19}N_2O_2S$	$[(CH_2)_5NSO_2N(CH_3)_3]Cl$	S:	S-N-Verb.1-172
$C_8ClH_{19}OSn$	$((CH_3)_3CO)(C_2H_5)_2SnCl$	Sn:	Org.Verb.5-105
–	$(C_2H_5)_2ClSnOCH(CH_3)C_2H_5$	Sn:	Org.Verb.6-54
–	$(C_4H_9)_2Sn(OH)Cl$	Sn:	Org.Verb.6-89/91
–	$(t\text{-}C_4H_9)_2Sn(OH)Cl$	Sn:	Org.Verb.6-107
$C_8ClH_{19}O_2Sn$	$C_2H_5SnCl(OC_3H_7)_2$	Sn:	Org.Verb.6-230
–	$C_8H_{17}Sn(OH)_2Cl$	Sn:	Org.Verb.6-252
$C_8ClH_{19}Sn$	$(CH_3)_2(C_6H_{13})SnCl$	Sn:	Org.Verb.5-194, 198
–	$(C_2H_5)_2(C_4H_9)SnCl$	Sn:	Org.Verb.5-201, 204
–	$(C_4H_9)_2SnHCl$	Sn:	Org.Verb.6-289/90
–	$(i\text{-}C_4H_9)_2SnHCl$	Sn:	Org.Verb.6-290
$C_8ClH_{20}I_2NSn$	$[(C_2H_5)_4N]SnClI_2$	Sn:	MVol.C3-120
$C_8ClH_{20}NOSn$	$[(CH_3)_3SnCH_2NH(CH_2CH_2)_2O]Cl$	Sn:	Org.Verb.2-16
$C_8ClH_{20}NO_4S_2$	$[((C_2H_5)_4N)(SOCl)]SO_3$	S:	SVol.1-54
$C_8ClH_{21}NiOP_2$	$CH_3C(O)Ni(P(CH_3)_3)_2Cl$	Ni:	Org.Verb.1-14, 22
$C_8ClH_{21}O_2Sn_2$	$Cl(C_2H_5)_2SnOSn(C_2H_5)_2OH$	Sn:	Org.Verb.6-55

Formula	Compound		Reference
$C_8ClH_{22}NSn$	$(C_2H_5)_3SnCl \cdot (CH_3)_2NH$	Sn:	Org.Verb.5-107
$C_8ClH_{22}OPSSn$	$(CH_3)_3SnCl \cdot (C_2H_5)_2P(O)SCH_3$	Sn:	Org.Verb.5-87
$C_8Cl_2FH_9N_2O$	$(NH_2)((CH_3)_2CHO)C_5FCl_2N$	F:	PerFHalOrg.5-214
–	$(NH_2)(C_3H_7O)C_5FCl_2N$	F:	PerFHalOrg.5-214
$C_8Cl_2FH_{15}Si$	$CFClCClSi(C_2H_5)_3$	F:	PerFHalOrg.4-19
$C_8Cl_2F_2HNO_2$	$C_6F_2Cl_2C(O)NHC(O)$	F:	PerFHalOrg.6-124
$C_8Cl_2F_2H_5NOS$	$CH_3COCH_2SC_5Cl_2F_2N$	F:	PerFHalOrg.5-216/7
$C_8Cl_2F_2H_8NO_3PS$	$NC_5F_2Cl_2(OPS(OCH_3)OC_2H_5)$	F:	PerFHalOrg.5-212
$C_8Cl_2F_2H_8NO_4P$	$NC_5F_2Cl_2(OPO(OCH_3)OC_2H_5)$	F:	PerFHalOrg.5-212
$C_8Cl_2F_3H_5N_2$	$ClCF_2CFClNNC_6H_5$	F:	PerFHalOrg.7-192
$C_8Cl_2F_4H_8N_2O_2S$	$SC(CF_2Cl)_2NHC(O)CH(NHCOCH_3)CH_2$	F:	PerFHalOrg.7-66, 68
$C_8Cl_2F_4N_2$	$C(CFNCFCCl)_2C$	F:	PerFHalOrg.6-135, 148
–	$C_6F_4NCClCClN$	F:	PerFHalOrg.6-133/4, 147
$C_8Cl_2F_5H_3O$	$CHCl_2CH(OH)C_6F_5$	F:	PerFHalOrg.4-101
$C_8Cl_2F_6$	$Cl_2CCFC_6F_5$	F:	PerFHalOrg.4-17
$C_8Cl_2F_7HN_4$	$C_3F_7C_5Cl_2HN_4$	F:	PerFHalOrg.6-127, 139, 162
$C_8Cl_2F_8Hg$	$Hg(C_4F_4Cl)_2$	F:	PerFHalOrg.4-33
$C_8Cl_2F_{12}MoO_4P_2$	$Mo(CO)_4[(CF_3)_2PCl]_2$	F:	PerFHalOrg.3-110/1, 120
$C_8Cl_2F_{18}N_2O_2$	$ClCF_2CF_2[N(CF_3)OCF_2CF_2]_2Cl$	F:	PerFHalOrg.7-95, 112
$C_8Cl_2FeGeH_8O_6$	$(CO)_4Fe(GeCl_2)(C_4H_8O_2)$	Fe:	Org.Verb.B3-161
$C_8Cl_2FeH_4O_4$	$(CH_2CClCClCH_2)Fe(CO)_4$	Fe:	Org.Verb.C2-31
$C_8Cl_2FeH_9Hg_2NO_5$	$(CO)_4Fe(HgCl)_2(NH(CH_2CH_2)_2O)$	Fe:	Org.Verb.B2-153/5, 158
$C_8Cl_2FeH_9NO_3$	$(t\text{-}C_4H_9NC)Fe(CO)_3Cl_2$	Fe:	Org.Verb.B4-2/3
$C_8Cl_2FeH_{14}O_2S_2$	$(CO)_2Fe[(C_2H_5)SCH_2CH_2S(C_2H_5)]Cl_2$	Fe:	Org.Verb.B1-101/2, 107
$C_8Cl_2FeH_{16}O_2P_2$	$(CO)_2Fe[(CH_3)_2PCH_2CH_2P(CH_3)_2]Cl_2$	Fe:	Org.Verb.B1-101/2, 106
$C_8Cl_2FeH_{18}HgI_2O_2P_2$	$(CO)_2Fe(P(CH_3)_3)_2I_2 \cdot HgCl_2$	Fe:	Org.Verb.B1-109
$C_8Cl_2FeH_{18}O_2P_2$	$(CO)_2Fe(P(CH_3)_3)_2Cl_2$	Fe:	Org.Verb.B1-101/3, 109
$C_8Cl_2FeH_{18}O_8P_2$	$(CO)_2Fe(P(OCH_3)_3)_2Cl_2$	Fe:	Org.Verb.B1-101/2, 105
$C_8Cl_2Fe_2H_2O_6$	$(CO)_3Fe(CHCHCl)ClFe(CO)_3$	Fe:	Org.Verb.C2-79/81
$C_8Cl_2H_6N_2O_8S_2$	$C_6H_4(OCONHSO_2Cl)_2$	S:	S-N-Verb.1-124/5
$C_8Cl_2H_6O_4Sn$	$SnCl_2(O_2C_6H_2(OH)COCH_3)$	Sn:	MVol.C5-89
$C_8Cl_2H_6S_2Sn$	$(C_4H_3S)_2SnCl_2$	Sn:	Org.Verb.6-184/5, 190
$C_8Cl_2H_7NSn$	$(C_6H_5)(NCCH_2)SnCl_2$	Sn:	Org.Verb.6-200
$C_8Cl_2H_8Ni$	$(CH_2C_6H_4CH_2NiCl_2)_x$	Ni:	Org.Verb.2-400/1
$C_8Cl_2H_8Sn$	$(C_6H_5)(CH_2CH)SnCl_2$	Sn:	Org.Verb.6-200
$C_8Cl_2H_{10}N_4O_4Sn$	$[SnCl_2(NNCCOOC_2H_5)_2]$	Sn:	MVol.C6-31
$C_8Cl_2H_{10}Sn$	$(C_2H_5)(C_6H_5)SnCl_2$	Sn:	Org.Verb.6-198, 204
$C_8Cl_2H_{11}NS_2Ti$	$C_5H_5Ti(S_2CN(CH_3)_2)Cl_2$	Ti:	Org.Verb.1-198/9
$C_8Cl_2H_{12}N_4$	$NCClC(N(CH_3)_2)NCClC(N(CH_3)_2)$	F:	PerFHalOrg.6-60, 67
$C_8Cl_2H_{12}Ni$	$C_4(CH_3)_4NiCl_2 \cdot C_{12}H_8N_2 \cdot 2\ H_2O$	Ni:	Org.Verb.2-75
–	$C_4(CH_3)_4NiCl_2 \cdot P(C_6H_5)_3$	Ni:	Org.Verb.2-75
$C_8Cl_2H_{12}NiO_8$	$(CH_3)_4C_4Ni(ClO_4)_2$	Ni:	Org.Verb.2-329
$C_8Cl_2H_{12}OTi$	$CH_3C_5H_4Ti(OC_2H_5)Cl_2$	Ti:	Org.Verb.1-180, 183
–	$C_5H_5Ti(OC_3H_7\text{-}n)Cl_2$	Ti:	Org.Verb.1-180/1
–	$C_5H_5Ti(OC_3H_7\text{-}i)Cl_2$	Ti:	Org.Verb.1-180/1
$C_8Cl_2H_{12}SSn$	$(C_4H_9)(C_4H_3S)SnCl_2$	Sn:	Org.Verb.6-199
$C_8Cl_2H_{13}NSn$	$(CH_3)_2SnCl_2 \cdot CH_3C_5H_4N$	Sn:	Org.Verb.6-35, 38
$C_8Cl_2H_{14}N_2O_8S_2$	$C_2H_4(CH_2CH_2OCONHSO_2Cl)_2$	S:	S-N-Verb.1-125
$C_8Cl_2H_{14}NiO$	$[(CH_3)_4C_4Ni(H_2O)Cl_2]$	Ni:	Org.Verb.2-325

Formula	Compound	Reference
$C_8Cl_2H_{14}Ni_2$	$(CH_2C(CH_3)CH_2NiCl)_2$	Ni: Org.Verb.2-305
–	$(CH_3CHCHCH_2NiCl)_2$	Ni: Org.Verb.2-289/93
$C_8Cl_2H_{14}Ni_2O_8$	$(CH_3CHCHCH_2NiOClO_3)_2$	Ni: Org.Verb.2-296
$C_8Cl_2H_{14}O_2Sn$	$[CH_3C(O)CH_2CH_2]_2SnCl_2$	Sn: Org.Verb.6-130
$C_8Cl_2H_{14}O_4S_2Sn$	$SnCl_2(SCH_2COOC_2H_5)_2$	Sn: MVol.C6-89
$C_8Cl_2H_{14}O_4Sn$	$(CH_3OCOCH_2CH_2)_2SnCl_2$	Sn: Org.Verb.6-130/1
–	$[HOCOCH(CH_3)CH_2]_2SnCl_2$	Sn: Org.Verb.6-130
–	$SnCl_2(CH_3CH_2CH_2COO)_2$	Sn: MVol.C2-223
–	$SnCl_2((CH_3)_2CHCOO)_2$	Sn: MVol.C2-223
$C_8Cl_2H_{14}O_7U$	$UO_2Cl_2 \cdot (CH_3CO)_2O \cdot C_4H_8O_2$	U: SVol.E1-123
$C_8Cl_2H_{14}Sn$	$[CH_3CHC(CH_3)]_2SnCl_2$	Sn: Org.Verb.6-149
$C_8Cl_2H_{16}MnO_{12}$	$Mn(ClO_4)_2 \cdot 2\ C_4H_8O_2 \cdot 4\ H_2O$	Mn: MVol.D1-154
$C_8Cl_2H_{16}N_2O_2Sn$	$[H_2NC(O)CH(CH_3)CH_2]_2SnCl_2$	Sn: Org.Verb.6-131
–	$[H_2NC(O)CH_2CH(CH_3)]_2SnCl_2$	Sn: Org.Verb.6-131
$C_8Cl_2H_{16}N_4S_2Te$	$Te(SC(NHCH_2)_2CH_2)_2Cl_2$	Te: SVol.B3-159
$C_8Cl_2H_{16}O_4Sn$	$SnCl_2 \cdot 2\ CH_3COOC_2H_5$	Sn: MVol.C5-9
$C_8Cl_2H_{16}O_4Sn^{2+}$	$[SnCl_2(CH_3COOC_2H_5)_2]^{2+}$	Sn: MVol.C5-130
$C_8Cl_2H_{16}O_6U$	$UO_2Cl_2 \cdot [CH_2OCH_2]_4$	U: SVol.E1-73, 78
–	$UO_2Cl_2 \cdot [CH_2OCH_2]_4 \cdot 2\ H_2O$	U: SVol.E1-73, 78
–	$UO_2Cl_2 \cdot 2\ CH_3COOC_2H_5$	U: SVol.E1-120/2
$C_8Cl_2H_{16}O_8P_2Sn$	$SnCl_2(CH_2COP(O)(OCH_3)_2)_2$	Sn: MVol.C6-164
$C_8Cl_2H_{16}Sn$	$(CH_3)_2(C_3H_4Cl(CH_2)_3)SnCl$	Sn: Org.Verb.5-194, 198
–	$(CH_3)_2SnC_5H_9CHCl_2$	Sn: Org.Verb.3-106
–	$(CH_3)_3SnCH_2CH(CCl_2)CHCH_3$	Sn: Org.Verb.2-27, 29
–	$(CH_3)_3SnCH_2CHCHCH_2CHCl_2$	Sn: Org.Verb.2-79, 86
–	$(CH_3)_3SnCH_2CH_2CH(CCl_2)CH_2$	Sn: Org.Verb.2-37, 39
$C_8Cl_2H_{17}NO_4Sn$	$SnCl_2 \cdot (CH_3)_2C_3H_4(NH_2)COOH \cdot CH_3COOH$	Sn: MVol.C5-31/2
$C_8Cl_2H_{17}Ti$	$C_8H_{17}TiCl_2$	Ti: Org.Verb.1-8
$C_8Cl_2H_{18}N_2O_2S$	$SO_2[N(C_4H_9)Cl]_2$	S: S-N-Verb.1-177
–	$SO_2[N(i\text{-}C_4H_9)Cl]_2$	S: S-N-Verb.1-177
$C_8Cl_2H_{18}N_2O_2Sn$	$SnCl_2 \cdot 2\ HN(CH_2CH_2)_2O$	Sn: MVol.C5-29
$C_8Cl_2H_{18}N_2O_4U$	$UO_2Cl_2 \cdot 2\ CH_3CON(CH_3)_2$	U: SVol.E1-103/4, 106
$C_8Cl_2H_{18}OSn$	$(C_4H_9O)(C_4H_9)SnCl_2$	Sn: Org.Verb.6-94
–	$C_8H_{17}Sn(OH)Cl_2 \cdot H_2O$	Sn: Org.Verb.6-252
$C_8Cl_2H_{18}O_2S_2Sn$	$(CH_2CH)_2SnCl_2 \cdot 2\ (CH_3)_2SO$	Sn: Org.Verb.6-145
$C_8Cl_2H_{18}S_2Sn$	$SnCl_2(SC(CH_3)_3)_2$	Sn: MVol.C6-79/80
–	$SnCl_2(SC_4H_9)_2$	Sn: MVol.C6-79/80
$C_8Cl_2H_{18}Sn$	$(CH_3)(C_7H_{15})SnCl_2$	Sn: Org.Verb.6-197, 204
–	$(CH_3)_3SnCH(CH_3)CH(CH_3)CHCl_2$	Sn: Org.Verb.2-53, 55
–	$(CH_3)_3SnCH_2C(CH_3)_2CHCl_2$	Sn: Org.Verb.2-55
–	$(CH_3)_3SnCH_2CH(CHCl_2)C_2H_5$	Sn: Org.Verb.2-54
–	$(C_2H_5)_2Sn(CHClCH_3)_2$	Sn: Org.Verb.3-26
–	$(C_4H_9)_2SnCl_2$	
	Analysis	Sn: Org.Verb.6-74/5
	Chemical reactions	Sn: Org.Verb.6-85/100
	Formation and preparation	Sn: Org.Verb.6-71/8
	Molecule	Sn: Org.Verb.6-78/9
	Physical properties	Sn: Org.Verb.6-83/4
	Physiology	Sn: Org.Verb.6-101/2

Formula	Compound	Reference
$C_8Cl_2H_{18}Sn$	$(C_4H_9)_2SnCl_2$	
	Polarography	Sn: Org.Verb.6-84
	Spectra	Sn: Org.Verb.6-79/82
	Uses	Sn: Org.Verb.6-102/3
–	$(i\text{-}C_4H_9)_2SnCl_2$	Sn: Org.Verb.6-104/5
–	$(s\text{-}C_4H_9)_2SnCl_2$	Sn: Org.Verb.6-106
–	$(t\text{-}C_4H_9)_2SnCl_2$	Sn: Org.Verb.6-106/8
$C_8Cl_2H_{18}Ti$	$(C_4H_9)_2TiCl_2$	Ti: Org.Verb.1-72
–	$(i\text{-}C_4H_9)_2TiCl_2$	Ti: Org.Verb.1-72
$C_8Cl_2H_{19}NSn$	$(C_4H_9)[(C_2H_5)_2N]SnCl_2$	Sn: Org.Verb.6-245
$C_8Cl_2H_{19}O_2PS_2Sn$	$C_4H_9Sn[SP(S)(OC_2H_5)_2]Cl_2$	Sn: Org.Verb.6-247
$C_8Cl_2H_{20}INSn$	$[(C_2H_5)_4N]SnCl_2I$	Sn: MVol.C3-120
$C_8Cl_2H_{20}MnO_2$	$MnCl_2 \cdot 2\ C_4H_9OH$	Mn: MVol.D1-33
$C_8Cl_2H_{20}MnO_{12}$	$[Mn(C_2H_4(OCH_3)_2)_2 \cdot (H_2O)_2](ClO_4)_2$	Mn: MVol.D1-141
$C_8Cl_2H_{20}N_2O_2Sn$	$(CH_3)_2SnCl_2 \cdot 2\ HCON(CH_3)_2$	Sn: Org.Verb.6-36
$C_8Cl_2H_{20}OSSn$	$(CH_3)_2SnCl_2 \cdot (C_3H_7)_2SO$	Sn: Org.Verb.6-37
$C_8Cl_2H_{20}OSn_2$	$[(C_2H_5)_2ClSn]_2O$	Sn: Org.Verb.5-204
		Sn: Org.Verb.6-52, 54, 56
$C_8Cl_2H_{20}O_2P_2Sn$	$SnCl_2[OP(C_2H_5)_2]_2$	Sn: MVol.C6-140
$C_8Cl_2H_{20}O_4U$	$UO_2Cl_2 \cdot 2\ (C_2H_5)_2O \cdot H_2O$	U: SVol.E1-71, 75/6
$C_8Cl_2H_{20}O_6P_2Sn$	$SnCl_2(C_2H_5(C_2H_5O)P(O)O)_2$	Sn: MVol.C6-162
$C_8Cl_2H_{20}O_8P_2Sn$	$SnCl_2[(C_2H_5O)_2PO_2]_2$	Sn: MVol.C2-267
		Sn: MVol.C6-168/9
$C_8Cl_2H_{20}P_2S_4Sn$	$SnCl_2((C_2H_5)_2P(S)S)_2$	Sn: MVol.C6-157
$C_8Cl_2H_{20}Sn_2$	$[(C_2H_5)_2ClSn]_2$	Sn: Org.Verb.6-51
$C_8Cl_2H_{21}O_3PSn$	$(C_2H_5)_2SnCl_2 \cdot (C_2H_5O)_2HPO$	Sn: Org.Verb.6-62
$C_8Cl_2H_{22}N_2O_2U$	$UO_2Cl_2 \cdot 2\ C_4H_9NH_2$	U: SVol.E1-18, 21
$C_8Cl_2H_{22}N_2Sn$	$SnCl_2 \cdot 2\ (C_2H_5)_2NH$	Sn: MVol.C5-11/2
–	$SnCl_2 \cdot 2\ C_4H_9NH_2$	Sn: MVol.C5-11/2
$C_8Cl_2H_{22}N_2Ti$	$(CH_3)_2TiCl_2 \cdot C_2H_4(N(CH_3)_2)_2$	Ti: Org.Verb.1-75
$C_8Cl_2H_{22}O_2S_2Sn$	$(C_2H_5)_2SnCl_2 \cdot 2\ (CH_3)_2SO$	Sn: Org.Verb.6-61/2
$C_8Cl_2H_{22}O_2Se_2Sn$	$(C_2H_5)_2SnCl_2 \cdot 2\ (CH_3)_2SeO$	Sn: Org.Verb.6-62
$C_8Cl_2H_{22}O_2Si_2Sn$	$[CH_3O(CH_3)_2SiCH_2]_2SnCl_2$	Sn: Org.Verb.6-131
$C_8Cl_2H_{22}O_2Si_3Sn$	$(CH_3)_2Si(OSi(CH_3)_2CH_2)_2SnCl_2$	Sn: Org.Verb.6-207
$C_8Cl_2H_{22}O_4Si_2Sn$	$[CH_3(CH_3O)_2SiCH_2]_2SnCl_2$	Sn: Org.Verb.6-131
$C_8Cl_2H_{22}O_6Si_2Sn$	$[(CH_3O)_3SiCH_2]_2SnCl_2$	Sn: Org.Verb.6-132
$C_8Cl_2H_{22}Si_2Sn$	$[(CH_3)_3SiCH_2]_2SnCl_2$	Sn: Org.Verb.6-132
$C_8Cl_2H_{22}Si_2Ti$	$[(CH_3)_3SiCH_2]_2TiCl_2$	Ti: Org.Verb.1-73, 75
$C_8Cl_2H_{24}MnO_4$	$Mn(C_2H_5OH)_4Cl_2$	Mn: MVol.C5-77
		Mn: MVol.D1-27
$C_8Cl_2H_{24}MnO_{14}$	$[Mn(C_2H_4(OCH_3)_2)_2 \cdot (H_2O)_2](ClO_4)_2$	Mn: MVol.D1-141
$C_8Cl_2H_{24}MnO_{16}$	$[Mn(C_4H_8O_2)_2 \cdot (H_2O)_4](ClO_4)_2$	Mn: MVol.D1-154
$C_8Cl_2H_{24}N_2Sn$	$(C_4H_9)_2SnCl_2 \cdot 2\ NH_3$	Sn: Org.Verb.6-99
–	$(i\text{-}C_4H_9)_2SnCl_2 \cdot 2\ NH_3$	Sn: Org.Verb.6-105
$C_8Cl_2H_{24}N_3OPSn$	$(CH_3)_2SnCl_2 \cdot [(CH_3)_2N]_3PO$	Sn: Org.Verb.6-36
$C_8Cl_2H_{24}N_4Sn$	$SnCl_2 \cdot 2\ NH_2(CH_2)_4NH_2$	Sn: MVol.C5-14/5
$C_8Cl_2H_{24}N_6O_2P_2S$	$SO_2[NPCl(N(CH_3)_2)_2]_2$	S: S-N-Verb.1-192
$C_8Cl_2H_{24}Ni_2P_2$	$(CH_3Ni(P(CH_3)_3)Cl)_2$	Ni: Org.Verb.2-250/1
$C_8Cl_2H_{24}O_2P_2S_2Sn$	$(CH_3)_2SnCl_2 \cdot 2\ (CH_3)_2(CH_3S)PO$	Sn: Org.Verb.6-37
$C_8Cl_2H_{24}O_4P_2Sn$	$(CH_3)_2SnCl_2 \cdot 2\ (CH_3)_2(CH_3O)PO$	Sn: Org.Verb.6-37

Formula	Compound	Reference
$C_8Cl_2H_{24}O_{14}S_4U$	$UO_2(ClO_4)_2 \cdot 4\ (CH_3)_2SO$	U: SVol.E1-204, 208
$C_8Cl_2H_{24}S_2Sb_2Sn$	$(CH_3)_2SnCl_2 \cdot 2\ (CH_3)_3SbS$	Sn: Org.Verb.6-23
$C_8Cl_2H_{26}IrN_2S_2^+$	$[Ir(S(C_2H_5)_2)_2(NH_3)_2Cl_2]^+$	Ir: SVol.2-165
$C_8Cl_2S_{19}$	$C_8S_{19}Cl_2$	C: MVol.D4-215
$C_8Cl_3FH_5NOS$	$CH_3COCH_2SC_5FCl_3N$	F: PerFHalOrg.5-216/7
$C_8Cl_3F_3HN_3O_2$	$C_6Cl_3(NO_2)NC(CF_3)NH$	F: PerFHalOrg.6-125/6, 139
$C_8Cl_3F_3H_2N_2O_3S$	$C_6Cl_3(SO_3H)NC(CF_3)NH$	F: PerFHalOrg.6-125/6, 162
$C_8Cl_3F_5H_2O$	$CCl_3CH(OH)C_6F_5$	F: PerFHalOrg.4-101
$C_8Cl_3F_{13}H_4Sn$	$CF_3(CF_2)_5CH_2CH_2SnCl_3$	Sn: Org.Verb.6-258
$C_8Cl_3FeH_{12}NO_4Si$	$[N(CH_3)_4][(CO)_4FeSiCl_3]$	Fe: Org.Verb.B2-142
$C_8Cl_3H_4MnO_3Sn$	$(CO)_3MnC_5H_4SnCl_3$	Sn: Org.Verb.6-258
$C_8Cl_3H_7O_2Sn$	$CH_3OCOC_6H_4SnCl_3$	Sn: Org.Verb.6-285
–	$SnCl_3(OC_6H_4COCH_3)$	Sn: MVol.C5-88/9
$C_8Cl_3H_7O_3Sn$	$SnCl_3(OC_6H_3(OH)COCH_3)$	Sn: MVol.C5-89
–	$SnCl_3(OC_6H_4COOCH_3)$	Sn: MVol.C5-146/7
$C_8Cl_3H_7Sn$	$C_6H_5CHCHSnCl_3$	Sn: Org.Verb.6-265
$C_8Cl_3H_8IrN_4$	$[Ir(C_4H_4N_2)_2Cl_3]_n$	Ir: SVol.2-77
$C_8Cl_3H_9OSn$	$C_2H_5OC_6H_4SnCl_3$	Sn: Org.Verb.6-285
$C_8Cl_3H_9Sn$	$(CH_2Cl)_2(C_6H_5)SnCl$	Sn: Org.Verb.5-222
$C_8Cl_3H_{10}IrN_4O$	$Ir[N(CH)N(CH)_3]_2(H_2O)Cl_3 \cdot H_2O$	Ir: SVol.2-80
–	$Ir[N(CH)_2N(CH)_2]_2(H_2O)Cl_3 \cdot H_2O$	Ir: SVol.2-78
–	$Ir[NN(CH)_4]_2(H_2O)Cl_3$	Ir: SVol.2-79
–	$Ir[NN(CH)_4]_2(H_2O)Cl_3 \cdot H_2O$	Ir: SVol.2-79
$C_8Cl_3H_{10}NO_2Sn$	$CH_3SnCl_3 \cdot CH_3C_6H_4NO_2$	Sn: Org.Verb.6-222
$C_8Cl_3H_{10}NSn$	$(CH_3)_2NC_6H_4SnCl_3$	Sn: Org.Verb.6-285
$C_8Cl_3H_{13}O_4Sn$	$C_2H_5OCOCH_2CH(COOC_2H_5)SnCl_3$	Sn: Org.Verb.6-254, 258
$C_8Cl_3H_{13}Sn$	$C_6H_9CH_2CH_2SnCl_3$	Sn: Org.Verb.6-265
$C_8Cl_3H_{16}MnO_4$	$MnCl_3 \cdot 2\ C_4H_8O_2$	Mn: MVol.D1-153/4
$C_8Cl_3H_{16}O_2Ti$	$TiCl_3 \cdot 2\ C_4H_8O$	Ti: Org.Verb.1-39
$C_8Cl_3H_{17}O_2Sn$	$C_4H_9SnCl_3 \cdot CH_3COOC_2H_5$	Sn: Org.Verb.6-247/8
$C_8Cl_3H_{17}Sn$	$C_8H_{17}SnCl_3$	Sn: Org.Verb.6-251/2
–	$(i\text{-}C_8H_{17})SnCl_3$	Sn: Org.Verb.6-252
$C_8Cl_3H_{18}N_2O_2PS$	$(C_4H_9)_2NSO_2NPCl_3$	S: S-N-Verb.1-188
$C_8Cl_3H_{18}O_2PS_2Sn$	$SnCl_3((C_4H_9O)_2P(S)S)$	Sn: MVol.C6-175/6
$C_8Cl_3H_{18}O_3U^-$	$[UO_2Cl_3((C_4H_9)_2O)]^-$	U: SVol.E1-76
$C_8Cl_3H_{19}OSn_2$	$Cl_2(C_4H_9)SnOSn(C_2H_5)_2Cl$	Sn: Org.Verb.6-245
$C_8Cl_3H_{19}O_2Sn$	$SnCl_3OC_4H_9 \cdot C_4H_9OH$	Sn: MVol.C5-73/4
$C_8Cl_3H_{19}O_3SiSn$	$(CH_3)_2Si[O(CH_2CH_2O)_2CH_3]CH_2SnCl_3$	Sn: Org.Verb.6-259
$C_8Cl_3H_{19}SiSn$	$(C_2H_5)_3SnCH_2CH_2SiCl_3$	Sn: Org.Verb.2-163, 164
$C_8Cl_3H_{20}NSn$	$[(C_2H_5)_4N]SnCl_3$	Sn: MVol.C3-95/6
$C_8Cl_3H_{20}O_2Ti$	$TiCl_3 \cdot 2\ (C_2H_5)_2O$	Ti: Org.Verb.1-20, 24, 38
$C_8Cl_3H_{20}O_3PSn$	$C_2H_5SnCl_3 \cdot OP(OC_2H_5)_2C_2H_5$	Sn: Org.Verb.6-232
–	$C_4H_9SnCl_3 \cdot (C_2H_5O)_2P(O)H$	Sn: Org.Verb.6-247/8
$C_8Cl_3H_{21}O_2Sn$	$HSnCl_3 \cdot 2\ (C_2H_5)_2O$	Sn: MVol.C5-7/8
$C_8Cl_3H_{23}IrNS_2$	$Ir(S(C_2H_5)_2)_2(NH_3)Cl_3$	Ir: SVol.2-165
$C_8Cl_3H_{23}N_3OPSn$	$C_2H_5SnCl_3 \cdot [(CH_3)_2N]_3PO$	Sn: Org.Verb.6-233
$C_8Cl_3H_{23}O_4P_2Sn$	$C_2H_5SnCl_3 \cdot 2\ CH_3O(CH_3)_2PO$	Sn: Org.Verb.6-232
$C_8Cl_3H_{24}O_7P_3S_2Sn$	$SnCl_3(CH_3(CH_3O)PS_2) \cdot 2\ CH_3PO(OCH_3)_2$	Sn: MVol.C6-168
$C_8Cl_3H_{24}O_8P_3S_2Sn$	$SnCl_3(CH_3O)_2P(S)S \cdot 2\ CH_3PO(OCH_3)_2$	Sn: MVol.C6-176
$C_8Cl_3H_{32}IN_{16}O_8U$	$UICl_3 \cdot 8\ (NH_2)_2CO$	U: SVol.E1-82/3

$C_8Cl_4CoH_{11}OSn_2$ .... $(CH_3Cl_2Sn)_2Co(CO)C_5H_5$ .... Sn: Org.Verb.6-219, 221
$C_8Cl_4F_3HN_2$ .... $C_6Cl_4NC(CF_3)NH$ .... F: PerFHalOrg.6-125/6, 139, 160/2
$C_8Cl_4F_3HN_2O$ ...... $C_6Cl_4NC(CF_3)N(OH)$ .... F: PerFHalOrg.6-125/6, 159, 161/2
$C_8Cl_4F_3HN_2O_2S$ ..... $C_6Cl_3(SO_2Cl)NC(CF_3)NH$ .... F: PerFHalOrg.6-125/6, 139, 160
$C_8Cl_4F_4N_6$ .... $N_3C_3F_2CCl_2CCl_2C_3F_2N_3$ .... F: PerFHalOrg.6-89
$C_8Cl_4F_8HN_3O$ ...... $OCNCNC(CF_2Cl)_2NHC(CF_2Cl)_2$ .... F: PerFHalOrg.5-63/4, 87
$C_8Cl_4F_9N_3$ .... $(CF_2ClCFCl)_2(CF_3)C_3N_3$ .... F: PerFHalOrg.6-111
$C_8Cl_4FeH_{18}HgO_2P_2$ .. $(CO)_2Fe(P(CH_3)_3)_2Cl_2 \cdot HgCl_2$ .... Fe: Org.Verb.B1-109
$C_8Cl_4Fe_2Ge_2O_8$ ..... $[(CO)_4FeGeCl_2]_2$ .... Fe: Org.Verb.C1-42, 45
$C_8Cl_4Fe_2O_8Si_2$ ...... $[(CO)_4FeSiCl_2]_2$ .... Fe: Org.Verb.C1-42/5, 48
$C_8Cl_4Fe_2O_8Sn_2$ ..... $[(CO)_4FeSnCl_2]_2$ .... Fe: Org.Verb.C1-42/3, 46
$C_8Cl_4H_4S_2Sn$ .... $[C_4H_2(Cl)S]_2SnCl_2$ .... Sn: Org.Verb.6-184/5, 190
$C_8Cl_4H_5NO_2Sn$ ..... $SnCl_4 \cdot C_6H_4(CO)_2NH$ .... Sn: MVol.C6-54
– .... $SnCl_4 \cdot C_8H_5NO_2$ .... Sn: MVol.C5-192
$C_8Cl_4H_6O_2Sn$ .... $SnCl_4 \cdot C_6H_4(CHO)_2$ .... Sn: MVol.C5-83/4
$C_8Cl_4H_7NOSn$ ...... $SnCl_4 \cdot CH_3OC_6H_4CN$ .... Sn: MVol.C6-71/2
$C_8Cl_4H_7NSn$ .... $SnCl_4 \cdot CH_3C_6H_4CN$ .... Sn: MVol.C6-71/2
$C_8Cl_4H_8IrKN_4$ ...... $K[Ir(N(CH)N(CH)_3)_2Cl_4]$ .... Ir: SVol.2-80
– .... $K[Ir(N(CH)N(CH)_3)_2Cl_4] \cdot 0.5\ H_2O$ .... Ir: SVol.2-80
– .... $K[Ir(N(CH)_2N(CH)_2)_2Cl_4]$ .... Ir: SVol.2-77
– .... $K[Ir(N(CH)_2N(CH)_2)_2Cl_4] \cdot H_2O$ .... Ir: SVol.2-77
– .... $K[Ir(NN(CH)_4)_2Cl_4]$ .... Ir: SVol.2-79
$C_8Cl_4H_8IrN_4^-$ .... $[Ir(N(CH)N(CH)_3)_2Cl_4]^-$ .... Ir: SVol.2-80, 138
– .... $[Ir(N(CH)_2N(CH)_2)_2Cl_4]^-$ .... Ir: SVol.2-77, 123, 138
$C_8Cl_4H_8IrN_4Na$ ..... $Na[Ir(N(CH)_2N(CH)_2)_2Cl_4] \cdot 0.5\ H_2O$ ...... Ir: SVol.2-77
– .... $Na[Ir(N(CH)_2N(CH)_2)_2Cl_4] \cdot H_2O$ .... Ir: SVol.2-77
$C_8Cl_4H_8N_2O_2Sn$ ..... $SnCl_4 \cdot C_6H_4(CONH_2)_2$ .... Sn: MVol.C6-52
– .... $SnCl_4 \cdot C_6H_5CONHCONH_2$ .... Sn: MVol.C6-58/9
$C_8Cl_4H_8N_4Sn$ .... $SnCl_4 \cdot 2\ N(CHCH)_2N$ .... Sn: MVol.C5-225/6
$C_8Cl_4H_8OSn$ .... $SnCl_4 \cdot CH_3COC_6H_5$ .... Sn: MVol.C5-88
– .... $SnCl_4 \cdot CH_3C_6H_4CHO$ .... Sn: MVol.C5-81
$C_8Cl_4H_8OTe$ .... $TeCl_4 \cdot CH_3COC_6H_5$ .... Te: SVol.B3-166
$C_8Cl_4H_8O_2Sn$ .... $SnCl_4 \cdot CH_3OC_6H_4CHO$ .... Sn: MVol.C5-81
$C_8Cl_4H_9IrN_4$ .... $H[Ir(N(CH)_2N(CH)_2)_2Cl_4] \cdot 0.5\ H_2O$ .... Ir: SVol.2-77
– .... $IrH[N(CH)N(CH)_3]_2Cl_4$ .... Ir: SVol.2-80
– .... $IrH[N(CH)N(CH)_3]_2Cl_4 \cdot H_2O$ .... Ir: SVol.2-80
– .... $IrH[NN(CH)_4]_2Cl_4$ .... Ir: SVol.2-79
– .... $Ir[N(CH)_2N(CH)_2][N(CH)_2N(CH)_2H]Cl_4$ ..... Ir: SVol.2-78
$C_8Cl_4H_9NOSn$ ...... $SnCl_4 \cdot NC_5H_4CH_2OC_2H_3$ .... Sn: MVol.C5-203/4
$C_8Cl_4H_9NSSn$ ...... $SnCl_4 \cdot CH_2CHSC_6H_4NH_2$ .... Sn: MVol.C6-99
$C_8Cl_4H_9N_3O_4Sn$ ..... $SnCl_4 \cdot (NO_2)_2C_6H_3N(CH_3)_2$ .... Sn: MVol.C5-180
$C_8Cl_4H_{10}N_2O_2Sn$ .... $SnCl_4 \cdot NO_2C_6H_2(CH_3)_2NH_2$ .... Sn: MVol.C5-178
$C_8Cl_4H_{10}N_2Sn$ ...... $SnCl_4 \cdot 2\ CH_2C(CH_3)CN$ .... Sn: MVol.C6-69/70
– .... $SnCl_4 \cdot 2\ C_4H_4NH$ .... Sn: MVol.C5-190
$C_8Cl_4H_{10}OSn$ .... $SnCl_4 \cdot CH_3C_6H_4OCH_3$ .... Sn: MVol.C5-152
– .... $SnCl_4 \cdot (CH_3)_2C_6H_3OH$ .... Sn: MVol.C5-76
$C_8Cl_4H_{10}SSn$ .... $SnCl_4 \cdot C_6H_5SC_2H_5$ .... Sn: MVol.C6-94/5

Formula	Compound	Element	Reference
$C_8Cl_4H_{11}NO_2Sn$	$SnCl_4 \cdot C_6H_{10}(CO)_2NH$	Sn:	MVol.C6-53
$C_8Cl_4H_{12}IrN_5$	$NH_4[Ir(N(CH)_2N(CH)_2)_2Cl_4] \cdot H_2O$	Ir:	SVol.2-77
–	$NH_4[Ir(NN(CH)_4)_2Cl_4]$	Ir:	SVol.2-79
$C_8Cl_4H_{12}N_2Te$	$TeCl_4 \cdot (CH_3)_2NC_6H_4NH_2$	Te:	SVol.B3-169
$C_8Cl_4H_{12}N_4Pa$	$PaCl_4 \cdot 4\ CH_3CN$	Pa:	SVol.2-88/9
$C_8Cl_4H_{12}N_4Sn$	$SnCl_4 \cdot NCC(CH_3)_2NNC(CH_3)_2CN$	Sn:	MVol.C6-73/5
$C_8Cl_4H_{12}N_4Ti$	$[CH_3NC(Cl)]_4Ti$	Ti:	Org.Verb.1-82, 94
$C_8Cl_4H_{12}N_4U$	$[UCl_4(CH_3CN)_4]$	U:	SVol.E1-48/50
$C_8Cl_4H_{12}Ni_2O$	$(CH_3)_2C(OH)CCCH_2CHCH_2 \cdot 2\ NiCl_2$	Ni:	Org.Verb.2-333
–	$(CH_3)_2C(OH)CCCH_2CHCH_2 \cdot 2\ NiCl_2 \cdot 6\ H_2O$	Ni:	Org.Verb.2-333
$C_8Cl_4H_{12}O_4Sn$	$SnCl_4 \cdot C_2H_2(COOC_2H_5)_2$	Sn:	MVol.C5-143/4
$C_8Cl_4H_{12}O_5Sn$	$SnCl_4 \cdot C_4H_6O(OCOCH_3)_2$	Sn:	MVol.C5-133
$C_8Cl_4H_{12}O_6Sn$	$SnCl_4 \cdot 2\ (CH_3CO)_2O$	Sn:	MVol.C5-124/5
$C_8Cl_4H_{12}O_6Te$	$TeCl_4 \cdot 2\ (CH_3CO)_2O$	Te:	SVol.B3-166
$C_8Cl_4H_{12}Sn$	$(CH_3CClCHCH_2)_2SnCl_2$	Sn:	Org.Verb.6-148
$C_8Cl_4H_{14}N_2Sn$	$SnCl_4 \cdot 2\ C_3H_7CN$	Sn:	MVol.C6-68/9
$C_8Cl_4H_{14}N_2U$	$[UCl_4(C_3H_7CN)_2]$	U:	SVol.E1-48/50
$C_8Cl_4H_{14}O_4Sn$	$SnCl_4 \cdot C_2H_4(COOC_2H_5)_2$	Sn:	MVol.C5-141/3
–	$SnCl_4 \cdot C_4H_8(COOCH_3)_2$	Sn:	MVol.C5-141/3
$C_8Cl_4H_{15}NOSn$	$SnCl_4 \cdot NC_5H_9(CH_3)OCHCH_2$	Sn:	MVol.C5-193/4
$C_8Cl_4H_{15}NO_2Sn$	$SnCl_4 \cdot CH_3CONHCOC_5H_{11}$	Sn:	MVol.C6-50/1
$C_8Cl_4H_{16}N_2O_{10}U$	$UO_2(NO_3)_2 \cdot 2\ (ClC_2H_4)_2O$	U:	SVol.E1-71, 76
$C_8Cl_4H_{16}N_4O_2Sn$	$SnCl_4 \cdot 2\ CH_2CHCH_2NHCONH_2$	Sn:	MVol.C6-57
$C_8Cl_4H_{16}N_4S_2Sn$	$SnCl_4 \cdot 2\ NH_2CSNHCH_2CHCH_2$	Sn:	MVol.C6-121
$C_8Cl_4H_{16}O_2S_2Sn$	$SnCl_4 \cdot 2\ (CH_2)_4SO$	Sn:	MVol.C6-110
–	$SnCl_4 \cdot 2\ S(CH_2CH_2)_2O$	Sn:	MVol.C6-113/4
$C_8Cl_4H_{16}O_2Se_2Sn$	$SnCl_4 \cdot 2\ Se(CH_2CH_2)_2O$	Sn:	MVol.C6-124
$C_8Cl_4H_{16}O_2Sn$	$SnCl_4 \cdot 2\ CH_3COC_2H_5$	Sn:	MVol.C5-87
–	$SnCl_4 \cdot 2\ C_4H_8O$	Sn:	MVol.C5-156
–	$[SnCl_4 \cdot C_5H_{11}COOC_2H_5]_2$	Sn:	MVol.C5-136/7
$C_8Cl_4H_{16}O_4S_2Sn$	$SnCl_4 \cdot 2\ (CH_2)_4SO_2$	Sn:	MVol.C6-110
$C_8Cl_4H_{16}O_4Sn$	$SnCl_4 \cdot 2\ CH_3COOC_2H_5$	Sn:	MVol.C5-128/30
–	$SnCl_4 \cdot 2\ C_2H_5COOCH_3$	Sn:	MVol.C5-134
–	$SnCl_4 \cdot 2\ C_3H_7COOH$	Sn:	MVol.C5-120/1
–	$SnCl_4 \cdot 2\ C_4H_8O_2$	Sn:	MVol.C5-158
–	$SnCl_4 \cdot 2\ C_4H_8O_2 \cdot 2\ H_2O$	Sn:	MVol.C5-159
–	$SnCl_4 \cdot 2\ HCOOC_3H_7$	Sn:	MVol.C5-128
$C_8Cl_4H_{16}O_4U$	$UCl_4 \cdot 2\ O(CH_2CH_2)_2O$	U:	SVol.E1-70, 74/5
$C_8Cl_4H_{16}O_8Sn$	$SnCl_4 \cdot 4\ CH_3COOH$	Sn:	MVol.C5-117/8
$C_8Cl_4H_{16}S_2Sn$	$SnCl_4 \cdot 2\ (CH_2)_4S$	Sn:	MVol.C6-109/10
$C_8Cl_4H_{16}Sn$	$(C_3H_7CHCl)_2SnCl_2$	Sn:	Org.Verb.6-131
–	$Sn(CHClCH_3)_4$	Sn:	Org.Verb.1-123
$C_8Cl_4H_{17}NO_2Sn$	$SnCl_4 \cdot (CH_3)_2C_3H_4(NH_2)COOC_2H_5$	Sn:	MVol.C6-12/3
$C_8Cl_4H_{17}O_3U$	$U(OC_3H_7)Cl_4 \cdot CH_3COOC_3H_7$	U:	SVol.E1-120/2
$C_8Cl_4H_{17}O_5PSn$	$SnCl_4 \cdot C_2H_5OCOCH_2P(O)(OC_2H_5)_2$	Sn:	MVol.C6-164/5
$C_8Cl_4H_{18}N_2O_2Sn$	$SnCl_4 \cdot 2\ CH_3CON(CH_3)_2$	Sn:	MVol.C6-43/4
–	$SnCl_4 \cdot 2\ HN(CH_2CH_2)_2O$	Sn:	MVol.C5-245
$C_8Cl_4H_{18}N_2O_2Te$	$TeCl_4 \cdot 2\ CH_3CON(CH_3)_2$	Te:	SVol.B3-174
–	$TeCl_4 \cdot 2\ NH(CH_2CH_2)_2O$	Te:	SVol.B3-172

Formula	Compound	Reference
$C_8Cl_4H_{18}N_2O_4Sn$	$SnCl_4 \cdot 2\,NH_2CH_2COOC_2H_5$	Sn: MVol.C6-12/3
$C_8Cl_4H_{18}N_2O_4U$	$UCl_4 \cdot 2\,CH_3NHCOOC_2H_5$	U: SVol.E1-120/3
$C_8Cl_4H_{18}N_2O_8Sn$	$SnCl_4 \cdot 2\,NH_2CH_2COOH \cdot 2\,CH_3COOH$	Sn: MVol.C6-11
$C_8Cl_4H_{18}N_2S_2Sn$	$SnCl_4 \cdot 2\,CH_3CSN(CH_3)_2$	Sn: MVol.C6-116/7
$C_8Cl_4H_{18}OSn$	$SnCl_4 \cdot (C_4H_9)_2O$	Sn: MVol.C5-152
$C_8Cl_4H_{18}OTe$	$TeCl_4 \cdot (C_4H_9)_2O$	Te: SVol.B3-166
$C_8Cl_4H_{18}O_2SSn$	$SnCl_4 \cdot (n\text{-}C_4H_9)_2SO_2$	Sn: MVol.C6-103
$C_8Cl_4H_{18}O_2Sn$	$SnCl_4 \cdot C_2H_5O(CH_2)_4OC_2H_5$	Sn: MVol.C5-154/5
$C_8Cl_4H_{18}O_3SiSn$	$SnCl_4 \cdot CH_2CHSi(OC_2H_5)_3$	Sn: MVol.C6-127/8
$C_8Cl_4H_{18}O_4U$	$UCl_4 \cdot C_2H_4(OC_2H_4OCH_3)_2$	U: SVol.E1-70, 74/5
$C_8Cl_4H_{18}O_5Sn$	$[(C_2H_5)_2OH][HSnCl_4(CH_3COO)_2]$	Sn: MVol.C5-118
$C_8Cl_4H_{18}SSn$	$SnCl_4 \cdot (C_4H_9)_2S$	Sn: MVol.C6-93/4
$C_8Cl_4H_{18}STe$	$TeCl_4 \cdot (C_4H_9)_2S$	Te: SVol.B3-175
$C_8Cl_4H_{18}Ti^{2-}$	$[(i\text{-}C_4H_9)_2TiCl_4]^{2-}$	Ti: Org.Verb.1-81
$C_8Cl_4H_{19}NSn_2$	$2\,SnCl_2 \cdot (C_4H_9)_2NH$	Sn: MVol.C5-11/2
$C_8Cl_4H_{20}IrS_2$	$Ir(S(C_2H_5)_2)_2Cl_4$	Ir: SVol.2-166
$C_8Cl_4H_{20}IrS_2^-$	$[Ir(S(C_2H_5)_2)_2Cl_4]^-$	Ir: SVol.2-166
$C_8Cl_4H_{20}MnN$	$(C_2H_5)_4NMnCl_4 \cdot 2\,CH_3COOH$	Mn: MVol.C5-212
$C_8Cl_4H_{20}MnN^-$	$[MnCl_4(C_2H_5)_4N]^-$	Mn: MVol.C5-87
$C_8Cl_4H_{20}N_4O_2P_2S$	$SO_2[NPCl_2N(C_2H_5)_2]_2$	S: S-N-Verb.1-192
–	$SO_2[NPCl_2NHC_4H_9]_2$	S: S-N-Verb.1-192
$C_8Cl_4H_{20}N_4O_4Sn$	$SnCl_4 \cdot 4\,CH_3CONH_2$	Sn: MVol.C6-44
$C_8Cl_4H_{20}N_4O_8Sn$	$SnCl_4 \cdot 4\,NH_2CH_2COOH$	Sn: MVol.C6-10
$C_8Cl_4H_{20}O_2S_2Sn$	$SnCl_4 \cdot 2\,(C_2H_5)_2SO$	Sn: MVol.C6-102
$C_8Cl_4H_{20}O_2Sn$	$SnCl_4 \cdot 2\,C_2H_5OC_2H_5$	Sn: MVol.C5-148/51
–	$SnCl_4 \cdot 2\,C_4H_9OH$	Sn: MVol.C5-72/3
$C_8Cl_4H_{20}O_3SiSn$	$SnCl_4 \cdot C_2H_5Si(OC_2H_5)_3$	Sn: MVol.C6-127/8
$C_8Cl_4H_{20}O_4SiSn$	$SnCl_4 \cdot Si(OC_2H_5)_4$	Sn: MVol.C6-126/7
$C_8Cl_4H_{20}O_4U$	$UCl_4 \cdot 2\,C_2H_4(OCH_3)_2$	U: SVol.E1-70, 74/5
$C_8Cl_4H_{20}O_6Sn$	$[C_4H_8O_2 \cdot H]_2[SnCl_4(OH)_2]$	Sn: MVol.C5-159
$C_8Cl_4H_{20}P_2S_2Sn$	$SnCl_4 \cdot (C_2H_5)_2PSPS(C_2H_5)_2$	Sn: MVol.C6-148
$C_8Cl_4H_{20}S_2Sn$	$SnCl_4 \cdot 2\,(CH_3)_3CSH$	Sn: MVol.C6-79/80
–	$SnCl_4 \cdot 2\,C_2H_5SC_2H_5$	Sn: MVol.C6-92/3
–	$SnCl_4 \cdot 2\,C_4H_9SH$	Sn: MVol.C6-79/80
$C_8Cl_4H_{20}S_4U$	$UCl_4 \cdot 2\,C_2H_4(SCH_3)_2$	U: SVol.E1-221
$C_8Cl_4H_{22}N_2Sn$	$SnCl_4 \cdot 2\,(C_2H_5)_2NH$	Sn: MVol.C5-171
–	$SnCl_4 \cdot 2\,C_4H_9NH_2$	Sn: MVol.C5-169
$C_8Cl_4H_{22}N_2Sn_2$	$[Sn_2Cl_3((C_2H_5)_2NH)_2]Cl$	Sn: MVol.C5-11/2
–	$[Sn_2Cl_3(C_4H_9NH_2)_2]Cl$	Sn: MVol.C5-11/2
$C_8Cl_4H_{22}N_2Te$	$TeCl_4 \cdot 2\,C_4H_9NH_2$	Te: SVol.B3-166
$C_8Cl_4H_{22}N_2U$	$UCl_4 \cdot 2\,(C_2H_5)_2NH$	U: SVol.E1-19, 22
$C_8Cl_4H_{22}O_2P_2S_2Sn$	$SnCl_4 \cdot 2\,(C_2H_5)_2POSH$	Sn: MVol.C6-156/7
$C_8Cl_4H_{22}O_4P_2S_2Sn$	$SnCl_4 \cdot 2\,(C_2H_5O)_2P(S)H$	Sn: MVol.C6-168
$C_8Cl_4H_{22}O_4P_2Sn$	$SnCl_4 \cdot 2\,C_2H_5(H)P(O)OC_2H_5$	Sn: MVol.C6-151/2
$C_8Cl_4H_{22}O_6P_2Sn$	$SnCl_4 \cdot 2\,HP(O)(OC_2H_5)_2$	Sn: MVol.C6-160/1
$C_8Cl_4H_{22}O_8P_2Sn$	$SnCl_4 \cdot 2\,(C_2H_5O)_2P(O)OH$	Sn: MVol.C6-168/9
$C_8Cl_4H_{24}IrNS_2$	$NH_4[Ir(S(C_2H_5)_2)_2Cl_4]$	Ir: SVol.2-166
$C_8Cl_4H_{24}MnN_2$	$[(CH_3)_4N]_2MnCl_4$	Mn: MVol.C5-84/6, 193/4
–	$(C_4H_9NH_3)_2MnCl_4$	Mn: MVol.C5-182/5
$C_8Cl_4H_{24}N_2O_2U$	$[N(CH_3)_4]_2UO_2Cl_4$	U: SVol.C9-93/101

Formula	Compound	Reference
$C_8Cl_4H_{24}N_4O_3P_2Sn$	$SnCl_4 \cdot O[PO(N(CH_3)_2)_2]_2$	Sn: MVol.C6-172
$C_8Cl_4H_{24}N_4Sn$	$SnCl_4 \cdot 2\ NH_2(CH_2)_4NH_2$	Sn: MVol.C5-185
$C_8Cl_4H_{24}N_4SnTi$	$SnCl_4 \cdot Ti(N(CH_3)_2)_4$	Sn: MVol.C6-183
$C_8Cl_4H_{24}O_2P_2U$	$[P(C_2H_5)_2H_2]_2UO_2Cl_4$	U: SVol.C9-59
$C_8Cl_4H_{24}O_4S_4U$	$UCl_4 \cdot 4\ (CH_3)_2SO$	U: SVol.E1-202/3
$C_8Cl_4H_{24}O_4Sn$	$SnCl_4 \cdot 4\ C_2H_5OH$	Sn: MVol.C5-70
$C_8Cl_4H_{24}O_4U$	$UCl_4 \cdot 4\ C_2H_5OH$	U: SVol.E1-58/9
$C_8Cl_4H_{28}N_4Sn$	$SnCl_4 \cdot 4\ C_2H_5NH_2$	Sn: MVol.C5-169
$C_8Cl_4H_{32}N_8Sn$	$SnCl_4 \cdot 4\ C_2H_8N_2$	Sn: MVol.C5-183
$C_8Cl_4H_{32}N_8U$	$UCl_4 \cdot 4\ C_2H_8N_2$	U: SVol.E1-23/4
$C_8Cl_5CoH_8OSn_2$	$C_5H_5Co(CO)(SnCl_3)(CH_2CHSnCl_2)$	Sn: Org.Verb.6-263
$C_8Cl_5Fe_2H_{18}O_2P_2$	$(CO)_2Fe(P(CH_3)_3)_2Cl(ClFeCl_3)$	Fe: Org.Verb.B1-109
$C_8Cl_5H_8N_4U$	$[UCl_5(N(CHCH)_2N)_2]$	U: SVol.E1-46, 47
$C_8Cl_5H_{11}OSn$	$SnCl_4 \cdot (CH_3)_2C_5H_5(Cl)CO$	Sn: MVol.C5-94
$C_8Cl_5H_{18}OU$	$UCl_5 \cdot (C_4H_9)_2O$	U: SVol.E1-70, 75
$C_8Cl_5H_{20}NOU$	$N(C_2H_5)_4UOCl_5$	U: SVol.C9-92
$C_8Cl_5H_{20}NSn$	$[(C_2H_5)_4N]SnCl_5$	Sn: MVol.C3-101
$C_8Cl_5H_{20}NTe$	$[(C_2H_5)_4N]TeCl_5$	Te: SVol.B2-113, 133
$C_8Cl_5H_{21}O_2P_2Sn$	$C_4H_9SnCl_3 \cdot 2\ (CH_3)_2P(O)Cl$	Sn: Org.Verb.6-247/8
$C_8Cl_5H_{24}MnN_2$	$[(CH_3)_4N]_2MnCl_5$	Mn: MVol.C5-211
–	$[(CH_3)_4N]_2MnCl_5 \cdot H_2O$	Mn: MVol.C5-91, 211
$C_8Cl_6F_{11}NO$	$Cl_3C(CF_2CFCl)_3CF_2NO$	F: PerFHalOrg.7-179, 186
$C_8Cl_6Fe_2O_8Si_2$	$Cl_3Si(CO)_4FeFe(CO)_4SiCl_3$	Fe: Org.Verb.C1-72
$C_8Cl_6H_8Ir_2N_4S_4$	$[Ir(C_2H_4(SCN)_2)Cl_3]_2$	Ir: SVol.2-172
$C_8Cl_6H_8Ir_2N_4Se_4$	$[Ir(C_2H_4(SeCN)_2)Cl_3]_2$	Ir: SVol.2-175
$C_8Cl_6H_{10}O_7U$	$UO_2(Cl_3CCOO)_2 \cdot (C_2H_5)_2O$	U: SVol.E1-71, 77
–	$UO_2(Cl_3CCOO)_2 \cdot (C_2H_5)_2O \cdot H_2O$	U: SVol.E1-71, 77
–	$UO_2(Cl_3CCOO)_2 \cdot (C_2H_5)_2O \cdot 2\ H_2O$	U: SVol.E1-71, 77
$C_8Cl_6H_{12}N_4Na_2Sn$	$[Na(CH_3CN)_2]_2[SnCl_6]$	Sn: MVol.C3-24
$C_8Cl_6H_{14}O_4Sn$	$SnCl_4 \cdot 2\ CH_2ClCOOC_2H_5$	Sn: MVol.C5-133/4
$C_8Cl_6H_{14}O_8Sn$	$[CH_2ClCOOH \cdot H]_2[SnCl_4(CH_3COO)_2]$	Sn: MVol.C5-118
$C_8Cl_6H_{20}NPa$	$[N(C_2H_5)_4]PaCl_6$	Pa: SVol.2-58
$C_8Cl_6H_{20}NU$	$[N(C_2H_5)_4]UCl_6$	U: SVol.C9-60/3
$C_8Cl_6H_{20}P_2Sn$	$SnCl_4 \cdot 2\ (C_2H_5)_2PCl$	Sn: MVol.C6-138
$C_8Cl_6H_{22}O_2Sn$	$[(C_2H_5)_2OH]_2[SnCl_6]$	Sn: MVol.C3-107
$C_8Cl_6H_{22}O_4Sn_2$	$(SnCl_3OC_2H_5 \cdot C_2H_5OH)_2$	Sn: MVol.C5-71
$C_8Cl_6H_{22}S_2Sn$	$((CH_3)_3CSH_2)_2[SnCl_6]$	Sn: MVol.C6-80
–	$(C_4H_9SH_2)_2[SnCl_6]$	Sn: MVol.C6-80
$C_8Cl_6H_{24}IrN_2$	$((CH_3)_4N)_2[IrCl_6]$	Ir: SVol.2-127
$C_8Cl_6H_{24}MnN_2$	$[(CH_3)_4N]_2MnCl_6$	Mn: MVol.C5-213/4
$C_8Cl_6H_{24}N_2Pa$	$[N(CH_3)_4]_2PaCl_6$	Pa: SVol.2-56/7
$C_8Cl_6H_{24}N_2Pr$	$[(CH_3)_4N]_2PrCl_6$	Sc: MVol.C5-223
$C_8Cl_6H_{24}N_2Sb$	$[(CH_3)_4N]_2[SbCl_6]$ solid solutions $[(CH_3)_4N]_2[(Sb,Sn)Cl_6]$	Sn: MVol.C3-106
$C_8Cl_6H_{24}N_2Sn$	$[(CH_3)_4N]_2[SnCl_6]$	Sn: MVol.C3-105/6 Sn: Org.Verb.5-190
–	$[(CH_3)_4N]_2[SnCl_6]$ solid solutions $[(CH_3)_4N]_2[(Sn,Sb)Cl_6]$	Sn: MVol.C3-106
–	$[(C_2H_5)_2NH_2]_2[SnCl_6]$	Sn: MVol.C3-104/5

Formula	Compound	Reference
$C_8Cl_6H_{24}N_2Sn$	$[C_4H_9NH_3]_2[SnCl_6]$	Sn: MVol.C3-103/4
$C_8Cl_6H_{24}N_2Te$	$[(CH_3)_4N]_2TeCl_6$	Te: SVol.B2-116, 119/21, 124, 137/8
–	$[(C_2H_5)_2NH_2]_2TeCl_6$	Te: SVol.B2-116, 137/8
$C_8Cl_6H_{24}N_2U$	$[N(CH_3)_4]_2UCl_6$	U: SVol.C9-52/9
$C_8Cl_6H_{24}N_4O_2P_2Sn$	$SnCl_4 \cdot 2\ ClPO(N(CH_3)_2)_2$	Sn: MVol.C6-173
$C_8Cl_6H_{24}O_5Sn_2$	$(SnCl_3)_2O \cdot 4\ C_2H_5OH$	Sn: MVol.C5-71
$C_8Cl_6H_{24}P_2U$	$(P(C_2H_5)_2H_2)_2UCl_6$	U: SVol.C9-59
$C_8Cl_6H_{26}O_4S_4Te$	$H_2TeCl_6 \cdot 4\ (CH_3)_2SO$	Te: SVol.B3-176
$C_8Cl_6H_{26}O_4Sn$	$[(CH_3)_2OH]_2[SnCl_6] \cdot 2\ (CH_3)_2O$	Sn: MVol.C3-107
$C_8Cl_7H_7Ni_2O_4$	$[CH_3CHCHCH_2Ni][(CCl_3COO)_2NiCl]$	Ni: Org.Verb.2-301
$C_8Cl_7H_{20}NOSSn$	$[((C_2H_5)_4N)(SOCl)][SnCl_6]$	S: SVol.1-54 Sn: MVol.C2-76
$C_8Cl_7H_{20}NOSTi$	$[((C_2H_5)_4N)(SOCl)][TiCl_6]$	S: SVol.1-54
$C_8Cl_7H_{32}LaN_4$	$LaCl_3 \cdot 4\ [(CH_3)_2NH_2]Cl$	Sc: MVol.C5-215
$C_8Cl_7H_{32}N_4Nd$	$NdCl_3 \cdot 4\ [(CH_3)_2NH_2]Cl$	Sc: MVol.C5-215
$C_8Cl_7H_{32}N_4Pr$	$PrCl_3 \cdot 4\ [(CH_3)_2NH_2]Cl$	Sc: MVol.C5-215
$C_8Cl_8FeH_{20}NO_2S_2$	$[((C_2H_5)_4N)(SOCl)_2][FeCl_6]$	S: SVol.1-54
$C_8Cl_8Fe_2O_8Si_2Sn$	$[Cl_3Si(CO)_4Fe]_2SnCl_2$	Fe: Org.Verb.C1-60
$C_8Cl_8Fe_3H_{18}O_2P_2$	$(CO)_2Fe(P(CH_3)_3)_2(ClFeCl_3)_2$	Fe: Org.Verb.B1-109
$C_8Cl_8H_{16}O_2Sn$	$SnCl_4 \cdot 2\ (ClC_2H_4)_2O$	Sn: MVol.C5-151
$C_8Cl_8H_{20}NiO_4P_4$	$Ni(P(OC_2H_5)Cl_2)_4$	Ni: Org.Verb.1-240
$C_8Cl_8H_{20}O_4Si_2Sn$	$SnCl_4 \cdot 2\ Cl_2Si(OC_2H_5)_2$	Sn: MVol.C6-128/9
$C_8Cl_8H_{20}SnTi_2$	$Sn(C_2H_5)_4 \cdot 2\ TiCl_4$	Sn: Org.Verb.1-76
$C_8Cl_{13}H_{20}NTe_3$	$[(C_2H_5)_4N]Te_3Cl_{13}$	Te: SVol.B2-144/5
$C_8CmH_8O_{12}^-$	$Cm(C_4H_4O_6)_2^-$	Np: TrU.D1-159
$C_8CmH_{14}O_6^+$	$Cm((CH_3)_2C(OH)COO)_2^+$	Np: TrU.D1-154
$C_8CmH_{18}N_2O_6P_2^-$	$Cm(C_2H_4(NHC_3H_6PO_3)_2)^-$	Np: TrU.D1-168
$C_8CmH_{19}N_2O_6P_2$	$Cm(C_2H_4(NHC_3H_6PO_3)_2H)$	Np: TrU.D1-168/9
$C_8CmH_{20}N_2O_6P_2^+$	$Cm(C_2H_4(NHC_3H_6PO_3)_2H_2)^+$	Np: TrU.D1-168/9
$C_8CmH_{21}N_2O_6P_2^{2+}$	$Cm(C_2H_4(NHC_3H_6PO_3)_2H_3)^{2+}$	Np: TrU.D1-168/9
$C_8CoFeO_8^-$	$[(CO)_4FeC(O)Co(CO)_3]^-$	Fe: Org.Verb.B2-189/90
$C_8CoFe_2H_{20}N_6O_8$	$[Co(NH_3)_6][HFe(CO)_4]_2$	Fe: Org.Verb.B2-18
$C_8CoH_{15}S_7$	$[Co(S_2CSC_2H_5)_2(SC_2H_5)]_2$	C: MVol.D4-266, 270
$C_8Co_2H_{24}N_6S_6$	$Co_2(C_2H_8N_2)_3(CS_3)_2$	C: MVol.D4-225
$C_8Co_2O_8$	$Co_2(CO)_8$ systems	
	$Co_2(CO)_8$-$Co_4(CO)_{12}$-$Fe(CO)_5$-$Ni(CO)_4$	Fe: Org.Verb.B3-64
	$Co_2(CO)_8$-$Fe(CO)_5$	Fe: Org.Verb.B3-64
	$Co_2(CO)_8$-$Fe(CO)_5$-$Ni(CO)_4$	Fe: Org.Verb.B3-64
$C_8Co_3FeH_{19}O_3P^+$	$[HFeCo_3P(OC_3H_7)_2OC_2H_4]^+$	Fe: Org.Verb.B1-204
$C_8Cs_4N_8S_8U$	$Cs_4U(NCS)_8$	U: SVol.C9-59
$C_8CuF_{12}H_4N_6$	$[N(C(CF_3)NH)_2]_2Cu$	F: PerFHalOrg.7-58
$C_8CuFeH_5N_6$	$[C_2H_5NCFe(CN)_5]Cu \cdot x\ H_2O$	Fe: Org.Verb.B4-10
$C_8CuH_{10}NOS_2$	$Cu[S_2COC_2H_5](C_5H_5N)$	C: MVol.D4-257
$C_8CuH_{20}N_6Se_2$	$Cu(NH_2C_3H_6NH_2)_2(SeCN)_2$	C: MVol.D6-230
$C_8FFeH_{10}N_2O_4P$	$(CO)_4Fe(PF(NCH_3CH_2)_2)$	Fe: Org.Verb.C2-49
$C_8FFeH_{10}O_6P$	$(CO)_4FePF(OC_2H_5)_2$	Fe: Org.Verb.B2-82, 113
$C_8FFeH_{15}O_8P_2$	$(CO)_3FeP_2F(OCH_3)_5$	Fe: Org.Verb.B1-161
$C_8FH_5Ni^+$	$C_5H_5NiC_3F^+$	Ni: Org.Verb.2-359
$C_8FH_6Ni^+$	$C_3FHNiC_5H_5^+$	Ni: Org.Verb.2-359

Formula	Compound		Reference
$C_8FH_8NO_3S$	$FSO_2NCHC_6H_4OCH_3$	S:	S-N-Verb.1-105
$C_8FH_{10}NO_2$	$(CH_3)_2NC_6H_2F(OH)_2$	F:	PerFHalOrg.5-144
$C_8FH_{11}Sn$	$(CH_3)_2(FC_6H_4)SnH$	Sn:	Org.Verb.4-90
$C_8FH_{16}N_5O_3S_3$	$[NS(O)F][NS(O)N(CH_2)_4]_2$	S:	S-N-Verb.1-13
$C_8FH_{16}N_5O_5S_3$	$[NS(O)N(CH_2CH_2)_2O]_2[NS(O)F]$	S:	S-N-Verb.1-13
$C_8FH_{18}Sn^+$	$(C_4H_9)_2SnF^+$	Sn:	Org.Verb.5-30
$C_8FH_{19}Sn$	$(C_2H_5)_2(C_4H_9)SnF$	Sn:	Org.Verb.5-40/1
–	$(C_4H_9)_2SnHF$	Sn:	Org.Verb.5-53
$C_8FH_{20}N_5O_3S_3$	$[NS(O)F][NS(O)N(C_2H_5)_2]_2$	S:	S-N-Verb.1-14
$C_8F_2FeH_{10}NO_4P$	$(CO)_4FePF_2N(C_2H_5)_2$	Fe:	Org.Verb.B2-85, 94, 115
$C_8F_2H_5NO_2$	$C_6H_5N(COF)_2$	C:	MVol.D3-46
$C_8F_2H_6Ni^+$	$C_5H_5NiC_3F_2H^+$	Ni:	Org.Verb.2-359
$C_8F_2H_6O_2$	$C_6H_5COOCF_2H$	F:	PerFHalOrg.5-18
$C_8F_2H_9NO_3$	$(CH_3O)_3C_5F_2N$	F:	PerFHalOrg.5-135
$C_8F_2H_{10}NNiO_3P$	$(CO)_3NiP(NC_5H_{10})F_2$	Ni:	Org.Verb.1-163
$C_8F_2H_{10}N_2O_2$	$NC(OC_2H_5)NC(OC_2H_5)CFCF$	F:	PerFHalOrg.6-58
$C_8F_2H_{10}N_2O_4$	$N(C(OCH_2CH_2OH))_2NCFCF$	F:	PerFHalOrg.6-58, 64
$C_8F_2H_{10}Sn$	$(C_2H_5)(C_6H_5)SnF_2$	Sn:	Org.Verb.5-50
$C_8F_2H_{12}N_4$	$((CH_3)_2N)_2C_4F_2N_2$	F:	PerFHalOrg.6-59
–	$FC(NCN(CH_3)_2)_2CF$	F:	PerFHalOrg.6-59, 66
$C_8F_2H_{12}Ni$	$(CH_3)_4C_4NiF_2$	Ni:	Org.Verb.2-329
$C_8F_2H_{18}N_2NiO_2P_2$	$(CO)_2Ni(P(CH_3)(N(CH_3)_2)F)_2$	Ni:	Org.Verb.1-127, 128
$C_8F_2H_{18}O_6S_2Sn$	$(C_4H_9)_2Sn(SO_3F)_2$	Sn:	Org.Verb.6-89, 95
$C_8F_2H_{18}S$	$(C_4H_9)_2SF_2$	S:	SVol.2-91
$C_8F_2H_{18}Sn$	$(C_4H_9)_2SnF_2$	Sn:	Org.Verb.5-47/8
–	$(i\text{-}C_4H_9)_2SnF_2$	Sn:	Org.Verb.5-49
–	$(t\text{-}C_4H_9)_2SnF_2$	Sn:	Org.Verb.5-49
$C_8F_2H_{24}N_6O_{10}P_2U$	$UO_2(NO_3)_2 \cdot 2\ [(CH_3)_2N]_2FPO$	U:	SVol.E1-166, 168
$C_8F_2H_{24}Ni_2P_2$	$(CH_3Ni(P(CH_3)_3)F)_2$	Ni:	Org.Verb.2-250
$C_8F_2MnN_3O_5$	$C_3N_3F_2Mn(CO)_5$	C:	MVol.D3-272
$C_8F_2N_3O_5Re$	$C_3N_3F_2Re(CO)_5$	C:	MVol.D3-272
$C_8F_3FeH_4O_4$	$(Fe(CO)_4C_4H_4F_3)_n$	Fe:	Org.Verb.B3-171
$C_8F_3FeH_5O_5$	$C_3H_5Fe(CO)_3OCOCF_3$	Fe:	Org.Verb.B5-39/40, 48, 51, 59/60
$C_8F_3FeH_{10}O_4PSi$	$(CO)_4FeP(CH_3)_2CH_2CH_2SiF_3$	Fe:	Org.Verb.B2-89, 110
$C_8F_3GeH_{15}$	$(C_2H_5)_3GeCFCF_2$	F:	PerFHalOrg.4-72
$C_8F_3GeH_{19}Sn$	$(CH_3)_3SnCF[Ge(CH_3)_3]CHF_2$	Sn:	Org.Verb.2-35, 37, 39
–	$(CH_3)_3SnCF_2CHFGe(CH_3)_3$	Sn:	Org.Verb.2-35, 37
$C_8F_3H_4O_2PuS^{3+}$	$Pu(SC_4H_3COCHCOCF_3)^{3+}$	Np:	TrU.D1-167/8
$C_8F_3H_5N_2O_2$	$CF_3NNC_6H_4COOH$	F:	PerFHalOrg.7-158/9, 160
$C_8F_3H_5NiO$	$C_5H_5Ni(CO)CFCF_2$	Ni:	Org.Verb.2-168
$C_8F_3H_7N_2$	$CF_3NNC_6H_4CH_3$	F:	PerFHalOrg.7-158/9, 160
$C_8F_3H_7N_2O$	$CF_3C(NOH)NHC_6H_5$	F:	PerFHalOrg.7-146
–	$CF_3NNC_6H_4OCH_3$	F:	PerFHalOrg.7-158/9
$C_8F_3H_7O_5PaS^+$	$Pa(OH)_3(C_4H_3SCOCHCOCF_3)^+$	Pa:	SVol.2-178
$C_8F_3H_8N_3O_2$	$(NO_2)(N(CH_3)_2)C_6F_3NH_2$	F:	PerFHalOrg.7-60
$C_8F_3H_8O_6PaS$	$Pa(OH)_4(C_4H_3SCOCHCOCF_3)$	Pa:	SVol.2-178/9
$C_8F_3H_9N_2$	$NCFCFNCFC(C_4H_9)$	F:	PerFHalOrg.6-60, 68
$C_8F_3H_9N_2O$	$NCFCFNCFC(OC(CH_3)_3)$	F:	PerFHalOrg.6-58, 64
$C_8F_3H_{10}N_3$	$NCFCFC(NHC_4H_9)CFN$	F:	PerFHalOrg.6-59, 65

$C_8F_3H_{10}N_3$	$NCFC(N(C_2H_5)_2)CFCFN$	F:	PerFHalOrg.6-59, 64
$C_8F_3H_{11}O$	$(CH_2)_5C(CFCF_2)OH$	F:	PerFHalOrg.4-45
$C_8F_3H_{12}N_5O_4$	$(CF_3)(N(CH_2OH)_2)_2C_3N_3$	F:	PerFHalOrg.6-112
$C_8F_3H_{12}NaO$	$(CH_3)_3CC(CH_3)(CFCF_2)ONa$	F:	PerFHalOrg.4-1
$C_8F_3H_{13}N_2O_2$	$C_2H_5OCOCH(NNCF_3)CH(CH_3)_2$	F:	PerFHalOrg.7-160
$C_8F_3H_{13}O$	$(CH_3)_3CC(CH_3)(CFCF_2)OH$	F:	PerFHalOrg.4-1
$C_8F_3H_{15}O_3Si$	$(C_2H_5O)_3SiCFCF_2$	F:	PerFHalOrg.4-72
$C_8F_3H_{15}Si$	$CF_2CFSi(C_2H_5)_3$	F:	PerFHalOrg.4-19, 72
$C_8F_3H_{15}Sn$	$(C_2H_5)_3SnCFCF_2$	F:	PerFHalOrg.4-28, 73
		Sn:	Org.Verb.2-192, 193
$C_8F_3H_{18}N_3OSSi_2$	$[(CH_3)_3SiN]_2SNC(O)CF_3$	S:	S-N-Verb.1-206
$C_8F_3H_{18}Sn^-$	$[(C_4H_9)_2SnF_3]^-$	Sn:	Org.Verb.5-47
$C_8F_3H_{19}SiSn$	$(CH_3)_3SnCF[Si(CH_3)_3]CHF_2$	Sn:	Org.Verb.2-37, 39
–	$(CH_3)_3SnCF_2CHFSi(CH_3)_3$	Sn:	Org.Verb.2-37, 39
$C_8F_4Fe_2H_3NO_7P_2$	$(CO)_7Fe_2(PF_2N(CH_3)PF_2)$	Fe:	Org.Verb.C1-200/2
$C_8F_4Fe_2O_6S_2$	$(CO)_3Fe(SC_2F_4S)Fe(CO)_3$	Fe:	Org.Verb.C1-115/6, 120
$C_8F_4Fe_2O_9P_2$	$(CO)_4Fe(F_2POPF_2)Fe(CO)_4$	Fe:	Org.Verb.C1-35, 40
$C_8F_4HNO_2$	$C_6F_4C(O)NHC(O)$	F:	PerFHalOrg.6-124, 138, 159
$C_8F_4H_2N_2$	$C_6F_4NCHCHN$	F:	PerFHalOrg.6-157, 158
$C_8F_4H_2N_2O_2$	$C_6F_4C(O)NHC(O)NH$	F:	PerFHalOrg.6-134, 148
–	$C_6F_4NHC(O)C(O)NH$	F:	PerFHalOrg.6-133/4, 147, 160
$C_8F_4H_2O_4$	$C_6F_4(COOH)_2$	F:	PerFHalOrg.4-103
$C_8F_4H_3NO$	$C_6F_4NC(CH_3)O$	F:	PerFHalOrg.7-60
$C_8F_4H_3N_3O$	$C_6F_4N(COCH_3)NN$	F:	PerFHalOrg.6-160
$C_8F_4H_4N_2$	$C_6F_4N(CH_3)CHN$	F:	PerFHalOrg.7-61
$C_8F_4H_4N_2O_2$	$C_6F_4[C(O)NH_2]_2$	F:	PerFHalOrg.7-19, 22, 48
$C_8F_4H_4NiO_4P_2$	$[(F_2POC_6H_4OPF_2)Ni(CO)_2]_x$	Ni:	Org.Verb.1-245
$C_8F_4H_4O_2$	$CH_3COOC_6F_4H$	F:	PerFHalOrg.4-53
$C_8F_4H_5N$	$CH_3CHCHC_5F_4N$	F:	PerFHalOrg.5-142
$C_8F_4H_5NO_2$	$NH_2C_6F_4C(O)OCH_3$	F:	PerFHalOrg.7-61
$C_8F_4H_5N_3O_2$	$O_2NCF_2CF_2NNC_6H_5$	F:	PerFHalOrg.7-192
$C_8F_4H_6N_2O$	$(CH_3)_2NC_6F_4NO$	F:	PerFHalOrg.7-193
$C_8F_4H_6N_2O_2$	$NO_2C_6F_4N(CH_3)_2$	F:	PerFHalOrg.7-60/1
$C_8F_4H_6N_6$	$C_6F_4(NC(NHNH_2))_2$	F:	PerFHalOrg.6-133/4, 147, 158
$C_8F_4H_6S_2$	$CH_3SC_6F_4SCH_3$	F:	PerFHalOrg.4-31
$C_8F_4H_7NO$	$(CH_3)_2CHOC_5F_4N$	F:	PerFHalOrg.5-143
$C_8F_4H_8N_2$	$(CH_3)_2NC_6F_4NH_2$	F:	PerFHalOrg.7-59
$C_8F_4H_8Si$	$HC_6F_4SiH(CH_3)_2$	F:	PerFHalOrg.4-24
$C_8F_4H_9NSi$	$(CH_3)_3SiC_5F_4N$	F:	PerFHalOrg.4-77
		F:	PerFHalOrg.5-143
$C_8F_4H_{10}NiO_2Si_2$	$(CO)_2Ni(SiF_2C(C_4H_9)CHSiF_2)$	Ni:	Org.Verb.1-147
$C_8F_4H_{10}O_4$	$C_2H_4(CH_2COOCHF_2)_2$	F:	PerFHalOrg.5-19
$C_8F_4H_{10}O_5$	$O(CF_2COOC_2H_5)_2$	F:	PerFHalOrg.6-12
$C_8F_4H_{14}NiO_4P_2$	$(CO)_2Ni(P(OC_3H_7\text{-}n)F_2)_2$	Ni:	Org.Verb.1-139
$C_8F_4H_{16}O_2Sn$	$SnF_4 \cdot 2\ C_4H_8O$	Sn:	MVol.C5-156
$C_8F_4H_{18}O_4P_2Sn$	$(C_4H_9)_2Sn(PO_2F_2)_2$	Sn:	Org.Verb.6-89, 95
$C_8F_4H_{20}N_3S_2^+$	$SF_4NS(N(C_2H_5)_2)_2^+$	S:	S-N-Verb.1-142

$C_8F_4H_{20}O_2Sn$	$SnF_4 \cdot 2\ C_2H_5OC_2H_5$	Sn:	MVol.C5-148
$C_8F_4H_{20}O_4Sn$	$SnF_4 \cdot 2\ C_2H_4(OCH_3)_2$	Sn:	MVol.C5-153/4
$C_8F_4Mn_2N_2O_{10}S_2$	$[Mn(CO)_4NS(O)F_2]_2$	S:	S-N-Verb.1-62
$C_8F_4N_2O_{10}Re_2S_2$	$[Re(CO)_4NS(O)F_2]_2$	S:	S-N-Verb.1-61
$C_8F_5HN_2O$	$C_6F_4CFNNC(OH)$	F:	PerFHalOrg.6-134/5, 148
–	$C_6F_4C(OH)CFNN$	F:	PerFHalOrg.6-134, 147
$C_8F_5H_2N_3$	$C_6F_4CFNNC(NH_2)$	F:	PerFHalOrg.6-134/5, 148
–	$C_6F_4C(NH_2)CFNN$	F:	PerFHalOrg.6-134, 147
–	$C_6F_4C(NH_2)NCFN$	F:	PerFHalOrg.6-134, 148
$C_8F_5H_3N_2O_2$	$C_6F_5C(O)NHC(O)NH_2$	F:	PerFHalOrg.7-24, 50
$C_8F_5H_3O$	$C_6F_5COCH_3$	F:	PerFHalOrg.4-99
$C_8F_5H_3O_2$	$C_6F_5COOCH_3$	F:	PerFHalOrg.4-98
$C_8F_5H_5NOP$	$CF_3CONP(F_2)C_6H_5$	F:	PerFHalOrg.7-63
$C_8F_5H_5N_2$	$(CH_3CHCH)(CF_3)C_4F_2N_2$	F:	PerFHalOrg.6-60, 68
$C_8F_5H_5NiO$	$C_5H_5Ni(CO)C_2F_5$	Ni:	Org.Verb.2-167
$C_8F_5H_5O$	$C_6F_5CH(CH_3)OH$	F:	PerFHalOrg.4-99, 101
–	$C_6F_5CH_2CH_2OH$	F:	PerFHalOrg.4-107
$C_8F_5H_6N$	$C_6F_5N(CH_3)_2$	F:	PerFHalOrg.7-59
$C_8F_5H_6NO$	$C_6F_5NHCH_2CH_2OH$	F:	PerFHalOrg.7-59
$C_8F_5H_6NO_2$	$(CH_3O)_2(CF_3)C_5F_2N$	F:	PerFHalOrg.5-177
$C_8F_5H_6N_5$	$C_6F_5(NHCNH)_2NH_2 \cdot HCl$	F:	PerFHalOrg.7-3/4
$C_8F_5H_6O_2P$	$C_6F_5P(OCH_3)_2$	F:	PerFHalOrg.3-129, 133
$C_8F_5H_6P$	$C_6F_5P(CH_3)_2$	F:	PerFHalOrg.3-129, 130
$C_8F_5H_7Si$	$C_6F_5Si(CH_3)_2H$	F:	PerFHalOrg.4-23, 24, 80
$C_8F_5H_8NO_2S$	$SF_5NHCOOCH_2C_6H_5$	S:	S-N-Verb.1-149
$C_8F_5H_8N_2P$	$C_6F_5P(NHCH_3)_2$	F:	PerFHalOrg.3-125, 131
$C_8F_5H_{10}N_5O_3$	$(C_2F_5)(N(CH_2OH)_2)(NHCH_2OH)C_3N_3$	F:	PerFHalOrg.6-112
$C_8F_5H_{15}Sn$	$(C_2H_5)_3SnC_2F_5$	Sn:	Org.Verb.2-163, 164
$C_8F_5H_{20}NTe$	$[(C_2H_5)_4N]TeF_5$	Te:	SVol.B2-15
$C_8F_5H_{20}N_3S_2$	$SF_5NS[N(C_2H_5)_2]_2$	S:	S-N-Verb.1-142
$C_8F_5H_{20}N_3S_2^+$	$SF_5NS(N(C_2H_5)_2)_2^+$	S:	S-N-Verb.1-142
C_8F_5Li	C_6F_5CCLi	F:	PerFHalOrg.4-2, 33
$C_8F_5N_2O_2P$	$C_6F_5P(NCO)_2$	F:	PerFHalOrg.3-60/1, 66, 68
$C_8F_5N_2P$	$C_6F_5P(CN)_2$	F:	PerFHalOrg.3-165, 170, 173/4
$C_8F_5N_2PS$	$C_6F_5P(S)(CN)_2$	F:	PerFHalOrg.3-69
$C_8F_5N_2PS_2$	$C_6F_5P(NCS)_2$	F:	PerFHalOrg.3-61, 66, 68/9
$C_8F_5N_2PS_3$	$C_6F_5P(S)(NCS)_2$	F:	PerFHalOrg.3-69
$C_8F_6FeH_7N_2O_4PS$	$[(CH_3)_2C_3HNSFe(CO)_3NO][PF_6]$	Fe:	Org.Verb.B4-133
$C_8F_6FeH_{18}NO_9P_3$	$[(CO)_2Fe(P(OCH_3)_3)_2NO]PF_6$	Fe:	Org.Verb.B1-130/3
$C_8F_6FeO_4$	$(CFCF_2)_2Fe(CO)_4$	Fe:	Org.Verb.B4-336, 345
–	$(CF_2CF_2CFCF)Fe(CO)_4$	Fe:	Org.Verb.B4-336, 345
$C_8F_6Fe_2O_6S_2$	$(CO)_3Fe(SCF_3)_2Fe(CO)_3$	Fe:	Org.Verb.C1-81, 86
$C_8F_6Fe_2O_6S_3$	$(CO)_3Fe(SCF_3)_2SFe(CO)_3$	Fe:	Org.Verb.C1-23/4
$C_8F_6Fe_2O_6Se_2$	$(CO)_3Fe(SeCF_3)_2Fe(CO)_3$	Fe:	Org.Verb.C1-83, 90
$C_8F_6GeH_6$	$(CF_3CC)_2Ge(CH_3)_2$	F:	PerFHalOrg.4-74
$C_8F_6H_4O$	$(CF_3CC)_2C(OH)CH_3$	F:	PerFHalOrg.4-47
$C_8F_6H_5P$	$(CF_3)_2PC_6H_5$	F:	PerFHalOrg.3-101, 105

Formula	Compound		Reference
$C_8F_6H_6NP$	$(CF_3)_2PNHC_6H_5$	F:	PerFHalOrg.3-108, 114
$C_8F_6H_6Si$	$(CF_3CC)_2Si(CH_3)_2$	F:	PerFHalOrg.4-74, 88
$C_8F_6H_6Sn$	$(CF_3CC)_2Sn(CH_3)_2$	F:	PerFHalOrg.4-74
		Sn:	Org.Verb.3-7, 16
–	$(CH_2CH)_2Sn(CFCF_2)_2$	F:	PerFHalOrg.4-72
		Sn:	Org.Verb.3-71
$C_8F_6H_7NO_4$	$(CF_3)_2CNOCOCH(CH_3)C(O)OCH_3$	F:	PerFHalOrg.7-141/2
$C_8F_6H_8NO_2P$	$(CF_3)_2P(O)OH \cdot C_6H_5NH_2$	F:	PerFHalOrg.3-50
$C_8F_6H_8N_2O_2S$	$SF(CF_3)_2NHC(O)CH(NHCOCH_3)CH_2$	F:	PerFHalOrg.7-66, 68
$C_8F_6H_9NO_3$	$(CF_3)_2CNOC(O)OCH(C_2H_5)CH_3$	F:	PerFHalOrg.7-141/2
$C_8F_6H_9NO_4$	$(CH_3OCO)_2CHC(CF_3)_2NH_2$	F:	PerFHalOrg.7-73
$C_8F_6H_{10}NP$	$(CF_2CF)_2PN(C_2H_5)_2$	F:	PerFHalOrg.4-106
$C_8F_6H_{10}N_2O_3S$	$[NS(O)N(C_2H_5)_2][OC(CF_3)_2CO]$	S:	S-N-Verb.1-41
$C_8F_6H_{10}Sn$	$(C_2H_5)_2Sn(CFCF_2)_2$	F:	PerFHalOrg.4-73
		Sn:	Org.Verb.3-26
$C_8F_6H_{11}N$	$(CH_3)_3CCH_2NC(CF_3)_2$	F:	PerFHalOrg.5-26
$C_8F_6H_{11}N_2O_3S$	$(C_2H_5)_2NHS(O)NC(O)C(CF_3)_2O$	F:	PerFHalOrg.5-98/9
$C_8F_6H_{12}$	$(CH_3)_3CCH_2C(CF_3)_2H$	F:	PerFHalOrg.5-26
$C_8F_6H_{12}N_4$	$CF_3NN(CH_2)_6NNCF_3$	F:	PerFHalOrg.7-159
$C_8F_6H_{12}OSi_2$	$[CF_2CFSi(CH_3)_2]_2O$	F:	PerFHalOrg.4-19
$C_8F_6H_{12}Sn$	$(CH_3)_3SnCH(CF_2)_3CH_2$	Sn:	Org.Verb.2-61
$C_8F_6H_{14}O_6S_2Sn$	$(C_3H_7)_2Sn(SO_3CF_3)_2$	Sn:	Org.Verb.6-68
$C_8F_6H_{14}Sn$	$(CH_3)_2Sn(CH_2CH_2CF_3)_2$	Sn:	Org.Verb.3-7, 17
$C_8F_6H_{16}N_2O$	$(C_2H_5)_3N \cdot (CF_3)_2NOH$	F:	PerFHalOrg.7-140
$C_8F_6H_{24}IrN_2$	$[(CH_3)_4N]_2[IrF_6]$	Ir:	SVol.2-108
$C_8F_6N_2$	$C(CFNCFCF)_2C$	F:	PerFHalOrg.6-135, 148
–	$C(NCFCFCF)_2C$	F:	PerFHalOrg.6-134/5, 148
–	$C_6F_4CFCFNN$	F:	PerFHalOrg.6-134, 147, 158
–	$C_6F_4CFNCFN$	F:	PerFHalOrg.6-134, 147, 158, 161
–	$C_6F_4CFNNCF$	F:	PerFHalOrg.6-134/5, 148, 158, 161
–	$C_6F_4NCFCFN$	F:	PerFHalOrg.6-133/4, 146, 156/7, 161
$C_8F_6N_4$	$FCN(CF)_2NCCN(CF)_2NCF$	F:	PerFHalOrg.6-31/2, 52
–	$NCFN(CF)_2CC(CF)_2NCFN$	F:	PerFHalOrg.6-24, 46
$C_8F_6N_6O_2S_2$	$[NC(CF_3)SC(NCO)N]_2$	F:	PerFHalOrg.5-67/8, 107
$C_8F_7GeH_9$	$(CH_3)_3GeCCCF(CF_3)_2$	F:	PerFHalOrg.4-76
$C_8F_7HN_2$	$C_6F_4NC(CF_3)NH$	F:	PerFHalOrg.6-124/5, 139
$C_8F_7HN_2O_3$	$(NO_2C_6F_4)NHC(O)CF_3$	F:	PerFHalOrg.7-23/4, 49
$C_8F_7H_7N_2O_2$	$C_3F_7C(NH_2)NOC(O)C(CH_3)CH_2$	F:	PerFHalOrg.7-58, 76
$C_8F_7H_9Sn$	$(CH_3)_3SnCCF_2CC_2F_5$	Sn:	Org.Verb.2-57/8
$C_8F_7H_{11}Si$	$CH_3(C_2H_5)Si(CHCH_2)C_3F_7$	F:	PerFHalOrg.4-18
$C_8F_7H_{11}Sn$	$(CH_3)_3SnCF(CF_2)_3CH_2$	Sn:	Org.Verb.2-61
C_8F_7I	C_6F_5CFCFI	F:	PerFHalOrg.4-17
$C_8F_7N_5$	$(i\text{-}C_3F_7)(NC)_2C_3N_3$	F:	PerFHalOrg.6-79/80, 101
C_8F_8	$C_6F_5CFCF_2$	F:	PerFHalOrg.4-38, 111
$C_8F_8FeH_2O_4$	$(CO)_4Fe(CF_2CF_2H)_2$	Fe:	Org.Verb.B2-192/3
$C_8F_8FeNO_5P$	$(CO)_4FePF_2(OC(CF_3)_2CN)$	Fe:	Org.Verb.B2-94, 115

Formula	Compound		Reference
$C_8F_8FeO_4$	$(CF_2)_4Fe(CO)_4$	Fe:	Org.Verb.B4-336, 344/5
–	$CF_3CFCFCF_3Fe(CO)_4$	Fe:	Org.Verb.B4-320, 325
–	$[(CO)_4FeC_4F_8]_n$	Fe:	Org.Verb.B2-31
$C_8F_8Fe_2H_6N_2O_6P_4$	$(CO)_3Fe(PF_2NCH_3PF_2)_2Fe(CO)_3$	Fe:	Org.Verb.C1-29/30
C_8F_8HNO	$C_6F_5NHC(O)CF_3$	F:	PerFHalOrg.7-23/4, 49
$C_8F_8H_2O$	$C_6F_5CH(OH)CF_3$	F:	PerFHalOrg.4-99, 101
$C_8F_8H_3NO$	$(CH_3O)(CF_3)_2C_5F_2N$	F:	PerFHalOrg.5-177
$C_8F_8H_6N_8$	$C_2F_4(CF_2CNNHC(NH_2)N)_2$	F:	PerFHalOrg.5-72/3, 93
$C_8F_8H_{10}N_8O_2$	$(CF_2)_4(C(O)NHNHC(NH)NH_2)_2$	F:	PerFHalOrg.7-6/7, 34
$C_8F_8H_{10}O_2$	$(CH_3CH(OH)CF_2)_2C_2F_4$	F:	PerFHalOrg.4-42
$C_8F_8H_{16}O_4Te_2$	$[TeF_3(C_4H_8O_2)_2][TeF_5]$	Te:	SVol.B3-166
$C_8F_8H_{24}N_4NiP_4$	$Ni(P(N(CH_3)_2)F_2)_4$	Ni:	Org.Verb.2-278
$C_8F_8Hg_2$	$Hg(C_4F_4)_2Hg$	F:	PerFHalOrg.4-33
C_8F_8O	$C_6F_5COCF_3$	F:	PerFHalOrg.4-99, 112
$C_8F_9HN_2$	$(CF_3)C_7F_6HN_2$	F:	PerFHalOrg.6-123, 138
$C_8F_9H_2N$	$(CF_3)_2C_6F_3NH_2$	F:	PerFHalOrg.7-10, 37, 60
–	$C_2F_5C_6F_4NH_2$	F:	PerFHalOrg.7-10, 37, 60
$C_8F_9H_3N_4$	$(CF_3)_3(CH_3)C_4N_4$	F:	PerFHalOrg.6-156, 159
$C_8F_9H_5NSb$	$(CF_3)_3Sb \cdot C_5H_5N$	F:	PerFHalOrg.3-231
$C_8F_9H_5O$	$FC(CF_2)_4CCH(OH)CH_3$	F:	PerFHalOrg.4-44
$C_8F_9H_6NO$	$(CF_3)_3CNC_4H_6O$	F:	PerFHalOrg.7-194/5
$C_8F_9H_9NO_3Sb$	$[C_5H_5NH][(CF_3)_3Sb(OH)_3]$	F:	PerFHalOrg.3-230
C_8F_9N	$NC_5F_3(CF_2)_3$	F:	PerFHalOrg.6-128/9, 140
C_8F_9NO	$[N(C_6F_5)OCF_2CF_2]_n$	F:	PerFHalOrg.7-208/9
$C_8F_9N_7$	$[(CF_3)_2CF](N_3)_2C_5F_2N$	F:	PerFHalOrg.5-185, 200
$C_8F_{10}FeO_4$	$(CO)_4Fe(C_2F_5)_2$	Fe:	Org.Verb.B2-191, 193
$C_8F_{10}H_2N_2$	$[(CF_3)_2CF](NH_2)C_5F_3N$	F:	PerFHalOrg.5-186, 203
$C_8F_{10}H_4N_4O$	$NC(C_2F_5)N(NHCOCH_3)C(C_2F_5)N$	F:	PerFHalOrg.5-101
$C_8F_{10}H_{10}Sn$	$(C_2H_5)_2Sn(C_2F_5)_2$	Sn:	Org.Verb.3-26
$C_8F_{10}H_{18}N_2Te_2$	$[CH_3N(C_2H_4)_3NCH_3][TeF_5]_2$	Te:	SVol.B2-15
$C_8F_{10}NO$	$(C_6F_5)(C_2F_5)NO$	F:	PerFHalOrg.7-86, 122/3
$C_8F_{10}N_2$	$(CF_3)_2C_6F_4N_2$	F:	PerFHalOrg.6-128/9, 142
–	$NCFC(C(CFCF_3)CF_3)CFCFN$	F:	PerFHalOrg.6-17, 37
–	$NCFC(C_2F_5)C(CFCF_2)CFN$	F:	PerFHalOrg.6-17, 38
–	$NCFNC(C(CF_3)CFCF_3)CFCF$	F:	PerFHalOrg.6-24, 46
$C_8F_{10}N_4$	$[(CF_3)_2CF](N_3)C_5F_3N$	F:	PerFHalOrg.5-185, 200, 211, 213/4
$C_8F_{11}H_3O$	$C_7F_{11}CH_2OH$	F:	PerFHalOrg.4-43
$C_8F_{11}H_9OSi$	$(CH_3)_3Si(CF_2)_2OCF(CF_3)_2$	F:	PerFHalOrg.4-86
$C_8F_{11}N$	$(CF_3)_2CFC_5F_4N$	F:	PerFHalOrg.5-151/3, 162, 173/4, 177/8
$C_8F_{12}Fe$	$Fe(CF_2CFCFCF_2)_2$	Fe:	Org.Verb.B3-171
$C_8F_{12}FeH_8P_4$	$C_6H_5CHCH_2Fe(PF_3)_4$	Fe:	Org.Verb.B4-178, 182
$C_8F_{12}FeO_4P_2S$	$(CO)_4FeP(CF_3)_2SP(CF_3)_2$	Fe:	Org.Verb.B2-97, 118
$C_8F_{12}FeO_5P_2$	$(CO)_4FeP(CF_3)_2OP(CF_3)_2$	Fe:	Org.Verb.B2-97, 118
$C_8F_{12}Ge$	$(CF_2CF)_4Ge$	F:	PerFHalOrg.4-168/9, 177
$C_8F_{12}GeH_6N_2$	$(CH_3)_2Ge[NC(CF_3)_2]_2$	F:	PerFHalOrg.7-64
$C_8F_{12}HNO_2$	$C(CF_3)_2CONHC(CF_2)_3O$	F:	PerFHalOrg.5-58/9, 80
–	$C(CF_3)_2NHCOC(CF_2)_3O$	F:	PerFHalOrg.5-58/9, 80
$C_8F_{12}HN_3O$	$OCNCNC(CF_3)_2NHC(CF_3)_2$	F:	PerFHalOrg.5-63/4, 86/7, 108/10

Formula	Compound		Reference
$C_8F_{12}H_2MoO_4P_2$	$Mo(CO)_4[(CF_3)_2PH]_2$	F:	PerFHalOrg.3-27
$C_8F_{12}H_2N_2O$	$(CF_2)_3COC(CF_3)_2NC(NH_2)$	F:	PerFHalOrg.5-59, 81
$C_8F_{12}H_2N_8$	$NHNNNC(CF_2)_6CNNNNH$	F:	PerFHalOrg.5-73/4, 93
$C_8F_{12}H_2O_4$	$HOCO(CF_2)_6COOH$	F:	PerFHalOrg.4-103
$C_8F_{12}H_3I_2N_3$	$I(CF_2)_3C(NH)NC(NH_2)(CF_2)_3I$	F:	PerFHalOrg.7-7, 34
$C_8F_{12}H_3NO_2$	$(CF_3)_2CC(OCH_3)NC(CF_3)_2O$	F:	PerFHalOrg.5-101/2
–	$(CF_3)_2CC(O)N(CH_3)C(CF_3)_2O$	F:	PerFHalOrg.5-101/2
$C_8F_{12}H_4N_2O$	$(CF_3)_2COC(CF_3)_2NC(NHCH_3)$	F:	PerFHalOrg.5-102
–	$(CF_3)_2CC(O)N(CH_3)C(CF_3)_2NH$	F:	PerFHalOrg.5-102/3
$C_8F_{12}H_4N_4O_2$	$(CF_3)_2CN(CH_3)C(CF_3)_2NC(NHNO_2)$	F:	PerFHalOrg.5-108
$C_8F_{12}H_4NiO_2P_2$	$(CO)_2Ni(CF_3)_2P(CH_2)_2P(CF_3)_2$	Ni:	Org.Verb.1-149, 152
–	$[(CF_3)_2PC_2H_4P(CF_3)_2Ni(CO)_2]_x$	Ni:	Org.Verb.2-395
$C_8F_{12}H_5N_3$	$(CF_3)_2CNHC(CF_3)_2N(CH_3)C(NH)$	F:	PerFHalOrg.5-104
–	$(CF_3)_2CNHC(CF_3)_2NC(NHCH_3)$	F:	PerFHalOrg.5-103/4
$C_8F_{12}H_5N_3O$	$(CF_3)_2CNHC(CF_3)_2NC(NHCH_2OH)$	F:	PerFHalOrg.5-106
$C_8F_{12}H_5N_4O_2$	$(CF_3)_2CNHC(CF_3)_2NHC(NN(O)OCH_3)$	F:	PerFHalOrg.5-108
$C_8F_{12}H_6N_2O_3$	$(CF_3)_2NOC(CH_3)_2C(O)ON(CF_3)_2$	F:	PerFHalOrg.7-130/1
$C_8F_{12}H_6N_2Si$	$(CH_3)_2Si[NC(CF_3)_2]_2$	F:	PerFHalOrg.7-64
$C_8F_{12}H_6N_2Sn$	$(CH_3)_2Sn[NC(CF_3)_2]_2$	F:	PerFHalOrg.7-64
		Sn:	Org.Verb.6-32
$C_8F_{12}H_6N_4O_2$	$NH_2(NH)CCF(CF_3)O(CF_2)_3(OCF_2)C(NH)NH_2$	F:	PerFHalOrg.7-5
$C_8F_{12}H_6NiO_2P_2$	$(CO)_2Ni(P(CF_3)_2CH_3)_2$	Ni:	Org.Verb.1-128
$C_8F_{12}H_8N_2O_2$	$(CF_3)_2NOC(CH_3)_2CH_2ON(CF_3)_2$	F:	PerFHalOrg.7-129, 131, 133
–	$(CF_3)_2NOCH_2C(CH_3)_2ON(CF_3)_2$	F:	PerFHalOrg.7-134/5
$C_8F_{12}H_8P_2$	$(CF_3)_2PCH(CH_3)CH(CH_3)P(CF_3)_2$	F:	PerFHalOrg.3-183
$C_8F_{12}H_{10}N_2O_4Si$	$[(CF_3)_2NOCH_2]_2Si(OCH_3)_2$	F:	PerFHalOrg.7-127
$C_8F_{12}H_{10}N_2O_5Si$	$[(CF_3)_2NOCH_2O]_2Si(CH_3)OCH_3$	F:	PerFHalOrg.7-127
$C_8F_{12}H_{10}Si_2$	$C_2F_4(CF_2CF_2SiH_2CH_3)_2$	F:	PerFHalOrg.4-87
$C_8F_{12}H_{12}N_4NiP_2$	$[(CH_3NC)_4Ni][PF_6]_2$	Ni:	Org.Verb.1-333
$C_8F_{12}H_{13}N_3O_2$	$(C_2H_5)_2NH \cdot [(CF_3)_2NOH]_2$	F:	PerFHalOrg.7-139
$C_8F_{12}ILi$	$IC(CF_2CF_2)_2(CFLiCF_2)CF$	F:	PerFHalOrg.4-4, 9, 15
$C_8F_{12}I_2MoO_4P_2$	$Mo(CO)_4[(CF_3)_2PI]_2$	F:	PerFHalOrg.3-110/1, 120
$C_8F_{12}N_2$	$(C_2F_5)_2C_4F_2N_2$	F:	PerFHalOrg.6-32/3, 53, 55/6
–	$(s\text{-}C_4F_9)C_4F_3N_2$	F:	PerFHalOrg.6-32/3, 53, 55/6
–	$NC(C(CF_3)_3)CFCFCFN$	F:	PerFHalOrg.6-17/8, 38
–	$NC(C_2F_5)CFNC(C_2F_5)CF$	F:	PerFHalOrg.6-30/2, 51
–	$NC(C_2F_5)CFNCFC(C_2F_5)$	F:	PerFHalOrg.6-30/2, 51/2
–	$NC(s\text{-}C_4F_9)CFNCFCF$	F:	PerFHalOrg.6-30/2, 51
–	$NCFC(C(CF_3)_3)CFCFN$	F:	PerFHalOrg.6-17/8, 38
–	$NCFC(C_2F_5)C(C_2F_5)CFN$	F:	PerFHalOrg.6-15/6, 35, 55
–	$NCFC(s\text{-}C_4F_9)CFCFN$	F:	PerFHalOrg.6-18, 38, 55/6
–	$NCFNC(C_2F_5)C(C_2F_5)CF$	F:	PerFHalOrg.6-22, 44
–	$NCFNC(s\text{-}C_4F_9)CFCF$	F:	PerFHalOrg.6-24, 46
$C_8F_{12}N_2O$	$[C_6F_{12}C_2N_2O]_n$	F:	PerFHalOrg.5-69/70
$C_8F_{12}N_4O_4$	$[CF_3NCOCONCF_3]_2$	F:	PerFHalOrg.6-166

Formula	Compound		Reference
$C_8F_{12}NiS_4^{2-}$	$[Ni(S_2C_2(CF_3)_2)_2]^{2-}$	Ni:	Org.Verb.2-114
$C_8F_{12}O_8Sn$	$Sn(CF_3COO)_4$	Sn:	MVol.C2-223
$C_8F_{12}Si$	$(CF_2CF)_4Si$	F:	PerFHalOrg.4-151, 158, 161
$C_8F_{12}Sn$	$(CF_2CF)_4Sn$	F:	PerFHalOrg.4-182, 187/90
		Sn:	Org.Verb.1-138
$C_8F_{13}H_2NO$	$C_6F_{11}CF_2C(O)NH_2$	F:	PerFHalOrg.7-19/21, 48
$C_8F_{13}H_9N_2O_2Si$	$(CH_3)_2SiFCH[ON(CF_3)_2]CH_2ON(CF_3)_2$	F:	PerFHalOrg.7-127/8
$C_8F_{13}I$	$CF_3C(CF_2)(CF_2CF_2)_2CI$	F:	PerFHalOrg.4-14
–	$IC(CF_2CF_2)_3CF$	F:	PerFHalOrg.4-14
$C_8F_{13}Li$	$CF_3C(CF_2CF_2)_2(CF_2)CLi$	F:	PerFHalOrg.4-4, 9, 13, 14, 16
–	$LiC(CF_2CF_2)_3CF$	F:	PerFHalOrg.4-4, 9, 14/6
$C_8F_{14}HNO_2$	$[C_3F_7C(O)]_2NH$	F:	PerFHalOrg.7-24, 50
$C_8F_{14}HN_3$	$NC(C_3F_7)NHC(C_3F_7)N$	F:	PerFHalOrg.5-71/2, 92, 97, 99
$C_8F_{14}H_2N_2O_2$	$C_3F_7C(NH_2)NOC(O)C_3F_7$	F:	PerFHalOrg.7-5/6, 32
–	$C_3F_7C(NOH)C(NOH)C_3F_7$	F:	PerFHalOrg.7-89, 105
$C_8F_{14}H_2N_4$	$NC(C_3F_7)NHNHC(C_3F_7)N$	F:	PerFHalOrg.6-122
–	$NC(C_3F_7)N(NH_2)C(C_3F_7)N$	F:	PerFHalOrg.5-71/2, 92, 101
$C_8F_{14}H_3N_3$	$C_3F_7C(NH)NC(NH_2)C_3F_7$	F:	PerFHalOrg.7-7, 34, 58
$C_8F_{14}H_3N_3O$	$C_3F_7C(NH)NC(NH_2)(CF_2)_2OCF_3$	F:	PerFHalOrg.7-7, 35
–	$C_3F_7C(NH_2)NNHCOC_3F_7$	F:	PerFHalOrg.7-6/7, 33, 58
$C_8F_{14}H_3N_3O_2$	$CF_3O(CF_2)_2C(NH)NC(NH_2)(CF_2)_2OCF_3$	F:	PerFHalOrg.7-7, 35
$C_8F_{14}H_4HgN_4$	$(C_3F_7C(NH)NH)_2Hg$	F:	PerFHalOrg.7-3, 31
$C_8F_{14}H_4N_2O_2$	$C_3F_7C(NH)NH_2 \cdot C_3F_7COOH$	F:	PerFHalOrg.7-3/4
$C_8F_{14}H_4O$	$(n\text{-}C_3F_7)_2C(CH_3)(OH)$	F:	PerFHalOrg.4-42, 93, 95
$C_8F_{14}H_5OP$	$(C_3F_7)_2POC_2H_5$	F:	PerFHalOrg.3-101, 103
$C_8F_{14}H_6NP$	$(C_3F_7)_2PN(CH_3)_2$	F:	PerFHalOrg.3-108, 115
$C_8F_{14}H_6Si$	$(CF_3CF_2CF_2)_2Si(CH_3)_2$	F:	PerFHalOrg.4-18, 19
$C_8F_{14}MoO_4P_2$	$Mo(CO)_4[(CF_3)_2PF]_2$	F:	PerFHalOrg.3-110/1, 121
$C_8F_{14}N_2O$	$C_3F_7CNC(C_3F_7)NO$	F:	PerFHalOrg.5-66, 88
–	$NC(C_3F_7)OC(C_3F_7)N$	F:	PerFHalOrg.5-64/5, 87
$C_8F_{14}N_2O_2$	$(CF_2ONCF_3)C(C(CF_2)CF_2ON(CF_3)CF_2)$	F:	PerFHalOrg.6-5
–	$(CF_2ONCF_3)C(C_2F_4C(CF_2)ONCF_3)$	F:	PerFHalOrg.6-5
–	$CF_3NCO(CF_2)_4CONCF_3$	F:	PerFHalOrg.6-166
–	$C_3F_7CNON(O)CC_3F_7$	F:	PerFHalOrg.5-66/7, 89
$C_8F_{14}N_2O_3$	$(CF_3NC(O)CF_2CF_2)_2O$	F:	PerFHalOrg.6-167
–	$O(CF_2CF_2)_2NOCNC_2F_4OCF_2$	F:	PerFHalOrg.6-2/3
$C_8F_{14}N_2S$	$NC(C_3F_7)SC(C_3F_7)N$	F:	PerFHalOrg.5-67/8, 89
$C_8F_{14}N_4$	$(C_3F_7)_2C_2N_4$	F:	PerFHalOrg.6-122
–	$NC_4F_7NNC_4F_7N$	F:	PerFHalOrg.6-24, 46
$C_8F_{15}HN_2O_2$	$(CF_3)_2CNOCF_2CFHONC(CF_3)_2$	F:	PerFHalOrg.7-146
$C_8F_{15}H_2NO$	$C_7F_{15}C(O)NH_2$	F:	PerFHalOrg.7-18, 20/1, 47, 76
$C_8F_{15}H_2N_3O$	$(CF_3)_2NCHC(CHN(CF_3)_2)N(CF_3)O$	F:	PerFHalOrg.7-163, 166
$C_8F_{15}H_3N_2$	$CF_3(CF_2)_6C(NH)NH_2$	F:	PerFHalOrg.7-3/4, 31
$C_8F_{15}H_3N_2O$	$C_7F_{15}C(NOH)NH_2$	F:	PerFHalOrg.7-5, 32, 58

Formula	Compound		Reference
$C_8F_{15}H_3NiO_2P_2$	$(CO)_2Ni(P(CF_3)_3)(P(CF_3)_2CH_3)$	Ni:	Org.Verb.1-145
$C_8F_{15}H_4N_3$	$C_7F_{15}C(NH_2)NNH_2$	F:	PerFHalOrg.7-6/7, 33
$C_8F_{15}N$	$NC(CF(CF_3)_2)(CF_2)_4$	F:	PerFHalOrg.5-151/3, 163
$C_8F_{15}NO$	$O(CF_2CF_2)_2NCF(CF_2)_3$	F:	PerFHalOrg.6-2/3, 7
C_8F_{16}	$C_6F_{13}CFCF_2$	F:	PerFHalOrg.4-68/9
$C_8F_{16}HgN_2O_4$	$(O(CF_2CF_2)_2NO)_2Hg$	F:	PerFHalOrg.6-2/3
$C_8F_{16}N_2$	$FN(CF_2CF_2)_2NCF(CF_2)_3$	F:	PerFHalOrg.6-30/2, 52
$C_8F_{16}N_2O_2$	$O(CF_2CF_2)_2NN(CF_2CF_2)_2O$	F:	PerFHalOrg.6-2/3, 7
$C_8F_{16}N_2O_3$	$O(CF_2CF_2)_2NOCF_2NCF_2CF_2OCF_2$	F:	PerFHalOrg.6-2/3, 8
–	$O(CF_2CF_2)_2NOCF_2OCF_2CFNCF_3$	F:	PerFHalOrg.6-2/3, 7
–	$O(CF_2CF_2)_2NOCF_2OC_2F_4NCF_2$	F:	PerFHalOrg.6-2/3, 8
–	$O(CF_2CF_2)_2NON(CF_2CF_2)_2O$	F:	PerFHalOrg.6-2/3, 7
$C_8F_{16}N_4$	$C_3F_3N_2(CF_3)_2(NCF_3)CFNCF_3$	F:	PerFHalOrg.5-61/2, 84
–	$C_3F_4N_2(CF_3)C_3F_6N_2(CF_3)$	F:	PerFHalOrg.5-62, 85
$C_8F_{16}NiO_2P_2$	$(CO)_2Ni(CF_3)_2P(CF_2)_2P(CF_3)_2$	F:	PerFHalOrg.3-174
		Ni:	Org.Verb.1-150, 153
$C_8F_{16}O_4S$	$SF_4 \cdot 2\ CF_3C(O)C(O)CF_3$	S:	SVol.2-81
$C_8F_{17}HO$	$(C_3F_7)_2C(OH)CF_3$	F:	PerFHalOrg.4-93, 95
$C_8F_{17}H_2O_3P$	$C_8F_{17}P(O)(OH)_2$	F:	PerFHalOrg.3-43, 46, 50
$C_8F_{17}IMg$	$(n\text{-}C_8F_{17})MgI$	F:	PerFHalOrg.4-68
$C_8F_{17}I_2P$	$C_8F_{17}PI_2$	F:	PerFHalOrg.3-76, 95
$C_8F_{17}N$	$CF_2N(C_3F_7)CF_2CF(CF_3)CF_2$	F:	PerFHalOrg.5-52/3
–	$(CF_2)_4NC(CF_3)_3$	F:	PerFHalOrg.5-52/3, 75
–	$(CF_2)_4NCF(CF_3)CF_2CF_3$	F:	PerFHalOrg.5-52/3, 75
–	$(CF_2)_4NCF_2CF(CF_3)_2$	F:	PerFHalOrg.5-52/3, 75
–	$(CF_2)_4NCF_2CF_2CF_2CF_3$	F:	PerFHalOrg.5-52/3, 75
–	$CF_3CF_2CF_2N(CF_2)_5$	F:	PerFHalOrg.5-157/8, 170, 175
–	$CF_3NC_5F_8(CF_3)_2$	F:	PerFHalOrg.5-158, 172
–	$(CF_3)_2CFN(CF_2)_5$	F:	PerFHalOrg.5-157/8, 170/1, 175
–	$FNC_5F_7(CF_3)_3$	F:	PerFHalOrg.5-150/1, 161
$C_8F_{17}NO$	$CF_3(CF_2)_7NO$	F:	PerFHalOrg.7-179, 185, 208/9
–	$O(CF_2CF_2)_2NC_4F_9$	F:	PerFHalOrg.6-2, 7
$C_8F_{17}NO_2$	$C_6F_{13}OCF_2CF_2NO$	F:	PerFHalOrg.7-179, 185
$C_8F_{18}HN_3O_4$	$[(CF_3)_2NO]_2CHC(O)ON(CF_3)_2$	F:	PerFHalOrg.7-135
$C_8F_{18}HO_2P$	$(C_4F_9)_2P(O)OH$	F:	PerFHalOrg.3-43, 50
$C_8F_{18}H_2P_2$	$(CF_3)_2PCH(CF_3)CH(CF_3)P(CF_3)_2$	F:	PerFHalOrg.3-29
$C_8F_{18}IP$	$(n\text{-}C_4F_9)_2PI$	F:	PerFHalOrg.3-76, 94
$C_8F_{18}N_2O_2$	$CF_3C(ON(CF_3)_2)C(ON(CF_3)_2)CF_3$	F:	PerFHalOrg.7-97/8, 115
–	$(CF_3)_2NOCFCF_2CF_2CFON(CF_3)_2$	F:	PerFHalOrg.7-95/6, 113, 145
–	$(CF_3)_2NOCF_2CFCFCF_2ON(CF_3)_2$	F:	PerFHalOrg.7-95/6, 114
–	$(CF_3)_2NOC(N(CF_3)_2)(CF_3)C(O)CF_3$	F:	PerFHalOrg.7-97/8, 116
$C_8F_{18}N_2O_3$	$[(CF_3)_2NO]_2C(CF_3)C(O)CF_3$	F:	PerFHalOrg.7-97/8, 115
$C_8F_{18}N_4O_3$	$[CF_3NC(ON(CF_3)_2)]_2O$	F:	PerFHalOrg.7-96
$C_8F_{18}NiO_2P_2$	$[(CF_3)_3P]_2Ni(CO)_2$	F:	PerFHalOrg.3-174
		Ni:	Org.Verb.1-128, 130
$C_8F_{20}N_2O_2$	$(CF_3)_2NOC(CF_3)_2CF_2ON(CF_3)_2$	F:	PerFHalOrg.7-95/6, 114

Formula	Compound		Reference
$C_8F_{20}P_4$	$(C_2F_5P)_4$	F:	PerFHalOrg.3-1/3
$C_8F_{22}HgN_4O_2$	$[(CF_3)_2NOCF_2N(CF_3)]_2Hg$	F:	PerFHalOrg.7-102, 124
$C_8F_{24}GeN_4O_4$	$[(CF_3)_2NO]_4Ge$	F:	PerFHalOrg.7-102, 119
$C_8F_{24}N_4O_4Se$	$[(CF_3)_2NO]_4Se$	F:	PerFHalOrg.7-92/3, 108, 126
$C_8F_{24}N_4O_4Si$	$[(CF_3)_2NO]_4Si$	F:	PerFHalOrg.7-100, 117
$C_8F_{24}N_4O_4Sn$	$[(CF_3)_2NO]_4Sn$	F:	PerFHalOrg.7-102
$C_8F_{24}N_4O_4Te$	$[(CF_3)_2NO]_4Te$	F:	PerFHalOrg.7-92/3, 108, 126
$C_8F_{24}N_4O_4Ti$	$[(CF_3)_2NO]_4Ti$	F:	PerFHalOrg.7-102
$C_8F_{24}N_4P_4$	$[(CF_3)_2PN]_4$	F:	PerFHalOrg.3-61/2, 66, 68
$C_8F_{24}N_8O_4S_4$	$[(CF_3)_2NO]_4S_4N_4$	F:	PerFHalOrg.7-92/3, 108, 121
$C_8F_{24}O_4P_4Si$	$[(CF_3)_2PO]_4Si$	F:	PerFHalOrg.3-40/2
$C_8F_{24}P_4Pt_2S_6$	$(CF_3)_2PS_2Pt(P(CF_3)_2S)_2PtS_2P(CF_3)_2$	F:	PerFHalOrg.3-159
$C_8F_{28}NiP_4$	$Ni[(CF_3)_2PF]_4$	F:	PerFHalOrg.3-110/1, 120
		Ni:	Org.Verb.1-236, 238
$C_8F_{28}P_4Pt$	$Pt[(CF_3)_2PF]_4$	F:	PerFHalOrg.3-110/1, 121
$C_8FeGeH_9NO_4$	$(CH_3)_3GeNCFe(CO)_4$	Fe:	Org.Verb.B4-6/7
$C_8FeH_2O_6$	$(CHCO)_2Fe(CO)_4$	Fe:	Org.Verb.B4-334, 340/1
$C_8FeH_2O_7$	$O(C(O)CH)_2Fe(CO)_4$	Fe:	Org.Verb.B4-302/3, 307/8
$C_8FeH_3NO_6$	$HN(C(O)CH)_2Fe(CO)_4$	Fe:	Org.Verb.B4-304, 309/10
$C_8FeH_4N_2O_4$	$(CO)_4Fe(N(CH)_2N(CH)_2)$	Fe:	Org.Verb.B2-71/2, 75, 77
–	$(CO)_4Fe(NN(CH)_4)$	Fe:	Org.Verb.B2-71, 76, 79
$C_8FeH_4NaO_7^-$	$[NaOCO(CH_2)_2C(O)Fe(CO)_4]^-$	Fe:	Org.Verb.B4-154, 162
$C_8FeH_4Na_2O_7$	$Na[NaOCO(CH_2)_2C(O)Fe(CO)_4]$	Fe:	Org.Verb.B4-154, 162
$C_8FeH_4O_6$	$(CH_2CO)_2Fe(CO)_4$	Fe:	Org.Verb.B4-341
$C_8FeH_4O_8$	$(HOCO)_2C_2H_2Fe(CO)_4$	Fe:	Org.Verb.B4-269/70, 280/4
$C_8FeH_5IO_5$	$ICHCHCOCH_3Fe(CO)_4$	Fe:	Org.Verb.B4-271, 289
$C_8FeH_5IO_6$	$CH_3OC(O)CHCHIFe(CO)_4$	Fe:	Org.Verb.B4-275, 293
$C_8FeH_5NO_4S$	$(SCHCHN(CH_3)C)Fe(CO)_4$	Fe:	Org.Verb.B4-128/9
$C_8FeH_5N_6^{2-}$	$[C_2H_5NCFe(CN)_5]^{2-}$	Fe:	Org.Verb.B4-10
$C_8FeH_5N_6Ni$	$[C_2H_5NCFe(CN)_5]Ni \cdot x\ H_2O$	Fe:	Org.Verb.B4-10
$C_8FeH_5NaO_5$	$Na[(CHO)_2C_3H_3Fe(CO)_3]$	Fe:	Org.Verb.B5-72
$C_8FeH_5O_2^+$	$[FeC_5H_4CCCOOH]^+$	Fe:	Org.Verb.A3-79
$C_8FeH_5O_3$	$C_5H_5Fe(CO)_3$	Fe:	Org.Verb.B3-151
$C_8FeH_5O_5^-$	$[CH_3C(O)C_3H_2(O)Fe(CO)_3]^-$	Fe:	Org.Verb.B5-72, 75
–	$[(CHO)_2C_3H_3Fe(CO)_3]^-$	Fe:	Org.Verb.B5-73, 76
$C_8FeH_5O_6^-$	$[CH_3OC(O)C_3H_2(O)Fe(CO)_3]^-$	Fe:	Org.Verb.B5-72, 74/5
$C_8FeH_6^+$	$[FeC_8H_6]^+$	Fe:	Org.Verb.A1-93/4
$C_8FeH_6Hg_2O_8$	$(CO)_4FeHg \cdot Hg(OOCCH_3)_2$	Fe:	Org.Verb.B2-154
$C_8FeH_6N_2O_4$	$(CO)_4Fe(C(CH_3)NCHCHNH)$	Fe:	Org.Verb.B2-71, 74
$C_8FeH_6O_3$	$C_5H_6Fe(CO)_3$	Fe:	Org.Verb.B3-151
$C_8FeH_6O_4$	$CH(O)C_4H_5Fe(CO)_3$	Fe:	Org.Verb.B5-76
–	$CH_2CHCHCH_2Fe(CO)_4$	Fe:	Org.Verb.B4-238, 250/1
$C_8FeH_6O_5$	$(C(CH_3)_2OC)Fe(CO)_4$	Fe:	Org.Verb.B4-128, 131, 135
–	$CH_3COCHCH_2Fe(CO)_4$	Fe:	Org.Verb.B4-237

Formula	Compound	Reference
$C_8FeH_6O_5$	$O(CH_2CH)_2Fe(CO)_4$	Fe: Org.Verb.B4-303, 307
$C_8FeH_6O_5S$	$(CO)_4Fe(OS(CH_2CH)_2)$	Fe: Org.Verb.B2-128/31, 133
$C_8FeH_6O_6$	$CH_3COOCHCH_2Fe(CO)_4$	Fe: Org.Verb.B4-234, 241
–	$CH_3OC(O)CHCH_2Fe(CO)_4$	Fe: Org.Verb.B4-232/3, 236, 246/7
$C_8FeH_6O_6S$	$CH_2CHSO_2CHCH_2Fe(CO)_4$	Fe: Org.Verb.B4-234, 241
$C_8FeH_7^+$	$[FeC_8H_7]^+$	Fe: Org.Verb.A1-93/4
$C_8FeH_7KO_5$	$K[C_3H_7C(O)Fe(CO)_4]$	Fe: Org.Verb.B4-150, 160
$C_8FeH_7KO_6$	$K[CH_3CH(OH)CH_2C(O)Fe(CO)_4]$	Fe: Org.Verb.B4-153, 162
$C_8FeH_7KO_7$	$K[HOCH_2CH(OH)CH_2C(O)Fe(CO)_4]$	Fe: Org.Verb.B4-154, 162
$C_8FeH_7NO_4S$	$(SCH_2CH_2N(CH_3)C)Fe(CO)_4$	Fe: Org.Verb.B4-128/9
$C_8FeH_7NO_5$	$CH_3NHCOCHCH_2Fe(CO)_4$	Fe: Org.Verb.B4-237, 248/9
$C_8FeH_7NO_6$	$CH_3NHCOOCHCH_2Fe(CO)_4$	Fe: Org.Verb.B4-234, 241
$C_8FeH_7N_3O_4$	$(N(CH_3)NCHN(CH_3)C)Fe(CO)_4$	Fe: Org.Verb.B4-128/9
$C_8FeH_7N_6$	$[C_2H_5NCFe(CN)_5]H_2$	Fe: Org.Verb.B4-10
$C_8FeH_7NaO_5$	$Na[C_3H_7C(O)Fe(CO)_4]$	Fe: Org.Verb.B4-150
$C_8FeH_7O^+$	$[HFeC_5H_4CHCHCO]^+$	Fe: Org.Verb.A3-123
$C_8FeH_7O_2^+$	$[FeC_5H_4CHCHCOOH]^+$	Fe: Org.Verb.A3-77
$C_8FeH_7O_4^+$	$[CH_3C_3H_4Fe(CO)_4]^+$	Fe: Org.Verb.B5-99/100, 102/4, 106
$C_8FeH_7O_5$	$[CH_3OCOC_3H_4Fe(CO)_3]$	Fe: Org.Verb.B5-69, 71
$C_8FeH_7O_5^-$	$[(CHO)(HOCH_2)C_3H_3Fe(CO)_3]^-$	Fe: Org.Verb.B5-73, 76
–	$[C_3H_7C(O)Fe(CO)_4]^-$	Fe: Org.Verb.B4-150, 156, 160, 167
$C_8FeH_7O_6^-$	$[CH_3CH(OH)CH_2C(O)Fe(CO)_4]^-$	Fe: Org.Verb.B4-153, 162
$C_8FeH_7O_7^-$	$[HOCH_2CH(OH)CH_2C(O)Fe(CO)_4]^-$	Fe: Org.Verb.B4-154, 162
$C_8FeH_8^+$	$[C_5H_5FeC_3H_3]^+$	Fe: Org.Verb.A1-93/4
$C_8FeH_8NO_4$	$(CO)_4FeN(CH_2)_4$	Fe: Org.Verb.B2-70/1
$C_8FeH_8N_2O_3$	$CH_3NC(C_2H_5NC)Fe(CO)_3$	Fe: Org.Verb.B4-20
$C_8FeH_8N_6$	$[C_2H_5NCFe(CN)_5]H_3$	Fe: Org.Verb.B4-10
$C_8FeH_8OS_3$	$(C_5H_5)Fe(CO)[S_2CSCH_3]$	C: MVol.D4-269
$C_8FeH_8O_3S_2$	$(C_2H_5SCCHCH_2S)Fe(CO)_3$	Fe: Org.Verb.B4-213, 220/1
$C_8FeH_8O_4$	$(CH_3CHCHC(CH_3)O)Fe(CO)_3$	Fe: Org.Verb.B5-132, 141
–	$(CH_3)_2CCH_2Fe(CO)_4$	Fe: Org.Verb.B4-258, 260
$C_8FeH_8O_4S$	$(CO)_4FeSC_4H_8$	Fe: Org.Verb.B2-129
$C_8FeH_8O_4S_2$	$(CO)_4Fe(CH_2(SCH_2)_2CH_2)$	Fe: Org.Verb.B2-128/30, 132
$C_8FeH_8O_5$	$(CO)_4FeOC_4H_8$	Fe: Org.Verb.B2-129 Fe: Org.Verb.B3-85
–	$C_2H_5OCHCH_2Fe(CO)_4$	Fe: Org.Verb.B4-233/4, 240/1
–	$C_3H_5Fe(CO)_3OCOCH_3$	Fe: Org.Verb.B5-39, 48, 51, 59
$C_8FeH_8O_5S$	$(CO)_4Fe(OS(CH_2CH_2)_2)$	Fe: Org.Verb.B2-128/31, 133
$C_8FeH_9I_2NO_3$	$(t\text{-}C_4H_9NC)Fe(CO)_3I_2$	Fe: Org.Verb.B4-2/3
$C_8FeH_9NO_3$	$C_6H_9Fe(CO)_2NO$	Fe: Org.Verb.B5-14/5, 17, 23
$C_8FeH_9NO_4$	$CH_2CHC(O)N(CH_3)_2Fe(CO)_3$	Fe: Org.Verb.B5-158
–	$CH_3COCH_2C_3H_4Fe(CO)_2NO$	Fe: Org.Verb.B5-14/6, 22
–	$(CO)_4FeNH(CH_2)_4$	Fe: Org.Verb.B2-71, 73
$C_8FeH_9NO_4S$	$(CO)_4FeSC(CH_3)N(CH_3)_2$	Fe: Org.Verb.B2-128/32
$C_8FeH_9NO_4Si$	$(CH_3)_3SiNCFe(CO)_4$	Fe: Org.Verb.B4-6/7
$C_8FeH_9NO_4Sn$	$(CH_3)_3SnNCFe(CO)_4$	Fe: Org.Verb.B4-6/7

Formula	Compound	Reference
$C_8FeH_9O_3^+$	$[(CH_3)_2C_3H_3Fe(CO)_3]^+$	Fe: Org.Verb.B5-82/6
$C_8FeH_9O_4^-$	$[C_4H_9C(O)Fe(CO)_3]^-$	Fe: Org.Verb.B4-109
$C_8FeH_9O_5^-$	$[(CH_2OH)_2C_3H_3Fe(CO)_3]^-$	Fe: Org.Verb.B5-73
–	$[C_2H_5CH(OH)CH_2C(O)Fe(CO)_3]^-$	Fe: Org.Verb.B4-109
$C_8FeH_{10}Hg_2O_4$	$(CO)_4Fe(HgC_2H_5)_2$	Fe: Org.Verb.B2-153, 156
$C_8FeH_{10}N_2Na_2O_6S_2$	$(CO)_2Fe(SCH_2CH(NH_2)COONa)_2 \cdot 2\ H_2O$	Fe: Org.Verb.B1-116
$C_8FeH_{10}N_6O_2$	$[C_3H_7NCFe(CN)_4NO_2]H_3 \cdot 2\ H_2O$	Fe: Org.Verb.B4-10
$C_8FeH_{10}O_2$	$(C_3H_5)_2Fe(CO)_2$	Fe: Org.Verb.B5-107
$C_8FeH_{10}O_4$	$[Fe(CO)_3(CH_3C_4H_7O)]$	Fe: Org.Verb.B3-88
$C_8FeH_{10}O_4S_4$	$(CO)_2Fe[S_2COC_2H_5]_2$	C: MVol.D4-259
$C_8FeH_{11}NO_2S_2$	$C_3H_5Fe(CO)_2(S_2CN(CH_3)_2)$	Fe: Org.Verb.B5-24/7
$C_8FeH_{11}NO_4$	$[C_4H_8NH_2][HFe(CO)_4]$	Fe: Org.Verb.B4-165
$C_8FeH_{11}N_5O$	$[(CH_3NC)_3Fe(CN)_2H_2O]$	Fe: Org.Verb.B4-44
$C_8FeH_{11}NiOP^+$	$NiP(CH_3)_2Fe(C_5H_5)(CO)^+$	Ni: Org.Verb.1-167
$C_8FeH_{11}O_4P$	$(CO)_4FeP(C_2H_5)_2H$	Fe: Org.Verb.B2-88, 109
$C_8FeH_{12}I_2N_4$	$(CH_3NC)_4FeI_2$	Fe: Org.Verb.B4-48/51, 56
$C_8FeH_{12}NO_4P$	$(CO)_4FeP(CH_3)_2N(CH_3)_2$	Fe: Org.Verb.B2-97, 118
$C_8FeH_{12}N_2O_2S_4$	$(CO)_2Fe(S_2CN(CH_3)_2)_2$	Fe: Org.Verb.B1-113/4
		Fe: Org.Verb.C2-43
$C_8FeH_{12}N_2O_6S_2$	$(CO)_2Fe(SCH_2CH(NH_2)COOH)_2 \cdot H_2O$	Fe: Org.Verb.B1-115
$C_8FeH_{12}N_4O_2S_2^+$	$[(CO)_2Fe(SC(NHCH_2)_2)_2]^+$	Fe: Org.Verb.B1-165
$C_8FeH_{12}O_4P_2$	$(CO)_4FeP(CH_3)_2P(CH_3)_2$	Fe: Org.Verb.B2-28
$C_8FeH_{14}IO_5P$	$C_3H_5Fe(CO)_2P(OCH_3)_3I$	Fe: Org.Verb.B5-24/6
$C_8FeH_{14}I_2O_2S_2$	$(CO)_2Fe[(C_2H_5)SCH_2CH_2S(C_2H_5)]I_2$	Fe: Org.Verb.B1-101/2, 107
$C_8FeH_{14}O_2P$	$[C_3H_5Fe(CO)_2P(CH_3)_3]$	Fe: Org.Verb.B5-69/70
$C_8FeH_{14}O_5P$	$[C_3H_5Fe(CO)_2P(OCH_3)_3]$	Fe: Org.Verb.B5-69/70
$C_8FeH_{15}N_3O_4$	$((C_2H_5)_2N)(C_2H_5O)CFe(NO)_2CO$	Fe: Org.Verb.B4-109/11
$C_8FeH_{15}N_4OS_4^+$	$[CH_3NCFe(S_2CN(CH_3)_2)_2NO]^+$	Fe: Org.Verb.B4-10
$C_8FeH_{15}S_7$	$[Fe(S_2CSC_2H_5)_2(SC_2H_5)]_2$	C: MVol.D4-266, 269
$C_8FeH_{16}I_2N_4O_4$	$(CO)_4Fe(C_2H_8N_2)_2I_2$	Fe: Org.Verb.B2-67
$C_8FeH_{16}N_4O_2^{2+}$	$[(CH_3NC)_4Fe(H_2O)_2]^{2+}$	Fe: Org.Verb.B4-70
$C_8FeH_{16}O_3Si_2$	$(CO)_3Fe(H)(Si(CH_3)_3)Si(CH_3)_2$	Fe: Org.Verb.C2-52
$C_8FeH_{18}I_2O_2P_2$	$(CO)_2Fe(P(CH_3)_3)_2I_2$	Fe: Org.Verb.B1-101/3, 109
$C_8FeH_{18}I_2O_8P_2$	$(CO)_2Fe(P(OCH_3)_3)_2I_2$	Fe: Org.Verb.B1-101/2, 105, 110
$C_8FeH_{18}I_6O_2P_2$	$(CO)_2Fe(P(CH_3)_3)_2(I_3)_2$	Fe: Org.Verb.B1-109
$C_8FeH_{18}NO_9P_2^+$	$[(CO)_2Fe(P(OCH_3)_3)_2NO]^+$	Fe: Org.Verb.B1-130/3
$C_8FeH_{22}Si_2$	$((CH_3)_3SiCH_2)_2Fe$	Fe: Org.Verb.B1-6/7
$C_8FeH_{24}N_4P^+$	$[FeP(N(CH_3)_2)_4]^+$	Fe: Org.Verb.B1-164
$C_8Fe_2Ge_2I_4O_8$	$[(CO)_4FeGeI_2]_2$	Fe: Org.Verb.C1-42, 45, 50
$C_8Fe_2HLiO_8$	$Li[HFe_2(CO)_8]$	Fe: Org.Verb.C2-10
$C_8Fe_2HNaO_8$	$Na[HFe_2(CO)_8]$	Fe: Org.Verb.C2-10/1
$C_8Fe_2HNa_2O_8^+$	$[Na_2HFe_2(CO)_8]^+$	Fe: Org.Verb.C2-10
$C_8Fe_2HO_8^-$	$[HFe_2(CO)_8]^-$	Fe: Org.Verb.C2-3/6, 11/5
–	$[DFe_2(CO)_8]^-$	Fe: Org.Verb.C2-4
$C_8Fe_2HO_8Tl$	$(CO)_4FeTl \cdot HFe(CO)_4$	Fe: Org.Verb.C1-249
$C_8Fe_2H_2I_2O_6$	$(CO)_3Fe(CHCHI)IFe(CO)_3$	Fe: Org.Verb.C2-79/80, 82, 84
$C_8Fe_2H_2O_6$	$C_2H_2Fe_2(CO)_6$	Fe: Org.Verb.C2-73

$C_8Fe_2H_2O_6S$ .... $(CO)_3Fe(CHCHS)Fe(CO)_3$ .... Fe: Org.Verb.C2-95/6, 98/100
$C_8Fe_2H_2O_6S_2$ .... $(CO)_3FeSC_2H_2SFe(CO)_3$ .... Fe: Org.Verb.C1-115/6
$C_8Fe_2H_2O_8^{2-}$ .... $[H_2Fe_2(CO)_8]^{2-}$ .... Fe: Org.Verb.C2-3
$C_8Fe_2H_4O_6S_2$ .... $(CO)_3FeSC_2H_4SFe(CO)_3$ .... Fe: Org.Verb.C1-115/6, 119/20
Fe: Org.Verb.C2-44
$C_8Fe_2H_6I_2O_6S_2$ ..... $I(CO)_3Fe(SCH_3)_2Fe(CO)_3I$ .... Fe: Org.Verb.C1-6/7, 9
$C_8Fe_2H_6N_2O_6$ ...... $(CO)_3Fe(CH_3NNCH_3)Fe(CO)_3$ .... Fe: Org.Verb.C1-149/51, 153/4
$C_8Fe_2H_6O_6S_2$ .... $(CO)_3Fe(SCH_3)_2Fe(CO)_3$ .... Fe: Org.Verb.C1-79/81, 84/6
$C_8Fe_2H_6O_6Se_2$ ...... $(CO)_3Fe(SeCH_3)_2Fe(CO)_3$ .... Fe: Org.Verb.C1-83
$C_8Fe_2H_7O_6P$ .... $Fe_2(CO)_6(P(CH_3)_2)H$ .... Fe: Org.Verb.B2-108
$C_8Fe_2H_7O_7P$ .... $(CO)_3Fe(P(CH_3)_2)(OH)Fe(CO)_3$ .... Fe: Org.Verb.C1-185/7, 189
$C_8Fe_2H_8^{+}$ .... $[Fe_2(C_5H_5)(C_3H_3)]^{+}$ .... Fe: Org.Verb.A1-93/4
$C_8Fe_2H_8N_2O_6$ ...... $(CO)_3Fe(CH_3NH)_2Fe(CO)_3$ .... Fe: Org.Verb.C1-130, 135
$C_8Fe_2H_8N_2O_8$ ...... $(NH_4)_2[Fe_2(CO)_8]$ .... Fe: Org.Verb.C2-7
$C_8Fe_2H_8O_6P_2$ .... $(CO)_3Fe(P(CH_3)H)_2Fe(CO)_3$ .... Fe: Org.Verb.C1-164/6, 170, 172
$C_8Fe_2H_{20}MnN_6O_8$ ... $[Mn(NH_3)_6][HFe(CO)_4]_2$ .... Fe: Org.Verb.B2-18
$C_8Fe_2H_{20}N_6NiO_8$ .... $[Ni(NH_3)_6][HFe(CO)_4]_2$ .... Fe: Org.Verb.B2-18
$C_8Fe_2H_{24}N_4O_4P_4$ .... $(NO)_2Fe(P(CH_3)_2P(CH_3)_2)_2Fe(NO)_2$ .... Fe: Org.Verb.B1-143
$C_8Fe_2HgK_2O_8$ ...... $K_2[(CO)_4FeHgFe(CO)_4]$ .... Fe: Org.Verb.C1-58/9
$C_8Fe_2HgO_8^{2-}$ .... $[(CO)_4FeHgFe(CO)_4]^{2-}$ .... Fe: Org.Verb.C1-58/9
$C_8Fe_2I_2O_8$ .... $I(CO)_4FeFe(CO)_4I$ .... Fe: Org.Verb.C1-71/2
$C_8Fe_2I_4O_8$ .... $[(CO)_4FeI_2]_2$ .... Fe: Org.Verb.B2-64
$C_8Fe_2K_2O_8$ .... $K_2[Fe_2(CO)_8]$ .... Fe: Org.Verb.C2-7
$C_8Fe_2K_2O_8Zn$ ...... $K_2[(CO)_4FeZnFe(CO)_4]$ .... Fe: Org.Verb.C1-58/9
$C_8Fe_2N_2O_8$ .... $Fe_2(CO)_8N_2$ .... Fe: Org.Verb.C2-1
$C_8Fe_2Na_2O_8$ .... $Na_2[Fe_2(CO)_8]$ .... Fe: Org.Verb.C2-6/7
$C_8Fe_2O_8$ .... $Fe_2(CO)_8$ .... Fe: Org.Verb.B3-80
Fe: Org.Verb.C2-1/3
$C_8Fe_2O_8^{+}$ .... $[Fe_2(CO)_8]^{+}$ .... Fe: Org.Verb.C2-26
$C_8Fe_2O_8^{2-}$ .... $[Fe_2(CO)_8]^{2-}$ .... Fe: Org.Verb.C2-3/5, 11/5
$C_8Fe_2O_8Sb$ .... $(CO)_4FeSbFe(CO)_4$ .... Fe: Org.Verb.C1-33
$C_8Fe_2O_8Zn^{2-}$ .... $[(CO)_4FeZnFe(CO)_4]^{2-}$ .... Fe: Org.Verb.C1-58/9
$C_8Fe_2O_{10}S$ .... $(CO)_4Fe(SO_2)Fe(CO)_4$ .... Fe: Org.Verb.C1-197/8
Fe: Org.Verb.C2-27
$C_8Fe_2O_{14}P_4$ .... $(CO)_4Fe(P_4O_6)Fe(CO)_4$ .... Fe: Org.Verb.C1-36, 40
Fe: Org.Verb.C2-28
$C_8Fe_2O_{15}P_4$ .... $[(CO)_4Fe]_2P_4O_7$ .... Fe: Org.Verb.C2-28
$C_8Fe_3H_4I_5O_6S_2$ ..... $Fe_3(CO)_6(SCH_2CH_2S)I_5$ .... Fe: Org.Verb.C1-18/20
$C_8Fe_3H_{18}N_6O_8$ ...... $[Fe(NH_3)_6][Fe_2(CO)_8]$ .... Fe: Org.Verb.C2-7/8
$C_8Fe_3H_{20}N_6O_8$ ...... $[Fe(NH_3)_6][HFe(CO)_4]_2$ .... Fe: Org.Verb.B2-18
Fe: Org.Verb.B3-113
$C_8Fe_4H_{18}N_6O_4S_2$ .... $Fe_4(NO)_4S_2(NC(CH_3)_3)_2$ .... Fe: Org.Verb.C1-64
$C_8FmH_8O_{12}^{-}$ .... $Fm(C_4H_4O_6)_2^{-}$ .... Np: TrU.D1-159
$C_8GeH_{18}OSn$ .... $(CH_3)_3SnC[Ge(CH_3)_3]CO$ .... Sn: Org.Verb.2-26, 29
$C_8GeH_{18}Sn$ .... $(CH_3)_3SnCCGe(CH_3)_3$ .... Sn: Org.Verb.2-114

Formula	Compound		Reference
$C_8H_4I_2NiO_2$	$(C_6H_4(CO)_2NiI_2)_x$	Ni:	Org.Verb.2-398/9
$C_8H_4I_2Ni_2O_6$	$(INi(CHCO)_2O)_2$	Ni:	Org.Verb.2-270
$C_8H_4K_2Ni$	$K_2[Ni(CCH)_4]$	Ni:	Org.Verb.1-105/6
$C_8H_4K_4Ni$	$K_4[Ni(CCH)_4]$	Ni:	Org.Verb.1-106, 107
$C_8H_4N_2Sn^+$	$[SnC_6H_4(CN)_2]^+$	Sn:	MVol.C5-28
$C_8H_4N_4Ni$	$Ni(NCCHCHCN)_2$	Ni:	Org.Verb.1-390/1
$C_8H_4Na_2Ni$	$Na_2[Ni(CCH)_4] \cdot 2\ NH_3$	Ni:	Org.Verb.1-105, 106
$C_8H_4NiO_6$	$Ni(C_4H_2O_3)_2$	Ni:	Org.Verb.1-388
$C_8H_4O_{16}Pu^-$	$Pu(HC_2O_4)_4^-$	Np:	TrU.D1-156
C_8H_4Sn	$Sn(CCH)_4$	Sn:	Org.Verb.1-139/40
		Sn:	Org.Verb.6-266
$C_8H_5MnO_3$	$C_5H_5Mn(CO)_3$ systems		
	$C_5H_5Mn(CO)_3$-$Fe(CO)_5$	Fe:	Org.Verb.B3-64
$C_8H_5NNiO_3$	$(CO)_3Ni(NC_5H_5)$	Ni:	Org.Verb.1-160
$C_8H_5NO_4Ti$	$C_5H_5Ti(CO)_3NO$	Ti:	Org.Verb.1-211
$C_8H_5N_3S_3Ti$	$C_5H_5Ti(NCS)_3$	Ti:	Org.Verb.1-141, 152
$C_8H_6MnNO_4^+$	$[Mn(OC_6H_2(CH_3)(NO_2)CHO)]^+$	Mn:	MVol.D1-53/4
$C_8H_6N_2O_2Se$	$NO_2C_6H_4CH_2SeCN$	C:	MVol.D6-227
$C_8H_6N_2S_2Sn$	$(CH_2CH)_2Sn(SC(CN)C(CN)S)$	Sn:	Org.Verb.6-145
$C_8H_6N_4S_3Sn$	$[C_5H_6N]Sn(SCN)_3$	Sn:	MVol.C3-126
$C_8H_6O_{13}Sn^{4-}$	$SnO(OCOCH(O)CH(OH)COO)_2^{4-}$	Sn:	MVol.C2-226
$C_8H_7MnO_2^+$	$[Mn(OC_6H_3(CH_3)CHO)]^+$	Mn:	MVol.D1-53/4
$C_8H_7MnO_3^+$	$[Mn(OC_6H_3(OCH_3)CHO)]^+$	Mn:	MVol.D1-53/4
C_8H_7NNi	$C_6H_5CH_2NiCN$	Ni:	Org.Verb.1-16, 22/3
C_8H_7NSe	$C_6H_5CH_2SeCN$	C:	MVol.D6-227
C_8H_7NTe	$C_6H_5CH_2TeCN$	C:	MVol.D6-260
$C_8H_7Ni_2^+$	$Ni_2(C_5H_5)C_3H_2^+$	Ni:	Org.Verb.2-200
$C_8H_7OS_2^-$	$[S_2COCH_2C_6H_5]^-$	C:	MVol.D4-244
$C_8H_7O_2S^-$	$[SOCOCH_2C_6H_5]^-$	C:	MVol.D4-237
$C_8H_7O_{12}Sn^{3-}$	$SnH(OCOCH(O)CH(OH)COO)_2^{3-}$	Sn:	MVol.C2-225/6
$C_8H_8I_6Ir_2N_4S_4$	$[Ir(S_2N_2C_4H_4)I_3]_2$	Ir:	SVol.2-172
$C_8H_8K_2O_{13}Sn$	$K_2SnO(C_4H_4O_6)_2 \cdot 2\ H_2O$	Sn:	MVol.C3-80
$C_8H_8Na_2O_8S_2Sn$	$(CH_3)_2Sn(OC_6H_2(SO_3Na)_2O)$	Sn:	Org.Verb.6-27
$C_8H_8Na_2O_{13}Sn$	$Na_2SnO(C_4H_4O_6)_2 \cdot 6\ H_2O$	Sn:	MVol.C3-35
C_8H_8Ni	$(C_4H_4)_2Ni$	Ni:	Org.Verb.2-324
–	$(NiC_8H_8)_x$	Ni:	Org.Verb.2-401
$C_8H_8Ni^+$	$(C_4H_4)_2Ni^+$	Ni:	Org.Verb.2-324
$C_8H_8NpO_{14}^{3-}$	$NpO_2(C_4H_4O_6)_2^{3-}$	Np:	TrU.D1-157, 159
$C_8H_8OS_2$	$SC(OCH_2C_6H_5)(SH)$	C:	MVol.D4-238/42
$C_8H_8O_{10}Sn^{2-}$	$Sn(OCOCH_2CH(OH)COO)_2^{2-}$	Sn:	MVol.C2-225
$C_8H_8O_{12}Pu^-$	$Pu(C_4H_4O_6)_2^-$	Np:	TrU.D1-159
$C_8H_8O_{12}Sn^{2-}$	$Sn(OCOCH(OH)CH(OH)COO)_2^{2-}$	Sn:	MVol.C2-225/6
C_8H_8Sn	$(CHC)_3SnC_2H_5$	Sn:	Org.Verb.2-351/3
$C_8H_9KNiO_5$	$K[(CO)_3NiO_2CC_4H_9$-t]	Ni:	Org.Verb.1-191, 192
$C_8H_9LiNiO_4$	$Li[(CO)_3NiCOC_4H_9]$	Ni:	Org.Verb.1-193, 194
$C_8H_9NNiO_3$	(n-$C_4H_9NC)Ni(CO)_3$	Ni:	Org.Verb.1-303
$C_8H_9NO_5S$	$CH_3OC(O)NHSO_2OC_6H_5$	S:	S-N-Verb.1-97
$C_8H_9NiO_6P$	$(CO)_3Ni(P(OCH_2)_3CCH_3)$	Ni:	Org.Verb.1-178, 179
$C_8H_{10}I_2Sn$	$(C_2H_5)(C_6H_5)SnI_2$	Sn:	Org.Verb.5-50
$C_8H_{10}K_2N_2Ni$	$K_2[Ni(CCH)_4] \cdot 2\ NH_3$	Ni:	Org.Verb.1-106/7

Formula	Compound	Reference
$C_8H_{10}LiNNiO_4$	$Li[(CO)_3NiC(O)N(C_2H_5)_2]$	Ni: Org.Verb.1-193, 194
$C_8H_{10}MoNO_2S_2$	$MoO[S_2COC_2H_5](C_5H_5N)$	C: MVol.D4-258
$C_8H_{10}N_2Na_2Ni$	$Na_2[Ni(CCH)_4] \cdot 2\ NH_3$	Ni: Org.Verb.1-105, 106
$C_8H_{10}N_2NiO_2$	$(C_2H_5NC)_2Ni(CO)_2$	Ni: Org.Verb.1-308, 309
$C_8H_{10}N_2Ni_2$	$(C_3H_5NiCN)_2$	Ni: Org.Verb.2-282
$C_8H_{10}N_2O_4S$	$C_6H_5NHSO_2NHCOOCH_3$	S: S-N-Verb.1-170
$C_8H_{10}N_2O_5SU$	$UO_2(OH)_2 \cdot NH_2CSNHCOC_6H_5$	U: SVol.E1-220/1
$C_8H_{10}N_2S_2Sn$	$(C_2H_5)_2Sn(SC(CN)C(CN)S)$	Sn: Org.Verb.6-59
$C_8H_{10}N_4OTe_2$	$2\ Te(CN)_2 \cdot (C_2H_5)_2O$	C: MVol.D6-260
$C_8H_{10}Ni$	$C_5H_5NiC_3H_5$	Ni: Org.Verb.2-175
$C_8H_{10}NiO$	$C_5H_5Ni(CO)C_2H_5$	Ni: Org.Verb.2-345
$C_8H_{10}NiS_2$	$C_5H_5NiSC(S)C_2H_5$	Ni: Org.Verb.2-114
$C_8H_{10}NiS_3$	$(C_5H_5)Ni[S_2CSC_2H_5]$	C: MVol.D4-270
		Ni: Org.Verb.2-115
$C_8H_{10}OSn$	$[(C_2H_5)(C_6H_5)SnO]_x$	Sn: Org.Verb.6-204
$C_8H_{10}O_{10}Sn$	$SnH_2(OCOCH_2CH(OH)COO)_2$	Sn: MVol.C2-225
$C_8H_{10}O_{12}Pa^{3+}$	$Pa(HC_4H_4O_6)_2{}^{3+}$	Pa: SVol.2-172
$C_8H_{10}O_{12}Sn$	$Sn(OCOCH(OH)CH(OH)COOH)_2$	Sn: MVol.C2-225/6
$C_8H_{10}Ti$	$(CH_3)(C_6H_5CH_2)Ti$	Ti: Org.Verb.1-68
$C_8H_{11}NNiO_3$	$(CO)_3Ni(NC_5H_{11})$	Ni: Org.Verb.1-160
$C_8H_{11}NNiO_4$	$(CH_3)_2N(C_2H_5O)CNi(CO)_3$	Ni: Org.Verb.1-334
$C_8H_{11}NO_3S$	$C_6H_5NHSO_2OC_2H_5$	S: S-N-Verb.1-97
$C_8H_{11}Sn^+$	$Sn(C_6H_5)(C_2H_5)H^+$	Sn: Org.Verb.3-25
$C_8H_{12}Hg_7Mn_2O_{20}$	$Hg(MnO_4)_2 \cdot 2\ [Hg(C_2H_3O_2)_2 \cdot 2\ HgO]$	Mn: MVol.C2-302
$C_8H_{12}INi$	$C_8H_{12}NiI$	Ni: Org.Verb.2-74
$C_8H_{12}I_4N_4Pa$	$PaI_4 \cdot 4\ CH_3CN$	Pa: SVol.2-88/9
$C_8H_{12}MnO_3{}^+$	$[Mn(CH(COCH_3)_2)CH_3COCH_2]^+$	Mn: MVol.D1-92/3
$C_8H_{12}Mo_2O_7S_8$	$Mo_2O_3[S_2COCH_3]_4$	C: MVol.D4-258
$C_8H_{12}N_2NiO_2P_2$	$(CO)_2Ni(P(C_2H_4CN)H_2)_2$	Ni: Org.Verb.1-129, 130
$C_8H_{12}N_2NiO_6$	$(CH_3)_4C_4Ni(NO_3)_2$	Ni: Org.Verb.2-329
$C_8H_{12}N_2O_2S$	$C_2H_5NHSO_2NHC_6H_5$	S: S-N-Verb.1-170
$C_8H_{12}N_2O_3S$	$C_2H_5OC_6H_4NHSO_2NH_2$	S: S-N-Verb.1-167
$C_8H_{12}N_2O_4Sn$	$Sn(CH_3COC(CH_3)NO)_2$	Sn: MVol.C5-41
$C_8H_{12}N_2O_{12}U$	$UO_2(NO_3)_2 \cdot 2\ O(CH_2)_3C(O)$	U: SVol.E1-119/20
$C_8H_{12}N_4Ni$	$(CH_3NC)_4Ni$	Ni: Org.Verb.1-328, 329
$C_8H_{12}N_4Ni_2O_2$	$[(CO)NiNCN(CH_3)_2]_2$	Ni: Org.Verb.2-256/7
$C_8H_{12}N_4O_2Sn$	$(CH_3)_2Sn[CH_3C(O)NCN]_2$	Sn: Org.Verb.6-32
$C_8H_{12}N_6Ni$	$(CH_3)_4C_4Ni(N_3)_2$	Ni: Org.Verb.2-329
$C_8H_{12}N_6S_2Se_2Te$	$Te(SC(NHCH_2)_2)_2(SeCN)_2$	Te: SVol.B3-154/5
$C_8H_{12}N_6S_4Te$	$Te(SC(NHCH_2)_2)_2(SCN)_2$	Te: SVol.B3-154
$C_8H_{12}Na_2O_4Sn$	$Na_2[(CH_3)_2(OH)_2Sn(OC_6H_4O)]$	Sn: Org.Verb.6-37/8
$C_8H_{12}Ni$	$C_8H_{12}Ni$	Ni: Org.Verb.2-226/9, 231
–	$Ni(C_4H_6)_2$	Ni: Org.Verb.2-86
$C_8H_{12}Ni^+$	$NiC_8H_{12}{}^+$	Ni: Org.Verb.2-88
$C_8H_{12}NiO_2$	$C_3H_5Ni(CH(COCH_3)_2)$	Ni: Org.Verb.2-7/8
$C_8H_{12}NiO_4S$	$((CH_3)_4C_4NiSO_4)$	Ni: Org.Verb.2-329
$C_8H_{12}O_8Pu$	$Pu(CH_3COO)_4$	Np: TrU.D1-153
$C_8H_{12}O_8Pu^-$	$Pu(CH_3COO)_4{}^-$	Np: TrU.D1-152
$C_8H_{12}O_8S_4Te$	$Te(SCH_2COOH)_4$	Te: SVol.B3-134
$C_8H_{12}O_8Sn$	$Sn(CH_3COO)_4$	Sn: MVol.C2-221

Formula	Compound	Ref.
$C_8H_{12}Sn$	$(CH_3)_2(C_6H_5)SnH$	Sn: Org.Verb.4-90
–	$(CH_3)_2Sn(CHCCH_2)_2$	Sn: Org.Verb.3-7, 16
–	$(CH_3)_2Sn(CH_2CCH)_2$	Sn: Org.Verb.3-7
–	$(CH_3)_3SnCCCCCH_3$	Sn: Org.Verb.2-113
–	$C_2H_5(C_6H_5)SnH_2$	Sn: Org.Verb.4-119/20
–	$(C_2H_5)_2Sn(CCH)_2$	Sn: Org.Verb.3-25/6
–	$Sn(CHCH_2)_4$	Sn: Org.Verb.1-129/33
$C_8H_{12}Ti$	$(CH_2CHCHCH_2)_2Ti$	Ti: Org.Verb.1-123, 131
–	$(CH_2CH)_4Ti$	Ti: Org.Verb.1-91, 99, 123
$C_8H_{13}MnO_2{}^+$	$[Mn(CH_3COCHCOCH_2CH(CH_3)_2)]^+$	Mn: MVol.D1-126
$C_8H_{13}NOSn$	$(CH_3)_3SnC_5H_4NO$	Sn: Org.Verb.2-153
$C_8H_{13}NSn$	$(CH_3)_3SnC_5H_4N$	Sn: Org.Verb.2-153, 156
$C_8H_{13}N_3O_4S_2$	$(CH_3)_2NSO_2NHNHSO_2C_6H_5$	S: S-N-Verb.1-222
$C_8H_{13}OTi^+$	$[C_5H_5Ti(CH_3)OC_2H_5]^+$	Ti: Org.Verb.1-160
$C_8H_{14}I_2Ni_2$	$(CH_2C(CH_3)CH_2NiI)_2$	Ni: Org.Verb.2-306
–	$(CH_3CHCHCH_2NiI)_2$	Ni: Org.Verb.2-294/6
$C_8H_{14}Mo_2O_4S_6$	$Mo_2O_2S_2[S_2COC_3H_7]_2$	C: MVol.D4-258
$C_8H_{14}N_2O_2Sn$	$[(C_2H_5)_2SnNCH_2C(O)NCH_2C(O)]_x$	Sn: Org.Verb.6-56
$C_8H_{14}Ni$	$Ni(C_3H_4CH_3\text{-}1)_2$	Ni: Org.Verb.2-61/3
–	$Ni(C_3H_4CH_3\text{-}2)_2$	Ni: Org.Verb.2-63/6
$C_8H_{14}NiO_2S_4$	$Ni[S_2COC_3H_7]_2$	C: MVol.D4-260
$C_8H_{14}NiS_6$	$Ni[S_2CSC_3H_7]_2$	C: MVol.D4-267, 270
$C_8H_{14}O_2PbS_4$	$Pb[S_2COC_3H_7]_2$	C: MVol.D4-254
$C_8H_{14}O_2PtS_4$	$Pt[S_2COC_3H_7]_2$	C: MVol.D4-261
$C_8H_{14}O_2S_4$	$C_3H_7OC(S)SSC(S)OC_3H_7$	C: MVol.D4-251
$C_8H_{14}O_2S_4Se$	$Se[S_2COC_3H_7]_2$	C: MVol.D4-255
$C_8H_{14}O_2S_4Te$	$Te[S_2COC_3H_7]_2$	C: MVol.D4-255
		Te: SVol.B3-136
$C_8H_{14}O_2S_4Zn$	$Zn[S_2COC_3H_7\text{-}i]_2$	C: MVol.D4-245, 257
$C_8H_{14}O_2Sn$	$(CH_3)_3SnCCCOOC_2H_5$	Sn: Org.Verb.2-113, 117
$C_8H_{14}O_3Ti$	$C_5H_5Ti(OCH_3)_3$	Ti: Org.Verb.1-156/7
$C_8H_{14}O_4S_2$	$C_3H_7OC(O)SS(O)COC_3H_7$	C: MVol.D4-236/7
$C_8H_{14}O_4S_4U$	$UO_2[S_2COC_3H_7]_2$	C: MVol.D4-261
$C_8H_{14}O_4Sn$	$[(CH_3)_2SnOCO(CH_2)_4COO]_x$	Sn: Org.Verb.6-25
$C_8H_{14}O_6S_3Ti$	$C_5H_5Ti(OS(O)CH_3)_3$	Ti: Org.Verb.1-170
$C_8H_{14}O_8S_4Te$	$H_2[Te(SCH_2COOH)_4]$	Te: SVol.B3-134
$C_8H_{14}O_8Sn$	$H_2Sn(CH_3COO)_4$	Sn: MVol.C2-220/1
$C_8H_{14}S_3Ti$	$C_5H_5Ti(SCH_3)_3$	Ti: Org.Verb.1-171
$C_8H_{14}Sn$	$(CH_2CH)_2Sn(CH_3)(CHCH_2CH_2)$	Sn: Org.Verb.3-92, 94
–	$(CH_2CH)_3SnC_2H_5$	Sn: Org.Verb.2-347
–	$(CH_3)_2SnC_6H_8$	Sn: Org.Verb.3-109
–	$(CH_3)_3SnCCC(CH_3)CH_2$	Sn: Org.Verb.2-113, 116
–	$(CH_3)_3SnCCCHCHCH_3$	Sn: Org.Verb.2-113, 117
–	$(CH_3)_3SnC_5H_5$	Sn: Org.Verb.2-93/4
$C_8H_{14}Ti$	$C_5H_5Ti(CH_3)_3$	Ti: Org.Verb.1-209/10
$C_8H_{15}KOS_2$	$K[S_2COC_7H_{15}]$	C: MVol.D4-253
$C_8H_{15}KOSe_2$	$K[Se_2COC_7H_{15}]$	C: MVol.D6-221
$C_8H_{15}NNi$	$(CH_3)_2C_3H_3Ni(NCCH_3)CH_3$	Ni: Org.Verb.2-34
$C_8H_{15}NO_8U$	$U(CH_3COO)_4 \cdot NH_3$	U: SVol.E1-12, 15
$C_8H_{15}NSn$	$(CH_3)_3SnCH_2N(CHCH)_2$	Sn: Org.Verb.2-12

Formula	Compound	Reference
$C_8H_{15}NSn$	$(CH_3)_3SnC_4H_3NCH_3$	Sn: Org.Verb.2-153, 156
$C_8H_{15}NaSn$	$(C_2H_5)_3SnCCNa$	Sn: Org.Verb.2-221
$C_8H_{15}OS_2^-$	$[S_2COC_7H_{15}]^-$	C: MVol.D4-251
$C_8H_{16}I_2MnO_4$	$MnI_2 \cdot 2\ C_4H_8O_2$	Mn: MVol.D1-155
$C_8H_{16}I_4N_4S_2Sn$	$SnI_4 \cdot 2\ NH_2CSNHCH_2CHCH_2$	Sn: MVol.C6-121
$C_8H_{16}I_4O_2S_4Sn$	$SnI_4 \cdot 2\ S(CH_2CH_2)_2SO$	Sn: MVol.C6-114
$C_8H_{16}I_4O_4Sn$	$SnI_4 \cdot 2\ C_4H_8O_2$	Sn: MVol.C5-159/60
$C_8H_{16}N_2O_4S$	$C_6H_{11}NHSO_2NHCOOCH_3$	S: S-N-Verb.1-170
–	$SO_2[N(CH_2CH_2)_2O]_2$	S: S-N-Verb.1-19
$C_8H_{16}N_2O_{10}U$	$UO_2(CH_3COO)_2 \cdot 2\ NH_2CH_2COOH$	U: SVol.E1-18, 21
–	$UO_2(NO_3)_2 \cdot 2\ (CH_2)_4O$	U: SVol.E1-71, 76
–	$UO_2(NO_3)_2 \cdot 2\ CH_3COC_2H_5$	U: SVol.E1-79/80
$C_8H_{16}N_2O_{12}S_2U$	$UO_2(NO_3)_2 \cdot 2\ SO(CH_2CH_2)_2O$	U: SVol.E1-216/7
$C_8H_{16}N_2O_{12}U$	$UO_2(NO_3)_2 \cdot [CH_2OCH_2]_4 \cdot 2\ H_2O$	U: SVol.E1-73, 78
–	$UO_2(NO_3)_2 \cdot 2\ CH_3COOC_2H_5$	U: SVol.E1-120/2
–	$UO_2(NO_3)_2 \cdot 2\ C_4H_8O_2$	U: SVol.E1-71, 76
$C_8H_{16}N_2O_{13}Sn$	$(NH_4)_2SnO(C_4H_4O_6)_2 \cdot 1.5\ H_2O$	Sn: MVol.C3-125/6
$C_8H_{16}N_4S_2Te^{2+}$	$[Te(SC(NHCH_2)_2CH_2)_2]^{2+}$	Te: SVol.B3-159
$C_8H_{16}N_6O_{10}U$	$UO_2(NO_3)_2 \cdot 2\ CO(NHCH_2)_2CH_2$	U: SVol.E1-92, 95
$C_8H_{16}N_6O_{12}U$	$UO_2(NO_3)_2 \cdot 2\ (CH_3CO)_2N_2H_2 \cdot 2\ H_2O$	U: SVol.E1-114/6
$C_8H_{16}NiO_2P_2$	$(CO)_2Ni(CH_3)_2P(CH_2)_2P(CH_3)_2$	Ni: Org.Verb.1-149
$C_8H_{16}Ni_2$	$(C_3H_5NiCH_3)_2$	Ni: Org.Verb.2-284/5
$C_8H_{16}Ni_2O_2$	$(C_3H_5NiOCH_3)_2$	Ni: Org.Verb.2-282
$C_8H_{16}O_2Sn$	$(CH_3)_3SnC(CH_2)COOC_2H_5$	Sn: Org.Verb.2-79, 87
–	$(CH_3)_3SnCHCHCOOC_2H_5$	Sn: Org.Verb.2-79, 87
$C_8H_{16}O_3S_5Zn$	$Zn[S_2COC_2H_5]_2((CH_3)_2SO)$	C: MVol.D4-257
$C_8H_{16}O_5Sn_2$	$[(CH_3)_2Sn(C_2H_4(COO)_2) \cdot (CH_3)_2SnO]_x$	Sn: Org.Verb.6-28
$C_8H_{16}Sn$	$(CH_2)_4Sn(CH_2)_4$	Sn: Org.Verb.3-127
–	$(CH_3)_2Sn(CH_2CHCH_2)_2$	Sn: Org.Verb.3-7, 17
–	$(CH_3)_3SnCCCH(CH_3)_2$	Sn: Org.Verb.2-113, 117
–	$(CH_3)_3SnCCC_3H_7$	Sn: Org.Verb.2-113
–	$(CH_3)_3SnC(CHCH_2)CHCH_3$	Sn: Org.Verb.2-79, 86
–	$(CH_3)_3SnC(CH_2)C(CH_3)CH_2$	Sn: Org.Verb.2-79, 86
–	$(CH_3)_3SnC(CH_3)CCHCH_3$	Sn: Org.Verb.2-79, 86
–	$(CH_3)_3SnC(CH_3)CHCHCH_2$	Sn: Org.Verb.2-79, 86
–	$(CH_3)_3SnCHCC(CH_3)_2$	Sn: Org.Verb.2-78, 86
–	$(CH_3)_3SnCHCCHC_2H_5$	Sn: Org.Verb.2-78
–	$(CH_3)_3SnCHCHC(CH_3)CH_2$	Sn: Org.Verb.2-78, 86
–	$(CH_3)_3SnCHCHCHCHCH_3$	Sn: Org.Verb.2-79
–	$(CH_3)_3Sn(CHCH_2CHCH_2CH)$	Sn: Org.Verb.2-98, 105
–	$(CH_3)_3Sn(CHCH_2CH_2CHCH)$	Sn: Org.Verb.2-98, 105
–	$(CH_3)_3SnCH_2C(CH_3)CCH_2$	Sn: Org.Verb.2-78
–	$(CH_3)_3SnCH_2CH(CH_3)CCH$	Sn: Org.Verb.2-113, 117
–	$(CH_3)_3SnCH_2CH_2CCCH_3$	Sn: Org.Verb.2-113, 117
–	$(C_2H_5)_2Sn(CHCH_2)_2$	Sn: Org.Verb.3-25/6
–	$(C_2H_5)_3SnCCH$	Sn: Org.Verb.2-206, 208
–	$(C_2H_5)_3SnCCD$	Sn: Org.Verb.2-208
$C_8H_{17}IrN_6O_8$	$H[Ir(NO_2)_2(CH_3C(NOH)C(NOH)CH_3)_2]$	Ir: SVol.2-61
$C_8H_{17}NNi$	$(CH_3)_2C_3H_3Ni(NH(CH_2)_2)CH_3$	Ni: Org.Verb.2-35
$C_8H_{17}NS_5$	$[(C_2H_5)_3NH](HC_2S_5)$	C: MVol.D4-162, 217

Formula	Compound	Reference
$C_8H_{17}NSn$	$(C_2H_5)_3SnCH_2CN$	Sn: Org.Verb.2-163, 164
$C_8H_{18}I_2N_2O_2Sn$	$SnI_2 \cdot 2\ HN(CH_2CH_2)_2O$	Sn: MVol.C5-29
$C_8H_{18}I_4N_2O_2Te$	$[TeI_3(CH_3CON(CH_3)_2)_2]I$	Te: SVol.B3-174
$C_8H_{18}MnO_8$	$Mn(OC_4H_6(OH)_3)_2$	Mn: MVol.D1-38/9
$C_8H_{18}N_2NiO_2S_4$	$Ni[S_2COC_2H_5]_2(C_2H_8N_2)$	C: MVol.D4-260
$C_8H_{18}N_2O_4Sn$	$Sn(HN(CH_2CH_2O)_2)_2$	Sn: MVol.C6-2
$C_8H_{18}N_2O_5S_3Se_2U$	$UO_2(NCSe)_2 \cdot 3\ (CH_3)_2SO$	U: SVol.E1-203, 207
$C_8H_{18}N_2O_5S_5U$	$UO_2(NCS)_2 \cdot 3\ (CH_3)_2SO$	U: SVol.E1-203, 207
$C_8H_{18}N_2Sn$	$(CH_3)_3SnCH(CN)CH_2N(CH_3)_2$	Sn: Org.Verb.2-38
$C_8H_{18}N_4O_2S$	$(C_4H_9)_2NSO_2N_3$	S: S-N-Verb.1-214
$C_8H_{18}N_4O_4S$	$SO_2[N(C_4H_9)NO]_2$	S: S-N-Verb.1-176
–	$SO_2[N(i\text{-}C_4H_9)NO]_2$	S: S-N-Verb.1-176
$C_8H_{18}N_4O_4S_6Te$	$Te(SC(NHCH_2)_2)_2(S_2O_2CH_3)_2$	Te: SVol.B3-155/6
$C_8H_{18}N_4O_6S$	$SO_2[N(C_4H_9)NO_2]_2$	S: S-N-Verb.1-176
–	$SO_2[N(i\text{-}C_4H_9)NO_2]_2$	S: S-N-Verb.1-176
$C_8H_{18}N_4O_{12}U$	$UO_2(NO_3)_2 \cdot 2\ CH_3NHCOOC_2H_5$	U: SVol.E1-120/3
$C_8H_{18}N_6O_9U$	$UO_2C_2O_4 \cdot 3\ NH_2CONHCH_3$	U: SVol.E1-93, 95
$C_8H_{18}N_8O_8S_4Te$	$Te(SC(NH_2)_2)_4(HC_2O_4)_2 \cdot 2\ H_2O$	Te: SVol.B3-146
$C_8H_{18}NiOS$	$(CH_3)_2C_3H_3Ni(OS(CH_3)_2)CH_3$	Ni: Org.Verb.2-34
$C_8H_{18}NiO_2$	$Ni(OC(CH_3)_3)_2$	Ni: Org.Verb.1-11
–	$[Ni(OC(CH_3)_3)_2]_n$	Ni: Org.Verb.2-388
$C_8H_{18}NiO_2P_2$	$(CO)_2Ni(P(CH_3)_3)_2$	Ni: Org.Verb.1-127, 128
$C_8H_{18}NiO_3^{2+}$	$[(CH_3)_4C_4Ni(H_2O)_3]^{2+}$	Ni: Org.Verb.2-325
$C_8H_{18}NiO_8P_2$	$(CO)_2Ni(P(OCH_3)_3)_2$	Ni: Org.Verb.1-139/40
$C_8H_{18}OSiSn$	$(CH_3)_3SnC[Si(CH_3)_3]CO$	Sn: Org.Verb.2-26, 29
$C_8H_{18}OSn$	$(CH_3)_3Sn(CH_2)_3COCH_3$	Sn: Org.Verb.2-43, 46
–	$[(C_4H_9)_2SnO]_x$	Sn: Org.Verb.6-91
–	$[(s\text{-}C_4H_9)_2SnO]_x$	Sn: Org.Verb.6-106
–	$[(t\text{-}C_4H_9)_2SnO]_x$	Sn: Org.Verb.6-107
$C_8H_{18}O_2Sn$	$(CH_3)_3SnCH(CH_3)CH_2COOCH_3$	Sn: Org.Verb.2-44
–	$(CH_3)_3SnCH_2CH_2COOC_2H_5$	Sn: Org.Verb.2-34
–	$(CH_3)_3SnCH_2CH_2OCH_2CH(O)CH_2$	Sn: Org.Verb.2-37
–	$(C_8H_{17}SnOOH)_x$	Sn: Org.Verb.6-252
$C_8H_{18}O_8U$	$UO_2(CH_3COO)_2 \cdot 2\ C_2H_5OH$	U: SVol.E1-59, 63
$C_8H_{18}SiSn$	$(CH_3)_3SnCCSi(CH_3)_3$	Sn: Org.Verb.2-114, 118
$C_8H_{18}Sn$	$(CH_3)_2Sn(CH_2)_6$	Sn: Org.Verb.3-108
–	$(CH_3)_3SnC(CH_2)CH(CH_3)_2$	Sn: Org.Verb.2-79, 87
–	$(CH_3)_3SnC(CH_3)C(CH_3)_2$	Sn: Org.Verb.2-79, 87
–	$(CH_3)_3SnC(CH_3)CHC_2H_5$	Sn: Org.Verb.2-79
–	$(CH_3)_3SnC(C_2H_5)CHCH_3$	Sn: Org.Verb.2-79, 87
–	$(CH_3)_3SnCHCHC_3H_7$	Sn: Org.Verb.2-79
–	$(CH_3)_3SnCH(CH_2)_4$	Sn: Org.Verb.2-61, 67
–	$(CH_3)_3SnCH(CH_3)CHCHCH_3$	Sn: Org.Verb.2-80
–	$(CH_3)_3SnCH_2C(CH_3)CHCH_3$	Sn: Org.Verb.2-80, 87
–	$(CH_3)_3SnCH_2CHC(CH_3)_2$	Sn: Org.Verb.2-80, 87
–	$(CH_3)_3SnCH_2CHCHC_2H_5$	Sn: Org.Verb.2-80, 87
–	$(CH_3)_3SnCH_2CH(CH_3)CHCH_2$	Sn: Org.Verb.2-80
–	$(CH_3)_3SnCH_2CH_2C(CH_3)CH_2$	Sn: Org.Verb.2-80
–	$(CH_3)_3SnCH_2CH_2CHCHCH_3$	Sn: Org.Verb.2-80, 88
–	$(CH_3)_3SnCH_2CH_2CH_2CHCH_2$	Sn: Org.Verb.2-80, 88

Formula	Compound	Reference
$C_8H_{18}Sn$	$(C_2H_5)_3SnCHCH_2$	Sn: Org.Verb.2-181, 183
–	$[(C_4H_9)_2Sn]_x$	Sn: Org.Verb.6-84, 88
$C_8H_{18}Sn_2$	$(CH_3)_3SnCCSn(CH_3)_3$	Sn: Org.Verb.5-80
$C_8H_{18}Ti$	$(C_4H_9)_2Ti$	Ti: Org.Verb.1-67
$C_8H_{19}NNi$	$(CH_3)_2C_3H_3Ni(NH_2C_2H_5)CH_3$	Ni: Org.Verb.2-35
$C_8H_{19}NOSn$	$(CH_3)_3SnCH_2N(CH_2CH_2)_2O$	Sn: Org.Verb.2-12/3
–	$(CH_3)_3SnCONH(t\text{-}C_4H_9)$	Sn: Org.Verb.2-17
$C_8H_{19}NO_3Sn$	$(i\text{-}C_3H_7O)_2Sn(NHCH_2CH_2O)$	Sn: MVol.C6-2
$C_8H_{19}Sn$	$(C_2H_5)_3SnCHCH_3$	Sn: Org.Verb.1-71
–	$(C_2H_5)_3SnCH_2CH_2$	Sn: Org.Verb.1-71
$C_8H_{20}I_2MnN_4O_4$	$MnI_2 \cdot 4\ CH_3CONH_2$	Mn: MVol.C5-321
$C_8H_{20}I_2O_4U$	$UO_2I_2 \cdot 2\ (C_2H_5)_2O$	U: SVol.E1-71, 76
$C_8H_{20}I_3NSn$	$[(C_2H_5)_4N]SnI_3$	Sn: MVol.C3-119/20
$C_8H_{20}I_5NTe$	$[(C_2H_5)_4N]TeI_5$	Te: SVol.B3-69/70
$C_8H_{20}I_6NU$	$[N(C_2H_5)_4]UI_6$	U: SVol.C9-176
$C_8H_{20}IrN_7O_8$	$NH_4[Ir(NO_2)_2(CH_3C(NOH)C(NOH)CH_3)_2]$	Ir: SVol.2-61
$C_8H_{20}MnNO_4$	$(C_2H_5)_4NMnO_4$	Mn: MVol.C2-202
$C_8H_{20}NO_{11}UV_3$	$C_8H_{17}NH_3[UO_2(VO_3)_3] \cdot n\ H_2O$	U: SVol.C3-302/3
$C_8H_{20}N_2O_2S$	$(C_2H_5)_2NSO_2N(C_2H_5)_2$	S: S-N-Verb.1-173
–	$(C_2H_5)_2NSO_2NHC_4H_9$	S: S-N-Verb.1-170
–	$(C_4H_9)_2NSO_2NH_2$	S: S-N-Verb.1-168
–	$(C_4H_9NH)_2SO_2$	S: S-N-Verb.1-169
–	$(i\text{-}C_4H_9NH)_2SO_2$	S: S-N-Verb.1-169
–	$(s\text{-}C_4H_9NH)_2SO_2$	S: S-N-Verb.1-169
$C_8H_{20}N_2O_2SSi$	$(CH_2)_5NSO_2NHSi(CH_3)_3$	S: S-N-Verb.1-172
$C_8H_{20}N_2O_2Sn$	$Sn(OC_2H_4N(CH_3)_2)_2$	Sn: MVol.C5-30
$C_8H_{20}N_2O_3Sn_2$	$(SnOC_2H_4N(CH_3)_2)_2O$	Sn: MVol.C5-30
$C_8H_{20}N_2O_{10}S_2U$	$UO_2(NO_3)_2 \cdot 2\ (C_2H_5)_2SO$	U: SVol.E1-211, 215
$C_8H_{20}N_2O_{10}U$	$UO_2(NO_3)_2 \cdot 2\ (C_2H_5)_2O$	U: SVol.E1-71, 76
–	$UO_2(NO_3)_2 \cdot 2\ (C_2H_5)_2O \cdot 2\ H_2O$	U: SVol.E1-71, 76
$C_8H_{20}N_4NpO_{11}$	$(C_2H_5)_4NNpO_2(NO_3)_3$	Np: TrU.D2-356
$C_8H_{20}N_4O_4S_2$	$(CH_3)_2NSO_2N(CH_2CH_2)_2NSO_2N(CH_3)_2$	S: S-N-Verb.1-174
$C_8H_{20}N_4O_8S_4$	$[C_2H_5NSO_2]_4$	S: S-N-Verb.1-4
$C_8H_{20}N_4O_{11}Pu$	$(C_2H_5)_4NPuO_2(NO_3)_3$	Np: TrU.D2-356
$C_8H_{20}N_4O_{12}S_2U$	$U(SO_4)_2 \cdot 4\ CH_3CONH_2$	U: SVol.E1-96/8
$C_8H_{20}N_7O_{18}Pa$	$[N(C_2H_5)_4]Pa(NO_3)_6$	Pa: SVol.2-35
$C_8H_{20}OSn$	$(CH_3)_3Sn(CH_2)_2C(CH_3)_2OH$	Sn: Org.Verb.2-55, 57
–	$(CH_3)_3Sn(CH_2)_5OH$	Sn: Org.Verb.2-55
–	$(C_2H_5)_3SnCH_2OCH_3$	Sn: Org.Verb.2-163, 164
$C_8H_{20}O_2S_6Sn$	$Sn(SCH_2CH_2S)_2 \cdot 2\ (CH_3)_2SO$	Sn: MVol.C6-101
$C_8H_{20}O_2Sn$	$[(CH_3)_3C]_2Sn(OH)_2$	Sn: Org.Verb.6-107
$C_8H_{20}O_2Ti$	$(CH_3)_4Ti \cdot C_4H_8O_2$	Ti: Org.Verb.1-113
$C_8H_{20}O_3SSn$	$(C_2H_5)_3SnOSO_2C_2H_5$	Sn: Org.Verb.5-22
$C_8H_{20}O_3Sn$	$C_2H_5Sn(OC_2H_5)_3$	Sn: Org.Verb.6-230/1
$C_8H_{20}O_4P_2S_4Te$	$Te((C_2H_5O)_2P(S)S)_2$	Te: SVol.B3-165
$C_8H_{20}O_4P_2Se_4Sn$	$Sn((C_2H_5O)_2P(Se)Se)_2$	Sn: MVol.C5-62
$C_8H_{20}O_4Sn$	$Sn(CH_2OCH_3)_4$	Sn: Org.Verb.1-123
$C_8H_{20}P_2S_2Se_2Te$	$Te((C_2H_5)_2P(S)Se)_2$	Te: SVol.B3-163
$C_8H_{20}P_2S_4Te$	$Te((C_2H_5)_2P(S)S)_2$	Te: SVol.B3-162/3
$C_8H_{20}P_2Se_4Te$	$Te((C_2H_5)_2P(Se)Se)_2$	Te: SVol.B3-164

Formula	Compound		Reference
$C_8H_{20}S_{2.5}Sn_2$	$[(C_2H_5)_2Sn]_2S_{2.5}$	Sn:	Org.Verb.6-59
$C_8H_{20}Sn$	$(CH_3)_2Sn(C_2H_5)(s\text{-}C_4H_9)$	Sn:	Org.Verb.3-75, 76
–	$(CH_3)_2Sn(C_3H_7)_2$	Sn:	Org.Verb.3-7
–	$(CH_3)_2Sn(C_3H_7)(i\text{-}C_3H_7)$	Sn:	Org.Verb.3-76
–	$(CH_3)_2Sn(i\text{-}C_3H_7)_2$	Sn:	Org.Verb.3-7
–	$(CH_3)_3SnC(CH_3)_2C_2H_5$	Sn:	Org.Verb.2-54
–	$(CH_3)_3SnCH(CH_3)CH(CH_3)_2$	Sn:	Org.Verb.2-54
–	$(CH_3)_3SnCH(CH_3)C_3H_7$	Sn:	Org.Verb.2-54
–	$(CH_3)_3SnCH(C_2H_5)_2$	Sn:	Org.Verb.2-54
–	$(CH_3)_3SnCH_2CH_2CH(CH_3)_2$	Sn:	Org.Verb.2-52/3
–	$(CH_3)_3SnC_5H_{11}$	Sn:	Org.Verb.2-52
–	$(C_2H_5)_2Sn(CH_3)(C_3H_7)$	Sn:	Org.Verb.3-86/7
–	$(C_4H_9)_2SnH_2$	Sn:	Org.Verb.4-104/10
–	$(C_4H_9)_2SnD_2$	Sn:	Org.Verb.4-104, 108
–	$(i\text{-}C_4H_9)_2SnH_2$	Sn:	Org.Verb.4-110/1
–	$(s\text{-}C_4H_9)_2SnH_2$	Sn:	Org.Verb.4-112
–	$(t\text{-}C_4H_9)_2SnH_2$	Sn:	Org.Verb.4-112
–	$C_8H_{17}SnH_3$	Sn:	Org.Verb.4-126/7
–	$Sn(C_2H_5)_4$		
	Chemical reactions	Sn:	Org.Verb.1-70/9
	Formation	Sn:	Org.Verb.1-57/62
	Molecule	Sn:	Org.Verb.1-62
	Physical properties	Sn:	Org.Verb.1-65/9
	Physiology	Sn:	Org.Verb.1-80/1
	Preparation	Sn:	Org.Verb.1-57/62
	Purity testing	Sn:	Org.Verb.1-59
	Spectra	Sn:	Org.Verb.1-62/5
	Uses	Sn:	Org.Verb.1-81/2
–	$Sn(C_2H_5)_4$ solutions		
	$Sn(C_2H_5)_4$-organic solvents	Sn:	Org.Verb.1-79
$C_8H_{20}Sn^+$	$Sn(C_2H_5)_4^+$	Sn:	Org.Verb.1-70/1
$C_8H_{20}Ti$	$(C_2H_5)_4Ti$	Ti:	Org.Verb.1-6, 90, 98
$C_8H_{21}INiOP_2$	$CH_3C(O)Ni(P(CH_3)_3)_2I$	Ni:	Org.Verb.1-14
$C_8H_{21}NNiP_2S$	$CH_3Ni(P(CH_3)_3)_2NCS$	Ni:	Org.Verb.1-18, 24
$C_8H_{21}N_3O_4S_2$	$HN(SO_2N(C_2H_5)_2)_2$	S:	S-N-Verb.1-161
–	$HN(SO_2NHCH_2CH(CH_3)_2)_2$	S:	S-N-Verb.1-161
–	$HN(SO_2NHC_4H_9)_2$	S:	S-N-Verb.1-161
$C_8H_{21}O_5U$	$UO(OC_2H_5)_3 \cdot C_2H_5OH$	U:	SVol.E1-58/9
$C_8H_{22}I_4O_4P_2Sn$	$SnI_4 \cdot 2\ C_2H_5(H)P(O)OC_2H_5$	Sn:	MVol.C6-151/2
$C_8H_{22}N_2O_3S$	$[(C_2H_5)_2NH_2][(C_2H_5)_2NSO_3]$	S:	S-N-Verb.1-97
$C_8H_{22}N_4O_8U$	$UO_2(NO_3)_2 \cdot 2\ (C_2H_5)_2NH$	U:	SVol.E1-19, 22
$C_8H_{22}N_8O_6S_4U$	$UO_2(CH_3COO)_2 \cdot 4\ (NH_2)_2CS$	U:	SVol.E1-220/1
$C_8H_{22}N_8O_{10}U$	$UO_2(CH_3COO)_2 \cdot 4\ (NH_2)_2CO$	U:	SVol.E1-84, 89
$C_8H_{22}NiSi_2$	$[(CH_3)_3SiCH_2]_2Ni$	Ni:	Org.Verb.1-71, 77
$C_8H_{22}OSi_2Sn$	$(CH_3)_2Sn(CH_2Si(CH_3)_2)_2O$	Sn:	Org.Verb.3-107
–	$[CH_2Si(CH_3)_2OSi(CH_3)_2CH_2Sn(CH_3)_2]_n$	Sn:	Org.Verb.6-19
$C_8H_{22}O_6U$	$UO_2(OC_2H_5)_2 \cdot 2\ C_2H_5OH$	U:	SVol.E1-59, 64
$C_8H_{23}IrN_6$	$Ir((NHC_2H_4)(NH_2C_2H_4)NH)((NHC_2H_4)_2NH)$	Ir:	SVol.2-95
$C_8H_{23}N_3Ti$	$C_2H_5Ti(N(CH_3)_2)_3$	Ti:	Org.Verb.1-52

Formula	Compound	Reference
$C_8H_{24}I_4N_2Sn$	$[(CH_3)_4N]_2SnI_4$	Sn: MVol.C3-119/20
$C_8H_{24}I_4N_4Sn$	$SnI_4 \cdot 2\ NH_2(CH_2)_4NH_2$	Sn: MVol.C5-185
$C_8H_{24}I_6N_2Sn$	$[(CH_3)_4N]_2[SnI_6]$	Sn: MVol.C3-120
$C_8H_{24}I_6N_2Te$	$[(CH_3)_4N]_2TeI_6$	Te: SVol.B3-67, 72
$C_8H_{24}MnN_6O_8$	$[(CH_3)_4N]_2[Mn(NO_2)_4]$	Mn: MVol.C3-266/7
$C_8H_{24}N_2O_2SSi_2$	$(CH_3)_2NSO_2N[Si(CH_3)_3]_2$	S: S-N-Verb.1-174
$C_8H_{24}N_2O_9S_3Sn$	$[(CH_3)_4N]_2Sn(SO_3)_3$	Sn: MVol.C3-121/2
$C_8H_{24}N_2O_{12}S_4U$	$UO_2(NO_3)_2 \cdot 4\ (CH_3)_2SO$	U: SVol.E1-204
–	$UO_2(NO_3)_2 \cdot 4\ (CH_3)_2SO \cdot H_2O$	U: SVol.E1-208
$C_8H_{24}N_2O_{13}U_4$	$[(C_2H_5)_2NH_2]_2O \cdot 4\ UO_3 \cdot n\ H_2O$	U: SVol.C3-65/6
$C_8H_{24}N_2O_{49}U_{16}$	$[NH_2(C_2H_5)_2]_2U_{16}O_{49}$	U: SVol.C3-65
$C_8H_{24}N_3O_7S_2$	$[(CH_3)_4N]_2[NO(SO_3)_2]$	S: S-N-Verb.1-57
$C_8H_{24}N_3SSi_2^+$	$[(CH_3)_3SiN]_2Sn(CH_3)_2^+$	S: S-N-Verb.1-210
$C_8H_{24}N_3SSi_3^+$	$S[NSi(CH_3)_3]_2NSi(CH_3)_2^+$	S: S-N-Verb.1-207
$C_8H_{24}N_4S_2Si_2$	$[NS(CH_3)_2NSi(CH_3)_2]_2$	S: S-N-Verb.1-7
$C_8H_{24}N_6Ni_2S_6$	$Ni_2(C_2H_8N_2)_3(CS_3)_2$	C: MVol.D4-225
$C_8H_{24}N_6O_{11}P_2U$	$UO_2(NO_3)_2 \cdot [((CH_3)_2N)_2PO]_2O$	U: SVol.E1-166/7
$C_8H_{24}N_{20}Sn$	$[(CH_3)_4N]_2[Sn(N_3)_6]$	Sn: MVol.C3-83
$C_8H_{24}NiP_2$	$(CH_3)_2Ni(P(CH_3)_3)_2$	Ni: Org.Verb.1-69, 72
$C_8H_{24}Si_2Sn$	$(CH_3)_2Sn[CH_2SiH(CH_3)_2]_2$	Sn: Org.Verb.3-8
$C_8H_{24}Ti_2$	$(CH_3)_3Ti(CH_3)_2Ti(CH_3)_3$	Ti: Org.Verb.1-96
$C_8H_{26}Ni_2O_2P_2$	$(CH_3Ni(P(CH_3)_3)OH)_2$	Ni: Org.Verb.2-250
–	$(CH_3Ni(P(CH_3)_3)OD)_2$	Ni: Org.Verb.2-250
$C_8H_{26}O_{10}Sn_3$	$[(C_2H_5Sn)_3(OCOCH_3)(OH)_8]$	Sn: Org.Verb.6-230
$C_8H_{28}I_4N_4Sn$	$SnI_4 \cdot 4\ (CH_3)_2NH$	Sn: MVol.C5-170
$C_8H_{28}N_2Ni_2P_2$	$(CH_3Ni(P(CH_3)_3)NH_2)_2$	Ni: Org.Verb.2-253
$C_8H_{32}I_4N_8Sn$	$SnI_4 \cdot 4\ C_2H_8N_2$	Sn: MVol.C5-183
$C_8H_{32}I_4N_{16}O_8U$	$UI_4 \cdot 8\ (NH_2)_2CO$	U: SVol.E1-82/3, 85
$C_8H_{32}N_8Na_4Sn_9$	$Na_4Sn_9 \cdot 4\ C_2H_8N_2$	Sn: MVol.C5-13
$C_8Ir_2O_8$	$Ir_2(CO)_8$	Ir: SVol.2-113
$C_8K_4N_6Ni_2O_2$	$K_4[Ni_2(CN)_6(CO)_2]$	Ni: Org.Verb.2-257
$C_8K_4O_{16}Sn$	$K_4Sn(C_2O_4)_4$	Sn: MVol.C3-79
$C_8N_4NiS_4^-$	$[Ni(S_2C_2(CN)_2)_2]^-$	Ni: Org.Verb.2-207
$C_8N_8PaS_8^{4-}$	$Pa(NCS)_8^{4-}$	Pa: SVol.2-35
$C_8Na_4O_{16}Sn$	$Na_4Sn(C_2O_4)_4 \cdot 3\ H_2O$	Sn: MVol.C3-34/5
$C_8Ni_2O_8$	$Ni_2(CO)_8$	Ni: Org.Verb.2-266
$C_8Ni_3O_8^{2-}$	$[Ni_3(CO)_8]^{2-}$	Ni: Org.Verb.2-373
$C_8NpO_{16}^{4-}$	$Np(C_2O_4)_4^{4-}$	Np: TrU.D1-158
$C_8O_{16}Pu^{4-}$	$Pu(C_2O_4)_4^{4-}$	Np: TrU.D1-158
$C_8O_{16}Pu^{5-}$	$Pu(C_2O_4)_4^{5-}$	Np: TrU.D1-156
$C_8O_{16}Sn^{4-}$	$Sn(C_2O_4)_4^{4-}$	Sn: MVol.C2-224/5
$C_{8.4}F_7O_{1.3}Sn$	$(C_6F_5)_{1.4}SnO_{1.3}$	F: PerFHalOrg.4-181, 186
$C_{8.5}H_{10.5}N_{1.5}O_{9.5}U$	$U(C_2O_4)_2 \cdot 1.5\ HCON(CH_3)_2 \cdot H_2O$	U: SVol.E1-99/100
$C_9CeCl_4H_{18}NO_2$	$CeCl_3 \cdot [C_5H_5NH]Cl \cdot 2\ C_2H_5OH$	Sc: MVol.C5-213/5, 217
$C_9CfCl_2H_4NO^{2+}$	$Cf(NC_9H_4Cl_2O)^{2+}$	Np: TrU.D1-168
$C_9CfH_8NO_3$	$Cf(OH)_2NC_9H_6O$	Np: TrU.D2-317
$C_9CfH_{15}O_9$	$Cf(CH_3CH(OH)COO)_3$	Np: TrU.D1-154
$C_9ClF_2H_{12}N_3$	$[(CH_3)_2N]_2C_5F_2ClN$	F: PerFHalOrg.5-138
$C_9ClF_3H_3N_5O_4S$	$C_2N_2S(CF_3)NHC_6H_2Cl(NO_2)_2$	F: PerFHalOrg.5-107
$C_9ClF_3H_4N_4$	$NC_5F_3ClNHC_4H_3N_2$	F: PerFHalOrg.5-215/6

Formula	Compound		Reference
$C_9ClF_3H_8N_2O$	$(O(CH_2CH_2)_2N)C_5F_3ClN$	F:	PerFHalOrg.5-138
$C_9ClF_3H_{12}N_4$	$(C_4H_9NH)(NH_2)(CF_3)C_4ClN_2$	F:	PerFHalOrg.6-59
C_9ClF_5HNO	HOC_9F_5ClN	F:	PerFHalOrg.6-130/1, 143, 160
$C_9ClF_6H_4NO_2$	$ClC_6H_4COON(CF_3)_2$	F:	PerFHalOrg.7-137/8
$C_9ClF_6H_6N$	$ClC_6H_4C(CF_3)_2NH_2$	F:	PerFHalOrg.7-69, 71
$C_9ClF_6H_6NO$	$ClC_6H_4CH_2ON(CF_3)_2$	F:	PerFHalOrg.7-137/8
$C_9ClF_6H_{10}O_3P$	$C_5F_6(Cl)PO(OC_2H_5)_2$	F:	PerFHalOrg.3-57, 59
C_9ClF_6N	$NCClC_6F_4CFCF$	F:	PerFHalOrg.6-132/3, 146
–	$NC_6F_4CFCFCCl$	F:	PerFHalOrg.6-130/1, 142, 153, 161
$C_9ClF_9H_{12}Sn$	$(CF_3CH_2CH_2)_3SnCl$	Sn:	Org.Verb.5-150
$C_9ClF_{10}HO$	$CF_3(C_6F_5)C(OH)CF_2Cl$	F:	PerFHalOrg.4-51
$C_9ClF_{12}H_4N_3O$	$(CF_3)_2CNHC(CF_3)_2NC(NHCOCH_2Cl)$	F:	PerFHalOrg.5-105
$C_9ClF_{12}H_{11}N_2O_2Si$	$(CH_3)_3SiCH[ON(CF_3)_2]CHClON(CF_3)_2$	F:	PerFHalOrg.7-127/8
$C_9ClF_{14}N_3$	$(C_3F_7)_2C_3ClN_3$	F:	PerFHalOrg.6-79/80, 100, 112
$C_9ClF_{21}Si$	$(C_3F_7)_3SiCl$	F:	PerFHalOrg.4-147
$C_9ClFeH_9O_3$	$C_6H_9Fe(CO)_3Cl$	Fe:	Org.Verb.B5-54
$C_9ClFeH_{10}O_6P$	$(CO)_4FePCl(OCH_2)_2C(CH_3)_2$	Fe:	Org.Verb.B2-96, 117/8
$C_9ClFeH_{13}NO_5P$	$ClC_3H_4Fe(CO)(NO)P(OCH_2)_3CCH_3$	Fe:	Org.Verb.B5-108/11
$C_9ClFeH_{18}HgO_9P_2^+$	$[(CO)_3Fe(P(OCH_3)_3)_2HgCl]^+$	Fe:	Org.Verb.B1-194
$C_9ClFeH_{21}O_2P_2$	$(CO)_2Fe(P(CH_3)_3)_2(CH_3)Cl$	Fe:	Org.Verb.B1-158
$C_9ClFe_2H_{12}O_5PS_2$	$(CO)_3Fe(SCH_3)_2Fe(CO)_2P(CH_3)_2Cl$	Fe:	Org.Verb.C1-93
$C_9ClFe_3N_3O_{12}Sn$	$ClSn(Fe(CO)_3NO)_3$	Fe:	Org.Verb.B1-198
$C_9ClH_5IrKNO_8$	$K[Ir(C_2O_4)_2(C_5H_5N)Cl]$	Ir:	SVol.2-35
$C_9ClH_5IrK_2NO_8$	$K_2[Ir(C_2O_4)_2(C_5H_5N)Cl]$	Ir:	SVol.2-35
C_9ClH_6NOSn	$SnCl(C_9H_6NO)$	Sn:	MVol.C5-20
$C_9ClH_7NOS^+$	$[C_9H_7NSOCl]^+$	S:	SVol.1-52
$C_9ClH_7O_2$	$C_6H_5CHCClCOOH$	F:	PerFHalOrg.4-46
C_9ClH_7Ti	C_9H_7TiCl	Ti:	Org.Verb.1-133
$C_9ClH_8NO_2S$	$C_6H_5CHCHCHNSO_2Cl$	S:	S-N-Verb.1-126
$C_9ClH_8NO_3S$	$ClSO_2N(COCH_2)CHC_6H_5$	S:	S-N-Verb.1-126
$C_9ClH_{12}N_3Sn$	$(CH_2CH_2CN)_3SnCl$	Sn:	MVol.C6-184
		Sn:	Org.Verb.5-151
$C_9ClH_{13}NOSSn$	$C_4H_9SnCl(SC_5H_4NO)$	Sn:	Org.Verb.6-246
$C_9ClH_{13}OSn$	$(CH_3)_2(CH_3OC_6H_4)SnCl$	Sn:	Org.Verb.5-194
$C_9ClH_{13}SSn$	$(CH_3)_3SnSC_6H_4Cl$	Sn:	Org.Verb.5-84
$C_9ClH_{13}Sn$	$(CH_3)_2(C_6H_5CH_2)SnCl$	Sn:	Org.Verb.5-194
–	$(CH_3)_2Sn(C_6H_5)(CH_2Cl)$	Sn:	Org.Verb.3-79
–	$(CH_3)_3SnC_6H_4Cl$	Sn:	Org.Verb.2-132/3
$C_9ClH_{15}O_2Ti$	$C_5H_5Ti(OC_2H_5)_2Cl$	Ti:	Org.Verb.1-184/7
$C_9ClH_{15}O_4S_2Sn$	$(CH_2CHCH_2)ClSn[OS(O)CH_2CHCH_2]_2$	Sn:	Org.Verb.5-157
$C_9ClH_{15}O_6Sn$	$(HOOCCH_2CH_2)_3SnCl$	Sn:	Org.Verb.5-151
$C_9ClH_{15}Sn$	$(CH_2C(CH_3))_3SnCl$	Sn:	Org.Verb.5-157
–	$(CH_2CHCH_2)_3SnCl$	Sn:	Org.Verb.5-157
–	$(CH_3)_2(C_7H_9)SnCl$	Sn:	Org.Verb.5-195, 198
–	$(C_2H_5)_2(C_5H_5)SnCl$	Sn:	Org.Verb.5-201
–	$(C_3H_5)_3SnCl$	Sn:	Org.Verb.5-150
$C_9ClH_{16}NSn$	$(CH_3)_3SnCl \cdot CH_3C_5H_4N$	Sn:	Org.Verb.5-87

Formula	Compound	Element	Reference
$C_9ClH_{16}NSn$	$(CH_3)_3SnCl \cdot C_6H_5NH_2$	Sn:	Org.Verb.5-87
$C_9ClH_{17}N_4NiS_2$	$C_3H_5Ni(SC(NHCH_2)_2)_2Cl$	Ni:	Org.Verb.2-21
$C_9ClH_{17}Sn$	$(C_2H_5)_3SnCCCH_2Cl$	Sn:	Org.Verb.2-221, 234
$C_9ClH_{19}MgNiO$	$C_3H_5NiCH_3 \cdot Mg(CH_3)Cl \cdot C_4H_8O$	Ni:	Org.Verb.2-285
$C_9ClH_{19}O_2Sn$	$(C_3H_7)_2(CH_3OCOCH_2)SnCl$	Sn:	Org.Verb.5-205/6
$C_9ClH_{19}Sn$	$(CH_3)_3Sn(CH_2)_3CH(CH_2)CHCl$	Sn:	Org.Verb.2-44
$C_9ClH_{21}ISn^-$	$(i\text{-}C_3H_7)_3SnClI^-$	Sn:	Org.Verb.5-116
$C_9ClH_{21}OSn$	$(C_2H_5)_3SnCl \cdot (CH_3)_2CO$	Sn:	Org.Verb.5-105, 107
–	$(C_4H_9)_2ClSnOCH_3$	Sn:	Org.Verb.6-21
$C_9ClH_{21}Sn$	$(CH_3)_2(C_7H_{15})SnCl$	Sn:	Org.Verb.5-195, 198
–	$(C_2H_5)_2(C_5H_{11})SnCl$	Sn:	Org.Verb.5-202
–	$(C_3H_7)_3SnCl$	Sn:	Org.Verb.5-110/6
–	$(i\text{-}C_3H_7)_3SnCl$	Sn:	Org.Verb.5-116/7
–	$(C_4H_9)_2(CH_3)SnCl$	Sn:	Org.Verb.5-207, 212
$C_9ClH_{21}Sn^+$	$(C_3H_7)_3SnCl^+$	Sn:	Org.Verb.5-112, 116
$C_9ClH_{23}O_3SiSn$	$(C_2H_5)_2[(CH_3O)_3SiCH_2CH_2]SnCl$	Sn:	Org.Verb.5-202
$C_9ClH_{24}LiSi_2Sn$	$(CH_3)_2[((CH_3)_3Si)_2CLi]SnCl$	Sn:	Org.Verb.5-196
$C_9ClH_{27}N_3OPSn$	$(CH_3)_3SnCl \cdot [(CH_3)_2N]_3PO$	Sn:	Org.Verb.5-87
$C_9ClH_{27}Ni_2OP_2$	$(CH_3Ni(P(CH_3)_3)Cl)(CH_3Ni(P(CH_3)_3)OCH_3)$	Ni:	Org.Verb.2-254
$C_9ClH_{29}O_2Sn_3$	$(CH_3)_3SnCl \cdot 2\ (CH_3)_3SnOH$	Sn:	Org.Verb.5-82
$C_9ClH_{30}O_9P_3Sn$	$(CH_3)_3SnCl \cdot 3\ (CH_3O)_2HPO$	Sn:	Org.Verb.5-87
$C_9Cl_2FH_{11}N_2O$	$(NH_2)(C_4H_9O)C_5FCl_2N$	F:	PerFHalOrg.5-214
$C_9Cl_2FH_{11}N_2O_2$	$(NH_2)(C_2H_5OC_2H_4O)C_5FCl_2N$	F:	PerFHalOrg.5-214/5
$C_9Cl_2FH_{12}N_3$	$[(CH_3)_2N]_2C_5FCl_2N$	F:	PerFHalOrg.5-138
$C_9Cl_2F_2H_8N_2$	$(CH_2)_4NC_5F_2Cl_2N$	F:	PerFHalOrg.5-138
$C_9Cl_2F_2H_8N_2O$	$(O(CH_2CH_2)_2N)C_5F_2Cl_2N$	F:	PerFHalOrg.5-138
$C_9Cl_2F_2H_{10}NO_3PS$	$NC_5F_2Cl_2(OPS(OC_2H_5)_2)$	F:	PerFHalOrg.5-212
$C_9Cl_2F_2H_{10}NO_4P$	$NC_5F_2Cl_2(OPO(OC_2H_5)_2)$	F:	PerFHalOrg.5-212
$C_9Cl_2F_3H_9Si$	$(CH_3)_3SiC_6F_3Cl_2$	F:	PerFHalOrg.4-26
$C_9Cl_2F_3H_{15}Si$	$CF_3C[Si(C_2H_5)_3]CCl_2$	F:	PerFHalOrg.4-163
$C_9Cl_2F_4H_2N_4$	$NC_5F_2Cl_2NHC_4HF_2N_2$	F:	PerFHalOrg.5-216
–	$NC_5F_4NHC_4HCl_2N_2$	F:	PerFHalOrg.5-215/6
$C_9Cl_2F_5N$	$C_9F_5Cl_2N$	F:	PerFHalOrg.6-130/1, 142
$C_9Cl_2F_9HO$	$(CF_2Cl)_2C(OH)C_6F_5$	F:	PerFHalOrg.4-51
$C_9Cl_2F_{11}N$	$(C_2F_5)_2C_5FCl_2N$	F:	PerFHalOrg.5-149/50
$C_9Cl_2F_{12}H_3N_3O$	$(CF_3)_2CNHC(CF_3)_2NC(NHCOCHCl_2)$	F:	PerFHalOrg.5-105
$C_9Cl_2F_{21}N_3O_3S$	$ClS[N(CF_3)OCF_2CF_2]_3Cl$	F:	PerFHalOrg.7-92/3, 108
$C_9Cl_2F_{21}P$	$(n\text{-}C_3F_7)_3PCl_2$	F:	PerFHalOrg.3-74, 92, 100
$C_9Cl_2FeH_5NO_4Sn$	$(CO)_3Fe(NO)Sn(C_6H_5)Cl_2$	Fe:	Org.Verb.B1-197/8
$C_9Cl_2FeH_7Hg_2NO_3$	$(CO)_3Fe(CH_3C_5H_4N)(HgCl)_2$	Fe:	Org.Verb.B1-192/3
–	$(CO)_3Fe(C_6H_5NH_2)(HgCl)_2$	Fe:	Org.Verb.B1-192/3
$C_9Cl_2FeH_{18}HgO_3P_2$	$(CO)_3Fe(P(CH_3)_3)_2 \cdot HgCl_2$	Fe:	Org.Verb.B1-192/4
$C_9Cl_2FeH_{18}HgO_9P_2$	$(CO)_3Fe(P(OCH_3)_3)_2 \cdot HgCl_2$	Fe:	Org.Verb.B1-192/4
$C_9Cl_2FeH_{23}N_5O_5$	$[COFe(NH(C_2H_4NHC_2H_4NH_2)_2)]Cl[ClO_4]$	Fe:	Org.Verb.B1-85
$C_9Cl_2FeH_{23}N_5O_9$	$[COFe(NH(C_2H_4NHC_2H_4NH_2)_2)][ClO_4]_2$	Fe:	Org.Verb.B1-85/6
$C_9Cl_2H_5NiO_3P$	$(CO)_3NiP(C_6H_5)Cl_2$	Ni:	Org.Verb.1-171, 172
$C_9Cl_2H_6NOTi^+$	$[Ti(C_9H_6NO)Cl_2]^+$	Ti:	Org.Verb.1-196
$C_9Cl_2H_8N_2Sn$	$SnCl_2 \cdot NH_2C_9H_6N$	Sn:	MVol.C5-21
$C_9Cl_2H_9N_3Ni$	$NiCl_2 \cdot 3\ CH_2CHCN$	Ni:	Org.Verb.1-419
$C_9Cl_2H_{12}OTi$	$C_6H_5Ti(OC_3H_7\text{-}i)Cl_2$	Ti:	Org.Verb.1-60

Formula	Compound		Reference
$C_9Cl_2H_{12}Sn$	$(C_2H_5)(CH_2Cl)(C_6H_5)SnCl$	Sn:	Org.Verb.5-226
–	$(C_3H_7)(C_6H_5)SnCl_2$	Sn:	Org.Verb.6-199, 204
$C_9Cl_2H_{14}N_2Sn$	$(CH_3)_2SnCl_2 \cdot NC_5H_4CHNCH_3$	Sn:	Org.Verb.6-36, 38
$C_9Cl_2H_{14}OTi$	$C_5H_5Ti(OC_4H_9\text{-t})Cl_2$	Ti:	Org.Verb.1-180
$C_9Cl_2H_{17}NNiO$	$((CH_3)_2C(OH)CH_2C_2CH_2N(CH_3)_2)NiCl_2$	Ni:	Org.Verb.1-339
$C_9Cl_2H_{17}O_2Ti$	$C_5H_5TiCl_2 \cdot 2\ C_2H_5OH$	Ti:	Org.Verb.1-136
$C_9Cl_2H_{18}N_2O_4U$	$UO_2Cl_2 \cdot (CH_3)_2C(CON(CH_3)_2)_2$	U:	SVol.E1-113/4
$C_9Cl_2H_{18}N_2SSn$	$(CH_2CH)_2SnCl_2 \cdot [(CH_3)_2N]_2CS$	Sn:	Org.Verb.6-146
$C_9Cl_2H_{18}N_6O_2U$	$[UO_2Cl_2(C_3N_3(N(CH_3)_2)_3)]$	U:	SVol.E1-46/8
$C_9Cl_2H_{18}N_6O_8S_3Te$	$Te(SC(NHCH_2)_2)_3(ClO_4)_2$	Te:	SVol.B3-155
$C_9Cl_2H_{20}N_2OS_2$	$[SOCl_2SCN]N(C_2H_5)_4$	S:	SVol.1-46
$C_9Cl_2H_{20}Sn$	$(CH_3)(C_8H_{17})SnCl_2$	Sn:	Org.Verb.6-197, 204
–	$(CH_3)_3SnCH_2CH(CHCl_2)CH(CH_3)_2$	Sn:	Org.Verb.2-56
–	$(CH_3)_3Sn(CH_2)_2C(CH_3)_2CHCl_2$	Sn:	Org.Verb.2-55
–	$(C_4H_9)_2(CH_2Cl)SnCl$	Sn:	Org.Verb.5-205, 207
$C_9Cl_2H_{21}N_3O_5U$	$UO_2Cl_2 \cdot 3\ HCON(CH_3)_2$	U:	SVol.E1-100/1
$C_9Cl_2H_{21}PSn$	$CH_3Cl_2SnP(t\text{-}C_4H_9)_2$	Sn:	Org.Verb.6-218
$C_9Cl_2H_{23}N_5O_2P_2S$	$[O_2S[NPClN(C_2H_5)_2]_2NCH_3]$	S:	S-N-Verb.1-34
$C_9Cl_2H_{23}OPSSn$	$(CH_3)_2SnCl_2 \cdot (C_3H_7)_2(CH_3S)PO$	Sn:	Org.Verb.6-37
$C_9Cl_2H_{24}N_6O_5U$	$UO_2Cl_2 \cdot 3\ NH_2CONHC_2H_5$	U:	SVol.E1-92/3
$C_9Cl_2H_{27}N_3O_2U$	$UO_2Cl_2 \cdot 3\ C_3H_7NH_2$	U:	SVol.E1-18, 21
$C_9Cl_3F_2H_3N_4$	$NC_5F_2Cl_2NHC_4H_2ClN_2$	F:	PerFHalOrg.5-216
$C_9Cl_3F_3H_2N_4$	$NC_5F_3ClNHC_4HCl_2N_2$	F:	PerFHalOrg.5-215
$C_9Cl_3F_3H_5N_3O_2S$	$(CH_3NHSO_2)(CF_3)C_7Cl_3HN_2$	F:	PerFHalOrg.6-160
$C_9Cl_3F_3H_7N$	$(CF_3)[(CH_3)_2CH]C_5Cl_3N$	F:	PerFHalOrg.5-177
$C_9Cl_3F_3H_8N_2$	$[(CH_3)_2CHNH](CF_3)C_5Cl_3N$	F:	PerFHalOrg.5-177
$C_9Cl_3F_8HO$	$C_6F_5(CF_2Cl)C(OH)CFCl_2$	F:	PerFHalOrg.4-51
$C_9Cl_3F_{12}N_3$	$(CF_3CFCl)_3C_3N_3$	F:	PerFHalOrg.6-82/3, 103
$C_9Cl_3FeH_{15}N_3$	$(C_2H_5NC)_3FeCl_3$	Fe:	Org.Verb.B4-44
$C_9Cl_3H_6NOSn$	$SnCl_3(C_9H_6NO)$	Sn:	MVol.C5-213
$C_9Cl_3H_6NSn$	$NC_9H_6SnCl_3$	Sn:	Org.Verb.6-287
$C_9Cl_3H_8MnN$	$(C_9H_7NH)MnCl_3$	Mn:	MVol.C5-208/9
$C_9Cl_3H_8NSn$	$SnCl_3(NC_8H_5CH_3)$	Sn:	MVol.C5-191
$C_9Cl_3H_9IrN_3S_2$	$Ir(S_2N_2C_4H_4)Cl_3 \cdot C_5H_5N$	Ir:	SVol.2-172
$C_9Cl_3H_9O_3Sn$	$SnCl_3(OC_6H_3(OCH_3)COCH_3)$	Sn:	MVol.C5-89
–	$SnCl_3(OC_6H_4COOC_2H_5)$	Sn:	MVol.C5-146/7
$C_9Cl_3H_{11}OSn$	$C_6H_5SnCl_3 \cdot (CH_3)_2CO$	Sn:	Org.Verb.6-280
$C_9Cl_3H_{11}Sn$	$(CH_2Cl)_3SnC_6H_5$	Sn:	Org.Verb.2-335
$C_9Cl_3H_{11}Ti$	$C_9H_{11}TiCl_3$	Ti:	Org.Verb.1-141, 152
$C_9Cl_3H_{12}NNi$	$(CH_3CClCHCH_3)NiCl_2 \cdot C_5H_5N$	Ni:	Org.Verb.1-339/40
$C_9Cl_3H_{12}NiO_6P$	$(CO)_3NiP(OCH_2CH_2Cl)_3$	Ni:	Org.Verb.1-177
$C_9Cl_3H_{13}SiSn$	$(C_6H_5)(CH_3)_2SiCH_2SnCl_3$	Sn:	Org.Verb.6-259
$C_9Cl_3H_{13}SiTi$	$(C_6H_5)(CH_3)_2SiCH_2TiCl_3$	Ti:	Org.Verb.1-16
$C_9Cl_3H_{14}NSn$	$[C_6H_5(CH_3)_3N]SnCl_3$	Sn:	MVol.C3-95
$C_9Cl_3H_{15}OSn$	$(CH_3)_2CCHC(O)CH_2C(CH_3)_2SnCl_3$	Sn:	Org.Verb.6-265
$C_9Cl_3H_{19}O_2S_2Ti$	$CH_3TiCl_3 \cdot 2\ S(CH_2CH_2)_2O$	Ti:	Org.Verb.1-30/1, 39
$C_9Cl_3H_{19}O_2Ti$	$CH_3TiCl_3 \cdot 2\ C_4H_8O$	Ti:	Org.Verb.1-30/1, 38/9
–	$CD_3TiCl_3 \cdot 2\ C_4H_8O$	Ti:	Org.Verb.1-34/5
$C_9Cl_3H_{19}S_2Ti$	$CH_3TiCl_3 \cdot 2\ S(CH_2)_4$	Ti:	Org.Verb.1-30/1, 39
$C_9Cl_3H_{19}Sn$	$(C_2H_5)_3SnCH_2CH_2CCl_3$	Sn:	Org.Verb.2-165

Formula	Compound	Reference
$C_9Cl_3H_{21}SSn_2$	$(CH_3)Cl_2SnSSnCl(C_4H_9)_2$	Sn: Org.Verb.6-218
$C_9Cl_3H_{22}N_3O_7Sn$	$SnCl_3OH \cdot 3\ CH_3CH(NH_2)COOH$	Sn: MVol.C6-12
$C_9Cl_3H_{23}O_2Ti$	$CH_3TiCl_3 \cdot 2\ (C_2H_5)_2O$	Ti: Org.Verb.1-30/1, 38
$C_9Cl_3H_{27}IrP_3$	$Ir(P(CH_3)_3)_3Cl_3$	Ir: SVol.2-199/200
$C_9Cl_4F_2H_2N_4$	$NC_5F_2Cl_2NHC_4HCl_2N_2$	F: PerFHalOrg.5-215/6
$C_9Cl_4F_3H_3N_2$	$C_6Cl_4NC(CF_3)N(CH_3)$	F: PerFHalOrg.6-160
$C_9Cl_4F_5HN_2O$	$C_6Cl_4NC(C_2F_5)N(OH)$	F: PerFHalOrg.6-125/6, 159, 161/2
$C_9Cl_4F_5HN_2O_3$	$(NO_2C_6Cl_4)NHC(O)C_2F_5$	F: PerFHalOrg.7-23/4
$C_9Cl_4FeH_{18}Hg_2O_9P_2$	$[(CO)_3Fe(P(OCH_3)_3)_2HgCl][HgCl_3]$	Fe: Org.Verb.B1-194
$C_9Cl_4H_7NOSn$	$SnCl_4 \cdot HOC_9H_6N$	Sn: MVol.C5-212
$C_9Cl_4H_7NSn_2$	$2\ SnCl_2 \cdot C_9H_7N$	Sn: MVol.C5-19
$C_9Cl_4H_8O_3Sn$	$SnCl_4 \cdot C_6H_4(OCOCH_3)CHO$	Sn: MVol.C5-81
$C_9Cl_4H_9NO_2Sn$	$SnCl_4 \cdot C_6H_5CONHCOCH_3$	Sn: MVol.C6-50/1
$C_9Cl_4H_{10}O_2Sn$	$SnCl_4 \cdot CH_3COC_6H_4OCH_3$	Sn: MVol.C5-89
–	$SnCl_4 \cdot C_6H_5COOC_2H_5$	Sn: MVol.C5-144/5
$C_9Cl_4H_{10}O_4Sn$	$SnCl_4 \cdot C_6H_5COOH \cdot CH_3COOH$	Sn: MVol.C5-122/3
$C_9Cl_4H_{10}Sn$	$(CH_3)_3SnC_6HCl_4$	Sn: Org.Verb.2-136
$C_9Cl_4H_{11}NOSn$	$SnCl_4 \cdot NC_5H_4C_2H_4OC_2H_3$	Sn: MVol.C5-203/4
$C_9Cl_4H_{11}O_4PSn$	$SnCl_4 \cdot C_6H_5COP(O)(OCH_3)_2$	Sn: MVol.C6-164
$C_9Cl_4H_{12}OSTe$	$TeCl_4 \cdot C_6H_5SOC_3H_7$	Te: SVol.B3-176
$C_9Cl_4H_{12}OSn$	$SnCl_4 \cdot C_6H_5CH_2OC_2H_5$	Sn: MVol.C5-152
$C_9Cl_4H_{13}NO_4Sn$	$SnCl_4 \cdot C_5H_5N \cdot 2\ CH_3COOH$	Sn: MVol.C5-197
$C_9Cl_4H_{14}O_3SiSn$	$SnCl_4 \cdot C_6H_5Si(OCH_3)_3$	Sn: MVol.C6-127/8
$C_9Cl_4H_{14}O_4Sn$	$SnCl_4 \cdot C_5H_8(OCOCH_3)_2$	Sn: MVol.C5-133
$C_9Cl_4H_{15}TiZr$	$Cl_3Ti(CH_2CHCH_2)(Cl)Zr(C_3H_5)_2$	Ti: Org.Verb.1-65/6
$C_9Cl_4H_{16}O_4Sn$	$SnCl_4 \cdot C_3H_6(COOC_2H_5)_2$	Sn: MVol.C5-141/3
–	$SnCl_4 \cdot C_5H_{10}(COOCH_3)_2$	Sn: MVol.C5-141/3
$C_9Cl_4H_{17}NOSn$	$SnCl_4 \cdot (CH_2)_5NC_2H_4OCHCH_2$	Sn: MVol.C5-193/4
$C_9Cl_4H_{17}NO_2Sn$	$SnCl_4 \cdot CH_3CONHCOC_6H_{13}$	Sn: MVol.C6-50/1
$C_9Cl_4H_{18}N_2O_2U$	$UCl_4 \cdot (CH_3)_2C(CON(CH_3)_2)_2$	U: SVol.E1-112/4
$C_9Cl_4H_{18}OSn$	$SnCl_4 \cdot CO(CH_2CH(CH_3)_2)_2$	Sn: MVol.C5-87
$C_9Cl_4H_{18}O_3U$	$UCl_4 \cdot 3\ (CH_3)_2CO$	U: SVol.E1-79/80
$C_9Cl_4H_{18}O_6Sn$	$SnCl_4 \cdot 3\ C_2H_5COOH$	Sn: MVol.C5-120/1
$C_9Cl_4H_{18}Sn$	$(C_2H_5)_3SnCHClCH_2CCl_3$	Sn: Org.Verb.2-165
$C_9Cl_4H_{19}NOSn$	$SnCl_4 \cdot (CH_2)_5NC_2H_4OC_2H_5$	Sn: MVol.C5-193/4
$C_9Cl_4H_{19}NO_4Sn$	$SnCl_4 \cdot 2\ CH_3COOH \cdot (CH_2)_5NH$	Sn: MVol.C5-193
$C_9Cl_4H_{19}O_4PSn$	$SnCl_4 \cdot CH_3COCH_2P(O)(OC_3H_7)_2$	Sn: MVol.C6-165/6
$C_9Cl_4H_{20}S_2Sn$	$SnCl_4 \cdot C_4H_9SCH_2SC_4H_9$	Sn: MVol.C6-97/8
$C_9Cl_4H_{21}N_3O_3U$	$UCl_4 \cdot 3\ HCON(CH_3)_2$	U: SVol.E1-99/100
$C_9Cl_4H_{24}Si_3Sn$	$[(CH_3)_2SiClCH_2]_3SnCl$	Sn: Org.Verb.5-154
$C_9Cl_4H_{27}O_3P_3U$	$UCl_4 \cdot 3\ (CH_3)_3PO$	U: SVol.E1-147/8
$C_9Cl_4H_{28}IrP_3$	$[P(CH_3)_3H][Ir(P(CH_3)_3)_2Cl_4]$	Ir: SVol.2-202/3
$C_9Cl_5FH_2N_4$	$NC_5FCl_3NHC_4HCl_2N_2$	F: PerFHalOrg.5-215/6
$C_9Cl_5FH_3N_5$	$NC_5FCl_2(NH_2)NHC_4Cl_3N_2$	F: PerFHalOrg.6-21/2
$C_9Cl_5F_2HN_4$	$NC_5F_2Cl_2NHC_4Cl_3N_2$	F: PerFHalOrg.5-215
$C_9Cl_5Fe_2H_{18}HgO_3P_2$	$[(CO)_3Fe(P(CH_3)_3)_2HgCl][FeCl_4]$	Fe: Org.Verb.B1-192/3
$C_9Cl_5Fe_2H_{18}HgO_9P_2$	$[(CO)_3Fe(P(OCH_3)_3)_2HgCl][FeCl_4]$	Fe: Org.Verb.B1-192/3
$C_9Cl_5Fe_2H_{18}O_3P_2$	$[(CO)_3Fe(P(CH_3)_3)_2Cl]FeCl_4$	Fe: Org.Verb.B1-184/5
$C_9Cl_5Fe_2H_{18}O_9P_2$	$[(CO)_3Fe(P(OCH_3)_3)_2Cl]FeCl_4$	Fe: Org.Verb.B1-184/5

Formula	Compound	Element	Reference
$C_9Cl_5H_7NSn^-$	$[SnCl_5(C_9H_7N)]^-$	Sn:	MVol.C5-210
$C_9Cl_5H_8NO_2Sn$	$SnCl_4 \cdot C_6H_5CONHCOCH_2Cl$	Sn:	MVol.C6-50/1
–	$SnCl_4 \cdot ClC_6H_4CONHCOCH_3$	Sn:	MVol.C6-50/1
$C_9Cl_5H_9Sn$	$(CH_3)_3SnC_6Cl_5$	Sn:	Org.Verb.2-135
$C_9Cl_6F_9N_3$	$(CF_2ClCFCl)_3C_3N_3$	F:	PerFHalOrg.6-82/3, 102
$C_9Cl_6F_{13}N$	$Cl_3CCF_2CF(C_2F_5)CF_2CClNCCl_2CF_3$	F:	PerFHalOrg.5-176
–	$[Cl_3C(CF_2)_3CClNCClC_3F_7]Cl$	F:	PerFHalOrg.5-175/6
–	$[Cl_3CC_2F_3(CF_3)CClNCClC_3F_7]Cl$	F:	PerFHalOrg.5-98
$C_9Cl_6H_{11}NiPt$	$[C_6H_6NiC_3H_5]PtCl_6$	Ni:	Org.Verb.2-220
$C_9Cl_6H_{30}LaN_3$	$LaCl_3 \cdot 3\ [(CH_3)_3NH]Cl$	Sc:	MVol.C5-213/6
$C_9Cl_6H_{30}N_3Nd$	$NdCl_3 \cdot 3\ [(CH_3)_3NH]Cl$	Sc:	MVol.C5-213/6
$C_9Cl_6H_{30}N_3Pr$	$PrCl_3 \cdot 3\ [(CH_3)_3NH]Cl$	Sc:	MVol.C5-213/4
$C_9Cl_7H_5O_4Ti$	$C_5H_5Ti(OCOCCl_3)_2Cl$	Ti:	Org.Verb.1-190
$C_9Cl_7H_{30}Mn_2N_3$	$[(CH_3)_3NH]_3Mn_2Cl_7$	Mn:	MVol.C5-190/1
$C_9Cl_8FeH_{18}Hg_4O_9P_2$	$(CO)_3Fe(P(OCH_3)_3)_2 \cdot (HgCl_2)_4$	Fe:	Org.Verb.B1-192/4
$C_9Cl_8H_7NO_2S_2Sn$	$[(C_9H_7NSOCl)(SOCl)][SnCl_6]$	S:	SVol.1-54
$C_9Cl_9H_6NiO_6P$	$(CO)_3NiP(OCH_2CCl_3)_3$	Ni:	Org.Verb.1-177
$C_9CmH_{15}O_9$	$Cm(CH_3CH(OH)COO)_3$	Np:	TrU.D1-154
$C_9CoF_4H_9O_4Sn$	$(CH_3)_3SnCF_2CF_2Co(CO)_4$	Sn:	Org.Verb.2-37, 39
$C_9CoH_{15}O_3S_6$	$Co[S_2COC_2H_5]_3$	C:	MVol.D4-245, 259
$C_9CoH_{15}O_3Se_6$	$Co[Se_2COC_2H_5]_3$	C:	MVol.D6-221
$C_9CoH_{15}S_9$	$Co[S_2CSC_2H_5]_3$	C:	MVol.D4-266/7, 270
$C_9CoH_{18}N_3O_2S_6$	$Co[S_2CNH_2]_3(CH_3COCH_3)_2$	C:	MVol.D6-140
$C_9CoH_{26}N_6S_3$	$Co(NH(C_2H_4NH_2)_2)_2CS_3$	C:	MVol.D4-225
$C_9CoH_{33}N_6O_3S_6$	$Co[S_2COC_2H_5]_3(NH_3)_6 \cdot 3\ H_2O$	C:	MVol.D4-259
$C_9Co_2FeO_9S$	$(CO)_3FeCo_2(CO)_6S$	Fe:	Org.Verb.B1-199
$C_9Co_2H_{24}N_6S_9$	$Co_2(C_2H_8N_2)_3(CS_3)_3$	C:	MVol.D4-225
$C_9Co_3FeH_{19}O_4P^+$	$[HFeCo_3(CO)P(OC_3H_7)_2OC_2H_4]^+$	Fe:	Org.Verb.B1-204
$C_9Co_3FeH_{22}O_3P^+$	$[HFeCo_3P(OC_3H_7)_3]^+$	Fe:	Org.Verb.B1-204
$C_9CrF_7FeH_{15}N_9O_2S_4$	$[(CO)_2Fe(C_3F_7)(NH_3)_3][Cr(NCS)_4(NH_3)_2]$	Fe:	Org.Verb.B1-121, 124, 128
$C_9CrFeNa_2O_9$	$Na_2[(CO)_4FeCr(CO)_5]$	Fe:	Org.Verb.B2-162/3
$C_9CrH_{15}O_3S_6$	$Cr[S_2COC_2H_5]_3$	C:	MVol.D4-245, 258
$C_9CrH_{15}O_3Se_6$	$Cr[Se_2COC_2H_5]_3$	C:	MVol.D6-221
$C_9CrH_{15}S_9$	$Cr[S_2CSC_2H_5]_3$	C:	MVol.D4-266/7, 269
$C_9Cr_2H_{24}N_{12}O_6S_9$	$[Cr(OC(NH_2)_2)_6][Cr(CS_3)_3]$	C:	MVol.D4-225
$C_9CsF_9FeO_3$	$Cs[C_6F_9Fe(CO)_3]$	Fe:	Org.Verb.B5-74, 76
$C_9EsH_{15}O_9$	$Es(CH_3CH(OH)COO)_3$	Np:	TrU.D1-154
$C_9FH_7O_2$	$C_6H_5CHCFCOOH$	F:	PerFHalOrg.4-89
$C_9FH_8NO_3S$	$FSO_2N(COCH_2)CH(C_6H_5)$	S:	S-N-Verb.1-104
$C_9FH_{11}Ti$	$C_9H_{11}TiF$	Ti:	Org.Verb.1-133
$C_9FH_{12}N_3Sn$	$(NCCH_2CH_2)_3SnF$	Sn:	Org.Verb.5-35/6
$C_9FH_{13}Sn$	$(CH_3)_3SnC_6H_4F$	Sn:	Org.Verb.2-131/2
$C_9FH_{15}O_2Ti$	$C_5H_5Ti(OC_2H_5)_2F$	Ti:	Org.Verb.1-184/7
$C_9FH_{15}Sn$	$(C_3H_5)_3SnF$	Sn:	Org.Verb.5-37
$C_9FH_{21}Sn$	$(C_3H_7)_3SnF$	Sn:	Org.Verb.5-22/4
$C_9FH_{27}Ni_2OP_2$	$(CH_3Ni(P(CH_3)_3)F)(CH_3Ni(P(CH_3)_3)OCH_3)$	Ni:	Org.Verb.2-254
$C_9F_2FeH_{10}NO_4P$	$(CO)_4FePF_2NC_5H_{10}$	Fe:	Org.Verb.B2-85, 94
$C_9F_2FeH_{14}O_3P_2$	$(CO)_3Fe(CH_3)_2PCF_2CH_2P(CH_3)_2$	Fe:	Org.Verb.B1-170/2, 176
$C_9F_2H_{11}NO_4Te$	$[C_9H_7NH][TeF_2(OH)_3O] \cdot HF$	Te:	SVol.B2-47/8

Formula	Compound	Reference
$C_9F_2H_{18}Sn$	$(C_2H_5)_3SnCHCF_2CH_2$	Sn: Org.Verb.2-181/2
$C_9F_3FeH_7O_5$	$CH_3C_3H_4Fe(CO)_3OCOCF_3$	Fe: Org.Verb.B5-42, 46, 48, 52/3, 61/3
$C_9F_3FeH_{15}O_3P_2$	$(CO)_3Fe(P(C_2H_5)_3)PF_3$	Fe: Org.Verb.B1-148/50, 156, 164
$C_9F_3GeH_{15}$	$(C_2H_5)_3Ge(CCCF_3)$	F: PerFHalOrg.4-76
$C_9F_3H_3N_6O_6S$	$NC(CF_3)SC(NHC_6H_2(NO_2)_3)N$	F: PerFHalOrg.5-107
$C_9F_3H_4N_5O_4S$	$NC(CF_3)SC(NHC_6H_3(NO_2)_2)N$	F: PerFHalOrg.5-107
$C_9F_3H_7N_2O_3$	$CF_3NHCOCOONHC_6H_5$	F: PerFHalOrg.7-63
$C_9F_3H_7O$	$C_6H_5CH(OH)CFCF_2$	F: PerFHalOrg.4-44
$C_9F_3H_8NO$	$C_6H_5CHCH_2N(CF_3)O$	F: PerFHalOrg.7-163, 168
$C_9F_3H_8N_5O_5$	$CF_3N(OH)CH_2CHNNHC_6H_3(NO_2)_2$	F: PerFHalOrg.7-164
$C_9F_3H_{11}N_2$	$(CH_3)_2(CF_3)(NC)C_5H_4NH$	F: PerFHalOrg.7-75
$C_9F_3H_{11}NiO_2$	$C_3H_5Ni(C_4H_6)O_2CCF_3$	Ni: Org.Verb.2-284
$C_9F_3H_{12}NO_4Te$	$[C_9H_7NH][TeF_2(OH)_3O] \cdot HF$	Te: SVol.B2-47/8
$C_9F_3H_{12}N_3$	$[(CH_3)_2N]_2C_5F_3N$	F: PerFHalOrg.5-136
$C_9F_3H_{12}N_5O_2$	$(C_2H_5NH)_2(NO_2)(CF_3)C_4N_2$	F: PerFHalOrg.6-60
$C_9F_3H_{13}N_2O_4$	$CF_3NNC_2H_3(COOC_2H_5)_2$	F: PerFHalOrg.7-160
$C_9F_3H_{13}N_4O_3S$	$C_2N_2S(CF_3)NHCON(CH_3)C_2H_3(OCH_3)_2$	F: PerFHalOrg.5-107
$C_9F_3H_{15}N_2O_2$	$C_2H_5OCOCH(NNCF_3)CH_2CH(CH_3)_2$	F: PerFHalOrg.7-160
$C_9F_3H_{15}Si$	$CF_3CCSi(C_2H_5)_3$	F: PerFHalOrg.4-19
$C_9F_3H_{15}Sn$	$(C_2H_5)_3SnCCCF_3$	Sn: Org.Verb.2-221
$C_9F_3H_{16}N_2P$	$CF_3P(N(CH_2)_4)_2$	F: PerFHalOrg.3-102, 116
$C_9F_3H_{18}NO_4P$	$CF_3N(O)P(O)(OC_4H_9)_2$	F: PerFHalOrg.7-169
$C_9F_3H_{18}P$	$CF_3P(n\text{-}C_4H_9)_2$	F: PerFHalOrg.3-101, 105
$C_9F_3H_{20}N_2P$	$CF_3P[N(C_2H_5)_2]_2$	F: PerFHalOrg.3-102, 115
$C_9F_3MnN_2O_5$	$NCFC[Mn(CO)_5]CFCFN$	F: PerFHalOrg.6-72
$C_9F_3N_2O_5Re$	$NCFC[Re(CO)_5]CFCFN$	F: PerFHalOrg.6-72
$C_9F_4Fe_2H_3NO_8P_2$	$(CO)_4Fe(F_2PN(CH_3)PF_2)Fe(CO)_4$	Fe: Org.Verb.C1-36, 40
$C_9F_4Fe_2O_7$	$(CO)_3Fe(CO)(CF_2)_2Fe(CO)_3$	Fe: Org.Verb.C1-226
$C_9F_4H_5N_3O_4$	$O_2NC_6H_4NN(CF_2)_2COOH$	F: PerFHalOrg.7-193
$C_9F_4H_6N_2O_2$	$C_6H_5NN(CF_2)_2COOH$	F: PerFHalOrg.7-193
$C_9F_4H_7NOSn$	$SnF_4 \cdot HOC_9H_6N$	Sn: MVol.C5-212
$C_9F_4H_7NO_2$	$(CH_3)_2NC_6F_4C(O)OH$	F: PerFHalOrg.7-61
–	$NH_2C_6H_4C(O)OC_2H_5$	F: PerFHalOrg.7-61
$C_9F_4H_8MnN$	$C_9H_7NHMnF_4 \cdot 3\ H_2O$	Mn: MVol.C4-212/3
$C_9F_4H_9LiSn$	$(CH_3)_3SnC_6F_4Li$	Sn: Org.Verb.2-136
$C_9F_4H_9N$	$C_4H_9C_5F_4N$	F: PerFHalOrg.5-142
$C_9F_4H_9NO$	$NC_5F_4C(C_2H_5)(CH_3)OH$	F: PerFHalOrg.4-97
		F: PerFHalOrg.5-143
$C_9F_4H_9NO_2Te$	$[C_9H_7NH][TeF_4(OH)O]$	Te: SVol.B2-47/8
$C_9F_4H_9NO_3$	$(CH_3O)_3(CF_3)C_5FN$	F: PerFHalOrg.5-177
$C_9F_4H_{10}Sn$	$(CH_3)_3SnC_6HF_4$	Sn: Org.Verb.2-136
$C_9F_5FeH_5O_2$	$C_5H_5Fe(CO)_2C_2F_5$	Fe: Org.Verb.B2-180
$C_9F_5FeH_5O_5$	$C_3H_5Fe(CO)_3OCOC_2F_5$	Fe: Org.Verb.B5-40, 51, 59/60
$C_9F_5GeH_9$	$C_6F_5Ge(CH_3)_3$	F: PerFHalOrg.4-180
$C_9F_5HI_4N_2O$	$C_6I_4NC(C_2F_5)N(OH)$	F: PerFHalOrg.6-125/6, 161/2
$C_9F_5H_2N$	$C_9F_5H_2N$	F: PerFHalOrg.6-153, 159

Formula	Structure		Reference
$C_9F_5H_2NO_2$	$(HO)_2C_9F_5N$	F:	PerFHalOrg.6-160
$C_9F_5H_3N_2O$	$C_6F_4NC(OCH_3)CFN$	F:	PerFHalOrg.6-156, 161
$C_9F_5H_3O_3$	$C_6F_5C(O)C(O)OCH_3$	F:	PerFHalOrg.4-52
$C_9F_5H_4NO$	$C_6F_5CNHC(O)CH_3$	F:	PerFHalOrg.7-59
$C_9F_5H_5NiO$	$C_5H_5Ni(CO)C_3F_5$	Ni:	Org.Verb.2-167/8
$C_9F_5H_7O$	$(CH_3)_2(C_6F_5)COH$	F:	PerFHalOrg.4-102
$C_9F_5H_9Pb$	$(CH_3)_3PbC_6F_5$	F:	PerFHalOrg.4-79
$C_9F_5H_9Si$	$C_6F_5Si(CH_3)_3$	F:	PerFHalOrg.4-22, 79, 86
$C_9F_5H_9Sn$	$(CH_3)_3SnC_6F_5$	F:	PerFHalOrg.4-81/2
		Sn:	Org.Verb.2-134/5
$C_9F_5H_{12}N_5O_4$	$(C_2F_5)(N(CH_2OH)_2)_2C_3N_3$	F:	PerFHalOrg.6-112
$C_9F_5I_2N$	$C_9F_5I_2N$	F:	PerFHalOrg.6-130/1, 143
$C_9F_6FeH_2O_9$	$[Fe(CO)_5H][CF_3CO_2HO_2CCF_3]$	Fe:	Org.Verb.B3-251
$C_9F_6FeH_5O_2P$	$C_5H_5Fe(CO)_2P(CF_3)_2$	F:	PerFHalOrg.3-27
$C_9F_6FeH_9O_4P$	$[CH_3COC_4H_6Fe(CO)_3]PF_6$	Fe:	Org.Verb.B5-94, 96
$C_9F_6FeH_{13}NO_3P_2$	$[C_4H_4P(CH_3)_3Fe(CO)_2NO][PF_6]$	Fe:	Org.Verb.B5-28/31
$C_9F_6FeH_{27}P_5$	$Fe(P(CH_3)_3)_3(PF_3)_2$	Fe:	Org.Verb.B1-6
$C_9F_6Fe_2H_9O_6PS_3$	$[(CO)_6Fe_2(SCH_3)_3][PF_6]$	Fe:	Org.Verb.C1-20/3
C_9F_6HN	C_9F_6HN	F:	PerFHalOrg.6-153/4, 158/9
C_9F_6HNO	$NC(OH)C_6F_4CFCF$	F:	PerFHalOrg.6-132/3, 146, 154/5, 159
–	$NC_6F_4CFCFC(OH)$	F:	PerFHalOrg.6-130/1, 142, 159
$C_9F_6H_2N_2$	$NC(NH_2)C_6F_4CFCF$	F:	PerFHalOrg.6-132/3, 146, 159, 161
–	$NC_6F_4CFCFC(NH_2)$	F:	PerFHalOrg.6-130/1, 142, 158
$C_9F_6H_3N_3$	$NC(NHNH_2)C_6F_4CFCF$	F:	PerFHalOrg.6-132/3, 146, 158
–	$NC_6F_4CFCFC(NHNH_2)$	F:	PerFHalOrg.6-130/1, 142, 159
$C_9F_6H_5NO$	$C_6H_5CFCF_2N(CF_3)O$	F:	PerFHalOrg.7-163, 168
$C_9F_6H_5NO_2$	$(CF_3)_2NOC(O)C_6H_5$	F:	PerFHalOrg.7-132, 136/8, 144
$C_9F_6H_5N_3O_2$	$O_2NCF_2CF(CF_3)NNC_6H_5$	F:	PerFHalOrg.7-192
$C_9F_6H_5NiS_2$	$C_5H_5NiS_2C_2(CF_3)_2$	Ni:	Org.Verb.2-114
$C_9F_6H_7N$	$C_6H_5C(CF_3)_2NH_2$	F:	PerFHalOrg.7-69/70
$C_9F_6H_7NO$	$C_6H_5CH_2ON(CF_3)_2$	F:	PerFHalOrg.7-137/8, 144
–	$HOC_6H_4C(CF_3)_2NH_2$	F:	PerFHalOrg.7-69/70
$C_9F_6H_7N_3O$	$CH_3(HONCF_3)C_6H_3NNCF_3$	F:	PerFHalOrg.7-160
$C_9F_6H_8N_2$	$NH_2C_6H_4C(CF_3)_2NH_2$	F:	PerFHalOrg.7-69, 71
$C_9F_6H_8N_{10}$	$(NH_2)_2C_3N_3(CF_2)_3C_3N_3(NH_2)_2$	F:	PerFHalOrg.6-92
$C_9F_6H_{10}$	$(CH_2)_4C_3H_2(CF_3)_2$	F:	PerFHalOrg.5-24/5
–	$C_3H_2(CF_3)_2(CH_3)C(CH_3)CH_2$	F:	PerFHalOrg.5-24
–	$C_6H_9CH(CF_3)_2$	F:	PerFHalOrg.5-24/5
$C_9F_6H_{10}O_4$	$C_2H_5OCO(CF_2)_3COOC_2H_5$	F:	PerFHalOrg.5-133
$C_9F_6H_{11}N$	$(CF_3)_2(CH_3)_2C_5H_4NH$	F:	PerFHalOrg.7-65
–	$C_6H_{11}NC(CF_3)_2$	F:	PerFHalOrg.5-25
$C_9F_6H_{11}NO$	$C_6H_9(O)C(CF_3)_2NH_2$	F:	PerFHalOrg.7-69, 73

Formula	Structure		Reference
$C_9F_6H_{11}NO$	$OC(CH_2)_4CHC(CF_3)_2NH_2$	F:	PerFHalOrg.7-65
$C_9F_6H_{11}NiP$	$[C_6H_6NiC_3H_5]PF_6$	Ni:	Org.Verb.2-220
$C_9F_6H_{12}$	$C_6H_{11}C(CF_3)_2H$	F:	PerFHalOrg.5-26
$C_9F_6H_{12}N_2$	$C_6H_{11}NNC(CF_3)_2H$	F:	PerFHalOrg.5-26
$C_9F_6H_{14}N_2$	$C_6H_{11}NHC(CF_3)_2NH_2$	F:	PerFHalOrg.7-65
$C_9F_6H_{19}NNiP_2$	$[(CH_3)_3P]_2NiC(CF_3)_2NH$	F:	PerFHalOrg.7-74
$C_9F_6H_{19}NP_2Pt$	$[(CH_3)_3P]_2PtC(CF_3)_2NH$	F:	PerFHalOrg.7-74
$C_9F_7^+$	$C_9F_7^+$	F:	PerFHalOrg.3-172
$C_9F_7H_5NiO$	$C_5H_5Ni(CO)C_3F_7$	Ni:	Org.Verb.2-167/8
$C_9F_7H_6N$	$FC_6H_4C(CF_3)_2NH_2$	F:	PerFHalOrg.7-69, 71
$C_9F_7H_6NO_2$	$(CH_3O)_2(CF_3)_2C_5FN$	F:	PerFHalOrg.5-177
$C_9F_7H_9NO_3P$	$C_3F_7P(O)(OH)_2 \cdot C_6H_5NH_2$	F:	PerFHalOrg.3-50
$C_9F_7H_9O$	$C_6H_8(n\text{-}C_3F_7)OH$	F:	PerFHalOrg.4-92
$C_9F_7H_{10}N_5O_3$	$(C_3F_7)(N(CH_2OH)_2)(NHCH_2OH)C_3N_3$	F:	PerFHalOrg.6-112
$C_9F_7H_{11}O$	$(CH_2)_5C(C_3F_7)OH$	F:	PerFHalOrg.4-91
$C_9F_7H_{15}Pb$	$(C_2H_5)_3PbC_3F_7$	F:	PerFHalOrg.4-127
$C_9F_7H_{15}Sn$	$(C_2H_5)_3SnC_3F_7$	Sn:	Org.Verb.2-163, 165
C_9F_7N	$NCFC_6F_4CFCF$	F:	PerFHalOrg.6-132/3, 145, 152, 154/5, 158/9, 161
–	$NC_6F_4CFCFCF$	F:	PerFHalOrg.6-130/1, 142, 152/4, 158/61
$C_9F_8FeO_4$	$C_5F_8Fe(CO)_4$	Fe:	Org.Verb.B4-322, 327
$C_9F_8H_8N_2O_7$	$2\ ON(CF_2)_2COOH \cdot (CH_3)_2CO$	F:	PerFHalOrg.7-192
$C_9F_9FeO_3^-$	$[C_6F_9Fe(CO)_3]^-$	Fe:	Org.Verb.B5-74
$C_9F_9GeH_9$	$(CH_3)_3GeCC(CF_3)_2C(CF_3)$	F:	PerFHalOrg.5-24
$C_9F_9H_2N$	$NH_2C_6F_3(CF_2)_3$	F:	PerFHalOrg.7-10/1, 38, 55, 62
$C_9F_9H_3N_4O$	$(N_3)(i\text{-}C_3F_7)(CH_3O)C_5F_2N$	F:	PerFHalOrg.5-213
$C_9F_9H_5Ni_2O_6$	$C_3H_5NiO_2CCF_3 \cdot Ni(O_2CCF_3)_2$	Ni:	Org.Verb.2-284
$C_9F_9H_5O$	$C_7F_9CH(OH)CH_3$	F:	PerFHalOrg.4-43, 97
$C_9F_9H_8NO$	$(CF_3)_3CNC_4H_5(CH_3)O$	F:	PerFHalOrg.7-194/5
$C_9F_9H_9O_3$	$CF_3OHCCH_2CCH_3CF_3OCCF_3OHCH_2$	F:	PerFHalOrg.4-44
$C_9F_9H_9Sn$	$(CH_3)_3SnC_3(CF_3)_3$	Sn:	Org.Verb.2-97/8
$C_9F_9H_{18}O_2PSi_2$	$(CF_3)_3P[OSi(CH_3)_3]_2$	F:	PerFHalOrg.3-36, 110
C_9F_9P	$(CF_3CC)_3P$	F:	PerFHalOrg.3-163, 166/7, 169, 172/3
C_9F_9Sb	$(CF_3CC)_3Sb$	F:	PerFHalOrg.3-227/31
C_9F_{10}	$C_6F_5CF_2CFCF_2$	F:	PerFHalOrg.4-16
$C_9F_{10}HNO$	$(CF_3C_6F_4)NHC(O)CF_3$	F:	PerFHalOrg.7-23/4, 49
$C_9F_{10}H_3NO$	$(CH_3O)[(CF_3)_2CF]C_5F_3N$	F:	PerFHalOrg.5-177
$C_9F_{10}H_4N_2$	$(NH_2)_2(CF_3)_3C_6F$	F:	PerFHalOrg.7-11/3, 39
$C_9F_{10}N_2$	$NC_5F_4C_3F_6CN$	F:	PerFHalOrg.5-197, 210
$C_9F_{11}HO$	$C_6F_5(CF_3)_2COH$	F:	PerFHalOrg.4-51, 102
$C_9F_{11}H_2N$	$(CF_3)_3C_6F_2NH_2$	F:	PerFHalOrg.7-10, 37, 60
$C_9F_{11}H_3O$	$C_7F_{11}C(O)CH_3$	F:	PerFHalOrg.4-43
$C_9F_{11}H_5O$	$C_7F_{11}CH(OH)CH_3$	F:	PerFHalOrg.4-43
$C_9F_{11}N$	$(CF_3)(C_5F_4N)CCFCF_3$	F:	PerFHalOrg.5-154/5, 167
–	$C_4F_7C_5F_4N$	F:	PerFHalOrg.5-155/6, 168
–	$NC_5F_3(CF_2)_4$	F:	PerFHalOrg.6-132/3, 145
$C_9F_{11}NO$	$NC_5F_4C(O)CF(CF_3)_2$	F:	PerFHalOrg.5-195, 208

Formula	Compound		Reference
$C_9F_{11}NO_2$	$(CF_3)_2NOC(O)C_6F_5$	F:	PerFHalOrg.7-94/5, 111, 126
$C_9F_{12}FeO_3$	$C_9F_{12}FeO_3$	Fe:	Org.Verb.B3-193
$C_9F_{12}HNO_2$	$(CF_2)_3CNHCOC(CF_2)_3O$	F:	PerFHalOrg.5-58/9, 81, 102
$C_9F_{12}H_3N_3O_9S_3$	$[CF_3CF(SO_3H)]_3C_3N_3$	F:	PerFHalOrg.6-112
$C_9F_{12}H_4N_2O_2$	$OC(CF_3)_2NC(NHCOCH_3)C(CF_3)_2$	F:	PerFHalOrg.5-101
$C_9F_{12}H_5NO_2$	$(CF_3)_2CC(OC_2H_5)NC(CF_3)_2O$	F:	PerFHalOrg.5-101/2
–	$(CF_3)_2CC(O)N(C_2H_5)C(CF_3)_2O$	F:	PerFHalOrg.5-101/2
$C_9F_{12}H_5N_3O_2$	$(CF_3)_2CNHC(CF_3)_2NC(NHCOOCH_3)$	F:	PerFHalOrg.5-108
–	$(CH_3CONHN)CC(CF_3)_2OC(CF_3)_2NH$	F:	PerFHalOrg.5-100/1
$C_9F_{12}H_6N_2O$	$(CF_3)_2COC(CF_3)_2NC(N(CH_3)_2)$	F:	PerFHalOrg.5-102/3
$C_9F_{12}H_6N_4O$	$(CH_3CONHN)CC(CF_3)_2NHC(CF_3)_2NH$	F:	PerFHalOrg.5-100
$C_9F_{12}H_7N_3$	$(CF_3)_2CNHC(CF_3)_2N(CH_3)C(NCH_3)$	F:	PerFHalOrg.5-104
–	$(CF_3)_2CNHC(CF_3)_2NC(N(CH_3)_2)$	F:	PerFHalOrg.5-104
$C_9F_{12}H_7N_3O_2$	$(CF_3)_2CNHC(CF_3)_2NC(NHCH_2OCH_2OH)$	F:	PerFHalOrg.5-106
$C_9F_{12}H_9N_3O$	$(CH_3)_3CNC(ON(CF_3)_2)N(CF_3)_2$	F:	PerFHalOrg.7-133, 143
$C_9F_{12}H_9N_3O_2$	$(CH_3)_3CNC(ON(CF_3)_2)_2$	F:	PerFHalOrg.7-133, 144
$C_9F_{12}H_{10}N_2O_2$	$(CF_3)_2NOC(CH_3)_2CH(CH_3)ON(CF_3)_2$	F:	PerFHalOrg.7-129, 132/3, 135
–	$(CF_3)_2NOCH_2C(CH_3)(C_2H_5)ON(CF_3)_2$	F:	PerFHalOrg.7-129, 132/3, 135
$C_9F_{12}H_{10}Sn$	$(CH_3)_3Sn(CF_2)_5CHF_2$	Sn:	Org.Verb.2-55
$C_9F_{12}H_{12}N_2OSi$	$(CH_3)_3SiCH[ON(CF_3)_2]CH_2N(CF_3)_2$	F:	PerFHalOrg.7-143
$C_9F_{12}H_{12}N_2O_5Si$	$(CH_3O)_3SiCH[ON(CF_3)_2]CH_2ON(CF_3)_2$	F:	PerFHalOrg.7-127/8
$C_9F_{12}NO$	$(C_6F_5)(i\text{-}C_3F_7)NO$	F:	PerFHalOrg.7-86, 122/3
$C_9F_{12}N_2$	$NCFC(C_2F_5)C(C(CF_3)CF_2)CFN$	F:	PerFHalOrg.6-17, 38
$C_9F_{12}N_4O_2$	$CF_2(CF_2CNOC(CF_3)N)_2$	F:	PerFHalOrg.5-69/71, 91
$C_9F_{13}H_2N_3O$	$HNC_5F_6(O)(NC(NH)C_3F_7)$	F:	PerFHalOrg.5-190, 205
$C_9F_{13}H_2N_3O_2$	$HNC_5F_6(O)(NC(NH)C_2F_4OCF_3)$	F:	PerFHalOrg.5-190, 205
$C_9F_{13}H_3N_4O_2$	$C_3F_7C_2N_2O(CF_2)_3C(NOH)NH_2$	F:	PerFHalOrg.7-5/6, 33
$C_9F_{13}N$	$(C_2F_5)_2C_5F_3N$	F:	PerFHalOrg.5-149/50, 160, 177/8
–	$C_4F_9C_5F_4N$	F:	PerFHalOrg.5-154, 166
$C_9F_{13}N_3$	$(C_2F_5)_2(CF_2CF)C_3N_3$	F:	PerFHalOrg.6-82/3
$C_9F_{14}H_6N_4$	$C_3F_7C(NCH_3)NHNHC(NH)C_3F_7$	F:	PerFHalOrg.7-58
$C_9F_{14}H_6N_4O$	$NH_2(NH)CCF(CF_3)O(CF_2)_5C(NH)NH_2$	F:	PerFHalOrg.7-5
$C_9F_{14}H_6O$	$(C_3F_7)_2C(OH)C_2H_5$	F:	PerFHalOrg.4-93, 96
$C_9F_{14}H_8Si$	$(C_3F_7)_2Si(C_2H_5)CH_3$	F:	PerFHalOrg.4-18
$C_9F_{14}N_4S$	$(C_3F_7)_2C_3N_4S$	F:	PerFHalOrg.6-123/4, 138
$C_9F_{15}H_4P_3$	$[(CF_3)_2PCHCH]_2PCF_3$	F:	PerFHalOrg.3-185
$C_9F_{15}H_5NiO_2P_2$	$(CO)_2Ni(P(CF_3)_3)(P(CF_3)_2C_2H_5)$	Ni:	Org.Verb.1-145
$C_9F_{15}H_5O$	$C_7F_{15}CH(OH)CH_3$	F:	PerFHalOrg.4-42
$C_9F_{15}H_7OSi$	$(CH_3)_2SiH(CF_2)_4OCF(CF_3)_2$	F:	PerFHalOrg.4-87
$C_9F_{15}N$	$C_9F_{15}N$	F:	PerFHalOrg.6-131/2
$C_9F_{15}N_3$	$(C_2F_5)_3C_3N_3$	F:	PerFHalOrg.6-82/3, 96, 102, 112, 114/5
–	$(n\text{-}C_3F_7)_2C_3FN_3$	F:	PerFHalOrg.6-78/80, 100
–	$(i\text{-}C_3F_7)_2C_3FN_3$	F:	PerFHalOrg.6-78/80, 100

Formula	Compound		Reference
$C_9F_{15}N_3O_6S_3$	$(CF_3CF(SO_2F))_3C_3N_3$	F:	PerFHalOrg.6-82/3, 103, 112
$C_9F_{17}HO_2$	$n-C_8F_{17}COOH$	F:	PerFHalOrg.4-104
$C_9F_{17}H_2NO_2$	$(CF_3)_2CFO(CF_2)_5C(O)NH_2$	F:	PerFHalOrg.7-18/20, 47, 63
$C_9F_{17}N$	$C_4F_7N(CF_2)_5$	F:	PerFHalOrg.5-157/8, 171, 175/6
–	$C_9F_{17}N$	F:	PerFHalOrg.6-132, 145
$C_9F_{18}H_2N_2$	$(C_3F_7)(C_5F_{11})C(NH)_2$	F:	PerFHalOrg.5-6
$C_9F_{18}N_2O$	$C_2F_5N(CF_2CF_2)_2NC_2F_4COF$	F:	PerFHalOrg.6-31/2
$C_9F_{18}N_6$	$[(CF_3)_2N]_3C_3N_3$	F:	PerFHalOrg.6-78, 98
$C_9F_{19}HO$	$(C_3F_7)_2C(OH)C_2F_5$	F:	PerFHalOrg.4-94, 96
$C_9F_{19}N$	$CF_2N(C_4F_9)CF_2CF(CF_3)CF_2$	F:	PerFHalOrg.5-52/3, 98
–	$C_2F_5NC_5F_9C_2F_5$	F:	PerFHalOrg.5-150/1, 162, 176
–	$C_4F_9N(CF_2)_5$	F:	PerFHalOrg.5-157/8, 171, 175/6
$C_9F_{19}NO$	$CF_2N(C_4F_9)CF_2OCF(C_2F_5)$	F:	PerFHalOrg.5-54/5, 77, 98
–	$CF_3(CF_2)_8NO$	F:	PerFHalOrg.7-179, 185
–	$(i-C_3F_7)NC_4F_6(CF_3)_2O$	F:	PerFHalOrg.6-4/5
–	$O(CF_2CF_2)_2NCF_2(CF_2)_3CF_3$	F:	PerFHalOrg.6-2, 7
$C_9F_{19}N_3S$	$(CF_3)_2CFNSNC(CF_3)_2NC(CF_3)_2$	S:	SVol.2-91
$C_9F_{21}MoO_3P_3$	$Mo(CO)_3[(CF_3)_2PF]_3$	F:	PerFHalOrg.3-110/1, 121
$C_9F_{21}NO$	$(C_3F_7)_2NO(C_3F_7)$	F:	PerFHalOrg.7-94, 109/10, 124
$C_9F_{21}P$	$(C_3F_7)_3P$	F:	PerFHalOrg.3-163, 169, 173
$C_9FeH_4NO_4$	$(N(CH)_4C)Fe(CO)_4$	Fe:	Org.Verb.B4-128, 130, 134
$C_9FeH_4O_6$	$(C(O)C(CH_3)CHCO)Fe(CO)_4$	Fe:	Org.Verb.B4-334
$C_9FeH_4O_7$	$(C(CH_3)C(O)OC(O)CH)Fe(CO)_4$	Fe:	Org.Verb.B4-317/8
$C_9FeH_5NO_4$	$(CO)_4Fe(NC_5H_5)$	Fe:	Org.Verb.B2-71, 75, 77/8
		Fe:	Org.Verb.B4-252
$C_9FeH_5N_3O_2$	$(CO)_2Fe(NC_5H_5)(CN)_2$	Fe:	Org.Verb.B1-100
$C_9FeH_6N_2O_4$	$(CO)_4Fe(NCHNCHCHC(CH_3))$	Fe:	Org.Verb.B2-71, 75, 77
$C_9FeH_6NaO_7^-$	$[NaOCO(CH_2)_3C(O)Fe(CO)_4]^-$	Fe:	Org.Verb.B4-154, 162
$C_9FeH_6Na_2O_7$	$Na[NaOCO(CH_2)_3C(O)Fe(CO)_4]$	Fe:	Org.Verb.B4-154, 162
$C_9FeH_6O_8$	$CH_3OC(O)CHCHCOOHFe(CO)_4$	Fe:	Org.Verb.B4-275, 293
–	$HOCOC(CH_3)CHCOOHFe(CO)_4$	Fe:	Org.Verb.B4-317
$C_9FeH_7^+$	$[FeC_9H_7]^+$	Fe:	Org.Verb.A1-93/4
$C_9FeH_7LiO_5S_2$	$Li[(CH_2S)_2C(CH_3)C(O)Fe(CO)_4]$	Fe:	Org.Verb.B4-157, 167/8
$C_9FeH_7NO_4S$	$(SCHC(CH_3)N(CH_3)C)Fe(CO)_4$	Fe:	Org.Verb.B4-128, 130, 132/3
$C_9FeH_7O_5^-$	$[(CHO)(CH_3CO)C_3H_3Fe(CO)_3]^-$	Fe:	Org.Verb.B5-74, 76
$C_9FeH_7O_5S_2^-$	$[(CH_2S)_2C(CH_3)C(O)Fe(CO)_4]^-$	Fe:	Org.Verb.B4-157, 167/8
$C_9FeH_7O_6^-$	$[(CHO)(CH_3OCO)C_3H_3Fe(CO)_3]^-$	Fe:	Org.Verb.B5-73, 76
–	$[(CH_3OCO)(CH_3)C_3H(O)Fe(CO)_3]^-$	Fe:	Org.Verb.B5-72, 75
$C_9FeH_8I_2O$	$(C_6H_5CHCHCHO)FeI_2$	Fe:	Org.Verb.B5-140

$C_9FeH_8NO_5^-$ .... $[(CH_2)_4NC(O)Fe(CO)_4]^-$ .... Fe: Org.Verb.B4-155, 165
$C_9FeH_8N_2O_3$ .... $(CO)_3Fe(CH_2CH_2CN)_2$ .... Fe: Org.Verb.B1-205
– .... $C_3H_3N_2CH_2CHCH_2Fe(CO)_3$ .... Fe: Org.Verb.B4-212, 217
$C_9FeH_8N_2O_4$ .... $(CHNCH_3)_2CFe(CO)_4$ .... Fe: Org.Verb.B4-128/30, 133
– .... $(NC(CH_3)CHC(CH_3)NH)Fe(CO)_4$ .... Fe: Org.Verb.B2-71/2, 74
– .... $(N(CH_3)N(CH_3)CHCHC)Fe(CO)_4$ .... Fe: Org.Verb.B4-128/30
– .... $(N_2C_5H_8)Fe(CO)_4$ .... Fe: Org.Verb.B2-71, 75, 79
$C_9FeH_8N_2O_5$ .... $(CH_3)_2C_4H_2N_2O_2Fe(CO)_3$ .... Fe: Org.Verb.B5-160/1
$C_9FeH_8NaO_6$ .... $Na[CH_2CHCOOC_2H_5(Fe(CO)_4)]$ .... Fe: Org.Verb.C2-14/5
$C_9FeH_8O_2S_3$ .... $(C_5H_5)Fe(CO)_2[S_2CSCH_3]$ .... C: MVol.D4-269
$C_9FeH_8O_3$ .... $(CO)_3FeC_4H_2(CH_3)_2$ .... Fe: Org.Verb.B2-31/2
– .... $C_5H_5(CH_3)Fe(CO)_3$ .... Fe: Org.Verb.B3-151
– .... $C_6H_8Fe(CO)_3$ .... Fe: Org.Verb.B5-144/5
$C_9FeH_8O_4$ .... $CH_2C(CH_3)CHCH_2Fe(CO)_4$ .... Fe: Org.Verb.B4-238, 251/2
– .... $(CH_2CHCHCHCOCH_3)Fe(CO)_3$ .... Fe: Org.Verb.B5-22
– .... $CH_3CHCHCHCH_2Fe(CO)_4$ .... Fe: Org.Verb.B4-238, 251
– .... $(CO)_3Fe(C(O)CHCHC(CH_3)_2)$ .... Fe: Org.Verb.C2-33
– .... $C_3H_5CHCH_2Fe(CO)_4$ .... Fe: Org.Verb.B4-237, 250
$C_9FeH_8O_5$ .... $(CO)_3Fe(C_4H_5COOCH_3)$ .... Fe: Org.Verb.C2-14
$C_9FeH_8O_6$ .... $(CH_2C(COOCH_3)COCH_3)Fe(CO)_3$ .... Fe: Org.Verb.B5-157/8
– .... $CH_3OC(O)CHCHCH_3Fe(CO)_4$ .... Fe: Org.Verb.B4-275, 293
– .... $C_2H_5OC(O)CHCH_2Fe(CO)_4$ .... Fe: Org.Verb.B4-232/3, 236, 247
– .... $(HO)(CH_3OCO)C_4H_4Fe(CO)_3$ .... Fe: Org.Verb.B5-76
$C_9FeH_9KNO_5$ .... $K[(CH_3)_2NCOCHCH_2Fe(CO)_4]$ .... Fe: Org.Verb.B4-248
$C_9FeH_9KO_6$ .... $K[C_2H_5CH(OH)CH_2C(O)Fe(CO)_4]$ .... Fe: Org.Verb.B4-154, 162
$C_9FeH_9LiO_5$ .... $Li[C_4H_9C(O)Fe(CO)_4]$ .... Fe: Org.Verb.B4-150, 160
$C_9FeH_9NO_4$ .... $(CH_3)_3CNCFe(CO)_4$ .... Fe: Org.Verb.B4-5/6
– .... $C_5H_6(COCH_3)Fe(CO)_2NO$ .... Fe: Org.Verb.B5-14/5, 17, 23
$C_9FeH_9NO_5$ .... $(CH_3)_2NCOCHCH_2Fe(CO)_4$ .... Fe: Org.Verb.B4-237, 248/9
– .... $(CO)_3FeN(C_4H_6)COOCH_3$ .... Fe: Org.Verb.B5-88, 91/2
$C_9FeH_9NO_5^-$ .... $[(CH_3)_2NCOCHCH_2Fe(CO)_4]^-$ .... Fe: Org.Verb.B4-248
$C_9FeH_9NO_6$ .... $(CO)_5Fe(NH(CH_2)_4O)$ .... Fe: Org.Verb.B3-249
$C_9FeH_9NaO_5$ .... $Na[C_4H_9C(O)Fe(CO)_4]$ .... Fe: Org.Verb.B4-150, 160
$C_9FeH_9O^+$ .... $[FeC_5H_4CHCHCOCH_3]^+$ .... Fe: Org.Verb.A3-10
$C_9FeH_9O_2^+$ .... $[FeC_5H_4CHCHCOOCH_3]^+$ .... Fe: Org.Verb.A3-123
$C_9FeH_9O_4^+$ .... $[CH_3COC_4H_6Fe(CO)_3]^+$ .... Fe: Org.Verb.B5-94, 96
– .... $[(CH_3)_2C_3H_3Fe(CO)_4]^+$ .... Fe: Org.Verb.B5-100/5
– .... $[C_2H_5C_3H_4Fe(CO)_4]^+$ .... Fe: Org.Verb.B5-100, 102/3
$C_9FeH_9O_4^-$ .... $[CH_2CH(CH_2)_3C(O)Fe(CO)_3]^-$ .... Fe: Org.Verb.B4-109
$C_9FeH_9O_5^+$ .... $[(CH_3COO)(CH_3)C_3H_3Fe(CO)_3]^+$ .... Fe: Org.Verb.B5-94/6
$C_9FeH_9O_5^-$ .... $[C_4H_9C(O)Fe(CO)_4]^-$ .... Fe: Org.Verb.B4-150, 156, 160, 167
$C_9FeH_9O_6^-$ .... $[C_2H_5CH(OH)CH_2C(O)Fe(CO)_4]^-$ .... Fe: Org.Verb.B4-154, 162
$C_9FeH_9O_7P$ .... $(CO)_4Fe(P(CH_2O)_3CCH_3)$ .... Fe: Org.Verb.B2-86
– .... $(CO)_4FeP(OCH_2)_3CCH_3$ .... Fe: Org.Verb.B2-95, 117
$C_9FeH_{10}LiNO_5$ .... $Li[(C_2H_5)_2NC(O)Fe(CO)_4]$ .... Fe: Org.Verb.B4-155, 165
$C_9FeH_{10}NO_5^+$ .... $[C_2H_5O(NH_2)CCHCH_2Fe(CO)_4]^+$ .... Fe: Org.Verb.B4-248/9
$C_9FeH_{10}NO_5^-$ .... $[(C_2H_5)_2NC(O)Fe(CO)_4]^-$ .... Fe: Org.Verb.B4-155, 165
– .... $[C_4H_9NHC(O)Fe(CO)_4]^-$ .... Fe: Org.Verb.B4-155, 165

Formula	Compound	Ref.
$C_9FeH_{10}N_2O_3$	$(C_2H_5NC)_2Fe(CO)_3$	Fe: Org.Verb.B4-18/9
$C_9FeH_{10}N_2O_4$	$CH_2CHC(N(CH_3)_2)NHCOFe(CO)_3$	Fe: Org.Verb.B5-159/60
–	$(CH_2NCH_3)_2CFe(CO)_4$	Fe: Org.Verb.B4-128, 130, 133/4
$C_9FeH_{10}N_2O_4^+$	$[(CH_2NCH_3)_2CFe(CO)_4]^+$	Fe: Org.Verb.B4-172
$C_9FeH_{10}N_2O_5$	$COFe(NCCOOC_2H_5)_2$	Fe: Org.Verb.B1-34
$C_9FeH_{10}N_6O$	$(C_2H_5NC)_2Fe(CN)_3NO$	Fe: Org.Verb.B4-11/2
$C_9FeH_{10}OS_3$	$(C_5H_5)Fe(CO)[S_2CSC_2H_5]$	C: MVol.D4-269
$C_9FeH_{10}O_3$	$C_3H_5Fe(CO)_3CH_2CHCH_2$	Fe: Org.Verb.B5-161
$C_9FeH_{10}O_3S$	$(CH_3CHC(C_2H_5)CHS)Fe(CO)_3$	Fe: Org.Verb.B5-122, 135, 148
–	$(SC(CH_3)C(C_2H_5)CH_2)Fe(CO)_3$	Fe: Org.Verb.B4-213, 220/1
$C_9FeH_{10}O_4$	$(CH_2C(CH_3)C(C_2H_5)O)Fe(CO)_3$	Fe: Org.Verb.B5-131, 141
$C_9FeH_{10}O_4S$	$(CH_3CHC(C_2H_5)CHSO)Fe(CO)_3$	Fe: Org.Verb.B5-136, 148/9
$C_9FeH_{10}O_5$	$(CO)_4Fe(O(CH_2)_3CHCH_3)$	Fe: Org.Verb.B2-128/31
$C_9FeH_{11}NO_4$	$CH_3(CH_3COCH_2)C_3H_3Fe(CO)_2NO$	Fe: Org.Verb.B5-14/5, 17, 23
–	$(CO)_4FeNH(CH_2)_5$	Fe: Org.Verb.B2-71/3
$C_9FeH_{11}NO_5$	$((CH_3)_2N)(C_2H_5O)CFe(CO)_4$	Fe: Org.Verb.B4-121/3, 125/6
$C_9FeH_{11}NiO_2P^+$	$NiP(CH_3)_2Fe(C_5H_5)(CO)_2^+$	Ni: Org.Verb.1-167
$C_9FeH_{11}O_4^-$	$[C_5H_{11}C(O)Fe(CO)_3]^-$	Fe: Org.Verb.B4-109
$C_9FeH_{12}^{2+}$	$[FeC_6H_3(CH_3)_3]^{2+}$	Fe: Org.Verb.A2-186, 278
$C_9FeH_{12}N_2O_4$	$CH_2CHC(O)NN(CH_3)_3Fe(CO)_3$	Fe: Org.Verb.B5-158
–	$[(CH_3)_2N]_2CFe(CO)_4$	Fe: Org.Verb.B4-120
$C_9FeH_{12}N_2O_4S$	$(CO)_4FeSC(N(CH_3)_2)_2$	Fe: Org.Verb.B2-128/30, 132
$C_9FeH_{12}N_2O_5Sn$	$Sn(OCN(CH_3)_2)_2Fe(CO)_3$	Fe: Org.Verb.B4-194, 196
$C_9FeH_{12}N_4O_3S_2$	$(CO)_3Fe(SC(NHCH_2)_2)_2$	Fe: Org.Verb.B1-148/50, 157, 165
$C_9FeH_{12}N_4O_3S_2^+$	$[(CO)_3Fe(SC(NHCH_2)_2)_2]^+$	Fe: Org.Verb.B1-165
$C_9FeH_{12}NiO_5P_2$	$(CO)_3Fe(P(CH_3)_2)_2Ni(CO)_2$	Fe: Org.Verb.B1-183
		Ni: Org.Verb.1-151, 154
$C_9FeH_{12}O_4Si$	$(CO)_4FeSi(CH_3)_2C_3H_6$	Fe: Org.Verb.B2-189
$C_9FeH_{12}O_5Si$	$(CH_3)_3SiOCHCH_2Fe(CO)_4$	Fe: Org.Verb.B4-234, 240/1
$C_9FeH_{12}O_9P_4$	$(CO)_3Fe(P(OCH_2)_3P)P(CH_2O)_3P$	Fe: Org.Verb.B1-148/50, 156, 165
$C_9FeH_{13}IO_3Si$	$Si(CH_3)_3C_3H_4Fe(CO)_3I$	Fe: Org.Verb.B5-44, 53, 63
$C_9FeH_{13}NO_3P^+$	$[C_4H_4P(CH_3)_3Fe(CO)_2NO]^+$	Fe: Org.Verb.B5-28/31
$C_9FeH_{13}NO_4$	$C(CH_3)_2OHCH_2C_3H_4Fe(CO)_2NO$	Fe: Org.Verb.B5-14/6, 22
–	$[C_5H_{10}NH_2][HFe(CO)_4]$	Fe: Org.Verb.B4-165
$C_9FeH_{14}NO_5P$	$C_3H_5Fe(CO)(NO)P(OCH_2)_3CCH_3$	Fe: Org.Verb.B5-6/9, 20/1
$C_9FeH_{14}N_5O^+$	$[(CH_3NC)_4Fe(H_2O)CN]^+$	Fe: Org.Verb.B4-70
$C_9FeH_{14}N_6O_4$	$[(CH_3NC)_4Fe(H_2O)CN]NO_3$	Fe: Org.Verb.B4-70
C_9FeH_{15}	$(C_3H_5)_3Fe$	Fe: Org.Verb.B5-108
$C_9FeH_{15}I_2O_3P$	$(CO)_3Fe(P(C_2H_5)_3)I_2$	Fe: Org.Verb.B1-180/2
$C_9FeH_{15}NO_4Sn$	$(CO)_3Fe(NO)Sn(C_2H_5)_3$	Fe: Org.Verb.B1-197/8
		Sn: Org. Verb.5-106
$C_9FeH_{15}O_3S_6$	$Fe[S_2COC_2H_5]_3$	C: MVol.D4-245, 259
$C_9FeH_{15}O_4PSi$	$(CO)_4FeP(CH_3)_2Si(CH_3)_3$	Fe: Org.Verb.B2-91
$C_9FeH_{15}O_5P$	$(C_4H_6)Fe(CO)_2(P(OCH_3)_3)$	Fe: Org.Verb.B1-162
$C_9FeH_{15}S_9$	$Fe[S_2CSC_2H_5]_3$	C: MVol.D4-267, 269
$C_9FeH_{16}IO_5P$	$CH_3C_3H_4Fe(CO)_2P(OCH_3)_3I$	Fe: Org.Verb.B5-24/5

Formula	Compound	Element	Reference
$C_9FeH_{16}O_3P_2$	$(CO)_3Fe(CH_3)_2PCH_2CH_2P(CH_3)_2$	Fe:	Org.Verb.B1-170/2, 176
$C_9FeH_{18}HgI_2O_3P_2$	$(CO)_3Fe(P(CH_3)_3)_2 \cdot HgI_2$	Fe:	Org.Verb.B1-192/3
$C_9FeH_{18}HgI_2O_9P_2$	$(CO)_3Fe(P(OCH_3)_3)_2 \cdot HgI_2$	Fe:	Org.Verb.B1-192/3
$C_9FeH_{18}Hg_2I_4O_9P_2$	$(CO)_3Fe(P(OCH_3)_3)_2 \cdot (HgI_2)_2$	Fe:	Org.Verb.B1-192/4
$C_9FeH_{18}I_4O_3P_2$	$[(CO)_3Fe(P(CH_3)_3)_2I]I_3$	Fe:	Org.Verb.B1-184/5
$C_9FeH_{18}O_2P_2S_2$	$CS_2Fe(CO)_2(P(CH_3)_3)_2$	Fe:	Org.Verb.B5-143
$C_9FeH_{18}O_3P_2$	$(CO)_3Fe(P(CH_3)_3)_2$	Fe:	Org.Verb.B1-148/51, 158
$C_9FeH_{18}O_3Sb_2$	$(CO)_3Fe(Sb(CH_3)_3)_2$	Fe:	Org.Verb.B1-148/50, 156, 164
$C_9FeH_{18}O_8P_2S_2$	$CS_2Fe(CO)_2(P(OCH_3)_3)_2$	Fe:	Org.Verb.B5-143
$C_9FeH_{18}O_9P_2$	$(CO)_3Fe(P(OCH_3)_3)_2$	Fe:	Org.Verb.B1-148/50, 153, 162/3
$C_9FeH_{21}IO_2P_2$	$(CO)_2Fe(CH_3)(P(CH_3)_3)_2I$	Fe:	Org.Verb.B1-121/2, 125
$C_9FeH_{23}I_2N_5O$	$[COFe(NH(C_2H_4NHC_2H_4NH_2)_2)][I]_2$	Fe:	Org.Verb.B1-86
$C_9FeH_{23}N_5O^{2+}$	$[COFe(NH(C_2H_4NHC_2H_4NH_2)_2)]^{2+}$	Fe:	Org.Verb.B1-85
$C_9FeMnNO_{10}$	$[Mn(CO)_6][Fe(CO)_3NO]$	Fe:	Org.Verb.B1-188/9
$C_9FeNO_{10}Re$	$[Re(CO)_6][Fe(CO)_3NO]$	Fe:	Org.Verb.B1-188
C_9FeNiO_9	$FeNi(CO)_9$	Fe:	Org.Verb.B3-49, 64
$C_9Fe_2H_2O_7S$	$(CO)_3Fe(CHCHCOS)Fe(CO)_3$	Fe:	Org.Verb.C2-109, 117
$C_9Fe_2H_6N_2O_7$	$(CO)_3Fe(CH_3NCONCH_3)Fe(CO)_3$	Fe:	Org.Verb.C1-143/4, 146
$C_9Fe_2H_6O_6S$	$(CO)_3Fe(CHCH_2)(SCH_3)Fe(CO)_3$	Fe:	Org.Verb.C2-85/6
$C_9Fe_2H_9NO_7$	$(NH_3)(CO)_2Fe(CH_3CO)_2Fe(CO)_3$	Fe:	Org.Verb.C1-232/3, 236
$C_9Fe_2H_9O_6S_3^+$	$[(CO)_6Fe_2(SCH_3)_3]^+$	Fe:	Org.Verb.C1-20/3
$C_9Fe_2H_{12}N_2O_7P_2$	$(CO)_4FeP(CH_3)_2P(CH_3)_2Fe(CO)(NO)_2$	Fe:	Org.Verb.C1-67, 69/70
$C_9Fe_2H_{16}N_4O_5$	$Fe_2(CO)_5(C_2H_8N_2)_2$	Fe:	Org.Verb.C2-8
$C_9Fe_2O_9$	$Fe_2(CO)_9$		
	Chemical reactions	Fe:	Org.Verb.C2-25/63
	Formation and preparation	Fe:	Org.Verb.C2-17/9
	Physical properties	Fe:	Org.Verb.C2-19/25
	Uses	Fe:	Org.Verb.C2-64/6
$C_9Fe_2O_9^+$	$[Fe_2(CO)_9]^+$	Fe:	Org.Verb.C2-26
$C_9Fe_2O_9^-$	$[Fe_2(CO)_9]^-$	Fe:	Org.Verb.C2-26
$C_9Fe_3H_2O_9S$	$H_2Fe_3(CO)_9S$	Fe:	Org.Verb.B3-126
$C_9Fe_3O_9SSe$	$Fe_3(CO)_9SSe$	Fe:	Org.Verb.B3-126
$C_9Fe_3O_9S_2$	$Fe_3(CO)_9S_2$	Fe:	Org.Verb.B3-126
		Fe:	Org.Verb.C2-27, 43, 118
$C_9Fe_3O_9Se_2$	$Fe_3(CO)_9Se_2$	Fe:	Org.Verb.B3-126
		Fe:	Org.Verb.C2-96
$C_9Fe_3O_9Te_2$	$Fe_3(CO)_9Te_2$	Fe:	Org.Verb.B3-126
		Fe:	Org.Verb.C2-129
$C_9GaH_{15}O_3S_6$	$Ga[S_2COC_2H_5]_3$	C:	MVol.D4-253
$C_9GeH_{14}NiO$	$C_5H_5Ni(CO)Ge(CH_3)_3$	Ni:	Org.Verb.2-346
$C_9GeH_{27}PSiSn$	$[(CH_3)_3Si][(CH_3)_3Ge][(CH_3)_3Sn]P$	Sn:	Org.Verb.5-82
$C_9GeH_{27}PSn_2$	$[(CH_3)_3Sn]_2PGe(CH_3)_3$	Sn:	Org.Verb.5-82
$C_9Ge_2H_{27}PSn$	$[(CH_3)_3Ge]_2PSn(CH_3)_3$	Sn:	Org.Verb.5-82
$C_9HNi_4O_9^-$	$[Ni_4(CO)_9H]^-$	Ni:	Org.Verb.2-384
$C_9HNi_5O_9^-$	$[Ni_5(CO)_9H]^-$	Ni:	Org.Verb.2-391
$C_9H_4INNpO_6S^-$	$NpO_2(NC_9H_4I(O)SO_3)^-$	Np:	TrU.D1-167/8
$C_9H_4INO_6PuS^-$	$PuO_2(NC_9H_4I(O)SO_3)^-$	Np:	TrU.D1-168
$C_9H_5NNpO_6S^-$	$NpO_2(NC_9H_5(O)SO_3)^-$	Np:	TrU.D1-167/8

Formula	Compound		Reference
$C_9H_5NO_6PuS^-$	$PuO_2(NC_9H_5(O)SO_3)^-$	Np:	TrU.D1-168
$C_9H_5N_2NiS_2$	$C_5H_5NiS_2C_2(CN)_2$	Ni:	Org.Verb.2-208
$C_9H_6KNiO_3P$	$K[(CO)_3NiP(H)C_6H_5]$	Ni:	Org.Verb.1-191, 192
$C_9H_6K_4MnN_6OS_4Se_2$	$K_4Mn(NCS)_4(NCSe)_2 \cdot (CH_3)_2CO$	Mn:	MVol.D1-56
C_9H_7NNiO	C_9H_7NiNO	Ni:	Org.Verb.2-349, 360
$C_9H_7NNiO_3$	$(CO)_3Ni(NC_5H_4CH_3)$	Ni:	Org.Verb.1-160
C_9H_7NNiS	$(CH_2CH)C_6H_4NiNCS$	Ni:	Org.Verb.1-42
$C_9H_7Ni^+$	$Ni(C_9H_7)^+$	Ni:	Org.Verb.2-68
$C_9H_7NiO_3P$	$(CO)_3NiP(C_6H_5)H_2$	Ni:	Org.Verb.1-172
$C_9H_8MoO_2S_3$	$(C_5H_5)Mo(CO)_2[S_2CSCH_3]$	C:	MVol.D4-269
$C_9H_8O_2S_3W$	$(C_5H_5)W(CO)_2[S_2CSCH_3]$	C:	MVol.D4-269
$C_9H_9MnO_2^+$	$[Mn(OC_6H_3(COCH_3)CH_3)]^+$	Mn:	MVol.D1-57/8
C_9H_9NOS	$HNCS \cdot C_6H_5COCH_3 \cdot n\ H_2O$	C:	MVol.D6-30
$C_9H_9NO_8U$	$[UO_2(O(CH_2COO)_2)(C_5H_5NO)]_n$	U:	SVol.E1-132, 135, 142/3
$C_9H_9NiO_3P$	$(CO)_3NiP(CHCH_2)_3$	Ni:	Org.Verb.1-166
$C_9H_9NiO_6P$	$(CO)_3Ni(PO_3C_6H_9)$	Ni:	Org.Verb.1-178, 179/80
$C_9H_{10}N_2NiO$	$[C_2H_5C_6H_5Ni(NO)]CN$	Ni:	Org.Verb.2-225
$C_9H_{10}O_2S_2Te$	$(CH_3OC_6H_4)Te[S_2COCH_3]$	C:	MVol.D4-255
$C_9H_{11}I_3Sn$	$(CH_2I)_3SnC_6H_5$	Sn:	Org.Verb.2-335
$C_9H_{11}MnO_4^+$	$[Mn(CH(COCH_3)_2)CH_3COCHCO]^+$	Mn:	MVol.D1-92/3
$C_9H_{11}N_5O_8U$	$UO_2(NC_5H_3(COO)_2) \cdot 2\ (NH_2)_2CO$	U:	SVol.E1-84, 86, 90
$C_9H_{11}NiO_6P$	$(CO)_3NiP(OCH_2)_3CC_2H_5$	Ni:	Org.Verb.1-178, 179
$C_9H_{12}N_2NiO_4$	$[(CH_3)_3C_6H_3Ni(NO)]NO_3$	Ni:	Org.Verb.2-222, 225
$C_9H_{12}N_2O_3S$	$C_6H_5C(O)NHSO_2NHC_2H_5$	S:	S-N-Verb.1-59
$C_9H_{12}N_2O_4S$	$C_6H_5CH_2NHSO_2NHCOOCH_3$	S:	S-N-Verb.1-170
$C_9H_{12}N_4Ni_5O_9$	$[Ni(NH_3)_4][Ni_4(CO)_9]$	Ni:	Org.Verb.1-160
$C_9H_{12}Ni$	$C_5H_5NiCH_2CHCHCH_3$	Ni:	Org.Verb.2-175/6
–	$C_9H_{12}Ni$	Ni:	Org.Verb.2-243, 244
$C_9H_{13}ISn$	$(CH_3)_2Sn(C_6H_5)(CH_2I)$	Sn:	Org.Verb.3-79
–	$(CH_3)_3SnC_6H_4I$	Sn:	Org.Verb.2-134
$C_9H_{13}I_2OTi$	$C_5H_5TiI_2 \cdot C_4H_8O$	Ti:	Org.Verb.1-134/6
$C_9H_{13}LiSn$	$(CH_3)_3SnC_6H_4Li$	Sn:	Org.Verb.2-140/1
$C_9H_{13}NS_4Sn$	$Sn(SCH_2CH_2S)_2 \cdot C_5H_5N$	Sn:	MVol.C6-82/3
$C_9H_{13}N_3O_2S$	$(CH_3)_2NSO_2NHNCHC_6H_5$	S:	S-N-Verb.1-222
$C_9H_{13}NiO_6P$	$(CO)_3Ni(CH_3OP(OCH_2)_2C(CH_3)_2)$	Ni:	Org.Verb.1-178
$C_9H_{13}Sn^+$	$Sn(C_6H_5)(C_2H_5)CH_3^+$	Sn:	Org.Verb.3-25
$C_9H_{14}N_2S_4Sn$	$Sn(SCH_2CH_2S)_2 \cdot NC_5H_4NH_2$	Sn:	MVol.C6-82/3
$C_9H_{14}NiOSn$	$C_5H_5Ni(CO)Sn(CH_3)_3$	Ni:	Org.Verb.2-164/6
$C_9H_{14}NiO_2$	$CH_3C_3H_4Ni(CH(COCH_3)_2)$	Ni:	Org.Verb.2-8
$C_9H_{14}OSn$	$(CH_3)_2(CH_3OC_6H_4)SnH$	Sn:	Org.Verb.4-90
$C_9H_{14}SSn$	$(CH_3)_3SnSC_6H_5$	Sn:	Org.Verb.5-82
$C_9H_{14}Sn$	$(CH_3)_2(CH_3C_6H_4)SnH$	Sn:	Org.Verb.4-90
–	$(CH_3)_3SnCCCCC_2H_5$	Sn:	Org.Verb.2-114
–	$(CH_3)_3SnC_6H_5$	Sn:	Org.Verb.2-120/4
$C_9H_{14}Sn^+$	$Sn(C_3H_4)(CH_2CHCH_2)_2^+$	Sn:	Org.Verb.1-134
$C_9H_{15}IO_2Ti$	$C_5H_5Ti(OC_2H_5)_2I$	Ti:	Org.Verb.1-184/7
$C_9H_{15}ISn$	$(C_3H_5)_3SnI$	Sn:	Org.Verb.5-150
$C_9H_{15}InO_3S_6$	$In[S_2COC_2H_5]_3$	C:	MVol.D4-253
$C_9H_{15}IrS_6$	$Ir(CH_3SCHCHS)_3$	Ir:	SVol.2-171
$C_9H_{15}KO_5S_6U$	$KUO_2[S_2COC_2H_5]_3$	C:	MVol.D4-261

$C_9H_{15}MnO_2^+$	$[Mn(CH(COCH(CH_3)_2)_2)]^+$	Mn: MVol.D1-125
$C_9H_{15}NSn$	$(CH_3)_3SnCH_2C_5H_4N$	Sn: Org.Verb.2-28, 30
–	$(CH_3)_3SnC_6H_4NH_2$	Sn: Org.Verb.2-140/1
–	$(C_2H_5)_3SnCCCN$	Sn: Org.Verb.2-221
$C_9H_{15}N_2NiO_3P$	$(NCCHCH_2)_2NiP(OCH_3)_3$	Ni: Org.Verb.1-412
$C_9H_{15}N_3NiO_2$	$Ni(O_2)(CNC_4H_9\text{-t})(CNCH_3)_2$	Ni: Org.Verb.2-397
$C_9H_{15}NiO_3P$	$(CO)_3NiP(C_2H_5)_3$	Ni: Org.Verb.1-166, 168
$C_9H_{15}NiO_3PS_3$	$(CO)_3NiP(SC_2H_5)_3$	Ni: Org.Verb.1-182
$C_9H_{15}NiO_3Sb$	$(CO)_3NiSb(C_2H_5)_3$	Ni: Org.Verb.1-188, 190
$C_9H_{15}NiO_6P$	$(CO)_3NiP(OC_2H_5)_3$	Ni: Org.Verb.1-177, 179
$C_9H_{15}ORhSn$	$(CH_3)_3Sn[RhH(C_5H_5)(CO)]$	Sn: Org.Verb.5-86
$C_9H_{15}O_2Ti^+$	$[C_5H_5Ti(OC_2H_5)_2]^+$	Ti: Org.Verb.1-160
$C_9H_{15}O_3PS_6$	$P[S_2COC_2H_5]_3$	C: MVol.D4-254
$C_9H_{15}O_3PSe_6$	$P[Se_2COC_2H_5]_3$	C: MVol.D6-221
$C_9H_{15}O_3S_6Sb$	$Sb[S_2COC_2H_5]_3$	C: MVol.D4-245, 248, 254
$C_9H_{15}O_3S_6Tl$	$Tl[S_2COC_2H_5]_3$	C: MVol.D4-253
$C_9H_{15}Sn^+$	$Sn(CH_2CHCH_2)_3^+$	Sn: Org.Verb.1-134
$C_9H_{15}Ti$	$(CH_2CHCH_2)_3Ti$	Ti: Org.Verb.1-83
$C_9H_{16}NO_6Sn$	$NH_4Sn(CH_3COO)_3 \cdot 0.5\ C_6H_6$	Sn: MVol.C3-124
$C_9H_{16}Sn$	$(CH_2CH_2CH)_2Sn(CH_3)(CHCH_2)$	Sn: Org.Verb.3-92, 94
–	$(CH_3)_3SnCCC(CH_3)CHCH_3$	Sn: Org.Verb.2-114
–	$(CH_3)_3SnC_5H_4CH_3$	Sn: Org.Verb.2-99, 105
$C_9H_{17}IN_4NiS_2$	$C_3H_5Ni(SC(NHCH_2)_2)_2I$	Ni: Org.Verb.2-21, 22
$C_9H_{17}IrN_6O_6S$	$H[Ir(NO_2)(CH_3C(NOH)C(NOH)CH_3)_2NCS]$	Ir: SVol.2-62
$C_9H_{17}KOS_2$	$K[S_2COC_8H_{17}]$	C: MVol.D4-253
$C_9H_{17}KOSe_2$	$K[Se_2COC_8H_{17}]$	C: MVol.D6-221
$C_9H_{17}NOSn$	$(C_2H_5)_3SnCCCONH_2$	Sn: Org.Verb.2-221
$C_9H_{17}NSn$	$(C_2H_5)_3SnC(CN)CH_2$	Sn: Org.Verb.2-192, 193
–	$(C_2H_5)_3SnC(CN)CHD$	Sn: Org.Verb.2-192
$C_9H_{17}NTi$	$(CH_3)_4Ti \cdot C_5H_5N$	Ti: Org.Verb.1-114, 117
$C_9H_{17}OS_2^-$	$[S_2COC_8H_{17}]^-$	C: MVol.D4-251
$C_9H_{18}IrN_3S_6$	$Ir(S_2CN(CH_3)_2)_3$	Ir: SVol.2-169
$C_9H_{18}MnO_9Sr$	$Sr[Mn(HOCH(CH_2O)_2)_3]$	Mn: MVol.D1-38
$C_9H_{18}N_2O_3S$	$C_5H_{10}NSO_2N(CH_2CH_2)_2O$	S: S-N-Verb.1-173
$C_9H_{18}N_2Sn$	$(C_2H_5)_3SnC_3H_3N_2$	Sn: Org.Verb.2-246
$C_9H_{18}N_3NiO_3$	$[(CH_3)_3C_6H_3Ni(NH_3)_2]NO_3$	Ni: Org.Verb.2-222, 225
$C_9H_{18}N_6NiO_3P_4$	$(CO)_3Ni(P_4(NCH_3)_6)$	Ni: Org.Verb.1-185, 186
$C_9H_{18}NiS_4$	$[Ni(S_2CSC_4H_9)(SC_4H_9)]_2$	C: MVol.D4-270
$C_9H_{18}OS_2$	$SC(OC_8H_{17})(SH)$	C: MVol.D4-239
$C_9H_{18}OSn$	$(C_2H_5)_3SnCCOCH_3$	Sn: Org.Verb.2-221
–	$(C_2H_5)_3SnOCH_2CCH$	Sn: Org.Verb.5-103
$C_9H_{18}O_2Sn$	$(CH_3)_3SnC(CHCH_3)COOC_2H_5$	Sn: Org.Verb.2-80, 88
–	$(CH_3)_3SnC(CH_3)CHCOOC_2H_5$	Sn: Org.Verb.2-80, 88
–	$(C_3H_7)_2Sn(CH_2CH_2COO)$	Sn: Org.Verb.2-257
$C_9H_{18}PdS_4$	$[Pd(S_2CSC_4H_9)(SC_4H_9)]_2$	C: MVol.D4-266, 270
$C_9H_{18}PtS_4$	$[Pt(S_2CSC_4H_9)(SC_4H_9)]_2$	C: MVol.D4-270
$C_9H_{18}SSn$	$(C_2H_5)_3SnCCSCH_3$	Sn: Org.Verb.2-222
$C_9H_{18}SeSn$	$(C_2H_5)_3SnCCSeCH_3$	Sn: Org.Verb.2-222
$C_9H_{18}Sn$	$(CH_3)_3SnCCC_4H_9$	Sn: Org.Verb.2-114
–	$(CH_3)_3SnCC(t\text{-}C_4H_9)$	Sn: Org.Verb.2-114, 118

$C_9H_{18}Sn$ .... $(CH_3)_3SnC(CH_3)CHCHCHCH_3$ .... Sn: Org.Verb.2-80, 88
– .... $(CH_3)_3SnCH(CHCH_2)CH_2CHCH_2$ .... Sn: Org.Verb.2-80, 88
– .... $(CH_3)_3SnCH_2CHCHCH_2CHCH_2$ .... Sn: Org.Verb.2-80, 88
– .... $(CH_3)_3SnC_6H_9$ .... Sn: Org.Verb.2-100/1, 105
Sn: Org.Verb.6-263, 266
– .... $(C_2H_5)_3SnCCCH_3$ .... Sn: Org.Verb.2-208/9
– .... $(C_2H_5)_3SnCHCCH_2$ .... Sn: Org.Verb.2-184/5
– .... $(C_2H_5)_3SnCDCCH_2$ .... Sn: Org.Verb.2-184
– .... $(C_2H_5)_3SnCH_2CCH$ .... Sn: Org.Verb.2-212/3
$C_9H_{19}I_4O_4PSn$ .... $SnI_4 \cdot CH_3COCH_2P(O)(OC_3H_7)_2$ .... Sn: MVol.C6-165/6
$C_9H_{19}NO_4Sn$ .... $(i\text{-}C_3H_7O)SnN(CH_2CH_2O)_3$ .... Sn: MVol.C6-4
$C_9H_{19}NSn$ .... $(C_2H_5)_3SnCH(CH_3)CN$ .... Sn: Org.Verb.2-165
– .... $(C_2H_5)_3SnCH_2CH_2CN$ .... Sn: Org.Verb.2-163, 165
$C_9H_{19}NiO_2P$ .... $CH_3Ni(P(CH_3)_3)(CH(COCH_3)_2)$ .... Ni: Org.Verb.1-12, 19
$C_9H_{20}INiP$ .... $C_3H_5Ni(P(C_2H_5)_3)I$ .... Ni: Org.Verb.2-281
$C_9H_{20}N_2Sn$ .... $(CH_3)_3SnC(CH_3)(CN)CH_2N(CH_3)_2$ .... Sn: Org.Verb.2-44
$C_9H_{20}N_2Te$ .... $(C_2H_5)_4NTeCN$ .... C: MVol.D6-258/9
$C_9H_{20}OSn$ .... $(C_2H_5)_3SnC(CH_2)OCH_3$ .... Sn: Org.Verb.2-193
– .... $(C_2H_5)_3SnC(CH_2OH)CH_2$ .... Sn: Org.Verb.2-192, 193
– .... $(C_2H_5)_3SnCHCHCH_2OH$ .... Sn: Org.Verb.2-192, 193
– .... $(C_2H_5)_3SnCH_2COCH_3$ .... Sn: Org.Verb.2-162
$C_9H_{20}O_2Sn$ .... $(CH_3)_3Sn(CH_2)_3OCH_2CH(O)CH_2$ .... Sn: Org.Verb.2-44
– .... $(CH_3)_3Sn(CH_2)_4COOCH_3$ .... Sn: Org.Verb.2-50
– .... $(C_2H_5)_2[CH_3OOCCH(CH_3)CH_2]SnH$ .... Sn: Org.Verb.4-91
– .... $(C_2H_5)_3SnCH_2COOCH_3$ .... Sn: Org.Verb.2-162/3
– .... $(C_2H_5)_3SnCH_2OOCCH_3$ .... Sn: Org.Verb.2-165, 176
$C_9H_{20}O_2Ti$ .... $CH_2Ti(OC_4H_9)_2$ .... Ti: Org.Verb.1-122
$C_9H_{20}SSn$ .... $(C_2H_5)_3SnCHCHSCH_3$ .... Sn: Org.Verb.2-192, 193
$C_9H_{20}Sn$ .... $(CH_3)_3SnC(CHCH_3)CH(CH_3)_2$ .... Sn: Org.Verb.2-81, 89
– .... $(CH_3)_3SnC(C_2H_5)C(CH_3)_2$ .... Sn: Org.Verb.2-81, 89
– .... $(CH_3)_3SnCHCHC_4H_9$ .... Sn: Org.Verb.2-81, 89
– .... $(CH_3)_3SnCH(CH_2)_5$ .... Sn: Org.Verb.2-58
– .... $(CH_3)_3SnCH_2C(CH_3)C(CH_3)_2$ .... Sn: Org.Verb.2-81, 88
– .... $(CH_3)_3SnCH_2CH(CH_3)C(CH_3)CH_2$ .... Sn: Org.Verb.2-81, 88/9
– .... $(C_2H_5)_2Sn(CH_2)_5$ .... Sn: Org.Verb.3-105
– .... $(C_2H_5)_3SnCHCH_2CH_2$ .... Sn: Org.Verb.2-181/2
– .... $(C_2H_5)_3SnCH_2CHCH_2$ .... Sn: Org.Verb.2-183/4
$C_9H_{21}NOSn$ .... $(CH_3)_3SnCH_2CON(C_2H_5)_2$ .... Sn: Org.Verb.2-25
– .... $(C_2H_5)_3SnCH_2CH_2CONH_2$ .... Sn: Org.Verb.2-165
$C_9H_{21}NO_2S$ .... $HNCS \cdot 2\ C_4H_9OH \cdot 2\ H_2O$ .... C: MVol. D6-30/1
$C_9H_{21}NO_2Sn$ .... $(C_2H_5)_2Sn(OCH_2CH_2)_2NCH_3$ .... Sn: Org.Verb.6-58
$C_9H_{21}NSn$ .... $(C_2H_5)_3SnCH_2NC_2H_4$ .... Sn: Org.Verb.2-165, 176
$C_9H_{21}N_3O_3S$ .... $(C_2H_5)_2NSO_2NHCON(C_2H_5)_2$ .... S: S-N-Verb.1-171
$C_9H_{21}N_3O_3Sn^{2+}$ .... $[Sn(OCHN(CH_3)_2)_3]^{2+}$ .... Sn: MVol.C5-43
$C_9H_{21}NiP$ .... $(CH_3)_2C_3H_3Ni(P(CH_3)_3)CH_3$ .... Ni: Org.Verb.2-36, 38
$C_9H_{21}Sn^+$ .... $Sn(C_3H_7)_3{}^+$ .... Sn: Org.Verb.1-86
$C_9H_{22}INOSn$ .... $[(CH_3)_3SnCH_2N(CH_3)(CH_2CH_2)_2O]I$ .... Sn: Org.Verb.2-13
$C_9H_{22}N_2O_{14}P_2U$ .... $UO_2(NO_3)_2 \cdot CH_2(PO(OC_2H_5)_2)_2$ .... U: SVol.E1-188/9
$C_9H_{22}OSn$ .... $(C_2H_5)_3SnCH_2CH_2CH_2OH$ .... Sn: Org.Verb.2-166
– .... $(C_2H_5)_3SnCH_2OC_2H_5$ .... Sn: Org.Verb.2-166

$C_9H_{22}OSn$	$(C_3H_7)_3SnOH$	Sn:	Org.Verb.5-114, 116
$C_9H_{22}O_2S_3$	$SC(SH)_2 \cdot 2\ (C_2H_5)_2O$	C:	MVol.D4-215
$C_9H_{22}O_5U$	$UO_2(OC_3H_7)_2 \cdot C_3H_7OH$	U:	SVol.E1-59, 64
$C_9H_{22}Sn$	$CH_3(C_2H_5)Sn(C_3H_7)(i\text{-}C_3H_7)$	Sn:	Org.Verb.3-96
–	$(CH_3)_2Sn(C_3H_7)(s\text{-}C_4H_9)$	Sn:	Org.Verb.3-75, 76
–	$(CH_3)_2Sn(i\text{-}C_3H_7)(s\text{-}C_4H_9)$	Sn:	Org.Verb.3-75, 77
–	$(CH_3)_3SnCH_2CH(CH_3)CH(CH_3)_2$	Sn:	Org.Verb.2-53/4
–	$(CH_3)_3SnCH_2CH(C_2H_5)_2$	Sn:	Org.Verb.2-54
–	$(CH_3)_3Sn(CH_2)_3CH(CH_3)_2$	Sn:	Org.Verb.2-53/4
–	$(C_2H_5)_3SnC_3H_7$	Sn:	Org.Verb.2-158
–	$(C_2H_5)_3Sn(i\text{-}C_3H_7)$	Sn:	Org.Verb.2-158
–	$(C_3H_7)_2Sn(CH_3)(C_2H_5)$	Sn:	Org.Verb.3-87
–	$(C_3H_7)_3SnH$	Sn:	Org.Verb.4-37/40
–	$(i\text{-}C_3H_7)_3SnH$	Sn:	Org.Verb.4-40/1
$C_9H_{23}NSn$	$(C_2H_5)_3SnCH_2CH_2CH_2NH_2$	Sn:	Org.Verb.2-166, 176
–	$(C_2H_5)_3SnCH_2N(CH_3)_2$	Sn:	Org.Verb.2-166, 176
$C_9H_{23}N_3Ti$	$CH_2CHCH_2Ti(N(CH_3)_2)_3$	Ti:	Org.Verb.1-53, 57
$C_9H_{24}IrN_9S_3$	$[Ir(C_2H_4(NH_2)_2)_3](CNS)_3$	Ir:	SVol.2-82
$C_9H_{24}N_2S_3$	$[N(CH_3)_4]_2CS_3$	C:	MVol.D4-218, 223
$C_9H_{24}N_2S_4$	$[(CH_3)_4N]_2CS_4$	C:	MVol.D4-231
$C_9H_{24}N_6O_9SU$	$UO_2SO_4 \cdot 3\ CO(NHCH_3)_2$	U:	SVol.E1-93, 95
–	$UO_2SO_4 \cdot 3\ NH_2CONHC_2H_5$	U:	SVol.E1-93, 95
$C_9H_{24}N_8O_{11}U$	$UO_2(NO_3)_2 \cdot 3\ NH_2CONHC_2H_5$	U:	SVol.E1-92, 95
$C_9H_{24}NiO_2P_2$	$CH_3Ni(P(CH_3)_3)_2O_2CCH_3$	Ni:	Org.Verb.1-12, 19
$C_9H_{26}N_2Ti$	$(CH_3)_4Ti \cdot (CH_3)_2NCH_2N(CH_3)_2$	Ti:	Org.Verb.1-113
$C_9H_{26}N_6NiS_3$	$Ni(NH(C_2H_4NH_2)_2)_2CS_3$	C:	MVol.D4-225
$C_9H_{26}NiP_2$	$(CH_3)_3PCH_2(CH_3)_2NiP(CH_3)_3$	Ni:	Org.Verb.1-103/4
$C_9H_{27}I_2NiP_3$	$Ni(P(CH_3)_3)_3I_2$	Ni:	Org.Verb.1-72
$C_9H_{27}NO_3SSi_3$	$[(CH_3)_3Si]_2NSO_2OSi(CH_3)_3$	S:	S-N-Verb.1-98
$C_9H_{27}N_3SSi_2Sn^+$	$[(CH_3)_3SiN]_2SNSn(CH_3)_3{}^+$	S:	S-N-Verb.1-211
$C_9H_{27}N_3SSi_3$	$S[NSi(CH_3)_3]_3$	S:	S-N-Verb.1-206/7
$C_9H_{27}N_3SSi_3{}^+$	$S[NSi(CH_3)_3]_3{}^+$	S:	S-N-Verb.1-207
$C_9H_{27}N_4O_{15}P_3U$	$U(NO_3)_4 \cdot 3\ (CH_3)_3PO$	U:	SVol.E1-148, 156
$C_9H_{27}N_4SSiSn_2{}^+$	$[(CH_3)_3SiN]NS[NSn(CH_3)_3]_2{}^+$	S:	S-N-Verb.1-211
$C_9H_{27}O_9Te_2{}^-$	$Te_2(OCH_3)_9{}^-$	Te:	SVol.B2-104
$C_9H_{27}PSiSn_2$	$[(CH_3)_3Sn]_2PSi(CH_3)_3$	Sn:	Org.Verb.5-81/2
$C_9H_{27}PSi_2Sn$	$[(CH_3)_3Si]_2PSn(CH_3)_3$	Sn:	Org.Verb.5-81/2
$C_9H_{27}PSn_3$	$[(CH_3)_3Sn]_3P$	Sn:	Org.Verb.5-81/2
$C_9H_{28}SiSn_3$	$[(CH_3)_3Sn]_3SiH$	Sn:	Org.Verb.5-84
$C_9H_{30}IrP_3$	$Ir(P(CH_3)_3)_3H_3$	Ir:	SVol.2-195
$C_9Ni_3O_{15}P_4$	$P_4O_6(Ni(CO)_3)_3$	Ni:	Org.Verb.2-373
$C_9Ni_4O_9{}^{2-}$	$[Ni_4(CO)_9]^{2-}$	Ni:	Org.Verb.2-384/5, 392
$C_9Ni_5O_9{}^{2-}$	$[Ni_5(CO)_9]^{2-}$	Ni:	Org.Verb.2-391
C_9S_{27}	$(CS_3)_9$	C:	MVol.D4-214, 221
$C_{10}CaH_{12}N_2O_8Sn$	$Ca[Sn(C_2H_4N_2(CH_2COO)_4)] \cdot 4\ H_2O$	Sn:	MVol.C5-37
$C_{10}CdF_8N_2$	$(NC_5F_4)_2Cd$	F:	PerFHalOrg.5-115/6, 125
$C_{10}CdF_{12}Fe_2O_{10}Sb_2$	$[Cd(Fe(CO)_5)_2][SbF_6]_2$	Fe:	Org.Verb.B3-130
$C_{10}CdFe_2O_{10}{}^{2+}$	$[(CO)_5FeCdFe(CO)_5]^{2+}$	Fe:	Org.Verb.C1-62
$C_{10}CdH_{18}O_2S_4$	$Cd[S_2COC_4H_9]_2$	C:	MVol.D4-245, 258
$C_{10}CeCl_6H_{12}N_2$	$[C_5H_5NH]_2CeCl_6$	Sc:	MVol.C5-223/4

$C_{10}CeCl_6H_{24}N_2$	$[C_5H_{10}NH_2]_2CeCl_6$	Sc: MVol.C5-224
$C_{10}CfH_{12}N_2O_8^-$	$CfC_2H_4N_2(CH_2COO)_4^-$	Np: TrU.D1-165
$C_{10}CfH_{15}N_2O_7$	$CfC_2H_4N_2(C_2H_4OH)(CH_2COO)_3$	Np: TrU.D1-164
$C_{10}ClF_3H_{10}N_2$	$(CH_2)_5NC_5F_3ClN$	F: PerFHalOrg.5-138
$C_{10}ClF_4H_{11}OSi$	$C_6F_4ClSi(OC_2H_5)(CH_3)_2$	F: PerFHalOrg.4-78
$C_{10}ClF_4H_{19}Sn$	$(C_4H_9)_2(C_2HF_4)SnCl$	Sn: Org.Verb.5-207
$C_{10}ClF_5H_3NO$	$C_6F_4CClCFC(O)NCH_3$	F: PerFHalOrg.6-154, 160
–	$C_6F_4C(OCH_3)CFCClN$	F: PerFHalOrg.6-153, 160/1
$C_{10}ClF_5H_9N$	$CH_3C_6H_4C(CF_3)(CF_2Cl)NH_2$	F: PerFHalOrg.7-69, 72, 75
$C_{10}ClF_5H_{11}H_3$	$(CF_3)(CF_2Cl)(C_5H_{11})C_3N_3$	F: PerFHalOrg.6-113
$C_{10}ClF_6Mn_2O_8P$	$Mn_2(CO)_8P(CF_3)_2Cl$	F: PerFHalOrg.3-110/1, 120
$C_{10}ClF_7Mg$	$C_{10}F_7MgCl$	F: PerFHalOrg.4-64/5
$C_{10}ClF_7N_2$	$NC_5F_3ClC_5F_4N$	F: PerFHalOrg.5-220/2, 224
$C_{10}ClF_{14}H_{10}NP_2$	$(C_3F_7)_2PNP(C_2H_5)_2Cl$	F: PerFHalOrg.3-69
$C_{10}ClF_{16}N_3$	$(C_3F_7)_2(CF_2Cl)C_3N_3$	F: PerFHalOrg.6-86/8, 106
$C_{10}ClF_{17}H_3NO_2$	$(CF_3)_2CFO(CF_2)_5C(O)NHCH_2Cl$	F: PerFHalOrg.7-63, 76
$C_{10}ClF_{17}N_2O_2$	$ClC_6F_5[ON(CF_3)_2]_2$	F: PerFHalOrg.7-98/9, 116
$C_{10}ClFeH_4NO_8S$	$NO_2C_6H_4SO_2Fe(CO)_4Cl$	Fe: Org.Verb.B3-187
$C_{10}ClFeH_5O_6S$	$C_6H_5SO_2Fe(CO)_4Cl$	Fe: Org.Verb.B3-187
$C_{10}ClFeH_8HgI$	$C_5H_5FeC_5H_3(HgCl)I$	Fe: Org.Verb.A6-212
$C_{10}ClFeH_9$	$C_5H_5FeC_5H_4Cl$	Fe: Org.Verb.A1-366/8, 372/80
$C_{10}ClFeH_9^+$	$[C_5H_5FeC_5H_4Cl]^+$	Fe: Org.Verb.A1-373
$C_{10}ClFeH_9Hg$	$C_5H_5FeC_5H_4HgCl$	Fe: Org.Verb.A1-175
$C_{10}ClFeH_{10}$	$[Fe(C_5H_5)_2Cl]$	Fe: Org.Verb.A1-206, 214/5
$C_{10}ClFeH_{10}O_4$	$[Fe(C_5H_5)_2ClO_4]$	Fe: Org.Verb.A1-206, 218/9
$C_{10}ClFeH_{11}MgO_5$	$MgCl[C_5H_{11}C(O)Fe(CO)_4]$	Fe: Org.Verb.B4-150, 160
$C_{10}ClFeH_{11}O_4$	$[Fe(C_5H_5)_2][HClO_4]$	Fe: Org.Verb.A1-219
$C_{10}ClFeH_{13}NO_6P$	$(CO)_2Fe(CH_2CClCH_2)(NO)P(OCH_2)_3CCH_3$	Fe: Org.Verb.B1-135/7
$C_{10}ClFeH_{15}NO_5P$	$ClC_3H_4Fe(CO)(NO)P(OCH_2)_3CC_2H_5$	Fe: Org.Verb.B5-108/11
$C_{10}ClFeH_{19}NO_5P$	$ClC_3H_4Fe(CO)(NO)P(OC_2H_5)_3$	Fe: Org.Verb.B5-108/11
$C_{10}ClH_8N_3NiO_3$	$Ni(C_{10}H_8N_2)(Cl)NO_3$	Ni: Org.Verb.1-40
$C_{10}ClH_9S_2Sn$	$(C_4H_3S)_2(CH_2CH)SnCl$	Sn: Org.Verb.5-223
$C_{10}ClH_{10}MnO_2^+$	$[Mn(OC_6H_3(Cl)COC_3H_7)]^+$	Mn: MVol.D1-58
$C_{10}ClH_{10}Ni$	$[Ni(C_5H_5)_2]Cl$	Ni: Org.Verb.2-213
$C_{10}ClH_{10}NiO_2$	$CH_3C_3H_4Ni \cdot C_6H_3ClO_2$	Ni: Org.Verb.2-6
$C_{10}ClH_{11}MoO_3Sn$	$(CH_3)_2ClSnMo(CO)_3C_5H_5$	Sn: Org.Verb.6-33
$C_{10}ClH_{11}Ni$	$[Ni(C_5H_5)(C_5H_6)]Cl$	Ni: Org.Verb.2-116
$C_{10}ClH_{11}Sn$	$(CH_2CH)_2(C_6H_5)SnCl$	Sn: Org.Verb.5-222
$C_{10}ClH_{13}IrKN_2O_8$	$K[IrH(C_2H_4N_2(CH_2COO)_4)Cl] \cdot 2\ H_2O$	Ir: SVol.2-32
$C_{10}ClH_{14}MnO_4$	$[MnCl(CH(COCH_3)_2)_2]$	Mn: MVol.D1-101/2
$C_{10}ClH_{15}NOS^+$	$[C_6H_5(C_2H_5)_2NSOCl]^+$	S: SVol.1-52
$C_{10}ClH_{15}Sn$	$(CH_3)(i-C_3H_7)(C_6H_5)SnCl$	Sn: Org.Verb.5-225, 227
–	$(CH_3)_3SnCH_2C_6H_4Cl$	Sn: Org.Verb.2-22/3
–	$(C_2H_5)_2(C_6H_5)SnCl$	Sn: Org.Verb.5-202
$C_{10}ClH_{17}O_2Sn_2$	$(C_6H_5O)(CH_3)_2SnOSn(CH_3)_2Cl$	Sn: Org.Verb.6-21
$C_{10}ClH_{17}O_2Ti$	$CH_3C_5H_4Ti(OC_2H_5)_2Cl$	Ti: Org.Verb.1-185, 187
$C_{10}ClH_{18}MnO_6$	$[Mn(CH(COCH_3)_2)_2(H_2O)_2]Cl \cdot 2\ H_2O$	Mn: MVol.D1-98/9
$C_{10}ClH_{18}MnO_{10}$	$[Mn(CH(COCH_3)_2)_2(H_2O)_2]ClO_4 \cdot 2.5\ H_2O$	Mn: MVol.D1-98/9
$C_{10}ClH_{18}N_2Ni$	$(Ni(CNC_4H_9\text{-}t)_2Cl)_x$	Ni: Org.Verb.2-397

Formula	Compound		Reference
$C_{10}ClH_{19}IrN_5^{2+}$	$[Ir(C_5H_5N)_2(NH_3)_3Cl]^{2+}$	Ir:	SVol.2-74
$C_{10}ClH_{19}N_4NiS_2$	$CH_3C_3H_4Ni(SC(NHCH_2)_2)_2Cl$	Ni:	Org.Verb.2-21
$C_{10}ClH_{19}Sn$	$(CH_3)_3SnC_7H_{10}Cl$	Sn:	Org.Verb.2-62, 67
–	$(C_2H_5)_3SnCCCH_2CH_2Cl$	Sn:	Org.Verb.2-223
$C_{10}ClH_{21}N_2NiS$	$C_3H_5Ni(SC(HNC_3H_7\text{-}i)_2)Cl$	Ni:	Org.Verb.2-10
$C_{10}ClH_{21}OSn$	$(C_2H_5)_2Sn(Cl)OCH(CH_2)_5$	Sn:	Org.Verb.6-54
$C_{10}ClH_{21}O_2Sn$	$(C_4H_9)_2SnCl(OCOCH_3)$	Sn:	Org.Verb.6-88, 290
$C_{10}ClH_{21}Sn$	$(C_2H_5)_2[CH_2CH(CH_2)_4]SnCl$	Sn:	Org.Verb.5-202
–	$(C_4H_9)_2(CH_2CH)SnCl$	Sn:	Org.Verb.5-207, 212
$C_{10}ClH_{23}MgNiO$	$CH_3C_3H_4NiCH_3 \cdot CH_3MgCl \cdot (C_2H_5)_2O$	Ni:	Org.Verb.2-40/1
$C_{10}ClH_{23}OSn$	$(C_3H_7)_2ClSnOC_4H_9$	Sn:	Org.Verb.5-114
$C_{10}ClH_{23}O_2Sn$	$(C_4H_9)(i\text{-}C_3H_7O)_2SnCl$	Sn:	Org.Verb.6-245
$C_{10}ClH_{23}Sn$	$(CH_3)_2(C_8H_{17})SnCl$	Sn:	Org.Verb.5-195, 199
–	$(C_2H_5)_2(C_6H_{13})SnCl$	Sn:	Org.Verb.5-202
–	$(C_3H_7)_2(C_4H_9)SnCl$	Sn:	Org.Verb.5-205/6
–	$(C_4H_9)_2(C_2H_5)SnCl$	Sn:	Org.Verb.5-207, 212
$C_{10}ClH_{24}OPS_2Sn$	$(CH_3)(CH_3O)PSSSn(C_4H_9)_2Cl$	Sn:	Org.Verb.6-89, 96
$C_{10}ClH_{24}O_2PS_2Sn$	$(CH_3O)_2PSSSn(C_4H_9)_2Cl$	Sn:	Org.Verb.6-89, 96
$C_{10}ClH_{24}O_3PSn$	$(C_2H_5)_2[(C_2H_5O)_2P(O)CH_2CH_2]SnCl$	Sn:	Org.Verb.5-202
$C_{10}ClH_{24}PSn$	$(CH_3)_2ClSnP[C(CH_3)_3]_2$	Sn:	Org.Verb.6-21
$C_{10}ClH_{26}NSn$	$(C_2H_5)_3SnCl \cdot (C_2H_5)_2NH$	Sn:	Org.Verb.5-107
$C_{10}ClH_{30}NNi_2P_2$	$(CH_3Ni(P(CH_3)_3)Cl)(CH_3Ni(P(CH_3)_3)N(CH_3)_2)$	Ni:	Org.Verb.2-254
$C_{10}Cl_{1.75}H_{13}IrN_2O_2$	$Ir(C_5H_5N)_2Cl_{1.75}(OH)(H_2O) \cdot 2\ H_2O$	Ir:	SVol.2-74
$C_{10}Cl_{1.88}H_{10}IrN_2$	$Ir(C_5H_5N)_2Cl_{1.88}$	Ir:	SVol.2-74
$C_{10}Cl_2CoH_5O_4Sn$	$(C_6H_5)Cl_2SnCo(CO)_4$	Sn:	Org.Verb.6-275, 278
$C_{10}Cl_2F_2H_3N_5O_4$	$NC_5F_2Cl_2NHC_5H_2(NO_2)_2N$	F:	PerFHalOrg.5-139
$C_{10}Cl_2F_2H_4N_4O_2$	$NC_5F_2Cl_2NHC_5H_3(NO_2)N$	F:	PerFHalOrg.5-139
$C_{10}Cl_2F_2H_9NO_2$	$NC_5F_2Cl_2(OCH_2CH(CH_2)_3O)$	F:	PerFHalOrg.5-136
		F:	PerFHalOrg.6-14
$C_{10}Cl_2F_2H_{10}N_2$	$(CH_2)_5NC_5F_2Cl_2N$	F:	PerFHalOrg.5-138
$C_{10}Cl_2F_4H_3N_3$	$NC_5F_2Cl_2NHC_5F_2H_2N$	F:	PerFHalOrg.5-138/9
$C_{10}Cl_2F_4H_5NOS$	$C_6H_4SC(CF_2Cl)_2NHC(O)$	F:	PerFHalOrg.7-66, 69
$C_{10}Cl_2F_4H_8N_2O_2Sn$	$SnF_4 \cdot 2\ ClC_5H_4NO$	Sn:	MVol.C5-208/9
$C_{10}Cl_2F_4H_{10}Ni_2O_4$	$(C_3H_5NiO_2CCF_2Cl)_2$	Ni:	Org.Verb.2-283
$C_{10}Cl_2F_4H_{11}N_3$	$(CF_2Cl)_2(C_5H_{11})C_3N_3$	F:	PerFHalOrg.6-113
$C_{10}Cl_2F_5H_8InO_2$	$C_6F_5InCl_2 \cdot C_4H_8O_2$	F:	PerFHalOrg.4-128/9
$C_{10}Cl_2F_6HN_3$	$NC_5F_2Cl_2NHC_5F_4N$	F:	PerFHalOrg.5-186/7
		F:	PerFHalOrg.6-13/4
–	$NC_5F_3ClNHC_5F_3ClN$	F:	PerFHalOrg.6-13/4
$C_{10}Cl_2F_6N_2$	$NC_5F_3ClC_5F_3ClN$	F:	PerFHalOrg.5-220/2, 224
$C_{10}Cl_2F_{12}Fe_2O_6P_2$	$Cl(CO)_3Fe(P(CF_3)_2)_2Fe(CO)_3Cl$	Fe:	Org.Verb.C1-6/8, 10/1
$C_{10}Cl_2F_{12}H_2O_4Sn$	$SnCl_2(CH(COCF_3)_2)_2$	Sn:	MVol.C5-109
$C_{10}Cl_2F_{12}Hg$	$Hg(C_5F_6Cl)_2$	F:	PerFHalOrg.4-34
$C_{10}Cl_2F_{13}N_3$	$(CF_2CFCF_2)(C_3F_5Cl_2)(CF_3)C_3N_3$	F:	PerFHalOrg.6-84/6, 104
$C_{10}Cl_2FeH_7Hg_2NO_4$	$(CO)_4Fe(HgCl)_2(CH_3C_5H_4N)$	Fe:	Org.Verb.B2-153/5, 158
–	$(CO)_4Fe(HgCl)_2(C_6H_5NH_2)$	Fe:	Org.Verb.B2-153/5, 158
$C_{10}Cl_2FeH_8Hg_2$	$Fe(C_5H_4HgCl)_2$	Fe:	Org.Verb.A1-175
$C_{10}Cl_2FeH_9P$	$C_5H_5FeC_5H_4PCl_2$	Fe:	Org.Verb.A1-170
$C_{10}Cl_2FeH_{10}$	$C_5H_5FeC_5H_5Cl_2$	Fe:	Org.Verb.A1-181
$C_{10}Cl_2FeH_{10}Hg$	$Fe(C_5H_5)_2 \cdot HgCl_2$	Fe:	Org.Verb.A1-234

Formula	Compound	Reference
$C_{10}Cl_2FeH_{10}O_8^-$	$[Fe(C_5H_5)_2(ClO_4)_2]^-$	Fe: Org.Verb.A1-206
$C_{10}Cl_2FeH_{18}Hg_2N_2O_4$	$(CO)_2Fe(NH(CH_2CH_2)_2O)_2(HgCl)_2$	Fe: Org.Verb.B1-116/7
$C_{10}Cl_2Fe_2GeH_{18}N_4O_4$	$((CH_3)_2C_3H_3Fe(NO)_2)_2GeCl_2$	Fe: Org.Verb.C2-156, 165
$C_{10}Cl_2Fe_2H_2O_{10}$	$[(CO)_5FeH]_2Cl_2$	Fe: Org.Verb.B3-252
$C_{10}Cl_2Fe_2H_{12}O_6P_2$	$Cl(CO)_3Fe(P(CH_3)_2)_2Fe(CO)_3Cl$	Fe: Org.Verb.C1-6/7, 10
$C_{10}Cl_2Fe_2H_{18}N_4O_4Pb$	$((CH_3)_2C_3H_3Fe(NO)_2)_2PbCl_2$	Fe: Org.Verb.C2-156, 165
$C_{10}Cl_2Fe_2H_{18}N_4O_4Sn$	$((CH_3)_2C_3H_3Fe(NO)_2)_2SnCl_2$	Fe: Org.Verb.C2-156, 165
$C_{10}Cl_2H_4O_4Sn$	$SnCl_2(C_{10}H_4O_4)$	Sn: MVol.C5-112
$C_{10}Cl_2H_6O_4Sn$	$[(CH_2CH)_2Sn(C_6Cl_2O_4)]_x$	Sn: Org.Verb.6-145
$C_{10}Cl_2H_8N_2O_2S_2Sn$	$SnCl_2(C_5H_4NOS)_2$	Sn: MVol.C6-106
–	$SnCl_2 \cdot ONC_5H_4SSC_5H_4NO$	Sn: MVol.C5-50
$C_{10}Cl_2H_8N_2O_2U$	$[UO_2Cl_2(NC_5H_4C_5H_4N)]$	U: SVol.E1-39/43
–	$[UO_2Cl_2(NC_5H_4C_5H_4N)] \cdot H_2O$	U: SVol.E1-39/42
–	$[UO_2Cl_2(NC_5H_4C_5H_4N)] \cdot 2\ H_2O$	U: SVol.E1-39/42
$C_{10}Cl_2H_8N_2O_4Sn$	$[SnCl_2(C_5H_4NO_2)_2]$	Sn: MVol.C5-208
$C_{10}Cl_2H_8N_2Sn$	$SnCl_2 \cdot NC_5H_4C_5H_4N$	Sn: MVol.C5-21/2
$C_{10}Cl_2H_8N_4O_8U$	$UO_2Cl_2 \cdot 2\ NO_2C_5H_4NO$	U: SVol.E1-133, 139, 142/3
$C_{10}Cl_2H_8N_4O_{10}U$	$UO_2(NO_3)_2 \cdot 2\ ClC_5H_4NO$	U: SVol.E1-133, 142/3
$C_{10}Cl_2H_9N_3O_2U$	$[UO_2Cl_2(NH(C_5H_4N)_2)]$	U: SVol.E1-41, 43
$C_{10}Cl_2H_{10}I_4Ti$	$C_5H_5Ti(C_5H_5I_4)Cl_2$	Ti: Org.Verb.1-206/7
$C_{10}Cl_2H_{10}MnN_2$	$MnCl_2 \cdot 2\ C_5H_5N$	Mn: MVol.C5-206
$C_{10}Cl_2H_{10}N_2O_2S$	$[C_5H_5NSO_2NC_5H_5]Cl_2$	S: S-N-Verb.1-168
$C_{10}Cl_2H_{10}N_2O_2Sn$	$SnCl_2 \cdot 2\ C_5H_5NO$	Sn: MVol.C5-19
$C_{10}Cl_2H_{10}N_2O_2U$	$UO_2Cl_2 \cdot 2\ C_5H_5N$	U: SVol.E1-28, 33/4
$C_{10}Cl_2H_{10}N_2O_3U$	$[UO_2Cl(H_2O)(C_{10}H_8N_2)]Cl$	U: SVol.E1-42
$C_{10}Cl_2H_{10}N_2O_4U$	$UO_2Cl_2 \cdot 2\ C_5H_5NO$	U: SVol.E1-132, 135, 142/3
$C_{10}Cl_2H_{10}N_2S$	$SCl_2 \cdot 2\ C_5H_5N$	S: SVol.2-273
$C_{10}Cl_2H_{10}N_2Sn$	$SnCl_2 \cdot 2\ C_5H_5N$	Sn: MVol.C5-16/7
$C_{10}Cl_2H_{10}Ni$	$C_5H_5NiC_5H_5Cl_2$	Ni: Org.Verb.2-205
$C_{10}Cl_2H_{10}S_9Ti$	$(C_5H_5)_2Ti(Cl)S_9Cl$	S: SVol.2-271
$C_{10}Cl_2H_{10}Sn$	$(C_5H_5)_2SnCl_2$	Sn: Org.Verb.6-143
$C_{10}Cl_2H_{12}IrN_2O_2^-$	$[Ir(C_5H_5N)_2Cl_2(OH)_2]^-$	Ir: SVol.2-74
$C_{10}Cl_2H_{12}N_4O_2U$	$UO_2Cl_2 \cdot 2\ NH_2C_5H_4N$	U: SVol.E1-29, 34
$C_{10}Cl_2H_{13}IrN_2O_2$	$Ir(C_5H_5N)_2Cl_2(OH)(H_2O)$	Ir: SVol.2-71, 74
–	$Ir(C_5H_5N)_2Cl_2(OH)(H_2O) \cdot 0.5\ H_2O$	Ir: SVol.2-74
$C_{10}Cl_2H_{13}NS_2Sn$	$CH_3C_6H_4SnCl_2(SCSN(CH_3)_2)$	Sn: Org.Verb.6-282
$C_{10}Cl_2H_{14}IrKN_2O_8$	$K[IrH_2(C_2H_4N_2(CH_2COO)_4)Cl_2]$	Ir: SVol.2-33
$C_{10}Cl_2H_{14}IrN_2O_2^+$	$[Ir(C_5H_5N)_2Cl_2(H_2O)_2]^+$	Ir: SVol.2-71
$C_{10}Cl_2H_{14}Ni_2$	$(CH_2CHCHCHCH_2NiCl)_2$	Ni: Org.Verb.2-297
–	$(C_5H_7NiCl)_2$	Ni: Org.Verb.2-319
$C_{10}Cl_2H_{14}Ni_2O_4$	$(C_3H_5NiO_2CCH_2Cl)_2$	Ni: Org.Verb.2-283
$C_{10}Cl_2H_{14}OSn$	$(CH_3)(C_6H_5)SnCl_2 \cdot (CH_3)_2CO$	Sn: Org.Verb.6-195
$C_{10}Cl_2H_{14}O_4Sn$	$[SnCl_2(CH(COCH_3)_2)_2]$	Sn: MVol.C5-101/6
$C_{10}Cl_2H_{14}O_4Te$	$(CH_3C(OH)CHCOCH_2)_2TeCl_2$	Te: SVol.B2-98
$C_{10}Cl_2H_{14}Sn$	$(CH_2Cl)_2Sn(C_2H_5)(C_6H_5)$	Sn: Org.Verb.3-92/3
–	$(CH_3)_3SnCCl_2C_6H_5$	Sn: Org.Verb.2-28, 30
–	$(C_4H_9)(C_6H_5)SnCl_2$	Sn: Org.Verb.6-199, 205
$C_{10}Cl_2H_{15}Mg_3Ti$	$Mg(CHCH_2)_2(Mg(CHCH_2)Cl)_2TiCHCH_2$	Ti: Org.Verb.1-131
–	$(Mg(CHCH_2)_2)_2(MgCl_2)TiCHCH_2$	Ti: Org.Verb.1-131
$C_{10}Cl_2H_{15}NS_2Ti$	$C_5H_5Ti(S_2CN(C_2H_5)_2)Cl_2$	Ti: Org.Verb.1-198/9

Formula	Compound	Reference
$C_{10}Cl_2H_{15}NSn$	$SnCl_2 \cdot C_6H_5N(C_2H_5)_2$	Sn: MVol.C5-13
$C_{10}Cl_2H_{16}IrN_4^+$	$[Ir(C_5H_5N)_2(NH_3)_2Cl_2]^+$	Ir: SVol.2-74
$C_{10}Cl_2H_{16}N_2OSn$	$(CH_3)_2SnCl_2 \cdot (CH_3)_2NCOC_5H_4N$	Sn: Org.Verb.6-36
$C_{10}Cl_2H_{17}NSn$	$(C_2H_5)_2SnCl_2 \cdot C_6H_5NH_2$	Sn: Org.Verb.6-62
$C_{10}Cl_2H_{18}N_2Ni$	$(t\text{-}C_4H_9NC)_2NiCl_2$	Ni: Org.Verb.1-315, 316
$C_{10}Cl_2H_{18}Ni_2$	$(CH_3CHC(CH_3)CH_2NiCl)_2$	Ni: Org.Verb.2-312
–	$(CH_3CHCHCHCH_3NiCl)_2$	Ni: Org.Verb.2-313/4
–	$(CH_3CHCHCHCH_3NiCl)_2 \cdot n\ C_6Cl_4O_2$	Ni: Org.Verb.2-317
–	$(CH_3CHCHCHCH_3NiCl)_2 \cdot x\ C_6Cl_4O_2 \cdot y\ C_6H_4O_2$	Ni: Org.Verb.2-317
–	$((CH_3)_2CCHCH_2NiCl)_2$	Ni: Org.Verb.2-310
$C_{10}Cl_2H_{18}O_2Sn$	$[CH_3C(O)(CH_2)_3]_2SnCl_2$	Sn: Org.Verb.6-132
–	$[C_2H_5C(O)CH_2CH_2]_2SnCl_2$	Sn: Org.Verb.6-132
$C_{10}Cl_2H_{18}O_4Sn$	$[CH_3OCOCH(CH_3)CH_2]_2SnCl_2$	Sn: Org.Verb.6-132
–	$[CH_3OCOCH_2CH(CH_3)]_2SnCl_2$	Sn: Org.Verb.6-133
–	$(C_2H_5OCOCH_2CH_2)_2SnCl_2$	Sn: Org.Verb.6-133
–	$[HOCO(CH_2)_4]_2SnCl_2$	Sn: Org.Verb.6-132
–	$SnCl_2((CH_3)_3CCOO)_2$	Sn: MVol.C2-223
$C_{10}Cl_2H_{18}Sn$	$[CH_2CHCH(CH_3)CH_2]_2SnCl_2$	Sn: Org.Verb.6-149
$C_{10}Cl_2H_{20}N_2O_2Sn$	$[CH_3NHC(O)CH(CH_3)CH_2]_2SnCl_2$	Sn: Org.Verb.6-133
–	$[CH_3NHC(O)CH_2CH(CH_3)]_2SnCl_2$	Sn: Org.Verb.6-133
$C_{10}Cl_2H_{20}N_2S_4Sn$	$[SnCl_2((C_2H_5)_2NCSS)_2]$	Sn: MVol.C6-119/20
$C_{10}Cl_2H_{20}NiO_2Sb_2$	$(CO)_2Ni(Sb(C_2H_5)_2Cl)_2$	Ni: Org.Verb.1-144
$C_{10}Cl_2H_{20}O_7U$	$UO_2Cl_2 \cdot (CH_2OCH_2)_5 \cdot 3\ H_2O$	U: SVol.E1-73, 78
$C_{10}Cl_2H_{20}Sn$	$(CH_3)_3SnC_6H_{10}CHCl_2$	Sn: Org.Verb.2-61
$C_{10}Cl_2H_{22}N_2O_2Sn$	$SnCl_2 \cdot 2\ CH_3N(CH_2CH_2)_2O$	Sn: MVol.C5-29
$C_{10}Cl_2H_{22}N_2O_4U$	$UO_2Cl_2 \cdot 2\ CH_3CONHC_3H_7$	U: SVol.E1-96/7, 99
$C_{10}Cl_2H_{22}N_2Sn$	$SnCl_2 \cdot 2\ C_5H_{10}NH$	Sn: MVol.C5-15/6
$C_{10}Cl_2H_{22}O_2S_2Sn$	$(CH_3)_2SnCl_2 \cdot 2\ (CH_2)_4SO$	Sn: Org.Verb.6-37
$C_{10}Cl_2H_{22}O_2Sn$	$(C_4H_9OCH_2)_2SnCl_2$	Sn: Org.Verb.6-133
$C_{10}Cl_2H_{22}O_4Si_2Sn$	$[(CH_3)_2(CH_3COO)SiCH_2]_2SnCl_2$	Sn: Org.Verb.6-133
$C_{10}Cl_2H_{22}Sn$	$[(CH_3)_2CHCH_2CH_2]_2SnCl_2$	Sn: Org.Verb.6-109
–	$[(CH_3)_3CCH_2]_2SnCl_2$	Sn: Org.Verb.6-109
–	$(CH_3)_3SnCH_2C(CH_3)(CHCl_2)CH(CH_3)_2$	Sn: Org.Verb.2-56
–	$(CH_3)_3SnCH_2CH(CHCl_2)CH_2CH(CH_3)_2$	Sn: Org.Verb.2-56
–	$(CH_3)_3Sn(CH_2)_3C(CH_3)_2CHCl_2$	Sn: Org.Verb.2-56
–	$[C_2H_5C(CH_3)_2]_2SnCl_2$	Sn: Org.Verb.6-109
–	$(C_5H_{11})_2SnCl_2$	Sn: Org.Verb.6-108
$C_{10}Cl_2H_{24}MnO_2$	$MnCl_2 \cdot 2\ C_2H_5CH(CH_3)CH_2OH$	Mn: MVol.D1-33
–	$MnCl_2 \cdot 2\ C_5H_{11}OH$	Mn: MVol.D1-33
$C_{10}Cl_2H_{24}N_4NiO_2P_2$	$(CO)_2Ni(P(N(CH_3)_2)_2Cl)_2$	Ni: Org.Verb.1-126
$C_{10}Cl_2H_{24}N_4O_4U$	$UO_2Cl_2 \cdot 2\ CO(N(CH_3)_2)_2$	U: SVol.E1-92/3
$C_{10}Cl_2H_{24}N_4S_2Te$	$Te((CH_3)_2NCSN(CH_3)_2)_2Cl_2$	Te: SVol.B3-148
$C_{10}Cl_2H_{24}OSSn$	$(CH_3)_2SnCl_2 \cdot (C_4H_9)_2SO$	Sn: Org.Verb.6-37
$C_{10}Cl_2H_{24}OSn_2$	$[(C_4H_9)(CH_3)ClSn]_2O$	Sn: Org.Verb.6-88
–	$(C_4H_9)_2SnO \cdot (CH_3)_2SnCl_2$	Sn: Org.Verb.6-21
$C_{10}Cl_2H_{25}OPSn$	$(C_2H_5)_3SnCl \cdot (C_2H_5)_2ClPO$	Sn: Org.Verb.5-107
$C_{10}Cl_2H_{26}N_2Sn$	$SnCl_2 \cdot 2\ C_5H_{11}NH_2$	Sn: MVol.C5-11/2
$C_{10}Cl_2H_{26}O_2S_2Sn$	$(CH_3)_2SnCl_2 \cdot 2\ (C_2H_5)_2SO$	Sn: Org.Verb.6-37
$C_{10}Cl_2H_{26}O_2Si_2Sn$	$[(CH_3)_2(C_2H_5O)SiCH_2]_2SnCl_2$	Sn: Org.Verb.6-133

Formula	Compound		Reference
$C_{10}Cl_2H_{26}O_8S_2Sn$	$(CH_3)_2SnCl_2 \cdot 2\ (C_2H_5)_2SO_4$	Sn:	Org.Verb.6-37
$C_{10}Cl_2H_{28}N_3OPSn$	$(C_2H_5)_2SnCl_2 \cdot [(CH_3)_2N]_3PO$	Sn:	Org.Verb.6-61
$C_{10}Cl_2H_{28}N_4Sn$	$SnCl_2 \cdot 2\ CH_3NH(CH_2)_3NHCH_3$	Sn:	MVol.C5-14
$C_{10}Cl_2H_{28}O_4P_2Sn$	$(C_2H_5)_2SnCl_2 \cdot 2\ (CH_3)_2(CH_3O)PO$	Sn:	Org.Verb.6-61/2
$C_{10}Cl_2H_{28}S_2Sb_2Sn$	$(C_2H_5)_2SnCl_2 \cdot 2\ (CH_3)_3SbS$	Sn:	Org.Verb.6-62
$C_{10}Cl_2H_{29}N_5Ti$	$C_5H_5Ti(NHCH_3)Cl_2 \cdot 4\ NH_2CH_3$	Ti:	Org.Verb.1-190
$C_{10}Cl_2H_{30}O_{15}S_5U$	$UO_2(ClO_4)_2 \cdot 5\ (CH_3)_2SO$	U:	SVol.E1-204, 208
$C_{10}Cl_2H_{30}O_{15}Se_5U$	$UO_2(ClO_4)_2 \cdot 5\ (CH_3)_2SeO$	U:	SVol.E1-217
$C_{10}Cl_2S_{19}$	$C_{10}S_{19}Cl_2$	C:	MVol.D4-215
		C:	MVol.D6-160
$C_{10}Cl_3F_3H_7N_3O_2S$	$((CH_3)_2NSO_2)(CF_3)C_7Cl_3HN_2$	F:	PerFHalOrg.6-162
–	$(C_2H_5NHSO_2)(CF_3)C_7Cl_3HN_2$	F:	PerFHalOrg.6-160
$C_{10}Cl_3F_4HN_4O_2$	$N_2C_4Cl_3NHC_6F_4NO_2$	F:	PerFHalOrg.7-15/6, 43, 76
$C_{10}Cl_3F_5HN_3$	$NC_5F_2Cl_2NHC_5F_3ClN$	F:	PerFHalOrg.5-186/7
		F:	PerFHalOrg.6-13/4
–	$N_2C_4Cl_3NHC_6F_5$	F:	PerFHalOrg.7-15/6, 43, 76
$C_{10}Cl_3F_5H_2N_4$	$NC_5F_2Cl_2NHC_4HCl(CF_3)N_2$	F:	PerFHalOrg.5-216
$C_{10}Cl_3F_{16}H_5Sn$	$CHF_2(CF_2)_7CH_2CH_2SnCl_3$	Sn:	Org.Verb.6-259
$C_{10}Cl_3FeH_8N_2$	$Fe(NC_5H_4C_5H_4N)Cl_3$	Fe:	Org.Verb.B1-8
$C_{10}Cl_3FeH_9Si$	$C_5H_5FeC_5H_4SiCl_3$	Fe:	Org.Verb.A1-170
$C_{10}Cl_3FeH_{10}$	$[Fe(C_5H_5)_2]Cl_3$	Fe:	Org.Verb.A1-215
$C_{10}Cl_3FeH_{10}OV$	$[Fe(C_5H_5)_2][VOCl_3]$	Fe:	Org.Verb.A1-225
$C_{10}Cl_3FeH_{16}NO_4Si$	$[N(C_2H_5)_3H][(CO)_4FeSiCl_3]$	Fe:	Org.Verb.B2-142, 149
$C_{10}Cl_3Fe_2H_{10}$	$[Fe(C_5H_5)_2]FeCl_3$	Fe:	Org.Verb.A1-221
$C_{10}Cl_3H_7Sn$	$C_{10}H_7SnCl_3$	Sn:	Org.Verb.6-283, 287
$C_{10}Cl_3H_8N_3OsS$	$[OsCl_3(NS)(NC_5H_4C_5H_4N)]$	S:	SVol.2-261
$C_{10}Cl_3H_{10}IrN_2$	$[Ir(C_5H_5N)_2Cl_3]_n$	Ir:	SVol.2-70, 71
$C_{10}Cl_3H_{10}MnN_2O$	$[Mn(NC_5H_4C_5H_4N)(H_2O)]Cl_3$	Mn:	MVol.C5-212/3
$C_{10}Cl_3H_{11}IrN_2O^-$	$[Ir(C_5H_5N)_2Cl_3(OH)]^-$	Ir:	SVol.2-71
$C_{10}Cl_3H_{12}IrN_2O$	$Ir(C_5H_5N)_2Cl_3(H_2O)$	Ir:	SVol.2-69, 70, 73
–	$Ir(C_5H_5N)_2Cl_3(H_2O) \cdot H_2O$	Ir:	SVol.2-74
–	$Ir(C_5H_5N)_2Cl_3(H_2O) \cdot 2\ H_2O$	Ir:	SVol.2-70, 74
$C_{10}Cl_3H_{13}N_4Sn$	$SnCl_3N_2H_3 \cdot 2\ C_5H_5N$	Sn:	MVol.C5-197
$C_{10}Cl_3H_{13}O_2S_2Sn$	$C_6H_5SnCl_3 \cdot OS(CH_2CH_2)_2SO$	Sn:	Org.Verb.6-280
$C_{10}Cl_3H_{13}O_2Sn$	$C_6H_5SnCl_3 \cdot CH_3COOC_2H_5$	Sn:	Org.Verb.6-280
$C_{10}Cl_3H_{13}Sn$	$(t\text{-}C_4H_9)C_6H_4SnCl_3$	Sn:	Org.Verb.6-283, 285
$C_{10}Cl_3H_{14}IrN_2O_2$	$[Ir(C_5H_5N)_2Cl_2(H_2O)_2]Cl$	Ir:	SVol.2-73
–	$[Ir(C_5H_5N)_2Cl_2(H_2O)_2]Cl \cdot H_2O$	Ir:	SVol.2-74
$C_{10}Cl_3H_{14}O_4Pa$	$Pa(CH_3COCHCOCH_3)_2Cl_3$	Pa:	SVol.2-86
$C_{10}Cl_3H_{15}N_2O_2Sn$	$C_4H_9SnCl_3 \cdot NO_2C_6H_4NH_2$	Sn:	Org.Verb.6-247/8
$C_{10}Cl_3H_{15}Ti$	$(CH_3)_5C_5TiCl_3$	Ti:	Org.Verb.1-141, 150/1
$C_{10}Cl_3H_{16}NOTi$	$CH_3TiCl_3 \cdot CH_3OC_6H_4N(CH_3)_2$	Ti:	Org.Verb.1-34/5
$C_{10}Cl_3H_{16}NSn$	$C_4H_9SnCl_3 \cdot CH_3C_5H_4N$	Sn:	Org.Verb.6-247/8
–	$[(C_6H_5)((CH_3)_2NHCH_2CH_2)SnCl_2]Cl$	Sn:	Org.Verb.6-201
$C_{10}Cl_3H_{16}N_3O_2Sn$	$C_4H_9SnCl_3 \cdot (NH_2)_2C_6H_3NO_2$	Sn:	Org.Verb.6-247/8
$C_{10}Cl_3H_{19}O_2Ti$	$CH_2CHTiCl_3 \cdot 2\ C_4H_8O$	Ti:	Org.Verb.1-36/7, 41
$C_{10}Cl_3H_{20}NO_2Sn$	$C_7H_{15}CONHC_2H_4OSnCl_3$	Sn:	MVol.C6-46
$C_{10}Cl_3H_{21}Sn$	$C_{10}H_{21}SnCl_3$	Sn:	Org.Verb.6-252

Formula	Compound		Reference
$C_{10}Cl_3H_{22}IrS_3$	$IrCl_3(S(CH_2CH_2S(n\text{-}C_3H_7))_2)$	Ir:	SVol.2-167
$C_{10}Cl_3H_{23}OSn_2$	$(C_2H_5)Cl_2SnOSnCl(C_4H_9)_2$	Sn:	Org.Verb.6-230
$C_{10}Cl_3H_{23}O_2Sn$	$SnCl_3OC_5H_{11} \cdot C_5H_{11}OH$	Sn:	MVol.C5-73/4
$C_{10}Cl_3H_{23}SiSn$	$(CH_3)(C_4H_9)_2SiCH_2SnCl_3$	Sn:	Org.Verb.6-259
$C_{10}Cl_3H_{24}MnN$	$C_{10}H_{21}NH_3MnCl_3$	Mn:	MVol.C5-188
$C_{10}Cl_3H_{27}O_6P_2Sn$	$C_4H_9SnCl_3 \cdot 2\ (CH_3O)_2P(O)CH_3$	Sn:	Org.Verb.6-247/8
$C_{10}Cl_3H_{28}O_8P_3S_2Sn$	$SnCl_3(C_2H_5O)_2P(S)S \cdot 2\ CH_3PO(OCH_3)_2$	Sn:	MVol.C6-176
$C_{10}Cl_4F_2HN_9$	$NC_5F_2Cl_2NHC_5Cl_2(N_3)_2N$	F:	PerFHalOrg.5-186/7
		F:	PerFHalOrg.6-13/4
$C_{10}Cl_4F_2H_{10}Ni_2O_4$	$(C_3H_5NiO_2CCFCl_2)_2$	Ni:	Org.Verb.2-283
$C_{10}Cl_4F_3H_2N_3O$	$C_6Cl_4NC(CF_3)N(OCH_2CN)$	F:	PerFHalOrg.6-159
$C_{10}Cl_4F_3H_5N_2OS$	$C_6Cl_4NC(CF_3)N(OCH_2SCH_3)$	F:	PerFHalOrg.6-159
$C_{10}Cl_4F_3H_5N_2O_2$	$C_6Cl_4NC(CF_3)N(OC_2H_4OH)$	F:	PerFHalOrg.6-159
$C_{10}Cl_4F_4$	$C_{10}Cl_4F_4$	F:	PerFHalOrg.4-70
$C_{10}Cl_4F_4HN_3$	$NC_5Cl_4NHC_5F_4N$	F:	PerFHalOrg.5-186/7
		F:	PerFHalOrg.6-13/4
–	$NC_5F_2Cl_2NHC_5F_2Cl_2N$	F:	PerFHalOrg.5-186/7
		F:	PerFHalOrg.6-13/4
$C_{10}Cl_4F_4N_2$	$NC_5F_2Cl_2C_5F_2Cl_2N$	F:	PerFHalOrg.5-220/2
$C_{10}Cl_4F_4N_2S_2$	$NC_5F_2Cl_2SSC_5F_2Cl_2N$	F:	PerFHalOrg.5-192, 206
$C_{10}Cl_4F_{13}N_3$	$(CF_2ClCFClCF_2)_2(CF_3)C_3N_3$	F:	PerFHalOrg.6-84/6, 105
$C_{10}Cl_4F_{16}H_4Sn$	$CClF_2(CF_2)_7CH_2CH_2SnCl_3$	Sn:	Org.Verb.6-259
$C_{10}Cl_4FeGaH_{10}$	$[Fe(C_5H_5)_2]GaCl_4$	Fe:	Org.Verb.A1-225
$C_{10}Cl_4FeGaH_{11}$	$[Fe(C_5H_5)_2][HGaCl_4]$	Fe:	Org.Verb.A1-225
$C_{10}Cl_4FeH_{10}Hg_2$	$[Fe(C_5H_5)_2][HgCl_2]_2$	Fe:	Org.Verb.A1-224
$C_{10}Cl_4FeH_{10}Sb$	$[Fe(C_5H_5)_2]SbCl_4$	Fe:	Org.Verb.A1-224
$C_{10}Cl_4Fe_2H_{10}$	$[Fe(C_5H_5)_2]FeCl_4$	Fe:	Org.Verb.A1-221/2
$C_{10}Cl_4Fe_2H_{11}$	$[Fe(C_5H_5)_2][HFeCl_4]$	Fe:	Org.Verb.A1-222
$C_{10}Cl_4H_6O_2Sn$	$SnCl_4 \cdot C_{10}H_6O_2$	Sn:	MVol.C5-112
$C_{10}Cl_4H_7NiO_2$	$CH_3C_3H_4Ni \cdot C_6Cl_4O_2$	Ni:	Org.Verb.2-6
$C_{10}Cl_4H_8N_2Sn$	$SnCl_4 \cdot NC_5H_4C_5H_4N$	Sn:	MVol.C5-217/9
$C_{10}Cl_4H_8N_2Te$	$TeCl_4 \cdot NC_5H_4C_5H_4N$	Te:	SVol.B3-171
$C_{10}Cl_4H_8N_2Ti$	$TiCl_4 \cdot NC_5H_4C_5H_4N$	Ti:	Org.Verb.1-41
$C_{10}Cl_4H_8N_4O_6Sn$	$SnCl_4 \cdot 2\ NO_2C_5H_4NO$	Sn:	MVol.C5-208/9
$C_{10}Cl_4H_8O_2S_3Sn$	$SnCl_4 \cdot CH_3OC_6H_3(OH)CSSC(S)CH$	Sn:	MVol.C6-112/3
$C_{10}Cl_4H_8O_4Sn$	$SnCl_4 \cdot 2\ C_4H_3OCHO$	Sn:	MVol.C5-84
$C_{10}Cl_4H_{10}IrKN_2$	$K[Ir(C_5H_5N)_2Cl_4]$	Ir:	SVol.2-70
–	$K[Ir(C_5H_5N)_2Cl_4] \cdot H_2O$	Ir:	SVol.2-70, 71
$C_{10}Cl_4H_{10}IrN_2$	$Ir(C_5H_5N)_2Cl_4$	Ir:	SVol.2-70, 71
$C_{10}Cl_4H_{10}IrN_2^-$	$[Ir(C_5H_5N)_2Cl_4]^-$	Ir:	SVol.2-70
$C_{10}Cl_4H_{10}IrN_2Na$	$Na[Ir(C_5H_5N)_2Cl_4]$	Ir:	SVol.2-70
$C_{10}Cl_4H_{10}N_2O_2Sn$	$SnCl_4 \cdot 2\ C_5H_5NO$	Sn:	MVol.C5-206/7
$C_{10}Cl_4H_{10}N_2O_2Te$	$TeCl_4 \cdot 2\ C_5H_5NO$	Te:	SVol.B3-171
$C_{10}Cl_4H_{10}N_2O_2U$	$[C_{10}H_8N_2H_2]UO_2Cl_4 \cdot 3\ H_2O$	U:	SVol.C9-93
$C_{10}Cl_4H_{10}N_2S_2Sn$	$SnCl_4 \cdot 2\ NC_5H_4SH$	Sn:	MVol.C6-105
$C_{10}Cl_4H_{10}N_2Sn$	$SnCl_4 \cdot C_3H_2N_2(CH_3)C_6H_5$	Sn:	MVol.C5-221
–	$SnCl_4 \cdot 2\ C_5H_5N$	Sn:	MVol.C5-195/6
$C_{10}Cl_4H_{10}N_2Te$	$TeCl_4 \cdot 2\ C_5H_5N$	Te:	SVol.B3-169/70
$C_{10}Cl_4H_{10}N_2U$	$UCl_4 \cdot 2\ C_5H_5N$	U:	SVol.E1-28, 33
$C_{10}Cl_4H_{10}Ni$	$C_5H_5NiC_5H_5Cl_4$	Ni:	Org.Verb.2-205

Formula	Compound	Reference
$C_{10}Cl_4H_{10}NiTi$	$[Ni(C_5H_5)_2][TiCl_4]$	Ni: Org.Verb.2-213
$C_{10}Cl_4H_{10}O_4Sn$	$SnCl_4 \cdot C_6H_4(COOCH_3)_2$	Sn: MVol.C5-147
$C_{10}Cl_4H_{10}Ti$	$C_5H_5Ti(C_5H_5Cl_2)Cl_2$	Ti: Org.Verb.1-206
$C_{10}Cl_4H_{11}IrN_2$	$H[Ir(C_5H_5N)_2Cl_4]$	Ir: SVol.2-69
–	$H[Ir(C_5H_5N)_2Cl_4] \cdot 2\ H_2O$	Ir: SVol.2-70
–	$H[Ir(C_5H_5N)_2Cl_4] \cdot 4\ H_2O$	Ir: SVol.2-70
–	$H[Ir(C_5H_5N)_2Cl_4] \cdot 5\ H_2O$	Ir: SVol.2-70
$C_{10}Cl_4H_{11}NO_2Sn$	$SnCl_4 \cdot CH_3COCH_2CONHC_6H_5$	Sn: MVol.C6-49/50
–	$SnCl_4 \cdot (CH_3CO)_2NC_6H_5$	Sn: MVol.C6-51
–	$SnCl_4 \cdot C_6H_5CONHCOC_2H_5$	Sn: MVol.C6-50/1
$C_{10}Cl_4H_{12}MnN_2$	$(C_5H_5NH)_2MnCl_4$	Mn: MVol.C5-84/6, 206/7
$C_{10}Cl_4H_{12}N_2O_2U$	$[C_5H_5NH]_2UO_2Cl_4$	U: SVol.C9-93, 101
$C_{10}Cl_4H_{12}N_4Sn$	$SnCl_4 \cdot 2\ NC_5H_4NH_2$	Sn: MVol.C5-203
$C_{10}Cl_4H_{12}N_4Te$	$TeCl_4 \cdot 2\ NH_2C_5H_4N$	Te: SVol.B3-171
$C_{10}Cl_4H_{12}Ni_2O_4$	$(C_3H_5NiO_2CCHCl_2)_2$	Ni: Org.Verb.2-283
$C_{10}Cl_4H_{12}O_4Sn$	$[SnCl_2(ClC(COCH_3)_2)_2]$	Sn: MVol.C5-107
$C_{10}Cl_4H_{13}NO_2Sn$	$C_5H_5NH[SnCl_4(CH(COCH_3)_2)]$	Sn: MVol.C5-107
$C_{10}Cl_4H_{13}N_3O_4Sn$	$SnCl_4 \cdot (NO_2)_2C_6H_3N(C_2H_5)_2$	Sn: MVol.C5-180
$C_{10}Cl_4H_{13}O_4PSn$	$SnCl_4 \cdot C_6H_5COCH_2P(O)(OCH_3)_2$	Sn: MVol.C6-166
$C_{10}Cl_4H_{14}IrN_3$	$NH_4[Ir(C_5H_5N)_2Cl_4]$	Ir: SVol.2-70
$C_{10}Cl_4H_{14}MnS_4$	$(CH(CSCH_3)_2)_2MnCl_4$	Mn: MVol.C5-209/10
$C_{10}Cl_4H_{14}N_2O_2Sn$	$SnCl_4 \cdot (CH_3)_2C(OH)CONHNHC_6H_5$	Sn: MVol.C6-61
$C_{10}Cl_4H_{14}N_2Sn$	$SnCl_4 \cdot NC_5H_4C_4H_7NCH_3$	Sn: MVol.C5-203
$C_{10}Cl_4H_{14}OSn$	$SnCl_4 \cdot (CH_3)_4C_6HOH$	Sn: MVol.C5-76
–	$SnCl_4 \cdot C_2H_5CH(CH_3)C_6H_4OH$	Sn: MVol.C5-76
$C_{10}Cl_4H_{14}O_5Sn$	$[C_6H_5OH \cdot H][HSnCl_4(CH_3COO)_2]$	Sn: MVol.C5-118
$C_{10}Cl_4H_{15}NO_4Sn$	$SnCl_4 \cdot C_6H_5NH_2 \cdot 2\ CH_3COOH$	Sn: MVol.C5-175/6
$C_{10}Cl_4H_{16}N_2Sn$	$SnCl_4 \cdot C_6H_4(N(CH_3)_2)_2$	Sn: MVol.C5-187/8
–	$SnCl_4 \cdot NC(CH_2)_8CN$	Sn: MVol.C6-73/5
$C_{10}Cl_4H_{16}N_2Te$	$TeCl_4 \cdot (CH_3)_2NC_6H_4N(CH_3)_2$	Te: SVol.B3-169
–	$TeCl_4 \cdot (C_2H_5)_2NC_6H_4NH_2$	Te: SVol.B3-169
$C_{10}Cl_4H_{16}N_4Sn$	$SnCl_4 \cdot 2\ C_2H_5C_3H_3N_2$	Sn: MVol.C5-223
$C_{10}Cl_4H_{16}O_4Sn$	$SnCl_4 \cdot 2\ CH_2C(CH_3)COOCH_3$	Sn: MVol.C5-137/8
–	$SnCl_4 \cdot 2\ CH_2(COCH_3)_2$	Sn: MVol.C5-101
–	$SnCl_4 \cdot C_6H_{10}(OCOCH_3)_2$	Sn: MVol.C5-133
$C_{10}Cl_4H_{18}N_2Ni_2O_2$	$(t\text{-}C_4H_9NCl_2NiCO)_2$	Ni: Org.Verb.2-394
$C_{10}Cl_4H_{18}NiO_4P_2$	$(CO)_2Ni(P(OC_4H_9\text{-}n)Cl_2)_2$	Ni: Org.Verb.1-139
$C_{10}Cl_4H_{18}O_4Sn$	$SnCl_4 \cdot C_4H_8(COOC_2H_5)_2$	Sn: MVol.C5-141/3
–	$SnCl_4 \cdot C_6H_{12}(COOCH_3)_2$	Sn: MVol.C5-141/3
$C_{10}Cl_4H_{19}NSn_2$	$2\ (CH_3)_2SnCl_2 \cdot CH_3C_5H_4N$	Sn: Org.Verb.6-35, 38
$C_{10}Cl_4H_{20}N_2S_2Sn$	$SnCl_4 \cdot (C_2H_5)_2NCSCSN(C_2H_5)_2$	Sn: MVol.C6-116/7
$C_{10}Cl_4H_{20}N_4Sn$	$SnCl_4 \cdot 2\ CNN(C_2H_5)_2$	Sn: MVol.C6-77/8
$C_{10}Cl_4H_{20}O_2Sn$	$SnCl_4 \cdot 2\ CH_3C_4H_7O$	Sn: MVol.C5-157
–	$SnCl_4 \cdot 2\ C_5H_{10}O$	Sn: MVol.C5-158
$C_{10}Cl_4H_{20}O_4Sn$	$SnCl_4 \cdot 2\ CH_3COOC_3H_7$	Sn: MVol.C5-131/2
–	$SnCl_4 \cdot 2\ C_2H_5COOC_2H_5$	Sn: MVol.C5-134
–	$SnCl_4 \cdot 2\ C_3H_7COOCH_3$	Sn: MVol.C5-135
–	$SnCl_4 \cdot 2\ HCOOC_4H_9$	Sn: MVol.C5-128
$C_{10}Cl_4H_{20}O_6Sn$	$SnCl_4 \cdot 2\ OC(OC_2H_5)_2$	Sn: MVol.C5-138/9
$C_{10}Cl_4H_{20}O_9U_2$	$(UO_2Cl_2)_2 \cdot (CH_2OCH_2)_5$	U: SVol.E1-73, 78

Formula	Compound	Reference
$C_{10}Cl_4H_{20}S_2Sn$	$SnCl_4 \cdot 2\ (CH_2)_5S$	Sn: MVol.C6-113
$C_{10}Cl_4H_{21}NO_2Sn$	$2\ SnCl_4 \cdot 2\ CH_3(CH_2)_6CONHC_2H_4OH$	Sn: MVol.C6-46
$C_{10}Cl_4H_{21}O_5PSn$	$SnCl_4 \cdot C_2H_5OCOCH_2P(O)(OC_3H_7)_2$	Sn: MVol.C6-165/6
$C_{10}Cl_4H_{22}N_2O_2U$	$UCl_4(CH_3CONH(i\text{-}C_3H_7))_2$	U: SVol.E1-9
$C_{10}Cl_4H_{22}N_2O_4Sn$	$SnCl_4 \cdot 2\ (CH_3)_2CHCH(NH_2)COOH$	Sn: MVol.C6-11
$C_{10}Cl_4H_{22}N_2O_4U$	$UCl_4 \cdot 2\ (CH_3)_2NCOOC_2H_5$	U: SVol.E1-120/3
$C_{10}Cl_4H_{22}N_2O_8Sn$	$SnCl_4 \cdot 2\ CH_3CH(NH_2)COOH \cdot 2\ CH_3COOH$	Sn: MVol.C6-12
–	$SnCl_4 \cdot 2\ NH_2C_2H_4COOH \cdot 2\ CH_3COOH$	Sn: MVol.C6-12
$C_{10}Cl_4H_{22}N_2Sn$	$SnCl_4 \cdot 2\ (CH_2)_5NH$	Sn: MVol.C5-193
$C_{10}Cl_4H_{22}N_2Te$	$TeCl_4 \cdot 2\ NH(CH_2)_5$	Te: SVol.B3-172
$C_{10}Cl_4H_{22}O_2Sn$	$SnCl_4 \cdot C_2H_4(OC_4H_9)_2$	Sn: MVol.C5-154/5
$C_{10}Cl_4H_{22}STe$	$TeCl_4 \cdot (C_5H_{11})_2S$	Te: SVol.B3-175
–	$TeCl_4 \cdot (i\text{-}C_5H_{11})_2S$	Te: SVol.B3-175
$C_{10}Cl_4H_{22}S_2Sn$	$SnCl_4 \cdot C_4H_9S(CH_2)_2SC_4H_9$	Sn: MVol.C6-97/8
$C_{10}Cl_4H_{22}S_2Te$	$TeCl_4 \cdot C_4H_9S(CH_2)_2SC_4H_9$	Te: SVol.B3-175
$C_{10}Cl_4H_{22.5}N_{2.5}O_{2.5}Pa$	$PaCl_4 \cdot 2.5\ CH_3CON(CH_3)_2$	Pa: SVol.2-89/90
$C_{10}Cl_4H_{22.5}N_{2.5}O_{2.5}Te$	$TeCl_4 \cdot 2.5\ CH_3CON(CH_3)_2$	Te: SVol.B3-174
$C_{10}Cl_4H_{22.5}N_{2.5}O_{2.5}U$	$UCl_4 \cdot 2.5\ CH_3CON(CH_3)_2$	U: SVol.E1-102/4
$C_{10}Cl_4H_{24}MnN_2$	$(C_5H_{10}NH_2)_2MnCl_4$	Mn: MVol.C5-204/5
$C_{10}Cl_4H_{24}N_4O_2Sn$	$SnCl_4 \cdot 2\ (CH_3)_2NCON(CH_3)_2$	Sn: MVol.C6-57
$C_{10}Cl_4H_{24}N_4O_2U$	$UCl_4 \cdot 2\ CO(N(CH_3)_2)_2$	U: SVol.E1-91/2, 94
$C_{10}Cl_4H_{24}N_4S_2Sn$	$SnCl_4 \cdot 2\ (CH_3)_2NCSN(CH_3)_2$	Sn: MVol.C6-121
$C_{10}Cl_4H_{24}N_4S_2Te$	$Te(SC(N(CH_3)_2)_2)_2Cl_4$	Te: SVol.B3-180/1
$C_{10}Cl_4H_{24}O_2Sn$	$SnCl_4 \cdot 2\ C_3H_7OC_2H_5$	Sn: MVol.C5-151/2
–	$SnCl_4 \cdot 2\ C_5H_{11}OH$	Sn: MVol.C5-72/3
$C_{10}Cl_4H_{26}N_2Sn_2$	$[Sn_2Cl_3(C_5H_{11}NH_2)_2]Cl$	Sn: MVol.C5-11/2
$C_{10}Cl_4H_{26}O_2P_2S_2Sn$	$SnCl_4 \cdot 2\ (C_2H_5)_2P(O)SCH_3$	Sn: MVol.C6-158/60
$C_{10}Cl_4H_{26}O_4P_2Sn$	$SnCl_4 \cdot 2\ C_2H_5(H)P(O)OCH(CH_3)_2$	Sn: MVol.C6-151/2
–	$SnCl_4 \cdot 2\ C_2H_5(H)P(O)OC_3H_7$	Sn: MVol.C6-151/2
$C_{10}Cl_4H_{26}O_6P_2S_2Sn$	$SnCl_4 \cdot 2\ (C_2H_5O)_2(CH_3S)PO$	Sn: MVol.C6-174/5
$C_{10}Cl_4H_{26}P_2S_4Sn$	$SnCl_4 \cdot 2\ (C_2H_5)_2P(S)SCH_3$	Sn: MVol.C6-158/60
$C_{10}Cl_4H_{28}MnN_2$	$(C_5H_{11}NH_3)_2MnCl_4$	Mn: MVol.C5-182/5
$C_{10}Cl_4H_{28}N_2O_4U$	$[HOC_2H_4N(CH_3)_3]_2UO_2Cl_4$	U: SVol.C9-154/5
$C_{10}Cl_4H_{30}N_2Si_2Sn$	$SnCl_4 \cdot 2\ (CH_3)_3SiN(CH_3)_2$	Sn: MVol.C6-130
$C_{10}Cl_4H_{30}O_5PaS_5$	$PaCl_4 \cdot 5\ (CH_3)_2SO$	Pa: SVol.2-91/2
$C_{10}Cl_4H_{30}O_5S_5U$	$UCl_4 \cdot 5\ (CH_3)_2SO$	U: SVol.E1-202/3, 207
$C_{10}Cl_4H_{30}O_9S_5U$	$[UCl_3((CH_3)_2SO)_5]ClO_4$	U: SVol.E1-202/3
$C_{10}Cl_5F_2H_2N_3$	$NC_5F_2Cl_2NHC_5Cl_3HN$	F: PerFHalOrg.5-138/9
$C_{10}Cl_5F_3HN_3$	$NC_5Cl_4NHC_5F_3ClN$	F: PerFHalOrg.5-186/7
		F: PerFHalOrg.6-13/4
–	$NC_5F_2Cl_2NHC_5FCl_3N$	F: PerFHalOrg.5-186/7
		F: PerFHalOrg.6-13/4
$C_{10}Cl_5FeH_5$	$C_5H_5FeC_5Cl_5$	Fe: Org.Verb.A1-374
$C_{10}Cl_5FeH_5O_4SiSn$	$(CO)_4Fe(Sn(C_6H_5)Cl_2)(SiCl_3)$	Fe: Org.Verb.B2-141
$C_{10}Cl_5FeH_5O_4Sn_2$	$(CO)_4Fe(SnCl_3)Sn(C_6H_5)Cl_2$	Fe: Org.Verb.B3-138
$C_{10}Cl_5Fe_3H_{10}O_6S_2$	$Fe_3(CO)_6(SC_2H_5)_2Cl_5$	Fe: Org.Verb.C1-18/20
$C_{10}Cl_5H_8N_2U$	$[UCl_5(NC_5H_4C_5H_4N)]$	U: SVol.E1-39/41
$C_{10}Cl_5H_{10}MnN_2$	$(NHC_5H_4C_5H_4NH)MnCl_5$	Mn: MVol.C5-212/3
$C_{10}Cl_5H_{10}NO_2Sn$	$SnCl_4 \cdot ClC_6H_4CONHCOC_2H_5$	Sn: MVol.C6-50/1

Formula	Compound	Reference
$C_{10}Cl_5H_{10}N_2S_2U$	$UCl_5 \cdot 2\ HSC_5H_4N$	U: SVol.E1-28, 33
$C_{10}Cl_5H_{10}N_2U$	$UCl_5 \cdot 2\ C_5H_5N$	U: SVol.E1-28, 33/4
$C_{10}Cl_5H_{12}MnN_2$	$(C_5H_5NH)_2MnCl_5$	Mn: MVol.C5-212
$C_{10}Cl_5H_{12}N_2OU$	$[C_5H_5NH]_2UOCl_5$	U: SVol.C9-90/2
$C_{10}Cl_5H_{15}LiN_5Sn$	$[Li(CH_3CN)_4][SnCl_5(CH_3CN)]$	Sn: MVol.C6-67/8
$C_{10}Cl_5H_{15}N_5NaSn$	$[Na(CH_3CN)_4][SnCl_5(CH_3CN)]$	Sn: MVol.C6-67/8
$C_{10}Cl_5H_{15}Ni_2O$	$(CH_3CHCHCH_2NiCl)_2 \cdot CCl_3CHO$	Ni: Org.Verb.2-300
$C_{10}Cl_5H_{15}Ni_2O_2$	$(CH_3CHCHCH_2NiCl)_2 \cdot CCl_3COOH$	Ni: Org.Verb.2-300
$C_{10}Cl_5H_{17}OSn$	$SnCl_4 \cdot CH_3COC_8H_{14}Cl$	Sn: MVol.C5-87/8
$C_{10}Cl_5H_{22}OSn^-$	$[SnCl_5 \cdot C_{10}H_{21}OH]^-$	Sn: MVol.C5-72
$C_{10}Cl_5H_{22}OU$	$UCl_5 \cdot (C_5H_{11})_2O$	U: SVol.E1-70, 75
$C_{10}Cl_5H_{23}N_2Sn$	$[(C_2H_5)_4N][SnCl_5(CH_3CN)]$	Sn: MVol.C6-67/8
$C_{10}Cl_6F_2HN_3$	$(NC_5FCl_3)_2NH$	F: PerFHalOrg.5-186/7
–	$NC_5F_2Cl_2NHC_5Cl_4N$	F: PerFHalOrg.5-186/7 F: PerFHalOrg.6-13/4
$C_{10}Cl_6F_{12}HNO_2$	$[CCl_3(CF_2)_3C(O)]_2NH$	F: PerFHalOrg.7-24, 50
$C_{10}Cl_6F_{12}N_2O$	$NC(C_3F_6CCl_3)OC(C_3F_6CCl_3)N$	F: PerFHalOrg.5-64/5, 87
$C_{10}Cl_6F_{15}N$	$CCl_3(CF_2)_3CClNCCl_2C_4F_9$	F: PerFHalOrg.5-175
–	$CCl_3(CF_2)_3CCl_2NCClC_4F_9$	F: PerFHalOrg.5-175
$C_{10}Cl_6FeH_8Si_2$	$Fe(C_5H_4SiCl_3)_2$	Fe: Org.Verb.A1-170
$C_{10}Cl_6FeO_4$	$((CClCCl)_2CCCl_2)Fe(CO)_4$	Fe: Org.Verb.C2-32
$C_{10}Cl_6Fe_3H_{10}$	$[Fe(C_5H_5)_2]Fe_2Cl_6 \cdot 3\ H_2O$	Fe: Org.Verb.A1-222
$C_{10}Cl_6Ge_2H_{24}N_2NiO_2$	$[N(CH_3)_4]_2[(CO)_2Ni(GeCl_3)_2]$	Ni: Org.Verb.1-157/8
$C_{10}Cl_6H_4O_2Sn$	$SnCl_4 \cdot C_{10}H_4Cl_2O_2 \cdot n\ C_6H_6$	Sn: MVol.C5-112
$C_{10}Cl_6H_8N_2NiSi_2$	$(Cl_3Si)_2Ni(NC_5H_4C_5H_4N)$	Ni: Org.Verb.1-76
$C_{10}Cl_6H_8N_2O_2Sn$	$SnCl_4 \cdot 2\ ClC_5H_4NO$	Sn: MVol.C5-208/9
$C_{10}Cl_6H_{10}N_2OSSn$	$SnCl_4 \cdot 2\ C_5H_5N \cdot SOCl_2$	Sn: MVol.C5-197
$C_{10}Cl_6H_{10}N_2OSTi$	$(C_5H_5N)_2TiCl_4 \cdot SOCl_2$	S: SVol.1-55, 57
$C_{10}Cl_6H_{10}N_2OSZr$	$(C_5H_5N)_2ZrCl_4 \cdot SOCl_2$	S: SVol.1-55
$C_{10}Cl_6H_{10}N_2O_2SSn$	$SnCl_4 \cdot 2\ C_5H_5N \cdot SO_2Cl_2$	Sn: MVol.C5-197
$C_{10}Cl_6H_{10}Ni_2O_4$	$(C_3H_5NiO_2CCCl_3)_2$	Ni: Org.Verb.2-283
$C_{10}Cl_6H_{10}Ti$	$C_5H_5Ti(C_5H_5Cl_4)Cl_2$	Ti: Org.Verb.1-206
$C_{10}Cl_6H_{12}IrN_2$	$(C_5H_5NH)_2[IrCl_6]$	Ir: SVol.2-127
$C_{10}Cl_6H_{12}N_2Sb$	$[C_5H_6N]_2[SbCl_6]$ solid solutions $[C_5H_6N]_2[SbCl_6]$-$[C_5H_6N]_2[SnCl_6]$	Sn: MVol.C3-106/7
$C_{10}Cl_6H_{12}N_2Sn$	$[C_5H_6N]_2[SnCl_6]$	Sn: MVol.C3-106/7 Sn: MVol.C5-196
–	$[C_5H_6N]_2[SnCl_6]$ solid solutions $[C_5H_6N]_2[SnCl_6]$-$[C_5H_6N]_2[SbCl_6]$	Sn: MVol.C3-106/7
$C_{10}Cl_6H_{12}N_2Te$	$[C_5H_5NH]_2TeCl_6$	Te: SVol.B2-116/7, 119, 137/8
$C_{10}Cl_6H_{12}N_2U$	$[C_5H_5NH]_2UCl_6$	U: SVol.C9-52
$C_{10}Cl_6H_{14}Ni_2O$	$(CH_3CHCHCH_2NiCl)_2 \cdot CCl_3COCl$	Ni: Org.Verb.2-300
$C_{10}Cl_6H_{14}S_4Sn$	$(CH(CSCH_3)_2)_2[SnCl_6]$	Sn: MVol.C3-109
$C_{10}Cl_6H_{28}N_2Sn$	$[(CH_3)_3(C_2H_5)N]_2[SnCl_6]$	Sn: MVol.C3-106
–	$[i\text{-}C_5H_{11}NH_3]_2[SnCl_6]$	Sn: MVol.C3-103/4
$C_{10}Cl_6H_{30}IrN_3$	$(C_5H_{10}NH_3)_2(NH_4)[IrCl_6] \cdot H_2O$	Ir: SVol.2-127
$C_{10}Cl_7H_{10}N_2PSSn$	$SnCl_4 \cdot 2\ C_5H_5N \cdot PSCl_3$	Sn: MVol.C5-197
$C_{10}Cl_8DyH_{40}N_5$	$DyCl_3 \cdot 5\ [(CH_3)_2NH_2]Cl$	Sc: MVol.C5-220

Formula	Compound	Reference
$C_{10}Cl_8FeH_{10}Sn_2$	$[Fe(C_5H_5)_2][SnCl_4]_2$	Fe: Org.Verb.A1-225
$C_{10}Cl_8H_{10}N_2O_2Se_2Sn$	$[C_5H_5NSeOCl]_2[SnCl_6]$	Sn: MVol.C2-126
$C_{10}Cl_8H_{10}N_2Te$	$[C_5H_4ClNH]_2TeCl_6$	Te: SVol.B2-116
$C_{10}Cl_8H_{40}LaN_5$	$LaCl_3 \cdot 5\ [(CH_3)_2NH_2]Cl \cdot 4\ H_2O$	Sc: MVol.C5-213/5, 219/20
$C_{10}Cl_{10}H_{16}N_2O_8Sn$	$SnCl_4 \cdot 2\ CH_3CH(NH_2)COOH \cdot 2\ CCl_3COOH$	Sn: MVol.C6-12
–	$SnCl_4 \cdot 2\ NH_2C_2H_4COOH \cdot 2\ CCl_3COOH$	Sn: MVol.C6-12
$C_{10}Cl_{12}FeH_{10}Ti_3$	$[Fe(C_5H_5)_2][TiCl_4]_3$	Fe: Org.Verb.A1-225
$C_{10}Cl_{14}FeH_{10}Hg_7$	$[Fe(C_5H_5)_2][HgCl_2]_7$	Fe: Org.Verb.A1-224/5
–	$Fe(C_5H_5)_2 \cdot 7\ HgCl_2$	Fe: Org.Verb.A1-234
$C_{10}CmH_{10}N_2O_4^+$	$Cm(NC_5H_4CH_2N(CH_2COO)_2)^+$	Np: TrU.D1-163
$C_{10}CmH_{12}N_2O_8^-$	$CmC_2H_4N_2(CH_2COO)_4^-$	Np: TrU.D1-165
$C_{10}CmH_{13}N_2O_8$	$CmC_2H_4N_2(CH_2COOH)(CH_2COO)_3$	Np: TrU.D1-165
$C_{10}CmH_{15}N_2O_7$	$CmC_2H_4N_2(C_2H_4OH)(CH_2COO)_3$	Np: TrU.D1-164
$C_{10}CoF_{12}H_{10}IrP_4$	$(C_5H_5)_2Co[Ir(PF_3)_4]$	Ir: SVol.2-193
$C_{10}CoH_{12}N_2O_8Sn$	$Co[Sn(C_2H_4N_2(CH_2COO)_4)] \cdot 4\ H_2O$	Sn: MVol.C5-37
$C_{10}Co_2F_{12}O_6P_2$	$[Co(CO)_3P(CF_3)_2]_2$	F: PerFHalOrg.3-27, 110/1, 121
$C_{10}Co_2H_6O_8Sn$	$(CH_3)_2Sn[Co(CO)_4]_2$	Sn: Org.Verb.6-33
$C_{10}Co_3FeH_{19}O_5P^+$	$[HFeCo_3(CO)_2P(OC_3H_7)_2OC_2H_4]^+$	Fe: Org.Verb.B1-204
$C_{10}Co_3FeH_{22}O_4P^+$	$[HFeCo_3(CO)P(OC_3H_7)_3]^+$	Fe: Org.Verb.B1-204
$C_{10}CrFeH_9^+$	$[C_5H_5FeC_5H_4Cr]^+$	Fe: Org.Verb.A1-346
$C_{10}CrH_8N_2O_6U$	$[UO_2CrO_4(C_{10}H_8N_2)] \cdot H_2O$	U: SVol.E1-39/42
$C_{10}CrH_{24}N_4O_8U$	$UO_2CrO_4 \cdot 2\ (CH_3)_2NCON(CH_3)_2$	U: SVol.E1-93/5
$C_{10}Cr_2H_{10}N_2O_9U$	$UO_2Cr_2O_7 \cdot 2\ C_5H_5N$	U: SVol.E1-29, 35
$C_{10}CuF_5H_8O_2$	$CuC_6F_5 \cdot C_4H_8O_2$	F: PerFHalOrg.4-111
$C_{10}CuH_{12}N_2O_8Sn$	$Cu[Sn(C_2H_4N_2(CH_2COO)_4)] \cdot 2\ H_2O$	Sn: MVol.C5-37
–	$Cu[Sn(C_2H_4N_2(CH_2COO)_4)] \cdot 4\ H_2O$	Sn: MVol.C5-37
$C_{10}CuH_{19}S_3$	$Cu[S_2CSC_9H_{19}]$	C: MVol.D4-269
$C_{10}FFeH_9$	$C_5H_5FeC_5H_4F$	Fe: Org.Verb.A1-365, 372/3
$C_{10}FFeH_{10}$	$[Fe(C_5H_5)_2]F$	Fe: Org.Verb.A1-214
$C_{10}FH_9O_2$	$C_6H_5C(CH_3)CFCOOH$	F: PerFHalOrg.4-89
$C_{10}FH_{10}Ni_2^+$	$FNi_2(C_5H_5)_2^+$	Ni: Org.Verb.2-359
$C_{10}FH_{11}Ni$	$[C_5H_5NiC_5H_6]F$	Ni: Org.Verb.2-189
$C_{10}FH_{11}Sn$	$(CH_2CH)_2(C_6H_5)SnF$	Sn: Org.Verb.5-40/1
$C_{10}FH_{14}NO_2$	$(C_2H_5)_2NC_6H_2F(OH)_2$	F: PerFHalOrg.5-144
$C_{10}FH_{15}Sn$	$(CH_3)_3SnCH_2C_6H_4F$	Sn: Org.Verb.2-22, 27, 29
$C_{10}FH_{17}N_2O_7Sn$	$[SnF(C_2H_4N_2(C_2H_4OH)(CH_2COO)_3H_2)]$	Sn: MVol.C5-32
$C_{10}FH_{20}N_5O_3S_3$	$[NS(O)F][NS(O)N(CH_2)_5]_2$	S: S-N-Verb.1-13
$C_{10}FH_{23}Sn$	$(C_4H_9)_2(C_2H_5)SnF$	Sn: Org.Verb.5-40
$C_{10}FH_{30}NNi_2P_2$	$(CH_3Ni(P(CH_3)_3)F)(CH_3Ni(P(CH_3)_3)N(CH_3)_2)$	Ni: Org.Verb.2-254
$C_{10}F_2FeH_5N_3O_2$	$C_3N_3F_2Fe(CO)_2(C_5H_5)$	C: MVol.D3-272
$C_{10}F_2FeH_5O_4P$	$(CO)_4FePF_2C_6H_5$	Fe: Org.Verb.B2-94
$C_{10}F_2H_8N_2O_2S_2Sn$	$SnF_2(C_5H_4NOS)_2$	Sn: MVol.C6-106
$C_{10}F_2H_{12}Ni$	$[C_5H_5NiC_5H_6]HF_2$	Ni: Org.Verb.2-189
$C_{10}F_2H_{14}O_4Sn$	$[SnF_2(CH(COCH_3)_2)_2]$	Sn: MVol.C5-101/6
$C_{10}F_2H_{14}Sn$	$(C_4H_9)(C_6H_5)SnF_2$	Sn: Org.Verb.5-50
$C_{10}F_2H_{18}N_2O_7Sn$	$[SnF(C_2H_4N_2(C_2H_4OH)(CH_2COO)_3H_2)] \cdot HF$	Sn: MVol.C5-32
$C_{10}F_2H_{22}Sn$	$(C_5H_{11})_2SnF_2$	Sn: Org.Verb.5-49
–	$(t\text{-}C_5H_{11})_2SnF_2$	Sn: Org.Verb.5-49

Formula	Compound	Element	Reference
$C_{10}F_2H_{24}N_4NiO_2P_2$	$(CO)_2Ni(P(N(CH_3)_2)_2F)_2$	Ni:	Org.Verb.1-125, 126
$C_{10}F_3FeH_5O_2$	$C_5H_5Fe(CO)_2CCCF_3$	F:	PerFHalOrg.4-33
$C_{10}F_3H_6N_3$	$NCFCFC(NHC_6H_5)CFN$	F:	PerFHalOrg.6-59, 65
–	$NCFNC(NHC_6H_5)CFCF$	F:	PerFHalOrg.6-59, 65
$C_{10}F_3H_6N_3O_2S$	$SNC(CF_3)NC(NHOCOC_6H_5)$	F:	PerFHalOrg.5-101
$C_{10}F_3H_6N_3O_3$	$NC(NHOC(O)C_6H_5)ONC(CF_3)$	F:	PerFHalOrg.5-101
$C_{10}F_3H_6N_5O_4S$	$C_2N_2S(CF_3)NHC_6H_2(NO_2)_2CH_3$	F:	PerFHalOrg.5-107
$C_{10}F_3H_8NaO$	$C_6H_5(CH_3)C(CFCF_2)ONa$	F:	PerFHalOrg.4-1
$C_{10}F_3H_9O$	$C_6H_5(CH_3)C(CFCF_2)OH$	F:	PerFHalOrg.4-1
$C_{10}F_3H_{13}Sn$	$(CH_3)_3SnC_6H_4CF_3$	Sn:	Org.Verb.2-126/7
$C_{10}F_3H_{14}N_3O$	$((C_2H_5)_2NC_2H_4O)C_4F_3N_2$	F:	PerFHalOrg.6-58
$C_{10}F_3H_{17}O_2$	$C_8H_{17}OC(O)CF_3$	F:	PerFHalOrg.7-63
$C_{10}F_3H_{19}N_2O_7Sn$	$[C_2H_4N_2(C_2H_4OH)(CH_2COOH)_3H][SnF_3]$	Sn:	MVol.C5-32
$C_{10}F_3H_{20}N_2P$	$CF_2CFP[N(C_2H_5)_2]_2$	F:	PerFHalOrg.4-106
$C_{10}F_4FeH_{18}O_8P_2$	$F_2CCF_2Fe(CO)_2(P(OCH_3)_3)_2$	Fe:	Org.Verb.B4-186, 189
$C_{10}F_4H_5N_2NaO_2$	$NC_5F_4CNa(CN)COOC_2H_5$	F:	PerFHalOrg.5-142
$C_{10}F_4H_6N_2O_2$	$C_6F_4NC(OCH_3)C(OCH_3)N$	F:	PerFHalOrg.6-156, 161
–	$NC_5F_4CH(CN)COOC_2H_5$	F:	PerFHalOrg.5-142
$C_{10}F_4H_6O_4$	$(CH_3COO)_2C_6F_4$	F:	PerFHalOrg.4-56
$C_{10}F_4H_8N_2Sn$	$SnF_4 \cdot NC_5H_4C_5H_4N$	Sn:	MVol.C5-217
$C_{10}F_4H_8N_4O_6Sn$	$SnF_4 \cdot 2\ NO_2C_5H_4NO$	Sn:	MVol.C5-208/9
$C_{10}F_4H_9MnO_5Sn$	$(CH_3)_3SnCF_2CF_2Mn(CO)_5$	Sn:	Org.Verb.2-37, 39
$C_{10}F_4H_{10}N_2$	$(CH_2)_5NC_5F_4N$	F:	PerFHalOrg.5-136
$C_{10}F_4H_{10}N_2O_2Sn$	$SnF_4 \cdot 2\ C_5H_5NO$	Sn:	MVol.C5-206
$C_{10}F_4H_{10}N_2Sn$	$SnF_4 \cdot 2\ C_5H_5N$	Sn:	MVol.C5-194/5
$C_{10}F_4H_{10}NiO_4Si_2$	$C_4H_9C_2HF_4Si_2Ni(CO)_4$	Ni:	Org.Verb.1-302
$C_{10}F_4H_{12}N_2$	$[(CH_3)_2N]_2C_6F_4$	F:	PerFHalOrg.7-61
$C_{10}F_4H_{12}Ni_2O_4$	$(C_3H_5NiO_2CCHF_2)_2$	Ni:	Org.Verb.2-283
$C_{10}F_4H_{12}OSn$	$(CH_3)_3SnC_6F_4OCH_3$	Sn:	Org.Verb.2-137
$C_{10}F_4H_{14}Si_2$	$C_6F_4[Si(CH_3)_2H]_2$	F:	PerFHalOrg.4-24
$C_{10}F_4H_{20}N_2NiO_2P_2$	$(CO)_2Ni(P(N(C_2H_5)_2)F_2)_2$	Ni:	Org.Verb.1-125, 126
$C_{10}F_4H_{20}O_9U_2$	$(UO_2F_2)_2 \cdot (CH_2OCH_2)_5$	U:	SVol.E1-73, 78
–	$(UO_2F_2)_2 \cdot (CH_2OCH_2)_5 \cdot 4\ H_2O$	U:	SVol.E1-73, 78
$C_{10}F_4H_{20}Sn$	$(C_4H_9)_2(CHF_2CF_2)SnH$	Sn:	Org.Verb.4-91
$C_{10}F_4MnNO_5$	$NC_5F_4Mn(CO)_5$	F:	PerFHalOrg.5-145
$C_{10}F_4NO_5Re$	$NC_5F_4Re(CO)_5$	F:	PerFHalOrg.5-145
$C_{10}F_5FeIO_4$	$(CO)_4Fe(C_6F_5)I$	Fe:	Org.Verb.B2-179, 182
$C_{10}F_5H_5N_2O$	$(NH_2)(CH_3O)C_9F_5N$	F:	PerFHalOrg.6-161
$C_{10}F_5H_5NiO$	$C_5H_5Ni(CO)C_4F_5$	Ni:	Org.Verb.2-167/8
$C_{10}F_5H_5O_3$	$C_6F_5C(O)C(O)OC_2H_5$	F:	PerFHalOrg.4-52
$C_{10}F_5H_8N$	$C_6F_5NC_2H_2(CH_3)_2$	F:	PerFHalOrg.7-195/6
–	$C_6F_5NHCH(CH_3)CHCH_2$	F:	PerFHalOrg.7-198
$C_{10}F_5H_8NO$	$C_6F_5N(OH)CH(CH_3)CHCH_2$	F:	PerFHalOrg.7-198
$C_{10}F_5H_{10}O_2P$	$C_6F_5P(OC_2H_5)_2$	F:	PerFHalOrg.3-129, 133
$C_{10}F_5H_{10}P$	$C_6F_5P(C_2H_5)_2$	F:	PerFHalOrg.3-129, 130
$C_{10}F_5H_{11}O_2Si$	$C_6F_5(C_2H_5O)_2SiH$	F:	PerFHalOrg.4-81
$C_{10}F_5H_{12}N_2P$	$C_6F_5P(N(CH_3)_2)_2$	F:	PerFHalOrg.3-125, 131
$C_{10}F_6FeH_{10}P$	$[Fe(C_5H_5)_2]PF_6$	Fe:	Org.Verb.A1-220
$C_{10}F_6FeH_{27}NO_{11}P_4$	$[COFe(NO)(P(OCH_3)_3)_3]PF_6$	Fe:	Org.Verb.B1-38/40
$C_{10}F_6FeO_4$	$C_6F_6Fe(CO)_4$	Fe:	Org.Verb.B4-321, 327

Formula	Compound		Reference
$C_{10}F_6Fe_2O_6S_2$	$(CO)_3FeSC(CF_3)C(CF_3)SFe(CO)_3$	Fe:	Org.Verb.C1-115, 117, 120
$C_{10}F_6H_3NO$	$CH_3OC_9F_6N$	F:	PerFHalOrg.6-153/4, 159/61
–	$C_6F_4CFCFC(O)NCH_3$	F:	PerFHalOrg.6-154, 159/60
–	$C_6F_4CFCFN(CH_3)C(O)$	F:	PerFHalOrg.6-155, 159
$C_{10}F_6H_3N_5O_4S$	$C_2N_2S(CF_3)NHC_6H_2(NO_2)_2CF_3$	F:	PerFHalOrg.5-107
$C_{10}F_6H_5NOS$	$C_6H_4SC(CF_3)_2NHC(O)$	F:	PerFHalOrg.7-66, 68
$C_{10}F_6H_5NO_2$	$(CF_3)_2CNOC(O)C_6H_5$	F:	PerFHalOrg.7-141
$C_{10}F_6H_5N_3O_4$	$O_2NC_6H_4NN(CF_2)_3COOH$	F:	PerFHalOrg.7-193
$C_{10}F_6H_6N_2O_2$	$(CF_3)_2CNOC(O)NHC_6H_5$	F:	PerFHalOrg.7-143
–	$C_6H_5NN(CF_2)_3COOH$	F:	PerFHalOrg.7-193
$C_{10}F_6H_7NO_2$	$CH_3C_6H_4C(O)ON(CF_3)_2$	F:	PerFHalOrg.7-132, 136
$C_{10}F_6H_7NO_3S$	$(CF_3)_2CNOSO_2C_6H_4CH_3$	F:	PerFHalOrg.7-143
$C_{10}F_6H_8MnO_4$	$[Mn(CF_3COCHCOCH_3)_2]$	Mn:	MVol.D1-108/9
$C_{10}F_6H_8MnO_4^-$	$[Mn(CF_3COCHCOCH_3)_2]^-$	Mn:	MVol.D1-108/9
$C_{10}F_6H_8N_2O_2$	$CF_3N(O)CH(C_6H_5)CH_2ON(CF_3)$	F:	PerFHalOrg.7-165, 168
$C_{10}F_6H_9MnO_4^+$	$[Mn(CF_3COCHCOCH_3)_2H]^+$	Mn:	MVol.D1-109
$C_{10}F_6H_9N$	$CH_3C_6H_4C(CF_3)_2NH_2$	F:	PerFHalOrg.7-69/70
$C_{10}F_6H_9NO$	$CH_3OC_6H_4C(CF_3)_2NH_2$	F:	PerFHalOrg.7-69/70
–	$C_6H_5CH[ON(CF_3)_2]CH_3$	F:	PerFHalOrg.7-136
$C_{10}F_6H_9NO_3$	$(CH_3O)_3(CF_3)_2C_5N$	F:	PerFHalOrg.5-177
$C_{10}F_6H_{10}NiO_4$	$Ni(CH_2CHCH_2OOCCF_3)_2$	Ni:	Org.Verb.2-90
$C_{10}F_6H_{10}Ni_2O_4$	$(C_3H_5NiO_2CCF_3)_2$	Ni:	Org.Verb.2-283/4
$C_{10}F_6H_{11}N_3$	$(CF_3)_2(C_5H_{11})C_3N_3$	F:	PerFHalOrg.6-113
$C_{10}F_6H_{12}MnO_6$	$[Mn(CF_3COCHCOCH_3)_2(H_2O)_2]$	Mn:	MVol.D1-109
$C_{10}F_6H_{12}N_2Ni$	$C_8H_{12}NiN_2(CF_3)_2$	Ni:	Org.Verb.2-74/5
$C_{10}F_6H_{12}N_2Te$	$[C_5H_5NH]_2TeF_6$	Te:	SVol.B2-18
$C_{10}F_6H_{14}N_2O_2$	$(CF_3)_2CNOC(O)N[CH(CH_3)_2]_2$	F:	PerFHalOrg.7-141/2
$C_{10}F_6H_{14}N_{10}O_4S$	$[C_4N_2(CF_3)(NH_2)_3H]_2SO_4$	F:	PerFHalOrg.6-27
$C_{10}F_6H_{16}Sn$	$(C_2H_5)_3SnC(CF_3)CHCF_3$	Sn:	Org.Verb.2-192, 194
–	$(C_2H_5)_3SnCF(CF_2)_2CHF$	Sn:	Org.Verb.2-181/2
$C_{10}F_6H_{18}NO_4P$	$(CF_3)_2NOP(O)(OC_4H_9)_2$	F:	PerFHalOrg.7-127
$C_{10}F_6H_{18}O_6S_2Sn$	$(C_4H_9)_2Sn(SO_3CF_3)_2$	Sn:	Org.Verb.6-89, 95
$C_{10}F_6H_{19}NNiP_2$	$(CF_3)_2CN(CH_3)Ni[(CH_3)_2P(CH_2)_2P(CH_3)_2]$	Ni:	Org.Verb.1-62, 64
$C_{10}F_6H_{20}NO_5P$	$(CF_3)_2NOP(OC_2H_5)_4$	F:	PerFHalOrg.7-126
$C_{10}F_6H_{22}N_4O_2$	$2\ CF_3NO \cdot 2\ (C_2H_5)_2NH$	F:	PerFHalOrg.7-160
$C_{10}F_6H_{25}NiO_6P_3$	$[CH_3C_3H_4Ni(P(OCH_3)_3)_2]PF_6$	Ni:	Org.Verb.2-24
$C_{10}F_6IMn_2O_8P$	$Mn_2(CO)_8P(CF_3)_2I$	F:	PerFHalOrg.3-110/1, 120
$C_{10}F_6N_2$	$NC_5F_3C_5F_3N$	F:	PerFHalOrg.6-135, 148/9, 152
$C_{10}F_7FeH_5O_2$	$C_5H_5Fe(CO)_2C_3F_7$	Fe:	Org.Verb.B2-181
$C_{10}F_7H_2N$	$NH_2C_{10}F_7$	F:	PerFHalOrg.7-11, 38
$C_{10}F_7H_2N_3$	$NC_5F_3(NH_2)C_5F_4N$	F:	PerFHalOrg.5-220/2
$C_{10}F_7H_5O$	$C_3F_7C(O)C_6H_5$	F:	PerFHalOrg.4-41
$C_{10}F_7H_7N_2O_2$	$CH_3CONHC(CF_2)_3C(NCOCH_3)CF$	F:	PerFHalOrg.7-58
$C_{10}F_7H_7O$	$C_3F_7CH(OH)C_6H_5$	F:	PerFHalOrg.4-41
$C_{10}F_7H_{12}N_5O_4$	$(C_3F_7)(N(CH_2OH)_2)_2C_3N_3$	F:	PerFHalOrg.6-112
$C_{10}F_7N_3O_2$	$NC_5F_3(NO_2)C_5F_4N$	F:	PerFHalOrg.5-220/2

Formula	Compound	Element	Reference
$C_{10}F_8HN_3$	$[C_5F_4N]_2NH$	F:	PerFHalOrg.5-186/7
		F:	PerFHalOrg.6-13/4
$C_{10}F_8H_6Si$	$CF_2CFSi(CH_3)_2(C_6F_5)$	F:	PerFHalOrg.4-81
$C_{10}F_8H_8N_2Te_2$	$2\ TeF_4 \cdot NC_5H_4C_5H_4N$	Te:	SVol.B3-171
$C_{10}F_8H_{10}N_2$	$NC(N(CH_2)_5)(CF_2)_4$	F:	PerFHalOrg.5-143
$C_{10}F_8H_{10}N_2Te_2$	$[TeF_3(C_5H_5N)_2][TeF_5]$	Te:	SVol.B3-169
$C_{10}F_8H_{15}N_2O_5P$	$CF_3N[OP(O)(OC_2H_5)_2]CF_2C(OC_2H_5)NCF_3$	F:	PerFHalOrg.7-126
$C_{10}F_8H_{18}NiO_6P_2$	$C_4F_8Ni(P(OCH_3)_3)_2$	Ni:	Org.Verb.1-345, 346
$C_{10}F_8HgN_2$	$(NC_5F_4)_2Hg$	F:	PerFHalOrg.5-115/6, 125/6
$C_{10}F_8N_2$	$NC_5F_4C_5F_4N$	F:	PerFHalOrg.5-219/24
$C_{10}F_8N_2OS$	$NC_5F_4S(O)C_5F_4N$	F:	PerFHalOrg.5-191/2
$C_{10}F_8N_2O_2S$	$NC_5F_4S(O_2)C_5F_4N$	F:	PerFHalOrg.5-191/2
$C_{10}F_8N_2S$	$S(C_5F_4N)_2$	F:	PerFHalOrg.5-191/2
$C_{10}F_8N_2S_2$	$NC_5F_4SSC_5F_4N$	F:	PerFHalOrg.5-191/2
$C_{10}F_8N_2Zn$	$(NC_5F_4)_2Zn$	F:	PerFHalOrg.5-115/6, 125
$C_{10}F_8N_4$	$NC_5F_4NNC_5F_4N$	F:	PerFHalOrg.5-190, 204
$C_{10}F_9GeH_3$	$(CF_3CC)_3GeCH_3$	F:	PerFHalOrg.4-75, 76
$C_{10}F_9H_3Si$	$(CF_3CC)_3SiCH_3$	F:	PerFHalOrg.4-88
$C_{10}F_9H_4NO$	$(CF_3)_2C_6F_3NHCOCH_3$	F:	PerFHalOrg.7-60
–	$C_2F_5C_6F_4NHC(O)CH_3$	F:	PerFHalOrg.7-60
$C_{10}F_9H_5NOP$	$C_3F_7CONP(F_2)C_6H_5$	F:	PerFHalOrg.7-63
$C_{10}F_9H_5NO_2$	$CF_3N(O)N(O)(CF_3)CF_2CF(C_6H_5)$	F:	PerFHalOrg.7-165, 168
$C_{10}F_9H_6N$	$(CF_3)_2C_6F_3N(CH_3)_2$	F:	PerFHalOrg.7-61
$C_{10}F_9H_6NO$	$(CF_3)_3CN(OH)C_6H_5$	F:	PerFHalOrg.7-192
$C_{10}F_9H_6NO_2$	$(CH_3O)_2[(CF_3)_2CF]C_5F_2N$	F:	PerFHalOrg.5-177
$C_{10}F_9H_7O$	$C_7F_9C(CH_3)_2OH$	F:	PerFHalOrg.4-44, 97
$C_{10}F_9H_9Sn$	$(CF_2CF)_3SnC_4H_9$	Sn:	Org.Verb.2-346, 349
$C_{10}F_9H_{10}NO$	$(CF_3)_3CNC_4H_4(CH_3)_2O$	F:	PerFHalOrg.7-194/5
$C_{10}F_9H_{15}NO_3P$	$(CF_3)_3CNP(OC_2H_5)_3$	F:	PerFHalOrg.7-200
$C_{10}F_9H_{15}Sn$	$(CF_3CH_2CH_2)_3SnCH_3$	Sn:	Org.Verb.2-336, 338
$C_{10}F_{10}FeO_4$	$C_6F_{10}Fe(CO)_4$	Fe:	Org.Verb.B4-322, 327
$C_{10}F_{10}Fe_2I_2O_6$	$C_2F_5(CO)_3FeI_2Fe(CO)_3C_2F_5$	Fe:	Org.Verb.C1-222
$C_{10}F_{10}Fe_2O_6Se_2$	$(CO)_3Fe(SeC_2F_5)_2Fe(CO)_3$	Fe:	Org.Verb.C1-83, 90
$C_{10}F_{10}GeH_6$	$(CF_3CF_2CC)_2Ge(CH_3)_2$	F:	PerFHalOrg.4-75, 76
$C_{10}F_{10}H_2N_2O_2$	$[CF_3C(O)NH]_2C_6F_4$	F:	PerFHalOrg.7-23/4, 49
$C_{10}F_{10}H_6N_3O_3P$	$(CF_2N(O)CF_3)_2P(O)NHC_6H_5$	F:	PerFHalOrg.5-99
$C_{10}F_{10}H_6Sn$	$(CH_3)_2Sn(C_3F_2CF_3)_2$	Sn:	Org.Verb.3-8
$C_{10}F_{10}H_{10}Sn$	$(CH_2CHCH_2)_2Sn(C_2F_5)_2$	Sn:	Org.Verb.3-72
$C_{10}F_{10}Ni_2O_6P_2$	$[(CF_3PCF_2)_2(Ni(CO)_3)_2]_x$	Ni:	Org.Verb.2-395
$C_{10}F_{11}H_2N$	$NH_2C_6F_3(CF_2)_4$	F:	PerFHalOrg.7-10/1, 38
$C_{10}F_{11}N$	$(CF_3)_2CFCCC_5F_4N$	F:	PerFHalOrg.5-155/6, 167
$C_{10}F_{12}FeO_4$	$(CF_3)_2CC(CF_3)_2Fe(CO)_4$	Fe:	Org.Verb.B4-321, 325
–	$(CO)_4Fe(C_2F_5)C(CF_3)CFCF_3$	Fe:	Org.Verb.B2-192
$C_{10}F_{12}FeO_4P_2$	$(CO)_4FeP_2C_2(CF_3)_4$	Fe:	Org.Verb.B2-97, 118
$C_{10}F_{12}Fe_2H_2O_6P_2$	$H(CO)_3Fe(P(CF_3)_2)_2Fe(CO)_3H$	Fe:	Org.Verb.C1-6/7, 9/10
–	$D(CO)_3Fe(P(CF_3)_2)_2Fe(CO)_3D$	Fe:	Org.Verb.C1-10
$C_{10}F_{12}Fe_2I_2O_6P_2$	$I(CO)_3Fe(P(CF_3)_2)_2Fe(CO)_3I$	Fe:	Org.Verb.C1-6/8, 11
$C_{10}F_{12}Fe_2NiO_{10}Sb_2$	$[Ni(Fe(CO)_5)_2][SbF_6]_2$	Fe:	Org.Verb.B3-130
$C_{10}F_{12}Fe_2O_6P_2$	$(CO)_3Fe(P(CF_3)_2)_2Fe(CO)_3$	Fe:	Org.Verb.C1-164/7, 173/4

Formula	Compound	Reference
$C_{10}F_{12}Fe_2O_6P_2Se$	$(CO)_6Fe_2(P(CF_3)_2)(SeP(CF_3)_2)$	Fe: Org.Verb.C1-191/2, 194/5
$C_{10}F_{12}Fe_2O_6P_4$	$(CO)_3Fe(PCF_3)_4Fe(CO)_3$	Fe: Org.Verb.C1-164, 168, 177
$C_{10}F_{12}Fe_2O_{10}Sb_2Sn$	$[Sn(Fe(CO)_5)_2][SbF_6]_2$	Fe: Org.Verb.B3-130
$C_{10}F_{12}Fe_3O_{10}Sb_2$	$[Fe(Fe(CO)_5)_2][SbF_6]_2$	Fe: Org.Verb.B3-130
		Fe: Org.Verb.C1-62
$C_{10}F_{12}H_2MnO_4$	$[Mn(CH(COCF_3)_2)_2]$	Mn: MVol.D1-110
$C_{10}F_{12}H_2O$	$C_3F_7CH(OH)C_6F_5$	F: PerFHalOrg.4-101
$C_{10}F_{12}H_3NO$	$(CH_3O)(C_2F_5)_2C_5F_2N$	F: PerFHalOrg.5-177/8
$C_{10}F_{12}H_3NO_2$	$(CF_2)_3CC(OCH_3)NC(CF_2)_3O$	F: PerFHalOrg.5-102
–	$(CF_2)_3CC(O)N(CH_3)C(CF_2)_3O$	F: PerFHalOrg.5-102
$C_{10}F_{12}H_4N_4O_2$	$HNC_5F_6(O)(NC(NH)C_3F_6CONH_2)$	F: PerFHalOrg.5-190, 205
$C_{10}F_{12}H_7N_3$	$(CF_3)_2CNHC(CF_3)_2NC(NHC_3H_5)$	F: PerFHalOrg.5-104/5
$C_{10}F_{12}H_7N_3OS$	$(CF_3)_2CNHC(CF_3)_2NC(NHC(O)SC_2H_5)$	F: PerFHalOrg.5-108
$C_{10}F_{12}H_7N_3O_2$	$(CF_3)_2CNHC(CF_3)_2NC(NHCOOC_2H_5)$	F: PerFHalOrg.5-108
$C_{10}F_{12}H_8N_4$	$(CF_3)_2CNHC(CF_3)_2NC(NCHN(CH_3)_2)$	F: PerFHalOrg.5-109
$C_{10}F_{12}H_9N_3$	$(CF_3)_2CN(CH_3)C(CF_3)_2NC(N(CH_3)_2)$	F: PerFHalOrg.5-104
$C_{10}F_{12}H_{10}N_2$	$((CF_3)_2C(NH_2)CH_2)_2C(CH_2)$	F: PerFHalOrg.7-65, 69, 72
$C_{10}F_{12}H_{10}N_4$	$C_3HN_2(CF_3)_4(NHCH_2N(CH_3)_2)$	F: PerFHalOrg.5-106
$C_{10}F_{12}H_{10}NiO_2P_2$	$(CO)_2Ni(P(CF_3)_3)(P(C_2H_5)_2CF_3)$	Ni: Org.Verb.1-145
–	$(CO)_2Ni(P(C_2H_5)(CF_3)_2)_2$	Ni: Org.Verb.1-128
$C_{10}F_{12}H_{12}N_2O_2$	$(CF_3)_2NOC(CH_3)_2C(CH_3)_2ON(CF_3)_2$	F: PerFHalOrg.7-132/3, 135
$C_{10}F_{12}H_{14}Si_2$	$C_2F_4(CF_2CF_2SiH(CH_3)_2)_2$	F: PerFHalOrg.4-87
$C_{10}F_{12}Hg_2$	$Hg(C_5F_6)_2Hg$	F: PerFHalOrg.4-34
$C_{10}F_{12}Ni_2O_6P_2$	$(CO)_3Ni(CF_3)_2PP(CF_3)_2Ni(CO)_3$	Ni: Org.Verb.1-238, 243/4
$C_{10}F_{12}Ni_2O_6P_4$	$(PCF_3)_4[Ni(CO)_3]_2$	Ni: Org.Verb.1-246
$C_{10}F_{12}O$	$C_6F_5COC_3F_7$	F: PerFHalOrg.4-102
$C_{10}F_{13}HN_2O$	$NCFC(C_6F_{11})CFC(O)NH$	F: PerFHalOrg.6-18/9, 39
–	$NCFCFC(C_6F_{11})C(O)NH$	F: PerFHalOrg.6-18/9, 39
$C_{10}F_{13}HO$	$(CF_3C_6F_4)(CF_3)_2COH$	F: PerFHalOrg.4-53
$C_{10}F_{13}H_3N_2O_2$	$[(CF_3)_2NO]_2C_6H_3F$	F: PerFHalOrg.7-136
$C_{10}F_{13}N$	$C_5F_9C_5F_4N$	F: PerFHalOrg.5-155/6, 168
$C_{10}F_{13}N_3$	$(CF_2CFCF_2)_2(CF_3)C_3N_3$	F: PerFHalOrg.6-84/6, 104
$C_{10}F_{14}FeNO_5P$	$(CO)_3Fe(PF_2NCO(C(CF_3)_2)_2O)$	Fe: Org.Verb.B1-170/1, 175, 179
$C_{10}F_{14}FeO_4$	$(CO)_4Fe(C_3F_7)_2$	Fe: Org.Verb.B2-191/3
$C_{10}F_{14}HNO_2$	$(CF_2)_4CCONHC(CF_2)_3O$	F: PerFHalOrg.5-58/9, 81
–	$(CF_2)_4CNHCOC(CF_2)_3O$	F: PerFHalOrg.5-58/9, 81
$C_{10}F_{14}H_2N_2O_2$	$HNC(O)(CC_3F_7\text{-}i)_2C(O)NH$	F: PerFHalOrg.6-17, 37
$C_{10}F_{14}H_4N_4O$	$NC(C_3F_7)N(NHCOCH_3)C(C_3F_7)N$	F: PerFHalOrg.5-101
$C_{10}F_{14}H_4N_4O_4$	$C_3F_7C(NH_2)NOCOCOONC(NH_2)C_3F_7$	F: PerFHalOrg.7-5/6, 32
$C_{10}F_{14}H_6O_2$	$C_3F_7C(CH_3)(OH)CH_2COC_3F_7$	F: PerFHalOrg.4-41
$C_{10}F_{14}H_8O$	$(C_3F_7)_2C(OH)C_3H_7$	F: PerFHalOrg.4-93, 96
$C_{10}F_{14}H_9P$	$(n\text{-}C_3F_7)_2P(n\text{-}C_4H_9)$	F: PerFHalOrg.3-101, 105
$C_{10}F_{14}H_{10}Si$	$(C_3F_7)_2Si(C_2H_5)_2$	F: PerFHalOrg.4-19
$C_{10}F_{14}N_2$	$NCFC(C_6F_{11})CFCFN$	F: PerFHalOrg.6-18/9, 39
$C_{10}F_{14}N_4$	$(i\text{-}C_3F_7)_2(NC)C_3N_3$	F: PerFHalOrg.6-79/80, 101
$C_{10}F_{14}N_4O_2$	$C_2F_4(COC(C_2F_5)NN)_2$	F: PerFHalOrg.5-69/70, 90
–	$NC(C_3F_7)NOCCONC(C_3F_7)N$	F: PerFHalOrg.5-69/71, 91
–	$NNC(C_3F_7)OCCOC(C_3F_7)NN$	F: PerFHalOrg.5-69/70, 90

Formula	Compound		Reference
$C_{10}F_{15}HN_2O$	$NCFC(i-C_3F_7)C(i-C_3F_7)C(O)NH$	F:	PerFHalOrg.6-17, 37
$C_{10}F_{15}H_2N_3$	$NC(i-C_3F_7)CFNC(i-C_3F_7)C(NH_2)$	F:	PerFHalOrg.6-30/2, 51/2
–	$NC(i-C_3F_7)C(NH_2)C(i-C_3F_7)CFN$	F:	PerFHalOrg.6-17, 37
–	$NC(i-C_3F_7)NC(NH_2)C(i-C_3F_7)CF$	F:	PerFHalOrg.6-27, 48
–	$NC(NH_2)C(i-C_3F_7)C(i-C_3F_7)CFN$	F:	PerFHalOrg.6-17, 37
$C_{10}F_{15}H_3N_4O_2$	$C_3F_7C_2N_2O(CF_2)_4C(NOH)NH_2$	F:	PerFHalOrg.7-5/6, 33
$C_{10}F_{15}H_4N_5$	$(C_7F_{15})(NH_2)_2C_3N_3$	F:	PerFHalOrg.6-79/80, 101
$C_{10}F_{15}H_4N_5O$	$(C_2F_5)_2(OH)C_3N_3 \cdot NH_2C(NH)C_2F_5$	F:	PerFHalOrg.6-79/80, 101
$C_{10}F_{15}H_9OSi$	$(CH_3)_3Si(CF_2)_4OCF(CF_3)_2$	F:	PerFHalOrg.4-86
$C_{10}F_{15}H_{10}N_5O_5$	$[CF_3N(O)CH_2]_5$	F:	PerFHalOrg.7-164, 170
$C_{10}F_{15}H_{28}N_6O_5U$	$U(OCH_2CF_3)_5 \cdot 6\ NH_3$	U:	SVol.E1-13, 16
$C_{10}F_{15}H_{31}N_7O_5U$	$U(OCH_2CF_3)_5 \cdot 7\ NH_3$	U:	SVol.E1-13, 16
$C_{10}F_{15}H_{34}N_8O_5U$	$U(OCH_2CF_3)_5 \cdot 8\ NH_3$	U:	SVol.E1-13, 16
$C_{10}F_{15}H_{37}N_9O_5U$	$U(OCH_2CF_3)_5 \cdot 9\ NH_3$	U:	SVol.E1-13, 16
$C_{10}F_{15}H_{40}N_{10}O_5U$	$U(OCH_2CF_3)_5 \cdot 10\ NH_3$	U:	SVol.E1-13, 16
$C_{10}F_{15}H_{43}N_{11}O_5U$	$U(OCH_2CF_3)_5 \cdot 11\ NH_3$	U:	SVol.E1-13, 16
$C_{10}F_{15}H_{46}N_{12}O_5U$	$U(OCH_2CF_3)_5 \cdot 12\ NH_3$	U:	SVol.E1-13, 16
$C_{10}F_{15}HgN_3$	$[HgC_2F_3(CF_3)C_3N_3(CF_3)CF_2CF(CF_3)]_n$	F:	PerFHalOrg.6-85/6
$C_{10}F_{15}I_2N_3$	$(CF_2IC_2F_4)_2(CF_3)C_3N_3$	F:	PerFHalOrg.6-86/8, 107
–	$(CF_3CFICF_2)_2(CF_3)C_3N_3$	F:	PerFHalOrg.6-84/6, 105
$C_{10}F_{15}N$	$[(CF_3)_2CF](C_2F_5)C_5F_3N$	F:	PerFHalOrg.5-153/4, 165
$C_{10}F_{15}NO_5$	$[N(CF_3)OC_4F_7(OCF_3)C_4F_2O_3]_n$	F:	PerFHalOrg.7-210
$C_{10}F_{16}H_4N_2O_4$	$NH_2COCF(CF_3)O(CF_2)_5OCF_2CONH_2$	F:	PerFHalOrg.7-19/20, 48
$C_{10}F_{16}H_6N_4O_2$	$NH_2(NH)CCF(CF_3)O(CF_2)_5(OCF_2)C(NH)NH_2$	F:	PerFHalOrg.7-5
–	$NH_2(NOH)C(CF_2)_8C(NOH)NH_2$	F:	PerFHalOrg.7-5/6, 32
$C_{10}F_{16}N_2$	$(i-C_3F_7)_2C_4F_2N_2$	F:	PerFHalOrg.6-32/3, 53/6
–	$NC(C_2F_5)C(C_2F_5)C(C_2F_5)CFN$	F:	PerFHalOrg.6-15/6, 35
–	$NC(i-C_3F_7)CFC(i-C_3F_7)CFN$	F:	PerFHalOrg.6-16, 36/7, 55/6
–	$NC(i-C_3F_7)CFNC(i-C_3F_7)CF$	F:	PerFHalOrg.6-30/2, 51
–	$NC(i-C_3F_7)CFNCFC(i-C_3F_7)$	F:	PerFHalOrg.6-30/2
–	$NCFC(i-C_3F_7)C(i-C_3F_7)CFN$	F:	PerFHalOrg.6-16, 36, 55/6
–	$NCFNC(i-C_3F_7)C(i-C_3F_7)CF$	F:	PerFHalOrg.6-23, 45
–	$NCFNC(i-C_3F_7)CFC(i-C_3F_7)$	F:	PerFHalOrg.6-23, 44/5
$C_{10}F_{17}HN_2O_3$	$HOC_6F_5[ON(CF_3)_2]_2$	F:	PerFHalOrg.7-98/9, 116
$C_{10}F_{17}H_2N_3O_2$	$NH_2C_6F_5[ON(CF_3)_2]_2$	F:	PerFHalOrg.7-98/9, 116
$C_{10}F_{17}H_4NO_3$	$(CF_3)_2CFO(CF_2)_5C(O)NHCH_2OH$	F:	PerFHalOrg.7-63
$C_{10}F_{17}N_3$	$(C_3F_7)_2(CF_3)C_3N_3$	F:	PerFHalOrg.6-83/4, 104
$C_{10}F_{17}N_3O$	$(C_2F_5)_2(CF_3OC_2F_4)C_3N_3$	F:	PerFHalOrg.6-86/8, 105
$C_{10}F_{17}N_3O_2$	$(CF_3OC_2F_4)_2(CF_3)C_3N_3$	F:	PerFHalOrg.6-86/8, 105
$C_{10}F_{18}GeH_3N_3$	$CH_3Ge[NC(CF_3)_2]_3$	F:	PerFHalOrg.7-64
$C_{10}F_{18}H_3N_3O_2$	$C_2F_5O(CF_2)_2C(NH)NC(NH_2)(CF_2)_2OC_2F_5$	F:	PerFHalOrg.7-7, 35
$C_{10}F_{18}H_3N_3Si$	$CH_3Si[NC(CF_3)_2]_3$	F:	PerFHalOrg.7-64
$C_{10}F_{18}H_3N_3Sn$	$CH_3Sn[NC(CF_3)_2]_3$	F:	PerFHalOrg.7-64
		Sn:	Org.Verb.6-219/20
$C_{10}F_{18}H_4N_2O_2$	$CF_3(CF_2)_4C(NH)NH_2 \cdot C_3F_7COOH$	F:	PerFHalOrg.7-3/4, 31
$C_{10}F_{18}H_4N_4$	$C(CF_3)_2NHC(CF_3)_2NCNHC(CF_3)_2NH_2$	F:	PerFHalOrg.5-63/4, 86
$C_{10}F_{18}N_2O$	$NC(C_4F_9)OC(C_4F_9)N$	F:	PerFHalOrg.5-64/5, 87
$C_{10}F_{18}N_2O_2$	$C(O)FC_2F_4N(CF_2CF_2)_2NC_2F_4C(O)F$	F:	PerFHalOrg.6-31/2, 52

Formula	Compound		Reference
$C_{10}F_{18}N_2O_2$	$C_6F_6[ON(CF_3)_2]_2$	F:	PerFHalOrg.7-98/9, 116, 123
$C_{10}F_{19}HN_2O_2$	$C_3F_7N(CF_2CF_2)_2NC_2F_4COOH$	F:	PerFHalOrg.6-31/2, 52
$C_{10}F_{19}H_2NO$	$C_9F_{19}C(O)NH_2$	F:	PerFHalOrg.7-18, 20/1, 47
$C_{10}F_{19}H_3N_2$	$CF_3(CF_2)_8C(NH)NH_2$	F:	PerFHalOrg.7-3/4, 31
$C_{10}F_{19}N$	$C_5F_{11}NC_5F_8$	F:	PerFHalOrg.5-158, 172
$C_{10}F_{20}N_2$	$(CF_2)_5NN(CF_2)_5$	F:	PerFHalOrg.5-220/2
$C_{10}F_{20}N_2O_4$	$(O(CF_2CF_2)_2NO)_2C_2F_4$	F:	PerFHalOrg.6-2/3, 8
$C_{10}F_{21}HN_4O$	$(CF_3)_2NCHON(CF_3)C[C(N(CF_3)_2)_2]$	F:	PerFHalOrg.7-166
$C_{10}F_{21}HO$	$(C_3F_7)_3COH$	F:	PerFHalOrg.4-41, 94
$C_{10}F_{21}H_2O_3P$	$C_{10}F_{21}P(O)(OH)_2$	F:	PerFHalOrg.3-43, 50
$C_{10}F_{21}H_3Si$	$(C_3F_7)_3SiCH_3$	F:	PerFHalOrg.4-18
$C_{10}F_{21}IMg$	$(n\text{-}C_{10}F_{21})MgI$	F:	PerFHalOrg.4-68
$C_{10}F_{21}I_2P$	$C_{10}F_{21}PI_2$	F:	PerFHalOrg.3-76, 95
$C_{10}F_{21}N$	$C_5F_{11}N(CF_2)_5$	F:	PerFHalOrg.5-157/8, 171/2, 175
$C_{10}F_{21}NO$	$CF_3(CF_2)_9NO$	F:	PerFHalOrg.7-179, 185
–	$[N(C_8F_{17})OCF_2CF_2]_n$	F:	PerFHalOrg.7-208/9
$C_{10}F_{21}NO_2$	$C_8F_{17}OCF_2CF_2NO$	F:	PerFHalOrg.7-179, 185
$C_{10}F_{22}N_2$	$(i\text{-}C_3F_7)N(CF_2CF_2)_2N(C_3F_7\text{-}i)$	F:	PerFHalOrg.6-31/2, 52
$C_{10}F_{22}N_2O_2$	$C_6F_{10}(ON(CF_3)_2)_2$	F:	PerFHalOrg.7-98
$C_{10}F_{24}N_2NiO_2P_2$	$(CO)_2Ni(P(CF_3)_2N(CF_3)_2)_2$	Ni:	Org.Verb.1-238
$C_{10}F_{25}P_5$	$(C_2F_5P)_5$	F:	PerFHalOrg.3-1/8
$C_{10}F_{27}NiOP_3$	$CONi(P(CF_3)_3)_3$	Ni:	Org.Verb.1-114
$C_{10}F_{30}N_5O_5P$	$[(CF_3)_2NO]_5P$	F:	PerFHalOrg.7-101, 119
$C_{10}FeH_4NNaO_5$	$Na[NC_5H_4C(O)Fe(CO)_4]$	Fe:	Org.Verb.B4-155, 165
$C_{10}FeH_4NO_5^-$	$[NC_5H_4C(O)Fe(CO)_4]^-$	Fe:	Org.Verb.B4-155, 165
$C_{10}FeH_4O_7$	$C_5H_2O_3CH_2Fe(CO)_4$	Fe:	Org.Verb.B4-259, 263
$C_{10}FeH_5I_3O_4Sn$	$(CO)_4Fe(Sn(C_6H_5)I_2)I$	Fe:	Org.Verb.B2-139
$C_{10}FeH_5O_4^-$	$[C_6H_5C(O)Fe(CO)_3]^-$	Fe:	Org.Verb.B4-109
$C_{10}FeH_6LiNO_5$	$Li[NC_5H_6C(O)Fe(CO)_4]$	Fe:	Org.Verb.B4-157, 168
$C_{10}FeH_6NO_5^-$	$[NC_5H_6C(O)Fe(CO)_4]^-$	Fe:	Org.Verb.B4-157, 168
$C_{10}FeH_6N_4O_4$	$Fe(OC(OCH_3)C(CN)_2)_2$	Fe:	Org.Verb.B3-189
$C_{10}FeH_6O_4$	$C_5H_5CHFe(CO)_4$	Fe:	Org.Verb.B4-119
–	1) $C_6H_6Fe(CO)_4$	Fe:	Org.Verb.B4-259, 264/5
	2) $C_6H_6Fe(CO)_4$	Fe:	Org.Verb.B4-303, 306
$C_{10}FeH_6O_6$	$(C(CH_3)CO)_2Fe(CO)_4$	Fe:	Org.Verb.B4-334, 341
–	$(C(O)C(C_2H_5)CHCO)Fe(CO)_4$	Fe:	Org.Verb.B4-334
$C_{10}FeH_7LiO_3$	$Li[C_7H_7Fe(CO)_3]$	Fe:	Org.Verb.B5-77
$C_{10}FeH_7NO_4$	$(CO)_4Fe(NC_5H_4CH_3)$	Fe:	Org.Verb.B2-71, 75
$C_{10}FeH_7NO_5$	$(CO)_3Fe(ONC_6H_4OCH_3)$	Fe:	Org.Verb.B1-170/1, 175, 179
$C_{10}FeH_7NaO_3$	$Na[C_7H_7Fe(CO)_3]$	Fe:	Org.Verb.B5-77/8
$C_{10}FeH_7O_3^-$	$[C_7H_7Fe(CO)_3]^-$	Fe:	Org.Verb.B5-74, 77/9
$C_{10}FeH_7O_4P$	$(CO)_4FeP(C_6H_5)H_2$	Fe:	Org.Verb.B2-88, 108
$C_{10}FeH_7O_5^-$	$[(CHO)(CH_2CHCO)C_3H_3Fe(CO)_3]^-$	Fe:	Org.Verb.B5-74
$C_{10}FeH_7O_8^-$	$[(CH_3OCO)_2C_3H(O)Fe(CO)_3]^-$	Fe:	Org.Verb.B5-72, 75, 79
$C_{10}FeH_8^+$	$[FeC_{10}H_8]^+$	Fe:	Org.Verb.A1-93/4
$C_{10}FeH_8I_2$	$C_5H_5FeC_5H_3I_2$	Fe:	Org.Verb.A6-212

Formula	Compound	Reference
$C_{10}FeH_8Li_2$	$C_5H_5FeC_5H_3Li_2$	Fe: Org.Verb.A6-212
–	$Fe(C_5H_4Li)_2$	Fe: Org.Verb.A1-172
$C_{10}FeH_8N_2O_4$	$(CO)_4Fe(N_2C_4H_2(CH_3)_2)$	Fe: Org.Verb.B2-71, 76, 79
$C_{10}FeH_8N_2O_6$	$(N(CH_3)C(O)CH)_2Fe(CO)_4$	Fe: Org.Verb.B4-305, 312
$C_{10}FeH_8N_2O_8$	$HOC(O)C_2H_2C(O)N_2H_2C(O)CH_3Fe(CO)_4$	Fe: Org.Verb.B4-274
$C_{10}FeH_8N_4O_2$	$Fe(NO)_2(NC_5H_4C_5H_4N)$	Fe: Org.Verb.B1-142
$C_{10}FeH_8Na_2$	$Fe(C_5H_4Na)_2$	Fe: Org.Verb.A1-173/4
$C_{10}FeH_8O_3$	$C_7H_8Fe(CO)_3$	Fe: Org.Verb.B5-79
$C_{10}FeH_8O_4$	$CH_2(CH)_3CHCH_2Fe(CO)_4$	Fe: Org.Verb.B4-238, 252
–	$C_6H_8Fe(CO)_4$	Fe: Org.Verb.B4-304, 310
$C_{10}FeH_8O_5S$	$(CO)_3FeOS(O)C_7H_8$	Fe: Org.Verb.B5-91, 93
$C_{10}FeH_8O_6$	$(CH_3COO)(CHO)C_4H_4Fe(CO)_3$	Fe: Org.Verb.B5-76
$C_{10}FeH_8O_8$	$(CH_3OCO)_2C_2H_2Fe(CO)_4$	Fe: Org.Verb.B4-270, 284/6
$C_{10}FeH_9$	$C_5H_5FeC_5H_4$	Fe: Org.Verb.A1-232
$C_{10}FeH_9^+$	$[FeC_{10}H_9]^+$	Fe: Org.Verb.A1-93/4
$C_{10}FeH_9^-$	$[C_5H_5FeC_5H_4]^-$	Fe: Org.Verb.A1-230
$C_{10}FeH_9I$	$C_5H_5FeC_5H_4I$	Fe: Org.Verb.A1-371/80
$C_{10}FeH_9I^+$	$[C_5H_5FeC_5H_4I]^+$	Fe: Org.Verb.A1-271, 273, 380
$C_{10}FeH_9I_3$	$C_5H_5FeC_5H_4I \cdot I_2$	Fe: Org.Verb.A1-370
$C_{10}FeH_9I_{2x+1}$	$C_5H_5FeC_5H_4I \cdot n\ I_2$	Fe: Org.Verb.A1-370, 374
$C_{10}FeH_9KO_6$	$K[C_3H_5CH(OH)CH_2C(O)Fe(CO)_4]$	Fe: Org.Verb.B4-154, 162
$C_{10}FeH_9KO_7$	$K[C_2H_5OC(O)CH(CH_3)C(O)Fe(CO)_4]$	Fe: Org.Verb.B4-154, 163
–	$K[C_2H_5OC(O)CH_2CH_2C(O)Fe(CO)_4]$	Fe: Org.Verb.B4-154, 163
$C_{10}FeH_9Li$	$C_5H_5FeC_5H_4Li$	Fe: Org.Verb.A1-172, 374
$C_{10}FeH_9LiO_5S_2$	$Li[(CH_2S)_2C(C_2H_5)C(O)Fe(CO)_4]$	Fe: Org.Verb.B4-157, 167/8
$C_{10}FeH_9N_3$	$C_5H_5FeC_5H_4N_3$	Fe: Org.Verb.A1-375
$C_{10}FeH_9Na$	$C_5H_5FeC_5H_4Na$	Fe: Org.Verb.A1-174
$C_{10}FeH_9O_5S_2^-$	$[(CH_2S)_2C(C_2H_5)C(O)Fe(CO)_4]^-$	Fe: Org.Verb.B4-157, 167/8
$C_{10}FeH_9O_6^-$	$[C_3H_5CH(OH)CH_2C(O)Fe(CO)_4]^-$	Fe: Org.Verb.B4-154, 162
$C_{10}FeH_9O_7^-$	$[C_2H_5OC(O)C_2H_4C(O)Fe(CO)_4]^-$	Fe: Org.Verb.B4-154, 163
$C_{10}FeH_9O_7P$	$(CO)_4FeP(OCHCH_2)_3$	Fe: Org.Verb.B2-96, 117
$C_{10}FeH_9W^+$	$[C_5H_5FeC_5H_4W]^+$	Fe: Org.Verb.A1-347
$C_{10}FeH_{10}$	$Fe(C_5H_5)_2$	
	Absorption of electromagnetic radiation	Fe: Org.Verb.A1-40/51
	Basicity	Fe: Org.Verb.A1-96/8
	Catalysis	Fe: Org.Verb.A1-190/2
	Chemical reactions	
	Basicity	Fe: Org.Verb.A1-96/7
	Complex formation	Fe: Org.Verb.A1-98/102
	General	Fe: Org.Verb.A1-90/1
	Oxidation to ferrocenium ion	Fe: Org.Verb.A1-102/11
	Physical effects	Fe: Org.Verb.A1-91/6
	Protonation	Fe: Org.Verb.A1-96/7
	Reactions with halogens and halogenating agents	Fe: Org.Verb.A1-104/5, 180/90
	Reactions with pseudohalogenating agents	Fe: Org.Verb.A1-186/7

$C_{10}FeH_{10}$ .... $Fe(C_5H_5)_2$
Chemical reactions
Reactions with rupture of
iron-cyclopentadienyl bond ... Fe: Org.Verb.A1-112/9
Substitution reactions
in cyclopentadienyl rings ..... Fe: Org.Verb.A1-119/80
Complex formation .... Fe: Org.Verb.A1-98/102
Crystallographic properties .... Fe: Org.Verb.A1-54/6
Decomposition .... Fe: Org.Verb.A1-91/6
Electrical properties .... Fe: Org.Verb.A1-37/9
Electrochemical behavior .... Fe: Org.Verb.A1-80/90
Formation .... Fe: Org.Verb.A1-15, 30/1
History .... Fe: Org.Verb.A1-5
Hydration .... Fe: Org.Verb.A1-113/4
Inhibition .... Fe: Org.Verb.A1-190/2
Ionization .... Fe: Org.Verb.A1-61/2
Isotope exchange .... Fe: Org.Verb.A1-29/30, 124/6
Magnetic properties .... Fe: Org.Verb.A1-37/9
Mechanical properties .... Fe: Org.Verb.A1-31/2
Molecule .... Fe: Org.Verb.A1-56/76
Nomenclature .... Fe: Org.Verb.A1-6/7
Nuclear magnetic resonance ..... Fe: Org.Verb.A1-40
Optical properties .... Fe: Org.Verb.A1-39/54
Physiological behavior .... Fe: Org.Verb.A1-192
Preparation
in the laboratory .... Fe: Org.Verb.A1-15/28
of special forms .... Fe: Org.Verb.A1-28/30
Protonation .... Fe: Org.Verb.A1-96/8
Purification .... Fe: Org.Verb.A1-28/9
Separation methods
analytical .... Fe: Org.Verb.A1-11/2
preparative .... Fe: Org.Verb.A1-11/2, 28
Solubility .... Fe: Org.Verb.A1-76/80
Solvolysis .... Fe: Org.Verb.A1-114/7
Thermal properties .... Fe: Org.Verb.A1-31/7
Uses .... Fe: Org.Verb.A1-193/5

– .... $Fe(C_5D_5)_2$
Absorption .... Fe: Org.Verb.A1-42/3
Crystallographic properties .... Fe: Org.Verb.A1-55/6
Molecule .... Fe: Org.Verb.A1-71/5
Preparation .... Fe: Org.Verb.A1-29/30
Raman scattering .... Fe: Org.Verb.A1-51/2

– .... $Fe(C_5H_5)_2$ solutions
$Fe(C_5H_5)_2$-H_2O .... Fe: Org.Verb.A1-76/7
$Fe(C_5H_5)_2$-organic solvents .... Fe: Org.Verb.A1-77/80, 187/90

$C_{10}FeH_{10}^+$ .... 1) $[Fe(C_5H_5)_2]^+$ .... Fe: Org.Verb.A1-230/1

$C_{10}FeH_{10}^+$ .... 2) $[Fe(C_5H_5)_2]^+$
Chemical reactions .... Fe: Org.Verb.A1-204/13
Diffusion .... Fe: Org.Verb.A1-204
Electrochemical behavior .... Fe: Org.Verb.A1-80/9, 204
Formation .... Fe: Org.Verb.A1-91/6, 102/13, 123, 135/6, 180/90, 196/7
Molecule .... Fe: Org.Verb.A1-197/204

– .... $[(C_5D_5)_2Fe]^+$ .... Fe: Org.Verb.A1-203
$C_{10}FeH_{10}^{2+}$ .... $[Fe(C_5H_5)_2]^{2+}$ .... Fe: Org.Verb.A1-93
$C_{10}FeH_{10}^-$ .... $[Fe(C_5H_5)_2]^-$ .... Fe: Org.Verb.A1-93/4, 231
$C_{10}FeH_{10}I$ .... $[Fe(C_5H_5)_2]I$ .... Fe: Org.Verb.A1-216
$C_{10}FeH_{10}I_2$ .... $Fe(C_5H_5)_2 \cdot I_2$ .... Fe: Org.Verb.A1-184, 216
$C_{10}FeH_{10}I_2O$ .... $(C_6H_5CHCHC(CH_3)O)FeI_2$ .... Fe: Org.Verb.B5-143
$C_{10}FeH_{10}I_3$ .... $[Fe(C_5H_5)_2]I_3$ .... Fe: Org.Verb.A1-216/7
– .... $[Fe(C_5H_5)_2I_3]$ .... Fe: Org.Verb.A1-206
$C_{10}FeH_{10}I_4$ .... $[Fe(C_5H_5)_2]I_4$ .... Fe: Org.Verb.A1-217
$C_{10}FeH_{10}I_5$ .... $[Fe(C_5H_5)_2]I_5$ .... Fe: Org.Verb.A1-217
$C_{10}FeH_{10}I_6$ .... $[Fe(C_5H_5)_2]I_6$ .... Fe: Org.Verb.A1-217
$C_{10}FeH_{10}I_{20}$ .... $[Fe(C_5H_5)_2]I_{20}$ .... Fe: Org.Verb.A1-217
$C_{10}FeH_{10}NO_3$ .... $[Fe(C_5H_5)_2]NO_3$ .... Fe: Org.Verb.A1-219
$C_{10}FeH_{10}NO_5^-$ ...... $[(CH_2)_5NC(O)Fe(CO)_4]^-$ .... Fe: Org.Verb.B4-156, 165
$C_{10}FeH_{10}N_2O_4$ ...... $(CO)Fe(NO)_2C(OC_2H_5)C_6H_5$ .... Fe: Org.Verb.B1-143
– .... $(CO)_4Fe(N_2C_6H_{10})$ .... Fe: Org.Verb.B2-71, 76
– .... $(C_6H_5)(C_2H_5O)CFe(NO)_2CO$ .... Fe: Org.Verb.B4-109/11
$C_{10}FeH_{10}N_4O_2$ ...... $Fe(NO)_2(NC_5H_5)_2$ .... Fe: Org.Verb.C1-63
$C_{10}FeH_{10}Ni^+$ .... $(C_5H_5)_2NiFe^+$ .... Ni: Org.Verb.2-168, 170, 351
$C_{10}FeH_{10}Ni_2^+$ ...... $(C_5H_5)_2Ni_2Fe^+$ .... Ni: Org.Verb.2-351
$C_{10}FeH_{10}O$ .... $C_5H_5FeC_5H_4OH$ .... Fe: Org.Verb.A2-5
$C_{10}FeH_{10}O_2S_3$ ...... $(C_5H_5)Fe(CO)_2[S_2CSC_2H_5]$ .... C: MVol.D4-269
$C_{10}FeH_{10}O_3$ .... $C_7H_{10}Fe(CO)_3$ .... Fe: Org.Verb.B5-144
$C_{10}FeH_{10}O_4$ .... $(CH_3)_2CCHCHCH_2Fe(CO)_4$ .... Fe: Org.Verb.B4-239, 254
– .... $C_6H_{10}Fe(CO)_4$ .... Fe: Org.Verb.B4-305, 312
$C_{10}FeH_{10}O_6$ .... $(CH_3O)(CH_3OCO)C_4H_4Fe(CO)_3$ .... Fe: Org.Verb.B5-76
$C_{10}FeH_{11}$ .... $HFe(C_5H_5)_2$ .... Fe: Org.Verb.A1-232
$C_{10}FeH_{11}^+$ .... $[HFe(C_5H_5)_2]^+$ .... Fe: Org.Verb.A1-94, 96/7, 102, 126, 231
$C_{10}FeH_{11}LiO_5$ ...... $Li[C_5H_{11}C(O)Fe(CO)_4]$ .... Fe: Org.Verb.B4-150, 160
$C_{10}FeH_{11}NO_4$ ...... $C_6H_8(COCH_3)Fe(CO)_2NO$ .... Fe: Org.Verb.B5-14/5, 18, 23
$C_{10}FeH_{11}NO_5$ ...... $(CO)_3Fe(C_2H_4CN)C_2H_4COOCH_3$ .... Fe: Org.Verb.B1-205
$C_{10}FeH_{11}N_2O_4$ ...... $HFe(C_5H_5)_2(NO_2)_2 \cdot 2\ H_2O$ .... Fe: Org.Verb.A1-219
$C_{10}FeH_{11}NaO_5$ ..... $Na[C_5H_{11}C(O)Fe(CO)_4]$ .... Fe: Org.Verb.B4-150, 160
$C_{10}FeH_{11}NiO_3P^+$ .... $NiP(CH_3)_2Fe(C_5H_5)(CO)_3^+$ .... Ni: Org.Verb.1-167
$C_{10}FeH_{11}O$ .... $[Fe(C_5H_5)_2OH]$ .... Fe: Org.Verb.A1-206, 214
$C_{10}FeH_{11}O_2P$ .... $C_5H_5FeC_5H_4PH(O)OH$ .... Fe: Org.Verb.A1-170
$C_{10}FeH_{11}O_3$ .... $C_3H_5Fe(CO)_3C_4H_6$ .... Fe: Org.Verb.C2-169
$C_{10}FeH_{11}O_4^+$ .... $[(CH_3)_3C_3H_2Fe(CO)_4]^+$ .... Fe: Org.Verb.B5-101
$C_{10}FeH_{11}O_5^-$ .... $[C_5H_{11}C(O)Fe(CO)_4]^-$ .... Fe: Org.Verb.B4-150, 160

Formula	Compound	Reference
$C_{10}FeH_{11}O_7P$	$(CO)_4FeP(OCH_2)_3CC_2H_5$	Fe: Org.Verb.B2-95, 117
$C_{10}FeH_{12}NO_5^+$	$[C_2H_5O(NHCH_3)CCHCH_2Fe(CO)_4]^+$	Fe: Org.Verb.B4-248/9
$C_{10}FeH_{12}N_2O_4$	$CH_2CHC(N(CH_3)_2)N(CH_3)COFe(CO)_3$	Fe: Org.Verb.B5-159/60
$C_{10}FeH_{12}N_3O_4P$	$(CO)_4FeP(CH_2NCH_2)_3$	Fe: Org.Verb.B2-86
$C_{10}FeH_{12}N_6$	$(CH_3NC)_4Fe(CN)_2$	Fe: Org.Verb.B4-48/51, 56/8
–	$(CH_3NC)_4Fe(CN)_2 \cdot H_2O$	Fe: Org.Verb.B4-67
–	$(CH_3NC)_4Fe(CN)_2 \cdot 2\,H_2O$	Fe: Org.Verb.B4-67
–	$[(C_2H_5NC)_2Fe(CN)_4]H_2$	Fe: Org.Verb.B4-28
$C_{10}FeH_{12}Na_2O_7$	$Na_2[Fe(CO)_4] \cdot 1.5\,C_4H_8O_2$	Fe: Org.Verb.B2-16, 20/1, 29
$C_{10}FeH_{12}O_4$	$(CH_2C(CH_3)C(C_3H_7)O)Fe(CO)_3$	Fe: Org.Verb.B5-131, 141
–	$CH_3(CH_2)_3CHCH_2Fe(CO)_4$	Fe: Org.Verb.B4-233, 237, 250
$C_{10}FeH_{12}O_8$	$(CH_3O)_2CC(OCH_3)_2Fe(CO)_4$	Fe: Org.Verb.B4-321, 325
$C_{10}FeH_{13}NO_4$	$CH_3COCH_2(CH_3)_2C_3H_2Fe(CO)_2NO$	Fe: Org.Verb.B5-14/5, 18, 23
$C_{10}FeH_{13}O_9P$	$HOC(O)CHCH_2Fe(CO)_4 \cdot P(OCH_3)_3$	Fe: Org.Verb.B4-245/6
$C_{10}FeH_{13}O_{10}P$	$HOC(O)CHCH_2Fe(CO)_4 \cdot OP(OCH_3)_3$	Fe: Org.Verb.B4-245/6
$C_{10}FeH_{14}N_2O_2S_4$	$Fe(CH(CSCH_3)_2)_2(NO)_2$	Fe: Org.Verb.B1-115
$C_{10}FeH_{14}O_2$	$(CH_3C_3H_4)_2Fe(CO)_2$	Fe: Org.Verb.B5-107
$C_{10}FeH_{14}O_2^+$	$[Fe(C_5H_5)_2(H_2O)_2]^+$	Fe: Org.Verb.A1-205, 206
$C_{10}FeH_{14}O_4$	$Fe(CH(COCH_3)_2)_2$	Fe: Org.Verb.B3-181
$C_{10}FeH_{14}O_4Si$	$(CO)_4Fe(CH_2CH_2CH(CH_3)Si(CH_3)_2)$	Fe: Org.Verb.C2-52
$C_{10}FeH_{15}NO_4$	$CH_3(C(CH_3)_2OHCH_2)C_3H_3Fe(CO)_2NO$	Fe: Org.Verb.B5-14/5, 17, 23
–	$(CO)_4FeN(C_2H_5)_3$	Fe: Org.Verb.B2-71/3
$C_{10}FeH_{15}NO_5$	$[N(CH_3)_4][CH_3C(O)Fe(CO)_4]$	Fe: Org.Verb.B4-149, 159
$C_{10}FeH_{15}NO_5Sn$	$((CH_3)_2N)((CH_3)_3SnO)CFe(CO)_4$	Fe: Org.Verb.B4-122, 124, 126
$C_{10}FeH_{15}O_4P$	$(CO)_4FeP(C_2H_5)_3$	Fe: Org.Verb.B2-87, 104
$C_{10}FeH_{15}O_7P$	$(CO)_4FeP(OC_2H_5)_3$	Fe: Org.Verb.B2-84/5, 95, 116
$C_{10}FeH_{15}O_7PS_2$	$((CH_3O)_3P)(CO)_2FeS_2C_2HCOOC_2H_5$	Fe: Org.Verb.B4-138
$C_{10}FeH_{16}O_4Si_2$	$(CO)_4Fe(Si(CH_3)_2CH_2CH_2Si(CH_3)_2)$	Fe: Org.Verb.B2-144, 150
$C_{10}FeH_{18}N_3O_4P$	$(CO)_4FeP(N(CH_3)_2)_3$	Fe: Org.Verb.B2-94, 114/5
$C_{10}FeH_{18}O_3Si_2$	$(CH_3)_2SiCHCH_2Fe(CO)_3Si(CH_3)_3$	Fe: Org.Verb.B4-211, 217
$C_{10}FeH_{18}O_4Si_2$	$(CO)_4Fe(Si(CH_3)_3)_2$	Fe: Org.Verb.B2-139, 147
$C_{10}FeH_{18}O_4Sn_2$	$(CO)_4Fe(Sn(CH_3)_3)_2$	Fe: Org.Verb.B2-136, 141, 148
$C_{10}FeH_{20}NO_5P$	$C_3H_5Fe(CO)(NO)P(OC_2H_5)_3$	Fe: Org.Verb.B5-6/9, 20/1
$C_{10}FeH_{20}N_3OS_4$	$(NO)Fe(S_2CN(C_2H_5)_2)_2$	Fe: Org.Verb.C1-85
$C_{10}FeH_{20}N_6O_2$	$((CH_3NCH_2)_2C)_2Fe(NO)_2$	Fe: Org.Verb.B4-109/10, 112
$C_{10}FeH_{21}IO_3P_2$	$(CO)_2Fe(COCH_3)(P(CH_3)_3)_2I$	Fe: Org.Verb.B1-121/2, 125
$C_{10}FeH_{21}I_3O_3P_2$	$[(CO)_3Fe(P(CH_3)_3)_2CH_3]I_3$	Fe: Org.Verb.B1-205
$C_{10}FeH_{22}N_2O_4Si$	$[SiH_3 \cdot 2\,N(CH_3)_3][HFe(CO)_4]$	Fe: Org.Verb.B2-20
$C_{10}FeH_{24}N_2O_4Si_2$	$(CO)_4Fe(SiH_3 \cdot N(CH_3)_3)_2$	Fe: Org.Verb.B2-136
$C_{10}FeH_{24}O_2P_2$	$(CO)_2Fe(CH_3)_2(P(CH_3)_3)_2$	Fe: Org.Verb.B1-121, 124, 128
$C_{10}FeH_{25}N_5O_5$	$(CO)_5Fe(NH_2CH_3)_5$	Fe: Org.Verb.B3-249
$C_{10}FeH_{27}I_2OP_3$	$COFe(P(CH_3)_3)_3I_2$	Fe: Org.Verb.B1-32
$C_{10}FeH_{27}NO_{11}P_3^+$	$[COFe(NO)(P(OCH_3)_3)_3]^+$	Fe: Org.Verb.B1-38/9
$C_{10}FeH_{28}IOP_3$	$COFeH(P(CH_3)_3)_3I$	Fe: Org.Verb.B1-32
$C_{10}FeH_{30}I_2O_5S_5$	$Fe(OS(CH_3)_2)_5I_2$	Fe: Org.Verb.B5-58

Formula	Compound	Reference
$C_{10}FeH_{30}I_3O_5S_5$	$Fe(OS(CH_3)_2)_5I_3$	Fe: Org.Verb.B5-58
$C_{10}FeH_{30}N_5P_2^+$	$[FeP_2(N(CH_3)_2)_5]^+$	Fe: Org.Verb.B1-164
$C_{10}Fe_2GeH_6O_8$	$(CO)_4FeGe(CH_3)_2Fe(CO)_4$	Fe: Org.Verb.B2-28
$C_{10}Fe_2GeH_{18}I_2N_4O_4$	$((CH_3)_2C_3H_3Fe(NO)_2)_2GeI_2$	Fe: Org.Verb.C2-156, 165
$C_{10}Fe_2H_2O_{14}S$	$[(CO)_5FeH]_2SO_4$	Fe: Org.Verb.B3-251
$C_{10}Fe_2H_4N_2O_6$	$(CO)_6Fe_2(C_4H_4N_2)$	Fe: Org.Verb.B3-146
$C_{10}Fe_2H_4O_6S$	$C_4H_4SFe_2(CO)_6$	Fe: Org.Verb.C2-126
$C_{10}Fe_2H_4O_6Te$	$C_4H_4TeFe_2(CO)_6$	Fe: Org.Verb.C2-122, 129
$C_{10}Fe_2H_4O_7$	$C_3H_4Fe_2(CO)_7$	Fe: Org.Verb.C2-156/7
$C_{10}Fe_2H_5IO_7$	$(CO)_3Fe(CHCHCOCH_3)IFe(CO)_3$	Fe: Org.Verb.C2-79/80, 82, 84
$C_{10}Fe_2H_5IO_8$	$(CO)_3Fe(CHCHCOOCH_3)IFe(CO)_3$	Fe: Org.Verb.C2-79/80, 82, 84
$C_{10}Fe_2H_6N_2O_7$	$(CO)_3Fe(NC(CH_3)_2)(OCN)Fe(CO)_3$	Fe: Org.Verb.C1-133, 141
$C_{10}Fe_2H_6O_6S$	$(CO)_3Fe(CHCH_2)(SCHCH_2)Fe(CO)_3$	Fe: Org.Verb.C2-85/6
$C_{10}Fe_2H_6O_6S_2$	$(CO)_3FeSC(CH_3)C(CH_3)SFe(CO)_3$	Fe: Org.Verb.C1-115, 117
$C_{10}Fe_2H_6O_8$	$(CO)_3Fe(CH_3CO)_2Fe(CO)_3$	Fe: Org.Verb.C1-232/3, 235
$C_{10}Fe_2H_8N_2O_6$	$(CO)_3Fe(CH_3NNC(CH_3)CH_2)Fe(CO)_3$	Fe: Org.Verb.C1-149/51, 154/5
$C_{10}Fe_2H_8O_4S_2$	$(C_3H_4SFe(CO)_2)_2$	Fe: Org.Verb.C2-156, 168
$C_{10}Fe_2H_8O_5S_2$	$C_3H_4SOFe_2(CO)_4C_3H_4S$	Fe: Org.Verb.C2-169
$C_{10}Fe_2H_8O_6S$	$(CO)_3Fe(CHCH_2)(SC_2H_5)Fe(CO)_3$	Fe: Org.Verb.C2-85/6
$C_{10}Fe_2H_9NO_6S$	$(CO)_3Fe(C_4H_9NS)Fe(CO)_3$	Fe: Org.Verb.C1-185/6, 188
–	$(CO)_6Fe_2((CH_3)_2NC(CH_3)S)$	Fe: Org.Verb.C1-238/40
$C_{10}Fe_2H_{10}^+$	$[Fe_2(C_5H_5)_2]^+$	Fe: Org.Verb.A1-93/4
$C_{10}Fe_2H_{10}I_2O_6S_2$	$I(CO)_3Fe(SC_2H_5)_2Fe(CO)_3I$	Fe: Org.Verb.C1-6/7, 9, 12
$C_{10}Fe_2H_{10}I_4$	$[Fe(C_5H_5)_2]FeI_4$	Fe: Org.Verb.A1-223
$C_{10}Fe_2H_{10}N_2O_6$	$(CO)_3Fe(C_2H_5NNC_2H_5)Fe(CO)_3$	Fe: Org.Verb.C1-149/51, 153
–	$(CO)_3Fe(NCHCH_3)(C_2H_5NH)Fe(CO)_3$	Fe: Org.Verb.C1-133
$C_{10}Fe_2H_{10}N_2O_7$	$(CO)_6Fe_2(CH_3CHN)(C_2H_5NHO)$	Fe: Org.Verb.C1-191/2, 194
–	$(CO)_6Fe_2(C_2H_5NH)(CH_3CHNO)$	Fe: Org.Verb.C1-191/3
$C_{10}Fe_2H_{10}O_6S_2$	$(CO)_3Fe(SC_2H_5)_2Fe(CO)_3$	Fe: Org.Verb.C1-79/81, 86/7
$C_{10}Fe_2H_{10}O_6Se_2$	$(CO)_3Fe(SeC_2H_5)_2Fe(CO)_3$	Fe: Org.Verb.C1-83
$C_{10}Fe_2H_{11}O_6P$	$Fe_2(CO)_6(P(C_2H_5)_2)H$	Fe: Org.Verb.B2-109
$C_{10}Fe_2H_{11}O_7P$	$(CO)_3Fe(P(C_2H_5)_2)(OH)Fe(CO)_3$	Fe: Org.Verb.C1-185/7
$C_{10}Fe_2H_{12}I_2O_6P_2$	$I(CO)_3Fe(P(CH_3)_2)_2Fe(CO)_3I$	Fe: Org.Verb.C1-6/8, 10/1
$C_{10}Fe_2H_{12}O_6P_2$	$(CO)_3Fe(P(CH_3)_2)_2Fe(CO)_3$	Fe: Org.Verb.C1-164/6, 170/3
$C_{10}Fe_2H_{12}O_6P_2^{2-}$	$[Fe_2(CO)_6(P(CH_3)_2)_2]^{2-}$	Fe: Org.Verb.C1-171
$C_{10}Fe_2H_{12}O_6P_4$	$(CO)_3Fe(PCH_3)_4Fe(CO)_3$	Fe: Org.Verb.C1-164, 168, 176/7
$C_{10}Fe_2H_{12}O_{12}S_3$	$[Fe(C_5H_5)_2][Fe(SO_4)(HSO_4)_2]$	Fe: Org.Verb.A1-223
$C_{10}Fe_2H_{15}O_8PS_2$	$(CO)_3Fe(SCH_3)_2Fe(CO)_2P(OCH_3)_3$	Fe: Org.Verb.C1-95
$C_{10}Fe_2H_{18}HgN_2O_{12}P_2$	$[NO(CO)_2FeP(OCH_3)_3]_2Hg$	Fe: Org.Verb.C1-64/5
$C_{10}Fe_2H_{18}I_2N_4O_4Pb$	$((CH_3)_2C_3H_3Fe(NO)_2)_2PbI_2$	Fe: Org.Verb.C2-156, 165
$C_{10}Fe_2H_{18}I_2N_4O_4Sn$	$((CH_3)_2C_3H_3Fe(NO)_2)_2SnI_2$	Fe: Org.Verb.C2-156, 165
$C_{10}Fe_2H_{24}N_6O_4$	$[Fe(C_2H_8N_2)_3][Fe(CO)_4]$	Fe: Org.Verb.B2-22
$C_{10}Fe_2O_{10}$	$Fe_2(CO)_{10}$	Fe: Org.Verb.B3-22/3, 68
$C_{10}Fe_3H_{10}I_5O_6S_2$	$Fe_3(CO)_6(SC_2H_5)_2I_5$	Fe: Org.Verb.C1-18/20
$C_{10}Fe_3H_{18}N_3O_7P_3$	$(CO)_4Fe_3(P(CH_3)_2)_3(NO)_3$	Fe: Org.Verb.C1-221

$C_{10}Fe_3O_{10}^{2+}$ .... $[(CO)_5FeFeFe(CO)_5]^{2+}$ .... Fe: Org.Verb.C1-62
$C_{10}H_5LiNiO_4$ .... $Li[(CO)_3NiC(O)C_6H_5]$ .... Ni: Org.Verb.1-193, 194
$C_{10}H_5NNiO_3$ .... $C_6H_5NCNi(CO)_3$ .... Ni: Org.Verb.1-303
$C_{10}H_5N_{10}NbS_{12}$ ..... $(C_2H_2N_2S_3)_2 \cdot H[Nb(CNS)_6]$ .... C: MVol.D6-41
$C_{10}H_6N_2Ni$ .... $Ni(CN)_2CCHC_6H_5$ .... Ni: Org.Verb.1-337/8
$C_{10}H_6N_4O_8Sn$ ..... $Sn(C_5H_3N_2O_4)_2$ .... Sn: MVol.C5-15
$C_{10}H_6N_6Sn$ .... $(CH_3)_2Sn[C(CN)_3]_2$ .... Sn: Org.Verb.6-31
$C_{10}H_6O_6PuS_2$ ..... $PuO_2(SC_4H_3COO)_2$ .... Np: TrU.D1-156
$C_{10}H_6O_8Pu$ .... $PuO_2(OC_4H_3COO)_2$ .... Np: TrU.D1-156
$C_{10}H_7KO_4Ti$ .... $K[CH_3C_5H_4Ti(CO)_4]$ .... Ti: Org.Verb.1-211
$C_{10}H_8I_2N_2O_2S_2Sn$ ... $SnI_2(C_5H_4NOS)_2$ .... Sn: MVol.C6-106
$C_{10}H_8I_2N_2Sn$ .... $SnI_2 \cdot NC_5H_4C_5H_4N$ .... Sn: MVol.C5-21/2
$C_{10}H_8I_4N_2Sn$ .... $SnI_4 \cdot NC_5H_4C_5H_4N$ .... Sn: MVol.C5-217/9
$C_{10}H_8MnN_5O_9$ ...... $Mn(NO_3)_3 \cdot NC_5H_4C_5H_4N$ .... Mn: MVol.C3-306
$C_{10}H_8N_2O_6SU$ ...... $[UO_2SO_4(C_{10}H_8N_2)]$ .... U: SVol.E1-39/43
– .... $[UO_2SO_4(C_{10}H_8N_2)] \cdot 4\ H_2O$ .... U: SVol.E1-39/42
$C_{10}H_8N_2O_8SU$ ...... $UO_2SO_4 \cdot C_{10}H_8N_2O_2$ .... U: SVol.E1-134, 141, 144
$C_{10}H_8N_4O_8SU$ ...... $[UO_2(NO_3)_2(S(C_5H_4N)_2)]$ .... U: SVol.E1-41, 43
$C_{10}H_8N_4O_8U$ .... $[UO_2(NO_3)_2(C_{10}H_8N_2)]$ .... U: SVol.E1-39/42
– .... $[UO_2(NO_3)_2(C_{10}H_8N_2)] \cdot 0.5\ H_2O$ .... U: SVol.E1-39/42
– .... $[UO_2(NO_3)_2(C_{10}H_8N_2)] \cdot 4\ H_2O$ .... U: SVol.E1-39/42
$C_{10}H_8N_4O_{10}U$ ...... $UO_2(NO_3)_2 \cdot C_{10}H_8N_2O_2$ .... U: SVol.E1-134, 141, 144
$C_{10}H_8Ni^+$ .... $(C_5H_4)_2Ni^+$ .... Ni: Org.Verb.2-345
$C_{10}H_8Ni_2^+$ .... $(C_5H_4)_2Ni_2^+$ .... Ni: Org.Verb.2-345
$C_{10}H_8O_2Sn$ .... $(C_{10}H_7SnOOH)_x$ .... Sn: Org.Verb.6-283
$C_{10}H_8Ti$ .... $C_{10}H_8Ti$ .... Ti: Org.Verb.1-122
$C_{10}H_9N_2NaO_3S$ ..... $Na[SO_3N(C_{10}H_7)NH_2]$ .... S: S-N-Verb.1-219
$C_{10}H_9N_3O_6U$ ...... $[UO_2(OH)(NO_3)(C_{10}H_8N_2)]$ .... U: SVol.E1-39/42
$C_{10}H_{10}I_2N_2Sn$ ...... $SnI_2 \cdot 2\ C_5H_5N$ .... Sn: MVol.C5-16/7
$C_{10}H_{10}I_2Ni$ .... $C_5H_5NiC_5H_5I_2$ .... Ni: Org.Verb.2-205
$C_{10}H_{10}I_4N_2O_2Sn$ .... $SnI_4 \cdot 2\ C_5H_5NO$ .... Sn: MVol.C5-206/7
$C_{10}H_{10}I_4N_2Sn$ ...... $SnI_4 \cdot 2\ C_5H_5N$ .... Sn: MVol.C5-200
$C_{10}H_{10}I_4N_2Te$ ...... $TeI_4 \cdot 2\ C_5H_5N$ .... Te: SVol.B3-170
$C_{10}H_{10}I_4Ni$ .... $C_5H_5NiC_5H_5I_4$ .... Ni: Org.Verb.2-205
$C_{10}H_{10}I_6Ti$ .... $C_5H_5Ti(C_5H_5I_4)I_2$ .... Ti: Org.Verb.1-206/7
$C_{10}H_{10}NNi_2O^+$ ...... $(C_5H_5)_2Ni_2NO^+$ .... Ni: Org.Verb.2-111
$C_{10}H_{10}N_2NpO_6^-$ .... $NpO_2(NC_5H_4CH_2N(CH_2COO)_2)^-$ .... Np: TrU.D1-163
$C_{10}H_{10}N_2O_2S$ ...... $C_{10}H_7NHSO_2NH_2$ .... S: S-N-Verb.1-167
$C_{10}H_{10}N_2O_6SU$ ..... $UO_2SO_4 \cdot 2\ C_5H_5N$ .... U: SVol.E1-29, 35
$C_{10}H_{10}N_2O_7U$ ...... $UO_2C_2O_4 \cdot (CH_3)_2NC_6H_4NO$ .... U: SVol.E1-144/5
$C_{10}H_{10}N_2O_8SU$ ..... $UO_2SO_4 \cdot 2\ C_5H_5NO$ .... U: SVol.E1-132, 136, 142/3
$C_{10}H_{10}N_2O_8U$ ...... $[UO_2(C_5H_3N(COO)_2)(NH_2COOC_2H_5)]_n$ .... U: SVol.E1-120/3
$C_{10}H_{10}N_4O_8U$ ...... $UO_2(NO_3)_2 \cdot 2\ C_5H_5N$ .... U: SVol.E1-25, 29, 34/5
$C_{10}H_{10}N_4O_{10}U$ ..... $UO_2(NO_3)_2 \cdot 2\ C_5H_5NO$ .... U: SVol.E1-132, 135, 142/3
$C_{10}H_{10}N_6O_{12}Sn$ .... $Sn(NO_3)_4 \cdot 2\ C_5H_5N$ .... Sn: MVol.C5-194
$C_{10}H_{10}Ni$ .... $Ni(C_5H_5)_2$
Catalytic properties .... Ni: Org.Verb.2-210/12
Chemical reactions .... Ni: Org.Verb.2-204/10
Formation and preparation .... Ni: Org.Verb.2-190/2
Molecule .... Ni: Org.Verb.2-202/4

$C_{10}H_{10}Ni$ .... $Ni(C_5H_5)_2$
Physical properties .... Ni: Org.Verb.2-192/202
Uses .... Ni: Org.Verb.2-210/2
$C_{10}H_{10}Ni^+$ .... $[Ni(C_5H_5)_2]^+$ .... Ni: Org.Verb.2-201, 213, 351, 381
$C_{10}H_{10}Ni^{2+}$ .... $[Ni(C_5H_5)_2]^{2+}$ .... Ni: Org.Verb.2-201, 214
$C_{10}H_{10}Ni_2^+$ .... $Ni_2(C_5H_5)_2^+$ .... Ni: Org.Verb.2-200/1, 351, 359, 381
$C_{10}H_{10}Ni_3^+$ .... $(C_5H_5)_2Ni_3^+$ .... Ni: Org.Verb.2-381
$C_{10}H_{10}OPa^+$ .... $Pa(C_5H_5)_2O^+$ .... Pa: SVol.2-99
$C_{10}H_{10}OPa_2^+$ .... $C_5H_5PaOPaC_5H_5^+$ .... Pa: SVol.2-99
$C_{10}H_{10}O_4Sn$ .... $(CH_3)_2Sn(OCOC_6H_4COO)$ .... Sn: Org.Verb.6-28
$C_{10}H_{10}O_8S_4Sn_2$ .... $Sn_2(C_2H_2(SCH_2COO)_4)$ .... Sn: MVol.C5-48
$C_{10}H_{10}Pa^+$ .... $Pa(C_5H_5)_2^+$ .... Pa: SVol.2-99
$C_{10}H_{11}INi$ .... $[Ni(C_5H_5)(C_5H_6)]I$ .... Ni: Org.Verb.2-116
$C_{10}H_{11}MnO_2^+$ .... $[Mn(OC_6H_3(CH_3)COC_2H_5)]^+$ .... Mn: MVol.D1-58
– .... $[Mn(OC_6H_4COC_3H_7)]^+$ .... Mn: MVol.D1-58
$C_{10}H_{11}MnO_4^+$ .... $[Mn(OC_6H_2(OCH_3)_2COCH_3)]^+$ .... Mn: MVol.D1-57/8
$C_{10}H_{11}N_2NpO_6$ .... $NpO_2(NC_5H_4CH_2N(CH_2COO)_2H)$ .... Np: TrU.D1-163
$C_{10}H_{11}Na_4O_8S_2Sn$ .... $C_2H_5Sn[SC_2H_3(COONa)_2]_2 \cdot H_2O$ .... Sn: Org.Verb.6-230/1
$C_{10}H_{11}Ni^+$ .... $[C_5H_5NiC_5H_6]^+$ .... Ni: Org.Verb.2-188/9, 201
$C_{10}H_{12}I_6N_2Sn$ .... $[C_5H_6N]_2[SnI_6]$ .... Sn: MVol.C3-120
$C_{10}H_{12}I_6N_2Te$ .... $(C_5H_5NH)_2TeI_6$ .... Te: SVol.B3-67, 72
$C_{10}H_{12}IrN_2O_8$ .... $IrC_2H_4N_2(CH_2COO)_4 \cdot n\ H_2O$ .... Ir: SVol.2-32
$C_{10}H_{12}MgN_2O_8Sn$ .... $Mg[Sn(C_2H_4N_2(CH_2COO)_4)] \cdot 4\ H_2O$ .... Sn: MVol.C5-37
$C_{10}H_{12}MnN_2O_8Sn$ .... $Mn[Sn(C_2H_4N_2(CH_2COO)_4)] \cdot 4\ H_2O$ .... Sn: MVol.C5-37
$C_{10}H_{12}N_2Na_2O_8Sn$ .... $Na_2[Sn(C_2H_4N_2(CH_2COO)_4)] \cdot 2\ H_2O$ .... Sn: MVol.C5-36
$C_{10}H_{12}N_2NiS_2$ .... $(CH_3)_4C_4Ni(SCN)_2$ .... Ni: Org.Verb.2-329
$C_{10}H_{12}N_2NpO_8$ .... $NpC_2H_4N_2(CH_2COO)_4$ .... Np: TrU.D1-165
$C_{10}H_{12}N_2NpO_8^-$ .... $NpC_2H_4N_2(CH_2COO)_4^-$ .... Np: TrU.D1-160
$C_{10}H_{12}N_2NpO_{10}^{3-}$ .... $NpO_2(C_2H_4N_2(CH_2COO)_4)^{3-}$ .... Np: TrU.D1-165
$C_{10}H_{12}N_2Np_2O_{12}^{2-}$ .... $(NpO_2)_2C_2H_4N_2(CH_2COO)_4^{2-}$ .... Np: TrU.D1-162
$C_{10}H_{12}N_2O_8Pu$ .... $PuC_2H_4N_2(CH_2COO)_4$ .... Np: TrU.D1-161, 165
$C_{10}H_{12}N_2O_8Pu^-$ .... $PuC_2H_4N_2(CH_2COO)_4^-$ .... Np: TrU.D1-160, 164
$C_{10}H_{12}N_2O_8Pu_2^{2+}$ .... $Pu_2C_2H_4N_2(CH_2COO)_4^{2+}$ .... Np: TrU.D1-160
$C_{10}H_{12}N_2O_8Pu_2^{4+}$ .... $Pu_2C_2H_4N_2(CH_2COO)_4^{4+}$ .... Np: TrU.D1-161
$C_{10}H_{12}N_2O_8Sn^{2-}$ .... $[Sn(C_2H_4(N(CH_2COO)_2)_2)]^{2-}$ .... Sn: MVol.C5-32/3
$C_{10}H_{12}N_2O_8SnSr$ .... $Sr[Sn(C_2H_4N_2(CH_2COO)_4)] \cdot 4\ H_2O$ .... Sn: MVol.C5-37
$C_{10}H_{12}N_2O_8SnZn$ .... $Sn[Zn(C_2H_4N_2(CH_2COO)_4)] \cdot 2\ H_2O$ .... Sn: MVol.C5-37
$C_{10}H_{12}N_2O_8Sn_2$ .... $[Sn_2(C_2H_4(N(CH_2COO)_2)_2)]$ .... Sn: MVol.C5-33
– .... $[Sn_2(C_2H_4(N(CH_2COO)_2)_2)] \cdot H_2O$ .... Sn: MVol.C5-35
– .... $[Sn_2(C_2H_4(N(CH_2COO)_2)_2)] \cdot 2\ H_2O$ .... Sn: MVol.C5-33/5
$C_{10}H_{12}N_2O_9Pa^{2-}$ .... $PaO(C_2H_4N_2(CH_2COO)_4)^{2-}$ .... Pa: SVol.2-175
$C_{10}H_{12}N_2O_{10}Pu^{2-}$ .... $PuO_2(C_2H_4N_2(CH_2COO)_4)^{2-}$ .... Np: TrU.D1-165
$C_{10}H_{12}N_2O_{10}Pu^{3-}$ .... $PuO_2(C_2H_4N_2(CH_2COO)_4)^{3-}$ .... Np: TrU.D1-165
$C_{10}H_{12}N_6NiO_2P_2$ .... $(CO)_2Ni(P(N(CH_3)_2)(CN)_2)_2$ .... Ni: Org.Verb.1-126
$C_{10}H_{12}N_6O_{10}U$ .... $UO_2(NO_3)_2 \cdot 2\ NH_2C_5H_4NO$ .... U: SVol.E1-133, 142/3
$C_{10}H_{12}N_8O_{18}Pu$ .... $(C_5H_5NH)_2Pu(NO_3)_6$ .... Np: TrU.D1-140
$C_{10}H_{12}Na_2O_8S_2Sn$ .... $(C_2H_5)_2Sn(OC_6H_2(SO_3Na)_2O)$ .... Sn: Org.Verb.6-58
$C_{10}H_{12}Ni$ .... $C_5H_5NiC_5H_7$ .... Ni: Org.Verb.2-179/80

Formula	Compound	Reference
$C_{10}H_{12}Ni$	$C_5H_5NiC_5H_5D_2$	Ni: Org.Verb.2-180
–	$Ni(C_5H_6)_2$	Ni: Org.Verb.2-87
$C_{10}H_{12}NiO_4$	$(CH_3)_4C_4Ni(C_2O_4)$	Ni: Org.Verb.2-329
$C_{10}H_{12}Ti$	$(C_5H_6)_2Ti$	Ti: Org.Verb.1-131
$C_{10}H_{13}I_4NO_2Sn$	$C_5H_5NH[SnI_4(CH(COCH_3)_2)]$	Sn: MVol.C5-107
$C_{10}H_{13}N_2NaO_9Sn$	$Na[Sn(OH)(C_2H_4N_2(CH_2COO)_4)] \cdot H_2O$	Sn: MVol.C6-16
$C_{10}H_{13}N_2NpO_{10}{}^{2-}$	$NpO_2(C_2H_4N_2(CH_2COOH)(CH_2COO)_3)^{2-}$	Np: TrU.D1-165
$C_{10}H_{13}N_2NpO_{11}{}^{4-}$	$NpO_2(OH)(C_2H_4N_2(CH_2COO)_4)^{4-}$	Np: TrU.D1-165
$C_{10}H_{13}N_2O_8Pu$	$PuC_2H_4N_2(CH_2COOH)(CH_2COO)_3$	Np: TrU.D1-160, 164
$C_{10}H_{13}N_2O_8Sn^-$	$[Sn(C_2H_4(N(CH_2COO)_2)_2H)]^-$	Sn: MVol.C5-32/3
$C_{10}H_{13}N_2O_9Pa^-$	$Pa(OH)(C_2H_4N_2(CH_2COO)_4)^-$	Pa: SVol.2-175
$C_{10}H_{13}N_2O_9Pu^-$	$Pu(OH)(C_2H_4N_2(CH_2COO)_4)^-$	Np: TrU.D1-165
$C_{10}H_{13}N_2O_9Pu^{2-}$	$Pu(OH)C_2H_4N_2(CH_2COO)_4{}^{2-}$	Np: TrU.D1-160
$C_{10}H_{13}N_2O_9Sn^-$	$[Sn(OH)(C_2H_4N_2(CH_2COO)_4)]^-$	Sn: MVol.C6-14
$C_{10}H_{13}N_2O_9Sn^{3-}$	$[Sn(OH)(C_2H_4(N(CH_2COO)_2)_2)]^{3-}$	Sn: MVol.C5-33
$C_{10}H_{13}N_2O_{10}Pu^{2-}$	$PuO_2(C_2H_4N_2(CH_2COOH)(CH_2COO)_3)^{2-}$	Np: TrU.D1-165
$C_{10}H_{13}N_5O_6S_3$	$(C_5H_5NH)_2H[NSO_2]_3$	S: S-N-Verb.1-9
$C_{10}H_{13}Na_2O_8S_2Sn$	$C_2H_5Sn[SCH(COONa)CH_2COOH]_2$	Sn: Org.Verb.6-230/1
$C_{10}H_{13}Ni$	$C_3H_5Ni(C_6H_5)CH_3$	Ni: Org.Verb.2-285
$C_{10}H_{13}NiO$	$C_3H_5Ni(OC_6H_5)CH_3$	Ni: Org.Verb.2-285
$C_{10}H_{13}NiO_6P$	$(CO)_3Ni(P(OCH_2)_3CC_3H_7\text{-}n)$	Ni: Org.Verb.1-178
$C_{10}H_{14}IMnO_4$	$[MnI(CH(COCH_3)_2)_2]$	Mn: MVol.D1-101/2
$C_{10}H_{14}I_2O_4Sn$	$[SnI_2(CH(COCH_3)_2)_2]$	Sn: MVol.C5-101/6
$C_{10}H_{14}I_2Sn$	$(C_4H_9)(C_6H_5)SnI_2$	Sn: Org.Verb.5-50
$C_{10}H_{14}IrN_2O_2{}^+$	$[Ir(C_5H_5N)_2(H_2O)_2]^+$	Ir: SVol.2-74
$C_{10}H_{14}MnN_3O_4$	$[MnN_3(CH(COCH_3)_2)_2]$	Mn: MVol.D1-99/101
$C_{10}H_{14}MnO_4$	$Mn(CH(COCH_3)_2)_2$	Mn: MVol.D1-65/71
$C_{10}H_{14}MnO_4{}^+$	$[Mn(CH(COCH_3)_2)_2]^+$	Mn: MVol.D1-69, 85, 92/3
$C_{10}H_{14}N_2Ni_2S_2$	$(CH_3CHCHCH_2NiCNS)_2$	Ni: Org.Verb.2-296
$C_{10}H_{14}N_2NpO_8{}^{2+}$	$NpC_2H_4N_2(CH_2COOH)_2(CH_2COO)_2{}^{2+}$	Np: TrU.D1-161, 165
$C_{10}H_{14}N_2O_8Sn$	$H_2[Sn(C_2H_4(N(CH_2COO)_2)_2)]$	Sn: MVol.C5-32/3, 35/6
$C_{10}H_{14}N_2O_9Sn$	$[Sn(H_2O)(C_2H_4N_2(CH_2COO)_4)]$	Sn: MVol.C6-14/5
$C_{10}H_{14}N_2O_9Sn_2$	$Sn[Sn(C_2H_4N_2(CH_2COO)_4)H_2O] \cdot H_2O$	Sn: MVol.C5-33
$C_{10}H_{14}N_2O_{10}Pa^-$	$Pa(OH)_2(C_2H_4N_2(CH_2COO)_4)^-$	Pa: SVol.2-175
$C_{10}H_{14}N_4O_8S_3$	$SO_2(NHSO_3NHC_5H_5)_2$	S: S-N-Verb.1-178
$C_{10}H_{14}N_6O_8S_4$	$(C_5H_5NH)_2H_2[NSO_2]_4$	S: S-N-Verb.1-4
$C_{10}H_{14}Ni$	$Ni(C_3H_4CHCH_2)_2$	Ni: Org.Verb.2-63
$C_{10}H_{14}Ni_2$	$(C_5H_7)_2Ni_2$	Ni: Org.Verb.2-336/7
$C_{10}H_{14}NpO_4{}^{2+}$	$Np(CH_3COCHCOCH_3)_2{}^{2+}$	Np: TrU.D2-242
$C_{10}H_{14}O_2Sn$	$(CH_3)_3SnC_6H_4COOH$	Sn: Org.Verb.2-140/1
–	$(C_2H_5)_2Sn(OC_6H_4O)$	Sn: Org.Verb.6-56
$C_{10}H_{14}O_4Pu^{2+}$	$Pu(CH_3COCHCOCH_3)_2{}^{2+}$	Np: TrU.D1-167/8
$C_{10}H_{14}O_4Sn$	$[Sn(CH(COCH_3)_2)_2]$	Sn: MVol.C5-8/9
$C_{10}H_{14}O_5Pa^+$	$PaO((CH_3CO)_2CH)_2{}^+$	Pa: SVol.2-177/8
$C_{10}H_{14}O_6Pa^-$	$PaO_2((CH_3CO)_2CH)_2{}^-$	Pa: SVol.2-177/8
$C_{10}H_{15}MnO_4{}^+$	$[Mn(CH(COCH_3)_2)_2H]^+$	Mn: MVol.D1-70
$C_{10}H_{15}NNiO_4$	$(C_2H_5)_2N(C_2H_5O)CNi(CO)_3$	Ni: Org.Verb.1-334
$C_{10}H_{15}NO_3S$	$C_6H_5NSO(OC_2H_5)_2$	S: S-N-Verb.1-97
$C_{10}H_{15}NO_4U$	$UO_2(C_6H_4O_2) \cdot (C_2H_5)_2NH$	U: SVol.E1-19, 22
$C_{10}H_{15}NS_4Sn$	$Sn(SCH_2CH_2S)_2 \cdot NC_5H_4CH_3$	Sn: MVol.C6-82/3

Formula	Compound	Reference
$C_{10}H_{15}N_2NpO_7^+$	$NpC_2H_4N_2(C_2H_4OH)(CH_2COO)_3^+$	Np: TrU.D1-164
$C_{10}H_{15}N_2NpO_9^{2-}$	$NpO_2(C_2H_4N_2(C_2H_4OH)(CH_2COO)_3)^{2-}$	Np: TrU.D1-164
$C_{10}H_{15}N_2O_{11}Sn^{3-}$	$[Sn(OH)_3(C_2H_4N_2(CH_2COO)_4)]^{3-}$	Sn: MVol.C6-14
$C_{10}H_{15}N_3O_8Sn$	$[Sn(NH_3)(C_2H_4N_2(CH_2COO)_4)]$	Sn: MVol.C6-16
–	$[Sn(NH_3)(C_2H_4N_2(CH_2COO)_4)] \cdot 2\ H_2O$	Sn: MVol.C6-15/6
$C_{10}H_{15}NaSn$	$(C_2H_5)_3SnCCCCNa$	Sn: Org.Verb.2-222
$C_{10}H_{15}O_5Pa^+$	$Pa(OH)((CH_3CO)_2CH)_2^+$	Pa: SVol.2-176
$C_{10}H_{15}O_{10}Sn_2^-$	$Sn_2(CH_3COO)_5^-$	Sn: MVol.C2-219/20
$C_{10}H_{15}Sn^+$	$Sn(C_2H_5)_2C_6H_5^+$	Sn: Org.Verb.3-25
$C_{10}H_{16}N_2NpO_9^-$	$NpO_2(C_2H_4N_2(C_2H_4OH)(CH_2COO)_3H)^-$	Np: TrU.D1-164
$C_{10}H_{16}N_2NpO_{10}^{3-}$	$NpO_2(OH)(C_2H_4N_2(C_2H_4OH)(CH_2COO)_3)^{3-}$	Np: TrU.D1-164
$C_{10}H_{16}N_2O_2S$	$(C_2H_5)_2NSO_2NHC_6H_5$	S: S-N-Verb.1-170
$C_{10}H_{16}N_2O_2S_4Te$	$Te(OC_4H_8NCSS)_2$	Te: SVol.B3-139
$C_{10}H_{16}N_2O_9Pu^-$	$PuO_2(C_2H_4N_2(C_2H_4OH)(CH_2COO)_3H)^-$	Np: TrU.D1-164
$C_{10}H_{16}N_2S_4Sn$	$Sn(NC_4H_8CSS)_2$	Sn: MVol.C5-54
$C_{10}H_{16}N_2Sn$	$(C_2H_5)_3SnC(CN)CHCN$	Sn: Org.Verb.2-192, 194
–	$[NHC_6H_4NHSn(C_2H_5)_2]_x$	Sn: Org.Verb.6-56
$C_{10}H_{16}N_4O_4Sn$	$(CH_3)_2Sn[C(N_2)COOC_2H_5]_2$	Sn: Org.Verb.3-8, 17
$C_{10}H_{16}N_6S_2Se_2Te$	$Te(SC(NHCH_2)_2CH_2)_2(SeCN)_2$	Te: SVol.B3-160
$C_{10}H_{16}N_6S_4Te$	$Te(SC(NHCH_2)_2CH_2)_2(SCN)_2$	Te: SVol.B3-160
$C_{10}H_{16}NiP_2$	$Ni(C_5H_5)_2(PH_3)_2$	Ni: Org.Verb.2-206
$C_{10}H_{16}Ni_2O_4$	$(C_3H_5NiO_2CCH_3)_2$	Ni: Org.Verb.2-283
$C_{10}H_{16}OSSn$	$(CH_3)_3SnSC_6H_4OCH_3$	Sn: Org.Verb.5-84
$C_{10}H_{16}OSn$	$(CH_3)_3SnC_6H_4OCH_3$	Sn: Org.Verb.2-139, 140/1, 144
$C_{10}H_{16}O_2Sn$	$(C_2H_5)(C_6H_5)Sn(OCH_3)_2$	Sn: Org.Verb.5-50
$C_{10}H_{16}O_2Ti^+$	$[CH_2C_5H_4Ti(OC_2H_5)_2]^+$	Ti: Org.Verb.1-164
$C_{10}H_{16}O_6Pa$	$Pa(OH)_2((CH_3CO)_2CH)_2$	Pa: SVol.2-175/6
$C_{10}H_{16}SSn$	$(CH_3)_3SnC_6H_4SCH_3$	Sn: Org.Verb.2-141, 144
–	$(CH_3)_3SnSC_6H_4CH_3$	Sn: Org.Verb.5-84
$C_{10}H_{16}Sn$	$(CH_3)_2Sn(C_2H_5)(C_6H_5)$	Sn: Org.Verb.3-76
–	$(CH_3)_3SnCCC_5H_7$	Sn: Org.Verb.2-114
–	$(CH_3)_3SnCH_2C_6H_5$	Sn: Org.Verb.2-21/2
–	$(CH_3)_3SnC_6H_4CH_3$	Sn: Org.Verb.2-124/5
–	$(CH_3)_3SnC_7H_7$	Sn: Org.Verb.2-102, 106
–	$(C_2H_5)_2Sn(CCCH_3)_2$	Sn: Org.Verb.3-26
–	$(C_2H_5)_3SnCCCCH$	Sn: Org.Verb.2-212/3
–	$(C_3H_7)_2Sn(CCH)_2$	Sn: Org.Verb.3-37
–	$(C_4H_9)(C_6H_5)SnH_2$	Sn: Org.Verb.4-120
$C_{10}H_{17}LiTi$	$Li[(CH_3)_4TiC_6H_5] \cdot 2\ C_4H_8O_2$	Ti: Org.Verb.1-121
$C_{10}H_{17}MnNO_2S_2$	$[Mn(CH(COCH_3)_2)((C_2H_5)_2NCSS)]$	Mn: MVol.D1-84
$C_{10}H_{17}MnO_6$	$[Mn(CH(COCH_3)_2)_2(H_2O)OH]$	Mn: MVol.D1-94, 98
$C_{10}H_{17}NNi$	$(CH_3)_2C_3H_3Ni(NH(CH)_4)CH_3$	Ni: Org.Verb.2-35, 37
$C_{10}H_{17}NO_6U$	$UO_2(CH(COCH_3)_2)_2 \cdot NH_3$	U: SVol.E1-13, 16
$C_{10}H_{17}N_3O_4S_2$	$(C_2H_5)_2NSO_2NHNHSO_2C_6H_5$	S: S-N-Verb.1-222
$C_{10}H_{17}NiO_3P$	$(CO)_3Ni(CH_2P(C_2H_5)_3)$	Ni: Org.Verb.1-183
$C_{10}H_{17}Ti^-$	$[(CH_3)_4TiC_6H_5]^-$	Ti: Org.Verb.1-121
$C_{10}H_{18}I_2N_2Ni$	$(t\text{-}C_4H_9NC)_2NiI_2$	Ni: Org.Verb.1-315, 316
$C_{10}H_{18}I_2Ni_2$	$(CH_3CHCHCHCH_3NiI)_2$	Ni: Org.Verb.2-314
$C_{10}H_{18}IrN_2O_4^{3+}$	$[Ir(C_5H_5N)_2(H_2O)_4]^{3+}$	Ir: SVol.2-71, 74

$C_{10}H_{18}MgO_6$ .... $[Mg(CH(COCH_3)_2)_2(H_2O)_2]$ solid solutions
$[(Mg,Mn)(CH(COCH_3)_2)_2(H_2O)_2]$ .... Mn: MVol.D1-74
$C_{10}H_{18}MnO_6$ .... $[Mn(CH(COCH_3)_2)_2(H_2O)_2]$ .... Mn: MVol.D1-71/4
– .... $[Mn(CH(COCH_3)_2)_2(H_2O)_2]$ solid solutions
$[(Mg,Mn)(CH(COCH_3)_2)_2(H_2O)_2]$ .... Mn: MVol.D1-74
$C_{10}H_{18}MnO_6^+$ ..... $[Mn(CH(COCH_3)_2)_2(H_2O)_2]^+$ .... Mn: MVol.D1-94, 98
$C_{10}H_{18}Mo_2O_4S_6$ ..... $Mo_2O_2S_2[S_2COC_4H_9]_2$ .... C: MVol.D4-258
$C_{10}H_{18}N_2Ni$ .... $(Ni(CNC_4H_9\text{-}t)_2)_x$ .... Ni: Org.Verb.2-396
$C_{10}H_{18}N_2NiO_2$ ..... $(Ni(O_2)(CNC_4H_9\text{-}t)_2)_x$ .... Ni: Org.Verb.2-396/7
$C_{10}H_{18}N_2NiO_4S$ ..... $(t\text{-}C_4H_9NC)_2NiSO_4$ .... Ni: Org.Verb.1-315, 317
$C_{10}H_{18}N_2Ni_2$ .... $(C_3H_5NiNC_2H_4)_2$ .... Ni: Org.Verb.2-282
$C_{10}H_{18}N_2O_2Sn$ ...... $(C_4H_9)_2Sn(NCO)_2$ .... Sn: Org.Verb.6-89, 95
$C_{10}H_{18}N_2O_3Sn$ ...... $(CH_3)_3Sn(CH_2)_3CH(CONH)_2CO$ .... Sn: Org.Verb.2-45/6
$C_{10}H_{18}N_2O_{14}Sn_2^{2-}$ ... $[(OH)_3Sn(C_2H_4N_2(CH_2COO)_4)Sn(OH)_3]^{2-}$ ... Sn: MVol.C6-14
$C_{10}H_{18}N_2S_2Sn$ ...... $(C_4H_9)_2Sn(NCS)_2$ .... Sn: Org.Verb.6-89, 97
$C_{10}H_{18}N_2S_4Sn$ ...... $(CH_2CH)_2Sn[S_2CN(CH_3)_2]_2$ .... Sn: Org.Verb.6-145
$C_{10}H_{18}N_2Sn$ .... $(C_2H_5)_2Sn(CH_2CH_2CN)_2$ .... Sn: Org.Verb.3-26
$C_{10}H_{18}N_4NiO_2$ ..... $(CO)_2Ni(C_2N_2(CH_3)_2(N(CH_3)_2)_2)$ .... Ni: Org.Verb.1-146/7
$C_{10}H_{18}N_4NiO_6$ ..... $(t\text{-}C_4H_9NC)_2Ni(NO_3)_2$ .... Ni: Org.Verb.1-316
$C_{10}H_{18}N_4O_4S_2$ ...... $(CH_3)_2NSO_2NHC_6H_4NHSO_2N(CH_3)_2$ .... S: S-N-Verb.1-171
$C_{10}H_{18}N_4O_6S_2U$ .... $UO_2(CH_3COO)_2 \cdot 2\ CS(NHCH_2)_2$ .... U: SVol.E1-220/1
$C_{10}H_{18}N_4O_6U$ .... $UO_2(CH_3COO)_2 \cdot (CH_2)_6N_4$ .... U: SVol.E1-23/6
$C_{10}H_{18}Ni$ .... $CH_3CHCHCH_2NiCH_2CHCHCH_2C_2H_5$ .... Ni: Org.Verb.2-299
– .... $Ni(C_3H_3(CH_3)_2)_2$ .... Ni: Org.Verb.2-63, 66
$C_{10}H_{18}NiO_2$ .... $Ni(OC(CH_3)_2CH_2CH)_2$ .... Ni: Org.Verb.2-92
$C_{10}H_{18}NiO_2S_4$ ..... $Ni[S_2COC_4H_9]_2$ .... C: MVol.D4-260
$C_{10}H_{18}NiS_6$ .... $Ni[S_2CSC_4H_9]_2$ .... C: MVol.D4-267, 270
$C_{10}H_{18}Ni_2O_{10}P_6$ ..... $(CH_3)_3P(CO)_2Ni[P_4O_6]Ni(CO)_2P(CH_3)_3$ ..... Ni: Org.Verb.2-262
$C_{10}H_{18}OSn$ .... $(CH_3)_3SnC_7H_9O$ .... Sn: Org.Verb.2-64, 67
– .... $(C_2H_5)_3SnCCCOCH_3$ .... Sn: Org.Verb.2-222
$C_{10}H_{18}O_2PbS_4$ ...... $Pb[S_2COC_4H_9]_2$ .... C: MVol.D4-245, 254
$C_{10}H_{18}O_2PdS_4$ ...... $Pd[S_2COC_4H_9]_2$ .... C: MVol.D4-261
$C_{10}H_{18}O_2S_4$ .... $C_4H_9OC(S)SSC(S)OC_4H_9$ .... C: MVol.D4-251
$C_{10}H_{18}O_2S_4Zn$ ...... $Zn[S_2COC_4H_9]_2$ .... C: MVol.D4-257
$C_{10}H_{18}O_2Sn$ .... $(C_2H_5)_3SnCCCOOCH_3$ .... Sn: Org.Verb.2-222, 234
$C_{10}H_{18}O_2Ti$ .... $C_5H_5Ti(CH_3)(OC_2H_5)_2$ .... Ti: Org.Verb.1-205
$C_{10}H_{18}O_4OsSiSn$ .... $[(CH_3)_3Si][(CH_3)_3Sn]Os(CO)_4$ .... Sn: Org.Verb.5-82
$C_{10}H_{18}O_4OsSn_2$ ..... $[(CH_3)_3Sn]_2Os(CO)_4$ .... Sn: Org.Verb.5-86
$C_{10}H_{18}O_4RuSiSn$ .... $[(CH_3)_3Si][(CH_3)_3Sn]Ru(CO)_4$ .... Sn: Org.Verb.5-82
$C_{10}H_{18}O_4RuSn_2$ ..... $[(CH_3)_3Sn]_2Ru(CO)_4$ .... Sn: Org.Verb.5-86
$C_{10}H_{18}O_4STi$ .... $CH_3Ti(SO_2C_5H_5)(OC_2H_5)_2$ .... Ti: Org.Verb.1-60, 190, 205
$C_{10}H_{18}O_4S_2$ .... $C_4H_9OC(O)SS(O)COC_4H_9$ .... C: MVol.D4-236/7
$C_{10}H_{18}O_4S_4U$ .... $UO_2[S_2COC_4H_9]_2$ .... C: MVol.D4-261
$C_{10}H_{18}O_4Sn$ .... $[(C_2H_5)_2SnOCO(CH_2)_4COO]_x$ .... Sn: Org.Verb.6-58
$C_{10}H_{18}O_6S_2Ti$ ...... $CH_3SO_2Ti(SO_2C_5H_5)(OC_2H_5)_2$ .... Ti: Org.Verb.1-60
$C_{10}H_{18}O_6S_8Sn^{4-}$ .... $[Sn(SCH_2CH(S)CH_2SC_2H_4SO_3)_2]^{4-}$ .... Sn: MVol.C5-49/50
$C_{10}H_{18}O_8S_6Sn^{4-}$ .... $[Sn(SCH_2CH(S)CH_2OC_2H_4SO_3)_2]^{4-}$ .... Sn: MVol.C5-49/50
$C_{10}H_{18}O_{10}S_8Sn^{4-}$ .... $[Sn(SCH_2CH(S)CH_2SO_2C_2H_4SO_3)_2]^{4-}$ ...... Sn: MVol.C5-49/50
$C_{10}H_{18}PdS_6$ .... $Pd[S_2CSC_4H_9]_2$ .... C: MVol.D4-270

$C_{10}H_{18}Sn$ .... $(CH_2CHCH_2)_3SnCH_3$ .... Sn: Org.Verb.2-348
– .... $(CH_2CH)_3SnC_4H_9$ .... Sn: Org.Verb.2-346/7
– .... $((CH_2)_2CH)_3SnCH_3$ .... Sn: Org.Verb.2-340
– .... 1) $(CH_3)_3SnC_7H_9$ .... Sn: Org.Verb.2-64, 67
2) $(CH_3)_3SnC_7H_9$ .... Sn: Org.Verb.2-94/6
3) $(CH_3)_3SnC_7H_9$ .... Sn: Org.Verb.2-101, 106
– .... $(C_2H_5)_2SnC_6H_8$ .... Sn: Org.Verb.3-109
– .... $(C_2H_5)_3SnCCCHCH_2$ .... Sn: Org.Verb.2-209/10
– .... $(C_2H_5)_3SnCHCHCCH$ .... Sn: Org.Verb.2-113
$C_{10}H_{19}IN_4NiS_2$ .... $CH_3C_3H_4Ni(SC(NHCH_2)_2)_2I$ .... Ni: Org.Verb.2-21
$C_{10}H_{19}KOS_2$ .... $K[S_2COC_9H_{19}]$ .... C: MVol.D4-253
$C_{10}H_{19}KOSe_2$ .... $K[Se_2COC_9H_{19}]$ .... C: MVol.D6-221
$C_{10}H_{19}NSn$ .... $(CH_3)_3SnCHCH(CN)CH_2C(CH_3)_2$ .... Sn: Org.Verb.2-60
– .... $(CH_3)_3SnC_6H_{10}CN$ .... Sn: Org.Verb.2-62
$C_{10}H_{19}Ni$ .... $C_3H_5Ni(C_6H_{11})CH_3$ .... Ni: Org.Verb.2-285
$C_{10}H_{19}OS_2^-$ .... $[S_2COC_9H_{19}]^-$ .... C: MVol.D4-251
$C_{10}H_{20}I_2N_2S_4Sn$ .... $[SnI_2((C_2H_5)_2NCSS)_2]$ .... Sn: MVol.C6-119/20
$C_{10}H_{20}IrN_5O^{2+}$ .... $[Ir(C_5H_5N)_2(NH_3)_3(OH)]^{2+}$ .... Ir: SVol.2-74
$C_{10}H_{20}Li_2NiO_2P_2$ .... $Li_2[(CO)_2Ni(P(C_2H_5)_2)_2] \cdot 3\ C_4H_8O$ .... Ni: Org.Verb.1-157/8
$C_{10}H_{20}MnN_2O_4$ .... $[Mn(CH(COCH_3)_2)_2(NH_3)_2]$ .... Mn: MVol.D1-75/6
$C_{10}H_{20}N_2O_2Sn$ .... $(C_2H_5)_3SnC(N_2)COOC_2H_5$ .... Sn: Org.Verb.2-166, 176
$C_{10}H_{20}N_2O_4S$ .... $(CH_3CO)(C_3H_7)NSO_2N(C_3H_7)(COCH_3)$ .... S: S-N-Verb.1-173/4
– .... $(CH_3CO)(i\text{-}C_3H_7)NSO_2N(i\text{-}C_3H_7)(COCH_3)$ .. S: S-N-Verb.1-173/4
$C_{10}H_{20}N_2O_8U$ .... $UO_2(CH_3COO)_2 \cdot 2\ HCON(CH_3)_2$ .... U: SVol.E1-100, 102
$C_{10}H_{20}N_2O_{10}U$ .... $UO_2(NO_3)_2 \cdot 2\ CH_3COCH(CH_3)_2$ .... U: SVol.E1-79/80
$C_{10}H_{20}N_2O_{13}U$ .... $UO_2(NO_3)_2 \cdot (CH_2OCH_2)_5 \cdot 2\ H_2O$ .... U: SVol.E1-73, 78
$C_{10}H_{20}N_2S_4Se$ .... $Se(S_2CN(C_2H_5)_2)_2$ .... Se: SVol.A1-193
$C_{10}H_{20}N_2S_4Sn$ .... $Sn((C_2H_5)_2NCSS)_2$ .... Sn: MVol.C5-54
$C_{10}H_{20}N_2S_4Te$ .... $Te((C_2H_5)_2NCSS)_2$ .... Te: SVol.B3-137/8
$C_{10}H_{20}N_4NiO_7$ .... $(t\text{-}C_4H_9NC)_2Ni(H_2O)(NO_3)_2$ .... Ni: Org.Verb.1-315, 317
$C_{10}H_{20}N_4O_8Sn$ .... $(NH_4)_2[Sn(C_2H_4N_2(CH_2COO)_4)] \cdot 2\ H_2O$ ... Sn: MVol.C5-36
$C_{10}H_{20}N_4O_{14}U_2$ .... $[UO_2(CH_3COO)_2((NH_2)_2CO)]_2$ .... U: SVol.E1-84, 86, 89
$C_{10}H_{20}NiO_2S_2$ .... $(CO)_2Ni(S(C_2H_5)_2)_2$ .... Ni: Org.Verb.1-123
$C_{10}H_{20}NiO_6P_2$ .... $C_5H_5Ni(P(OCH_3)_3)PO(OCH_3)_2$ .... Ni: Org.Verb.2-134
$C_{10}H_{20}Ni_2$ .... $(CH_2C(CH_3)CH_2NiCH_3)_2$ .... Ni: Org.Verb.2-307
– .... $(CH_3CHCHCH_2NiCH_3)_2$ .... Ni: Org.Verb.2-298/9
– .... $(C_3H_5NiC_2H_5)_2$ .... Ni: Org.Verb.2-285
– .... $(C_3H_5NiCH_2CD_3)_2$ .... Ni: Org.Verb.2-285
$C_{10}H_{20}Ni_2O_2$ .... $(C_3H_5NiOC_2H_5)_2$ .... Ni: Org.Verb.2-282
$C_{10}H_{20}Ni_2S_8$ .... $[Ni(S_2CSC_2H_5)(SC_2H_5)]_2$ .... C: MVol.D4-266, 270
$C_{10}H_{20}OSn$ .... $(CH_3)_3SnC_7H_{10}OH$ .... Sn: Org.Verb.2-63, 67
– .... $(C_2H_5)_3SnCCCH_2OCH_3$ .... Sn: Org.Verb.2-218
– .... $(C_2H_5)_3SnCCOC_2H_5$ .... Sn: Org.Verb.2-218/9
– .... $(C_2H_5)_3SnOCH(CH_3)CCH$ .... Sn: Org.Verb.5-103
– .... $(C_2H_5)_3SnO(CH_2)_2CCH$ .... Sn: Org.Verb.5-103
$C_{10}H_{20}O_2Sn$ .... $(C_2H_5)_3SnC(CH_2)COOCH_3$ .... Sn: Org.Verb.2-194, 203
– .... $(C_2H_5)_3SnC(CHD)COOCH_3$ .... Sn: Org.Verb.2-203
– .... $(C_2H_5)_3SnCHCHCOOCH_3$ .... Sn: Org.Verb.2-194, 203
$C_{10}H_{20}O_4Sn$ .... $(CH_3)_3SnCH(COOC_2H_5)_2$ .... Sn: Org.Verb.2-25
– .... $(C_2H_5)_2Sn(CH_2COOCH_3)_2$ .... Sn: Org.Verb.3-25/6

Formula	Compound	Reference
$C_{10}H_{20}O_6S_8Sn$	$[Sn(SCH_2CH(SH)CH_2SC_2H_4SO_3)_2]$	Sn: MVol.C5-49/50
$C_{10}H_{20}O_8S_6Sn$	$[Sn(SCH_2CH(SH)CH_2OC_2H_4SO_3)_2]$	Sn: MVol.C5-49/50
$C_{10}H_{20}O_{10}S_8Sn$	$[Sn(SCH_2CH(SH)CH_2SO_2C_2H_4SO_3)_2]$	Sn: MVol.C5-49/50
$C_{10}H_{20}SSn$	$(C_2H_5)_3SnCCSC_2H_5$	Sn: Org.Verb.2-219
$C_{10}H_{20}Sn$	$(CH_2)_5Sn(CH_2)_5$	Sn: Org.Verb.3-127
–	$(CH_3)_2Sn(CH_2CHCHCH_3)_2$	Sn: Org.Verb.3-8, 17
–	$(CH_3)_3SnCCCH_2CH_2CH(CH_3)_2$	Sn: Org.Verb.2-114
–	$(CH_3)_3SnC_7H_{11}$	Sn: Org.Verb.2-62/3, 67
–	$(C_2H_5)_2Sn(CH_2CHCH_2)_2$	Sn: Org.Verb.3-25/6
–	$(C_2H_5)_3SnCCC_2H_5$	Sn: Org.Verb.2-212/3
–	$(C_2H_5)_3SnC(CH_2)CHCH_2$	Sn: Org.Verb.2-184/5, 189
–	$(C_2H_5)_3SnCHCCHCH_3$	Sn: Org.Verb.2-184/5
–	$(C_2H_5)_3SnCHCHCHCH_2$	Sn: Org.Verb.2-185, 189
–	$(C_2H_5)_3SnCH(CH_3)CCH$	Sn: Org.Verb.2-213
–	$(C_2H_5)_3SnCH_2CCCH_3$	Sn: Org.Verb.2-213
–	$(C_2H_5)_3SnCH_2CHCCH_2$	Sn: Org.Verb.2-185
–	$(C_2H_5)_3SnCH_2CH_2CCH$	Sn: Org.Verb.2-213
$C_{10}H_{21}IN_2NiS$	$C_3H_5Ni(SC(NHCH(CH_3)_2)_2)I$	Ni: Org.Verb.2-281
$C_{10}H_{21}I_4O_5PSn$	$SnI_4 \cdot C_2H_5OCOCH_2P(O)(OC_3H_7)_2$	Sn: MVol.C6-165/6
$C_{10}H_{21}IrN_5O^{3+}$	$[Ir(C_5H_5N)_2(NH_3)_3(H_2O)]^{3+}$	Ir: SVol.2-74
$C_{10}H_{21}NNi$	$(CH_3)_2C_3H_3Ni(NH(CH_2)_4)CH_3$	Ni: Org.Verb.2-35
$C_{10}H_{21}NNiO$	$(CH_3)_2C_3H_3Ni(NH(CH_2CH_2)_2O)CH_3$	Ni: Org.Verb.2-35
$C_{10}H_{21}NO_4Sn$	$(t\text{-}C_4H_9O)SnN(CH_2CH_2O)_3$	Sn: MVol.C6-4
$C_{10}H_{21}NSn$	$(C_2H_5)_3SnC(CN)(CH_3)_2$	Sn: Org.Verb.2-166
$C_{10}H_{21}N_3O_4S_2$	$HN(SO_2N(CH_2)_5)_2$	S: S-N-Verb.1-161
$C_{10}H_{21}N_3O_{12}U$	$UO_2(CH_3COO)_2 \cdot 3\ NH_2CH_2COOH$	U: SVol.E1-18, 21
$C_{10}H_{21}NiO_3PSn_2$	$(CO)_3NiP(CH_3)(Sn(CH_3)_3)_2$	Ni: Org.Verb.1-164/5
$C_{10}H_{21}Sn^+$	$(C_4H_9)_2SnCHCH_2{}^+$	Sn: Org.Verb.3-41
$C_{10}H_{21}Ti^+$	$[((CH_3)_3CCH_2)Ti(C_5H_{10})]^+$	Ti: Org.Verb.1-99
$C_{10}H_{22}I_2N_2O_2Sn$	$SnI_2 \cdot 2\ CH_3N(CH_2CH_2)_2O$	Sn: MVol.C5-29
$C_{10}H_{22}I_3IrS_3$	$IrI_3(S(CH_2CH_2SC_3H_7\text{-}n)_2)$	Ir: SVol.2-167
$C_{10}H_{22}N_2O_2S$	$(C_2H_5)_2NSO_2NHC_6H_{11}$	S: S-N-Verb.1-170
–	$C_4H_9NHSO_2NHC_6H_{11}$	S: S-N-Verb.1-170
–	$(i\text{-}C_4H_9)NHSO_2NHC_6H_{11}$	S: S-N-Verb.1-170
$C_{10}H_{22}N_2O_4Sn$	$Sn(CH_3N(CH_2CH_2O)_2)_2$	Sn: MVol.C6-2
$C_{10}H_{22}N_4O_4S_6Te$	$Te(SC(NHCH_2)_2CH_2)_2(S_2O_2CH_3)_2$	Te: SVol.B3-160
$C_{10}H_{22}N_4O_6S_2$	$(O(CH_2CH_2)_2NSO_2NH)_2C_2H_4$	S: S-N-Verb.1-171
$C_{10}H_{22}N_4O_{10}U$	$UO_2(NO_3)_2 \cdot 2\ C_2H_5CON(CH_3)_2$	U: SVol.E1-110/1
$C_{10}H_{22}N_4O_{12}U$	$UO_2(NO_3)_2 \cdot 2\ (CH_3)_2NCOOC_2H_5$	U: SVol.E1-120/3
$C_{10}H_{22}NiO_4S_6$	$Ni[S_2COC_2H_5]_2((CH_3)_2SO)_2$	C: MVol.D4-260
$C_{10}H_{22}OSn$	$(C_2H_5)_3SnC(CH_2)OC_2H_5$	Sn: Org.Verb.2-194, 203
–	$(C_2H_5)_3SnCHCHCH_2CH_2OH$	Sn: Org.Verb.2-194
–	$(C_2H_5)_3SnCHCHOC_2H_5$	Sn: Org.Verb.2-194, 203
–	$(C_2H_5)_3SnCH_2CH_2COCH_3$	Sn: Org.Verb.2-166, 176
–	$(C_4H_9)_2(CH_2CH)SnOH$	Sn: Org.Verb.5-212
$C_{10}H_{22}O_2Sn$	$(C_2H_5)_2Sn(CH_3)(C_2H_4COOC_2H_5)$	Sn: Org.Verb.3-86/7
–	$(C_2H_5)_3SnCH_2CH_2COOCH_3$	Sn: Org.Verb.2-167, 176
–	$(C_2H_5)_3SnCH_2CH_2OOCCH_3$	Sn: Org.Verb.2-166, 176
–	$(C_2H_5)_3SnCH_2COOC_2H_5$	Sn: Org.Verb.2-166, 176
$C_{10}H_{22}O_2Ti$	$C_2H_4 \cdot Ti(OC_4H_9)_2$	Ti: Org.Verb.1-123

Formula	Compound	Element	Reference
$C_{10}H_{22}O_8U$	$UO_2(CH_3COO)_2 \cdot 2\ C_3H_7OH$	U:	SVol.E1-59, 63
$C_{10}H_{22}SSn$	$(C_2H_5)_3SnCHCHSC_2H_5$	Sn:	Org.Verb.2-194, 203
$C_{10}H_{22}Sn$	$(CH_3)_3SnC(CH_3)CH(CH_2)_3CH_3$	Sn:	Org.Verb.2-81
–	$(CH_3)_3SnC(C_2H_5)CHC_3H_7$	Sn:	Org.Verb.2-81
–	$(CH_3)_3SnCHCH(CH_2)_4CH_3$	Sn:	Org.Verb.2-81
–	$(CH_3)_3SnCH(CH_2)_6$	Sn:	Org.Verb.2-62
–	$(CH_3)_3SnCH_2C_6H_{11}$	Sn:	Org.Verb.2-27
–	$(CH_3)_3SnC_6H_{10}CH_3$	Sn:	Org.Verb.2-61
–	$(C_2H_5)_2(CH_2CHCH_2CH_2CH_2CH_2)SnH$	Sn:	Org.Verb.4-91
–	$(C_2H_5)_2Sn(CH_2)_6$	Sn:	Org.Verb.3-108
–	$(C_2H_5)_3SnCH_2C(CH_3)CH_2$	Sn:	Org.Verb.2-185, 189
–	$(C_2H_5)_3SnCH_2CHCHCH_3$	Sn:	Org.Verb.2-185, 189
–	$(C_2H_5)_3SnCH_2CH_2CHCH_2$	Sn:	Org.Verb.2-185, 189
$C_{10}H_{22.5}I_4N_{2.5}O_{2.5}Te$	$TeI_4 \cdot 2.5\ CH_3CON(CH_3)_2$	Te:	SVol.B3-174
$C_{10}H_{22.5}N_{6.5}O_{14.5}U$	$U(NO_3)_4 \cdot 2.5\ CH_3CON(CH_3)_2$	U:	SVol.E1-104/5
–	$U(NO_3)_4 \cdot 2.5\ CH_3CONHC_2H_5$	U:	SVol.E1-96/9
$C_{10}H_{23}ISn$	$(C_4H_9)_2Sn(CH_3)(CH_2I)$	Sn:	Org.Verb.3-88/9
$C_{10}H_{23}NNi$	$(CH_3)_2C_3H_3Ni(NH(C_2H_5)_2)CH_3$	Ni:	Org.Verb.2-35
$C_{10}H_{23}NNiP_2$	$C_3H_5Ni(P(CH_3)_3)_2CN$	Ni:	Org.Verb.2-282
$C_{10}H_{23}NOSn$	$(C_2H_5)_3SnCH_2CH(CH_3)CONH_2$	Sn:	Org.Verb.2-167
–	$(C_2H_5)_3SnCH_2CON(CH_3)_2$	Sn:	Org.Verb.2-167, 176
$C_{10}H_{23}NO_4Sn$	$(i\text{-}C_3H_7O)_2SnHN(CH_2CH_2O)_2$	Sn:	MVol.C6-2
$C_{10}H_{23}NS_4Sn$	$Sn(SCH_2CH_2S)_2 \cdot (C_2H_5)_3N$	Sn:	MVol.C6-82/3
$C_{10}H_{23}NiP$	$C_3H_5Ni(P(C_2H_5)_3)CH_3$	Ni:	Org.Verb.2-30
$C_{10}H_{23}O_2Ti$	$C_2H_5Ti(OC_4H_9\text{-}n)_2$	Ti:	Org.Verb.1-9
–	$C_2H_5Ti(OC_4H_9\text{-}s)_2$	Ti:	Org.Verb.1-9
$C_{10}H_{24}I_2N_4S_2Te$	$Te((CH_3)_2NCSN(CH_3)_2)_2I_2$	Te:	SVol.B3-148
$C_{10}H_{24}Li_2NiO$	$C_2H_4Ni(LiC_2H_5)_2 \cdot (C_2H_5)_2O$	Ni:	Org.Verb.1-340/1
$C_{10}H_{24}NO_{11}UV_3$	$C_{10}H_{21}NH_3[UO_2(VO_3)_3] \cdot n\ H_2O$	U:	SVol.C3-302/3
$C_{10}H_{24}N_2NiS_6$	$[(CH_3)_4N]_2[Ni(CS_3)_2]$	C:	MVol.D4-225
$C_{10}H_{24}N_2NiS_7$	$[(CH_3)_4N]_2[Ni(CS_3)(CS_4)]$	C:	MVol.D4-225, 231
$C_{10}H_{24}N_2O_2S$	$(CH_3)(C_4H_9)NSO_2N(CH_3)C_4H_9$	S:	S-N-Verb.1-173
–	$(CH_3)(i\text{-}C_4H_9)NSO_2NCH_3(i\text{-}C_4H_9)$	S:	S-N-Verb.1-173
–	$(C_2H_5)(C_3H_7)NSO_2N(C_2H_5)C_3H_7$	S:	S-N-Verb.1-173
$C_{10}H_{24}N_2O_{13}W_4$	$(C_5H_{12}N)_2W_4O_{13}$	W:	SVol.B3-244
$C_{10}H_{24}N_2O_{16}W_5$	$(C_5H_{12}N)_2W_5O_{16}$	W:	SVol.B3-244
$C_{10}H_{24}N_2S_6$	$[(CH_3)_4N]_2C_2S_6 \cdot 0.5\ CS_2$	C:	MVol.D4-229
$C_{10}H_{24}N_3O_7PU$	$UO_2(CH_3COO)_2 \cdot ((CH_3)_2N)_3PO$	U:	SVol.E1-163, 165
$C_{10}H_{24}N_4O_2S_2Ti$	$(CO)_2Ti(S(CH_2CH_2NH_2)_2)_2$	Ti:	Org.Verb.1-122
$C_{10}H_{24}N_4O_8SU$	$UO_2SO_4 \cdot 2\ (CH_3)_2NCON(CH_3)_2$	U:	SVol.E1-93/5
–	$UO_2SO_4 \cdot 2\ CO(NHC_2H_5)_2$	U:	SVol.E1-93/5
$C_{10}H_{24}N_4S_2Te^{2+}$	$[Te((CH_3)_2NCSN(CH_3)_2)_2]^{2+}$	Te:	SVol.B3-148/50
$C_{10}H_{24}N_6NpO_{20}$	$(NH_4)_4[Np(C_2O_4)_4] \cdot (NH_4)_2C_2O_4 \cdot x\ H_2O$	Np:	TrU.D1-118
$C_{10}H_{24}N_6O_{10}U$	$UO_2(NO_3)_2 \cdot 2\ CO(N(CH_3)_2)_2$	U:	SVol.E1-92, 95
–	$UO_2(NO_3)_2 \cdot 2\ CO(NHC_2H_5)_2$	U:	SVol.E1-92, 95
$C_{10}H_{24}OSn$	$(C_2H_5)_3SnCH_2CH_2OC_2H_5$	Sn:	Org.Verb.2-167
$C_{10}H_{24}O_2S_6Sn$	$Sn(SCH_2CH(S)CH_3)_2 \cdot 2\ (CH_3)_2SO$	Sn:	MVol.C6-101
$C_{10}H_{24}O_2Sn$	$(C_2H_5)_2Sn(CH_2OC_2H_5)_2$	Sn:	Org.Verb.3-27
–	$(C_4H_9)_2Sn(OCH_3)_2$	Sn:	Org.Verb.6-91
$C_{10}H_{24}O_2Ti$	$(CH_3)_2Ti(OC_4H_9\text{-}t)_2$	Ti:	Org.Verb.1-97

Formula	Compound		Reference
$C_{10}H_{24}O_3Sn$	$C_4H_9Sn(OC_2H_5)_3$	Sn:	Org.Verb.6-246
$C_{10}H_{24}O_3Ti$	$CH_3Ti(OC_3H_7\text{-}i)_3$	Ti:	Org.Verb.1-44, 48
–	$(n\text{-}C_4H_9)Ti(OC_2H_5)_3$	Ti:	Org.Verb.1-6, 45
$C_{10}H_{24}O_4Sn$	$(CH_3)_2Sn[OOC(CH_3)_3]_2$	Sn:	Org.Verb.6-27
$C_{10}H_{24}SSn$	$(C_2H_5)_3SnSC_4H_9$	Sn:	Org.Verb.5-104
$C_{10}H_{24}Sn$	$(CH_3)_2Sn(C_4H_9)(s\text{-}C_4H_9)$	Sn:	Org.Verb.3-75, 77
–	$(CH_3)_2Sn(i\text{-}C_4H_9)(s\text{-}C_4H_9)$	Sn:	Org.Verb.3-75, 78
–	$(CH_3)_2Sn(C_4H_9)_2$	Sn:	Org.Verb.3-8, 17
		Sn:	Org.Verb.6-18
–	$(CH_3)_2Sn(i\text{-}C_4H_9)_2$	Sn:	Org.Verb.3-9
–	$(CH_3)_2Sn(s\text{-}C_4H_9)(t\text{-}C_4H_9)$	Sn:	Org.Verb.3-75, 78
–	$(CH_3)_2Sn(s\text{-}C_4H_9)_2$	Sn:	Org.Verb.3-9, 17
–	$(CH_3)_2Sn(t\text{-}C_4H_9)_2$	Sn:	Org.Verb.3-9, 17
–	$(C_2H_5)_2Sn(C_3H_7)_2$	Sn:	Org.Verb.3-27
–	$(C_2H_5)_2Sn(i\text{-}C_3H_7)_2$	Sn:	Org.Verb.3-27
–	$(C_2H_5)_3SnC_4H_9$	Sn:	Org.Verb.2-158/9
–	$(C_2H_5)_3Sn(i\text{-}C_4H_9)$	Sn:	Org.Verb.2-159
–	$(C_2H_5)_3Sn(t\text{-}C_4H_9)$	Sn:	Org.Verb.2-159
–	$(C_3H_7)_3SnCH_3$	Sn:	Org.Verb.2-250
–	$(i\text{-}C_3H_7)_3SnCH_3$	Sn:	Org.Verb.2-270/1
–	$(C_5H_{11})_2SnH_2$	Sn:	Org.Verb.4-112
$C_{10}H_{24}Ti$	$(C_4H_9)(C_2H_5)_3Ti$	Ti:	Org.Verb.1-90
$C_{10}H_{25}NSn$	$(C_2H_5)_3SnN(C_2H_5)_2$	Sn:	Org.Verb.5-104
$C_{10}H_{25}N_3Ti$	$H_2CC(CH_3)CH_2Ti(N(CH_3)_2)_3$	Ti:	Org.Verb.1-53, 57
$C_{10}H_{25}NiO_6P_2^+$	$[CH_3C_3H_4Ni(P(OCH_3)_3)_2]^+$	Ni:	Org.Verb.2-24
$C_{10}H_{25}O_3PSn$	$(C_2H_5)_3SnOP(OC_2H_5)_2$	Sn:	Org.Verb.5-105
$C_{10}H_{25}O_5Pa$	$Pa(OC_2H_5)_5$	Pa:	SVol.2-82
$C_{10}H_{26}I_4O_4P_2Sn$	$SnI_4 \cdot 2\ C_2H_5(H)P(O)OCH(CH_3)_2$	Sn:	MVol.C6-151/2
–	$SnI_4 \cdot 2\ C_2H_5(H)P(O)OC_3H_7$	Sn:	MVol.C6-151/2
$C_{10}H_{26}N_2O_2S_4Sn$	$Sn(SCH_2CH_2S)_2 \cdot 2\ (CH_3)_3NO$	Sn:	MVol.C6-82/3
$C_{10}H_{26}N_4O_4S_2$	$(CH_3)_2NSO_2NH(CH_2)_6NHSO_2N(CH_3)_2$	S:	S-N-Verb.1-171
–	$(C_2H_5)_2NSO_2NH(CH_2)_2NHSO_2N(C_2H_5)_2$	S:	S-N-Verb.1-171
$C_{10}H_{26}N_4O_6U$	$UO_2(OH)_2 \cdot 2\ (CH_3)_2NCON(CH_3)_2$	U:	SVol.E1-94
$C_{10}H_{26}Ni_2O_4P_2$	$(CH_3Ni(P(CH_3)_3)O_2CH)_2$	Ni:	Org.Verb.2-252
$C_{10}H_{26}OSi_2Sn$	$[CH_2Si(CH_3)_2OSi(CH_3)_2CH_2Sn(C_2H_5)_2]_x$	Sn:	Org.Verb.6-53
–	$O(Si(CH_3)_2CH_2)_2Sn(C_2H_5)_2$	Sn:	Org.Verb.6-207
$C_{10}H_{26}P_2Sn$	$(CH_3)_2Sn[CHP(CH_3)_3]_2$	Sn:	Org.Verb.3-9
$C_{10}H_{26}SiSn$	$(CH_3)_3SnCH_2Si(CH_3)_2C_4H_9$	Sn:	Org.Verb.2-14/5
$C_{10}H_{27}N_3Ti$	$C_4H_9Ti(N(CH_3)_2)_3$	Ti:	Org.Verb.1-52
$C_{10}H_{27}NiOP_3$	$CONi(P(CH_3)_3)_3$	Ni:	Org.Verb.1-114/5
$C_{10}H_{27}NiO_{10}P_3$	$CONi(P(OCH_3)_3)_3$	Ni:	Org.Verb.1-116, 117
$C_{10}H_{27}PSiSn$	$(CH_3)_3SnC[Si(CH_3)_3]P(CH_3)_3$	Sn:	Org.Verb.2-13/4
$C_{10}H_{28}N_2OS_4$	$[(CH_3)_4N]_2CS_4 \cdot CH_3OH$	C:	MVol.D4-231
$C_{10}H_{28}N_2Ti$	$(CH_3)_4Ti \cdot (CH_3)_2N(CH_2)_2N(CH_3)_2$	Ti:	Org.Verb.1-113, 117
$C_{10}H_{28}N_6Ni$	$[Ni(NH_3)_6][C_5H_5]_2$	Ni:	Org.Verb.1-123
$C_{10}H_{28}O_7U$	$UO_2(OC_2H_5)_2 \cdot 3\ C_2H_5OH$	U:	SVol.E1-59, 64
$C_{10}H_{28}P_2Ti$	$(CH_3)_4Ti \cdot (CH_3)_2PCH_2CH_2P(CH_3)_2$	Ti:	Org.Verb.1-114, 117
$C_{10}H_{28}Si_2Sn$	$(CH_3)_2Sn[CH_2Si(CH_3)_3]_2$	Sn:	Org.Verb.3-9, 17
$C_{10}H_{30}N_2O_{13}S_5U$	$UO_2(NO_3)_2 \cdot 5\ (CH_3)_2SO$	U:	SVol.E1-204, 208
$C_{10}H_{30}N_2Ti$	$(CH_3)_4Ti \cdot 2\ N(CH_3)_3$	Ti:	Org.Verb.1-117

Formula	Compound		Reference
$C_{10}H_{30}N_3SSi_2Sn^+$	$[(CH_3)_3SiN]_2SN(CH_3)Sn(CH_3)_3{}^+$	S:	S-N-Verb.1-211
$C_{10}H_{30}N_4SSi_2$	$[(CH_3)_3SiN]_2S[N(CH_3)_2]_2$	S:	S-N-Verb.1-209/10
$C_{10}H_{30}N_4SSi_2{}^+$	$[(CH_3)_3SiN]_2S[N(CH_3)_2]_2{}^+$	S:	S-N-Verb.1-210
$C_{10}H_{30}N_6NiS_3$	$Ni(CH_3CH(NH_2)CH_2NH_2)_3CS_3$	C:	MVol.D4-225
$C_{10}H_{30}N_6O_{20}U_2$	$[UO_2(NO_3)_2(HOC_2H_4N(CH_3)_3OH)]_2$	U:	SVol.E1-23
$C_{10}H_{30}N_{12}O_{12}U$	$UO_2(CH_3COO)_2 \cdot 6\ (NH_2)_2CO$	U:	SVol.E1-84, 89
$C_{10}H_{30}Ni_2O_2P_2$	$(CH_3Ni(P(CH_3)_3)OCH_3)_2$	Ni:	Org.Verb.2-251/2
$C_{10}H_{30}O_3Si_2Sn_2$	$[(CH_3)_3SiOSn(CH_3)_2]_2O$	Sn:	Org.Verb.6-21
$C_{10}H_{30}P_2Ti$	$(CH_3)_4Ti \cdot 2\ P(CH_3)_3$	Ti:	Org.Verb.1-113, 117
$C_{10}H_{31}NO_8Sn_2$	$2\ Sn(OCH_3)_4 \cdot (CH_3)_2NH$	Sn:	MVol.C5-170/1
$C_{10}K_6O_{20}Sn_2$	$K_6Sn_2(C_2O_4)_5 \cdot 3\ H_2O$	Sn:	MVol.C3-78
$C_{10}Mn_2O_{10}$	$Mn_2(CO)_{10}$ systems		
	$Mn_2(CO)_{10}$-$Fe(CO)_5$	Fe:	Org.Verb.B3-64
$C_{10}O_9Re_4S_3$	$Re_4(CO)_9(CS_3)$	C:	MVol.D4-225
$C_{10}O_{20}Pu^{6-}$	$Pu(C_2O_4)_5{}^{6-}$	Np:	TrU.D1-157
$C_{10.5}Cl_2H_{21}N_3O_5U$	$UO_2Cl_2 \cdot 1.5\ CH_2(CON(CH_3)_2)_2$	U:	SVol.E1-113/4
$C_{10.5}Cl_4H_{21}N_3O_3U$	$UCl_4 \cdot 1.5\ CH_2(CON(CH_3)_2)_2$	U:	SVol.E1-112/4
$C_{10.5}H_{24}N_2S_7$	$[(CH_3)_4N]_2C_2S_6 \cdot 0.5\ CS_2$	C:	MVol.D4-229
$C_{11}ClF_3H_5NS$	$(C_6H_5S)C_5F_3ClN$	F:	PerFHalOrg.5-141
$C_{11}ClF_3H_9N_3OS$	$N(CH_2COC_6H_5)C(NH \cdot HCl)SC(CF_3)N$	F:	PerFHalOrg.5-100
$C_{11}ClF_3H_9N_3OSe$	$N(CH_2COC_6H_5)C(NH \cdot HCl)SeC(CF_3)N$	F:	PerFHalOrg.5-100
$C_{11}ClF_4H_{13}O_2Si$	$C_6F_4ClSi(OC_2H_5)_2CH_3$	F:	PerFHalOrg.4-78
$C_{11}ClF_5H_{11}N_3$	$(CF_3)(CF_2Cl)(C_6H_{11})C_3N_3$	F:	PerFHalOrg.6-113
$C_{11}ClF_6H_3N_2O$	$NC_5F_2Cl(OCH_3)C_5F_4N$	F:	PerFHalOrg.5-224
–	$NC_5F_3ClC_5F_3(OCH_3)N$	F:	PerFHalOrg.5-224
$C_{11}ClF_6H_9NO_3P$	$C_5F_6(Cl)P(O)(OH)_2 \cdot C_6H_5NH_2$	F:	PerFHalOrg.3-50/1
$C_{11}ClF_8H_{12}N_3$	$(CF_3)_2(CF_2Cl)(C_5H_{11})C_3HN_3$	F:	PerFHalOrg.6-113
$C_{11}ClF_{12}H_5N_2O_2$	$ClC_6H_4CH(ON(CF_3)_2)_2$	F:	PerFHalOrg.7-137/8
$C_{11}ClFeH_4NaO_5$	$Na[ClC_6H_4C(O)Fe(CO)_4]$	Fe:	Org.Verb.B4-152, 162
$C_{11}ClFeH_4O_5{}^-$	$[ClC_6H_4C(O)Fe(CO)_4]^-$	Fe:	Org.Verb.B4-152, 162
$C_{11}ClFeH_7O_6S$	$CH_3C_6H_4SO_2Fe(CO)_4Cl$	Fe:	Org.Verb.B3-187
$C_{11}ClFeH_7O_7S$	$CH_3OC_6H_4SO_2Fe(CO)_4Cl$	Fe:	Org.Verb.B3-187
$C_{11}ClFeH_8N$	$ClC_5H_4FeC_5H_4CN$	Fe:	Org.Verb.A1-374
$C_{11}ClFeH_9O$	$C_5H_5FeC_5H_4COCl$	Fe:	Org.Verb.A3-152/7
$C_{11}ClFeH_9O^+$	$[C_5H_5FeC_5H_4COCl]^+$	Fe:	Org.Verb.A3-152/3
$C_{11}ClFeH_9O_2$	$C_5H_5FeC_5H_3(Cl)COOH$	Fe:	Org.Verb.A3-59
–	$ClC_5H_4FeC_5H_4COOH$	Fe:	Org.Verb.A3-58/9
$C_{11}ClFeH_9O_3$	$C_8H_9Fe(CO)_3Cl$	Fe:	Org.Verb.B5-46, 54, 64
$C_{11}ClFeH_{11}$	$C_5H_5FeC_5H_4CH_2Cl$	Fe:	Org.Verb.A1-359, 381/2
$C_{11}ClFeH_{11}O$	$[C_5H_5FeC_5H_4CHOH]Cl$	Fe:	Org.Verb.A2-151
–	$[C_5H_5FeC_5H_4CHOD]Cl$	Fe:	Org.Verb.A2-151
$C_{11}ClFeH_{12}NaO_5$	$Na[ClCH_2(CH_2)_5C(O)Fe(CO)_4]$	Fe:	Org.Verb.B4-151, 160
$C_{11}ClFeH_{12}O_5{}^-$	$[ClCH_2(CH_2)_5C(O)Fe(CO)_4]^-$	Fe:	Org.Verb.B4-151, 160
$C_{11}ClFeH_{13}MgO_5$	$MgCl[C_6H_{13}C(O)Fe(CO)_4]$	Fe:	Org.Verb.B4-151, 160
$C_{11}ClFeH_{15}N_6O_4$	$[(CH_3NC)_5FeCN]ClO_4$	Fe:	Org.Verb.B4-78
$C_{11}ClFeH_{19}NO_6P$	$ClC_3H_4C(O)Fe(CO)(NO)P(OC_2H_5)_3$	Fe:	Org.Verb.B5-108/11
$C_{11}ClFeH_{21}NO_5P$	$CH_3C_3H_3ClFe(CO)(NO)P(OC_2H_5)_3$	Fe:	Org.Verb.B5-108/11
$C_{11}ClFeH_{22}NO_9P_2$	$COFe(NO)(P(OCH_3)_3)_2COCH_2CClCH_2$	Fe:	Org.Verb.B1-40/2
$C_{11}ClFe_2GeH_{19}N_4O_4$	$(C_3H_5Fe(NO)_2)_2Ge(Cl)C_5H_9$	Fe:	Org.Verb.C2-165

$C_{11}ClFe_2GeH_{21}N_4O_4$. $((CH_3)_2C_3H_3Fe(NO)_2)_2Ge(Cl)CH_3$ .... Fe: Org.Verb.C2-165
$C_{11}ClFe_2H_{10}O_2$ ..... $[C_5H_5FeC_5H_4COOH]FeCl$ .... Fe: Org.Verb.A3-63/5
$C_{11}ClFe_2H_{19}N_4O_4Pb$. $(C_3H_5Fe(NO)_2)_2Pb(Cl)C_5H_9$ .... Fe: Org.Verb.C2-165
$C_{11}ClFe_2H_{19}N_4O_4Sn$. $(C_3H_5Fe(NO)_2)_2Sn(Cl)C_5H_9$ .... Fe: Org.Verb.C2-165
$C_{11}ClFe_2H_{21}N_4O_4Pb$. $((CH_3)_2C_3H_3Fe(NO)_2)_2Pb(Cl)CH_3$ .... Fe: Org.Verb.C2-165
$C_{11}ClFe_2H_{21}N_4O_4Sn$. $((CH_3)_2C_3H_3Fe(NO)_2)_2Sn(Cl)CH_3$ .... Fe: Org.Verb.C2-165
$C_{11}ClH_8NO_2S$ .... $C_{10}H_7CHNSO_2Cl$ .... S: S-N-Verb.1-126
$C_{11}ClH_{10}NO_2Sn$ ..... $C_5H_5NH[SnClC_6H_4O_2]$ .... Sn: MVol.C5-7/8
$C_{11}ClH_{12}NOSn$ ..... $(CH_3)_2SnCl(OC_9H_6N)$ .... Sn: Org.Verb.6-21
$C_{11}ClH_{13}Sn$ .... $(CH_2CH)_2(C_6H_5CH_2)SnCl$ .... Sn: Org.Verb.5-222
– .... $(C_5H_5)_2(CH_3)SnCl$ .... Sn: Org.Verb.5-223, 224
$C_{11}ClH_{15}OSn$ .... $(CH_3)_2(C_6H_5COCH_2CH_2)SnCl$ .... Sn: Org.Verb.5-195, 199
$C_{11}ClH_{16}IrN_2O_8$ ..... $[IrH_2(C_3H_6N_2(CH_2COO)_4)Cl] \cdot H_2O$ .... Ir: SVol.2-33
$C_{11}ClH_{16}OPPdS_2$ .... $[(CH_3)_2P(C_6H_5)]PdCl[S_2COC_2H_5]$ .... C: MVol.D4-261
$C_{11}ClH_{17}Ni$ .... $CH_3C_3H_4Ni(C_7H_{10})Cl$ .... Ni: Org.Verb.2-46
$C_{11}ClH_{17}Sn$ .... $(CH_3)(i\text{-}C_4H_9)(C_6H_5)SnCl$ .... Sn: Org.Verb.5-225, 227
$C_{11}ClH_{18}NiO_5P$ ..... $(CO)_3NiP(OC_4H_9\text{-}n)_2Cl$ .... Ni: Org.Verb.1-178, 179
$C_{11}ClH_{19}O_2Sn$ ...... $[SnCl(CH(COC(CH_3)_3)_2)]$ .... Sn: MVol.C5-9
$C_{11}ClH_{19}O_2Ti$ ...... $C_2H_5C_5H_4Ti(OC_2H_5)_2Cl$ .... Ti: Org.Verb.1-187/8
– .... $C_5H_5Ti(OC_3H_7\text{-}n)_2Cl$ .... Ti: Org.Verb.1-185/6
– .... $C_5H_5Ti(OC_3H_7\text{-}i)_2Cl$ .... Ti: Org.Verb.1-185/6
$C_{11}ClH_{19}O_3Ti$ ...... $(CH_2CHC_2H_4O)(C_2H_5)Ti(CH(COCH_3)_2)Cl$ ... Ti: Org.Verb.1-124/5
– .... $C_5H_5TiCl \cdot (CH_3OC_2H_4)_2O$ .... Ti: Org.Verb.1-132
$C_{11}ClH_{20}NSn$ .... $(C_2H_5)_3SnCl \cdot C_5H_5N$ .... Sn: Org.Verb.5-107
– .... $(i\text{-}C_4H_9)_2(NCCHCH)SnCl$ .... Sn: Org.Verb.5-213/4
$C_{11}ClH_{20}N_3O_9SU$ .... $UO_2[(C_{10}H_{16}N_3O_6S)Cl] \cdot CH_3OH \cdot 2\ H_2O$ .. U: SVol.E1-59, 63
$C_{11}ClH_{20}NiP$ .... $C_5H_5Ni(P(C_2H_5)_3)Cl$ .... Ni: Org.Verb.2-117, 119
$C_{11}ClH_{22}NSn$ .... $(i\text{-}C_4H_9)_2(NCCH_2CH_2)SnCl$ .... Sn: Org.Verb.5-213/4
$C_{11}ClH_{23}MnN_4O_6$ .... $[Mn(CH(COCH_3)_2)(C_2H_4(NC_2H_4NH_2)_2)]ClO_4$. Mn: MVol.D1-108
$C_{11}ClH_{23}N_2OSn$ ..... $[CH_2CHC(NH_2)NOSn(C_4H_9)_2Cl]_x$ .... Sn: Org.Verb.6-89, 95
$C_{11}ClH_{23}O_2Sn$ ...... $(C_4H_9)_2(CH_2Cl)SnOCOCH_3$ .... Sn: Org.Verb.5-205
– .... $(C_4H_9)_2(CH_3OCOCH_2)SnCl$ .... Sn: Org.Verb.5-208
$C_{11}ClH_{23}Sn$ .... $(C_4H_9)_2(CH_2CHCH_2)SnCl$ .... Sn: Org.Verb.5-207
$C_{11}ClH_{25}MgNiO$ .... $(CH_3)_2C_3H_3Ni(CH_3)ClMgCH_3 \cdot (C_2H_5)_2O$ ... Ni: Org.Verb.2-315
$C_{11}ClH_{25}OSn$ .... $(i\text{-}C_4H_9)_2ClSnOCH_2C_2H_5$ .... Sn: Org.Verb.6-105
– .... $(i\text{-}C_4H_9)_2[HO(CH_2)_3]SnCl$ .... Sn: Org.Verb.5-213/4
$C_{11}ClH_{25}Sn$ .... $(C_4H_9)_2(C_3H_7)SnCl$ .... Sn: Org.Verb.5-208, 212
$C_{11}ClH_{26}O_3PSn$ ..... $(C_2H_5)_2[(C_2H_5O)_2PO(CH_2)_3]SnCl$ .... Sn: Org.Verb.5-202
$C_{11}ClH_{27}OSn_2$ ...... $(CH_3)_3SnOSnCl(C_4H_9)_2$ .... Sn: Org.Verb.5-82
$C_{11}ClH_{28}OPSn$ ..... $(C_2H_5)_3SnCl \cdot (C_2H_5)_2(CH_3O)P$ .... Sn: Org.Verb.5-107
$C_{11}ClH_{31}O_6P_2Sn$ .... $(CH_3)_3SnCl \cdot 2\ (C_2H_5O)_2HPO$ .... Sn: Org.Verb.5-87
$C_{11}Cl_2FH_7N_2O$ ...... $(NH_2)(C_6H_5O)C_5FCl_2N$ .... F: PerFHalOrg.5-214/5
$C_{11}Cl_2FH_7N_2S$ ...... $(NH_2)(C_6H_5S)C_5FCl_2N$ .... F: PerFHalOrg.5-214/5
$C_{11}Cl_2F_2H_3N_5O_2$ .... $NC_5F_2Cl_2NHC_5H_2(CN)(NO_2)N$ .... F: PerFHalOrg.5-139
$C_{11}Cl_2F_2H_3N_5O_6$ .... $NC_5F_2Cl_2NHC_6H_2(NO_2)_3$ .... F: PerFHalOrg.5-215
$C_{11}Cl_2F_2H_4N_2O_3$ .... $NO_2C_6H_4OC_5F_2Cl_2N$ .... F: PerFHalOrg.5-142/3
$C_{11}Cl_2F_2H_4N_4O_4$ .... $NC_5F_2Cl_2NHC_6H_3(NO_2)_2$ .... F: PerFHalOrg.5-140, 215
$C_{11}Cl_2F_2H_5N$ .... $C_6H_5C_5F_2Cl_2N$ .... F: PerFHalOrg.5-216
$C_{11}Cl_2F_2H_5NS$ ...... $(C_6H_5S)C_5F_2Cl_2N$ .... F: PerFHalOrg.5-141
$C_{11}Cl_2F_2H_5N_3O_2$ .... $NC_5F_2Cl_2NHC_6H_4NO_2$ .... F: PerFHalOrg.5-140

Formula	Compound	Reference
$C_{11}Cl_2F_2H_5N_5$	$NC_5F_2Cl_2NHC_4HN_2(CN)CH_3$	F: PerFHalOrg.5-216
$C_{11}Cl_2F_2H_6N_2$	$NC_5F_2Cl_2NHC_6H_5$	F: PerFHalOrg.5-140
$C_{11}Cl_2F_2H_6N_2O$	$NC_5F_2Cl_2(OCH_2C_5H_4N)$	F: PerFHalOrg.5-136
		F: PerFHalOrg.6-14
$C_{11}Cl_2F_2H_{14}NO_3PS$	$NC_5F_2Cl_2(OPS(OC_3H_7)_2)$	F: PerFHalOrg.5-212
$C_{11}Cl_2F_2H_{14}NO_4P$	$NC_5F_2Cl_2(OPO(OC_3H_7)_2)$	F: PerFHalOrg.5-212
$C_{11}Cl_2F_2H_{15}N_3$	$N(CFCCl)_2CNH(CH_2)_6NH_2$	F: PerFHalOrg.5-140
$C_{11}Cl_2F_3H_3N_4$	$NC_5F_2Cl_2NHC_5H_2F(CN)N$	F: PerFHalOrg.5-139
$C_{11}Cl_2F_3H_3N_4O_4$	$NC_5F_2Cl_2NHC_6H_2F(NO_2)_2$	F: PerFHalOrg.5-215
$C_{11}Cl_2F_3H_5N_2$	$NC_5F_2Cl_2NHC_6H_4F$	F: PerFHalOrg.5-140
$C_{11}Cl_2F_4H_7NOS$	$SC(CF_2Cl)_2NHC(O)CH(C_6H_5)$	F: PerFHalOrg.7-66/7
$C_{11}Cl_2F_4H_{11}N_3$	$(CF_2Cl)_2(C_6H_{11})C_3N_3$	F: PerFHalOrg.6-113
$C_{11}Cl_2F_5HN_4$	$NC_5F_2Cl_2NHC_5F_3(CN)N$	F: PerFHalOrg.5-186/7
		F: PerFHalOrg.6-13/4
$C_{11}Cl_2F_5H_2N_5$	$NC_5F_2Cl_2NHC_4HN_2(CN)CF_3$	F: PerFHalOrg.5-216
$C_{11}Cl_2F_5H_3N_2O$	$NC_5F_2Cl(OCH_3)C_5F_3ClN$	F: PerFHalOrg.5-224
$C_{11}Cl_2F_5H_4N_3O$	$NC_5F_4NHC_5FCl_2(OCH_3)N$	F: PerFHalOrg.5-138/9
$C_{11}Cl_2F_6HN_3O_2$	$NC_5F_2Cl_2NHC_6F_4NO_2$	F: PerFHalOrg.5-188/9
		F: PerFHalOrg.6-13
$C_{11}Cl_2F_7HN_2$	$NC_5F_2Cl_2NHC_6F_5$	F: PerFHalOrg.5-188
		F: PerFHalOrg.6-13
$C_{11}Cl_2F_7H_{12}N_3$	$(CF_2Cl)_2(CF_3)(C_5H_{11})C_3HN_3$	F: PerFHalOrg.6-113
$C_{11}Cl_2F_{25}N_3O_3$	$ClCF_2CF_2[N(CF_3)OCF_2CF_2]_3Cl$	F: PerFHalOrg.7-95, 112
$C_{11}Cl_2FeGeH_5NO_4S$	$(CO)_4Fe(GeCl_2)(C_6H_4SCHN)$	Fe: Org.Verb.B3-161
$C_{11}Cl_2FeH_7NO_3$	$CH_3C_6H_4NCFe(CO)_3Cl_2$	Fe: Org.Verb.B4-2/3
$C_{11}Cl_2H_9NOSn$	$SnCl_2 \cdot C_9H_6N(OCHCH_2)$	Sn: MVol.C5-21
$C_{11}Cl_2H_{10}N_4Sn$	$SnCl_2 \cdot NC_5H_4CHNNHC_5H_4N$	Sn: MVol.C5-39
$C_{11}Cl_2H_{10}OTi$	$C_5H_5Ti(OC_6H_5)Cl_2$	Ti: Org.Verb.1-180/1
$C_{11}Cl_2H_{10}STi$	$C_5H_5Ti(SC_6H_5)Cl_2$	Ti: Org.Verb.1-189
$C_{11}Cl_2H_{10}Sn$	$(C_6H_5)(C_5H_5)SnCl_2$	Sn: Org.Verb.6-201
$C_{11}Cl_2H_{14}OSn$	$(C_6H_5)[CH_3CO(CH_2)_3]SnCl_2$	Sn: Org.Verb.6-201
$C_{11}Cl_2H_{15}NS_2Ti$	$C_5H_5Ti(S_2CN(CH_2)_5)Cl_2$	Ti: Org.Verb.1-198/9
$C_{11}Cl_2H_{16}OSn$	$(C_2H_5)(C_6H_5)SnCl_2 \cdot (CH_3)_2CO$	Sn: Org.Verb.6-204
$C_{11}Cl_2H_{17}O_{1.5}Ti$	$C_5H_5TiCl_2 \cdot 1.5\ C_4H_8O$	Ti: Org.Verb.1-134/6
$C_{11}Cl_2H_{20}O_2S_2Sn$	$(CH_3)(C_6H_5)SnCl_2 \cdot 2\ (CH_3)_2SO$	Sn: Org.Verb.6-195
$C_{11}Cl_2H_{21}P_2Ti$	$C_5H_5TiCl_2 \cdot (CH_3)_2PCH_2CH_2P(CH_3)_2$	Ti: Org.1-134/5
$C_{11}Cl_2H_{22}N_2O_4U$	$UO_2Cl_2 \cdot (CH_3)_2C(CH_2CON(CH_3)_2)_2$	U: SVol.E1-113
$C_{11}Cl_2H_{24}OSn$	$(C_4H_9)_2SnCl_2 \cdot (CH_3)_2CO$	Sn: Org.Verb.6-100
$C_{11}Cl_2H_{24}Sn$	$(CH_3)(C_{10}H_{21})SnCl_2$	Sn: Org.Verb.6-197
$C_{11}Cl_2H_{27}O_2PSn$	$(C_2H_5)_2SnCl_2 \cdot (C_2H_5)_2(C_3H_7O)PO$	Sn: Org.Verb.6-61/2
$C_{11}Cl_2H_{29}NSn$	$[(C_2H_5)_4N][(CH_3)_3SnCl_2]$	Sn: Org.Verb.5-87
$C_{11}Cl_2H_{30}Si_2Sn_2$	$[(CH_3)_3Si]_2C[Sn(CH_3)_2Cl]_2$	Sn: Org.Verb.6-18
$C_{11}Cl_2H_{31}NiP_3$	$[((CH_3)_3PCH_2)_2Ni(P(CH_3)_3)Cl]Cl$	Ni: Org.Verb.1-71, 77/8
$C_{11}Cl_3F_2H_3N_4O_4$	$NC_5F_2Cl_2NHC_6H_2Cl(NO_2)_2$	F: PerFHalOrg.5-140
$C_{11}Cl_3F_3H_9N_3O_2S$	$(C_3H_7NHSO_2)(CF_3)C_7Cl_3HN_2$	F: PerFHalOrg.6-160
$C_{11}Cl_3F_4HN_4$	$NC_5F_2Cl_2NHC_5F_2Cl(CN)N$	F: PerFHalOrg.5-186/7
		F: PerFHalOrg.6-13/4
–	$N_2C_4Cl_3NHC_6F_4CN$	F: PerFHalOrg.7-15/6, 43, 76
$C_{11}Cl_3F_6H_{12}N_3$	$(CF_2Cl)_3(C_5H_{11})C_3HN_3$	F: PerFHalOrg.6-113

$C_{11}Cl_3F_7HN_3$	$CF_3C_6F_4NHCCClCClNCClN$	F:	PerFHalOrg.6-21/2
$C_{11}Cl_3FeH_9$	$C_5H_5FeC_5H_4CCl_3$	Fe:	Org.Verb.A1-359, 381/2
$C_{11}Cl_3FeH_{11}$	$Fe(C_5H_5)_2 \cdot CHCl_3$	Fe:	Org.Verb.A1-189
$C_{11}Cl_3FeH_{13}N_6$	$(CH_3NC)_4Fe(CN)_2 \cdot CHCl_3$	Fe:	Org.Verb.B4-68
$C_{11}Cl_3GeH_{20}NNiO_3$	$[N(C_2H_5)_4][(CO)_3NiGeCl_3]$	Ni:	Org.Verb.1-191, 192
$C_{11}Cl_3GeH_{20}NiP$	$C_5H_5Ni(P(C_2H_5)_3)GeCl_3$	Ni:	Org.Verb.2-130
$C_{11}Cl_3H_8NTi$	$NC_5H_4C_6H_4TiCl_3$	Ti:	Org.Verb.1-42
$C_{11}Cl_3H_9Sn$	$CH_3C_{10}H_6SnCl_3$	Sn:	Org.Verb.6-287
$C_{11}Cl_3H_{11}NOSSn^-$	$[SnCl_3(SC_6H_4NC(CH_3)CHCOCH_3)]^-$	Sn:	MVol.C6-90/1
$C_{11}Cl_3H_{11}N_2Sn$	$CH_3SnCl_3 \cdot NC_5H_4C_5H_4N$	Sn:	Org.Verb.6-222
$C_{11}Cl_3H_{11}N_2Ti$	$CH_3TiCl_3 \cdot NC_5H_4C_5H_4N$	Ti:	Org.Verb.1-34/5, 41
$C_{11}Cl_3H_{12}N_3NiO_2P_2$	$(CO)_2Ni(PCl_3)(P(C_2H_4CN)_3)$	Ni:	Org.Verb.1-146
$C_{11}Cl_3H_{13}IrN_3S_2$	$Ir(S_2N_2C_4H_4)Cl \cdot CH_3C_6H_4NH_2$	Ir:	SVol.2-172
$C_{11}Cl_3H_{13}IrN_3Se_2$	$Ir(Se_2N_2C_4H_4)(CH_3C_6H_4NH_2)Cl_3$	Ir:	SVol.2-175
$C_{11}Cl_3H_{13}N_2Sn$	$CH_3SnCl_3 \cdot 2\ C_5H_5N$	Sn:	Org.Verb.6-222
$C_{11}Cl_3H_{13}N_2Ti$	$CH_3TiCl_3 \cdot 2\ C_5H_5N$	Ti:	Org.Verb.1-30/1, 39/40
$C_{11}Cl_3H_{14}IrN_2O$	$Ir(C_5H_5N)_2Cl_3(CH_3OH)$	Ir:	SVol.2-74
$C_{11}Cl_3H_{17}N_2O_2Sn$	$C_4H_9SnCl_3 \cdot (CH_3)(NO_2)C_6H_3NH_2$	Sn:	Org.Verb.6-247/8
$C_{11}Cl_3H_{19}N_2Ti$	$CH_3TiCl_3 \cdot C_6H_4(N(CH_3)_2)_2$	Ti:	Org.Verb.1-34/5
$C_{11}Cl_3H_{20}NiPSi$	$C_5H_5Ni(P(C_2H_5)_3)SiCl_3$	Ni:	Org.Verb.2-130/1
$C_{11}Cl_3H_{20}NiPSn$	$C_5H_5Ni(P(C_2H_5)_3)SnCl_3$	Ni:	Org.Verb.2-130, 132
$C_{11}Cl_3H_{24}IrS_3$	$IrCl_3(CH_3C(CH_2SC_2H_5)_3)$	Ir:	SVol.2-167
$C_{11}Cl_3H_{24}NOSn$	$[CH_2(CH_2C(CH_3)_2)_2NO][(CH_3)_2SnCl_3]$	Sn:	Org.Verb.6-35
$C_{11}Cl_3H_{26}O_3PSSn$	$C_2H_5SnCl_3 \cdot CH_3S(C_4H_9O)_2PO$	Sn:	Org.Verb.6-233
$C_{11}Cl_3H_{29}O_2P_2S_2Sn$	$CH_3SnCl_3 \cdot 2\ (C_2H_5)_2P(O)SCH_3$	Sn:	Org.Verb.6-222/3
$C_{11}Cl_3H_{33}N_2Sn$	$[(CH_3)_4N]_2[(CH_3)_3SnCl_3]$	Sn:	Org.Verb.5-87
$C_{11}Cl_4F_2H_3N_3O_2$	$NC_5F_2Cl_2NHC_6H_2Cl_2(NO_2)$	F:	PerFHalOrg.5-140
$C_{11}Cl_4F_2H_4N_2$	$NC_5F_2Cl_2NHC_6H_3Cl_2$	F:	PerFHalOrg.5-140
$C_{11}Cl_4F_3HN_4$	$NC_5F_2Cl_2NHC_5FCl_2(CN)N$	F:	PerFHalOrg.5-186/7
		F:	PerFHalOrg.6-13/4
$C_{11}Cl_4F_3H_4N_3O$	$NC_5F_2Cl_2NHC_5FCl_2(OCH_3)N$	F:	PerFHalOrg.5-138/9
$C_{11}Cl_4F_3H_4N_3O_2S$	$NC_5F_2Cl_2NHC_5FCl_2(SO_2CH_3)N$	F:	PerFHalOrg.5-138/9
$C_{11}Cl_4F_3H_4N_3S$	$NC_5F_2Cl_2NHC_5FCl_2(SCH_3)N$	F:	PerFHalOrg.5-138/9
$C_{11}Cl_4F_3H_5N_4$	$NC_5F_2Cl_2NHC_5FCl_2(NHCH_3)N$	F:	PerFHalOrg.5-138/9
$C_{11}Cl_4F_5H_3N_2O$	$C_6Cl_4NC(C_2F_5)N(OCHCH_2)$	F:	PerFHalOrg.6-159
$C_{11}Cl_4FeH_{10}$	$Fe(C_5H_5)_2 \cdot CCl_4$	Fe:	Org.Verb.A1-188
$C_{11}Cl_4H_{13}NO_2Sn$	$SnCl_4 \cdot CH_3COCH_2CONHC_6H_4CH_3$	Sn:	MVol.C6-49/50
–	$SnCl_4 \cdot C_6H_5CONHCOC_3H_7$	Sn:	MVol.C6-50/1
$C_{11}Cl_4H_{13}NSn$	$SnCl_4 \cdot (CH_3)_3CC_6H_4CN$	Sn:	MVol.C6-71/2
$C_{11}Cl_4H_{14}O_6Sn$	$SnCl_4 \cdot C_6H_5COOH \cdot 2\ CH_3COOH$	Sn:	MVol.C5-122/3
$C_{11}Cl_4H_{16}INOSn$	$SnCl_4 \cdot CH_2CHOC_6H_4N(CH_3)_3I$	Sn:	MVol.C6-9/10
$C_{11}Cl_4H_{20}O_4Sn$	$SnCl_4 \cdot C_7H_{14}(COOCH_3)_2$	Sn:	MVol.C5-141/3
$C_{11}Cl_4H_{23}NO_2Sn$	$(C_2H_5)_3NH[SnCl_4(CH(COCH_3)_2)]$	Sn:	MVol.C5-107
$C_{11}Cl_4H_{24}S_2Sn$	$SnCl_4 \cdot C_4H_9S(CH_2)_3SC_4H_9$	Sn:	MVol.C6-97/8
$C_{11}Cl_4H_{24}S_2Te$	$TeCl_4 \cdot C_4H_9S(CH_2)_3SC_4H_9$	Te:	SVol.B3-175
$C_{11}Cl_5F_2H_2N_3O_2$	$NC_5F_2Cl_2NHC_5Cl_3(COOH)N$	F:	PerFHalOrg.5-186/7
		F:	PerFHalOrg.6-13/4
$C_{11}Cl_5F_2H_3N_2$	$NC_5F_2Cl_2NHC_6H_2Cl_3$	F:	PerFHalOrg.5-140
$C_{11}Cl_5F_2H_3N_4O$	$NC_5F_2Cl_2NHC_5Cl_3(CONH_2)N$	F:	PerFHalOrg.5-186/7
		F:	PerFHalOrg.6-13/4

Formula	Compound		Reference
$C_{11}Cl_5F_2H_4N_3O$	$NC_5FCl_3NHC_5FCl_2(OCH_3)N$	F:	PerFHalOrg.5-138/9
–	$NC_5F_2Cl_2NHC_5Cl_3(OCH_3)N$	F:	PerFHalOrg.5-138/9
$C_{11}Cl_5F_4HN_2$	$NC_5F_2Cl_2NHC_6Cl_3F_2$	F:	PerFHalOrg.5-188/9
$C_{11}Cl_5F_5HN_3$	$NC_5F_2Cl_2NHC_5Cl_3(CF_3)N$	F:	PerFHalOrg.5-186/7
		F:	PerFHalOrg.6-13/4
$C_{11}Cl_5FeLiO_5$	$Li[C_6Cl_5C(O)Fe(CO)_4]$	Fe:	Org.Verb.B3-132
$C_{11}Cl_5H_8NSn$	$C_6H_5SnCl_3 \cdot C_5H_3Cl_2N$	Sn:	Org.Verb.6-279
$C_{11}Cl_5H_{12}NO_2Sn$	$SnCl_4 \cdot ClC_6H_4CONHCOC_3H_7$	Sn:	MVol.C6-50/1
$C_{11}Cl_5H_{15}N_2Sn$	$[C_5H_5NH]_2[CH_3SnCl_5]$	Sn:	Org.Verb.6-222
$C_{11}Cl_6FH_4N_3O$	$NC_5FCl_3NHC_5Cl_3(OCH_3)N$	F:	PerFHalOrg.5-138/9
$C_{11}Cl_6F_{17}NO$	$ClCF_2CFCl(CF_2CFCl)_4CF_2NO$	F:	PerFHalOrg.7-179, 185
$C_{11}Cl_6H_{12}Sn$	1) $(CH_3)_3SnC_7Cl_6CH_3$	Sn:	Org.Verb.2-64, 67
	2) $(CH_3)_3SnC_7Cl_6CH_3$	Sn:	Org.Verb.2-102, 106
$C_{11}Cl_7F_2HN_2$	$NC_5F_2Cl_2NHC_6Cl_5$	F:	PerFHalOrg.5-188
		F:	PerFHalOrg.6-13
$C_{11}Cl_7H_{17}N_4Sn$	$SnCl_4 \cdot 2\ C_2H_5C_3H_3N_2 \cdot CHCl_3$	Sn:	MVol.C5-223
$C_{11}Cl_8H_{24}S_2Te_2$	$2\ TeCl_4 \cdot C_4H_9S(CH_2)_3SC_4H_9$	Te:	SVol.B3-176
$C_{11}CmH_8NO_6$	$CmN(C_6H_4COO)(CH_2COO)_2$	Np:	TrU.D1-163
$C_{11}CoH_{21}S_7$	$[Co(S_2CSC_3H_7)_2(SC_3H_7)]_2$	C:	MVol.D4-270
$C_{11}Co_2F_6H_5O_4P$	$C_5H_5Co(CO)P(CF_3)_2Co(CO)_3$	F:	PerFHalOrg.3-27
$C_{11}Co_2H_{10}O_6Sn$	$(CH_3)_3SnC_2HCo_2(CO)_6$	Sn:	Org.Verb.2-154, 156
$C_{11}Co_2H_{26}N_6S_9$	$Co_2(NH(C_2H_4NH_2)_2)_2(CS_3)_3$	C:	MVol.D4-225
$C_{11}Co_3FeH_{19}O_6P^+$	$[HFeCo_3(CO)_3P(OC_3H_7)_2OC_2H_4]^+$	Fe:	Org.Verb.B1-204
$C_{11}Co_3FeH_{22}O_5P^+$	$[HFeCo_3(CO)_2P(OC_3H_7)_3]^+$	Fe:	Org.Verb.B1-204
$C_{11}CrFeH_9O^+$	$[C_5H_5FeC_5H_4C(O)Cr]^+$	Fe:	Org.Verb.A1-346
$C_{11}CrFeH_{11}N^+$	$[C_5H_5FeC_5H_4C(NH_2)Cr]^+$	Fe:	Org.Verb.A1-346
$C_{11}CrFeH_{12}O_7P_2$	$(CO)_3Fe(P(CH_3)_2)_2Cr(CO)_4$	Fe:	Org.Verb.B1-183
$C_{11}FFeH_9O_4S$	$FSO_2C_5H_4FeC_5H_4COOH$	Fe:	Org.Verb.A3-58/9
$C_{11}FH_{15}N_2OS$	$C_5H_{10}NS(O)FNC_6H_5$	S:	S-N-Verb.1-195
$C_{11}FH_{18}NiO_3P$	$(CO)_3NiP(C_4H_9\text{-}t)_2F$	Ni:	Org.Verb.1-166, 168
$C_{11}FH_{20}N_5$	$C_3N_3F(N(C_2H_5)_2)_2$	C:	MVol.D3-272
$C_{11}FH_{25}OSn$	$(C_4H_9)_2(HOCH_2CH_2CH_2)SnF$	Sn:	Org.Verb.5-40
$C_{11}F_2FeH_{15}N_6O_6S_2$	$[(CH_3NC)_5FeCN][SO_3F]_2$	Fe:	Org.Verb.B4-96
$C_{11}F_2H_5MoN_3O_3$	$C_3N_3F_2Mo(CO)_3(C_5H_5)$	C:	MVol.D3-272
$C_{11}F_2H_{12}N_2O_3U$	$UO_2F_2 \cdot C_6H_5C_3HN_2O(CH_3)_2$	U:	SVol.E1-118
–	$UO_2F_2 \cdot C_6H_5C_3HN_2O(CH_3)_2 \cdot 2\ H_2O$	U:	SVol.E1-116/8
$C_{11}F_3FeH_5N_2O_2$	$[C_5H_5Fe(CO)_2]C_4F_3N_2$	F:	PerFHalOrg.6-72
$C_{11}F_3H_7N_4S$	$(CF_3)(C_6H_5CH_2)C_3N_4S$	F:	PerFHalOrg.5-100
$C_{11}F_3H_8N_3$	$NCFCFC(NHCH_2C_6H_5)CFN$	F:	PerFHalOrg.6-59, 65
$C_{11}F_3H_9N_2O_3$	$(CH_3O)_3C_8F_3N_2$	F:	PerFHalOrg.6-161
–	$C_6F_3(OCH_3)N(COCH_3)_2N$	F:	PerFHalOrg.6-157, 161
$C_{11}F_3H_9O$	$CF_3CCC(OH)(CH_3)C_6H_5$	F:	PerFHalOrg.4-47
$C_{11}F_3H_{10}N$	$(CH_3CHCH)_2C_5F_3N$	F:	PerFHalOrg.5-142
$C_{11}F_3H_{10}N_5O_6P$	$NH_2C_5N_4(CF_3)C_4H_4O(OH)CH_2PO_3OH$	F:	PerFHalOrg.7-63
$C_{11}F_3H_{15}Sn$	$(CH_3)_3SnCH_2C_6H_4CF_3$	Sn:	Org.Verb.2-24
$C_{11}F_3H_{17}N_4O_3S$	$C_2N_2S(CF_3)NHCON(CH_3)C_2H_3(OC_2H_5)_2$	F:	PerFHalOrg.5-107
$C_{11}F_3H_{18}NaO$	$((CH_3)_2CHCH_2)_2C(CFCF_2)ONa$	F:	PerFHalOrg.4-1
$C_{11}F_3H_{19}O$	$((CH_3)_2CHCH_2)_2C(CFCF_2)OH$	F:	PerFHalOrg.4-1
$C_{11}F_3H_{20}N_2P$	$CF_3P[N(CH_2)_5]_2$	F:	PerFHalOrg.3-102, 116
$C_{11}F_4H_3N_5O_6$	$NC_5F_4NHC_6H_2(NO_2)_3$	F:	PerFHalOrg.5-215

Formula	Compound	Element	Reference
$C_{11}F_4H_4N_4$	$C_6H_4N_3C_5F_4N$	F:	PerFHalOrg.5-212/3
$C_{11}F_4H_4N_4O_4$	$NC_5F_4NHC_6H_3(NO_2)_2$	F:	PerFHalOrg.5-215
$C_{11}F_4H_5N$	$C_6H_5C_5F_4N$	F:	PerFHalOrg.5-142, 216
$C_{11}F_4H_5Ni^+$	$NiC_5H_5C_6F_4^+$	Ni:	Org.Verb.2-181
$C_{11}F_4H_6N_2$	$C_6H_5NHC_5F_4N$	F:	PerFHalOrg.5-137, 212
$C_{11}F_4H_8N_2O$	$[C_6H_5NH_3][OC_5F_4N]$	F:	PerFHalOrg.5-212
$C_{11}F_4H_{11}N_3O_2Sn$	$SnF_4 \cdot NC_5H_4C_5H_4N \cdot CH_3NO_2$	Sn:	MVol.C5-217
$C_{11}F_4H_{12}N_2$	$(CH_2)_6NC_5F_4N$	F:	PerFHalOrg.5-213
–	$(CH_3)(i\text{-}C_3H_7)C_2H_2NC_5F_4N$	F:	PerFHalOrg.5-213
–	$CH_3C_5H_9NC_5F_4N$	F:	PerFHalOrg.5-213
–	$C_6H_{11}NHC_5F_4N$	F:	PerFHalOrg.5-137, 213
–	$NC_5F_4N(CH(CH_3)CHCH(CH_3)_2)$	F:	PerFHalOrg.5-137
$C_{11}F_5^+$	$C_{11}F_5^+$	F:	PerFHalOrg.3-172
$C_{11}F_5FeH_4^+$	$[FeC_5H_4C_6F_5]^+$	Fe:	Org.Verb.A1-390
$C_{11}F_5FeH_5^+$	$[C_5H_5FeC_6F_5]^+$	Fe:	Org.Verb.A2-287
$C_{11}F_5FeLiO_5$	$Li[C_6F_5C(O)Fe(CO)_4]$	Fe:	Org.Verb.B4-152, 161
$C_{11}F_5FeO_5^-$	$[C_6F_5C(O)Fe(CO)_4]^-$	Fe:	Org.Verb.B4-152, 161
$C_{11}F_5H_3N_4O_4$	$NC_5F_4NHC_6H_2F(NO_2)_2$	F:	PerFHalOrg.5-215
$C_{11}F_5H_4NO$	$C_6F_5N(CH)_4C(O)$	F:	PerFHalOrg.7-197
$C_{11}F_5H_5NiS$	$(C_5H_5NiSC_6F_5)_x$	Ni:	Org.Verb.2-400
$C_{11}F_5H_6NO_2$	$(CH_3O)_2C_9F_5N$	F:	PerFHalOrg.6-154/5, 159/61
–	$C_6F_4C(OCH_3)CFC(O)NCH_3$	F:	PerFHalOrg.6-154, 159/60
$C_{11}F_5H_{15}Si_2$	$(CH_3)_3SiSi(CH_3)_2C_6F_5$	F:	PerFHalOrg.4-23
$C_{11}F_5H_{15}Sn$	$(CH_2CHCH_2)_3SnC_2F_5$	Sn:	Org.Verb.2-348
$C_{11}F_5MnO_5$	$C_6F_5Mn(CO)_5$	F:	PerFHalOrg.4-36, 111
$C_{11}F_5N_2^+$	$[C_{11}F_5N_2]^+$	F:	PerFHalOrg.6-152
$C_{11}F_6^+$	$C_{11}F_6^+$	F:	PerFHalOrg.3-216
$C_{11}F_6FeH_9NO_2S_2$	$(t\text{-}C_4H_9NC)Fe(CO)_2S_2C_2(CF_3)_2$	Fe:	Org.Verb.B4-2
$C_{11}F_6FeH_{11}P$	$[C_6H_6FeC_5H_5]PF_6$	Fe:	Org.Verb.A1-306
$C_{11}F_6FeH_{13}O_4P$	$[(CH_3CO)(CH_3)_2C_4H_4Fe(CO)_3]PF_6$	Fe:	Org.Verb.B5-94/7
$C_{11}F_6FeH_{14}NO_5P$	$[C_2H_5O((CH_3)_2N)CCHCH_2Fe(CO)_4]PF_6$	Fe:	Org.Verb.B4-249
$C_{11}F_6FeH_{18}N_3O_2PS_3$	$[S(CN(CH_3)_2)_2Fe(CO)_2S_2CN(CH_3)_2]PF_6$	Fe:	Org.Verb.B4-189, 192
$C_{11}F_6FeH_{18}O_8P_2$	$F_2CCFCF_3Fe(CO)_2(P(OCH_3)_3)_2$	Fe:	Org.Verb.B4-187, 189
$C_{11}F_6FeH_{20}O_3P_2Si_2$	$(CO)_3Fe(P(CH_3)_2CH_2CH_2SiF_3)_2$	Fe:	Org.Verb.B1-148/50, 152
$C_{11}F_6FeH_{21}O_5P_3$	$(CO)_2Fe(PF_2OC_3H_7)_3$	Fe:	Org.Verb.B1-95, 97
$C_{11}F_6H_6N_4$	$(NH_2)_2(NH_2C_6F_4)C_5F_2N$	F:	PerFHalOrg.5-188, 203
$C_{11}F_6H_7NOS$	$SC(CF_3)_2NHC(O)CH(C_6H_5)$	F:	PerFHalOrg.7-66/7
$C_{11}F_6H_{11}NO$	$(CH_3)_2(OH)C_6H_2C(CF_3)_2NH_2$	F:	PerFHalOrg.7-69/70
$C_{11}F_6H_{11}N_3$	$(CF_3)_2(C_6H_{11})C_3N_3$	F:	PerFHalOrg.6-113
$C_{11}F_6H_{11}O_{6.5}U$	$UO_2(CF_3COCHCOCH_3)_2 \cdot 0.5\ C_2H_5OH$	U:	SVol.E1-60, 64/5
$C_{11}F_6H_{12}N_2$	$(CH_3)_2NC_6H_4C(CF_3)_2NH_2$	F:	PerFHalOrg.7-69/70
$C_{11}F_6H_{12}Ni$	$(C_8H_{12})Ni(CF(CF_3)CF_2)$	Ni:	Org.Verb.2-84/5
$C_{11}F_6H_{12}NiO$	$C_8H_{12}Ni(C(CF_3)_2O)$	Ni:	Org.Verb.2-83
$C_{11}F_6H_{12}NiS$	$C_8H_{12}Ni(C(CF_3)_2S)$	Ni:	Org.Verb.2-83
$C_{11}F_6H_{12}Sn$	$(CH_3)_3SnC_6H_3(CF_3)_2$	Sn:	Org.Verb.2-127
$C_{11}F_6H_{13}NNi$	$C_8H_{12}Ni(C(CF_3)_2NH)$	Ni:	Org.Verb.2-83
$C_{11}F_6H_{17}N_3O_3$	$NHC(O)NHC(CF_3)_2OC(O) \cdot N(C_2H_5)_3$	F:	PerFHalOrg.6-60
$C_{11}F_6N_2^+$	$[C_{11}F_6N_2]^+$	F:	PerFHalOrg.6-152

$C_{11}F_7FeH_{16}IO_2P_2$	$(CO)_2Fe(C_3F_7)[(CH_3)_2PC_2H_4P(CH_3)_2]I$	Fe:	Org.Verb.B1-121/2, 126
$C_{11}F_7H_2NO_2$	$(HO)(HOC_6F_4)C_5F_3N$	F:	PerFHalOrg.5-184, 199/200
$C_{11}F_7H_3N_2$	$NC_5F_3(CH_3)C_5F_4N$	F:	PerFHalOrg.5-223/4
$C_{11}F_7H_3N_2O$	$NC_5F_3(OCH_3)C_5F_4N$	F:	PerFHalOrg.5-222/4
$C_{11}F_7H_4N_3$	$(NH_2C_6F_4)(NH_2)C_5F_3N$	F:	PerFHalOrg.5-188, 203
$C_{11}F_7H_4N_3O$	$NC_5F_4NHC_5F_3(OCH_3)N$	F:	PerFHalOrg.5-138/9
$C_{11}F_7H_8N_3$	$C_3F_7C(NNH_2)NCHC_6H_5$	F:	PerFHalOrg.7-58
$C_{11}F_7H_9O$	$C_3F_7C(CH_3)(C_6H_5)OH$	F:	PerFHalOrg.4-92
$C_{11}F_7N^+$	$[C_{11}F_7N]^+$	F:	PerFHalOrg.6-152
$C_{11}F_7N_2{}^+$	$[C_{11}F_7N_2]^+$	F:	PerFHalOrg.6-152
$C_{11}F_8HNO$	$(HO)(C_6F_5)C_5F_3N$	F:	PerFHalOrg.5-184, 199
$C_{11}F_8H_2N_2$	$(C_6F_5)(NH_2)C_5F_3N$	F:	PerFHalOrg.5-188, 203
$C_{11}F_8H_2N_4O$	$NC_5F_4NHNHC(O)C_5F_4N$	F:	PerFHalOrg.5-189/90, 204
$C_{11}F_8HgN_2O_2$	$NC_5F_4HgOC(O)C_5F_4N$	F:	PerFHalOrg.5-192/3, 207
$C_{11}F_8N_2O$	$NC_5F_4COC_5F_4N$	F:	PerFHalOrg.5-195, 208, 217
$C_{11}F_8Si^+$	$(C_5F_3)Si(C_6F_5)^+$	F:	PerFHalOrg.4-152
$C_{11}F_9Fe_2HO_6S_2$	$(CO)_3Fe(CF_3CCH)(SCF_3)_2Fe(CO)_3$	Fe:	Org.Verb.C2-67/8, 70
$C_{11}F_9HN_2O$	$CF_3C(O)NHC_9F_6N$	F:	PerFHalOrg.6-132/3, 146
$C_{11}F_9H_2N_3$	$NC_5F_4NHNHC_6F_5$	F:	PerFHalOrg.5-190, 204
$C_{11}F_9H_5O_6Ti$	$C_5H_5Ti(OCOCF_3)_3$	Ti:	Org.Verb.1-168/9
$C_{11}F_9H_{12}N_3$	$(CF_3)_3(C_5H_{11})C_3HN_3$	F:	PerFHalOrg.6-113
$C_{11}F_9H_{12}N_5$	$[(CH_3)_4N][C_4N_4(CF_3)_3]$	F:	PerFHalOrg.6-156, 160
$C_{11}F_9N$	$C_6F_5C_5F_4N$	F:	PerFHalOrg.5-155/6, 168/9, 176, 178
$C_{11}F_9NS$	$NC_5F_4SC_6F_5$	F:	PerFHalOrg.5-191/2, 206
$C_{11}F_9N_3$	$NC_5F_4NNC_6F_5$	F:	PerFHalOrg.5-190, 204
$C_{11}F_9Si^+$	$(C_6F_5)Si(C_5F_4)^+$	F:	PerFHalOrg.4-152, 153
$C_{11}F_{10}H_3N_3$	$NCC_2H_3NC_5F_3(i\text{-}C_3F_7)N$	F:	PerFHalOrg.5-214
$C_{11}F_{10}H_5NO_2$	$[(C_2H_5OCO)(CF_3)_2C]C_5F_4N$	F:	PerFHalOrg.5-142
$C_{11}F_{11}H_4NO$	$(CF_3)_3C_6F_2NHCOCH_3$	F:	PerFHalOrg.7-60
$C_{11}F_{11}H_5O_2$	$C_7F_{11}C(CH_2)OCOCH_3$	F:	PerFHalOrg.4-43
$C_{11}F_{12}H_6N_2O_2$	$C_6H_5CH(ON(CF_3)_2)_2$	F:	PerFHalOrg.7-137/8, 144
$C_{11}F_{12}H_7N_3O$	$(CF_3)_2CNHC(CF_3)_2NC(NHCOC_2H_2CH_3)$	F:	PerFHalOrg.5-105
$C_{11}F_{12}H_{10}N_4$	$C_3HN_2(CF_3)_4NC(CH_3)N(CH_3)_2$	F:	PerFHalOrg.5-109/10
$C_{11}F_{12}H_{12}N_4$	$(CF_3)_2CNHC(CF_3)_2NC(NHC_2H_4N(CH_3)_2)$	F:	PerFHalOrg.5-104/5
$C_{11}F_{13}FeH_9N_3O_2P$	$[(CO)_2Fe(C_3F_7)(CH_3CN)_3]PF_6$	Fe:	Org.Verb.B2-181
$C_{11}F_{13}H_8N_3O_2$	$C_3F_7C(NH)NC(NH_2)C_3F_6COOC_2H_5$	F:	PerFHalOrg.7-57/8
$C_{11}F_{13}H_8N_3O_3$	$CF_3OC_2F_4C(NH)NC(NH_2)C_3F_6COOC_2H_5$	F:	PerFHalOrg.7-57/8
$C_{11}F_{14}FeNO_6P$	$(CO)_4FePF_2(OC_3NO(CF_3)_4)$	Fe:	Org.Verb.B2-94, 115
$C_{11}F_{14}FeO_4$	$(CO)_4Fe(C_3F_7)C(CF_3)CFCF_3$	Fe:	Org.Verb.B2-192
$C_{11}F_{14}N_2$	$(CF_2)_2C_5F_4(C_3F_6CN)N$	F:	PerFHalOrg.5-156/7, 170
$C_{11}F_{15}FeIO_4$	$(CO)_4Fe(C_7F_{15})I$	Fe:	Org.Verb.B2-179, 181
$C_{11}F_{15}H_9NiO_2P_2$	$(CO)_2Ni(P(CF_3)_3)(P(CF_3)_2C_4H_9)$	Ni:	Org.Verb.1-145
$C_{11}F_{15}N$	$C_6F_{11}C_5F_4N$	F:	PerFHalOrg.5-155/6, 168, 178
$C_{11}F_{16}HNO_2$	$(CF_2)_4CNHCOC(CF_2)_4O$	F:	PerFHalOrg.5-59, 81
$C_{11}F_{16}H_3N_3$	$(CF_2)_4CNHC(CF_2)_4NC(NH_2)$	F:	PerFHalOrg.5-63/4, 86

$C_{11}F_{16}N_4O_2$	$CF_2(CF_2COC(C_2F_5)NN)_2$	F:	PerFHalOrg.5-69/70, 90
–	$CF_2(CF_2CONC(C_2F_5)N)_2$	F:	PerFHalOrg.5-69/71, 90
$C_{11}F_{17}HN_2O_4$	$HOC(O)C_6F_5[ON(CF_3)_2]_2$	F:	PerFHalOrg.7-98/9, 116
$C_{11}F_{17}H_5O_2$	$(n-C_8F_{17})COOC_2H_5$	F:	PerFHalOrg.4-105
$C_{11}F_{17}H_7O$	$(n-C_8F_{17})C(CH_3)_2OH$	F:	PerFHalOrg.4-103
$C_{11}F_{17}H_9Si$	$(n-C_8F_{17})Si(CH_3)_3$	F:	PerFHalOrg.4-86
$C_{11}F_{17}N$	$((CF_3)_2CF)_2C_5F_3N$	F:	PerFHalOrg.5-151/3, 163, 173/4, 176/8
–	$(C_2F_5)_3C_5F_2N$	F:	PerFHalOrg.5-149/50, 160, 176
$C_{11}F_{18}HN_3O_5$	$[OC(CF_3)_2C(NHNO_2)N]C[(OC(CF_3)_2)_2]$	F:	PerFHalOrg.5-59, 94
$C_{11}F_{18}H_2N_2O_3$	$[OC(CF_3)_2C(NH_2)N]C[(OC(CF_3)_2)_2]$	F:	PerFHalOrg.5-59, 94
$C_{11}F_{19}H_7OSi$	$(CH_3)_2SiH(CF_2)_6OCF(CF_3)_2$	F:	PerFHalOrg.4-87
$C_{11}F_{19}N_3$	$(C_3F_7)_2(C_2F_5)C_3N_3$	F:	PerFHalOrg.6-83/4, 104
$C_{11}F_{19}N_3O_2$	$(CF_3OC_2F_4)_2(C_2F_5)C_3N_3$	F:	PerFHalOrg.6-86/8, 105
$C_{11}F_{20}HNO_2$	$C_5F_{11}NCF_2CF(COOH)(CF_2)_3$	F:	PerFHalOrg.5-193/4, 208, 217
$C_{11}F_{20}H_3N_3O$	$C_6F_{13}C(NH)NC(NH_2)(CF_2)_2OCF_3$	F:	PerFHalOrg.7-7, 35
$C_{11}F_{23}HO$	$(n-C_8F_{17})C(CF_3)_2OH$	F:	PerFHalOrg.4-104
$C_{11}F_{23}N$	$(CF_3)_2CFCF_2CF(CF_3)N(CF_2)_5$	F:	PerFHalOrg.5-157/8, 171, 175
–	$C_6F_{13}N(CF_2)_5$	F:	PerFHalOrg.5-157/8, 172, 175
$C_{11}F_{23}NO$	$O(CF_2CF_2)_2NCF_2(CF_2)_5CF_3$	F:	PerFHalOrg.6-2, 7
$C_{11}F_{24}H_2N_2Ni_2O_3P_4$	$Ni_2(CO)_3[(CF_3)_2PNHP(CF_3)_2]_2$	Ni:	Org.Verb.2-258
$C_{11}F_{24}Ni_2O_3P_4S_2$	$Ni_2(CO)_3[(CF_3)_2PSP(CF_3)_2]_2$	Ni:	Org.Verb.2-259
$C_{11}FeH_4O_6$	$C_6H_4O_2CFe(CO)_4$	Fe:	Org.Verb.B4-128, 132, 135
$C_{11}FeH_5LiO_5$	$Li[C_6H_5C(O)Fe(CO)_4]$	Fe:	Org.Verb.B4-151, 161
$C_{11}FeH_5NO_4$	$(CO)_4Fe(NCC_6H_5)$	Fe:	Org.Verb.B2-71, 76
–	$C_6H_5NCFe(CO)_4$	Fe:	Org.Verb.B4-6/8
$C_{11}FeH_5NaO_5$	$Na[C_6H_5C(O)Fe(CO)_4]$	Fe:	Org.Verb.B4-151, 161
$C_{11}FeH_5O_4S_2$	$(CO)_4FeS_2CC_6H_5$	Fe:	Org.Verb.B2-65
$C_{11}FeH_5O_5^-$	$[C_6H_5C(O)Fe(CO)_4]^-$	Fe:	Org.Verb.B4-151/2, 157, 161, 167
$C_{11}FeH_5O_5Tl$	$Tl[(CO)_4FeCOC_6H_5]$	Fe:	Org.Verb.B2-173
$C_{11}FeH_6NO_5^-$	$[C_6H_5NHC(O)Fe(CO)_4]^-$	Fe:	Org.Verb.B4-156, 165/6
$C_{11}FeH_6N_2O_4$	$(CO)_4Fe(C_6H_4NHCHN)$	Fe:	Org.Verb.B2-71, 74
$C_{11}FeH_7IO_2$	$C_9H_7Fe(CO)_2I$	Fe:	Org.Verb.B2-66
$C_{11}FeH_7I_2NO_3$	$CH_3C_6H_4NCFe(CO)_3I_2$	Fe:	Org.Verb.B4-2/3
$C_{11}FeH_7I_2NO_4$	$CH_3OC_6H_4NCFe(CO)_3I_2$	Fe:	Org.Verb.B4-2/4
$C_{11}FeH_7NO_4$	$(C_6H_5)(NH_2)CFe(CO)_4$	Fe:	Org.Verb.B4-121/2
$C_{11}FeH_7O_4^-$	$[CH_3C_6H_4C(O)Fe(CO)_3]^-$	Fe:	Org.Verb.B4-109
$C_{11}FeH_8LiO_4P$	$(CO)_4FeP(C_6H_5)(CH_3)Li$	Fe:	Org.Verb.B2-50
$C_{11}FeH_8O_4$	$C_7H_8Fe(CO)_4$	Fe:	Org.Verb.B4-304, 310/1
–	$C_8H_8OFe(CO)_3$	Fe:	Org.Verb.B5-144/5
$C_{11}FeH_8O_6$	$C_7H_8O_2Fe(CO)_4$	Fe:	Org.Verb.B4-259, 263/4
–	$C_7H_6D_2O_2Fe(CO)_4$	Fe:	Org.Verb.B4-263/4
–	$C_7H_4D_4O_2Fe(CO)_4$	Fe:	Org.Verb.B4-263/4
$C_{11}FeH_9^+$	$[FeC_5H_4C_6H_5]^+$	Fe:	Org.Verb.A1-325

Formula	Compound	Reference
$C_{11}FeH_9^+$	$[FeHC_5H_4C_6H_4]^+$	Fe: Org.Verb.A3-126
$C_{11}FeH_9IO_3$	$C_8H_9Fe(CO)_3I$	Fe: Org.Verb.B5-54, 64
$C_{11}FeH_9N$	$C_5H_5FeC_5H_4CN$	Fe: Org.Verb.A1-186/7
–	$[C(C_5H_5FeC_5H_4)N]_n$	Fe: Org.Verb.A3-64
$C_{11}FeH_9N^+$	$[C_5H_5FeC_5H_4CN]^+$	Fe: Org.Verb.A1-346
$C_{11}FeH_9NO_3$	$C_6H_5C_3H_4Fe(CO)_2NO$	Fe: Org.Verb.B5-14/6, 22
$C_{11}FeH_9NO_4$	$(CO)_4Fe(NC_5H_3(CH_3)_2)$	Fe: Org.Verb.B2-71, 75
$C_{11}FeH_9NSe$	$C_5H_5FeC_5H_4SeCN$	Fe: Org.Verb.A6-185
$C_{11}FeH_9N_3O$	$C_5H_5FeC_5H_4CON_3$	Fe: Org.Verb.A3-156
$C_{11}FeH_9NaO_2$	$[C_5H_5FeC_5H_4COO]Na$	Fe: Org.Verb.A3-63/5
$C_{11}FeH_9O^+$	$[C_5H_5FeC_5H_4CO]^+$	Fe: Org.Verb.A2-208
		Fe: Org.Verb.A3-59, 152
$C_{11}FeH_9OW^+$	$[C_5H_5FeC_5H_4COW]^+$	Fe: Org.Verb.A1-347
$C_{11}FeH_9O_2$	$[C_5H_5FeC_5H_4COO]$	Fe: Org.Verb.A3-57
$C_{11}FeH_9O_2^-$	$[C_5H_5FeC_5H_4COO]^-$	Fe: Org.Verb.A3-57, 63
$C_{11}FeH_9O_4P$	$(CO)_4FeP(CH_3)(C_6H_5)H$	Fe: Org.Verb.B2-89, 110
$C_{11}FeH_{10}$	$(C_5H_4FeC_5H_4CH_2)_n$	Fe: Org.Verb.A6-234
–	$C_5H_5FeC_5H_4CH$	Fe: Org.Verb.A1-344
$C_{11}FeH_{10}^+$	$[C_5H_5FeC_6H_5]^+$	Fe: Org.Verb.A2-278
–	$[FeC_{11}H_{10}]^+$	Fe: Org.Verb.A2-17
–	$[FeC_{11}H_8D_2]^+$	Fe: Org.Verb.A2-17
$C_{11}FeH_{10}I_2N_2O$	$COFe(NC_5H_5)_2I_2$	Fe: Org.Verb.B1-31
$C_{11}FeH_{10}KNO_2$	$C_5H_5FeC_5H_4CONHOK$	Fe: Org.Verb.A3-110
$C_{11}FeH_{10}N^+$	$[(C_5H_5)_2FeCN]^+$	Fe: Org.Verb.A1-186/7
$C_{11}FeH_{10}NS$	$[Fe(C_5H_5)_2]SCN$	Fe: Org.Verb.A1-219
–	$[Fe(C_5H_5)_2SCN]$	Fe: Org.Verb.A1-206
$C_{11}FeH_{10}N_2O_4$	$(CO)_4Fe(C_7H_{10}N_2)$	Fe: Org.Verb.B2-183/4, 187
$C_{11}FeH_{10}NiO^+$	$(C_5H_5)_2NiFe(CO)^+$	Ni: Org.Verb.2-351
$C_{11}FeH_{10}Ni_2O^+$	$(C_5H_5)_2Ni_2Fe(CO)^+$	Ni: Org.Verb.2-351
$C_{11}FeH_{10}O$	$C_5H_5FeC_5H_4CHO$	
	Analysis	Fe: Org.Verb.A2-171
	Chemical reactions	Fe: Org.Verb.A2-147/70, 188
	Electrochemical behavior	Fe: Org.Verb.A2-146/7
	Formation and preparation	Fe: Org.Verb.A2-140/3
	Physical properties	Fe: Org.Verb.A2-143/6, 182/4
	Separation methods	Fe: Org.Verb.A2–171
	Solubility	Fe: Org.Verb.A2-144/6
	Uses	Fe: Org.Verb.A2-171/2
–	$C_5H_5FeC_5H_4CDO$	Fe: Org.Verb.A2-141, 144, 183/4
$C_{11}FeH_{10}O^+$	$[C_5H_5FeC_5H_4CHO]^+$	Fe: Org.Verb.A2-146, 148, 171
		Fe: Org.Verb.A3-165
$C_{11}FeH_{10}O^-$	$[C_5H_5FeC_5H_4CHO]^-$	Fe: Org.Verb.A2-146
$C_{11}FeH_{10}OS$	$C_5H_5FeC_5H_4COSH$	Fe: Org.Verb.A3-156
$C_{11}FeH_{10}O_2$	$C_5H_5FeC_5H_4COOH$	Fe: Org.Verb.A3-50/62
–	$C_5H_5{}^{59}FeC_5H_4COOH$	Fe: Org.Verb.A3-60
$C_{11}FeH_{10}O_2^+$	$[C_5H_5FeC_5H_4COOH]^+$	Fe: Org.Verb.A3-56, 58, 60, 62, 111
$C_{11}FeH_{10}O_3$	$C_8H_{10}Fe(CO)_3$	Fe: Org.Verb.B5-144/5

Formula	Compound	Reference
$C_{11}FeH_{10}O_4S$	$HO_3SC_5H_4FeC_5H_4CHO$	Fe: Org.Verb.A2-150
$C_{11}FeH_{10}O_5S$	$HO_3SC_5H_4FeC_5H_4COOH$	Fe: Org.Verb.A3-60
$C_{11}FeH_{10}O_6$	$(CH_3COO)(CH_3CO)C_4H_4Fe(CO)_3$	Fe: Org.Verb.B5-76
$C_{11}FeH_{10}O_7$	$(CH_3COO)_2C_4H_4Fe(CO)_3$	Fe: Org.Verb.B5-76
$C_{11}FeH_{10}O_8$	$(CH_3OCO)_2C_2HC(OCH_3)Fe(CO)_3$	Fe: Org.Verb.B5-157/8
$C_{11}FeH_{11}{}^+$	$[C_5H_5FeC_5H_4CH_2]^+$	Fe: Org.Verb.A1-249, 350/2, 356/63
		Fe: Org.Verb.A2-18, 71
–	$[C_5H_5FeC_6H_6]^+$	Fe: Org.Verb.A1-118, 268, 325, 389
		Fe: Org.Verb.A2-186, 208
$C_{11}FeH_{11}KO_7$	$K[C_2H_5OC(O)CH(C_2H_5)C(O)Fe(CO)_4]$	Fe: Org.Verb.B4-155, 163
–	$K[C_2H_5OCO(CH_2)_3C(O)Fe(CO)_4]$	Fe: Org.Verb.B4-155, 163
–	$K[C_2H_5OC(O)C_2H_3(CH_3)C(O)Fe(CO)_4]$	Fe: Org.Verb.B4-155, 163
$C_{11}FeH_{11}Li$	$CH_3C_5H_4FeC_5H_4Li$	Fe: Org.Verb.A1-257
–	$C_5H_5FeC_5H_4CH_2Li$	Fe: Org.Verb.A2-112, 114
$C_{11}FeH_{11}LiO$	$C_5H_5FeC_5H_4CH_2OLi$	Fe: Org.Verb.A2-114
$C_{11}FeH_{11}LiO_5S_2$	$Li[(CH_2S)_2C(C_3H_7)C(O)Fe(CO)_4]$	Fe: Org.Verb.B4-157, 167/8
$C_{11}FeH_{11}N^+$	$[C_5H_5FeC_5H_4CH_2N]^+$	Fe: Org.Verb.A1-346
$C_{11}FeH_{11}NO$	$C_5H_5FeC_5H_4CHNOH$	Fe: Org.Verb.A2-165
–	$C_5H_5FeC_5H_4CONH_2$	Fe: Org.Verb.A3-154
$C_{11}FeH_{11}NO_4$	$C_6H_{11}NCFe(CO)_4$	Fe: Org.Verb.B4-6, 8/9
$C_{11}FeH_{11}NO_4S$	$NH_2SO_2C_5H_4FeC_5H_4COOH$	Fe: Org.Verb.A3-58/9
$C_{11}FeH_{11}NO_6$	$O(CH_2CH_2)_2NCHCH_2COFe(CO)_4$	Fe: Org.Verb.B4-237, 249
$C_{11}FeH_{11}NiO_4P^+$	$NiP(CH_3)_2Fe(C_5H_5)(CO)_4{}^+$	Ni: Org.Verb.1-167
$C_{11}FeH_{11}O^+$	$[C_5H_5FeC_5H_4CHOH]^+$	Fe: Org.Verb.A2-146, 151
$C_{11}FeH_{11}O^-$	$[C_5H_5FeC_5H_4CHOH]^-$	Fe: Org.Verb.A2-146
$C_{11}FeH_{11}O_2{}^+$	$[C_5H_5FeC_5H_4C(OH)_2]^+$	Fe: Org.Verb.A3-59
$C_{11}FeH_{11}O_5{}^-$	$[C_4H_9C(O)C_3H_2(O)Fe(CO)_3]^-$	Fe: Org.Verb.B5-72, 75
$C_{11}FeH_{11}O_5S$	$[C_5H_5FeC_5H_4CHO]HSO_4$	Fe: Org.Verb.A2-172
$C_{11}FeH_{11}O_5S_2{}^-$	$[(CH_2S)_2C(C_3H_7)C(O)Fe(CO)_4]^-$	Fe: Org.Verb.B4-157, 167/8
$C_{11}FeH_{11}O_7{}^-$	$[C_2H_5OC(O)C_3H_6C(O)Fe(CO)_4]^-$	Fe: Org.Verb.B4-155, 163
$C_{11}FeH_{12}$	$C_5H_5FeC_5H_4CH_3$	Fe: Org.Verb.A1-253/9
$C_{11}FeH_{12}{}^+$	$[C_5H_5FeC_5H_4CH_3]^+$	Fe: Org.Verb.A1-249, 254, 257, 277
$C_{11}FeH_{12}I_3$	$[C_5H_5FeC_5H_4CH_3]I_3$	Fe: Org.Verb.A1-277/8
$C_{11}FeH_{12}I_3O$	$[C_5H_5FeC_5H_4CH_2OH]I_3$	Fe: Org.Verb.A2-70
$C_{11}FeH_{12}N^+$	$[C_5H_5FeC_5H_4CH_3N]^+$	Fe: Org.Verb.A1-346
$C_{11}FeH_{12}NOS_2$	$COFe(NC_5H_5)(CH(CSCH_3)_2)$	Fe: Org.Verb.B1-34
$C_{11}FeH_{12}NO_5{}^-$	$[C_6H_{11}NHC(O)Fe(CO)_4]^-$	Fe: Org.Verb.B4-156, 165/6
$C_{11}FeH_{12}N_2O$	$C_5H_5FeC_5H_4CONHNH_2$	Fe: Org.Verb.A3-110, 155
$C_{11}FeH_{12}N_2O_3$	$(CH_3)_2C_3HN_2CH_2CHCH_2Fe(CO)_3$	Fe: Org.Verb.B4-216, 225
$C_{11}FeH_{12}O$	$C_5H_5FeC_5H_4CH_2OH$	
	Catalysis	Fe: Org.Verb.A2-23
	Chemical reactions	Fe: Org.Verb.A2-3, 17/23
	Electrochemical behavior	Fe: Org.Verb.A2-16/7
	Formation and preparation	Fe: Org.Verb.A2-11/4
	Physical properties	Fe: Org.Verb.A2-4, 14/6
	Solubility	Fe: Org.Verb.A2-15/6
	Uses	Fe: Org.Verb.A2-23

Formula	Compound	Reference
$C_{11}FeH_{12}O$	$C_5H_5FeC_5H_4CD_2OH$	Fe: Org.Verb.A2-13, 17
–	$C_5H_5FeC_5H_4OCH_3$	Fe: Org.Verb.A2-108/9
$C_{11}FeH_{12}O^+$	$[C_5H_5FeC_5H_4CH_2OH]^+$	Fe: Org.Verb.A2-16/8, 70
–	$[C_5H_5FeC_5H_4OCH_3]^+$	Fe: Org.Verb.A2-109
$C_{11}FeH_{12}O^{2+}$	$[HC_5H_5FeC_5H_4CHOH]^{2+}$	Fe: Org.Verb.A2-151
$C_{11}FeH_{12}O_3$	$C_4(CH_3)_4Fe(CO)_3$	Fe: Org.Verb.B2-31
		Fe: Org.Verb.B3-150, 153
–	$(CO)_3FeC_4H_3C_4H_9$	Fe: Org.Verb.B2-32
$C_{11}FeH_{12}O_3S$	$((CH_2)_5C_3H_2S)Fe(CO)_3$	Fe: Org.Verb.B4-213, 220/1
–	$(C_7H_{11}CSH)Fe(CO)_3$	Fe: Org.Verb.B5-135
$C_{11}FeH_{12}O_4$	$(CH_3)_2CCC(CH_3)_2Fe(CO)_4$	Fe: Org.Verb.B4-321, 325
$C_{11}FeH_{12}O_6$	$(C(CH_3)_2O)_2CFe(CO)_4$	Fe: Org.Verb.B4-128, 130, 134
$C_{11}FeH_{12}O_6S$	$(CO)_4FeSC(OC(CH_3)_2)_2$	Fe: Org.Verb.B2-128/32
$C_{11}FeH_{12}O_7$	$(CH_3OC(O)CHCH_2)_2Fe(CO)_3$	Fe: Org.Verb.B4-357/8
$C_{11}FeH_{13}{}^+$	$[HC_5H_5FeC_5H_4CH_3]^+$	Fe: Org.Verb.A1-277
$C_{11}FeH_{13}NO_2$	$[C_5H_5FeC_5H_4COO][NH_4]$	Fe: Org.Verb.A3-63/5
$C_{11}FeH_{13}NO_5$	$(C_2H_5)_2NCOCHCH_2Fe(CO)_4$	Fe: Org.Verb.B4-237, 249
$C_{11}FeH_{13}NO_6$	$NH_2C_4H_8[OC(O)CHCH_2Fe(CO)_4]$	Fe: Org.Verb.B4-245/6
$C_{11}FeH_{13}NaO_5$	$Na[C_6H_{13}C(O)Fe(CO)_4]$	Fe: Org.Verb.B4-150, 160
$C_{11}FeH_{13}Ni^+$	$(C_5H_5)_2NiFeCH_3{}^+$	Ni: Org.Verb.2-168
$C_{11}FeH_{13}O_4{}^+$	$[(CH_3CO)(CH_3)_2C_4H_4Fe(CO)_3]^+$	Fe: Org.Verb.B5-94/7
–	$[((CH_3)_2CH)(CH_3)C_3H_3Fe(CO)_4]^+$	Fe: Org.Verb.B5-101, 105/6
–	$[(CH_3)_4C_3HFe(CO)_4]^+$	Fe: Org.Verb.B5-102, 105/6
$C_{11}FeH_{13}O_4{}^-$	$[C_6H_{11}CH_2C(O)Fe(CO)_3]^-$	Fe: Org.Verb.B4-109
$C_{11}FeH_{13}O_5{}^-$	$[C_6H_{13}C(O)Fe(CO)_4]^-$	Fe: Org.Verb.B4-150/1, 160
$C_{11}FeH_{13}O_7P$	$(CO)_4FeP(OCH_2)_3CC_3H_7$	Fe: Org.Verb.B2-95, 117
$C_{11}FeH_{14}NO_5{}^+$	$[C_2H_5O(N(CH_3)_2)CCHCH_2Fe(CO)_4]^+$	Fe: Org.Verb.B4-248/9
$C_{11}FeH_{14}N_2O_4$	$(CH_2NC_2H_5)_2CFe(CO)_4$	Fe: Org.Verb.B4-128, 131
$C_{11}FeH_{14}N_2O_5$	$CH_3NH(CH_2)_2N(CH_3)COCHCH_2Fe(CO)_4$	Fe: Org.Verb.B4-237, 249
$C_{11}FeH_{14}O_4$	$(CH_2C(CH_3)C(C_4H_9)O)Fe(CO)_3$	Fe: Org.Verb.B5-132, 141
$C_{11}FeH_{15}IN_6$	$[(CH_3NC)_5FeCN]I \cdot H_2O$	Fe: Org.Verb.B4-76/8
$C_{11}FeH_{15}I_3N_6$	$[(CH_3NC)_5FeCN]I_3$	Fe: Org.Verb.B4-78
$C_{11}FeH_{15}NO_3$	$(CH_3CHCHCHNC_4H_9)Fe(CO)_3$	Fe: Org.Verb.B5-127, 136
$C_{11}FeH_{15}NO_5$	$((C_2H_5)_2N)(C_2H_5O)CFe(CO)_4$	Fe: Org.Verb.B4-121/2, 124, 127
$C_{11}FeH_{15}N_2O_4{}^+$	$[C_3H_3(N(CH_3)_2)NHC(OC_2H_5)Fe(CO)_3]^+$	Fe: Org.Verb.B5-161
$C_{11}FeH_{15}N_6{}^+$	$[(CH_3NC)_5FeCN]^+$	Fe: Org.Verb.B4-78, 96
$C_{11}FeH_{16}N_2O_3$	$(CO)_3Fe(C_2H_2(NCH(CH_3)_2)_2)$	Fe: Org.Verb.B1-168
$C_{11}FeH_{16}N_2O_7S_2$	$(CO)_3Fe(S_2(CH_2CH(COOCH_3)NH_2)_2)$	Fe: Org.Verb.B1-170/1, 175
$C_{11}FeH_{16}N_4O$	$[(CH_3)_2C_3H_2N_2]_2FeCO$	Fe: Org.Verb.B5-108
$C_{11}FeH_{16}N_6O$	$(CH_3NC)_4Fe(CN)_2 \cdot CH_3OH$	Fe: Org.Verb.B4-67/8
$C_{11}FeH_{16}O_2Sn$	$(CH_3)_3SnCH_2Fe(CO)_2C_5H_5$	Sn: Org.Verb.2-16
$C_{11}FeH_{17}NO_4$	$C(C_2H_5)_2OHCH_2C_3H_4Fe(CO)_2NO$	Fe: Org.Verb.B5-14/6, 22
$C_{11}FeH_{17}N_2O_4{}^+$	$[C_2H_3C(OC_2H_5)NN(CH_3)_3Fe(CO)_3]^+$	Fe: Org.Verb.B5-151
$C_{11}FeH_{18}N_2O_5Sn$	$(CH_3)_2Sn(OCN(CH_3)_2)_2Fe(CO)_3$	Fe: Org.Verb.B4-194, 196
$C_{11}FeH_{18}N_3O_2S_3{}^+$	$[S(CN(CH_3)_2)_2Fe(CO)_2S_2CN(CH_3)_2]^+$	Fe: Org.Verb.B4-189, 192
$C_{11}FeH_{18}O_3Si_2$	$(CH_3)_2SiCHCH_2Fe(CO)_3Si(CH_3)_2CHCH_2$	Fe: Org.Verb.B4-212, 217
$C_{11}FeH_{19}N_2O_5P$	$NCC_3H_4Fe(CO)(NO)P(OC_2H_5)_3$	Fe: Org.Verb.B5-108/11
$C_{11}FeH_{19}O_4PSi$	$(CO)_4FeP(CH_3)_2CH_2CH_2Si(CH_3)_3$	Fe: Org.Verb.B2-89, 110

Formula	Compound	Reference
$C_{11}FeH_{19}O_5PS$	$(CH_2CHCHS)Fe(CO)_2P(OC_2H_5)_3$	Fe: Org.Verb.B5-117, 120
$C_{11}FeH_{20}IO_5P$	$(C_3H_5)Fe(CO)_2P(OC_2H_5)_3I$	Fe: Org.Verb.B5-24/6
$C_{11}FeH_{20}I_3NO_3$	$[N(C_2H_5)_4][(CO)_3FeI_3]$	Fe: Org.Verb.B1-147
$C_{11}FeH_{20}N_2O_4$	$[N(C_2H_5)_4][Fe(CO)_3NO]$	Fe: Org.Verb.B1-190/1
$C_{11}FeH_{20}O_2P$	$[C_3H_5Fe(CO)_2P(C_2H_5)_3]$	Fe: Org.Verb.B5-69/70
$C_{11}FeH_{21}S_7$	$[Fe(S_2CSC_3H_7)_2(SC_3H_7)]_2$	C: MVol.D4-269
$C_{11}FeH_{22}NO_5P$	$CH_3C_3H_4Fe(CO)(NO)P(OC_2H_5)_3$	Fe: Org.Verb.B5-6/8, 10
$C_{11}FeH_{27}O_{11}P_3$	$(CO)_2Fe(P(OCH_3)_3)_3$	Fe: Org.Verb.B1-95, 98
$C_{11}FeH_{30}N_5OP_2^+$	$[(CO)FeP_2(N(CH_3)_2)_5]^+$	Fe: Org.Verb.B1-164
$C_{11}Fe_2GeH_{19}IN_4O_4$	$(C_3H_5Fe(NO)_2)_2Ge(I)C_5H_9$	Fe: Org.Verb.C2-165
$C_{11}Fe_2GeH_{21}IN_4O_4$	$((CH_3)_2C_3H_3Fe(NO)_2)_2Ge(I)CH_3$	Fe: Org.Verb.C2-165
$C_{11}Fe_2Ge_2H_{12}O_7$	$(CO)_7Fe_2(Ge(CH_3)_2)_2$	Fe: Org.Verb.C1-205/6
$C_{11}Fe_2H_4N_2O_7$	$(CO)_7Fe_2(N(CH)_4N)$	Fe: Org.Verb.C1-200/3
$C_{11}Fe_2H_6N_2O_7$	$(CO)_3Fe((CH_2CHN)_2CO)Fe(CO)_3$	Fe: Org.Verb.C2-47
$C_{11}Fe_2H_6N_2O_8$	$(CO)_4Fe((CH_3)_2CN_2)Fe(CO)_4$	Fe: Org.Verb.C1-34/7
$C_{11}Fe_2H_6O_6S$	$C_4H_3(CH_3)SFe_2(CO)_6$	Fe: Org.Verb.C2-126/7
$C_{11}Fe_2H_6O_{11}P_2$	$(CO)_4Fe(P(OCH_2)_3P)Fe(CO)_4$	Fe: Org.Verb.C1-36, 40
$C_{11}Fe_2H_8N_2O_6$	$(CO)_3Fe(C_5H_8N_2)Fe(CO)_3$	Fe: Org.Verb.C1-149/50, 152, 156/7
$C_{11}Fe_2H_8O_7S_2$	$Fe_2(CO)_7(C_4H_8S_2)$	Fe: Org.Verb.C2-44
$C_{11}Fe_2H_9NO_6$	$(HCCHCHN(CH_3)_2)Fe_2(CO)_6$	Fe: Org.Verb.C2-156, 159
$C_{11}Fe_2H_{10}N_2O_7$	$(CO)_3Fe(C_2H_5NCONC_2H_5)Fe(CO)_3$	Fe: Org.Verb.C1-143/4
$C_{11}Fe_2H_{10}N_2O_8$	$(CO)_6Fe_2((CH_3)_2CN)(NHC(OCH_3)O)$	Fe: Org.Verb.C1-191/2, 195
$C_{11}Fe_2H_{10}O_6S$	$(CO)_3Fe(CHCH_2)(SC_3H_7)Fe(CO)_3$	Fe: Org.Verb.C2-85/6
$C_{11}Fe_2H_{12}N_2O_6S$	$(CO)_6Fe_2((CH_3)_2NC(N(CH_3)_2)S)$	Fe: Org.Verb.C1-238/40
$C_{11}Fe_2H_{12}O_7Si_2$	$(CO)_7Fe_2(Si(CH_3)_2)_2$	Fe: Org.Verb.C1-205/6
$C_{11}Fe_2H_{18}NO_5PS_2$	$(CO)_5Fe_2(SCH_3)_2(P(CH_3)_2N(CH_3)_2)$	Fe: Org.Verb.C1-93
$C_{11}Fe_2H_{19}IN_4O_4Pb$	$(C_3H_5Fe(NO)_2)_2Pb(I)C_5H_9$	Fe: Org.Verb.C2-165
$C_{11}Fe_2H_{19}IN_4O_4Sn$	$(C_3H_5Fe(NO)_2)_2Sn(I)C_5H_9$	Fe: Org.Verb.C2-165
$C_{11}Fe_2H_{19}N_2O_5P$	$(CO)_3Fe(NH_2)_2Fe(CO)_2P(C_2H_5)_3$	Fe: Org.Verb.C1-158/9, 161
$C_{11}Fe_2H_{21}IN_4O_4Pb$	$((CH_3)_2C_3H_3Fe(NO)_2)_2Pb(I)CH_3$	Fe: Org.Verb.C2-165
$C_{11}Fe_2H_{21}IN_4O_4Sn$	$((CH_3)_2C_3H_3Fe(NO)_2)_2Sn(I)CH_3$	Fe: Org.Verb.C2-165
$C_{11}Fe_3HO_{11}^-$	$[HFe_3(CO)_{11}]^-$	Fe: Org.Verb.B3-81
$C_{11}Fe_3H_4NO_9^-$	$[HN(CH_3)CFe_3(CO)_9]^-$	Fe: Org.Verb.B3-189
$C_{11}Fe_3H_6N_2O_9$	$Fe_3(CO)_9(NCH_3)_2$	Fe: Org.Verb.C1-135
$C_{11}GeH_{22}Sn$	$(CH_3)_3SnC_5H_4Ge(CH_3)_3$	Sn: Org.Verb.2-99
$C_{11}GeH_{24}OSn$	$(C_2H_5)_3SnC(CO)Ge(CH_3)_3$	Sn: Org.Verb.2-196
$C_{11}GeH_{30}NPSn$	$(CH_3)_3SnN(Ge(CH_3)_3)P(CH_3)C_4H_9$	Sn: Org.Verb.5-84
$C_{11}H_5N_3Ni$	$Ni(CN)_2CC(CN)C_6H_5$	Ni: Org.Verb.1-337/8
$C_{11}H_7LiNiO_4$	$Li[(CO)_3NiC(O)C_6H_4CH_3]$	Ni: Org.Verb.1-193, 194
$C_{11}H_7LiNiO_5$	$Li[(CO)_3NiC(O)C_6H_4OCH_3]$	Ni: Org.Verb.1-193, 194
$C_{11}H_7MnO_3S^+$	$[Mn(OC_4H_3COCHCOC_4H_3S)]^+$	Mn: MVol.D1-115
$C_{11}H_8N_4O_9U$	$[UO_2(NO_3)_2(CO(C_5H_4N)_2)]$	U: SVol.E1-40/1, 43
$C_{11}H_9NO_3Sn$	$Sn(CH_3COO)C_9H_6NO$	Sn: MVol.C5-20
$C_{11}H_9NiO_5P$	$(CO)_3NiP(OCH_2)_2C_6H_5$	Ni: Org.Verb.1-177
$C_{11}H_9NiO_6P$	$(CO)_3NiP(OC_6H_5)(OCH_2)_2$	Ni: Org.Verb.1-181
$C_{11}H_9Ti$	$C_5H_5TiC_6H_4$	Ti: Org.Verb.1-212
$C_{11}H_{10}N_6O_9SU$	$UO_2(NO_3)_2 \cdot C_4H_3OCHNNHCSNHC_5H_4N \cdot H_2O$	U: SVol.E1-71, 76
$C_{11}H_{10}Ni_2O^+$	$(C_5H_5)_2Ni_2(CO)^+$	Ni: Org.Verb.2-345, 351

$C_{11}H_{10}S_3V$ .... $(C_5H_5)_2V(CS_3)$ .... C: MVol.D4-224
$C_{11}H_{10}Ti$ .... $C_5H_5TiC_6H_5$ .... Ti: Org.Verb.1-204
$C_{11}H_{11}NO_4Ti$ .... $C_3H_7C_5H_4Ti(CO)_3NO$ .... Ti: Org.Verb.1-211
$C_{11}H_{11}NO_8U$ .... $[UO_2(C_6H_4(COO)_2)(NH_2COOC_2H_5)]_n$ .... U: SVol.E1-121, 123
$C_{11}H_{11}NiO_3P$ .... $(CO)_3NiP(C_6H_5)(CH_3)_2$ .... Ni: Org.Verb.1-172
$C_{11}H_{12}MnO_3$ .... $[Mn(C_4H_9COC_6H_3O_2)]$ .... Mn: MVol.D1-58
$C_{11}H_{12}N_2NpO_6^-$ .... $NpO_2(CH_3C_5H_3NCH_2N(CH_2COO)_2)^-$ .... Np: TrU.D1-163
$C_{11}H_{12}Ni$ .... $C_5H_5NiC_5H_4CH_3$ .... Ni: Org.Verb.2-214
$C_{11}H_{13}MnO_2^+$ .... $[Mn(OC_6H_3(CH_3)COC_3H_7)]^+$ .... Mn: MVol.D1-58
$C_{11}H_{13}N_2NpO_6$ .... $NpO_2(CH_3C_5H_3NCH_2N(CH_2COO)_2H)$ .... Np: TrU.D1-163
$C_{11}H_{14}MnNO_4S$ .... $[MnSCN(CH(COCH_3)_2)_2]$ .... Mn: MVol.D1-102/3
$C_{11}H_{14}MoO_3Sn$ .... $(CH_3)_3SnMo(CO)_3C_5H_5$ .... Sn: Org.Verb.5-86
$C_{11}H_{14}Ni$ .... $C_8H_9NiCH_2CHCH_2$ .... Ni: Org.Verb.2-320
$C_{11}H_{14}NiO$ .... $C_5H_5NiC_5H_6OCH_3$ .... Ni: Org.Verb.2-180
$C_{11}H_{14}O_3SnW$ .... $(CH_3)_3SnW(CO)_3C_5H_5$ .... Sn: Org.Verb.5-86
$C_{11}H_{14}O_6Sn$ .... $(C_5H_5)Sn(OCOCH_3)_3$ .... Sn: Org.Verb.6-143
$C_{11}H_{14}O_6Ti$ .... $C_5H_5Ti(OCOCH_3)_3$ .... Ti: Org.Verb.1-168/9
$C_{11}H_{14}Sn$ .... $(CH_3)_3SnCCC_6H_5$ .... Sn: Org.Verb.2-109/11
– .... $(CH_3)_3SnC_6H_4CCH$ .... Sn: Org.Verb.2-126/7
$C_{11}H_{15}IO_2Sn$ .... $(CH_3)_3SnC_6H_3ICOOCH_3$ .... Sn: Org.Verb.2-137
$C_{11}H_{15}MnO_6$ .... $[MnHCOO(CH(COCH_3)_2)_2]$ .... Mn: MVol.D1-103/4
$C_{11}H_{15}NO_2S_4Zn$ .... $Zn[S_2COC_2H_5]_2(C_5H_5N)$ .... C: MVol.D4-257
$C_{11}H_{15}NO_8S_2U$ .... $UO_2(NC_5H_3(COO)_2) \cdot 2\ (CH_3)_2SO$ .... U: SVol.E1-204, 208
$C_{11}H_{15}N_3OSn$ .... $(CH_3)_3SnC_2HN_3(O)C_6H_5$ .... Sn: Org.Verb.2-153
$C_{11}H_{15}Ni^+$ .... $C_5H_5NiC_6H_{10}^+$ .... Ni: Org.Verb.2-111
$C_{11}H_{16}N_2O_2S$ .... $C_5H_{10}NSO_2NHC_6H_5$ .... S: S-N-Verb.1-171
$C_{11}H_{16}N_2O_8U$ .... $UO_2(NO_3)(HOC_6H_4COO) \cdot C_4H_9NH_2$ .... U: SVol.E1-18, 21
$C_{11}H_{16}N_2Ti$ .... $C_5H_5TiC_6H_5 \cdot 2\ NH_3$ .... Ti: Org.Verb.1-204
$C_{11}H_{16}OSn$ .... $(CH_3)_3SnCH_2COC_6H_5$ .... Sn: Org.Verb.2-25, 29
$C_{11}H_{16}Sn$ .... $(CH_3)_3SnC(C_6H_5)CH_2$ .... Sn: Org.Verb.2-81
– .... $(CH_3)_3SnCHCHC_6H_5$ .... Sn: Org.Verb.2-76
– .... $(CH_3)_3SnC_6H_4CHCH_2$ .... Sn: Org.Verb.2-126
– .... 1) $(CH_3)_3SnC_8H_7$ .... Sn: Org.Verb.2-64
2) $(CH_3)_3SnC_8H_7$ .... Sn: Org.Verb.2-146/7
$C_{11}H_{17}IO_4Sn$ .... $CH_3SnI(CH(COCH_3)_2)_2$ .... Sn: MVol.C5-9
$C_{11}H_{17}MnNO_2S_2$ .... $[Mn(CH(COCH_3)_2)(C_5H_{10}NCSS)]$ .... Mn: MVol.D1-84
$C_{11}H_{17}NNi$ .... $(CH_3)_2C_3H_3Ni(C_5H_5N)CH_3$ .... Ni: Org.Verb.2-35
$C_{11}H_{17}NNiO$ .... $C_6H_{13}C_5H_4NiNO$ .... Ni: Org.Verb.2-349
$C_{11}H_{17}NSSn$ .... $(CH_3)_3SnSC(CH_3)NC_6H_5$ .... Sn: Org.Verb.5-84
$C_{11}H_{17}NS_4Sn$ .... $Sn(SCH_2CH_2CH_2S)_2 \cdot C_5H_5N$ .... Sn: MVol.C6-84/5
$C_{11}H_{17}N_2S_4Ti$ .... $C_5H_5Ti(S_2CN(CH_3)_2)_2$ .... Ti: Org.Verb.1-197
$C_{11}H_{17}N_3O_2S$ .... $(C_2H_5)_2NSO_2NHNCHC_6H_5$ .... S: S-N-Verb.1-222
$C_{11}H_{17}N_3O_4S_2$ .... $(CH_2)_5NSO_2NHNHSO_2C_6H_5$ .... S: S-N-Verb.1-222
$C_{11}H_{17}Ni^+$ .... $C_5H_5NiC_6H_{12}^+$ .... Ni: Org.Verb.2-111
$C_{11}H_{17}NiO_5P$ .... $(CH_2CHCHO)_2NiP(CH_2O)_3CCH_3$ .... Ni: Org.Verb.1-412
$C_{11}H_{17}PSn$ .... $(CH_3)_2SnCH_2CH_2CH_2PC_6H_5$ .... Sn: Org.Verb.5-197
$C_{11}H_{18}N_2NiS_2$ .... $(t\text{-}C_4H_9NC)_2NiCS_2$ .... Ni: Org.Verb.1-308, 309
$C_{11}H_{18}N_2O_2S$ .... $(C_2H_5)_2NSO_2NCH_3(C_6H_5)$ .... S: S-N-Verb.1-173
$C_{11}H_{18}N_2Sn$ .... $(C_2H_5)_3SnCCC_3H_3N_2$ .... Sn: Org.Verb.2-223, 234
$C_{11}H_{18}N_8O_5S_2U$ .... $UO_2(NCS)_2 \cdot 3\ CO(NHCH_2)_2$ .... U: SVol.E1-92/3

Formula	Compound	Reference
$C_{11}H_{18}OSn$	$(CH_3)_3SnCH_2C_6H_4OCH_3$	Sn: Org.Verb.2-27
–	$(CH_3)_3SnC_6H_4OC_2H_5$	Sn: Org.Verb.2-141, 144
$C_{11}H_{18}O_4Ti$	$C_5H_5Ti(OC_2H_5)_2OCOCH_3$	Ti: Org.Verb.1-190
$C_{11}H_{18}Sn$	$(CH_3)_2Sn(C_3H_7)(C_6H_5)$	Sn: Org.Verb.3-76
–	$(CH_3)_3SnCCC_6H_9$	Sn: Org.Verb.2-114
–	$(CH_3)_3SnCH(C_6H_5)CH_3$	Sn: Org.Verb.2-36
–	$(CH_3)_3SnCH_2CH_2C_6H_5$	Sn: Org.Verb.2-34
–	$(CH_3)_3SnCH_2C_6H_4CH_3$	Sn: Org.Verb.2-23/4
–	$(CH_3)_3SnC_6H_3(CH_3)_2$	Sn: Org.Verb.2-126/8
–	$(CH_3)_3SnC_6H_4C_2H_5$	Sn: Org.Verb.2-127
–	$(C_2H_5)_2Sn(CH_3)(C_6H_5)$	Sn: Org.Verb.3-86/7
–	$(C_2H_5)_3SnCCCCCH_3$	Sn: Org.Verb.2-214
$C_{11}H_{19}LiTi$	$Li[(CH_3)_4TiCH_2C_6H_5]$	Ti: Org.Verb.1-121
$C_{11}H_{19}MnO_2^+$	$[Mn((CH_3)_3CCOCHCOC(CH_3)_3)]^+$	Mn: MVol.D1-125
$C_{11}H_{19}NSn$	$(CH_3)_3SnC_6H_4N(CH_3)_2$	Sn: Org.Verb.2-140
$C_{11}H_{19}NiO_3P$	$(CO)_3NiP(C_8H_{17}\text{-n})H_2$	Ni: Org.Verb.1-166, 169
$C_{11}H_{19}Ti^-$	$[(CH_3)_4TiCH_2C_6H_5]^-$	Ti: Org.Verb.1-120/1
$C_{11}H_{20}INiP$	$C_5H_5Ni(P(C_2H_5)_3)I$	Ni: Org.Verb.2-117, 119
$C_{11}H_{20}MnN_{10}O_2S_6$	$(NH_4)_4[Mn(SCN)_6] \cdot C_4H_3OCHO$	Mn: MVol.D1-54
$C_{11}H_{20}N_2NiO_4$	$(t\text{-}C_4H_9NC)_2Ni(H_2O)CO_3$	Ni: Org.Verb.1-315, 316
$C_{11}H_{20}N_2O_2Sn$	$(C_2H_5)_3SnC_3H_2N_2COOCH_3$	Sn: Org.Verb.2-246, 249
$C_{11}H_{20}N_2O_3Sn$	$(CH_3)_3Sn(CH_2)_3C(CH_3)(CONH)_2CO$	Sn: Org.Verb.2-45/6
$C_{11}H_{20}N_2Sn$	$(C_2H_5)_3SnCCC_3H_5N_2$	Sn: Org.Verb.2-223
$C_{11}H_{20}N_4O_{12}SU$	$UO_2[(C_{10}H_{16}N_3O_6S)NO_3] \cdot CH_3OH \cdot H_2O$	U: SVol.E1-59, 63
$C_{11}H_{20}N_4S_3Sn$	$[(C_2H_5)_4N]Sn(SCN)_3$	Sn: MVol.C3-126
$C_{11}H_{20}OSn$	$(C_2H_5)_3SnCCCHCHOCH_3$	Sn: Org.Verb.2-223
$C_{11}H_{20}O_2Sn$	$(C_2H_5)_3SnCCCH_2OOCCH_3$	Sn: Org.Verb.2-223
–	$(C_2H_5)_3SnCCCOOC_2H_5$	Sn: Org.Verb.2-224
–	$(C_2H_5)_3SnCH_2COOCH_2CCH$	Sn: Org.Verb.2-167
$C_{11}H_{20}O_3Ti$	$C_5H_5Ti(OC_2H_5)_3$	Ti: Org.Verb.1-156/7, 160/1
$C_{11}H_{20}O_4Sn$	$(CH_3)_3SnC(COOC_2H_5)CHCOOC_2H_5$	Sn: Org.Verb.2-81, 89
$C_{11}H_{20}SSn$	$(CH_3)_3SnCCCHCHSC_4H_9$	Sn: Org.Verb.2-115
$C_{11}H_{20}Sn$	$(CH_2CHCH_2)_3SnC_2H_5$	Sn: Org.Verb.2-348
–	$(CH_3)_3SnCH(CHCH)_2C(CH_3)_2$	Sn: Org.Verb.2-101
–	$(C_2H_5)_3SnCCC(CH_3)CH_2$	Sn: Org.Verb.2-211/2
–	$(C_2H_5)_3SnCCCHCHCH_3$	Sn: Org.Verb.2-211
–	$(C_2H_5)_3SnCCC_3H_5$	Sn: Org.Verb.2-223, 234
–	$(C_2H_5)_3SnC_5H_5$	Sn: Org.Verb.2-206/7
$C_{11}H_{21}KOS_2$	$K[S_2COC_{10}H_{21}]$	C: MVol.D4-253
$C_{11}H_{21}KOSe_2$	$K[Se_2COC_{10}H_{21}]$	C: MVol.D6-221
$C_{11}H_{21}NNi$	$(CH_3)_2C_3H_3Ni(CNC_4H_9\text{-t})CH_3$	Ni: Org.Verb.2-45
$C_{11}H_{21}NSn$	$(C_2H_5)_3SnCH_2NC_4H_4$	Sn: Org.Verb.2-167, 176
$C_{11}H_{21}N_3O_9U$	$UO_2(C_2O_4) \cdot 3\ HCON(CH_3)_2$	U: SVol.E1-100, 102
$C_{11}H_{21}NaSn$	$(C_2H_5)_3SnC(CH_3)C(Na)CHCH_2$	Sn: Org.Verb.2-189
$C_{11}H_{21}OS_2^-$	$[S_2COC_{10}H_{21}]^-$	C: MVol.D4-251
$C_{11}H_{21}O_3PS_4$	$CH_3(O)P[S_2COC_4H_9]_2$	C: MVol.D4-254
$C_{11}H_{22}N_2O_2S$	$(CH_2)_5NSO_2NHC_6H_{11}$	S: S-N-Verb.1-171
$C_{11}H_{22}OSn$	$(C_2H_5)_3SnCCCH(CH_3)OCH_3$	Sn: Org.Verb.2-224
–	$(C_2H_5)_3SnCCCH_2CH_2OCH_3$	Sn: Org.Verb.2-224
–	$(C_2H_5)_3SnCCCH_2OC_2H_5$	Sn: Org.Verb.2-224

$C_{11}H_{22}OSn$ .... $(C_2H_5)_3SnCHCHCHCHOCH_3$ .... Sn: Org.Verb.2-194
– .... $(C_2H_5)_3SnOC(CH_3)_2CCH$ .... Sn: Org.Verb.5-103
$C_{11}H_{22}O_2Sn$ .... $(CH_3)_3SnC(COOC_2H_5)(CHCH_3)_2$ .... Sn: Org.Verb.2-59/60
– .... $(CH_3)_3SnC(COOC_2H_5)CH_2C(CH_3)_2$ .... Sn: Org.Verb.2-60
– .... $(CH_3)_3SnCHCH(COOCH_3)CH_2C(CH_3)_2$ ..... Sn: Org.Verb.2-61
– .... $(C_2H_5)_3SnCCCH_2OCH_2OCH_3$ .... Sn: Org.Verb.2-224
– .... $(C_2H_5)_3SnC(CH_2)COOC_2H_5$ .... Sn: Org.Verb.2-195, 204
– .... $(C_2H_5)_3SnC(CHD)COOC_2H_5$ .... Sn: Org.Verb.2-204
– .... $(C_2H_5)_3SnC(CH_3)CHCOOCH_3$ .... Sn: Org.Verb.2-195, 203
– .... $(C_2H_5)_3SnCHCHCOOC_2H_5$ .... Sn: Org.Verb.2-195, 203/4
– .... $(C_2H_5)_3SnCH_2COOCH_2CHCH_2$ .... Sn: Org.Verb.2-167, 176
– .... $(C_4H_9)_2Sn(CH_2CH_2COO)$ .... Sn: Org.Verb.2-299
$C_{11}H_{22}SSn$ .... $(C_2H_5)_3SnCCCH_2SC_2H_5$ .... Sn: Org.Verb.2-224
$C_{11}H_{22}SiSn$ .... $(CH_3)_3SnC_5H_4Si(CH_3)_3$ .... Sn: Org.Verb.2-99, 105
$C_{11}H_{22}Sn$ .... $(CH_3)_3SnCH_2CH_2C_6H_9$ .... Sn: Org.Verb.2-37, 39
– .... 1) $(CH_3)_3SnC_8H_{13}$ .... Sn: Org.Verb.2-64, 68
2) $(CH_3)_3SnC_8H_{13}$ .... Sn: Org.Verb.2-65, 68
3) $(CH_3)_3SnC_8H_{13}$ .... Sn: Org.Verb.2-101, 105/6
– .... $(C_2H_5)_3SnCCCH(CH_3)_2$ .... Sn: Org.Verb.2-212, 214
– .... $(C_2H_5)_3SnCCC_3H_7$ .... Sn: Org.Verb.2-214
– .... $(C_2H_5)_3SnC(CHCH_3)CHCH_2$ .... Sn: Org.Verb.2-186
– .... $(C_2H_5)_3SnC(CH_2)C(CH_3)CH_2$ .... Sn: Org.Verb.2-186
– .... $(C_2H_5)_3SnC(CH_2)CHCHCH_3$ .... Sn: Org.Verb.2-186
– .... $(C_2H_5)_3SnC(CH_3)CCHCH_3$ .... Sn: Org.Verb.2-186
– .... $(C_2H_5)_3SnC(CH_3)CHCHCH_2$ .... Sn: Org.Verb.2-186, 189
– .... $(C_2H_5)_3SnC(CH_3)_2CCH$ .... Sn: Org.Verb.2-214
– .... $(C_2H_5)_3SnCHCC(CH_3)_2$ .... Sn: Org.Verb.2-186
– .... $(C_2H_5)_3SnCHCCHCH_2CH_3$ .... Sn: Org.Verb.2-185
– .... $(C_2H_5)_3SnCHCHC(CH_3)CH_2$ .... Sn: Org.Verb.2-186
– .... $(C_2H_5)_3SnCHCHCHCHCH_3$ .... Sn: Org.Verb.2-185
– .... $(C_2H_5)_3SnCH_2C(CH_3)CCH_2$ .... Sn: Org.Verb.2-186
– .... $(C_2H_5)_3SnCH_2CH(CH_3)CCH$ .... Sn: Org.Verb.2-214
– .... $(C_2H_5)_3SnCH_2CH_2CCCH_3$ .... Sn: Org.Verb.2-214
– .... $(C_2H_5)_3SnC_5H_7$ .... Sn: Org.Verb.2-207
– .... $(C_3H_7)_3SnCCH$ .... Sn: Org.Verb.2-263/4
$C_{11}H_{23}I_4NO_2Sn$ ..... $(C_2H_5)_3NH[SnI_4(CH(COCH_3)_2)]$ .... Sn: MVol.C5-107
$C_{11}H_{23}NO_2Sn$ ...... $(C_2H_5)_3SnCH_2CON(CH_3)COCH_3$ .... Sn: Org.Verb.2-167, 176
$C_{11}H_{23}NSn$ .... $(C_2H_5)_3SnCCC(CH_3)_2NH_2$ .... Sn: Org.Verb.2-225
$C_{11}H_{23}N_3Ti$ .... $C_5H_5Ti(N(CH_3)_2)_3$ .... Ti: Org.Verb.1-172, 174/8
$C_{11}H_{23}O_2Ti$ .... $C_3H_5Ti(OC_4H_9)_2$ .... Ti: Org.Verb.1-127
$C_{11}H_{24}N_2O_2S$ .... $(C_2H_5)_2NSO_2NCH_3(C_6H_{11})$ .... S: S-N-Verb.1-173
$C_{11}H_{24}N_2Sn$ .... $(C_2H_5)_3SnCH(CN)CH_2N(CH_3)_2$ .... Sn: Org.Verb.2-167
$C_{11}H_{24}N_3O_{12}PU$ ..... $UO_2(NO_3)_2 \cdot (C_3H_7O)_2POCON(C_2H_5)_2$ ..... U: SVol.E1-188/9
$C_{11}H_{24}N_6O_5S_3U$ ..... $UO_2(NCS)NO_3 \cdot 2\ CS(N(CH_3)_2)_2$ .... U: SVol.E1-220/1
$C_{11}H_{24}N_6O_9U$ .... $UO_2C_2O_4 \cdot 3\ CO(NHCH_3)_2$ .... U: SVol.E1-93, 95
– .... $UO_2C_2O_4 \cdot 3\ NH_2CONHC_2H_5$ .... U: SVol.E1-93, 95
$C_{11}H_{24}N_8O_2S_5U$ ..... $UO_2(NCS)_2 \cdot 3\ CS(NHCH_3)_2$ .... U: SVol.E1-220/1
$C_{11}H_{24}OSiSn$ .... $(C_2H_5)_3SnC(CO)Si(CH_3)_3$ .... Sn: Org.Verb.2-195
$C_{11}H_{24}OSn$ .... $(C_2H_5)_3SnC(CH_2)CH(OH)C_2H_5$ .... Sn: Org.Verb.2-195
– .... $(C_2H_5)_3SnC(CH_2)C(OH)(CH_3)_2$ .... Sn: Org.Verb.2-195

Formula	Compound	Reference
$C_{11}H_{24}OSn$	$(C_2H_5)_3SnCHCHCH_2OC_2H_5$	Sn: Org.Verb.2-195, 204
–	$(C_2H_5)_3SnCHCHC(OH)(CH_3)_2$	Sn: Org.Verb.2-195
–	$(C_3H_7)_3SnCH_2CHO$	Sn: Org.Verb.2-251/2
–	$(C_4H_9)_2(HOCH_2CHCH)SnH$	Sn: Org.Verb.4-91
$C_{11}H_{24}O_2Sn$	$(CH_3)_3SnCH(COOC_2H_5)CH(CH_3)C_2H_5$	Sn: Org.Verb.2-55, 57
–	$(CH_3)_3SnCH(C_4H_9)COOC_2H_5$	Sn: Org.Verb.2-55, 57
–	$(C_2H_5)_3SnCH_2CH(CH_3)COOCH_3$	Sn: Org.Verb.2-168
–	$(C_2H_5)_3SnCH_2CH_2COOCH_2CH_3$	Sn: Org.Verb.2-168, 177
–	$(C_2H_5)_3SnCH_2CH_2OCH_2CH(O)CH_2$	Sn: Org.Verb.2-168
–	$(C_2H_5)_3SnCH_2COOC_3H_7$	Sn: Org.Verb.2-168, 176
–	$(C_2H_5)_3Sn(CH_2)_3OOCCH_3$	Sn: Org.Verb.2-168
–	$(C_3H_7)_2Sn(CH_3)(CH_2CH_2COOCH_3)$	Sn: Org.Verb.3-87
–	$(i\text{-}C_3H_7)_2Sn(CH_3)(C_2H_4COOCH_3)$	Sn: Org.Verb.3-92/3
–	$[Sn(C_4H_9)_2O(CH_2)_3O]_x$	Sn: Org.Verb.6-89, 94
$C_{11}H_{24}PbSn$	$(CH_3)_3SnCCPb(C_2H_5)_3$	Sn: Org.Verb.2-115
–	$(C_2H_5)_3SnCCPb(CH_3)_3$	Sn: Org.Verb.2-225
$C_{11}H_{24}SSn$	$(C_2H_5)_3SnC(CH_3)CHSC_2H_5$	Sn: Org.Verb.2-196
–	$(C_2H_5)_3SnCHCHCH_2SC_2H_5$	Sn: Org.Verb.2-196
$C_{11}H_{24}SiSn$	$(C_2H_5)_3SnCCSi(CH_3)_3$	Sn: Org.Verb.2-219
$C_{11}H_{24}Sn$	$(CH_3)_2Sn(i\text{-}C_3H_7)(C_6H_{11})$	Sn: Org.Verb.3-77
–	$(CH_3)_2Sn(s\text{-}C_4H_9)(CH(CH_2)_4)$	Sn: Org.Verb.3-75, 78
–	$(CH_3)_3SnC(C_4H_9)CHC_2H_5$	Sn: Org.Verb.2-82
–	$(CH_3)_3SnCHCHC_6H_{13}$	Sn: Org.Verb.2-81
–	$(C_2H_5)_3SnCH(CH_3)CHCHCH_3$	Sn: Org.Verb.2-187
–	$(C_2H_5)_3SnCH_2C(CH_3)CHCH_3$	Sn: Org.Verb.2-187
–	$(C_2H_5)_3SnCH_2CHC(CH_3)_2$	Sn: Org.Verb.2-187
–	$(C_2H_5)_3SnCH_2CHCHCH_2CH_3$	Sn: Org.Verb.2-187, 190
–	$(C_2H_5)_3SnCH_2CH(CH_3)CHCH_2$	Sn: Org.Verb.2-187
–	$(C_2H_5)_3SnCH_2CH_2C(CH_3)CH_2$	Sn: Org.Verb.2-187
–	$(C_2H_5)_3SnCH_2CH_2CHCHCH_3$	Sn: Org.Verb.2-187, 190
–	$(C_2H_5)_3SnCH_2CH_2CH_2CHCH_2$	Sn: Org.Verb.2-187, 189
–	$(C_3H_7)_2Sn(CH_2)_5$	Sn: Org.Verb.3-105
–	$(C_3H_7)_3SnCHCH_2$	Sn: Org.Verb.2-259/60
–	$(i\text{-}C_3H_7)_3SnCHCH_2$	Sn: Org.Verb.2-270/1
$C_{11}H_{25}NOSn$	$(C_2H_5)_3SnCH(CHO)CH_2N(CH_3)_2$	Sn: Org.Verb.2-168
–	$(C_2H_5)_3SnCH_2N(CH_2CH_2)_2O$	Sn: Org.Verb.2-169, 177
$C_{11}H_{25}NO_2S$	$[(C_3H_7)_3NH][SOCOCH_3]$	C: MVol.D4-237
$C_{11}H_{25}NSn$	$(C_2H_5)_3SnNC_5H_{10}$	Sn: Org.Verb.5-102
$C_{11}H_{25}N_2O_{11}PU$	$UO_2(NO_3)_2 \cdot (C_5H_{11}O)_2(CH_3)PO$	U: SVol.E1-186/7
$C_{11}H_{25}Ti^+$	$[((CH_3)_3CCH_2)_2TiCH_3]^+$	Ti: Org.Verb.1-99
$C_{11}H_{26}N_2O_{14}P_2U$	$UO_2(NO_3)_2 \cdot (C_2H_5O)_2POCH_2PO(OC_3H_7)_2$	U: SVol.E1-188/9
$C_{11}H_{26}OSiSn$	$(C_2H_5)_3SnCHCHSi(CH_3)_2OCH_3$	Sn: Org.Verb.2-196
$C_{11}H_{26}OSn$	$(C_2H_5)_3SnOCH_2CH(CH_3)C_2H_5$	Sn: Org.Verb.5-105
–	$(C_4H_9)_2(HOCH_2CH_2CH_2)SnH$	Sn: Org.Verb.4-91
$C_{11}H_{26}O_2SiSn$	$(C_2H_5)_3SnCH_2COOSi(CH_3)_3$	Sn: Org.Verb.2-169, 177
$C_{11}H_{26}O_2Sn$	$(C_2H_5)_3SnCH_2CH_2CH_2OCH_2CH_2OH$	Sn: Org.Verb.2-169
$C_{11}H_{26}O_3Sn$	$C_2H_5Sn(OC_3H_7)_3$	Sn: Org.Verb.6-230/1
$C_{11}H_{26}SSn$	$(CH_3CH_2)_3SnCH_2CH_2SCH(CH_3)_2$	Sn: Org.Verb.2-169, 177
$C_{11}H_{26}SiSn$	$(C_2H_5)_3SnCHCHSi(CH_3)_3$	Sn: Org.Verb.2-196
$C_{11}H_{26}Sn$	$(CH_3)(C_3H_7)Sn(i\text{-}C_3H_7)C_4H_9$	Sn: Org.Verb.3-96

Formula	Compound	Reference
$C_{11}H_{26}Sn$	$(CH_3)(C_3H_7)Sn(i\text{-}C_3H_7)(i\text{-}C_4H_9)$	Sn: Org.Verb.3-96
–	$(CH_3)(C_3H_7)Sn(i\text{-}C_3H_7)(s\text{-}C_4H_9)$	Sn: Org.Verb.3-96
–	$(CH_3)(C_3H_7)Sn(i\text{-}C_3H_7)(t\text{-}C_4H_9)$	Sn: Org.Verb.3-96
–	$(CH_3)_2Sn(s\text{-}C_4H_9)(CH(CH_3)C_3H_7)$	Sn: Org.Verb.3-75, 78
–	$(CH_3)_2Sn(s\text{-}C_4H_9)(CH(C_2H_5)_2)$	Sn: Org.Verb.3-75, 78
–	$(CH_3)_2Sn(s\text{-}C_4H_9)(CH_2C(CH_3)_3)$	Sn: Org.Verb.3-75, 78
–	$(CH_3)_2Sn(s\text{-}C_4H_9)(CH_2CH(CH_3)C_2H_5)$	Sn: Org.Verb.3-75, 78
–	$(CH_3)_2Sn(s\text{-}C_4H_9)(i\text{-}C_5H_{11})$	Sn: Org.Verb.3-75, 78
–	$(CH_3)_3SnC_8H_{17}$	Sn: Org.Verb.2-53
–	$(C_2H_5)_3SnCH_2C(CH_3)_3$	Sn: Org.Verb.2-159
–	$(C_2H_5)_3SnC_5H_{11}$	Sn: Org.Verb.2-159
–	$(C_2H_5)_3Sn(i\text{-}C_5H_{11})$	Sn: Org.Verb.2-159
–	$(C_3H_7)_3SnC_2H_5$	Sn: Org.Verb.2-252
$C_{11}H_{27}NSn$	$(i\text{-}C_4H_9)_2(NH_2CH_2CH_2CH_2)SnH$	Sn: Org.Verb.4-91
$C_{11}H_{27}PSn$	$(CH_3)_3SnC(CH_3)P(C_2H_5)_3$	Sn: Org.Verb.2-38
$C_{11}H_{28}N_3SSi_2^+$	$[(CH_3)_3SiN]_2SNC_5H_{10}^+$	S: S-N-Verb.1-211
$C_{11}H_{28}O_3SiSn$	$(C_2H_5)_3SnCH(CH_3)Si(OCH_3)_3$	Sn: Org.Verb.2-169
–	$(C_2H_5)_3SnCH_2CH_2Si(OCH_3)_3$	Sn: Org.Verb.2-169
$C_{11}H_{28}PbSn$	$(CH_3)_3Sn(CH_2)_5Pb(CH_3)_3$	Sn: Org.Verb.2-55
$C_{11}H_{28}SiSn$	$(C_2H_5)_3SnCH_2Si(C_2H_5)(CH_3)_2$	Sn: Org.Verb.2-169
$C_{11}H_{30}N_6Ni_2S_6$	$Ni_2(CH_3CH(NH_2)CH_2NH_2)_3(CS_3)_2$	C: MVol.D4-225
$C_{11}H_{33}NNi_2OP_2$	$(CH_3Ni(P(CH_3)_3)N(CH_3)_2)(CH_3Ni(P(CH_3)_3)OCH_3)$	Ni: Org.Verb.2-254
$C_{11}H_{33}N_4SSiSn_2^+$	$[(CH_3)_3SiN]NS[N(CH_3)Sn(CH_3)_3]_2^+$	S: S-N-Verb.1-210
$C_{11}H_{33}NiP_3$	$(CH_3)_2Ni(P(CH_3)_3)_3$	Ni: Org.Verb.1-69, 72
$C_{12}CaCl_6H_{12}O_{12}Sn_2$	$Ca[Sn(CH_2ClCOO)_3]_2$	Sn: MVol.C3-179
$C_{12}CaCl_6H_{18}N_6Sn$	$[Ca(CH_3CN)_6][SnCl_6]$	Sn: MVol.C3-176
$C_{12}CaH_{18}O_{12}Sn_2$	$Ca[Sn(CH_3COO)_3]_2$	Sn: MVol.C3-179
$C_{12}CaH_{24}N_2O_6S_2$	$Ca[C_6H_{11}NHSO_3]_2 \cdot 2\ H_2O$	S: S-N-Verb.1-97
$C_{12}Ca_3Ir_2O_{24}$	$Ca_3[Ir(C_2O_4)_3]_2 \cdot 14\ H_2O$	Ir: SVol.2-35
$C_{12}CdF_8N_2O_4$	$Cd[OC(O)C_5F_4N]_2$	F: PerFHalOrg.5-192/3, 207, 211
$C_{12}CdF_{10}$	$(C_6F_5)_2Cd$	F: PerFHalOrg.4-35, 112, 136
$C_{12}CdFe_2H_{24}N_8O_8$	$[Cd(C_2H_8N_2)_3][Fe(CO)_3NO]_2$	Fe: Org.Verb.B1-189
$C_{12}CdH_9N_9S_6Sn$	$Cd[Sn(NCS)_6] \cdot 3\ CH_3CN$	Sn: MVol.C4-56
$C_{12}Cd_3Ir_2N_{12}$	$Cd_3[Ir(CN)_6]_2 \cdot 12\ H_2O$	Ir: SVol.2-183, 184
$C_{12}CeCl_6H_{16}N_2$	$[CH_3C_5H_4NH]_2CeCl_6$	Sc: MVol.C5-224
$C_{12}CeCl_6H_{32}N_2$	$CeCl_4 \cdot 2\ [(C_2H_5)_3NH]Cl$	Sc: MVol.C5-223
$C_{12}CeCl_7H_{40}N_4$	$CeCl_3 \cdot 4\ [(CH_3)_3NH]Cl$	Sc: MVol.C5-213/6
$C_{12}CeCl_9H_{48}N_6$	$CeCl_3 \cdot 6\ [(CH_3)_2NH_2]Cl$	Sc: MVol.C5-213/5
–	$CeCl_3 \cdot 6\ [(CH_3)_2NH_2]Cl \cdot 2\ H_2O$	Sc: MVol.C5-213/5
$C_{12}CfH_{12}N_2O_{12}^{3-}$	$Cf(N(CH_2COO)_3)_2^{3-}$	Np: TrU.D1-163
$C_{12}CfH_{21}O_9$	$Cf((CH_3)_2C(OH)COO)_3$	Np: TrU.D1-154
$C_{12}ClFH_{18}N_2O_2$	$(C_4H_9O)_2C_4FClN_2$	F: PerFHalOrg.6-58
$C_{12}ClFH_{20}N_4$	$((C_2H_5)_2N)_2C_4FClN_2$	F: PerFHalOrg.6-59
$C_{12}ClF_3H_{10}N_4$	$(C_6H_5CH_2NH)(NH_2)(CF_3)C_4ClN_2$	F: PerFHalOrg.6-59
$C_{12}ClF_3H_{10}Ni$	$C_5H_5NiC_5H_5C_2F_3Cl$	Ni: Org.Verb.2-181
$C_{12}ClF_5H_6N_2O_2$	$NC_5F_2Cl(OCH_3)C_5F_3(OCH_3)N$	F: PerFHalOrg.5-224

$C_{12}ClF_6FeH_{20}N_3O_2PS_3$ $[C_6H_{12}N_2SFe(CO)_2S_2CN(CH_3)_2]PF_6 \cdot 0.5\ C_2H_4Cl_2$ Fe: Org.Verb.B4-192
$C_{12}ClF_8H_{12}N_3$ ...... $(CF_3)_2(CF_2Cl)(C_6H_{11})C_3HN_3$ ...... F: PerFHalOrg.6-113
$C_{12}ClF_8Li$ ...... $LiC_6F_4C_6F_4Cl$ ...... F: PerFHalOrg.4-7/8, 13
$C_{12}ClF_9Mg$ ...... $C_6F_5C_6F_4MgCl$ ...... F: PerFHalOrg.4-70
$C_{12}ClF_9MgO$ ...... $C_6F_5OC_6F_4MgCl$ ...... F: PerFHalOrg.4-63, 65, 71
$C_{12}ClF_{10}P$ ...... $(C_6F_5)_2PCl$ ...... F: PerFHalOrg.3-124/9
$C_{12}ClF_{10}PS$ ...... $(C_6F_5)_2P(S)Cl$ ...... F: PerFHalOrg.3-141, 148
$C_{12}ClF_{10}Sb$ ...... $(C_6F_5)_2SbCl$ ...... F: PerFHalOrg.3-226/7, 229
$C_{12}ClF_{10}Tl$ ...... $(C_6F_5)_2TlCl$ ...... F: PerFHalOrg.4-131/2, 134, 136/8
$C_{12}ClF_{20}H_5S_2Sn$ ...... $CF_2Cl(CF_2)_9CH_2CH_2Sn(S)SH$ ...... Sn: Org.Verb.6-264
$C_{12}ClFeH_7MgO_5$ ...... $MgCl[C_6H_5CH_2C(O)Fe(CO)_4]$ ...... Fe: Org.Verb.B4-152
$C_{12}ClFeH_7O_4$ ...... $(ClCHCHC(C_6H_5)O)Fe(CO)_3$ ...... Fe: Org.Verb.B5-134, 146
– ...... $(ClC_6H_4CHCHCHO)Fe(CO)_3$ ...... Fe: Org.Verb.B5-122/7, 130, 140
– ...... $ClC_6H_4CHCH_2Fe(CO)_4$ ...... Fe: Org.Verb.B4-238/9, 254
$C_{12}ClFeH_8NO_5$ ...... $[C_5H_5NCOCHCH_2Fe(CO)_4]Cl$ ...... Fe: Org.Verb.B4-249
$C_{12}ClFeH_9O_3$ ...... $C_6H_5C_3H_4Fe(CO)_3Cl$ ...... Fe: Org.Verb.B5-43, 48, 52/3, 62, 64
$C_{12}ClFeH_{11}$ ...... $C_5H_5FeC_5H_4CClCH_2$ ...... Fe: Org.Verb.A1-384, 386
– ...... $C_5H_5FeC_5H_4CHCHCl$ ...... Fe: Org.Verb.A1-384, 386
$C_{12}ClFeH_{11}O$ ...... $C_5H_5FeC_5H_4CH_2COCl$ ...... Fe: Org.Verb.A3-159
– ...... $C_5H_5FeC_5H_4COCH_2Cl$ ...... Fe: Org.Verb.A2-188, 242/3, 245
$C_{12}ClFeH_{11}O_2$ ...... $C_5H_5FeC_5H_4OOCCH_2Cl$ ...... Fe: Org.Verb.A2-93
$C_{12}ClFeH_{12}^+$ ...... $[C_5H_5FeC_5H_4CHCH_2Cl]^+$ ...... Fe: Org.Verb.A2-47
$C_{12}ClFeH_{12}O^+$ ...... $[C_5H_5FeC_5H_4COHCH_2Cl]^+$ ...... Fe: Org.Verb.A2-242
$C_{12}ClFeH_{13}$ ...... $C_5H_5FeC_5H_4CHClCH_3$ ...... Fe: Org.Verb.A1-381/2
– ...... $C_5H_5FeC_5H_4CH_2CH_2Cl$ ...... Fe: Org.Verb.A1-381
Fe: Org.Verb.A2-71
– ...... $C_5H_5FeC_5H_4CH_2CD_2Cl$ ...... Fe: Org.Verb.A2-72
$C_{12}ClFeH_{13}O$ ...... $C_5H_5FeC_5H_4CHClCH_2OH$ ...... Fe: Org.Verb.A2-74, 79
– ...... $C_5H_5FeC_5H_4CHOHCH_2Cl$ ...... Fe: Org.Verb.A2-44, 47/8
$C_{12}ClFeH_{13}O^{2+}$ ...... $[HC_5H_5FeC_5H_4COHCH_2Cl]^{2+}$ ...... Fe: Org.Verb.A2-242
$C_{12}ClFeH_{16}N$ ...... $C_5H_5FeC_5H_4CH(CH_3)NH_2 \cdot HCl$ ...... Fe: Org.Verb.A2-220
$C_{12}ClFeH_{17}N_2O_5$ ...... $[CH_3NH(CH_2)_2N(CH_3)_2COCHCH_2Fe(CO)_4]Cl$ Fe: Org.Verb.B4-249
$C_{12}ClFe_2GeH_{15}N_4O_4$. $(C_3H_5Fe(NO)_2)_2Ge(Cl)C_6H_5$ ...... Fe: Org.Verb.C2-165
$C_{12}ClFe_2H_{15}N_4O_4Pb$. $(C_3H_5Fe(NO)_2)_2Pb(Cl)C_6H_5$ ...... Fe: Org.Verb.C2-165
$C_{12}ClFe_2H_{15}N_4O_4Sn$. $(C_3H_5Fe(NO)_2)_2Sn(Cl)C_6H_5$ ...... Fe: Org.Verb.C2-165
$C_{12}ClH_7N_2O_7SSn$ ...... $[Sn(OH)(O_2C_6H_3NNC_6H_2Cl(OH)SO_3)]$ ...... Sn: MVol.C6-34
$C_{12}ClH_9N_2O_3U$ ...... $[UO_2(OH)Cl(C_{12}H_8N_2)] \cdot 2\ H_2O$ ...... U: SVol.E1-44/6
$C_{12}ClH_9S_3Sn$ ...... $(C_4H_3S)_3SnCl$ ...... Sn: Org.Verb.5-189
$C_{12}ClH_{10}ISn$ ...... $(C_6H_5)_2ClSnI$ ...... Sn: Org.Verb.6-162/4
$C_{12}ClH_{10}N_3O_3S_3$ ...... $[NS(O)Cl][NS(O)C_6H_5]_2$ ...... S: S-N-Verb.1-19
$C_{12}ClH_{10}Ti$ ...... $(C_6H_5)_2TiCl \cdot 3\ (C_2H_5)_2O$ ...... Ti: Org.Verb.1-70
$C_{12}ClH_{11}OSn$ ...... $(C_6H_5)_2ClSnOH$ ...... Sn: Org.Verb.6-162/4
$C_{12}ClH_{11}Sn$ ...... $(C_6H_5)_2SnHCl$ ...... Sn: Org.Verb.6-290
$C_{12}ClH_{12}NO_6U$ ...... $[UO_2(HOC_6H_4COO)(C_5H_5N)H_2O]Cl$ ...... U: SVol.E1-29, 34

Formula	Compound	Reference
$C_{12}ClH_{15}Sn$	$(C_5H_5)_2(C_2H_5)SnCl$	Sn: Org.Verb.5-223
$C_{12}ClH_{16}MnO_6$	$[MnCH_2ClCOO(CH(COCH_3)_2)_2]$	Mn: MVol.D1-103/4
$C_{12}ClH_{17}OSn$	$(CH_3)_2[C_6H_5CO(CH_2)_3]SnCl$	Sn: Org.Verb.5-195, 199
$C_{12}ClH_{19}Sn$	$(CH_3)(i\text{-}C_4H_9)(C_6H_5CH_2)SnCl$	Sn: Org.Verb.5-225, 227
–	$(CH_3)_2Sn(C_6H_5)(CH_2CH_2CHClCH_3)$	Sn: Org.Verb.3-80
–	$(C_2H_5)_3SnC_6H_4Cl$	Sn: Org.Verb.2-239
$C_{12}ClH_{21}Sn$	$[CH_3CHC(CH_3)]_3SnCl$	Sn: Org.Verb.5-157
–	$(CH_3)_2(C_{10}H_{15})SnCl$	Sn: Org.Verb.5-195
$C_{12}ClH_{21}Ti$	$(CH_2C(CH_3)CH_2)_3TiCl$	Ti: Org.Verb.1-128
$C_{12}ClH_{22}NO_2S$	$ClSO_2N(C_6H_{11})_2$	S: S-N-Verb.1-124
$C_{12}ClH_{22}NSn$	$[(CH_3)_3SnC_6H_4N(CH_3)_3]Cl$	Sn: Org.Verb.2-142, 144
–	$(C_2H_5)_3SnCl \cdot CH_3C_5H_4N$	Sn: Org.Verb.5-107
$C_{12}ClH_{22}N_3O_9SU$	$UO_2[(C_{10}H_{16}N_3O_6S)Cl] \cdot C_2H_5OH$	U: SVol.E1-59, 63
$C_{12}ClH_{23}N_6S_2Sn$	$(C_6H_5)ClSn[N(CH_3)S(NCH_3)_2]_2$	Sn: Org.Verb.6-275, 277
$C_{12}ClH_{23}Sn$	$(C_3H_7)_3SnCCCH_2Cl$	Sn: Org.Verb.2-263/4
–	$(C_6H_{11})_2SnHCl$	Sn: Org.Verb.6-290
$C_{12}ClH_{25}OSn$	$(C_4H_9)_2(CH_3COCH_2CH_2)SnCl$	Sn: Org.Verb.5-208
$C_{12}ClH_{25}O_2Sn$	$(i\text{-}C_4H_9)_2(CH_3OCOCH_2CH_2)SnCl$	Sn: Org.Verb.5-213/4
$C_{12}ClH_{25}Sn$	$(C_3H_7)_2(C_6H_{11})SnCl$	Sn: Org.Verb.5-205/6
–	$(C_4H_9)_2(C_2H_5CHCH)SnCl$	Sn: Org.Verb.5-208
–	$(i\text{-}C_4H_9)_2(CH_2CHCH_2CH_2)SnCl$	Sn: Org.Verb.5-213/4
–	$(i\text{-}C_4H_9)_2(CH_3CHCHCH_2)SnCl$	Sn: Org.Verb.5-213/4
$C_{12}ClH_{26}Sn$	$(C_4H_9)_2ClSnCH_2CHC_2H_5$	Sn: Org.Verb.5-128
$C_{12}ClH_{27}I_3N_3O_3U$	$UI_3Cl \cdot 3\ CH_3CON(CH_3)_2$	U: SVol.E1-103
$C_{12}ClH_{27}OSn$	$(C_4H_9)_2[CH_3CH(OH)CH_2CH_2]SnCl$	Sn: Org.Verb.5-208
–	$(C_4H_9)_2[C_2H_5CH(OH)CH_2]SnCl$	Sn: Org.Verb.5-208
–	$(C_4H_9)_2(C_3H_7CHOH)SnCl$	Sn: Org.Verb.5-208
–	$(C_4H_9)_2ClSnOC_4H_9$	Sn: Org.Verb.5-131/2
–	$(i\text{-}C_4H_9)_2ClSnOCH_2C_3H_7$	Sn: Org.Verb.6-105
–	$(i\text{-}C_4H_9)_2ClSn(OC_4H_9\text{-}i)$	Sn: Org.Verb.6-105
–	$(C_4H_9)_2[HO(CH_2)_4]SnCl$	Sn: Org.Verb.5-208
$C_{12}ClH_{27}O_3Sn$	$(C_2H_5OCH_2CH_2)_3SnCl$	Sn: Org.Verb.5-151
$C_{12}ClH_{27}SSn$	$(C_4H_9)_2ClSnSC_4H_9$	Sn: Org.Verb.5-132
		Sn: Org.Verb.6-88
$C_{12}ClH_{27}Sn$	$(CH_3)_2(C_{10}H_{21})SnCl$	Sn: Org.Verb.5-195
–	$(C_2H_5)_2(C_8H_{17})SnCl$	Sn: Org.Verb.5-203
–	$(C_4H_9)_2(i\text{-}C_4H_9)SnCl$	Sn: Org.Verb.5-208
–	$(C_4H_9)_2(t\text{-}C_4H_9)SnCl$	Sn: Org.Verb.5-208, 212
–	$(t\text{-}C_4H_9)_2(C_4H_9)SnCl$	Sn: Org.Verb.5-217, 221
–	$(C_4H_9)_3SnCl$	
	Analysis	Sn: Org.Verb.5-120
	Chemical reactions	Sn: Org.Verb.5-128/34
	Formation and preparation	Sn: Org.Verb.5-117/23
	Physical properties	Sn: Org.Verb.5-126/7
	Physiological action	Sn: Org.Verb.5-134/6
	Polarography	Sn: Org.Verb.5-128
	Spectra	Sn: Org.Verb.5-124/6
	Uses	Sn: Org.Verb.5-136/7
–	$(i\text{-}C_4H_9)_3SnCl$	Sn: Org.Verb.5-137/8
–	$(t\text{-}C_4H_9)_3SnCl$	Sn: Org.Verb.5-138/9

Formula	Structure		Reference
$C_{12}ClH_{27}Sn^+$	$(C_4H_9)_3SnCl^+$	Sn:	Org.Verb.5-124, 138
$C_{12}ClH_{27}Ti$	$(C_4H_9)_3TiCl$	Ti:	Org.Verb.1-84
$C_{12}ClH_{28}O_2PS_2Sn$	$(C_2H_5O)_2PSSSn(C_4H_9)_2Cl$	Sn:	Org.Verb.6-89, 96
$C_{12}ClH_{29}N_2Sn$	$(C_4H_9)[(C_2H_5)_2N]_2SnCl$	Sn:	Org.Verb.6-245
$C_{12}ClH_{30}OPSn$	$(C_2H_5)_3SnCl \cdot (C_2H_5)_2(C_2H_5O)P$	Sn:	Org.Verb.5-107
$C_{12}ClH_{33}O_2Si_4Sn$	$CH_2(Si(CH_3)_2OSi(CH_3)_2CH_2)_2Sn(CH_3)Cl$	Sn:	Org.Verb.5-230
$C_{12}ClH_{33}Si_3Sn$	$[(CH_3)_3SiCH_2]_3SnCl$	Sn:	Org.Verb.5-153/4
–	$(CH_3)_3SnC(Si(CH_3)_3)_2SiCl(CH_3)_2$	Sn:	Org.Verb.5-79
$C_{12}ClH_{33}Si_3Ti$	$[(CH_3)_3SiCH_2]_3TiCl$	Ti:	Org.Verb.1-85
$C_{12}ClH_{34}NiP_3$	$[CH_3[(CH_3)_3PCH_2]_2NiP(CH_3)_3]Cl$	Ni:	Org.Verb.1-104
$C_{12}Cl_2Cu_2H_{24}Sn$	$(C_4H_9)_2Sn(CHCH_2)_2 \cdot 2\ CuCl$	Sn:	Org.Verb.3-41, 43
$C_{12}Cl_2FH_9N_2O$	$(NH_2)(C_6H_5CH_2O)C_5FCl_2N$	F:	PerFHalOrg.5-214/5
$C_{12}Cl_2F_2H_3N_5O_4$	$NC_5F_2Cl_2NHC_6H_2(NO_2)_2CN$	F:	PerFHalOrg.5-215
$C_{12}Cl_2F_2H_4N_4O_2$	$NC_5F_2Cl_2NHC_6H_3(NO_2)CN$	F:	PerFHalOrg.5-215
$C_{12}Cl_2F_2H_6N_2O_3$	$NC_5F_2Cl_2(OCH_2C_6H_4NO_2)$	F:	PerFHalOrg.5-136
		F:	PerFHalOrg.6-14
$C_{12}Cl_2F_2H_7NO$	$NC_5F_2Cl_2(OCH_2C_6H_5)$	F:	PerFHalOrg.5-136
		F:	PerFHalOrg.6-14
$C_{12}Cl_2F_2H_7N_3O_2$	$NC_5F_2Cl_2NHC_6H_3(NO_2)CH_3$	F:	PerFHalOrg.5-140
$C_{12}Cl_2F_2H_7N_3O_3$	$NC_5F_2Cl_2NHC_6H_3(NO_2)OCH_3$	F:	PerFHalOrg.5-140
$C_{12}Cl_2F_2H_8O_2S$	$(ClC_6H_4O)_2SF_2$	S:	SVol.2-44, 83
$C_{12}Cl_2F_2H_8Sn$	$(FC_6H_4)_2SnCl_2$	Sn:	Org.Verb.6-183/6
$C_{12}Cl_2F_3H_6NO$	$NC_5F_2Cl_2(OCH_2C_6H_4F)$	F:	PerFHalOrg.5-136
		F:	PerFHalOrg.6-14
$C_{12}Cl_2F_4H_6N_2O_2$	$NC_5F_2Cl(OCH_3)C_5F_2Cl(OCH_3)N$	F:	PerFHalOrg.5-224
$C_{12}Cl_2F_5H_4N_3O_2$	$NC_5F_2Cl_2NHC_6H_3(NO_2)CF_3$	F:	PerFHalOrg.5-140
$C_{12}Cl_2F_5H_5N_2$	$NC_5F_2Cl_2NHC_6H_4CF_3$	F:	PerFHalOrg.5-140
$C_{12}Cl_2F_6HN_3$	$NC_5F_2Cl_2NHC_6F_4CN$	F:	PerFHalOrg.5-188/9
		F:	PerFHalOrg.6-13
$C_{12}Cl_2F_7H_2NO$	$NC_5F_2Cl_2(OCH_2C_6F_5)$	F:	PerFHalOrg.5-136
		F:	PerFHalOrg.6-14
$C_{12}Cl_2F_7H_{12}N_3$	$(CF_2Cl)_2(CF_3)(C_6H_{11})C_3HN_3$	F:	PerFHalOrg.6-113
$C_{12}Cl_2F_8$	$ClC_6F_4C_6F_4Cl$	F:	PerFHalOrg.4-14
$C_{12}Cl_2F_8MgO$	$ClC_6F_4OC_6F_4MgCl$	F:	PerFHalOrg.4-64/5
$C_{12}Cl_2F_8Mg_2O$	$(ClMgC_6F_4)_2O$	F:	PerFHalOrg.4-66/7, 85
$C_{12}Cl_2F_{10}H_2NP$	$(C_6F_5)_2P(Cl_2)NH_2$	F:	PerFHalOrg.3-61, 69
$C_{12}Cl_2F_{10}Si$	$(C_6F_5)_2SiCl_2$	F:	PerFHalOrg.4-149/50, 156, 162
$C_{12}Cl_2F_{10}Sn$	$(C_6F_5)_2SnCl_2$	F:	PerFHalOrg.4-181/2, 186, 189
		Sn:	Org.Verb.6-184/6
$C_{12}Cl_2F_{10}Ti$	$(C_6F_5)_2TiCl_2$	Ti:	Org.Verb.1-73, 75
$C_{12}Cl_2F_{10}Tl^-$	$[(C_6F_5)_2TlCl_2]^-$	F:	PerFHalOrg.4-137
$C_{12}Cl_2F_{12}GeH_6$	$(CH_3)_2Ge(C_5F_6Cl)_2$	F:	PerFHalOrg.4-27
$C_{12}Cl_2F_{16}Hg$	$Hg(C_6F_8Cl)_2$	F:	PerFHalOrg.4-34
$C_{12}Cl_2F_{17}N_3$	$(C_3F_7)(CF_2CFCF_2)(C_3F_5Cl_2)C_3N_3$	F:	PerFHalOrg.6-84/6, 104
$C_{12}Cl_2F_{18}IN_3$	$(C_3F_7)(C_3F_6I)(C_3F_5Cl_2)C_3N_3$	F:	PerFHalOrg.6-84/6, 105
$C_{12}Cl_2F_{19}N_3$	$(CF_2ClCFClCF_2)(C_3F_7)_2C_3N_3$	F:	PerFHalOrg.6-84/6, 105
$C_{12}Cl_2FeGeH_{10}O_3$	$C_3H_5Fe(CO)_3Ge(C_6H_5)Cl_2$	Fe:	Org.Verb.B5-67/8
$C_{12}Cl_2FeH_8Hg_2N_2O_2$	$(CO)_2Fe(NC_5H_4C_5H_4N)(HgCl)_2$	Fe:	Org.Verb.B1-116/7

$C_{12}Cl_2FeH_8O_3$ ...... $ClC_6H_4C_3H_4Fe(CO)_3Cl$ ...... Fe: Org.Verb.B5-52, 62
$C_{12}Cl_2FeH_{10}$ ...... $C_5H_5FeC_5H_4CHCCl_2$ ...... Fe: Org.Verb.A1-384/5, 386
$C_{12}Cl_2FeH_{10}Hg_2N_2O_2$ $(CO)_2Fe(C_5H_5N)_2(HgCl)_2$ ...... Fe: Org.Verb.B1-116/7
$C_{12}Cl_2FeH_{10}O$ ...... $C_5H_5FeC_5H_4COCHCl_2$ ...... Fe: Org.Verb.A2-242/3
$C_{12}Cl_2FeH_{12}$ ...... $C_5H_5FeC_5H_4CCl_2CH_3$ ...... Fe: Org.Verb.A1-381, 383
– ...... $C_5H_5FeC_5H_4CHClCH_2Cl$ ...... Fe: Org.Verb.A1-381, 383
– ...... $C_5H_5FeC_5H_4CH_2CHCl_2$ ...... Fe: Org.Verb.A1-381, 383
$C_{12}Cl_2FeH_{12}O$ ...... $C_5H_5FeC_5H_4CHOHCHCl_2$ ...... Fe: Org.Verb.A2-45, 48
$C_{12}Cl_2FeH_{12}O^+$ ...... $[C_5H_5FeC_5H_4CHOHCHCl_2]^+$ ...... Fe: Org.Verb.A2-48
$C_{12}Cl_2FeH_{18}Hg_2N_2O_6$ $(CO)_4Fe(HgCl)_2(NH(CH_2CH_2)_2O)_2$ ...... Fe: Org.Verb.B2-153/5, 158
$C_{12}Cl_2FeH_{18}N_2O_2$ ... $(t\text{-}C_4H_9NC)_2Fe(CO)_2Cl_2$ ...... Fe: Org.Verb.B4-16
$C_{12}Cl_2FeH_{18}N_6$ ...... $[(CH_3NC)_6Fe]Cl_2$ ...... Fe: Org.Verb.B4-97, 99/101
– ...... $[(CH_3NC)_6Fe]Cl_2 \cdot 3\ H_2O$ ...... Fe: Org.Verb.B4-99/101
$C_{12}Cl_2FeH_{18}N_6O_8$ ... $[(CH_3NC)_6Fe][ClO_4]_2$ ...... Fe: Org.Verb.B4-97, 101
$C_{12}Cl_2FeH_{20}O_4Sn_2$ .. $(CO)_4Fe(Sn(C_2H_5)_2Cl)_2$ ...... Fe: Org.Verb.B2-136, 141, 148
$C_{12}Cl_2FeH_{22}N_2$ ...... $(C_5H_{11}NC)_2FeCl_2$ ...... Fe: Org.Verb.B4-11/2
$C_{12}Cl_2Fe_2H_{12}O_8Sn_2$ .. $[Cl(CH_3)_2Sn(CO)_4]_2Fe_2$ ...... Fe: Org.Verb.C1-72
$C_{12}Cl_2Fe_2H_{24}HgO_4P_2S_2$
$(CO)_4(P(CH_3)_3)_2Fe_2(SCH_3)_2 \cdot HgCl_2$ ...... Fe: Org.Verb.C1-110/1
$C_{12}Cl_2Fe_2H_{24}HgO_{10}P_2S_2$
$(CO)_4(P(OCH_3)_3)_2Fe_2(SCH_3)_2 \cdot HgCl_2$ ...... Fe: Org.Verb.C1-110/1
$C_{12}Cl_2GeH_{10}Ni_2O_2$ ... $(C_5H_5NiCO)_2GeCl_2$ ...... Ni: Org.Verb.2-350
$C_{12}Cl_2H_6MnO_4^{2-}$ .... $[Mn(ClC_6H_3O_2)_2]^{2-}$ ...... Mn: MVol.D1-45
$C_{12}Cl_2H_8I_2Sn$ ...... $(ClC_6H_4)_2SnI_2$ ...... Sn: Org.Verb.6-187
– ...... $(IC_6H_4)_2SnCl_2$ ...... Sn: Org.Verb.6-184/6
$C_{12}Cl_2H_8N_2O_2U$ ...... $[UO_2Cl_2(C_{12}H_8N_2)]$ ...... U: SVol.E1-44/6
$C_{12}Cl_2H_8N_2O_4Sn$ .... $SnCl_2(NC_5H_4COO)_2$ ...... Sn: MVol.C5-205/6
$C_{12}Cl_2H_8N_2Sn$ ...... $SnCl_2 \cdot C_{12}H_8N_2$ ...... Sn: MVol.C5-22/3
$C_{12}Cl_2H_8N_4O_4U$ ...... $UO_2Cl_2 \cdot 2\ NCC_5H_4NO$ ...... U: SVol.E1-133, 139, 142/3
$C_{12}Cl_2H_8OSn$ ...... $O(C_6H_4)_2SnCl_2$ ...... Sn: Org.Verb.6-207/8
– ...... $[(ClC_6H_4)_2SnO]_x$ ...... Sn: Org.Verb.6-187
$C_{12}Cl_2H_8OTe$ ...... $Cl_2Te(C_6H_4)_2O$ ...... Te: SVol.B2-98
$C_{12}Cl_2H_8Te$ ...... $(C_6H_4)_2TeCl_2$ ...... Te: SVol.B2-98
$C_{12}Cl_2H_9N_2Sn$ ...... $SnCl_2 \cdot C_{12}H_9N_2$ ...... Sn: MVol.C5-24
$C_{12}Cl_2H_{10}IrKN_2O_4$ ... $K[Ir(C_2O_4)(C_5H_5N)_2Cl_2]$ ...... Ir: SVol.2-36
$C_{12}Cl_2H_{10}NO_2PS$ .... $ClSO_2NP(Cl)(C_6H_5)_2$ ...... S: S-N-Verb.1-127
$C_{12}Cl_2H_{10}Ni_2O_2Sn$ ... $(C_5H_5NiCO)_2SnCl_2$ ...... Ni: Org.Verb.2-350
$C_{12}Cl_2H_{10}O_8Sn$ ...... $[SnCl_2(C_5H_2O_2(CH_2OH)O)_2]$ ...... Sn: MVol.C5-160/1
$C_{12}Cl_2H_{10}S_2Sn$ ...... $SnCl_2(SC_6H_5)_2$ ...... Sn: MVol.C6-80/1
$C_{12}Cl_2H_{10}Sn$ ...... $(C_6H_5)_2SnCl_2$
Analysis ...... Sn: Org.Verb.6-153
Chemical reactions ...... Sn: Org.Verb.6-159/78
Formation and preparation ...... Sn: Org.Verb.6-151/4
Physical properties ...... Sn: Org.Verb.6-158/9
Physiology ...... Sn: Org.Verb.6-179
Polarography ...... Sn: Org.Verb.6-159
Spectra ...... Sn: Org.Verb.6-155/8
Structure ...... Sn: Org.Verb.6-155
Uses ...... Sn: Org.Verb.6-179/80

$C_{12}Cl_2H_{10}Sn$ .... $(C_6H_5)_2SnCl_2 \cdot n\ H_2O$ .... Sn: Org.Verb.6-174, 178
$C_{12}Cl_2H_{10}Ti$ .... $(C_6H_5)_2TiCl_2$ .... Ti: Org.Verb.1-73
$C_{12}Cl_2H_{11}NO_2U$ ..... $UO_2Cl_2 \cdot (C_6H_5)_2NH$ .... U: SVol.E1-19, 22
$C_{12}Cl_2H_{12}N_2O_2U$ .... $[UO_2Cl_2((CH_3)_2C_{10}H_6N_2)]$ .... U: SVol.E1-39/40, 43
$C_{12}Cl_2H_{12}N_4O_6U$ .... $UO_2Cl_2 \cdot 2\ NH_2C_6H_4NO_2$ .... U: SVol.E1-18, 21
$C_{12}Cl_2H_{14}N_2O_2Sn$ ... $(CH_3)_2SnCl_2 \cdot ONC_5H_4C_5H_4NO$ .... Sn: Org.Verb.6-36
$C_{12}Cl_2H_{14}N_2O_2U$ .... $UO_2Cl_2 \cdot 2\ CH_3C_5H_4N$ .... U: SVol.E1-28, 34
$C_{12}Cl_2H_{14}N_2O_4U$ .... $UO_2Cl_2 \cdot 2\ CH_3C_5H_4NO$ .... U: SVol.E1-133, 137, 142/3
– .... $UO_2Cl_2 \cdot 2\ NH_2C_6H_4OH$ .... U: SVol.E1-18, 21
$C_{12}Cl_2H_{14}N_2O_6U$ .... $UO_2Cl_2 \cdot 2\ CH_3OC_5H_4NO$ .... U: SVol.E1-133, 138, 142/3
$C_{12}Cl_2H_{14}N_2Sn$ ..... $(CH_3)_2SnCl_2 \cdot NC_5H_4C_5H_4N$ .... Sn: Org.Verb.6-35, 38
– .... $SnCl_2 \cdot 2\ CH_3C_5H_4N$ .... Sn: MVol.C5-18
– .... $SnCl_2 \cdot 2\ C_6H_5NH_2$ .... Sn: MVol.C5-12/3
$C_{12}Cl_2H_{15}MnO_6$ ..... $[MnCHCl_2COO(CH(COCH_3)_2)_2]$ .... Mn: MVol.D1-103/4
$C_{12}Cl_2H_{16}N_2O_2Sn$ ... $(CH_3)_2SnCl_2 \cdot 2\ C_5H_5NO$ .... Sn: Org.Verb.6-36
$C_{12}Cl_2H_{16}N_2Sn$ ..... $(CH_3)_2SnCl_2 \cdot (CH_3)_2C_8H_4N_2$ .... Sn: Org.Verb.6-36
– .... $(CH_3)_2SnCl_2 \cdot 2\ C_5H_5N$ .... Sn: Org.Verb.6-35, 38
$C_{12}Cl_2H_{16}N_4O_2U$ .... $UO_2Cl_2 \cdot 2\ C_6H_4(NH_2)_2$ .... U: SVol.E1-23/5
– .... $UO_2Cl_2 \cdot 2\ (NH_2)(CH_3)C_5H_3N$ .... U: SVol.E1-29, 34
$C_{12}Cl_2H_{18}N_2O_2Sn$ ... $[SnCl_2(C_2H_4(NC(CH_3)CHCOCH_3)_2)]$ .... Sn: MVol.C6-28
$C_{12}Cl_2H_{18}N_4Sn$ ..... $(CH_3)_2SnCl_2 \cdot 2\ NH_2C_5H_4N$ .... Sn: Org.Verb.6-35, 38
$C_{12}Cl_2H_{18}Ni_2$ .... $(C_6H_9NiCl)_2$ .... Ni: Org.Verb.2-319
$C_{12}Cl_2H_{18}O_2Sn$ ..... $(C_2H_5OC_2H_4)(C_2H_5OC_6H_4)SnCl_2$ .... Sn: Org.Verb.6-203
$C_{12}Cl_2H_{18}Sn$ .... $(CH_3)_3SnCH_2CH(CHCl_2)C_6H_5$ .... Sn: Org.Verb.2-43, 46
– .... $(C_6H_9)_2SnCl_2$ .... Sn: Org.Verb.6-149
$C_{12}Cl_2H_{20}N_2O_2Sn$ ... $SnCl_2(N(CH_2)_5CO)_2$ .... Sn: MVol.C5-221
$C_{12}Cl_2H_{20}OTi$ ...... $(CH_3)_5C_5Ti(OC_2H_5)Cl_2$ .... Ti: Org.Verb.1-180, 183
$C_{12}Cl_2H_{22}N_2O_2S$ .... $SO_2[N(C_6H_{11})Cl]_2$ .... S: S-N-Verb.1-177
$C_{12}Cl_2H_{22}O_2S_2Sn$ ... $(C_2H_5)(C_6H_5)SnCl_2 \cdot 2\ (CH_3)_2SO$ .... Sn: Org.Verb.6-204
$C_{12}Cl_2H_{22}O_2Sn$ ..... $[CH_3C(O)CH_2C(CH_3)_2]_2SnCl_2$ .... Sn: Org.Verb.6-133
– .... $[C_3H_7C(O)CH_2CH_2]_2SnCl_2$ .... Sn: Org.Verb.6-133
$C_{12}Cl_2H_{22}O_4SU$ ..... $UO_2Cl_2 \cdot (HOC_6H_{10})_2S$ .... U: SVol.E1-5/6, 8, 219/20
$C_{12}Cl_2H_{22}O_4Sn$ ..... $[CH_3OCO(CH_2)_4]_2SnCl_2$ .... Sn: Org.Verb.6-134
– .... $[C_2H_5OCOCH(CH_3)CH_2]_2SnCl_2$ .... Sn: Org.Verb.6-134
– .... $[C_2H_5OCOCH_2CH(CH_3)]_2SnCl_2$ .... Sn: Org.Verb.6-133
$C_{12}Cl_2H_{22}Sn$ .... $[(CH_2)_5CH]_2SnCl_2$ .... Sn: Org.Verb.6-125
– .... $[(CH_3)_2CHCH_2C(CH_2)]_2SnCl_2$ .... Sn: Org.Verb.6-149
$C_{12}Cl_2H_{24}N_2O_2Sn$ ... $[(CH_3)_2NC(O)CH(CH_3)CH_2]_2SnCl_2$ .... Sn: Org.Verb.6-134
– .... $[(CH_3)_2NC(O)CH_2CH(CH_3)]_2SnCl_2$ .... Sn: Org.Verb.6-134
– .... $[C_2H_5NHC(O)CH(CH_3)CH_2]_2SnCl_2$ .... Sn: Org.Verb.6-134
$C_{12}Cl_2H_{24}N_6O_8S_3Te$ .. $Te(SC(NHCH_2)_2CH_2)_3(ClO_4)_2$ .... Te: SVol.B3-161
$C_{12}Cl_2H_{24}N_8O_8S_4Te$ .. $Te(SC(NHCH_2)_2)_4(ClO_4)_2$ .... Te: SVol.B3-158
$C_{12}Cl_2H_{24}N_8S_4Te$ .... $Te(SC(NHCH_2)_2)_4Cl_2$ .... Te: SVol.B3-158
– .... $Te(SC(NHCH_2)_2)_4Cl_2 \cdot 2\ H_2O$ .... Te: SVol.B3-158
$C_{12}Cl_2H_{24}N_8Sn$ ..... $SnCl_2 \cdot 2\ (CH_2)_6N_4$ .... Sn: MVol.C5-24
$C_{12}Cl_2H_{24}O_8P_2Sn$ ... $SnCl_2((CH_3)_2CCOP(O)(OCH_3)_2)_2$ .... Sn: MVol.C6-164
$C_{12}Cl_2H_{24}O_8U$ ...... $UO_2Cl_2 \cdot (CH_2OCH_2)_6 \cdot x\ H_2O$ .... U: SVol.E1-71/2
$C_{12}Cl_2H_{26}N_2O_4Sn$ ... $SnCl_2 \cdot 2\ (CH_3)_2C_3H_4(NH_2)COOH$ .... Sn: MVol.C5-31/2
$C_{12}Cl_2H_{26}N_2O_4U$ .... $UO_2Cl_2 \cdot 2\ C_3H_7CON(CH_3)_2$ .... U: SVol.E1-110/1
$C_{12}Cl_2H_{26}Sn$ .... $[(CH_3)_3CCH_2CH_2]_2SnCl_2$ .... Sn: Org.Verb.6-110

$C_{12}Cl_2H_{26}Sn$ .... $(C_4H_9)(C_8H_{17})SnCl_2$ .... Sn: Org.Verb.6-200
– .... $(C_6H_{13})_2SnCl_2$ .... Sn: Org.Verb.6-109
$C_{12}Cl_2H_{27}I_2N_3O_3U$ ... $UI_2Cl_2 \cdot 3\ CH_3CON(CH_3)_2$ .... U: SVol.E1-103
$C_{12}Cl_2H_{27}N_3O_5U$ .... $UO_2Cl_2 \cdot 3\ CH_3CONHC_2H_5$ .... U: SVol.E1-96/9
$C_{12}Cl_2H_{27}O_6PU$ ..... $UO_2Cl_2 \cdot (C_4H_9O)_3PO$ .... U: SVol.E1-182
$C_{12}Cl_2H_{28}OSn_2$ ..... $[(C_3H_7)_2ClSn]_2O$ .... Sn: Org.Verb.6-68
$C_{12}Cl_2H_{28}O_{10}U$ ..... $[UO_2Cl_2(CH_2OCH_2)_6(H_2O)_2]$ .... U: SVol.E1-71/2, 77/8
$C_{12}Cl_2H_{28}P_2S_4Sn$ .... $SnCl_2((C_3H_7)_2P(S)S)_2$ .... Sn: MVol.C6-157
$C_{12}Cl_2H_{28}Sn_2$ .... $[(C_3H_7)_2ClSn]_2$ .... Sn: Org.Verb.6-68
$C_{12}Cl_2H_{29}O_3PSn$ .... $(C_4H_9)_2SnCl_2 \cdot (C_2H_5O)_2P(O)H$ .... Sn: Org.Verb.6-100
$C_{12}Cl_2H_{30}N_2O_4U$ .... $UO_2Cl_2 \cdot 2\ (C_2H_5)_2N(C_2H_4OH)$ .... U: SVol.E1-20, 23
$C_{12}Cl_2H_{30}N_2Sn$ ..... $SnCl_2 \cdot 2\ (C_3H_7)_2NH$ .... Sn: MVol.C5-11/2
$C_{12}Cl_2H_{30}NiP_2$ ...... $Ni(P(C_2H_5)_3)_2Cl_2$ .... Ni: Org.Verb.1-43
$C_{12}Cl_2H_{30}O_4P_2U$ .... $UO_2Cl_2 \cdot 2\ (C_2H_5)_3PO$ .... U: SVol.E1-149, 156/7
$C_{12}Cl_2H_{30}O_4Si_2Sn$ ... $[(C_2H_5O)_2(CH_3)SiCH_2]_2SnCl_2$ .... Sn: Org.Verb.6-134
$C_{12}Cl_2H_{30}O_{11}U$ ..... $[(UO_2Cl_2)(CH_2OCH_2)_6(H_2O)_3]$ .... U: SVol.E1-71/2, 77/8
$C_{12}Cl_2H_{30}O_{13}U$ ..... $UO_2(ClO_4)_2 \cdot 3\ (C_2H_5)_2O$ .... U: SVol.E1-71, 76/7
$C_{12}Cl_2H_{32}N_8O_6U$ .... $UO_2Cl_2 \cdot 4\ CO(NHCH_3)_2$ .... U: SVol.E1-92/3
$C_{12}Cl_2H_{32}N_8S_4Te$ .... $Te(SC(NH_2)_2)_2((CH_3)_2NCSN(CH_3)_2)_2Cl_2$ .... Te: SVol.B3-149/50
$C_{12}Cl_2H_{32}O_2P_2S_2Sn$ .. $(CH_3)_2SnCl_2 \cdot 2\ (C_2H_5)_2(CH_3S)PO$ .... Sn: Org.Verb.6-37
$C_{12}Cl_2H_{33}N_3O_2U$ .... $UO_2Cl_2 \cdot 3\ C_4H_9NH_2$ .... U: SVol.E1-18, 21
$C_{12}Cl_2H_{34}N_2Sn_2$ ..... $2\ (C_2H_5)_3SnCl \cdot N_2H_4$ .... Sn: Org.Verb.5-107
$C_{12}Cl_2H_{34}N_6Sn$ ..... $SnCl_2 \cdot 2\ NH(C_3H_6NH_2)_2$ .... Sn: MVol.C5-14
$C_{12}Cl_2H_{36}IrNO_3P_4$ ... $[Ir(P(CH_3)_3)_4Cl_2]NO_3$ .... Ir: SVol.2-199
$C_{12}Cl_2H_{36}MnO_{14}$ .... $[Mn(C_2H_5OH)_6](ClO_4)_2$ .... Mn: MVol.D1-27
$C_{12}Cl_2H_{36}N_4O_{14}U$ .... $UO_2(ClO_4)_2 \cdot 4\ (CH_3)_3NO$ .... U: SVol.E1-131/2, 135
$C_{12}Cl_2H_{36}N_6O_4P_2U$ .. $UO_2Cl_2 \cdot 2\ ((CH_3)_2N)_3PO$ .... U: SVol.E1-5, 153, 163/4
$C_{12}Cl_2H_{36}O_{14}P_4U$ .... $UO_2(ClO_4)_2 \cdot 4\ (CH_3)_3PO$ .... U: SVol.E1-148, 156
$C_{12}Cl_2H_{45}O_{15}Sn_6^+$ ... $[(C_2H_5Sn)_6(OH)_{15}Cl_2]^+$ .... Sn: Org.Verb.6-230
$C_{12}Cl_3FH_7NO$ ...... $NC_5FCl_3(OCH_2C_6H_5)$ .... F: PerFHalOrg.5-136
F: PerFHalOrg.6-14
$C_{12}Cl_3F_2H_6NO$ ...... $NC_5F_2Cl_2(OCH_2C_6H_4Cl)$ .... F: PerFHalOrg.5-136
F: PerFHalOrg.6-14
$C_{12}Cl_3F_3H_{11}N_3O_2S$ ... $(C_4H_9NHSO_2)(CF_3)C_7Cl_3HN_2$ .... F: PerFHalOrg.6-160, 162
$C_{12}Cl_3F_4H_4N_3O_2$ .... $NC_5FCl_3NHC_6H_3(CF_3)NO_2$ .... F: PerFHalOrg.5-140
$C_{12}Cl_3F_6H_{12}N_3$ ...... $(CF_2Cl)_3(C_6H_{11})C_3HN_3$ .... F: PerFHalOrg.6-113
$C_{12}Cl_3F_{10}P$ .... $(C_6F_5)_2PCl_3$ .... F: PerFHalOrg.3-124, 125, 127
$C_{12}Cl_3F_{21}H_3ISn$ ..... $CF_3(CF_2)_9CH_2CHISnCl_3$ .... Sn: Org.Verb.6-259
$C_{12}Cl_3F_{21}H_4Sn$ ..... $CF_3(CF_2)_9CH_2CH_2SnCl_3$ .... Sn: Org.Verb.6-259
$C_{12}Cl_3FeH_9O$ .... $C_5H_5FeC_5H_4COCCl_3$ .... Fe: Org.Verb.A2-242/3
$C_{12}Cl_3FeH_9O_2$ ...... $C_5H_5FeC_5H_4OOCCCl_3$ .... Fe: Org.Verb.A2-93
$C_{12}Cl_3FeH_{10}O_2$ ..... $[Fe(C_5H_5)_2][CCl_3COO]$ .... Fe: Org.Verb.A1-227
$C_{12}Cl_3FeH_{11}O$ ...... $C_5H_5FeC_5H_4CHOHCCl_3$ .... Fe: Org.Verb.A2-45, 48
$C_{12}Cl_3FeH_{20}Hg_2NO_4$. $[N(C_2H_5)_4][(CO)_4Fe(HgCl)_2Cl]$ .... Fe: Org.Verb.B2-153, 157
$C_{12}Cl_3FeH_{20}NO_4Si$ .. $[N(C_2H_5)_4][(CO)_4FeSiCl_3]$ .... Fe: Org.Verb.B2-142, 149
$C_{12}Cl_3FeH_{20}NO_4Sn$ .. $[N(C_2H_5)_4][(CO)_4FeSnCl_3]$ .... Fe: Org.Verb.B2-143, 150
$C_{12}Cl_3H_9Sn$ .... $C_6H_5C_6H_4SnCl_3$ .... Sn: Org.Verb.6-286
$C_{12}Cl_3H_{10}IrN_2O$ ..... $Ir(C_{12}H_8N_2)Cl_3(H_2O)$ .... Ir: SVol.2-92
$C_{12}Cl_3H_{10}MnN_2O$ .... $[Mn(C_{12}H_8N_2)(H_2O)]Cl_3$ .... Mn: MVol.C5-212/3

Formula	Compound	Reference
$C_{12}Cl_3H_{11}N_2O_2Sn$	$C_6H_5SnCl_3 \cdot NO_2C_6H_4NH_2$	Sn: Org.Verb.6-279
$C_{12}Cl_3H_{11}Ti$	$C_6H_5CH_2C_5H_4TiCl_3$	Ti: Org.Verb.1-140
$C_{12}Cl_3H_{12}IrN_6$	$Ir(N(CH)_2N(CH)_2)_3Cl_3$	Ir: SVol.2-77
–	$Ir(NN(CH)_4)_3Cl_3$	Ir: SVol.2-79
$C_{12}Cl_3H_{12}N_3O_2Sn$	$C_6H_5SnCl_3 \cdot (NH_2)_2C_6H_3NO_2$	Sn: Org.Verb.6-279
$C_{12}Cl_3H_{13}N_2Sn$	$C_2H_5SnCl_3 \cdot NC_5H_4C_5H_4N$	Sn: Org.Verb.6-232
$C_{12}Cl_3H_{13}N_2Ti$	$C_2H_5TiCl_3 \cdot NC_5H_4C_5H_4N$	Ti: Org.Verb.1-34/5, 41
$C_{12}Cl_3H_{14}MnO_6$	$[MnCCl_3COO(CH(COCH_3)_2)_2]$	Mn: MVol.D1-103/4
$C_{12}Cl_3H_{15}O_2Sn$	$CH_3SnCl_3 \cdot C_6H_5CHCHCOOC_2H_5$	Sn: Org.Verb.6-222
$C_{12}Cl_3H_{16}IrN_2O$	$[Ir(CH_3C_5H_4N)_2(H_2O)Cl_3] \cdot H_2O$	Ir: SVol.2-75
–	$Ir(C_5H_5N)_2Cl_3(C_2H_5OH)$	Ir: SVol.2-74
$C_{12}Cl_3H_{16}IrN_2OS$	$Ir(C_5H_5N)_2Cl_3((CH_3)_2SO)$	Ir: SVol.2-74
$C_{12}Cl_3H_{18}IrN_2O_2$	$Ir(C_5H_5N)_2Cl_3(CH_3OH) \cdot CH_3OH$	Ir: SVol.2-74
$C_{12}Cl_3H_{18}NOSn$	$[(C_6H_5)SnCl_2(C_2H_4NH(CH_2CH_2)_2O)]Cl$	Sn: Org.Verb.6-201
$C_{12}Cl_3H_{20}NSn$	$[(C_2H_5)_2NHCH_2CH_2Sn(C_6H_5)Cl_2]Cl$	Sn: Org.Verb.6-201
$C_{12}Cl_3H_{20}TiZr$	$Cl_3Ti(CH_2CHCH_2)_2Zr(C_3H_5)_2$	Ti: Org.Verb.1-80/1
$C_{12}Cl_3H_{21}N_2Ti$	$CH_3TiCl_3 \cdot (CH_3)_2NCH_2C_6H_4N(CH_3)_2$	Ti: Org.Verb.1-34/5
$C_{12}Cl_3H_{24}IrS_3$	$IrCl_3(C_4H_8S)_3$	Ir: SVol.2-166
$C_{12}Cl_3H_{24}NO_3Sn$	$C_7H_{15}CON(C_2H_4OH)C_2H_4OSnCl_3$	Sn: MVol.C6-46
$C_{12}Cl_3H_{24}O_6Sn^+$	$[SnCl_3(CH_3COOC_2H_5)_3]^+$	Sn: MVol.C5-130
$C_{12}Cl_3H_{25}Sn$	$C_{12}H_{25}SnCl_3$	Sn: Org.Verb.6-253
$C_{12}Cl_3H_{27}OSn_2$	$Cl_2(C_4H_9)SnOSn(C_4H_9)_2Cl$	Sn: Org.Verb.6-245
$C_{12}Cl_3H_{27}O_3PU^-$	$[UO_2Cl_3((C_4H_9)_3PO)]^-$	U: SVol.E1-158
$C_{12}Cl_3H_{27}O_6PU^-$	$[UO_2Cl_3((C_4H_9O)_3PO)]^-$	U: SVol.E1-182
$C_{12}Cl_3H_{27}SSn_2$	$(C_4H_9)_2ClSnSSn(C_4H_9)Cl_2$	Sn: Org.Verb.6-245
$C_{12}Cl_3H_{30}IrS_3$	$Ir(S(C_2H_5)_2)_3Cl_3$	Ir: SVol.2-165
$C_{12}Cl_3H_{30}O_3Sn^+$	$[SnCl_3 \cdot 3\ C_4H_9OH]^+$	Sn: MVol.C5-72
$C_{12}Cl_3H_{30}O_3Ti$	$TiCl_3 \cdot 3\ (C_2H_5)_2O$	Ti: Org.Verb.1-20, 38
$C_{12}Cl_3H_{31}O_2P_2S_2Sn$	$C_2H_5SnCl_3 \cdot 2\ CH_3S(C_2H_5)_2PO$	Sn: Org.Verb.6-232
$C_{12}Cl_4CsH_{14}IrN_2$	$Cs[Ir(CH_3C_5H_4N)_2Cl_4]$	Ir: SVol.2-75
$C_{12}Cl_4F_2H_5NO$	$NC_5F_2Cl_2(OCH_2C_6H_3Cl_2)$	F: PerFHalOrg.5-136
		F: PerFHalOrg.6-14
$C_{12}Cl_4F_2H_7N_3O_2$	$NC_5F_2Cl_2NHC_5Cl_2(OCH_3)_2N$	F: PerFHalOrg.5-138/9
$C_{12}Cl_4F_3H_6N_3O$	$NC_5F_2Cl_2NHC_5FCl_2(OC_2H_5)N$	F: PerFHalOrg.5-138/9
$C_{12}Cl_4F_3H_7N_4$	$NC_5F_2Cl_2NHC_5FCl_2(N(CH_3)_2)N$	F: PerFHalOrg.5-138/9
–	$NC_5F_2Cl_2NHC_5FCl_2(NHC_2H_5)N$	F: PerFHalOrg.5-138/9
$C_{12}Cl_4F_4HN_3$	$NCC_6F_4NHC_5Cl_4N$	F: PerFHalOrg.7-15/6, 43
$C_{12}Cl_4F_4H_6N_4$	$NC_5F_2Cl_2NH(CH_2)_2NHC_5F_2Cl_2N$	F: PerFHalOrg.5-141
$C_{12}Cl_4F_4H_{10}O_2P_2Sn$	$SnCl_4 \cdot 2\ C_6H_5POF_2$	Sn: MVol.C6-167
$C_{12}Cl_4F_5H_4N_2NaO_3$	$C_6Cl_4NC(C_2F_5)N(OC_2H_4COONa)$	F: PerFHalOrg.6-159
$C_{12}Cl_4F_8N_2$	$NC_5F_4(CCl_2)_2C_5F_4N$	F: PerFHalOrg.5-156/7, 169
$C_{12}Cl_4F_{17}N_3$	$(CF_2ClCFClCF_2)_2(C_3F_7)C_3N_3$	F: PerFHalOrg.6-84/6, 105
$C_{12}Cl_4FeH_{12}OTi$	$C_5H_5FeC_5H_4COCH_3 \cdot TiCl_4$	Fe: Org.Verb.A2-214
$C_{12}Cl_4FeH_{18}O_4Sn_2$	$(CO)_4Fe(Sn(C_4H_9)Cl_2)_2$	Fe: Org.Verb.B3-139
$C_{12}Cl_4FeNaO_7^-$	$[NaOC(O)C_6Cl_4C(O)Fe(CO)_4]^-$	Fe: Org.Verb.B4-154, 162
$C_{12}Cl_4FeNa_2O_7$	$Na[NaOC(O)C_6Cl_4C(O)Fe(CO)_4]$	Fe: Org.Verb.B4-154, 162
$C_{12}Cl_4Fe_2H_{12}O$	$[C_5H_5FeC_5H_4COCH_3]FeCl_4$	Fe: Org.Verb.A2-301
$C_{12}Cl_4Fe_2H_{20}N_4O$	$(C_2H_5NC)_4Fe_2OCl_4$	Fe: Org.Verb.B4-71
$C_{12}Cl_4Fe_2H_{24}Hg_2O_4P_2S_2$	$(CO)_4(P(CH_3)_3)_2Fe_2(SCH_3)_2 \cdot 2\ HgCl_2$	Fe: Org.Verb.C1-110/1

Formula	Compound		Reference
$C_{12}Cl_4Fe_2H_{24}Hg_2O_{10}P_2S_2$	$(CO)_4(P(OCH_3)_3)_2Fe_2(SCH_3)_2 \cdot 2\ HgCl_2$	Fe:	Org.Verb.C1-110/1
$C_{12}Cl_4H_6S_3Sn$	$(C_4H_2ClS)_3SnCl$	Sn:	Org.Verb.5-189
$C_{12}Cl_4H_7IrKN_3O_2$	$K[Ir(C_{12}H_7N_2(NO_2))Cl_4]$	Ir:	SVol.2-93
$C_{12}Cl_4H_8IrKN_2$	$K[Ir(C_{12}H_8N_2)Cl_4]$	Ir:	SVol.2-91
$C_{12}Cl_4H_8IrN_2$	$Ir(C_{12}H_8N_2)Cl_4$	Ir:	SVol.2-92
$C_{12}Cl_4H_8IrN_2^-$	$[Ir(C_{12}H_8N_2)Cl_4]^-$	Ir:	SVol.2-91/2
$C_{12}Cl_4H_8N_2Sn$	$SnCl_4 \cdot C_{12}H_8N_2$	Sn:	MVol.C5-220/1
$C_{12}Cl_4H_8N_4O_2Sn$	$SnCl_4 \cdot 2\ NCC_5H_4NO$	Sn:	MVol.C5-208/9
$C_{12}Cl_4H_8N_4Sn$	$SnCl_4 \cdot 2\ NC_5H_4CN$	Sn:	MVol.C5-202
$C_{12}Cl_4H_8Sn$	$(ClC_6H_4)_2SnCl_2$	Sn:	Org.Verb.6-184/5, 187
$C_{12}Cl_4H_9N_3O_2Sn$	$SnCl_4 \cdot C_6H_5NNC_6H_4NO_2$	Sn:	MVol.C6-31/2
$C_{12}Cl_4H_{10}N_2OSn_2$	$2\ SnCl_2 \cdot C_6H_5NNC_6H_4OH$	Sn:	MVol.C5-40
$C_{12}Cl_4H_{10}N_2O_2P_2S$	$SO_2[NPCl_2(C_6H_5)]_2$	S:	S-N-Verb.1-191
$C_{12}Cl_4H_{10}N_2O_2Sn$	$C_6H_5SnCl_3 \cdot NH_2C_6H_3(NO_2)Cl$	Sn:	Org.Verb.6-279
–	$ClC_6H_4SnCl_3 \cdot NO_2C_6H_4NH_2$	Sn:	Org.Verb.6-283
–	$SnCl_4 \cdot NO_2C_6H_4NHC_6H_5$	Sn:	MVol.C5-180/1
$C_{12}Cl_4H_{10}N_2O_2U$	$[C_{12}H_8N_2H_2]UO_2Cl_4$	U:	SVol.C9-93
–	$UCl_4 \cdot 2\ NC_5H_4(CHO)$	U:	SVol.E1-28, 33
$C_{12}Cl_4H_{10}N_2O_6U$	$UCl_4 \cdot 2\ HOCOC_5H_4NO$	U:	SVol.E1-132, 139, 142/3
$C_{12}Cl_4H_{10}N_2Sn$	$SnCl_4 \cdot C_6H_5NNC_6H_5$	Sn:	MVol.C6-31/2
$C_{12}Cl_4H_{10}OSTe$	$TeCl_4 \cdot (C_6H_5)_2SO$	Te:	SVol.B3-176
$C_{12}Cl_4H_{10}O_2SSn$	$SnCl_4 \cdot (C_6H_5)_2SO_2$	Sn:	MVol.C6-103
$C_{12}Cl_4H_{10}O_2Sn$	$SnCl_4 \cdot C_6H_4O_2 \cdot C_6H_6$	Sn:	MVol.C5-111/2
$C_{12}Cl_4H_{11}IrN_2O$	$H_3O[Ir(C_{12}H_8N_2)Cl_4] \cdot 4\ H_2O$	Ir:	SVol.2-91/2
$C_{12}Cl_4H_{11}N_3O_2Sn$	$ClC_6H_4SnCl_3 \cdot (NH_2)_2C_6H_3NO_2$	Sn:	Org.Verb.6-283
$C_{12}Cl_4H_{11}N_3Sn$	$SnCl_4 \cdot C_6H_5NNC_6H_4NH_2$	Sn:	MVol.C6-31/3
$C_{12}Cl_4H_{12}I_2N_2Sn$	$SnCl_4 \cdot 2\ IC_6H_4NH_2$	Sn:	MVol.C5-179/80
$C_{12}Cl_4H_{12}IrN_3$	$NH_4[Ir(C_{12}H_8N_2)Cl_4] \cdot H_2O$	Ir:	SVol.2-91/2
$C_{12}Cl_4H_{12}N_2Sn$	$SnCl_4 \cdot NH_2C_6H_4C_6H_4NH_2$	Sn:	MVol.C5-187
$C_{12}Cl_4H_{12}N_2Te$	$TeCl_4 \cdot NH_2C_6H_4C_6H_4NH_2$	Te:	SVol.B3-169
$C_{12}Cl_4H_{12}N_4O_2Sn$	$SnCl_4 \cdot 2\ NC_5H_4CONH_2$	Sn:	MVol.C5-206
$C_{12}Cl_4H_{12}N_4O_4Sn$	$SnCl_4 \cdot 2\ NO_2C_6H_4NH_2$	Sn:	MVol.C5-179
$C_{12}Cl_4H_{12}S_2Sn$	$SnCl_4 \cdot 2\ C_6H_5SH$	Sn:	MVol.C6-80
$C_{12}Cl_4H_{13}IrN_6$	$(C_4H_4N_2H)[Ir(C_4H_4N_2)_2Cl_4]$	Ir:	SVol.2-79, 80
$C_{12}Cl_4H_{14}IrKN_2$	$K[Ir(CH_3C_5H_4N)_2Cl_4]$	Ir:	SVol.2-75
–	$K[Ir(CH_3C_5H_4N)_2Cl_4] \cdot 2\ H_2O$	Ir:	SVol.2-75
$C_{12}Cl_4H_{14}IrN_2$	$Ir(CH_3C_5H_4N)_2Cl_4$	Ir:	SVol.2-75
$C_{12}Cl_4H_{14}IrN_2Na$	$Na[Ir(CH_3C_5H_4N)_2Cl_4]$	Ir:	SVol.2-75
–	$Na[Ir(CH_3C_5H_4N)_2Cl_4] \cdot 2\ H_2O$	Ir:	SVol.2-75
$C_{12}Cl_4H_{14}IrN_2Rb$	$Rb[Ir(CH_3C_5H_4N)_2Cl_4]$	Ir:	SVol.2-75
$C_{12}Cl_4H_{14}IrN_2Tl$	$Tl[Ir(CH_3C_5H_4N)_2Cl_4]$	Ir:	SVol.2-75
$C_{12}Cl_4H_{14}MnN_2$	$[HNC_5H_3(CH_3)C_5H_3(CH_3)NH]MnCl_4$	Mn:	MVol.C5-208/9
$C_{12}Cl_4H_{14}N_2NiSi_2$	$[(CH_3Cl_2)Si]_2Ni(NC_5H_4C_5H_4N)$	Ni:	Org.Verb.1-76
$C_{12}Cl_4H_{14}N_2O_2Sn$	$SnCl_4 \cdot 2\ CH_3COC_4H_3NH$	Sn:	MVol.C5-190
–	$SnCl_4 \cdot 2\ CH_3C_5H_4NO$	Sn:	MVol.C5-208/9
–	$SnCl_4 \cdot 2\ NH_2C_6H_4OH$	Sn:	MVol.C6-4
$C_{12}Cl_4H_{14}N_2O_4Sn$	$SnCl_4 \cdot 2\ CH_3OC_5H_4NO$	Sn:	MVol.C5-208/9
$C_{12}Cl_4H_{14}N_2Sn$	$SnCl_4 \cdot 2\ CH_3C_5H_4N$	Sn:	MVol.C5-201
–	$SnCl_4 \cdot 2\ C_6H_5NH_2$	Sn:	MVol.C5-175/6

Formula	Compound	Reference
$C_{12}Cl_4H_{14}N_2Te$	$TeCl_4 \cdot 2\ CH_3C_5H_4N$	Te: SVol.B3-171
–	$TeCl_4 \cdot 2\ C_6H_5NH_2$	Te: SVol.B3-166/7
$C_{12}Cl_4H_{14}N_2U$	$UCl_4 \cdot 2\ C_6H_5NH_2$	U: SVol.E1-18
$C_{12}Cl_4H_{14}O_4Sn$	$SnCl_4 \cdot C_6H_4(COOC_2H_5)_2$	Sn: MVol.C5-147
$C_{12}Cl_4H_{15}IrN_2$	$H[Ir(CH_3C_5H_4N)_2Cl_4]$	Ir: SVol.2-75
$C_{12}Cl_4H_{15}NO_2Sn$	$SnCl_4 \cdot C_6H_5CONHCOC_4H_9$	Sn: MVol.C6-50/1
$C_{12}Cl_4H_{15}N_3Sn$	$SnCl_4 \cdot (C_{12}H_{15}N_3)$	Sn: MVol.C5-190
$C_{12}Cl_4H_{16}MnN_2$	$[CH_3C_5H_4NH]_2MnCl_4$	Mn: MVol.C5-208/9
$C_{12}Cl_4H_{16}N_4Sn$	$SnCl_4 \cdot 2\ C_6H_4(NH_2)_2$	Sn: MVol.C5-186/7
–	$SnCl_4 \cdot 2\ C_6H_5NHNH_2$	Sn: MVol.C5-189
$C_{12}Cl_4H_{16}N_4Sn_2$	$SnCl_4 \cdot Sn(CH_2CH_2CN)_4$	Sn: MVol.C6-184
$C_{12}Cl_4H_{16}Ni_2$	$(H_2(CH_3)_2C_4NiCl_2)_2$	Ni: Org.Verb.2-324
$C_{12}Cl_4H_{16}O_2Sn$	$SnCl_4 \cdot CH_2C(CH_3)COOCH_3 \cdot C_6H_5CH_3$	Sn: MVol.C5-138
$C_{12}Cl_4H_{16}O_4Sn$	$Sn(CH_2CH_2COCl)_4$	Sn: Org.Verb.1-124
$C_{12}Cl_4H_{17}O_4PSn$	$SnCl_4 \cdot C_6H_5COCH_2P(O)(OC_2H_5)_2$	Sn: MVol.C6-166
$C_{12}Cl_4H_{18}IrN_3$	$NH_4[Ir(CH_3C_5H_4N)_2Cl_4]$	Ir: SVol.2-75
–	$NH_4[Ir(CH_3C_5H_4N)_2Cl_4] \cdot H_2O$	Ir: SVol.2-75
$C_{12}Cl_4H_{18}N_2O_2Sn$	$SnCl_4 \cdot 2\ CH_2CHCONHCH_2CHCH_2$	Sn: MVol.C6-45/6
$C_{12}Cl_4H_{18}N_2Sn$	$[C_5H_5NH]_2[(CH_3)_2SnCl_4]$	Sn: Org.Verb.6-35
–	$SnCl_4 \cdot 2\ (CH_3)_2C_4H_2NH$	Sn: MVol.C5-190
$C_{12}Cl_4H_{18}OSn$	$SnCl_4 \cdot [(CH_3)_2CH]_2C_6H_3OH$	Sn: MVol.C5-76
$C_{12}Cl_4H_{18}O_6Sn$	$SnCl_4 \cdot C_6H_9(OCOCH_3)_3$	Sn: MVol.C5-133
$C_{12}Cl_4H_{18}O_9Sn$	$SnCl_4 \cdot 3\ (CH_3CO)_2O$	Sn: MVol.C5-124/5
$C_{12}Cl_4H_{20}N_2Sn$	$SnCl_4 \cdot C_6H_{10}NNC_6H_{10}$	Sn: MVol.C6-30/1
$C_{12}Cl_4H_{20}N_4Te$	$TeCl_4 \cdot 2\ C_6H_5NH_2 \cdot 2\ NH_3$	Te: SVol.B3-167
$C_{12}Cl_4H_{20}N_4U$	$[UCl_4(C_2H_5CN)_4]$	U: SVol.E1-48/50
$C_{12}Cl_4H_{20}O_2Sn$	$SnCl_4 \cdot 2\ C_5H_{10}CO$	Sn: MVol.C5-94
$C_{12}Cl_4H_{20}O_4Sn$	$(C_4H_9)_2Sn(OCOCHCl_2)_2$	Sn: Org.Verb.6-89, 95
$C_{12}Cl_4H_{20}O_6Sn$	$SnCl_4 \cdot CH(CH_2COOC_2H_5)_2COOC_2H_5$	Sn: MVol.C5-144
$C_{12}Cl_4H_{20}O_8Sn$	$SnCl_4 \cdot 2\ C_2H_4(OCOCH_3)_2$	Sn: MVol.C5-133
$C_{12}Cl_4H_{20}S_2Sn$	$SnCl_4 \cdot 2\ (C_3H_5)_2S$	Sn: MVol.C6-93/4
$C_{12}Cl_4H_{22}N_2O_2Sn$	$SnCl_4 \cdot (C_4H_9)_2C_4H_4N_2O_2$	Sn: MVol.C5-227
–	$SnCl_4 \cdot 2\ HN(CH_2)_5CO$	Sn: MVol.C5-221
$C_{12}Cl_4H_{22}OSTe$	$TeCl_4 \cdot (C_6H_{11})_2SO$	Te: SVol.B3-176
$C_{12}Cl_4H_{22}O_4Sn$	$SnCl_4 \cdot C_8H_{16}(COOCH_3)_2$	Sn: MVol.C5-141/3
$C_{12}Cl_4H_{24}Mn_2O_6$	$2\ MnCl_2 \cdot (C_2H_4O)_6 \cdot 8\ H_2O$	Mn: MVol.D1-159
$C_{12}Cl_4H_{24}N_3PSSn$	$SnCl_4 \cdot (C_4H_8N)_3PS$	Sn: MVol.C6-149
$C_{12}Cl_4H_{24}O_2Sn$	$SnCl_4 \cdot 2\ CH_3C_5H_9O$	Sn: MVol.C5-158
–	$SnCl_4 \cdot 2\ (CH_3)_2C_4H_6O$	Sn: MVol.C5-157
$C_{12}Cl_4H_{24}O_3S_3Te$	$TeCl_4 \cdot 3\ (CH_2)_4SO$	Te: SVol.B3-176
$C_{12}Cl_4H_{24}O_3Sn$	$[SnCl_3(C_4H_8O)_3]Cl$	Sn: MVol.C5-156
$C_{12}Cl_4H_{24}O_3U$	$UCl_4 \cdot 3\ (CH_2)_4O$	U: SVol.E1-70, 74/5
$C_{12}Cl_4H_{24}O_4Sn$	$SnCl_4 \cdot 2\ CH_3COOC_4H_9$	Sn: MVol.C5-131/2
–	$SnCl_4 \cdot 2\ C_3H_7COOC_2H_5$	Sn: MVol.C5-135/6
–	$SnCl_4 \cdot 2\ C_5H_{11}COOH$	Sn: MVol.C5-120/1
$C_{12}Cl_4H_{24}O_6Sn$	$SnCl_4 \cdot 3\ C_3H_7COOH$	Sn: MVol.C5-120/1
$C_{12}Cl_4H_{24}O_6U$	$UCl_4 \cdot [CH_2OCH_2]_6$	U: SVol.E1-71/2, 77/8
–	$UCl_4 \cdot 3\ CH_3COOC_2H_5$	U: SVol.E1-120/2
–	$UCl_4 \cdot 3\ O(CH_2CH_2)_2O$	U: SVol.E1-70, 74/5
$C_{12}Cl_4H_{24}O_{10}U_2$	$[(UO_2Cl_2)_2(CH_2OCH_2)_6]$	U: SVol.E1-71/2, 77/8

Formula	Compound	Reference
$C_{12}Cl_4H_{24}S_2Sn$	$SnCl_4 \cdot 2\ S(CH(CH_3)CH_2)_2$	Sn: MVol.C6-110
$C_{12}Cl_4H_{26}N_2O_2U$	$UCl_4 \cdot 2\ C_3H_7CON(CH_3)_2$	U: SVol.E1-109/11
$C_{12}Cl_4H_{26}N_2O_4Sn$	$SnCl_4 \cdot 2\ (CH_3)_2CHCH_2CH(NH_2)COOH$	Sn: MVol.C6-12
$C_{12}Cl_4H_{26}S_2Sn$	$SnCl_4 \cdot C_4H_9S(CH_2)_4SC_4H_9$	Sn: MVol.C6-97/8
$C_{12}Cl_4H_{26}S_2Te$	$TeCl_4 \cdot C_4H_9S(CH_2)_4SC_4H_9$	Te: SVol.B3-175
$C_{12}Cl_4H_{27}N_3O_3Pa$	$PaCl_4 \cdot 3\ CH_3CON(CH_3)_2$	Pa: SVol.2-89/90
$C_{12}Cl_4H_{27}N_3U$	$UCl_4 \cdot 3\ (CH_2)_4NH$	U: SVol.E1-38/9
$C_{12}Cl_4H_{27}OPSn$	$SnCl_4 \cdot (C_4H_9)_3PO$	Sn: MVol.C6-144
$C_{12}Cl_4H_{27}O_4PSn$	$SnCl_4 \cdot (C_4H_9O)_3PO$	Sn: MVol.C6-170/1
$C_{12}Cl_4H_{27}O_{11.5}U_2$	$[(UO_2Cl_2)_2(CH_2OCH_2)_6(H_2O)_{1.5}]$	U: SVol.E1-71/2, 77/8
$C_{12}Cl_4H_{28}MnNO_4$	$(C_2H_5)_4NMnCl_4 \cdot 2\ CH_3COOH$	Mn: MVol.C5-212
$C_{12}Cl_4H_{28}N_4O_4U$	$UCl_4 \cdot 4\ CH_3CONHCH_3$	U: SVol.E1-96, 98/9
$C_{12}Cl_4H_{28}N_4O_8Sn$	$SnCl_4 \cdot 4\ CH_3CH(NH_2)COOH$	Sn: MVol.C6-11/2
–	$SnCl_4 \cdot 4\ NH_2CH_2CH_2COOH$	Sn: MVol.C6-11/2
$C_{12}Cl_4H_{28}O_2S_2Sn$	$SnCl_4 \cdot 2\ (n\text{-}C_3H_7)_2SO$	Sn: MVol.C6-102
$C_{12}Cl_4H_{28}O_2Sn$	$SnCl_4 \cdot 2\ (C_3H_7)_2O$	Sn: MVol.C5-151/2
$C_{12}Cl_4H_{28}O_4SiSn$	$SnCl_4 \cdot Si(OC_3H_7)_4$	Sn: MVol.C6-126/7
$C_{12}Cl_4H_{28}S_2Sn$	$SnCl_4 \cdot 2\ (C_3H_7)_2S$	Sn: MVol.C6-93/4
$C_{12}Cl_4H_{28}S_2Te$	$TeCl_4 \cdot 2\ (C_3H_7)_2S$	Te: SVol.B3-176
$C_{12}Cl_4H_{29}NO_3SiSn$	$SnCl_4 \cdot (C_4H_9O)_3SiNH_2$	Sn: MVol.C6-130/1
$C_{12}Cl_4H_{29}N_3O_2Sn$	$SnCl_4 \cdot 2\ OCHN(CH_3)_2 \cdot (C_2H_5)_3N$	Sn: MVol.C6-42
$C_{12}Cl_4H_{30}IrP_2$	$Ir(P(C_2H_5)_3)_2Cl_4$	Ir: SVol.2-203
$C_{12}Cl_4H_{30}NO_2Pa$	$[N(C_2H_5)_4]Pa(OC_2H_5)_2Cl_4$	Pa: SVol.2-61/2, 82
$C_{12}Cl_4H_{30}NO_2U$	$[N(C_2H_5)_4]U(OC_2H_5)_2Cl_4$	U: SVol.C9-63, 90
$C_{12}Cl_4H_{30}N_2Sn$	$[(CH_3)_4N]_2[(CH_2CH)_2SnCl_4]$	Sn: Org.Verb.6-145
–	$SnCl_4 \cdot 2\ (C_2H_5)_3N$	Sn: MVol.C5-173
$C_{12}Cl_4H_{30}N_2Sn_2$	$[Sn_2Cl_3((C_3H_7)_2NH)_2]Cl$	Sn: MVol.C5-11/2
$C_{12}Cl_4H_{30}N_2Te$	$TeCl_4 \cdot 2\ N(C_2H_5)_3$	Te: SVol.B3-168
$C_{12}Cl_4H_{30}N_6O_6U$	$UCl_4 \cdot 6\ CH_3CONH_2$	U: SVol.E1-96/7
$C_{12}Cl_4H_{30}O_2P_2S_2Sn$	$SnCl_4 \cdot 2\ (CH(CH_3)_2)_2POSH$	Sn: MVol.C6-156/7
–	$SnCl_4 \cdot 2\ (C_2H_5)_2P(O)SC_2H_5$	Sn: MVol.C6-158/60
–	$SnCl_4 \cdot 2\ (C_3H_7)_2POSH$	Sn: MVol.C6-156/7
$C_{12}Cl_4H_{30}O_2P_2Sn$	$SnCl_4 \cdot 2\ (C_2H_5)_2POC_2H_5$	Sn: MVol.C6-139/40
–	$SnCl_4 \cdot 2\ (C_2H_5)_3PO$	Sn: MVol.C6-143/4
$C_{12}Cl_4H_{30}O_2P_2U$	$UCl_4 \cdot 2\ (C_2H_5)_3PO$	U: SVol.E1-149, 156/7
$C_{12}Cl_4H_{30}O_3S_3Te$	$TeCl_4 \cdot 3\ (C_2H_5)_2SO$	Te: SVol.B3-176
$C_{12}Cl_4H_{30}O_3S_3U$	$UCl_4 \cdot 3\ (C_2H_5)_2SO$	U: SVol.E1-209/11
$C_{12}Cl_4H_{30}O_4P_2Sn$	$SnCl_4 \cdot 2\ C_2H_5(H)P(O)OC_4H_9$	Sn: MVol.C6-151/2
–	$SnCl_4 \cdot 2\ C_2H_5P(OC_2H_5)_2$	Sn: MVol.C6-141
–	$SnCl_4 \cdot 2\ (C_2H_5)_2P(O)OC_2H_5$	Sn: MVol.C6-153/4
$C_{12}Cl_4H_{30}O_6P_2S_2Sn$	$SnCl_4 \cdot 2\ (C_2H_5O)_2(C_2H_5S)PO$	Sn: MVol.C6-174/5
$C_{12}Cl_4H_{30}O_6P_2Sn$	$SnCl_4 \cdot 2\ C_2H_5P(O)(OC_2H_5)_2$	Sn: MVol.C6-162/3
–	$SnCl_4 \cdot 2\ HP(O)(OC_3H_7)_2$	Sn: MVol.C6-160/1
–	$SnCl_4 \cdot 2\ P(OC_2H_5)_3$	Sn: MVol.C6-141/2
$C_{12}Cl_4H_{30}O_8P_2Sn$	$SnCl_4 \cdot 2\ (C_2H_5O)_3PO$	Sn: MVol.C6-170
$C_{12}Cl_4H_{30}P_2S_2Sn$	$SnCl_4 \cdot 2\ (C_2H_5)_2PSC_2H_5$	Sn: MVol.C6-141
–	$SnCl_4 \cdot 2\ (C_2H_5)_3PS$	Sn: MVol.C6-148
$C_{12}Cl_4H_{30}P_2Sn$	$SnCl_4 \cdot 2\ P(C_2H_5)_3$	Sn: MVol.C6-133
$C_{12}Cl_4H_{30}P_2U$	$UCl_4[(C_2H_5)_3P]_2$	U: SVol.E1-51
$C_{12}Cl_4H_{32}MnN_2$	$[(C_2H_5)_3NH]_2MnCl_4$	Mn: MVol.C5-86

Formula	Compound	Reference
$C_{12}Cl_4H_{32}MnN_2$	$(C_6H_{13}NH_3)_2MnCl_4$	Mn: MVol.C5-182
$C_{12}Cl_4H_{32}N_2O_2U$	$[N(C_2H_5)_3H]_2UO_2Cl_4$	U: SVol.C9-93/8, 101
$C_{12}Cl_4H_{32}N_4Sn$	$SnCl_4 \cdot 2\ NH_2(CH_2)_6NH_2$	Sn: MVol.C5-185/6
$C_{12}Cl_4H_{32}N_8O_4U$	$UCl_4 \cdot 4\ CO(NHCH_3)_2$	U: SVol.E1-91/2, 94
$C_{12}Cl_4H_{32}O_2P_2U$	$[P(C_2H_5)_3H]_2UO_2Cl_4$	U: SVol.C9-93
		U: SVol.E1-51
$C_{12}Cl_4H_{32}O_2Si_4Sn_2$	$[Sn(Cl_2)CH_2Si(CH_3)_2OSi(CH_3)_2CH_2]_2$	Sn: Org.Verb.6-207
$C_{12}Cl_4H_{32}O_4Sn$	$SnCl_4 \cdot 4\ (CH_3)_2CHOH$	Sn: MVol.C5-72
$C_{12}Cl_4H_{32}O_4U$	$UCl_4 \cdot 4\ C_3H_7OH$	U: SVol.E1-58/9
$C_{12}Cl_4H_{33}N_3U$	$UCl_4 \cdot 3\ (CH_3)_3CNH_2$	U: SVol.E1-18, 20
$C_{12}Cl_4H_{36}N_4U$	$UCl_4 \cdot 4\ C_3H_7NH_2$	U: SVol.E1-18, 20
$C_{12}Cl_4H_{36}N_6O_2P_2Pa$	$PaCl_4 \cdot 2\ [(CH_3)_2N]_3PO$	Pa: SVol.2-93/4
$C_{12}Cl_4H_{36}N_6O_2P_2Sn$	$SnCl_4 \cdot 2\ ((CH_3)_2N)_3PO$	Sn: MVol.C6-171/2
$C_{12}Cl_4H_{36}N_6O_2P_2U$	$UCl_4 \cdot 2\ ((CH_3)_2N)_3PO$	U: SVol.E1-3, 153, 161/2, 164
$C_{12}Cl_4H_{36}N_6O_{4.5}P_3U$	$UCl_4 \cdot 1.5\ [((CH_3)_2N)_2PO]_2O$	U: SVol.E1-166/7
$C_{12}Cl_4H_{36}N_6P_2S_2Sn$	$SnCl_4 \cdot 2\ ((CH_3)_2N)_3PS$	Sn: MVol.C6-177
$C_{12}Cl_4H_{36}N_6Sn$	$SnCl_4 \cdot 3\ NH_2(CH_2)_4NH_2$	Sn: MVol.C5-185
$C_{12}Cl_4H_{36}O_6S_6U$	$UCl_4 \cdot 6\ (CH_3)_2SO$	U: SVol.E1-207
$C_{12}Cl_4H_{36}O_{14}S_6U$	$[UCl_2((CH_3)_2SO)_6](ClO_4)_2$	U: SVol.E1-3, 202/3
$C_{12}Cl_4H_{42}N_6Sn$	$SnCl_4 \cdot 6\ C_2H_5NH_2$	Sn: MVol.C5-168
$C_{12}Cl_5CoH_{10}OSn_2$	$C_5H_5(CO)Co(SnCl_3)(SnCl_2C_6H_5)$	Sn: Org.Verb.6-275, 277
$C_{12}Cl_5F_2H_4N_3O_2$	$NC_5F_2Cl_2NHC_5Cl_3(COOCH_3)N$	F: PerFHalOrg.5-138/9
$C_{12}Cl_5H_6N_3O_3S_3$	$[NS(O)Cl][NS(O)C_6H_3Cl_2]_2$	S: S-N-Verb.1-19
$C_{12}Cl_5H_7IrKN_2$	$K[Ir(C_{12}H_7ClN_2)Cl_4]$	Ir: SVol.2-93
$C_{12}Cl_5H_8N_2U$	$[UCl_5(C_{12}H_8N_2)]$	U: SVol.E1-46/8
$C_{12}Cl_5H_9N_2Sn$	$SnCl_4 \cdot C_6H_5NNC_6H_4Cl$	Sn: MVol.C6-31/2
$C_{12}Cl_5H_{10}MnN_2$	$(C_{12}H_8N_2H_2)MnCl_5$	Mn: MVol.C5-212/3
$C_{12}Cl_5H_{10}Se_2U$	$UCl_5 \cdot (C_6H_5)_2Se_2$	U: SVol.E1-221/2
$C_{12}Cl_5H_{10}Te_2U$	$UCl_5 \cdot (C_6H_5)_2Te_2$	U: SVol.E1-221/2
$C_{12}Cl_5H_{14}IrN_2$	$Ir(C_5H_5N)_2Cl_3 \cdot C_2H_4Cl_2$	Ir: SVol.2-69
$C_{12}Cl_5H_{14}NO_2Sn$	$SnCl_4 \cdot ClC_6H_4CONHCOC_4H_9$	Sn: MVol.C6-50/1
$C_{12}Cl_5H_{14}N_2U$	$UCl_5 \cdot 2\ CH_3C_5H_4N$	U: SVol.E1-28, 33/4
$C_{12}Cl_5H_{24}O_6U$	$UCl_5 \cdot 3\ C_4H_8O_2$	U: SVol.E1-71, 75
$C_{12}Cl_5H_{32}N_2OU$	$[N(C_2H_5)_3H]_2UOCl_5$	U: SVol.C9-90
$C_{12}Cl_6F_3H_6N_3$	$(C_6H_5)(CFCl_2)_3C_3HN_3$	F: PerFHalOrg.6-112
$C_{12}Cl_6F_{15}N_3$	$(CF_2ClCFClCF_2)_3C_3N_3$	F: PerFHalOrg.6-84/6, 105
$C_{12}Cl_6F_{16}H_3N_3$	$C_5F_8Cl_3C(NH)NC(NH_2)C_5F_8Cl_3$	F: PerFHalOrg.7-7
$C_{12}Cl_6FeH_{18}N_6Pt$	$[(CH_3NC)_6Fe][PtCl_6]$	Fe: Org.Verb.B4-98, 103
$C_{12}Cl_6Fe_3O_{12}Si_3$	$[(CO)_4FeSiCl_2]_3$	Fe: Org.Verb.B2-149
$C_{12}Cl_6H_6Sn$	$Cl_3C_6H_2C_6H_4SnCl_3$	Sn: Org.Verb.6-287
$C_{12}Cl_6H_8I_2Sn$	$(Cl_2IC_6H_4)_2SnCl_2$	Sn: Org.Verb.6-184/5, 187
$C_{12}Cl_6H_{10}N_4O_4Sn$	$SnCl_4 \cdot 2\ NO_2C_6H_3(Cl)NH_2$	Sn: MVol.C5-179
$C_{12}Cl_6H_{12}N_2Sn$	$SnCl_4 \cdot 2\ ClC_6H_4NH_2$	Sn: MVol.C5-179/80
$C_{12}Cl_6H_{12}NiO_4$	$(CH_3)_4C_4Ni(CCl_3COO)_2$	Ni: Org.Verb.2-329
$C_{12}Cl_6H_{12}O_{12}Sn_2Sr$	$Sr[Sn(CH_2ClCOO)_3]_2$	Sn: MVol.C3-190
$C_{12}Cl_6H_{14}N_2O_2SSn$	$SnCl_4 \cdot 2\ CH_3C_5H_4N \cdot SO_2Cl_2$	Sn: MVol.C5-201
$C_{12}Cl_6H_{14}Ni_2O_4$	$(CH_3CHCHCH_2NiO_2CCCl_3)_2$	Ni: Org.Verb.2-297
$C_{12}Cl_6H_{14}S_2Sn$	$[C_6H_5SH_2]_2[SnCl_6]$	Sn: MVol.C6-80
$C_{12}Cl_6H_{16}N_2Sn$	$[C_6H_5NH_3]_2[SnCl_6]$	Sn: MVol.C3-103/4

Formula	Compound	Reference
$C_{12}Cl_6H_{16}N_2Te$	$[CH_3C_5H_4NH]_2TeCl_6$	Te: SVol.B2-116
$C_{12}Cl_6H_{18}MgN_6Sn$	$[Mg(CH_3CN)_6][SnCl_6]$	Sn: MVol.C3-165
$C_{12}Cl_6H_{18}N_6Na_2Sn$	$[Na(CH_3CN)_3]_2[SnCl_6]$	Sn: MVol.C3-24
$C_{12}Cl_6H_{18}N_6SnSr$	$[Sr(CH_3CN)_6][SnCl_6]$	Sn: MVol.C3-187/8
$C_{12}Cl_6H_{20}O_6Sn_{1.5}$	$1.5\ SnCl_4 \cdot CH(CH_2CO_2C_2H_5)_2CO_2C_2H_5$	Sn: MVol.C5-144
$C_{12}Cl_6H_{20}O_8U$	$UO_2(Cl_3CCOO)_2 \cdot 2\ (C_2H_5)_2O \cdot 2\ H_2O$	U: SVol.E1-71, 77
$C_{12}Cl_6H_{21}O_6Sn^+$	$[SnCl_3(CH_2ClCOOC_2H_5)_3]^+$	Sn: MVol.C5-134
$C_{12}Cl_6H_{24}N_8S_4Te_2$	$Te(SC(NHCH_2)_2)_4[TeCl_6]$	Te: SVol.B3-158
$C_{12}Cl_6H_{26}O_8Sn_2$	$2\ SnCl_3OH \cdot 3\ C_4H_8O_2 \cdot 2\ H_2O$	Sn: MVol.C5-159
$C_{12}Cl_6H_{28}NU$	$[N(C_3H_7)_4]UCl_6$	U: SVol.C9-60, 62
$C_{12}Cl_6H_{30}HgP_2Sn$	$HgCl_2 \cdot SnCl_4 \cdot 2\ P(C_2H_5)_3$	Sn: MVol.C6-133
$C_{12}Cl_6H_{30}O_2Sn$	$[(C_2H_5)_3O]_2[SnCl_6]$	Sn: MVol.C3-107
$C_{12}Cl_6H_{30}O_6Si_2Sn$	$SnCl_4 \cdot 2\ ClSi(OC_2H_5)_3$	Sn: MVol.C6-128/9
$C_{12}Cl_6H_{32}N_2Sn$	$[(CH_3)_2(C_2H_5)_2N]_2[SnCl_6]$	Sn: MVol.C3-106
–	$[(CH_3)_3(C_3H_7)N]_2[SnCl_6]$	Sn: MVol.C3-106
–	$[(C_2H_5)_3NH]_2[SnCl_6]$	Sn: MVol.C3-105
–	$[(C_3H_7)_2NH_2]_2[SnCl_6]$	Sn: MVol.C3-104/5
$C_{12}Cl_6H_{32}N_2Te$	$[(C_2H_5)_3NH]_2TeCl_6$	Te: SVol.B2-116
$C_{12}Cl_6H_{32}O_5Sn_2$	$(SnCl_3)_2O \cdot 4\ C_3H_7OH$	Sn: MVol.C5-74
$C_{12}Cl_6H_{32}P_2U$	$[P(C_2H_5)_3H]_2UCl_6$	U: SVol.C9-52
$C_{12}Cl_7DyH_{40}N_4$	$DyCl_3 \cdot 4\ [(CH_3)_3NH]Cl$	Sc: MVol.C5-221
$C_{12}Cl_7H_{14}N_2PSSn$	$SnCl_4 \cdot 2\ CH_3C_5H_4N \cdot PSCl_3$	Sn: MVol.C5-201
$C_{12}Cl_7H_{21}O_6Sn$	$SnCl_4 \cdot 3\ CH_2ClCOOC_2H_5$	Sn: MVol.C5-133/4
$C_{12}Cl_7H_{40}LaN_4$	$LaCl_3 \cdot 4\ [(CH_3)_3NH]Cl$	Sc: MVol.C5-213/6
$C_{12}Cl_7H_{40}N_4Nd$	$NdCl_3 \cdot 4\ [(CH_3)_3NH]Cl$	Sc: MVol.C5-213/6
$C_{12}Cl_7H_{40}N_4Pr$	$PrCl_3 \cdot 4\ [(CH_3)_3NH]Cl$	Sc: MVol.C5-213/6
$C_{12}Cl_8F_2H_8O_2P_2Sn$	$SnCl_4 \cdot 2\ FC_6H_4POCl_2$	Sn: MVol.C6-167
$C_{12}Cl_8F_{17}NO$	$Cl_3C(CF_2CFCl)_5CF_2NO$	F: PerFHalOrg.7-179, 186
$C_{12}Cl_8Fe_2H_{24}MnO_6$	$Mn(CH_3CHO)_6(FeCl_4)_2$	Mn: MVol.D1-51/2
$C_{12}Cl_8Fe_3H_{18}N_6$	$[(CH_3NC)_6Fe][FeCl_4]_2$	Fe: Org.Verb.B4-98, 103
$C_{12}Cl_8H_{10}O_2P_2Sn$	$SnCl_4 \cdot 2\ C_6H_5POCl_2$	Sn: MVol.C6-167
$C_{12}Cl_8H_{10}O_4P_2Sn$	$SnCl_4 \cdot 2\ Cl_2PO(OC_6H_5)$	Sn: MVol.C6-171
$C_{12}Cl_8H_{10}S_2U_2$	$[UCl_4(C_6H_5S)]_2$	U: SVol.E1-222
$C_{12}Cl_8H_{14}N_2Sn$	$[ClC_6H_4NH_3]_2[SnCl_6]$	Sn: MVol.C3-103/4
$C_{12}Cl_8H_{14}Ni_2O_3$	$(CH_3CHCHCH_2NiCl)_2 \cdot (CCl_3CO)_2O$	Ni: Org.Verb.2-300
$C_{12}Cl_8H_{14}Ni_3O_4$	$(CH_3CHCHCH_2NiCl)_2 \cdot Ni(O_2CCCl_3)_2$	Ni: Org.Verb.2-300, 301
$C_{12}Cl_8H_{18}N_2O_4Sn_2$	$2\ SnCl_4 \cdot 2\ NCCH_2CH_2COOC_2H_5$	Sn: MVol.C6-75
$C_{12}Cl_8H_{24}In_2MnO_6$	$Mn(CH_3CHO)_6(InCl_4)_2$	Mn: MVol.D1-51/2
$C_{12}Cl_8H_{26}S_2Te_2$	$2\ TeCl_4 \cdot C_4H_9S(CH_2)_4SC_4H_9$	Te: SVol.B3-176
$C_{12}Cl_8H_{28}NiO_4P_4$	$Ni(P(OC_3H_7)Cl_2)_4$	Ni: Org.Verb.1-240, 242
$C_{12}Cl_8H_{30}Mn_4O_3$	$4\ MnCl_2 \cdot 3\ C_4H_9OH$	Mn: MVol.D1-33
$C_{12}Cl_8H_{36}N_3Pa$	$[N(CH_3)_4]_3PaCl_8$	Pa: SVol.2-58
$C_{12}Cl_8H_{36}N_3U$	$[N(CH_3)_4]_3UCl_8$	U: SVol.C9-60/1
$C_{12}Cl_8H_{36}O_6S_6U_2$	$[UCl_3((CH_3)_2SO)_5][UCl_5((CH_3)_2SO)]$	U: SVol.E1-202, 207
$C_{12}Cl_8NiO_4$	$Ni(C_6Cl_4O_2)_2$	Ni: Org.Verb.1-230
$C_{12}Cl_9H_{48}LaN_6$	$LaCl_3 \cdot 6\ [(CH_3)_2NH_2]Cl$	Sc: MVol.C5-213/5
–	$LaCl_3 \cdot 6\ [(CH_3)_2NH_2]Cl \cdot 2\ H_2O$	Sc: MVol.C5-213/5
$C_{12}Cl_9H_{48}N_6Nd$	$NdCl_3 \cdot 6\ [(CH_3)_2NH_2]Cl$	Sc: MVol.C5-213/5
–	$NdCl_3 \cdot 6\ [(CH_3)_2NH_2]Cl \cdot 2\ H_2O$	Sc: MVol.C5-213/5
$C_{12}Cl_9H_{48}N_6Pr$	$PrCl_3 \cdot 6\ [(CH_3)_2NH_2]Cl$	Sc: MVol.C5-213/5

Formula	Compound	Reference
$C_{12}Cl_9H_{48}N_6Pr$	$PrCl_3 \cdot 6\,[(CH_3)_2NH_2]Cl \cdot 2\,H_2O$	Sc: MVol.C5-213/5
$C_{12}CmH_{10}O_{14}{}^{3-}$	$Cm(C_6H_5O_7)_2{}^{3-}$	Np: TrU.D1-159
$C_{12}CmH_{11}O_{15}{}^{4-}$	$Cm(OH)(C_6H_5O_7)_2{}^{4-}$	Np: TrU.D1-157, 159
$C_{12}CmH_{12}N_2O_{12}{}^{3-}$	$Cm(N(CH_2COO)_3)_2{}^{3-}$	Np: TrU.D1-163
$C_{12}CmH_{18}N_2O_{10}{}^{-}$	$Cm(N(C_2H_4OH)(CH_2COO)_2)_2{}^{-}$	Np: TrU.D1-162
$C_{12}CmH_{21}O_9$	$Cm((CH_3)_2C(OH)COO)_3$	Np: TrU.D1-154
$C_{12}CoH_{21}O_3S_6$	$Co[S_2COC_3H_7]_3$	C: MVol.D4-259
$C_{12}CoH_{21}S_9$	$Co[S_2CSC_3H_7]_3$	C: MVol.D4-267, 270
$C_{12}Co_2FeO_{12}S_2$	$Co_2Fe(CO)_{12}S_2$	Fe: Org.Verb.C1-76
$C_{12}Co_3FeHO_{12}$	$HFeCo_3(CO)_{12}$	Fe: Org.Verb.B1-200/1
$C_{12}Co_3FeH_{19}O_7P^+$	$[HFeCo_3(CO)_4P(OC_3H_7)_2OC_2H_4]^+$	Fe: Org.Verb.B1-204
$C_{12}Co_3FeH_{22}O_6P^+$	$[HFeCo_3(CO)_3P(OC_3H_7)_3]^+$	Fe: Org.Verb.B1-204
$C_{12}Co_3FeO_{12}{}^{-}$	$[FeCo_3(CO)_{12}]^-$	Fe: Org.Verb.B1-200
$C_{12}Co_3Fe_2N_{10}O_2$	$Co_3[COFe(CN)_5]_2 \cdot 5.7\,H_2O$	Fe: Org.Verb.B1-26/7
$C_{12}Co_3Ir_2N_{12}$	$Co[Ir(CN)_6]_2 \cdot 12\,H_2O$	Ir: SVol.2-183, 184
–	$Co_3[Ir(CN)_6]_2 \cdot 15\,H_2O$	Ir: SVol.2-183, 184
$C_{12}Co_4O_{12}$	$Co_4(CO)_{12}$ systems	
	$Co_4(CO)_{12}$-$Co_2(CO)_8$-$Fe(CO)_5$-$Ni(CO)_4$	Fe: Org.Verb.B3-64
	$Co_4(CO)_{12}$-$Fe(CO)_5$	Fe: Org.Verb.B3-64
$C_{12}CrF_{12}H_{12}IrP_4$	$(C_6H_6)_2Cr[Ir(PF_3)_4]$	Ir: SVol.2-193
$C_{12}CrFeH_{11}NO^+$	$[C_5H_5FeC_5H_4C(NH_2)Cr(CO)]^+$	Fe: Org.Verb.A1-346
$C_{12}CrFeH_{12}O^+$	$[C_5H_5FeC_5H_4C(OCH_3)Cr]^+$	Fe: Org.Verb.A1-346
$C_{12}CrH_8N_2O_6U$	$[UO_2CrO_4(C_{12}H_8N_2)]$	U: SVol.E1-44/6
$C_{12}CrH_{14}O_3Sn$	$(CH_3)_3SnC_6H_5Cr(CO)_3$	Sn: Org.Verb.2-141, 144
$C_{12}Cr_3Ir_2N_{12}$	$Cr_3[Ir(CN)_6]_2 \cdot 12\,H_2O$	Ir: SVol.2-183, 184
$C_{12}CuF_{20}H_4N_6$	$[HNC(CF_3)NC(C_3F_7)NH]_2Cu$	F: PerFHalOrg.7-58
–	$[N(C(C_2F_5)NH)_2]_2Cu$	F: PerFHalOrg.7-58
$C_{12}CuFeH_9$	$C_5H_5FeC_5H_4CCCu$	Fe: Org.Verb.A1-302
$C_{12}Cu_3Ir_2N_{12}$	$Cu_3[Ir(CN)_6]_2 \cdot 10\,H_2O$	Ir: SVol.2-183, 184
–	$Cu_3[Ir(CN)_6]_2 \cdot 12\,H_2O$	Ir: SVol.2-183, 184
$C_{12}EsH_{21}O_9$	$Es((CH_3)_2C(OH)COO)_3$	Np: TrU.D1-154
$C_{12}FFeH_7O_4$	$FC_6H_4CHCH_2Fe(CO)_4$	Fe: Org.Verb.B4-238, 254
$C_{12}FFeH_{12}O_3S^+$	$[C_5H_5FeC_5H_4C(CH_3)OSO_2F]^+$	Fe: Org.Verb.A1-297, 349
$C_{12}FFeH_{20}N_2O_4P$	$(CO)_4FePF(N(C_2H_5)_2)_2$	Fe: Org.Verb.B2-85, 94, 115
$C_{12}FH_{10}ISn$	$(C_6H_5)_2SnFI$	Sn: Org.Verb.5-34
$C_{12}FH_{10}N_3O_3S_3$	$[NS(O)F][NS(O)C_6H_5]_2$	S: S-N-Verb.1-14
$C_{12}FH_{19}Sn$	$(C_2H_5)_3SnC_6H_4F$	Sn: Org.Verb.2-238/9
$C_{12}FH_{22}N_3O_7Sn$	$[SnF(C_4H_9N_3(C_2H_4OH)(CH_2COO)_3H_2)]$	Sn: MVol.C5-32
$C_{12}FH_{24}N_5O_5S_3$	$[NS(O)F][NS(O)N(CH_2CHCH_3)_2O]_2$	S: S-N-Verb.1-13
$C_{12}FH_{27}Sn$	$(C_4H_9)_3SnF$	Sn: Org.Verb.5-24/30
–	$(i\text{-}C_4H_9)_3SnF$	Sn: Org.Verb.5-30
–	$(t\text{-}C_4H_9)_3SnF$	Sn: Org.Verb.5-36
$C_{12}FH_{33}Si_3Sn$	$[(CH_3)_3SiCH_2]_3SnF$	Sn: Org.Verb.5-36
$C_{12}F_2H_8N_2O_2U$	$[UO_2F_2(C_{12}H_8N_2)]$	U: SVol.E1-44/6
$C_{12}F_2H_8N_2O_6S$	$(NO_2C_6H_4O)_2SF_2$	S: SVol.2-44, 83
$C_{12}F_2H_8O_2$	$C_{10}H_7COOCHF_2$	F: PerFHalOrg.5-18
$C_{12}F_2H_{10}O_2S$	$(C_6H_5O)_2SF_2$	S: SVol.2-44, 83
$C_{12}F_2H_{10}S$	$(C_6H_5)_2SF_2$	S: SVol.2-91
$C_{12}F_2H_{10}Sn$	$(C_6H_5)_2SnF_2$	Sn: Org.Verb.5-48
$C_{12}F_2H_{12}N_2O_4$	$(CH_3O)_4C_8F_2N_2$	F: PerFHalOrg.6-161

Formula	Compound		Reference
$C_{12}F_2H_{18}N_2$	$NCFC(C_4H_9)NCFC(C_4H_9)$	F:	PerFHalOrg.6-60, 69
$C_{12}F_2H_{20}N_4O_2$	$((CH_3)_2NC_2H_4O)_2C_4F_2N_2$	F:	PerFHalOrg.6-58
$C_{12}F_2H_{22}Sn$	$(C_6H_{11})_2SnF_2$	Sn:	Org.Verb.5-49
$C_{12}F_2H_{23}N_3O_7Sn$	$[SnF(C_4H_9N_3(C_2H_4OH)(CH_2COO)_3H_2)] \cdot HF$	Sn:	MVol.C5-32
$C_{12}F_2H_{26}Sn$	$(C_6H_{13})_2SnF_2$	Sn:	Org.Verb.5-49
$C_{12}F_2H_{27}P$	$(C_4H_9)_3PF_2$	F:	PerFHalOrg.5-22
$C_{12}F_2H_{27}PS_3$	$(C_4H_9S)_3PF_2$	F:	PerFHalOrg.5-23
$C_{12}F_2H_{36}N_6O_4P_2U$	$UO_2F_2 \cdot 2\ [(CH_3)_2N]_3PO$	U:	SVol.E1-163/4
$C_{12}F_3FeH_9O$	$C_5H_5FeC_5H_4COCF_3$	Fe:	Org.Verb.A2-186, 241/3
$C_{12}F_3FeH_{10}{}^+$	$[C_5H_5FeC_5H_4CHCF_3]^+$	Fe:	Org.Verb.A2-47
$C_{12}F_3FeH_{11}O$	$C_5H_5FeC_5H_4CHOHCF_3$	Fe:	Org.Verb.A2-44, 47
$C_{12}F_3H_5MoN_2O_3$	$[C_5H_5Mo(CO)_3]C_4F_3N_2$	F:	PerFHalOrg.6-72
$C_{12}F_3H_6N_5O_2S$	$NC(NHC_6H_5)NC(CF_3)C(NO_2)C(SCN)$	F:	PerFHalOrg.6-71
$C_{12}F_3H_7N_4S$	$(CF_3)(C_6H_5CHCH)C_3N_4S$	F:	PerFHalOrg.5-100
$C_{12}F_3H_8N_5S$	$(C_6H_5NH)(CF_3)(NH_2)C_5N_3S$	F:	PerFHalOrg.6-71
$C_{12}F_3H_{10}N_5O_4S$	$C_2N_2S(CF_3)NHC_6H_2(NO_2)_2CH(CH_3)_2$	F:	PerFHalOrg.5-107
$C_{12}F_3H_{13}N_2O_2$	$C_2H_5OCOCH(NNCF_3)CH_2C_6H_5$	F:	PerFHalOrg.7-160
$C_{12}F_3H_{14}MnO_6$	$[MnCF_3COO(CH(COCH_3)_2)_2]$	Mn:	MVol.D1-103/4
$C_{12}F_3H_{24}N_3O_7Sn$	$[C_4H_9N_3(C_2H_4OH)(CH_2COOH)_3H][SnF_3]$	Sn:	MVol.C5-32
$C_{12}F_3H_{25}Sn$	$C_{12}H_{25}SnF_3$	Sn:	Org.Verb.5-52
$C_{12}F_4FeH_5NO_2$	$NC_5F_4[Fe(CO)_2(C_5H_5)]$	F:	PerFHalOrg.5-145
$C_{12}F_4H_5NO$	$OC(CFCF)_2CNC_6H_5$	F:	PerFHalOrg.7-197/8
$C_{12}F_4H_7NO$	$C_6H_5CH(C_5F_4N)OH$	F:	PerFHalOrg.4-97
$C_{12}F_4H_7N_3$	$C_6H_5CHNNHC_5F_4N$	F:	PerFHalOrg.5-216
$C_{12}F_4H_7N_3O$	$C_6H_5NHNHC(O)C_5F_4N$	F:	PerFHalOrg.5-217
$C_{12}F_4H_8N_4O_2Sn$	$SnF_4 \cdot 2\ NCC_5H_4NO$	Sn:	MVol.C5-208/9
$C_{12}F_4H_8O_2S$	$(FC_6H_4O)_2SF_2$	S:	SVol.2-44, 83
$C_{12}F_4H_9NO_3$	$(CH_3O)_3C_9F_4N$	F:	PerFHalOrg.6-161
$C_{12}F_4H_{10}NNaO_4$	$NC_5F_4CNa(COOC_2H_5)_2$	F:	PerFHalOrg.5-142
$C_{12}F_4H_{10}N_2$	$C_7H_{10}NC_5F_4N$	F:	PerFHalOrg.5-213
$C_{12}F_4H_{10}Ni$	$C_5H_5NiC_5H_5C_2F_4$	Ni:	Org.Verb.2-181
$C_{12}F_4H_{11}NO_4$	$NC_5F_4CH(COOC_2H_5)_2$	F:	PerFHalOrg.5-142
$C_{12}F_4H_{14}N_2$	$(CH_2)_7NC_5F_4N$	F:	PerFHalOrg.5-213
–	$C_7H_{13}NHC_5F_4N$	F:	PerFHalOrg.5-137
$C_{12}F_4H_{14}N_2O_2Sn$	$SnF_4 \cdot 2\ CH_3C_5H_4NO$	Sn:	MVol.C5-208/9
$C_{12}F_4H_{14}N_2O_4Sn$	$SnF_4 \cdot 2\ CH_3OC_5H_4NO$	Sn:	MVol.C5-208/9
$C_{12}F_4H_{16}N_4Sn$	$SnF_4 \cdot 2\ C_6H_5NHNH_2$	Sn:	MVol.C5-189
$C_{12}F_4H_{18}Si_2$	$[(CH_3)_3Si]_2C_6F_4$	F:	PerFHalOrg.4-25
$C_{12}F_4H_{18}Sn_2$	$[(CH_3)_3Sn]_2C_6F_4$	F:	PerFHalOrg.4-29
$C_{12}F_4H_{20}N_2NiO_2P_2$	$(CO)_2Ni(P(NC_5H_{10})F_2)_2$	Ni:	Org.Verb.1-126
$C_{12}F_4H_{30}N_2Sn$	$SnF_4 \cdot 2\ (C_2H_5)_3N$	Sn:	MVol.C5-173
$C_{12}F_5H_5Hg$	$C_6F_5HgC_6H_5$	F:	PerFHalOrg.4-111
$C_{12}F_5H_5NiOS$	$C_5H_5Ni(CO)SC_6F_5$	Ni:	Org.Verb.2-164
$C_{12}F_5H_5S$	$C_6F_5SC_6H_5$	F:	PerFHalOrg.4-31
$C_{12}F_5H_8NO$	$OC_6H_8NC_6F_5$	F:	PerFHalOrg.7-198/9
$C_{12}F_5H_{10}N$	$C_6F_5N(CH(CH_2)_4CH)$	F:	PerFHalOrg.7-195/6
–	$C_6H_9NHC_6F_5$	F:	PerFHalOrg.7-198
$C_{12}F_5H_{12}N$	$C_6F_5NC_2(CH_3)_4$	F:	PerFHalOrg.7-195/6
$C_{12}F_5H_{12}NO$	$C_6F_5N(OH)C(CH_3)_2C(CH_3)CH_2$	F:	PerFHalOrg.7-198
$C_{12}F_5H_{13}O_2Si$	$(C_2H_5O)_2Si(C_2H_3)C_6F_5$	F:	PerFHalOrg.4-79

Formula	Compound		Reference
$C_{12}F_5H_{15}O_3Si$	$(C_2H_5O)_3SiC_6F_5$	F:	PerFHalOrg.4-81
$C_{12}F_5H_{15}Si$	$C_6F_5Si(C_2H_5)_3$	F:	PerFHalOrg.4-163
$C_{12}F_5H_{15}Sn$	$(C_2H_5)_3SnC_6F_5$	Sn:	Org.Verb.2-240
$C_{12}F_6FeH_{10}O_4$	$(CO)_3FeOC(CF_3)_2C_4H_4(CH_3)_2$	Fe:	Org.Verb.B5-90, 92/3
$C_{12}F_6FeH_{15}O_5PS_2$	$(CO)_2Fe(P(OC_2H_5)_3)S_2C_2(CF_3)_2$	Fe:	Org.Verb.B1-101/2, 108, 111
$C_{12}F_6FeH_{19}NO_3P_2$	$[C_4H_4P(C_2H_5)_3Fe(CO)_2NO][PF_6]$	Fe:	Org.Verb.B5-28/31
$C_{12}F_6FeH_{20}O_8P_2$	$(CO)_2Fe(P(OCH_3)_3)_2(CHFCF_2)_2$	Fe:	Org.Verb.B1-163
$C_{12}F_6Fe_2H_6O_6S_2$	$(CO)_3Fe(CF_3CCCF_3)(SCH_3)_2Fe(CO)_3$	Fe:	Org.Verb.C2-67/8, 70
$C_{12}F_6Fe_2H_{25}O_4P_3S_2$	$[(CO)_4(P(CH_3)_3)_2Fe_2(SCH_3)_2H]PF_6$	Fe:	Org.Verb.C1-126
$C_{12}F_6Fe_2I_2O_8$	$[(CO)_4Fe(CFCF_2)I]_2$	Fe:	Org.Verb.B2-181
$C_{12}F_6H_4N_4O_5$	$(NH_2)(NO_2)C_6F_3OC_6F_3(NO_2)NH_2$	F:	PerFHalOrg.7-12/3, 40
$C_{12}F_6H_6N_2O_2$	$NC_5F_2(OCH_3)_2C_5F_4N$	F:	PerFHalOrg.5-223
–	$NC_5F_3(OCH_3)C_5F_3(OCH_3)N$	F:	PerFHalOrg.5-223/4
$C_{12}F_6H_8N_4O$	$(NH_2)_2C_6F_3OC_6F_3(NH_2)_2$	F:	PerFHalOrg.7-12/3, 40
$C_{12}F_6H_9MnO_5Sn$	$(CH_3)_3SnC(CF_3)C(CF_3)Mn(CO)_5$	Sn:	Org.Verb.2-82, 89
$C_{12}F_6H_{10}Ni_2S_2$	$(C_5H_5NiSCF_3)_2$	Ni:	Org.Verb.2-338, 339
$C_{12}F_6H_{10}OPTi$	$(CF_3)_2POTi(C_5H_5)_2$	F:	PerFHalOrg.3-37
$C_{12}F_6H_{11}N$	$C_6H_5C(CH_2)CH_2C(CF_3)_2NH_2$	F:	PerFHalOrg.7-65, 69, 73
$C_{12}F_6H_{14}Ni_2O_4$	$(CH_2C(CH_3)CH_2NiO_2CCF_3)_2$	Ni:	Org.Verb.2-306
–	$(CH_3CHCHCH_2NiO_2CCF_3)_2$	Ni:	Org.Verb.2-296/7
$C_{12}F_6H_{15}NNi$	$C_8H_{12}Ni(C(CF_3)_2NCH_3)$	Ni:	Org.Verb.2-83/4
$C_{12}F_6H_{16}N_2O$	$NH_2C(CF_3)_2C_5H_6N(CH_2CH_2)_2O$	F:	PerFHalOrg.7-69, 73
$C_{12}F_6H_{18}Sn$	$(n\text{-}C_4H_9)_2Sn(CFCF_2)_2$	F:	PerFHalOrg.4-72, 73
		Sn:	Org.Verb.3-42/3
$C_{12}F_6H_{20}NiO_2P_2$	$(CO)_2Ni(P(C_2H_5)_2CF_3)_2$	Ni:	Org.Verb.1-128
$C_{12}F_6H_{24}O_{12}U_3$	$(UO_2F_2)_3 \cdot [CH_2OCH_2]_6$	U:	SVol.E1-71/2, 77/8
–	$(UO_2F_2)_3 \cdot [CH_2OCH_2]_6 \cdot 4\ H_2O$	U:	SVol.E1-71/2, 77/8
$C_{12}F_6H_{30}N_2Te$	$TeF_6 \cdot 2\ N(C_2H_5)_3$	Te:	SVol.B3-183
$C_{12}F_6N^+$	$[C_{12}F_6N]^+$	F:	PerFHalOrg.6-152
$C_{12}F_6N_4$	$C_6F_4NC(CN)C(CFCFCN)N$	F:	PerFHalOrg.6-133/4, 147
$C_{12}F_7^+$	$C_{12}F_7^+$	F:	PerFHalOrg.3-172, 179, 216
$C_{12}F_7HN_2$	$C_6F_4N_2C_6F_3H$	F:	PerFHalOrg.6-155, 158
$C_{12}F_7HN_2O$	$C_6F_3(OH)NNC_6F_4$	F:	PerFHalOrg.6-136/7, 149, 160
$C_{12}F_7H_5N_2O$	$NC_5F_3(OC_2H_5)C_5F_4N$	F:	PerFHalOrg.5-222/3
$C_{12}F_7H_6N_3O$	$NC_5F_4NHC_5F_3(OC_2H_5)N$	F:	PerFHalOrg.5-138/9
$C_{12}F_7H_{15}Sn$	$(CH_2CHCH_2)_3SnC_3F_7$	Sn:	Org.Verb.2-348
$C_{12}F_7Li$	$LiC_6F_3C_6F_4$	F:	PerFHalOrg.4-7/8, 34
$C_{12}F_7N_2^+$	$[C_{12}F_7N_2]^+$	F:	PerFHalOrg.6-152
$C_{12}F_8$	$C_6F_5C_6F_3$	F:	PerFHalOrg.4-9/10
$C_{12}F_8^+$	$(C_6F_4)_2^+$	F:	PerFHalOrg.3-179, 216
		F:	PerFHalOrg.5-95
$C_{12}F_8Ge_2H_{12}$	$(CH_3)_2Ge(C_4F_4)_2Ge(CH_3)_2$	F:	PerFHalOrg.4-26
$C_{12}F_8HNS$	$S(C_6F_4)_2NH$	F:	PerFHalOrg.6-137
$C_{12}F_8HN_3O_4$	$(NO_2C_6F_4)_2NH$	F:	PerFHalOrg.7-14/6, 42
$C_{12}F_8H_2N_2$	$NH(C_6F_4)_2NH$	F:	PerFHalOrg.6-136/7, 150
$C_{12}F_8H_2N_4$	$C_6F_4N_3C_6F_4NH_2$	F:	PerFHalOrg.6-127/8, 140

Formula	Structure		Reference
$C_{12}F_8H_2N_4O_2$	$[NC_5F_4C(O)NH]_2$	F:	PerFHalOrg.5-189/90, 204
$C_{12}F_8H_3NO$	$(CH_3O)(C_6F_5)C_5F_3N$	F:	PerFHalOrg.5-178
$C_{12}F_8H_3N_5$	$NH_2C_6F_3N(NHC_6F_5)NN$	F:	PerFHalOrg.6-128, 140, 159
$C_{12}F_8H_4N_2$	$NH_2C_6F_4C_6F_4NH_2$	F:	PerFHalOrg.7-11/3, 39
$C_{12}F_8H_4N_2O$	$NH_2C_6F_4OC_6F_4NH_2$	F:	PerFHalOrg.7-12/3, 40, 55, 62
$C_{12}F_8H_4N_2O_2$	$(NC_5F_4OC_2H_4O)C_5F_4N$	F:	PerFHalOrg.5-136
$C_{12}F_8H_4N_4$	$NH_2C_6F_4NNC_6F_4NH_2$	F:	PerFHalOrg.7-12/3
$C_{12}F_8H_8N_2$	$NC(N(CH_3)C_6H_5)(CF_2)_4$	F:	PerFHalOrg.5-144
$C_{12}F_8H_{12}Ni$	$C_8H_{12}Ni(CF_2)_4$	Ni:	Org.Verb.2-85
$C_{12}F_8H_{14}Ni$	$Ni(C_2F_4CH_2C(CH_3)CH_2)_2$	Ni:	Org.Verb.2-85
$C_{12}F_8H_{20}Sn$	$(C_4H_9)_2Sn(CF_2CHF_2)_2$	Sn:	Org.Verb.3-43
$C_{12}F_8HgN_2O_4$	$Hg[OC(O)C_5F_4N]_2$	F:	PerFHalOrg.5-192/3, 207, 211
$C_{12}F_8ILi$	$IC_6F_4C_6F_4Li$	F:	PerFHalOrg.4-6
$C_{12}F_8I_2$	$IC_6F_4C_6F_4I$	F:	PerFHalOrg.4-14
$C_{12}F_8Li_2$	$LiC_6F_4C_6F_4Li$	F:	PerFHalOrg.4-6/7, 12/4, 26/7, 29, 39, 53/5
$C_{12}F_8Li_2O$	$(LiC_6F_4)_2O$	F:	PerFHalOrg.4-6
$C_{12}F_8Li_2S$	$(LiC_6F_4)_2S$	F:	PerFHalOrg.4-6
$C_{12}F_8N_2$	$N(C_6F_4)_2N$	F:	PerFHalOrg.6-136, 149, 152, 155, 160
$C_{12}F_8N_2^+$	$[C_6F_4N_2C_6F_4]^+$	F:	PerFHalOrg.6-152
$C_{12}F_8N_2O_3$	$(NC_5F_4CO)_2O$	F:	PerFHalOrg.5-194, 208
$C_{12}F_8N_2O_4Zn$	$Zn[OC(O)C_5F_4N]_2$	F:	PerFHalOrg.5-192/3, 207, 211
$C_{12}F_8N_4O$	$NC(C_5F_4N)OC(C_5F_4N)N$	F:	PerFHalOrg.5-65, 88
$C_{12}F_8N_4S$	$NC(C_5F_4N)SC(C_5F_4N)N$	F:	PerFHalOrg.5-67/8, 90
$C_{12}F_8O$	$C_6F_5OC_6F_3$	F:	PerFHalOrg.4-12
$C_{12}F_9^+$	$C_{12}F_9^+$	F:	PerFHalOrg.3-172, 216
$C_{12}F_9HN_2$	$HC_6F_4NNC_6F_5$	F:	PerFHalOrg.7-59
$C_{12}F_9HN_2O$	$NC_5F_4NHC(O)C_6F_5$	F:	PerFHalOrg.5-188/9
$C_{12}F_9HN_2O_2$	$(NO_2C_6F_4)(C_6F_5)NH$	F:	PerFHalOrg.7-14/5, 42
$C_{12}F_9HN_4$	$C_6F_4N(NHC_6F_5)NN$	F:	PerFHalOrg.6-128, 140
$C_{12}F_9H_2N$	$NH_2C_6F_4C_6F_5$	F:	PerFHalOrg.7-11, 38, 55
$C_{12}F_9H_2NO$	$NH_2C_6F_4OC_6F_5$	F:	PerFHalOrg.7-62
$C_{12}F_9H_2N_3$	$NH_2C_6F_4NNC_6F_5$	F:	PerFHalOrg.7-11, 38/9
$C_{12}F_9H_2N_3O$	$NC_5F_4NHNHCOC_6F_5$	F:	PerFHalOrg.5-189/90, 204
–	$NH_2C_6F_4N(O)NC_6F_5$	F:	PerFHalOrg.7-11, 39, 55
$C_{12}F_9H_5Sn$	$C_6H_5Sn(CFCF_2)_3$	F:	PerFHalOrg.4-73
		Sn:	Org.Verb.2-346, 349
$C_{12}F_9H_7N_2O$	$C_4F_9C(NH)NHC(O)C_6H_5$	F:	PerFHalOrg.7-57
$C_{12}F_9H_7O_6Ti$	$CH_3C_5H_4Ti(OCOCF_3)_3$	Ti:	Org.Verb.1-168/9
$C_{12}F_9H_{12}N_3$	$(CF_3)_3(C_6H_{11})C_3HN_3$	F:	PerFHalOrg.6-113
$C_{12}F_9I$	$IC_6F_4C_6F_5$	F:	PerFHalOrg.4-16
$C_{12}F_9Li$	$LiC_6F_4C_6F_5$	F:	PerFHalOrg.4-6, 9/11, 16/7, 39

$C_{12}F_9LiO$ .... $LiC_6F_4OC_6F_5$ .... F: PerFHalOrg.4-6, 10, 12/7
$C_{12}F_9NO$ .... $NC_5F_4C(O)C_6F_5$ .... F: PerFHalOrg.5-195, 208
$C_{12}F_9NO_2$ .... $C_6F_5C_6F_4NO_2$ .... F: PerFHalOrg.4-106
$C_{12}F_{10}$ .... $C_6F_5C_6F_5$ .... F: PerFHalOrg.4-38, 70, 106, 111/2
$C_{12}F_{10}^+$ .... $(C_6F_5)_2^+$ .... F: PerFHalOrg.3-172
$C_{12}F_{10}GeH_2$ .... $(C_6F_5)_2GeH_2$ .... F: PerFHalOrg.4-165/6, 174, 179
$C_{12}F_{10}GeO$ .... $[(C_6F_5)_2GeO]_n$ .... F: PerFHalOrg.4-166
$C_{12}F_{10}HN$ .... $(C_6F_5)_2NH$ .... F: PerFHalOrg.7-14/5, 42, 60
$C_{12}F_{10}HOP$ .... $(C_6F_5)_2P(OH)$ .... F: PerFHalOrg.3-46, 48
$C_{12}F_{10}HOTl$ .... $(C_6F_5)_2TlOH$ .... F: PerFHalOrg.4-131/3
$C_{12}F_{10}HP$ .... $(C_6F_5)_2PH$ .... F: PerFHalOrg.3-15, 19, 24/7
$C_{12}F_{10}H_2NP$ .... $(C_6F_5)_2PNH_2$ .... F: PerFHalOrg.3-60, 67/9
$C_{12}F_{10}H_2O_2Si$ .... $(C_6F_5)_2Si(OH)_2$ .... F: PerFHalOrg.4-147/8, 154
$C_{12}F_{10}H_2Si$ .... $(C_6F_5)_2SiH_2$ .... F: PerFHalOrg.4-147, 154
$C_{12}F_{10}H_9N$ .... $CH_3C_6H_4C(C_2F_5)_2NH_2$ .... F: PerFHalOrg.7-69, 72, 75
$C_{12}F_{10}H_{18}Sn$ .... $(C_4H_9)_2Sn(C_2F_5)_2$ .... Sn: Org.Verb.3-43
$C_{12}F_{10}Hg$ .... $(C_6F_5)_2Hg$ .... F: PerFHalOrg.4-34, 110, 111, 136
$C_{12}F_{10}IIn$ .... $(C_6F_5)_2InI$ .... F: PerFHalOrg.4-128
$C_{12}F_{10}ITl$ .... $(C_6F_5)_2TlI$ .... F: PerFHalOrg.4-131/2, 135/7
$C_{12}F_{10}LiP$ .... $(C_6F_5)_2PLi$ .... F: PerFHalOrg.3-15, 24
$C_{12}F_{10}Mg$ .... $C_6F_5C_6F_4MgF$ .... F: PerFHalOrg.4-63, 65, 70, 71
– .... $(C_6F_5)_2Mg$ .... F: PerFHalOrg.4-67/8, 70, 86, 104
$C_{12}F_{10}NO_3Tl$ .... $(C_6F_5)_2TlNO_3$ .... F: PerFHalOrg.4-131/3, 136/8
$C_{12}F_{10}N_2$ .... $N(CF)_5CC(CF)_5N$ .... F: PerFHalOrg.6-165
$C_{12}F_{10}N_2O_2$ .... $(C_6F_5NO)_2$ .... F: PerFHalOrg.7-181, 188
$C_{12}F_{10}OSi$ .... $[(C_6F_5)_2SiO]_x$ .... F: PerFHalOrg.4-147/8, 154, 161
$C_{12}F_{10}OSn$ .... $(C_6F_5)_2SnO$ .... F: PerFHalOrg.4-181, 185, 188
$C_{12}F_{10}P^+$ .... $(C_6F_5)_2P^+$ .... F: PerFHalOrg.3-172, 179
$C_{12}F_{10}Si^+$ .... $(C_6F_5)_2Si^+$ .... F: PerFHalOrg.4-152/3
$C_{12}F_{10}Zn$ .... $(C_6F_5)_2Zn$ .... F: PerFHalOrg.4-112, 136
$C_{12}F_{11}Ge^+$ .... $(C_6F_5)_2GeF^+$ .... F: PerFHalOrg.4-167, 174
$C_{12}F_{11}OP$ .... $(C_6F_5)_2P(O)F$ .... F: PerFHalOrg.3-52, 55
$C_{12}F_{11}P$ .... $(C_6F_5)_2PF$ .... F: PerFHalOrg.3-123, 125/6
$C_{12}F_{11}PS$ .... $(C_6F_5)_2P(S)F$ .... F: PerFHalOrg.3-141, 148
$C_{12}F_{11}Sn^+$ .... $(C_6F_5)_2SnF^+$ .... F: PerFHalOrg.4-184/5
Sn: Org.Verb.1-168
$C_{12}F_{11}Tl$ .... $(C_6F_5)_2TlF$ .... F: PerFHalOrg.4-131/2, 134, 136
$C_{12}F_{12}FeH_{18}N_6P_2$ .... $[(CH_3NC)_6Fe][PF_6]_2$ .... Fe: Org.Verb.B4-98, 103

$C_{12}F_{12}FeH_{21}N_7P_2$ .... $[NH_2(CH_3NH)CFe(CNCH_3)_5][PF_6]_2$ .... Fe: Org.Verb.B4-173/4
$C_{12}F_{12}FeH_{22}N_8P_2$ .... $[(NHCNHCH_3)_2Fe(CNCH_3)_4][PF_6]_2$ .... Fe: Org.Verb.B4-353/4
$C_{12}F_{12}FeH_{26}N_{10}P_2$ ... $[((NHCNHCH_3)_2)_2Fe(CNCH_3)_2][PF_6]_2$ ...... Fe: Org.Verb.B4-359
$C_{12}F_{12}FeO_4$ .... $(CF_3C)_4Fe(CO)_4$ .... Fe: Org.Verb.B4-338, 348
Fe: Org.Verb.C1-86
$C_{12}F_{12}Fe_2O_6S_2$ ..... $(CO)_3Fe(CF_3CCCF_3)(SCF_3)_2Fe(CO)_3$ ...... Fe: Org.Verb.C2-67/8, 70
$C_{12}F_{12}Ge$ .... $(CF_3CC)_4Ge$ .... F: PerFHalOrg.4-168/9, 177, 179
– .... $(C_6F_5)_2GeF_2$ .... F: PerFHalOrg.4-166/7, 176
$C_{12}F_{12}H_4N_2O_4$ ...... $[(CF_3)_2NOC(O)]_2C_6H_4$ .... F: PerFHalOrg.7-132, 136
$C_{12}F_{12}H_8N_2O$ .... $(NH_2C(CF_3)_2)_2C_6H_3OH$ .... F: PerFHalOrg.7-69, 71
$C_{12}F_{12}H_8N_2O_2$ ...... $C_6H_5CH(ON(CF_3)_2)CH_2ON(CF_3)_2$ .... F: PerFHalOrg.7-136
$C_{12}F_{12}H_{10}N_4O_2$ ..... $C_3HN_2(CF_3)_4NHCON(CH_2CH_2)_2O$ .... F: PerFHalOrg.5-108/9
$C_{12}F_{12}H_{12}N_4$ .... $C_3HN_2(CF_3)_4(NHCH_2N(CH_2)_4)$ .... F: PerFHalOrg.5-106
$C_{12}F_{12}H_{12}N_4O$ ...... $C_3HN_2(CF_3)_4(NHCH_2N(CH_2CH_2)_2O)$ .... F: PerFHalOrg.5-106
$C_{12}F_{12}H_{12}N_4O_2$ ..... $C_3HN_2(CF_3)_4NHCOOC_2H_4N(CH_3)_2$ .... F: PerFHalOrg.5-108
$C_{12}F_{12}H_{13}N_5$ .... $(CF_3)_2CNHC(CF_3)_2NC(NC(N(CH_3)_2)_2)$ ...... F: PerFHalOrg.5-109/10
$C_{12}F_{12}H_{14}N_2$ .... $[NH_2C(CF_3)_2CH_2CH]_2C_2H_4$ .... F: PerFHalOrg.7-65
$C_{12}F_{12}H_{16}Sn$ .... $Sn(CH_2CH_2CF_3)_4$ .... Sn: Org.Verb.1-123
$C_{12}F_{12}H_{18}N_2SiSn$ .... $(CH_3)_2Sn[NC(CH_3)(CF_3)_2]_2Si(CH_3)_2$ .... Sn: Org.Verb.6-32
$C_{12}F_{12}Mn_2O_8P_2$ ..... $Mn_2(CO)_8[P(CF_3)_2]_2$ .... F: PerFHalOrg.3-181
$C_{12}F_{12}N_2$ .... $CF_3CF(C_5F_4N)_2$ .... F: PerFHalOrg.5-156/7, 169
$C_{12}F_{12}N_6$ .... $[C_3N_3(CF_2)_6C_3N_3]_n$ .... F: PerFHalOrg.6-119
$C_{12}F_{12}Si$ .... $(CF_3CC)_4Si$ .... F: PerFHalOrg.4-88, 151, 158
– .... $(C_6F_5)_2SiF_2$ .... F: PerFHalOrg.4-148/9, 155
$C_{12}F_{13}N$ .... $(i\text{-}C_3F_7)C_9F_6N$ .... F: PerFHalOrg.6-131/2, 143
$C_{12}F_{13}P$ .... $(C_6F_5)_2PF_3$ .... F: PerFHalOrg.3-123, 126
$C_{12}F_{14}$ .... $C_6F_5C(CF_2)_4CF$ .... F: PerFHalOrg.4-16
$C_{12}F_{14}Fe_2I_2O_6$ ...... $C_3F_7(CO)_3FeI_2Fe(CO)_3C_3F_7$ .... Fe: Org.Verb.C1-222
$C_{12}F_{14}H_2N_2O_2$ ...... $(CF_3)_3(C_2F_5)C_7H_2N_2O_2$ .... F: PerFHalOrg.6-129/30, 141
$C_{12}F_{14}H_3NO$ .... $(CH_3O)(C_6F_{11})C_5F_3N$ .... F: PerFHalOrg.5-178
$C_{12}F_{14}H_5P$ .... $(C_3F_7)_2PC_6H_5$ .... F: PerFHalOrg.3-101, 106
$C_{12}F_{14}H_8NO_2P$ ...... $(C_3F_7)_2P(O)OH \cdot C_6H_5NH_2$ .... F: PerFHalOrg.3-50
$C_{12}F_{14}N_4$ .... $NC(CN)NC(i\text{-}C_3F_7)C(CN)C(i\text{-}C_3F_7)$ .... F: PerFHalOrg.6-27, 47
$C_{12}F_{14}N_6$ .... $[CF_2C_3N_3(CF_3)C_2F_4C_3N_3(CF_3)CF_2]_n$ .... F: PerFHalOrg.6-120/1
$C_{12}F_{14}N_6O$ .... $[(CF_2C_3N_3(CF_3)CF_2)_2O]_n$ .... F: PerFHalOrg.6-120/1
$C_{12}F_{14}N_6S$ .... $[(CF_2C_3N_3(CF_3)CF_2)_2S]_n$ .... F: PerFHalOrg.6-120/1
$C_{12}F_{15}FeNaO_5$ ...... $Na[C_7F_{15}C(O)Fe(CO)_4]$ .... Fe: Org.Verb.B4-151, 160
$C_{12}F_{15}HN_2O$ .... $(CF_3)_3(C_2F_5)C_7FHN_2O$ .... F: PerFHalOrg.6-129/30, 141
$C_{12}F_{15}H_7N_2O_2$ ...... $C_7F_{15}C(NH_2)NOC(O)C(CH_3)CH_2$ .... F: PerFHalOrg.7-58, 76
$C_{12}F_{15}N_3$ .... $(CF_2CFCF_2)_3C_3N_3$ .... F: PerFHalOrg.6-84/6, 104
$C_{12}F_{16}H_3NO$ .... $[(CF_3)_2CF]_2(CH_3O)C_5F_2N$ .... F: PerFHalOrg.5-177/8
$C_{12}F_{16}N_2$ .... $[(CF_3)_2CF]_2(CN)C_5F_2N$ .... F: PerFHalOrg.5-197, 210
– .... $(CF_3)_3(C_2F_5)C_7F_2N_2$ .... F: PerFHalOrg.6-128/30, 141
– .... $C_4F_2N_2(C(CF_3)CFCF_3)_2$ .... F: PerFHalOrg.6-24, 46
$C_{12}F_{16}N_6$ .... $[(CF_3)_2C_3N_3]_2C_2F_4$ .... F: PerFHalOrg.6-89, 108

Formula	Structure		Reference
$C_{12}F_{17}N_3$	$(CF_2CFCF_2)_2(C_3F_7)C_3N_3$	F:	PerFHalOrg.6-84/6, 104
$C_{12}F_{18}H_3NO_7$	$(HOCOC_3F_6)_2NOC_3F_6COOH$	F:	PerFHalOrg.7-97, 115
$C_{12}F_{18}H_3N_3O_3$	$[(CF_3)_2NO]_3C_6H_3$	F:	PerFHalOrg.7-136
$C_{12}F_{18}H_6N_4O_4$	$[H_2NC(O)(CF_2)_3]_2NO(CF_2)_3C(O)NH_2$	F:	PerFHalOrg.7-18, 20, 48
$C_{12}F_{18}N_3$	$[(CF_2)_6C_3N_3(CF_2)_3]_x$	F:	PerFHalOrg.6-119
$C_{12}F_{18}N_4O$	$(NCC_3F_6)_2NOC_3F_6CN$	F:	PerFHalOrg.7-97, 115
$C_{12}F_{18}N_4O_2$	$C_2F_4(COC(C_3F_7)NN)_2$	F:	PerFHalOrg.5-69/70, 90
$C_{12}F_{18}Na_2O_{12}Te$	$Na_2Te(CF_3COO)_6$	Te:	SVol.B1-98
$C_{12}F_{18}P_2$	$C_6P_2(CF_3)_6$	F:	PerFHalOrg.3-9, 12
$C_{12}F_{19}H_4N_5$	$(C_9F_{19})(NH_2)_2C_3N_3$	F:	PerFHalOrg.6-79/80
$C_{12}F_{19}H_9OSi$	$(CH_3)_3Si(CF_2)_6OCF(CF_3)_2$	F:	PerFHalOrg.4-87
$C_{12}F_{19}HgN_3$	$[HgC_2F_3(CF_3)C_3N_3(C_3F_7)C_2F_3(CF_3)]_n$	F:	PerFHalOrg.6-85/6
$C_{12}F_{19}I_2N_3$	$(CF_2IC_2F_4)_2(C_3F_7)C_3N_3$	F:	PerFHalOrg.6-87/8, 107
–	$(CF_3CFICF_2)_2(C_3F_7)C_3N_3$	F:	PerFHalOrg.6-84/6, 105
$C_{12}F_{19}I_2N_3O$	$(CF_2IC_2F_4)_2(CF_3OC_2F_4)C_3N_3$	F:	PerFHalOrg.6-87/8, 107
$C_{12}F_{19}N$	$[(CF_3)_2CF](C_2F_5)_2C_5F_2N$	F:	PerFHalOrg.5-153/4, 164/5, 176
–	$[(CF_3)_2CF]_2(CF_3)C_5F_2N$	F:	PerFHalOrg.5-153/4, 164
$C_{12}F_{19}N_3$	$[(CF_2)_6C_3N_3(C_3F_7)]_x$	F:	PerFHalOrg.6-119
–	$[(CF_2)_8C_3N_3(CF_3)]_x$	F:	PerFHalOrg.6-118/20
–	$(C_3F_7)_2(CF_2CFCF_2)C_3N_3$	F:	PerFHalOrg.6-84/6
$C_{12}F_{20}H_3NO_2$	$C_5F_{11}NC_5F_9(COOCH_3)$	F:	PerFHalOrg.5-217
$C_{12}F_{20}N_2$	$(s\text{-}C_4F_9)_2C_4F_2N_2$	F:	PerFHalOrg.6-32/3, 53, 55/6
–	$N(CC_2F_5)_4N$	F:	PerFHalOrg.6-15, 36
–	$NC(s\text{-}C_4F_9)CFC(s\text{-}C_4F_9)CFN$	F:	PerFHalOrg.6-18, 38
–	$NC(t\text{-}C_4F_9)CFC(t\text{-}C_4F_9)CFN$	F:	PerFHalOrg.6-17/8, 38
–	$NC(t\text{-}C_4F_9)CFCFC(t\text{-}C_4F_9)N$	F:	PerFHalOrg.6-18, 38
–	$NC(s\text{-}C_4F_9)CFNC(s\text{-}C_4F_9)CF$	F:	PerFHalOrg.6-30/2
–	$NC(s\text{-}C_4F_9)CFNCFC(s\text{-}C_4F_9)$	F:	PerFHalOrg.6-30/2
–	$NCF(CC_4F_9\text{-}s)_2CFN$	F:	PerFHalOrg.6-18, 38, 55/6
–	$NCFNC(s\text{-}C_4F_9)CFC(s\text{-}C_4F_9)$	F:	PerFHalOrg.6-24, 46
$C_{12}F_{21}H_7Si$	$(CH_3)_2SiHC_{10}F_{21}$	F:	PerFHalOrg.4-86
$C_{12}F_{21}N_3$	$(n\text{-}C_3F_7)_3C_3N_3$	F:	PerFHalOrg.6-83/4, 96, 103, 114/5
–	$(i\text{-}C_3F_7)_3C_3N_3$	F:	PerFHalOrg.6-83/4, 103
$C_{12}F_{21}N_3O$	$(C_3F_7)_2(CF_3OC_2F_4)C_3N_3$	F:	PerFHalOrg.6-86/8, 106
$C_{12}F_{21}N_3O_2$	$(CF_3OC_2F_4)_2(C_3F_7)C_3N_3$	F:	PerFHalOrg.6-86/8, 105
$C_{12}F_{21}N_3O_3$	$(CF_3OC_2F_4)_3C_3N_3$	F:	PerFHalOrg.6-86/8, 106
$C_{12}F_{22}H_3N_3$	$C_7F_{15}C(NH)NC(NH_2)C_3F_7$	F:	PerFHalOrg.7-7
$C_{12}F_{23}N$	$CF_2C(C_8F_{17})NCF_2CF_2$	F:	PerFHalOrg.5-53/4
–	$CF_2CF(C_8F_{17})NCFCF_2$	F:	PerFHalOrg.5-53/4
$C_{12}F_{24}K_2MnO_4$	$K_2[Mn(OC(CF_3)_2C(CF_3)_2O)_2]$	Mn:	MVol.D1-37
$C_{12}F_{24}N_4S$	$[(CF_3)_2CNC(CF_3)_2N]_2S$	S:	SVol.2-91
$C_{12}F_{24}Ni_2O_6P_4$	$(CO)_2Ni[(CF_3)_2POP(CF_3)_2]_2Ni(CO)_2$	Ni:	Org.Verb.2-260
$C_{12}F_{25}N$	$CF_2CF(C_8F_{17})NFCF_2CF_2$	F:	PerFHalOrg.5-53/4
–	$C_7F_{15}N(CF_2)_5$	F:	PerFHalOrg.5-157/8, 172, 175
$C_{12}F_{25}NO$	$O(CF_2CF_2)_2NCF_2(CF_2)_6CF_3$	F:	PerFHalOrg.6-2, 7

$C_{12}F_{26}HOP$ .... $(C_6F_{13})_2POH$ .... F: PerFHalOrg.3-44, 46
$C_{12}F_{26}HO_2P$ .... $(C_6F_{13})_2P(O)OH$ .... F: PerFHalOrg.3-43, 46, 50
$C_{12}F_{26}IP$ .... $(C_6F_{13})_2PI$ .... F: PerFHalOrg.3-76, 95
$C_{12}F_{26}N_2$ .... $(n\text{-}C_4F_9)N(CF_2CF_2)_2N(C_4F_9\text{-}n)$ .... F: PerFHalOrg.6-31/2, 52
$C_{12}F_{28}P_4$ .... $(C_3F_7P)_4$ .... F: PerFHalOrg.3-1/3
$C_{12}F_{30}N_4O_4$ .... $[(CF_3)_2NO]_4C_4F_6$ .... F: PerFHalOrg.7-95/6, 114
$C_{12}F_{36}N_9O_6P_3$ .... $[((CF_3)_2NO)_2PN]_3$ .... F: PerFHalOrg.7-101, 119
$C_{12}FeH_4NaO_7^-$ .... $[NaOC(O)C_6H_4C(O)Fe(CO)_4]^-$ .... Fe: Org.Verb.B4-154, 162
$C_{12}FeH_4Na_2O_7$ .... $Na[NaOC(O)C_6H_4C(O)Fe(CO)_4]$ .... Fe: Org.Verb.B4-154, 162
$C_{12}FeH_4O_6$ .... $C_6H_4(CO)_2Fe(CO)_4$ .... Fe: Org.Verb.B4-335, 343
$C_{12}FeH_4O_{12}Ru_3$ .... $FeRu_3(CO)_{12}H_4$ .... Fe: Org.Verb.B2-197
$C_{12}FeH_5MoO_7$ .... $C_5H_5Mo(CO)_3Fe(CO)_4$ .... Fe: Org.Verb.C2-49
$C_{12}FeH_5N_5O^{2-}$ .... $[Fe(CN)_4(CCC_6H_5)NO]^{2-}$ .... Fe: Org.Verb.B1-17
$C_{12}FeH_5O_7W$ .... $C_5H_5W(CO)_3Fe(CO)_4$ .... Fe: Org.Verb.C2-49
$C_{12}FeH_6K_3$ .... $K_3[Fe(CCH)_6]$ .... Fe: Org.Verb.B1-16/7
$C_{12}FeH_6K_4$ .... $K_4[Fe(CCH)_6]$ .... Fe: Org.Verb.B1-16/7
$C_{12}FeH_6N_2O_4$ .... $(CO)_4Fe(C_6H_4CHNNCH)$ .... Fe: Org.Verb.B2-71, 76, 80
$C_{12}FeH_6Na_4$ .... $Na_4[Fe(CCH)_6]$ .... Fe: Org.Verb.B1-16/7
$C_{12}FeH_6O_6$ .... $(C(O)C(CCCH_3)C(CH_3)CO)Fe(CO)_4$ .... Fe: Org.Verb.B4-334, 341/2
– .... $C_8H_6O_2Fe(CO)_4$ .... Fe: Org.Verb.B4-339, 350
$C_{12}FeH_6O_6S$ .... $(CHC_6H_4SO_2CH)Fe(CO)_4$ .... Fe: Org.Verb.B4-304, 309
$C_{12}FeH_7LiO_5$ .... $Li[CH_3C_6H_4C(O)Fe(CO)_4]$ .... Fe: Org.Verb.B4-152/3, 162
– .... $Li[C_6H_5CH_2C(O)Fe(CO)_4]$ .... Fe: Org.Verb.B4-152
$C_{12}FeH_7LiO_6$ .... $Li[CH_3OC_6H_4C(O)Fe(CO)_4]$ .... Fe: Org.Verb.B4-153, 162
$C_{12}FeH_7NO_4$ .... $CH_3C_6H_4NCFe(CO)_4$ .... Fe: Org.Verb.B4-6, 8
$C_{12}FeH_7NO_4S$ .... $(SC_6H_4N(CH_3)C)Fe(CO)_4$ .... Fe: Org.Verb.B4-128, 130, 134
$C_{12}FeH_7NO_5$ .... $CH_3OC_6H_4NCFe(CO)_4$ .... Fe: Org.Verb.B4-6, 8
$C_{12}FeH_7NO_6$ .... $(O_2NC_6H_4CHCHCHO)Fe(CO)_3$ .... Fe: Org.Verb.B5-122/4, 130
$C_{12}FeH_7N_6$ .... $[C_6H_5NCFe(CN)_5]H_2$ .... Fe: Org.Verb.B4-10
$C_{12}FeH_7NaO_5$ .... $Na[CH_3C_6H_4C(O)Fe(CO)_4]$ .... Fe: Org.Verb.B4-152/3, 162
– .... $Na[C_6H_5CH_2C(O)Fe(CO)_4]$ .... Fe: Org.Verb.B4-152, 162
$C_{12}FeH_7NaO_6$ .... $Na[CH_3OC_6H_4C(O)Fe(CO)_4]$ .... Fe: Org.Verb.B4-153, 162
$C_{12}FeH_7O^+$ .... $[FeC_{11}H_7CO]^+$ .... Fe: Org.Verb.A3-126
$C_{12}FeH_7O_5^-$ .... $[CH_3C_6H_4C(O)Fe(CO)_4]^-$ .... Fe: Org.Verb.B4-152/3, 157, 162, 169
– .... $[C_6H_5CH_2C(O)Fe(CO)_4]^-$ .... Fe: Org.Verb.B4-152, 162
$C_{12}FeH_7O_5Rh$ .... $(CO)_2RhC_7H_7Fe(CO)_3$ .... Fe: Org.Verb.B5-152/5
$C_{12}FeH_7O_6^-$ .... $[CH_3OC_6H_4C(O)Fe(CO)_4]^-$ .... Fe: Org.Verb.B4-153, 162
$C_{12}FeH_8NO_5^+$ .... $[C_5H_5NCOCHCH_2Fe(CO)_4]^+$ .... Fe: Org.Verb.B4-237, 249
$C_{12}FeH_8NO_5^-$ .... $[CH_3C_6H_4NHC(O)Fe(CO)_4]^-$ .... Fe: Org.Verb.B4-156, 165/6
$C_{12}FeH_8N_2$ .... $Fe(C_5H_4CN)_2$ .... Fe: Org.Verb.A1-186/7
$C_{12}FeH_8N_4O_2$ .... $Fe(NO)_2(C_{12}H_8N_2)$ .... Fe: Org.Verb.B1-142
$C_{12}FeH_8O_4$ .... $(CH_2CHC(C_6H_5)O)Fe(CO)_3$ .... Fe: Org.Verb.B5-131
– .... $(C_6H_5CHCHCHO)Fe(CO)_3$ .... Fe: Org.Verb.B5-122/6, 129/30, 138/40
– .... $C_6H_5CHCH_2Fe(CO)_4$ .... Fe: Org.Verb.B4-238, 252/4
– .... 1) $C_8H_8Fe(CO)_4$ .... Fe: Org.Verb.B4-303, 306
2) $C_8H_8Fe(CO)_4$ .... Fe: Org.Verb.B4-304, 311/2
$C_{12}FeH_8O_5$ .... $(C_6H_5)(CH_3O)CFe(CO)_4$ .... Fe: Org.Verb.B4-121/2, 124

$C_{12}FeH_8O_5$ .... $C_7H_8C(O)Fe(CO)_4$ .... Fe: Org.Verb.B4-336, 344
– .... $C_8H_8OFe(CO)_4$ .... Fe: Org.Verb.B4-302/3, 308
$C_{12}FeH_8O_6$ .... $(CH_3)_2C_6H_2O_2Fe(CO)_4$ .... Fe: Org.Verb.B4-335, 343
$C_{12}FeH_9^+$ .... $[FeC_5H_4C_7H_5]^+$ .... Fe: Org.Verb.A3-127
– .... $[FeC_{12}H_9]^+$ .... Fe: Org.Verb.A2-38
– .... $[FeC_{12}H_8D]^+$ .... Fe: Org.Verb.A2-38
$C_{12}FeH_9Li$ .... $C_5H_5FeC_5H_4CCLi$ .... Fe: Org.Verb.A1-302
$C_{12}FeH_9Li_2N$ .... $C_5H_5FeC_5H_4CLi_2CN$ .... Fe: Org.Verb.A3-23
$C_{12}FeH_9NO_2$ .... $NCC_5H_4FeC_5H_4COOH$ .... Fe: Org.Verb.A3-58/9
$C_{12}FeH_9NO_6$ .... $C_6H_5C_3H_4Fe(CO)_3ONO_2$ .... Fe: Org.Verb.B5-52, 62
– .... $HOC(O)CHCH_2Fe(CO)_4 \cdot C_5H_5N$ .... Fe: Org.Verb.B4-245/6
$C_{12}FeH_9N_3O_3$ .... $C_6H_4N_3CH_2CHCH_2Fe(CO)_3$ .... Fe: Org.Verb.B4-216/7, 225/6
$C_{12}FeH_9NaO_5$ .... $Na[CH_3OC(O)C_7H_6Fe(CO)_3]$ .... Fe: Org.Verb.B5-77, 79
$C_{12}FeH_9O^+$ .... $[FeC_5H_4COC_6H_5]^+$ .... Fe: Org.Verb.A2-278
– .... $[FeHC_5H_4C_6H_4CO]^+$ .... Fe: Org.Verb.A3-126
$C_{12}FeH_9O_2^+$ .... $[FeC_5H_4C_6H_4COOH]^+$ .... Fe: Org.Verb.A3-80, 126
$C_{12}FeH_9O_3$ .... $[C_6H_5C_3H_4Fe(CO)_3]$ .... Fe: Org.Verb.B5-69/70
$C_{12}FeH_9O_4^-$ .... $[C_6H_5CH_2CH_2C(O)Fe(CO)_3]^-$ .... Fe: Org.Verb.B4-109
$C_{12}FeH_9O_5^-$ .... $[CH_3OC(O)C_7H_6Fe(CO)_3]^-$ .... Fe: Org.Verb.B5-74, 77/9
$C_{12}FeH_{10}$ .... $[C(C_5H_5FeC_5H_4)CH]_n$ .... Fe: Org.Verb.A2-220
– .... $C_5H_5FeC_5H_4CCH$ .... Fe: Org.Verb.A1-297/9
– .... $(C_6H_5)_2Fe \cdot 4\ C_4H_8O$ .... Fe: Org.Verb.B1-13
$C_{12}FeH_{10}^+$ .... $[C_5H_5FeC_5H_4CCH]^+$ .... Fe: Org.Verb.A3-79
– .... $[C_5H_5FeC_7H_5]^+$ .... Fe: Org.Verb.A1-384
$C_{12}FeH_{10}I_2N_2O_2$ .... $(CO)_2Fe(NC_5H_5)_2I_2$ .... Fe: Org.Verb.B1-101/3
$C_{12}FeH_{10}N_4O_4$ .... $Fe(OC(OC_2H_5)C(CN)_2)_2$ .... Fe: Org.Verb.B3-189
$C_{12}FeH_{10}NiO_2^+$ .... $(C_5H_5)_2NiFe(CO)_2^+$ .... Ni: Org.Verb.2-351
$C_{12}FeH_{10}Ni_2O_2^+$ .... $(C_5H_5)_2Ni_2Fe(CO)_2^+$ .... Ni: Org.Verb.2-351
$C_{12}FeH_{10}O_3$ .... $C_5H_5FeC_5H_4COCOOH$ .... Fe: Org.Verb.A3-87/8
– .... $C_9H_{10}Fe(CO)_3$ .... Fe: Org.Verb.B5-145
$C_{12}FeH_{10}O_4$ .... $C_5H_5FeC_5H_3(COOH)_2$ .... Fe: Org.Verb.A6-212
– .... $C_8H_{10}Fe(CO)_4$ .... Fe: Org.Verb.B4-303, 306
– .... $Fe(C_5H_4COOH)_2$ .... Fe: Org.Verb.A3-51
$C_{12}FeH_{10}O_5$ .... $CH_3OC(O)C_7H_7Fe(CO)_3$ .... Fe: Org.Verb.B5-77, 78/9
– .... $C_8H_{10}OFe(CO)_4$ .... Fe: Org.Verb.B4-339, 349
$C_{12}FeH_{10}O_6$ .... $(C(C_2H_5)CO)_2Fe(CO)_4$ .... Fe: Org.Verb.B4-334
– .... $(CH_3COO)(C_2H_3CO)C_4H_4Fe(CO)_3$ .... Fe: Org.Verb.B5-76
– .... $(C(O)C(C_4H_9)CHCO)Fe(CO)_4$ .... Fe: Org.Verb.B4-334
$C_{12}FeH_{10}O_8$ .... $(CH_3OCO)_2C_3H_2CH_2Fe(CO)_4$ .... Fe: Org.Verb.B4-259, 260/3
$C_{12}FeH_{11}^+$ .... $[C_5H_5FeC_5H_4CCH_2]^+$ .... Fe: Org.Verb.A1-349
Fe: Org.Verb.A6-61
– .... $[FeC_5H_4CH_2C_6H_5]^+$ .... Fe: Org.Verb.A1-306
$C_{12}FeH_{11}IO$ .... $C_5H_5FeC_5H_4COCH_2I$ .... Fe: Org.Verb.A2-243, 245
$C_{12}FeH_{11}I_3O$ .... $C_5H_5FeC_5H_4CHOHCI_3$ .... Fe: Org.Verb.A2-45
$C_{12}FeH_{11}NO$ .... $C_5H_5FeC_5H_4CHOHCN$ .... Fe: Org.Verb.A2-169
$C_{12}FeH_{11}NO_2$ .... $C_5H_5FeC_5H_4CHCHNO_2$ .... Fe: Org.Verb.A2-169
$C_{12}FeH_{11}NiO_5P$ .... $(CO)_3Ni(P(CH_3)_2(CO)_2FeC_5H_5)$ .... Ni: Org.Verb.1-165, 167
$C_{12}FeH_{11}O^+$ .... $[FeC_5H_4CHOHC_6H_5]^+$ .... Fe: Org.Verb.A2-38
– .... $[FeC_5H_4CDOHC_6H_5]^+$ .... Fe: Org.Verb.A2-38

Formula	Compound	Reference
$C_{12}FeH_{11}O_2$	$[C_5H_5FeC_5H_4CH_2COO]$	Fe: Org.Verb.A3-67
$C_{12}FeH_{11}O_4P$	$(CO)_4FeP(CH_3)_2C_6H_5$	Fe: Org.Verb.B2-84/5, 89, 111
$C_{12}FeH_{12}$	$C_5H_5FeC_5H_4CCH_3$	Fe: Org.Verb.A1-344
–	$C_5H_5FeC_5H_4CHCH_2$	Fe: Org.Verb.A1-279/80
$C_{12}FeH_{12}^+$	$[C_5H_5FeC_5H_4CHCH_2]^+$	Fe: Org.Verb.A1-286
		Fe: Org.Verb.A2-71
$C_{12}FeH_{12}HgO_2$	$C_5H_5FeC_5H_4Hg(OOCCH_3)$	Fe: Org.Verb.A1-174
$C_{12}FeH_{12}I_5O$	$[C_5H_5FeC_5H_4COCH_3]I_5$	Fe: Org.Verb.A2-300
$C_{12}FeH_{12}NNaO_3$	$[C_5H_5FeC_5H_4CHOHCHNO_2]Na$	Fe: Org.Verb.A2-169
$C_{12}FeH_{12}NNaO_5$	$Na[NC(CH_2)_6C(O)Fe(CO)_4]$	Fe: Org.Verb.B4-156, 166
$C_{12}FeH_{12}NO_5^-$	$[NC(CH_2)_6C(O)Fe(CO)_4]^-$	Fe: Org.Verb.B4-156, 166
$C_{12}FeH_{12}N_2$	$C_5H_5FeC_5H_4CH(CN)NH_2$	Fe: Org.Verb.A2-169
–	$(NH_2C_6H_4)_2Fe$	Fe: Org.Verb.B1-13
$C_{12}FeH_{12}O$	$[C(C_5H_5FeC_5H_4)(OH)CH_2]_n$	Fe: Org.Verb.A2-228
–	$C_5H_5FeC_5H_4CH_2CHO$	Fe: Org.Verb.A2-173/4
–	$C_5H_5FeC_5H_4COCH_3$	
	Catalysis	Fe: Org.Verb.A2-230/1
	Chemical reactions	Fe: Org.Verb.A2-184/9, 205/30
	Chromatography	Fe: Org.Verb.A2-230
	Electrochemical behavior	Fe: Org.Verb.A2-204/5
	Formation and preparation	Fe: Org.Verb.A2-181, 193/6
	Physical properties	Fe: Org.Verb.A2-182/4, 196/204
	Separation	Fe: Org.Verb.A2-230
	Solubility	Fe: Org.Verb.A2-201/4
	Uses	Fe: Org.Verb.A2-230/1
–	$C_5H_5FeC_5H_4COCD_3$	Fe: Org.Verb.A2-194
–	$C_5H_5{}^{59}FeC_5H_4COCH_3$	Fe: Org.Verb.A2-209
–	$C_5H_5FeC_5H_4C_2H_3O$	Fe: Org.Verb.A3-172/3
$C_{12}FeH_{12}O^+$	$[C_5H_5FeC_5H_4COCH_3]^+$	Fe: Org.Verb.A2-194, 204/6, 208, 300
–	$[C_5H_5FeC_6H_4OCH_3]^+$	Fe: Org.Verb.A2-296
$C_{12}FeH_{12}O_2$	$C_5H_5FeC_5H_4CH_2COOH$	Fe: Org.Verb.A3-66/8, 101
–	$C_5H_5FeC_5H_4COOCH_3$	Fe: Org.Verb.A3-108/11
–	$C_5H_5FeC_5H_4COOCD_3$	Fe: Org.Verb.A3-111
–	$C_5H_5{}^{59}FeC_5H_4COOCH_3$	Fe: Org.Verb.A3-110
–	$C_5H_5FeC_5H_4OOCCH_3$	Fe: Org.Verb.A2-94
$C_{12}FeH_{12}O_2^+$	$[C_5H_5FeC_5H_4CH_2COOH]^+$	Fe: Org.Verb.A3-66
–	$[C_5H_5FeC_5H_4COOCH_3]^+$	Fe: Org.Verb.A3-109
–	$[C_5H_5FeC_5H_4OOCCH_3]^+$	Fe: Org.Verb.A2-94
$C_{12}FeH_{12}O_3$	$CH_3OC_5H_4FeC_5H_4COOH$	Fe: Org.Verb.A3-58/9
–	$(CH_3)_2C_7H_6Fe(CO)_3$	Fe: Org.Verb.B5-144
–	$C_5H_5FeC_5H_3(OCH_3)COOH$	Fe: Org.Verb.A3-59
–	$C_5H_5FeC_5H_4C_2H_3O_3$	Fe: Org.Verb.A3-179
–	$C_5H_5FeC_5H_4OCH_2COOH$	Fe: Org.Verb.A3-84/6
$C_{12}FeH_{12}O_4$	$C_7H_{10}CH_2Fe(CO)_4$	Fe: Org.Verb.B4-259
–	$C_8H_{12}Fe(CO)_4$	Fe: Org.Verb.B4-302, 305, 313/4

Formula	Compound	Reference
$C_{12}FeH_{12}O_4S$	$CH_3COC_5H_4FeC_5H_4SO_3H$	Fe: Org.Verb.A2-211
$C_{12}FeH_{12}O_8$	$(CH_2CHCOOCH_3)_2Fe(CO)_4$	Fe: Org.Verb.B4-338, 347/8
–	$(C_2H_5OCO)_2C_2H_2Fe(CO)_4$	Fe: Org.Verb.B4-270, 286/7
$C_{12}FeH_{13}{}^+$	$[C_5H_5FeC_5H_4CHCH_3]^+$	Fe: Org.Verb.A1-349/53, 357/9, 362, 382
–	$[C_5H_5FeC_5H_4CH_2CH_2]^+$	Fe: Org.Verb.A1-349 Fe: Org.Verb.A2-79
–	$[C_6H_6FeC_5H_4CH_3]^+$	Fe: Org.Verb.A2-150
$C_{12}FeH_{13}N$	$C_5H_5FeC_5H_4C(CH_3)NH$	Fe: Org.Verb.A1-287
$C_{12}FeH_{13}NO$	$C_5H_5FeC_5H_4C(CH_3)NOH$	Fe: Org.Verb.A2-220
–	$C_5H_5FeC_5H_4CONHCH_3$	Fe: Org.Verb.A2-228
–	$C_5H_5FeC_5H_4NHCOCH_3$	Fe: Org.Verb.A2-228
$C_{12}FeH_{13}NO_2$	$C_5H_5FeC_5H_4CH_2CONHOH$	Fe: Org.Verb.A3-116
$C_{12}FeH_{13}NO_3$	$C_5H_5FeC_5H_4CHOHCH_2NO_2$	Fe: Org.Verb.A2-169
$C_{12}FeH_{13}NO_5$	$(CH_2)_5NCOCHCH_2Fe(CO)_4$	Fe: Org.Verb.B4-237, 249
$C_{12}FeH_{13}N_3$	$C_5H_5FeC_5H_4CH(CH_3)N_3$	Fe: Org.Verb.A1-287, 359, 362, 382
$C_{12}FeH_{13}N_3O$	$C_5H_5FeC_5H_4CHNNHCONH_2$	Fe: Org.Verb.A2-167
$C_{12}FeH_{13}N_3S$	$C_5H_5FeC_5H_4CHNNHCSNH_2$	Fe: Org.Verb.A2-167
$C_{12}FeH_{13}O^+$	$[C_5H_5FeC_5H_4COHCH_3]^+$	Fe: Org.Verb.A2-212/3
$C_{12}FeH_{13}O_2{}^+$	$[C_5H_5FeC_5H_4C(OH)OCH_3]^+$	Fe: Org.Verb.A3-111
$C_{12}FeH_{14}$	$C_5H_5FeC_5H_4C_2H_5$	Fe: Org.Verb.A1-259/67, 282
–	$C_5H_5FeC_5H_4CH_2CD_3$	Fe: Org.Verb.A1-260
–	$C_5H_5FeC_5H_4CD_2CH_3$	Fe: Org.Verb.A1-260
–	$C_5H_5FeC_5H_4C_2D_5$	Fe: Org.Verb.A1-260
–	$Fe(C_5H_4CH_3)_2$	Fe: Org.Verb.A1-278
$C_{12}FeH_{14}{}^+$	$[C_5H_5FeC_5H_4C_2H_5]^+$	Fe: Org.Verb.A1-249, 264, 278
–	$Fe(C_5H_4CH_3)_2{}^+$	Fe: Org.Verb.A1-278
$C_{12}FeH_{14}I_3$	$[C_5H_5FeC_5H_4C_2H_5]I_3$	Fe: Org.Verb.A1-278
$C_{12}FeH_{14}I_3O$	$[C_5H_5FeC_5H_4CHOHCH_3]I_3$	Fe: Org.Verb.A2-70
–	$[C_5H_5FeC_5H_4CH_2OCH_3]I_3$	Fe: Org.Verb.A2-112
$C_{12}FeH_{14}NOS$	$[Fe(C_5H_5)_2SCN] \cdot CH_3OH$	Fe: Org.Verb.A1-206
$C_{12}FeH_{14}NO_5{}^-$	$[(CHO)((C_2H_5)_2NCO)C_3H_3Fe(CO)_3]^-$	Fe: Org.Verb.B5-74, 76
$C_{12}FeH_{14}N_2$	$C_5H_5FeC_5H_4C(CH_3)NNH_2$	Fe: Org.Verb.A2-221
$C_{12}FeH_{14}N_2O$	$C_5H_5FeC_5H_4CH_2CONHNH_2$	Fe: Org.Verb.A3-116
–	$C_5H_5FeC_5H_4CH_2NHCONH_2$	Fe: Org.Verb.A2-20
$C_{12}FeH_{14}N_2O_3$	$(CH_3)_2C_3HN_2CH_2C(CH_3)CH_2Fe(CO)_3$	Fe: Org.Verb.B4-216, 225
–	$(CH_3)_3C_3N_2CH_2CHCH_2Fe(CO)_3$	Fe: Org.Verb.B4-216, 225
$C_{12}FeH_{14}N_2O_4$	$CH_2CHC(N(CH_2)_5)NHCOFe(CO)_3$	Fe: Org.Verb.B5-159/60
$C_{12}FeH_{14}O$	$C_5H_5FeC_5H_3(CH_3)CH_2OH$	Fe: Org.Verb.A2-70
–	$C_5H_5FeC_5H_4CHOHCH_3$ (Optically active form)	Fe: Org.Verb.A2-33/4
–	$C_5H_5FeC_5H_4CHOHCH_3$ (Racemic form)	Fe: Org.Verb.A2-4, 23/33
–	$C_5H_5FeC_5H_4CDOHCH_3$	Fe: Org.Verb.A2-24
–	$C_5H_5FeC_5H_4CDOHCH_2D$	Fe: Org.Verb.A2-24
–	$C_5H_5FeC_5H_4CH_2CH_2OH$	Fe: Org.Verb.A2-3/4, 70/2
–	$C_5H_5FeC_5H_4CH_2CHDOH$	Fe: Org.Verb.A2-71
–	$C_5H_5FeC_5H_4CH_2CD_2OH$	Fe: Org.Verb.A2-71/2
–	$C_5H_5FeC_5H_4CH_2OCH_3$	Fe: Org.Verb.A2-112/3

Formula	Compound	Reference
$C_{12}FeH_{14}O$	$C_5H_5FeC_5H_4OC_2H_5$	Fe: Org.Verb.A2-108/9
$C_{12}FeH_{14}O^+$	$[C_5H_5FeC_5H_4CHOHCH_3]^+$	Fe: Org.Verb.A2-27/8, 70
$C_{12}FeH_{14}O^{2+}$	$[HC_5H_5FeC_5H_4COHCH_3]^{2+}$	Fe: Org.Verb.A2-212/3
$C_{12}FeH_{14}O_2$	$C_5H_5FeC_5H_4CHOHCH_2OH$	Fe: Org.Verb.A2-87, 90
–	$FeC_{10}H_8(OCH_3)_2$	Fe: Org.Verb.A2-109
$C_{12}FeH_{14}O_2S_4$	$(CO)_2Fe(CH(CSCH_3)_2)_2$	Fe: Org.Verb.B1-115
$C_{12}FeH_{14}O_4$	$C_8H_{14}Fe(CO)_4$	Fe: Org.Verb.B4-302, 305, 313
$C_{12}FeH_{15}{}^+$	$[FeC_{12}H_{15}]^+$	Fe: Org.Verb.A1-94
–	$[HC_5H_5FeC_5H_4C_2H_5]^+$	Fe: Org.Verb.A1-277
$C_{12}FeH_{15}N$	$C_5H_5FeC_5H_4CH(CH_3)NH_2$	Fe: Org.Verb.A1-359, 361/2, 382
$C_{12}FeH_{15}NO_2$	$[C_5H_5FeC_5H_4CH_2COO]NH_4$	Fe: Org.Verb.A3-67
$C_{12}FeH_{15}NO_6$	$NH_2C_5H_{10}[OC(O)CHCH_2Fe(CO)_4]$	Fe: Org.Verb.B4-245/6
$C_{12}FeH_{15}N_7$	$(CH_3NC)_4Fe(CN)_2 \cdot NCCH_3$	Fe: Org.Verb.B4-67/8
$C_{12}FeH_{15}NaO_5$	$Na[C_7H_{15}C(O)Fe(CO)_4]$	Fe: Org.Verb.B4-151, 160
$C_{12}FeH_{15}O_4{}^+$	$[(CH_3)_2CC(CH_3)C(CH_3)_2Fe(CO)_4]^+$	Fe: Org.Verb.B4-325
$C_{12}FeH_{15}O_5{}^-$	$[C_7H_{15}C(O)Fe(CO)_4]^-$	Fe: Org.Verb.B4-151, 160
$C_{12}FeH_{16}N_6$	$[(C_2H_5NC)_3Fe(CN)_3]H$	Fe: Org.Verb.B4-45
$C_{12}FeH_{16}OSi$	$C_5H_5FeC_5H_4Si(CH_3)_2OH$	Fe: Org.Verb.A1-170
$C_{12}FeH_{16}O_4$	$(CH_2C(CH_3)C(C_5H_{11})O)Fe(CO)_3$	Fe: Org.Verb.B5-132, 141
$C_{12}FeH_{17}N_2O_4{}^+$	$[CH_2CHC(N(CH_3)_2)N(CH_3)C(OC_2H_5)Fe(CO)_3]^+$	Fe: Org.Verb.B5-161
$C_{12}FeH_{17}N_2O_5{}^+$	$[CH_3NHC_2H_4N(CH_3)_2COC_2H_3Fe(CO)_4]^+$	Fe: Org.Verb.B4-249
$C_{12}FeH_{17}N_3O_3{}^{2+}$	$[CH_3NCFe(H_2O)_3(NC_5H_4C_5H_4N)]^{2+}$	Fe: Org.Verb.B4-9
$C_{12}FeH_{17}O_3$	$C_3H_5Fe(CO)_3C_6H_{12}$	Fe: Org.Verb.C2-169
$C_{12}FeH_{17}O_4{}^-$	$[C_6H_{13}CH(CH_3)C(O)Fe(CO)_3]^-$	Fe: Org.Verb.B4-109
$C_{12}FeH_{17}O_5P$	$(C_7H_8)Fe(CO)_2(P(OCH_3)_3)$	Fe: Org.Verb.B1-162
$C_{12}FeH_{18}Hg_2I_6N_6$	$[(CH_3NC)_6Fe]I_2 \cdot 2\ HgI_2$	Fe: Org.Verb.B4-101
$C_{12}FeH_{18}Hg_2O_4$	$(CO)_4Fe(HgC_4H_9)_2$	Fe: Org.Verb.B2-153, 157, 160
$C_{12}FeH_{18}I_2N_2O_2$	$(t\text{-}C_4H_9NC)_2Fe(CO)_2I_2$	Fe: Org.Verb.B4-16
$C_{12}FeH_{18}I_2N_6$	$[(CH_3NC)_6Fe]I_2$	Fe: Org.Verb.B4-97, 101
$C_{12}FeH_{18}I_6N_6$	$[(CH_3NC)_6Fe]I_2 \cdot 2\ I_2$	Fe: Org.Verb.B4-101
$C_{12}FeH_{18}N_6{}^{2+}$	$[(CH_3NC)_6Fe]^{2+}$	Fe: Org.Verb.B4-97/103
$C_{12}FeH_{18}N_6OSn_3$	$[(CH_3)_2Sn]_2[Fe(CN)_6] \cdot (CH_3)_2SnO$	Sn: Org.Verb.6-29
$C_{12}FeH_{18}N_6O_2$	$(C_3H_7C(O)C_2H_2N_3)_2Fe$	Fe: Org.Verb.B4-358
–	$(C_3H_7C(O)C_2H_2N_3)_2Fe \cdot H_2O$	Fe: Org.Verb.B4-358
$C_{12}FeH_{18}N_6O_4S$	$[(CH_3NC)_6Fe]SO_4$	Fe: Org.Verb.B4-98, 103
$C_{12}FeH_{18}N_8O_6$	$[(CH_3NC)_6Fe][NO_3]_2$	Fe: Org.Verb.B4-97, 101
$C_{12}FeH_{18}O_4Si_2$	$(CH_3)_3SiCCSi(CH_3)_3Fe(CO)_4$	Fe: Org.Verb.B4-328
$C_{12}FeH_{19}NO_3P^+$	$[C_4H_4P(C_2H_5)_3Fe(CO)_2NO]^+$	Fe: Org.Verb.B5-28/31
$C_{12}FeH_{19}N_3O_5$	$[(N(CH_3)_2)_2CH][(CH_3)_2NC(O)Fe(CO)_4]$	Fe: Org.Verb.B4-155, 163/4
$C_{12}FeH_{20}Hg_2I_3NO_4$	$[N(C_2H_5)_4][(CO)_4Fe(HgI)_2I]$	Fe: Org.Verb.B2-153, 157
$C_{12}FeH_{20}INO_4$	$[N(C_2H_5)_4][(CO)_4FeI]$	Fe: Org.Verb.B2-51
$C_{12}FeH_{20}I_2N_4O_2$	$((CH_2NCH_3)_2C)_2Fe(CO)_2I_2$	Fe: Org.Verb.B4-172
$C_{12}FeH_{20}N_2O_2S_4$	$(CO)_2Fe(S_2CN(C_2H_5)_2)_2$	Fe: Org.Verb.B1-114
$C_{12}FeH_{20}N_6O_8S_2$	$(CH_3NC)_4Fe(CN)_2 \cdot (CH_3O)_2SO_2 \cdot H_2SO_4$	Fe: Org.Verb.B4-68
–	$[(CH_3NC)_6Fe][HSO_4]_2$	Fe: Org.Verb.B4-97, 101/2
$C_{12}FeH_{21}NO_4$	$[N(C_2H_5)_4][HFe(CO)_4]$	Fe: Org.Verb.B2-19

$C_{12}FeH_{21}N_7^{2+}$	$[(NH_2)(CH_3NH)CFe(CNCH_3)_5]^{2+}$	Fe:	Org.Verb.B4-172/4
$C_{12}FeH_{21}O_3S_6$	$Fe[S_2COC_3H_7]_3$	C:	MVol.D4-259
$C_{12}FeH_{21}S_9$	$Fe[S_2CSC_3H_7]_3$	C:	MVol.D4-267, 269
$C_{12}FeH_{22}N_2O_4S$	$(C_5H_{11}NC)_2FeSO_4$	Fe:	Org.Verb.B4-12
$C_{12}FeH_{22}N_8^{2+}$	$[(NHCNHCH_3)_2Fe(CNCH_3)_4]^{2+}$	Fe:	Org.Verb.B4-353/4
$C_{12}FeH_{22}N_8O$	$COFe(C_6H_7N_8(CH_3)_4)CH_3$	Fe:	Org.Verb.B1-92/3
$C_{12}FeH_{24}N_4S_2$	$(CH_3NC)_2Fe((SC_2H_4NCH_3)_2C_2H_4)$	Fe:	Org.Verb.B4-35/6
$C_{12}FeH_{24}O_9P_2S_2$	$OCFe(P(OCH_3)_3)_2S_2C_2H(COOC_2H_5)$	Fe:	Org.Verb.B4-138
$C_{12}FeH_{26}N_{10}^{2+}$	$[((NHCNHCH_3)_2)_2Fe(CNCH_3)_2]^{2+}$	Fe:	Org.Verb.B4-359
$C_{12}FeH_{27}Li$	$Li[Fe(C_4H_9\text{-}n)_3]$	Fe:	Org.Verb.B1-10
$C_{12}FeH_{27}NO_4P^+$	$[Fe(NO)P(OC_4H_9)_3]^+$	Fe:	Org.Verb.B1-47/8
$C_{12}FeH_{27}N_2O_2P$	$Fe(NO)_2(P(C_4H_9)_3)$	Fe:	Org.Verb.B1-43, 46
$C_{12}FeH_{27}N_2O_2P^+$	$[Fe(NO)_2P(C_4H_9)_3]^+$	Fe:	Org.Verb.B1-46
$C_{12}FeH_{27}N_2O_5P$	$Fe(NO)_2(P(OC_4H_9)_3)$	Fe:	Org.Verb.B1-43, 46
$C_{12}FeH_{27}N_2O_5P^+$	$[Fe(NO)_2P(OC_4H_9)_3]^+$	Fe:	Org.Verb.B1-47/8
$C_{12}FeH_{28}N_6O_6$	$[(CH_3NC)_4Fe(H_2O)_4(CH_3OH)_2][CN]_2$	Fe:	Org.Verb.B4-57
$C_{12}FeH_{30}N_2O_2P_2$	$Fe(NO)_2(P(C_2H_5)_3)_2$	Fe:	Org.Verb.B1-132
$C_{12}FeH_{30}N_5O_2P_2^+$	$[(CO)_2FeP_2(N(CH_3)_2)_5]^+$	Fe:	Org.Verb.B1-164
$C_{12}FeH_{30}O_2P_2S^+$	$[Fe(P(C_2H_5)_3)_2SO_2]^+$	Fe:	Org.Verb.B1-94
$C_{12}FeH_{30}P_2^+$	$[Fe(P(C_2H_5)_3)_2]^+$	Fe:	Org.Verb.B1-94
$C_{12}FeH_{32}O_9P_3$	$C_3H_5Fe(P(OCH_3)_3)_3$	Fe:	Org.Verb.B5-4
$C_{12}FeH_{33}Si_3$	$((CH_3)_3SiCH_2)_3Fe$	Fe:	Org.Verb.B1-9
$C_{12}FeH_{36}N_6P_2^+$	$[Fe(P(N(CH_3)_2)_3)_2]^+$	Fe:	Org.Verb.B1-164
$C_{12}FeH_{36}P_4$	$(CH_3)_2PCH_2Fe(P(CH_3)_3)_3H$	Fe:	Org.Verb.B1-6
$C_{12}FeO_{12}Os_2$	$FeOs_2(CO)_{12}$	Fe:	Org.Verb.B2-195
$C_{12}FeO_{12}Ru_2$	$FeRu_2(CO)_{12}$	Fe:	Org.Verb.B2-194/5
$C_{12}Fe_2GeH_{10}O_8$	$(CO)_4FeGe(C_2H_5)_2Fe(CO)_4$	Fe:	Org.Verb.B2-28
$C_{12}Fe_2GeH_{15}IN_4O_4$	$(C_3H_5Fe(NO)_2)_2Ge(I)C_6H_5$	Fe:	Org.Verb.C2-165
$C_{12}Fe_2Ge_2H_{12}O_8$	$[(CO)_4FeGe(CH_3)_2]_2$	Fe:	Org.Verb.C1-42/5, 49
$C_{12}Fe_2Ge_2H_{12}O_9$	$((CH_3)_2Ge)_2OFe_2(CO)_8$	Fe:	Org.Verb.B3-138
$C_{12}Fe_2Ge_3H_{18}O_6$	$(CO)_3Fe(Ge(CH_3)_2)_3Fe(CO)_3$	Fe:	Org.Verb.C1-195/6
$C_{12}Fe_2H_4O_6S_2$	$(CO)_3FeSC_6H_4SFe(CO)_3$	Fe:	Org.Verb.C1-115, 117, 122
$C_{12}Fe_2H_5NO_6S$	$(CO)_3Fe(C_6H_5NS)Fe(CO)_3$	Fe:	Org.Verb.C1-185/6, 188
–	$(CO)_3Fe(SC_6H_4NH)Fe(CO)_3$	Fe:	Org.Verb.C1-185/6, 188/9
$C_{12}Fe_2H_6N_2O_6$	$(CO)_3Fe(CH_2CHCN)_2Fe(CO)_3$	Fe:	Org.Verb.C2-148/9
–	$(CO)_3Fe(NHC_6H_4NH)Fe(CO)_3$	Fe:	Org.Verb.C1-131
$C_{12}Fe_2H_6N_4O_6$	$(CO)_3Fe(N(CH)_3N)_2Fe(CO)_3$	Fe:	Org.Verb.C1-134, 142
$C_{12}Fe_2H_6O_9$	$CH_3CHCH_2Fe(CO)_4 \cdot Fe(CO)_5$	Fe:	Org.Verb.B4-249/50
$C_{12}Fe_2H_8N_2O_7$	$(CO)_3Fe(CO)(N_2C_5H_8)Fe(CO)_3$	Fe:	Org.Verb.C1-200/1, 203
$C_{12}Fe_2H_8O_7S_2$	$(CO)_6Fe_2(CO)(S_2C_5H_8)$	Fe:	Org.Verb.C2-44
$C_{12}Fe_2H_8O_8$	$(CH_3)_2(OH)_2C_4(Fe(CO)_3)_2$	Fe:	Org.Verb.B2-31
$C_{12}Fe_2H_9O_9PPt$	$(CO)_8Fe_2Pt(CO)(P(CH_3)_3)$	Fe:	Org.Verb.C1-216/8
$C_{12}Fe_2H_{10}I_2O_6$	$I(CO)_3Fe(C_3H_5)_2Fe(CO)_3I$	Fe:	Org.Verb.C2-156, 164/5
$C_{12}Fe_2H_{10}O_6$	$(C_3H_5Fe(CO)_3)_2$	Fe:	Org.Verb.C2-156, 166/8
$C_{12}Fe_2H_{10}O_6S$	$(CO)_3Fe(C(C_2H_5)C(C_2H_5)S)Fe(CO)_3$	Fe:	Org.Verb.B3-150
$C_{12}Fe_2H_{10}O_6S_2$	$(CO)_3Fe(S(CC_2H_5)_2S)Fe(CO)_3$	Fe:	Org.Verb.B3-150
$C_{12}Fe_2H_{10}O_6S_4$	$(CO)_3Fe(SCSC_2H_5)_2Fe(CO)_3$	Fe:	Org.Verb.C1-236/7
$C_{12}Fe_2H_{11}NO_6$	$(CH_3CCHCHN(CH_3)_2)Fe_2(CO)_6$	Fe:	Org.Verb.C2-156, 159
$C_{12}Fe_2H_{12}N_2O_6$	$(CO)_3Fe((CH_3)_2CN)_2Fe(CO)_3$	Fe:	Org.Verb.C1-131, 138/9

Formula	Compound	Reference
$C_{12}Fe_2H_{12}N_2O_7$	$(CO)_6Fe_2((CH_3)_2CN)((CH_3)_2CNO)$	Fe: Org.Verb.C1-191/2, 194
$C_{12}Fe_2H_{12}O_8P_2$	$(CO)_4Fe((CH_3)_2PP(CH_3)_2)Fe(CO)_4$	Fe: Org.Verb.C1-33/4, 37/8
$C_{12}Fe_2H_{12}O_8Pb_2$	$[(CO)_4FePb(CH_3)_2]_2$	Fe: Org.Verb.C1-42/4, 47
$C_{12}Fe_2H_{12}O_8Sn_2$	$[(CO)_4FeSn(CH_3)_2]_2$	Fe: Org.Verb.C1-42/5, 50/1
$C_{12}Fe_2H_{12}O_9P_2$	$(CO)_4Fe((CH_3)_2POP(CH_3)_2)Fe(CO)_4$	Fe: Org.Verb.C1-35, 39
$C_{12}Fe_2H_{12}O_9Si_2$	$((CH_3)_2Si)_2OFe_2(CO)_8$	Fe: Org.Verb.B3-138
$C_{12}Fe_2H_{14}I_2O_6S_2$	$I(CO)_3Fe(SC_3H_7)_2Fe(CO)_3I$	Fe: Org.Verb.C1-6/7, 9
$C_{12}Fe_2H_{14}N_2O_6$	$(CO)_3Fe(C_3H_7NNC_3H_7)Fe(CO)_3$	Fe: Org.Verb.C1-149/51, 153/4
–	$(CO)_3Fe(NC(CH_3)_2)(C_3H_7NH)Fe(CO)_3$	Fe: Org.Verb.C1-133
$C_{12}Fe_2H_{14}N_2O_7$	$(CO)_6Fe_2(C_3H_7NH)((CH_3)_2CNO)$	Fe: Org.Verb.C1-191/3
$C_{12}Fe_2H_{14}O_6S_2$	$(CO)_3Fe(SC_3H_7)_2Fe(CO)_3$	Fe: Org.Verb.C1-78, 80/1, 87
$C_{12}Fe_2H_{14}O_6Se_2$	$(CO)_3Fe(SeC_3H_7)_2Fe(CO)_3$	Fe: Org.Verb.C1-83, 91
$C_{12}Fe_2H_{15}IN_4O_4Pb$	$(C_3H_5Fe(NO)_2)_2Pb(I)C_6H_5$	Fe: Org.Verb.C2-165
$C_{12}Fe_2H_{15}IN_4O_4Sn$	$(C_3H_5Fe(NO)_2)_2Sn(I)C_6H_5$	Fe: Org.Verb.C2-165
$C_{12}Fe_2H_{15}O_6S_3{}^+$	$[(CO)_6Fe_2(SC_2H_5)_3]^+$	Fe: Org.Verb.C1-20/3
$C_{12}Fe_2H_{16}N_2O_6$	$(CO)_3Fe(C_3H_7NH)_2Fe(CO)_3$	Fe: Org.Verb.C1-130, 136
$C_{12}Fe_2H_{18}O_6Si_3$	$(CO)_6Fe_2(Si(CH_3)_2)_3$	Fe: Org.Verb.C2-51
$C_{12}Fe_2H_{24}IO_{10}P_2S_2{}^+$	$[(CO)_4(P(OCH_3)_3)_2Fe_2I(SCH_3)_2]^+$	Fe: Org.Verb.C1-12/5
$C_{12}Fe_2H_{24}N_8NiO_8$	$[Ni(C_2H_8N_2)_3][Fe(CO)_3NO]_2$	Fe: Org.Verb.B1-190
$C_{12}Fe_2H_{24}O_4P_2S_2$	$(CO)_4(P(CH_3)_3)_2Fe_2(SCH_3)_2$	Fe: Org.Verb.C1-102, 108
$C_{12}Fe_2H_{24}O_{10}P_2S_2$	$(CO)_4(P(OCH_3)_3)_2Fe_2(SCH_3)_2$	Fe: Org.Verb.C1-103, 108
$C_{12}Fe_2H_{25}O_4P_2S_2{}^+$	$[(CO)_4(P(CH_3)_3)_2Fe_2(SCH_3)_2H]^+$	Fe: Org.Verb.C1-125/6
$C_{12}Fe_2MnO_{12}{}^-$	$[(CO)_8Fe_2Mn(CO)_4]^-$	Fe: Org.Verb.C1-213/6
$C_{12}Fe_2O_{12}Os$	$(CO)_8Fe_2Os(CO)_4$	Fe: Org.Verb.C1-212/3
$C_{12}Fe_2O_{12}Re^-$	$[(CO)_8Fe_2Re(CO)_4]^-$	Fe: Org.Verb.C1-213/6
$C_{12}Fe_2O_{12}Ru$	$(CO)_8Fe_2Ru(CO)_4$	Fe: Org.Verb.C1-212
$C_{12}Fe_2O_{12}Tc^-$	$[(CO)_8Fe_2Tc(CO)_4]^-$	Fe: Org.Verb.C1-213/6
$C_{12}Fe_3H_2O_{10}$	$C_2H_2Fe_3(CO)_{10}$	Fe: Org.Verb.C2-84
$C_{12}Fe_3H_2O_{12}$	$H_2Fe_3(CO)_{12}$	Fe: Org.Verb.B2-10
$C_{12}Fe_3H_{24}N_8O_8$	$[Fe(C_2H_8N_2)_3][Fe(CO)_3NO]_2$	Fe: Org.Verb.B1-188/9
$C_{12}Fe_3Ir_2N_{12}$	$Fe_3[Ir(CN)_6]_2 \cdot 12\ H_2O$	Ir: SVol.2-183, 184
–	$Fe_3[Ir(CN)_6]_2 \cdot 20\ H_2O$	Ir: SVol.2-183, 184
$C_{12}Fe_3O_{12}$	$Fe_3(CO)_{12}$	Fe: Org.Verb.B2-1, 3/4
$C_{12}Fe_3O_{12}P_4$	$P_4(Fe(CO)_4)_3$	Fe: Org.Verb.C2-27
$C_{12}Fe_3O_{12}Pb$	$PbFe_3(CO)_{12}$	Fe: Org.Verb.B2-27
$C_{12}Fe_3O_{12}Tl_2$	$Tl_2Fe_3(CO)_{12}$	Fe: Org.Verb.B2-27
$C_{12}Fe_3O_{18}P_4$	$[(CO)_4Fe]_3P_4O_6$	Fe: Org.Verb.B2-117
$C_{12}Fe_4GeO_{12}S_4$	$[(CO)_6Fe_2S_2]_2Ge$	Fe: Org.Verb.C1-76
$C_{12}Fe_4N_4O_{16}Sn$	$Sn(Fe(CO)_3NO)_4$	Fe: Org.Verb.C1-67
$C_{12}Fe_4O_{12}S_4Si$	$[(CO)_6Fe_2S_2]_2Si$	Fe: Org.Verb.C1-76
$C_{12}FmH_{21}O_9$	$Fm((CH_3)_2C(OH)COO)_3$	Np: TrU.D1-154
$C_{12}GeH_{20}NiO$	$C_5H_5Ni(CO)Ge(C_2H_5)_3$	Ni: Org.Verb.2-164/6
$C_{12}GeH_{20}O_4S_8$	$Ge[S_2COC_2H_5]_4$	C: MVol.D4-253
$C_{12}GeH_{23}ORhSn$	$Rh(C_5H_5)(CO)[Ge(CH_3)_3][Sn(CH_3)_3]$	Sn: Org.Verb.5-86
$C_{12}GeH_{30}Sn$	$(C_2H_5)_3GeSn(C_2H_5)_3$	Sn: Org.Verb.5-104
$C_{12}Ge_3H_{27}NiO_3P$	$(CO)_3NiP(Ge(CH_3)_3)_3$	Ni: Org.Verb.1-163/4
$C_{12}Ge_3H_{27}NiO_3Sb$	$(CO)_3NiSb(Ge(CH_3)_3)_3$	Ni: Org.Verb.1-187, 189
$C_{12}H_4MnN_6O_{14}$	$Mn(OC_6H_2(NO_2)_3)_2 \cdot 4\ H_2O$	Mn: MVol.D1-74
–	$Mn(OC_6H_2(NO_2)_3)_2 \cdot 6\ H_2O$	Mn: MVol.D1-41/2

Formula	Compound	Reference
$C_{12}H_5LiNiO_4$	$Li[(CO)_3NiC(O)CCC_6H_5]$	Ni: Org.Verb.1-193, 194
$C_{12}H_6K_4MnO_{16}S_4$	$K_4[Mn(OC_6H_2(SO_3)_2OH)_2]$	Mn: MVol.D1-46
$C_{12}H_6K_4MnO_{18}S_4$	$K_4[Mn(O_3C_6H(SO_3H)_2)_2]$	Mn: MVol.D1-47/8
$C_{12}H_6MnN_2O_8{}^{2-}$	$[Mn(NO_2C_6H_3O_2)_2]^{2-}$	Mn: MVol.D1-45
$C_{12}H_6MnN_4O_{10}$	$Mn(OC_6H_3(NO_2)_2)_2$	Mn: MVol.D1-41
–	$Mn(OC_6H_3(NO_2)_2)_2 \cdot 4\ H_2O$	Mn: MVol.D1-74
$C_{12}H_6Mn_2O_{10}Sn$	$(CH_3)_2Sn[Mn(CO)_5]_2$	F: PerFHalOrg.4-36
$C_{12}H_6O_{18}S_4Sn^{6-}$	$[Sn(OH)_2(O_2C_6H_2(SO_3)_2)_2]^{6-}$	Sn: MVol.C5-77
$C_{12}H_8I_2N_2O_4Sn$	$SnI_2(NC_5H_4COO)_2$	Sn: MVol.C5-205/6
$C_{12}H_8I_2N_2Sn$	$SnI_2 \cdot C_{12}H_8N_2$	Sn: MVol.C5-22/3
$C_{12}H_8I_2OSn$	$[(IC_6H_4)_2SnO]_x$	Sn: Org.Verb.6-187
$C_{12}H_8I_4N_2Sn$	$SnI_4 \cdot C_{12}H_8N_2$	Sn: MVol.C5-220/1
$C_{12}H_8I_4Sn$	$(IC_6H_4)_2SnI_2$	Sn: Org.Verb.6-187
$C_{12}H_8MnN_2O_4$	$Mn(OC_6H_4NO)_2$	Mn: MVol.D1-40
$C_{12}H_8MnN_2O_6$	$Mn(OC_6H_4NO_2)_2$	Mn: MVol.D1-41
$C_{12}H_8MnN_5O_9$	$Mn(NO_3)_3 \cdot C_{12}H_8N_2$	Mn: MVol.C3-306
$C_{12}H_8MnO_4{}^{2-}$	$[Mn(OC_6H_4O)_2]^{2-}$	Mn: MVol.D1-43/4
$C_{12}H_8MnO_{10}S_2{}^{2-}$	$[Mn(O_2C_6H_3SO_3H)_2]^{2-}$	Mn: MVol.D1-46
$C_{12}H_8MnO_{16}S_4{}^{2-}$	$[Mn(O_2C_6H_2(SO_3H)_2)_2]^{2-}$	Mn: MVol.D1-46
$C_{12}H_8N_2NiO_2$	$(CO)_2Ni(NC_5H_4C_5H_4N)$	Ni: Org.Verb.1-147/8
$C_{12}H_8N_2NpO_6$	$NpO_2(NC_5H_4COO)_2$	Np: TrU.D1-156
$C_{12}H_8N_2NpO_6{}^-$	$NpO_2(NC_5H_4COO)_2{}^-$	Np: TrU.D1-155
$C_{12}H_8N_2NpO_8$	$NpO_2(ONC_5H_4COO)_2$	Np: TrU.D1-156
$C_{12}H_8N_2NpO_8{}^-$	$NpO_2(ONC_5H_4COO)_2{}^-$	Np: TrU.D1-155
$C_{12}H_8N_2O_4Sn$	$Sn(NC_5H_4COO)_2$	Sn: MVol.C5-19
$C_{12}H_8N_2O_6Pu$	$PuO_2(NC_5H_4COO)_2$	Np: TrU.D1-156
$C_{12}H_8N_2O_6SU$	$[UO_2SO_4(C_{12}H_8N_2)]$	U: SVol.E1-44/6
–	$[UO_2SO_4(C_{12}H_8N_2)] \cdot 5\ H_2O$	U: SVol.E1-44/6
$C_{12}H_8N_2O_6U$	$[UO_2(C_2O_4)(C_{10}H_8N_2)] \cdot H_2O$	U: SVol.E1-39/42
$C_{12}H_8N_2O_8Pu$	$PuO_2(ONC_5H_4COO)_2$	Np: TrU.D1-156
$C_{12}H_8N_4O_8SU$	$UO_2SO_4 \cdot 2\ NCC_5H_4NO$	U: SVol.E1-139, 142/3
$C_{12}H_8N_4O_8U$	$[UO_2(NO_3)_2(C_{12}H_8N_2)]$	U: SVol.E1-44/6
–	$[UO_2(NO_3)_2(C_{12}H_8N_2)] \cdot 0.5\ H_2O$	U: SVol.E1-44/6
$C_{12}H_8N_6O_{10}U$	$UO_2(NO_3)_2 \cdot 2\ NCC_5H_4NO$	U: SVol.E1-133, 139, 142/3
$C_{12}H_8NiO_4$	$Ni(C_6H_4O_2)_2$	Ni: Org.Verb.1-230
$C_{12}H_8O_4Sn$	$Sn(C_6H_4O_2)_2$	Sn: MVol.C5-77
$C_{12}H_8S_4Sn$	$Sn(C_6H_4S_2)_2$	Sn: MVol.C6-85/6
$C_{12}H_9LiNiO_4$	$Li[(CO)_3NiC(O)C_6H_3(CH_3)_2]$	Ni: Org.Verb.1-193, 194
$C_{12}H_9N_3O_6U$	$[UO_2(OH)(NO_3)(C_{12}H_8N_2)] \cdot 2\ H_2O$	U: SVol.E1-44/6
–	$[UO_2(OH)(NO_3)(C_{12}H_8N_2)] \cdot 2.5\ H_2O$	U: SVol.E1-44/6
$C_{12}H_9N_5O_2S_2U$	$[UO_2(NCS)_2(NH(C_5H_4N)_2)]$	U: SVol.E1-41, 43
$C_{12}H_9Sn^+$	$C_6H_5SnC_6H_4{}^+$	Sn: Org.Verb.1-155
		Sn: Org.Verb.2-355
$C_{12}H_{10}LiTi$	$[(C_6H_5)_2Ti]Li$	Ti: Org.Verb.1-68, 78
$C_{12}H_{10}Li_2S_2Sn$	$(C_6H_5)_2Sn(SLi)_2$	Sn: Org.Verb.6-160
$C_{12}H_{10}Li_2Sn$	$(C_6H_5)_2SnLi_2$	Sn: Org.Verb.6-160
$C_{12}H_{10}MgNi_2O_2$	$(C_5H_5NiCO)_2Mg \cdot n\ C_2H_4(N(CH_3)_2)_2$	Ni: Org.Verb.2-349
–	$(C_5H_5NiCO)_2Mg \cdot n\ C_4H_8O$	Ni: Org.Verb.2-349
–	$(C_5H_5NiCO)_2Mg \cdot n\ C_5H_5N$	Ni: Org.Verb.2-349
$C_{12}H_{10}MnNO_6$	$[Mn(CH(COCH_3)_2)(NC_5H_3(COO)_2)] \cdot H_2O$	Mn: MVol.D1-106

Formula	Compound	Reference
$C_{12}H_{10}MnO_4$	$[Mn(OC_6H_4OH)_2]$	Mn: MVol.D1-43/4
$C_{12}H_{10}MnO_5S$	$[Mn(C_6H_5CH(CH_2CO)_2CSO_3)] \cdot H_2O$	Mn: MVol.D1-128
$C_{12}H_{10}MnO_6$	$[Mn(CH_3C_5H_2O_3)_2]$	Mn: MVol.D1-150
$C_{12}H_{10}MnO_8$	$[Mn(C_5H_2O_3CH_2OH)_2]$	Mn: MVol.D1-150/1
$C_{12}H_{10}N_2NiO_2$	$(CO)_2Ni(NC_5H_5)_2$	Ni: Org.Verb.1-123
$C_{12}H_{10}N_2Ni_2O_6$	$(CO)_3Ni(CH_3)_2C_4H_4N_2Ni(CO)_3$	Ni: Org.Verb.2-264
$C_{12}H_{10}N_2O_4U$	$[UO_2(OH)_2(C_{12}H_8N_2)] \cdot 2.5\ H_2O$	U: SVol.E1-44
$C_{12}H_{10}N_2O_8Pu$	$PuO_2(C_5H_4N(O)COOH)_2 \cdot 2\ H_2O$	Np: TrU.D1-125
$C_{12}H_{10}N_4O_4Sn$	$Sn(C_6H_5N(O)NO)_2$	Sn: MVol.C5-43
$C_{12}H_{10}N_4O_{12}U$	$UO_2(NO_3)_2 \cdot 2\ C_6H_5NO_2$	U: SVol.E1-144/5
$C_{12}H_{10}N_4S_2Sn$	$Sn(SCN)_2 \cdot 2\ C_5H_5N$	Sn: MVol.C5-16/7
$C_{12}H_{10}Na_2O_8S_2Sn$	$(CH_3)_2Sn(O_2C_{10}H_4(SO_3Na)_2)$	Sn: Org.Verb.6-27
$C_{12}H_{10}Ni$	$(C_6H_5)_2Ni$	Ni: Org.Verb.1-87, 88
$C_{12}H_{10}NiOSe$	$C_5H_5Ni(CO)SeC_6H_5$	Ni: Org.Verb.2-164
$C_{12}H_{10}NiS_2$	$C_5H_5NiSC(S)C_6H_5$	Ni: Org.Verb.2-114
$C_{12}H_{10}Ni_2O_2$	$(C_5H_5NiCO)_2$	Ni: Org.Verb.2-344/9
$C_{12}H_{10}Ni_2O_2^+$	$(C_5H_5)_2Ni_2(CO)_2^+$	Ni: Org.Verb.2-345, 351
$C_{12}H_{10}OSn$	$[(C_6H_5)_2SnO]_n$	Sn: Org.Verb.5-160, 174
		Sn: Org.Verb.6-162/4, 167
$C_{12}H_{10}O_4S_2Sn$	$Sn(C_6H_5SO_2)_2$	Sn: MVol.C5-45/6
$C_{12}H_{10}O_4S_4Te$	$Te(S_2O_2C_6H_5)_2$	Te: SVol.B3-149
$C_{12}H_{10}O_6Sn^{2-}$	$[Sn(OH)_2(C_6H_4O_2)_2]^{2-}$	Sn: MVol.C5-77
$C_{12}H_{10}O_8Sn^{2-}$	$[Sn(OH)_2(O_2C_6H_3OH)_2]^{2-}$	Sn: MVol.C5-77
$C_{12}H_{10}O_{14}Pu^{2-}$	$Pu(C_6H_5O_7)_2^{2-}$	Np: TrU.D1-159
$C_{12}H_{10}O_{14}Sn^{4-}$	$Sn(HOC_3H_4(COO)_3)_2^{4-}$	Sn: MVol.C2-226
$C_{12}H_{10}O_{19}Sn^{6-}$	$SnOH(OCOCH(O)CH(OH)COO)_3^{6-}$	Sn: MVol.C2-226
$C_{12}H_{10}SSn$	$[(C_6H_5)_2SnS]_3$	Sn: Org.Verb.6-162/3, 170
$C_{12}H_{10}S_2Sn$	$[Sn(C_6H_5S)_2]$	Sn: MVol.C5-46
$C_{12}H_{10}Sn$	$[(C_6H_5)_2Sn]_x$	Sn: Org.Verb.6-159, 171, 273
$C_{12}H_{10}Sn^+$	$Sn(C_6H_5)_2^+$	Sn: Org.Verb.1-155
		Sn: Org.Verb.2-386
$C_{12}H_{10}Te_2$	$C_6H_5TeTeC_6H_5$	C: MVol.D6-261
$C_{12}H_{10}Ti$	$(C_6H_5)_2Ti$	Ti: Org.Verb.1-67/8
$C_{12}H_{11}LiO_4Ti$	$Li[(CH_3)(C_2H_5)C_5H_3Ti(CO)_4]$	Ti: Org.Verb.1-211
$C_{12}H_{11}N_3O_8U$	$UO_2(NO_3)_2 \cdot C_6H_5CH_2C_5H_4N \cdot H_2O$	U: SVol.E1-29, 34
$C_{12}H_{11}Sn^+$	$(C_6H_5)_2SnH^+$	Sn: Org.Verb.2-355
$C_{12}H_{12}I_6N_2S_2Sn_2$	$[SnI_3(SCH_2C_5H_4N)]_2$	Sn: MVol.C6-105
$C_{12}H_{12}K_2Ni$	$K_2[Ni(CCCH_3)_4]$	Ni: Org.Verb.1-106, 107
$C_{12}H_{12}N_2NiS_2$	$((CH_3)_4C_4NiS_2C_2(CN)_2)$	Ni: Org.Verb.2-329
$C_{12}H_{12}N_2NpO_{12}^{2-}$	$Np(N(CH_2COO)_3)_2^{2-}$	Np: TrU.D1-163
$C_{12}H_{12}N_2O_2S$	$C_6H_5NHSO_2NHC_6H_5$	S: S-N-Verb.1-169
$C_{12}H_{12}N_2O_4Sn$	$Sn(HCOO)_2 \cdot 2\ C_5H_5N$	Sn: MVol.C5-16/7
$C_{12}H_{12}N_2O_5U$	$[UO_2(OH)(CH_3COO)(C_{10}H_8N_2)]$	U: SVol.E1-39/42
$C_{12}H_{12}N_2O_6SU$	$UO_2SO_4 \cdot NH_2C_6H_4C_6H_4NH_2$	U: SVol.E1-25/6
$C_{12}H_{12}N_2O_9U$	$[UO_2(HOC_6H_4COO)(C_5H_5N)(H_2O)]NO_3$	U: SVol.E1-29, 35
$C_{12}H_{12}N_3NaO_4S_2$	$Na[N(SO_2NHC_6H_5)_2]$	S: S-N-Verb.1-162
$C_{12}H_{12}N_3NiO_3P$	$(CO)_3NiP(C_2H_4CN)_3$	Ni: Org.Verb.1-166
$C_{12}H_{12}N_3NiO_6P$	$(CO)_3NiP(OCH_2CH_2CN)_3$	Ni: Org.Verb.1-178
$C_{12}H_{12}N_4Ni$	$Ni(C_6H_4(NH)_2)_2$	Ni: Org.Verb.1-230
$C_{12}H_{12}N_4O_2Sn$	$(CH_3)_2Sn[CH_3C(O)C(CN)_2]_2$	Sn: Org.Verb.6-32

Formula	Compound	Reference
$C_{12}H_{12}N_4O_8U$	$[UO_2(NO_3)_2((CH_3)_2C_{10}H_6N_2)]$	U: SVol.E1-39/40, 43
–	$[UO_2(NO_3)_2(C_2H_4(C_5H_4N)_2)]$	U: SVol.E1-41, 43
$C_{12}H_{12}Ni$	$(C_6H_6)_2Ni(AlBr_3)_x$	Ni: Org.Verb.2-248
–	$(C_6H_6)_2Ni(AlCl_3)_x$	Ni: Org.Verb.2-248, 249
$C_{12}H_{12}Ni_2$	$(C_5H_5Ni)_2HCCH$	Ni: Org.Verb.2-355, 357
$C_{12}H_{12}NpO_{20}{}^{5-}$	$NpO_2(C_4H_4O_6)_3{}^{5-}$	Np: TrU.D1-157, 159
$C_{12}H_{12}O_8Os_2Sn_2$	$[(CH_3)_2SnOs(CO)_4]_2$	Sn: Org.Verb.6-33
$C_{12}H_{12}O_8Ru_2Sn_2$	$[(CH_3)_2SnRu(CO)_4]_2$	Sn: Org.Verb.6-21
$C_{12}H_{12}Sn$	$(C_6H_5)_2SnH_2$	Sn: Org.Verb.4-113/8
–	$Sn(CCCH_3)_4$	Sn: Org.Verb.1-140
–	$Sn(CHCCH_2)_4$	Sn: Org.Verb.1-138
–	$Sn(CH_2CCH)_4$	Sn: Org.Verb.1-140/1
$C_{12}H_{13}NTi$	$(C_6H_5)_2Ti \cdot NH_3$	Ti: Org.Verb.1-69
$C_{12}H_{13}N_3O_4S_2$	$HN(SO_2NHC_6H_5)_2$	S: S-N-Verb.1-161
$C_{12}H_{13}Ni^+$	$[C_5H_5NiC_7H_8]^+$	Ni: Org.Verb.2-189
$C_{12}H_{13}O_{16}Pa^{2-}$	$PaO(OH)(HC_6H_5O_7)_2{}^{2-}$	Pa: SVol.2-173/4
$C_{12}H_{13}S_3V$	$(C_5H_5)_2V[S_2CSCH_3]$	C: MVol.D4-269
$C_{12}H_{14}I_2MnN_2$	$MnI_2 \cdot 2\ C_6H_5NH_2$	Mn: MVol.C5-317
$C_{12}H_{14}I_2N_2Sn$	$SnI_2 \cdot 2\ CH_3C_5H_4N$	Sn: MVol.C5-18
–	$SnI_2 \cdot 2\ C_6H_5NH_2$	Sn: MVol.C5-12/3, 174
$C_{12}H_{14}MnO_{12}$	$[Mn(C_5H_5O_2(OH)_2COO)_2]_n$	Mn: MVol.D1-148
$C_{12}H_{14}MoO_3Sn$	$(CH_3)_3SnC_6H_5Mo(CO)_3$	Sn: Org.Verb.2-142, 144
$C_{12}H_{14}N_2Ni$	$(CH_3)_2Ni(NC_5H_4C_5H_4N)$	Ni: Org.Verb.1-69, 73
$C_{12}H_{14}N_2O_8SU$	$UO_2SO_4 \cdot 2\ CH_3C_5H_4NO$	U: SVol.E1-137/8, 142/3
$C_{12}H_{14}N_4NiS_4$	$Ni[S_2CNH_2]_2(C_5H_5N)_2$	C: MVol.D6-140
$C_{12}H_{14}N_4O_8U$	$UO_2(NO_3)_2 \cdot 2\ C_6H_5NH_2$	U: SVol.E1-18, 21, 25
$C_{12}H_{14}N_4O_{10}U$	$UO_2(NO_3)_2 \cdot 2\ CH_3C_5H_4NO$	U: SVol.E1-133, 137/8, 142/3
–	$UO_2(NO_3)_2 \cdot 2\ NH_2C_6H_4OH$	U: SVol.E1-18, 21, 25
$C_{12}H_{14}Ni$	$Ni(C_5H_4CH_3)_2$	Ni: Org.Verb.2-215, 216
$C_{12}H_{14}Ni^+$	$[Ni(C_5H_4CH_3)_2]^+$	Ni: Org.Verb.2-216/7
$C_{12}H_{14}O_2SSn$	$(CH_2CH)_3SnOS(O)C_6H_5$	Sn: Org.Verb.5-156
$C_{12}H_{14}O_{14}Pu^+$	$Pu(C_6H_7O_7)_2{}^+$	Np: TrU.D1-159
$C_{12}H_{14}O_{15}Pa^+$	$PaO(H_2C_6H_5O_7)_2{}^+$	Pa: SVol.2-173/4
$C_{12}H_{14}O_{15}Sn_2$	$(SnH_2(HOC_3H_4(COO)_3))_2O$	Sn: MVol.C2-226
–	$Sn_2H_2(OCOCH_2CH(OH)COO)_3$	Sn: MVol.C2-225
$C_{12}H_{14}Sn$	$(CH_2CH)_3SnC_6H_5$	Sn: Org.Verb.2-347
–	$(CH_3)_2Sn(CHCH)_2C_6H_4$	Sn: Org.Verb.3-111, 113
$C_{12}H_{15}LiSn$	$(CH_3)_3SnCH_2C_6H_4CCLi$	Sn: Org.Verb.2-28
$C_{12}H_{15}Ni^+$	$Ni(C_5H_5)_2C_2H_5{}^+$	Ni: Org.Verb.2-201
$C_{12}H_{15}NiO_6P$	$(CO)_3NiP(OCH_2CHCH_2)_3$	Ni: Org.Verb.1-178
$C_{12}H_{15}O_{16}Pa$	$PaO(OH)(H_2C_6H_5O_7)_2$	Pa: SVol.2-173/4
$C_{12}H_{16}HgMnN_4O_2S_4$	$MnHg(NCS)_4 \cdot 2\ C_4H_8O$	Mn: MVol.D1-144
$C_{12}H_{16}I_4N_4Sn$	$SnI_4 \cdot 2\ C_6H_4(NH_2)_2$	Sn: MVol.C5-187
$C_{12}H_{16}MnN_2O_6S_2$	$Mn(HSO_3)_2 \cdot 2\ C_6H_5NH_2$	Mn: MVol.C6-79
$C_{12}H_{16}N_2Ni$	$(CH_2CHCH_2CH_2CHCN)_2Ni$	Ni: Org.Verb.2-278
–	$(Ni(CH(CN)CH_2CH_2CHCH_2)_2)_x$	Ni: Org.Verb.1-389
$C_{12}H_{16}N_2NiO_2$	$(CO)_2Ni(C_2N_2(CH_3)_2(C_3H_5)_2)$	Ni: Org.Verb.1-146
$C_{12}H_{16}N_2O_{12}S_3Sn$	$[C_6H_5NH_3]_2Sn(SO_4)_3$	Sn: MVol.C3-122
$C_{12}H_{16}N_4Ni_2O_2$	$[(CO)NiNCN(CH_2)_4]_2$	Ni: Org.Verb.2-256/7

Formula	Compound	Reference
$C_{12}H_{16}N_4Ni_2O_4$	$[(CO)NiNCN(CH_2CH_2)_2O]_2$	Ni: Org.Verb.2-256/7
$C_{12}H_{16}N_4Sn$	$Sn(CH_2CH_2CN)_4$	Sn: Org.Verb.1-123/4
$C_{12}H_{16}N_6O_8U$	$UO_2(NO_3)_2 \cdot 2\ (NH_2)_2C_6H_4 \cdot H_2O$	U: SVol.E1-23/4
$C_{12}H_{16}Na_4O_8Sn$	$Sn(CH_2CH_2COONa)_4$	Sn: Org.Verb.1-124
$C_{12}H_{16}Ni$	$CH_3C_5H_4NiC_5H_6CH_3$	Ni: Org.Verb.2-180/1, 216
–	$C_{12}H_{16}Ni$	Ni: Org.Verb.2-244, 246
–	$Ni(C_6H_8)_2$	Ni: Org.Verb.2-87
$C_{12}H_{16}NiO_2P_2$	$(CO)_2NiC_6H_4(P(CH_3)_2)_2$	Ni: Org.Verb.1-151
$C_{12}H_{16}Ni_2S_2$	$(C_5H_5NiSCH_3)_2$	Ni: Org.Verb.2-338, 339
$C_{12}H_{16}O_4S_8$	$S_2(C(S)OC_4H_8OCS_2)_2$	C: MVol.D4-251
$C_{12}H_{16}O_6S_8$	$SS(C(S)OC_2H_4OC_2H_4OCS_2)_2$	C: MVol.D4-251
$C_{12}H_{16}O_{17}Pa^-$	$PaO(OH)_2(H_2C_6H_5O_7)_2^-$	Pa: SVol.2-173/4
$C_{12}H_{16}Sn$	$(CH_3)_2Sn(C_5H_5)_2$	Sn: Org.Verb.3-9
–	$(CH_3)_3SnCHCCHC_6H_5$	Sn: Org.Verb.2-82
–	$(CH_3)_3SnCH_2CCC_6H_5$	Sn: Org.Verb.2-115
–	$(CH_3)_3SnCH_2C_6H_4CCH$	Sn: Org.Verb.2-28/30
–	1) $(CH_3)_3SnC_9H_7$	Sn: Org.Verb.2-96/7
	2) $(CH_3)_3SnC_9H_7$	Sn: Org.Verb.2-102
–	$(C_2H_5)_2Sn(CCCHCH_2)_2$	Sn: Org.Verb.3-27
$C_{12}H_{17}MnO_6$	$[MnCH_3COO(CH(COCH_3)_2)_2]$	Mn: MVol.D1-103/4
$C_{12}H_{17}NO_2S_4Zn$	$Zn[S_2COC_2H_5]_2(NC_5H_4CH_3)$	C: MVol.D4-257
$C_{12}H_{17}N_3O_2S$	$(CH_2)_5NSO_2NHNCHC_6H_5$	S: S-N-Verb.1-222
$C_{12}H_{18}K_2N_2Ni$	$K_2[Ni(CCCH_3)_4] \cdot 2\ NH_3$	Ni: Org.Verb.1-106, 107
$C_{12}H_{18}K_2O_{12}Sn$	$K_2[Sn(CH_3COO)_6]$	Sn: MVol.C3-78
$C_{12}H_{18}MgO_{12}Sn_2$	$Mg[Sn(CH_3COO)_3]_2$	Sn: MVol.C3-168
$C_{12}H_{18}MnNO_5S$	$[MnSCN(CH(COCH_3)_2)_2] \cdot CH_3OH$	Mn: MVol.D1-105
$C_{12}H_{18}MnN_6^{2+}$	$[Mn(CH_3CN)_6]^{2+}$	Mn: MVol.C5-254, 287
$C_{12}H_{18}MnO_6$	$[Mn(CH_3OCH_2COCHCOCH_3)_2]$	Mn: MVol.D1-112
$C_{12}H_{18}N_2NiO_2$	$(n-C_4H_9NC)_2Ni(CO)_2$	Ni: Org.Verb.1-308, 309
–	$(t-C_4H_9NC)_2Ni(CO)_2$	Ni: Org.Verb.1-308, 309
$C_{12}H_{18}N_2NpO_{10}$	$Np(N(C_2H_4OH)(CH_2COO)_2)_2$	Np: TrU.D1-162
$C_{12}H_{18}N_2O_2S$	$C_5H_{10}NSO_2NHCH_2C_6H_5$	S: S-N-Verb.1-171
–	$C_5H_{10}NSO_2NHC_6H_4CH_3$	S: S-N-Verb.1-171
$C_{12}H_{18}N_6Ni_2O_6P_4$	$(CO)_3NiP_4(NCH_3)_6Ni(CO)_3$	Ni: Org.Verb.2-264
$C_{12}H_{18}Na_2O_{12}Sn$	$Na_2[Sn(CH_3COO)_6]$	Sn: MVol.C3-34
$C_{12}H_{18}Ni$	$C_{12}H_{18}Ni$	Ni: Org.Verb.2-229/36, 238/40
–	$Ni(C_4H_6)_3$	Ni: Org.Verb.1-417
–	$Ni(C_6H_9)_2$	Ni: Org.Verb.2-67
$C_{12}H_{18}NiO_4$	$(CH_3)_4C_4Ni(CH_3COO)_2$	Ni: Org.Verb.2-329
–	$Ni(C_3H_4COOC_2H_5)_2$	Ni: Org.Verb.2-66
$C_{12}H_{18}NiO_8P_2$	$(CO)_2Ni(P(OCH_2)_3CCH_3)_2$	Ni: Org.Verb.1-139/40, 241/2
$C_{12}H_{18}NiO_{10}P_2$	$(CO)_2Ni(P(OCH_2)_3CCH_2OH)_2$	Ni: Org.Verb.1-139, 140
$C_{12}H_{18}OSn$	$(CH_3)_3SnCH_2CH_2COC_6H_5$	Sn: Org.Verb.2-36
$C_{12}H_{18}O_2Sn$	$(C_2H_5)_2C_6H_5SnOCOCH_3$	Sn: Org.Verb.6-53
$C_{12}H_{18}O_4Sn$	$[Sn(C_4H_9)_2OCOCCCOO]_x$	Sn: Org.Verb.6-89, 95
$C_{12}H_{18}O_6Sn$	$[Sn(CH_3COCHCOOC_2H_5)_2]$	Sn: MVol.C5-9/10
$C_{12}H_{18}O_{12}Sn_2$	$Sn_2(CH_3COO)_6$	Sn: MVol.C2-222
$C_{12}H_{18}O_{12}Sn_2Sr$	$Sr[Sn(CH_3COO)_3]_2$	Sn: MVol.C3-190
$C_{12}H_{18}Sn$	$(CH_3)_3SnCHCH_2CHC_6H_5$	Sn: Org.Verb.2-59

Formula	Compound	Element	Reference
$C_{12}H_{18}Sn$	$(CH_3)_3SnCH(C_6H_5)CHCH_2$	Sn:	Org.Verb.2-82, 89
–	$(CH_3)_3SnCH_2CHCHC_6H_5$	Sn:	Org.Verb.2-82, 89
–	$(CH_3)_3SnC_6H_4C(CH_3)CH_2$	Sn:	Org.Verb.2-128
–	$(CH_3)_3SnC_9H_9$	Sn:	Org.Verb.2-65
–	$(C_2H_5)_2SnC_8H_8$	Sn:	Org.Verb.3-112
$C_{12}H_{19}ISn$	$(C_2H_5)_3SnC_6H_4I$	Sn:	Org.Verb.2-240
$C_{12}H_{19}NNi$	$(CH_3)_2C_3H_3Ni(NH_2C_6H_5)CH_3$	Ni:	Org.Verb.2-35
$C_{12}H_{19}NO_2S$	$[(CH_3)_4N][O_2CSCH_2C_6H_5]$	C:	MVol.D4-237
–	$[(CH_3)_4N][SOCOCH_2C_6H_5]$	C:	MVol.D4-237
$C_{12}H_{19}N_3O_9U$	$UO_2(NO_3)_2 \cdot (C_4H_9)(CH_3)_3C_5HNO$	U:	SVol.E1-133, 142/3
$C_{12}H_{19}Na_3O_6Sn$	$(NaOOCCH_2CH_2)_3SnC_3H_7$	Sn:	Org.Verb.2-336, 338
$C_{12}H_{20}MnO_4PS_2$	$[(C_2H_5)_3P](CO)_3Mn[S_2COC_2H_5]$	C:	MVol.D4-259
$C_{12}H_{20}MnO_5S$	$[Mn(CH(COCH_3)_2)_2] \cdot (CH_3)_2SO$	Mn:	MVol.D1-84
$C_{12}H_{20}Mo_2O_4S_8$	$Mo_2[S_2COC_2H_5]_4$	C:	MVol.D4-258
$C_{12}H_{20}Mo_2O_7S_8$	$Mo_2O_3[S_2COC_2H_5]_4$	C:	MVol.D4-245, 258
$C_{12}H_{20}N_2O_{10}U$	$UO_2(NO_3)_2 \cdot 2\ (CH_2)_5CO$	U:	SVol.E1-79, 81
–	$UO_2(NO_3)_2 \cdot 2\ CH_3COCHC(CH_3)_2$	U:	SVol.E1-79/80
$C_{12}H_{20}N_2S_4Te$	$Te(C_5H_{10}NCSS)_2$	Te:	SVol.B3-139
$C_{12}H_{20}N_2Sn$	$CH_3NC(C_6H_5)N(CH_3)Sn(CH_3)_3$	Sn:	Org.Verb.5-84
–	$(CH_3)_3SnNC(C_6H_5)N(CH_3)_2$	Sn:	Org.Verb.5-84
$C_{12}H_{20}N_4Ni$	$(C_2H_5NC)_4Ni$	Ni:	Org.Verb.1-329
$C_{12}H_{20}N_4O_2S_4Sn$	$Sn(SCN)_4 \cdot 2\ C_2H_5OC_2H_5$	Sn:	MVol.C5-150/1
$C_{12}H_{20}N_4O_4Sn$	$(C_2H_5)_2Sn[C(N_2)COOC_2H_5]_2$	Sn:	Org.Verb.3-27, 33
$C_{12}H_{20}N_4Ti$	$(C_4H_4N)_2Ti(N(CH_3)_2)_2$	Ti:	Org.Verb.1-130
$C_{12}H_{20}NiO_4$	$(CO)_2Ni(C_6H_{10}(OC_2H_5)_2)$	Ni:	Org.Verb.1-146
$C_{12}H_{20}Ni_2$	$(CH_2CHCH_2NiCHCH_2CH_2)_2$	Ni:	Org.Verb.2-285
$C_{12}H_{20}Ni_2O_4$	$(CH_3CHCHCH_2NiO_2CCH_3)_2$	Ni:	Org.Verb.2-296
$C_{12}H_{20}OSn$	$(CH_3)_2Sn(C_6H_5)CH_2CH_2CH(OH)CH_3$	Sn:	Org.Verb.3-80
–	$(C_2H_5)_3SnC_6H_4OH$	Sn:	Org.Verb.2-240, 243
$C_{12}H_{20}O_2Sn$	$(C_4H_9)_2Sn(C(O)CH)_2$	Sn:	Org.Verb.3-104
$C_{12}H_{20}O_4Sn$	$[Sn(C_4H_9)_2OCOCHCHCOO]_x$	Sn:	Org.Verb.6-89, 95
$C_{12}H_{20}O_7U$	$UO_2(CH(COCH_3)_2)_2 \cdot C_2H_5OH$	U:	SVol.E1-60, 64/5
$C_{12}H_{20}O_8Sn$	$Sn(CH_2CH_2COOH)_4$	Sn:	Org.Verb.1-124
$C_{12}H_{20}O_{12}Pu$	$Pu(CH_3CH(OH)COO)_4$	Np:	TrU.D1-154
$C_{12}H_{20}SSn$	$(CH_3)_3Sn(CH_2)_3SC_6H_5$	Sn:	Org.Verb.2-45
$C_{12}H_{20}Sn$	$(CH_3)(C_2H_5)Sn(i\text{-}C_3H_7)C_6H_5$	Sn:	Org.Verb.3-97
–	$(CH_3)_2Sn(C_2H_5)(CH(CH_3)C_6H_5)$	Sn:	Org.Verb.3-75, 76
–	$(CH_3)_2Sn(C_4H_9)(C_6H_5)$	Sn:	Org.Verb.3-77
–	$(CH_3)_2Sn(s\text{-}C_4H_9)(C_6H_5)$	Sn:	Org.Verb.3-75, 78
–	$(CH_3)_3SnCH_2CH_2CH_2C_6H_5$	Sn:	Org.Verb.2-43
–	$(CH_3)_3SnC_6H_2(CH_3)_3$	Sn:	Org.Verb.2-126, 128
–	$(C_2H_5)_3SnCCCCC_2H_5$	Sn:	Org.Verb.2-214
–	$(C_2H_5)_3SnCCCH_2CH_2CCH$	Sn:	Org.Verb.2-214
–	$(C_2H_5)_3SnC_6H_5$	Sn:	Org.Verb.2-238
–	$(C_4H_9)_2Sn(CCH)_2$	Sn:	Org.Verb.3-43
–	$Sn(C(CH_3)CH_2)_4$	Sn:	Org.Verb.1-137
–	$Sn(CHCHCH_3)_4$	Sn:	Org.Verb.1-136/7
–	$Sn(CHCH_2CH_2)_4$	Sn:	Org.Verb.1-121
–	$Sn(CH_2CHCH_2)_4$	Sn:	Org.Verb.1-133/6
$C_{12}H_{20}Sn^+$	$Sn(CH_2CHCH_2)_4^+$	Sn:	Org.Verb.1-134

Formula	Compound	Reference
$C_{12}H_{20}Ti$	$(CH_2CHCH_2)_4Ti$	Ti: Org.Verb.1-91, 99
$C_{12}H_{21}KO_5S_6U$	$KUO_2[S_2COC_3H_7]_3$	C: MVol.D4-261
$C_{12}H_{21}NSn$	$(C_2H_5)_3SnCH_2C_5H_4N$	Sn: Org.Verb.2-169
$C_{12}H_{21}N_3O_4S_2$	$(C_3H_7)_2NSO_2NHNHSO_2C_6H_5$	S: S-N-Verb.1-222
$C_{12}H_{21}NiO_3P$	$(CO)_3NiP(C_3H_7\text{-}i)_3$	Ni: Org.Verb.1-166, 168
$C_{12}H_{21}NiO_6P$	$(CO)_3NiP(OC_3H_7\text{-}i)_3$	Ni: Org.Verb.1-177
$C_{12}H_{21}NiO_9P$	$(CO)_3NiP(OC_2H_4OCH_3)_3$	Ni: Org.Verb.1-177
$C_{12}H_{21}O_3PS_6$	$P[S_2COC_3H_7]_3$	C: MVol.D4-254
$C_{12}H_{21}O_3PSe_6$	$P[Se_2COC_3H_7]_3$	C: MVol.D6-221
$C_{12}H_{21}O_3S_6Sb$	$Sb[S_2COC_3H_7]_3$	C: MVol.D4-254
$C_{12}H_{21}O_3S_6Tl$	$Tl[S_2COC_3H_7]_3$	C: MVol.D4-253
$C_{12}H_{21}PSn$	$(CH_3)_3SnP(C_3H_7)C_6H_5$	Sn: Org.Verb.5-84
$C_{12}H_{21}Ti$	$(CH_3CHCHCH_2)_3Ti$	Ti: Org.Verb.1-127/8
$C_{12}H_{22}INSn$	$[(CH_3)_3SnC_6H_4N(CH_3)_3]I$	Sn: Org.Verb.2-142, 144
$C_{12}H_{22}MnN_2O_4$	$[Mn(CH(COCH_3)_2)_2(NH_2CH_2CH_2NH_2)]$	Mn: MVol.D1-78
–	$[Mn(CH(COCH_3)_2)_2(ND_2CH_2CH_2ND_2)]$	Mn: MVol.D1-78
$C_{12}H_{22}MnO_6$	$[Mn(CH(COCH_3)_2)_2(H_2O)(C_2H_5OH)]$	Mn: MVol.D1-75
$C_{12}H_{22}MnO_6^+$	$[Mn(CH(COCH_3)_2)_2(CH_3OH)_2]^+$	Mn: MVol.D1-105
$C_{12}H_{22}MnO_{12}$	$Mn(OCH(CHOH)_5)_2$	Mn: MVol.D1-38/9
$C_{12}H_{22}MnO_{18}P_2^{2-}$	$[Mn(C_5H_5O(OH)_4CH_2OPO_3)_2]^{2-}$	Mn: MVol.D1-147
$C_{12}H_{22}N_2O_2Sn$	$(C_2H_5)_3SnC_3HN_2(CH_3)COOCH_3$	Sn: Org.Verb.2-247
$C_{12}H_{22}N_2O_3Sn$	$(C_2H_5)_3SnCH_2CH_2CH(CONH)_2CO$	Sn: Org.Verb.2-170, 177
$C_{12}H_{22}N_2Sn$	$(C_2H_5)_3SnC_3H_2N_2C(CH_2)CH_3$	Sn: Org.Verb.2-246
–	$(C_3H_7)_2Sn(CH_2CH_2CN)_2$	Sn: Org.Verb.3-37
$C_{12}H_{22}N_4O_4S$	$SO_2[N(C_6H_{11})NO]_2$	S: S-N-Verb.1-176
$C_{12}H_{22}Na_2O_4Sn$	$(C_3H_7)_2Sn(CH_2CH_2COONa)_2$	Sn: Org.Verb.3-37
$C_{12}H_{22}NiO_2S_4$	$Ni[S_2COC_5H_{11}]_2$	C: MVol.D4-260
$C_{12}H_{22}OSn$	$(C_2H_5)_3SnCCCH_2CH_2OCHCH_2$	Sn: Org.Verb.2-225, 234
–	$[(C_6H_{11})_2SnO]_x$	Sn: Org.Verb.6-109, 125
$C_{12}H_{22}O_2PdS_4$	$Pd[S_2COC_5H_{11}]_2$	C: MVol.D4-261
$C_{12}H_{22}O_2S_4$	$C_5H_{11}OC(S)SSC(S)OC_5H_{11}$	C: MVol.D4-251
$C_{12}H_{22}O_2S_4Zn$	$Zn[S_2COC_5H_{11}]_2$	C: MVol.D4-257
$C_{12}H_{22}O_2Sn$	$(C_2H_5)_3SnCCCH(CH_3)OOCCH_3$	Sn: Org.Verb.2-225
–	$(C_2H_5)_3SnCCCH_2CH_2OOCCH_3$	Sn: Org.Verb.2-225
–	$(C_2H_5)_3SnCH_2COOCH_2CH_2CCH$	Sn: Org.Verb.2-170, 177
$C_{12}H_{22}O_2Ti$	$C_5H_5Ti(CH_3)(OC_3H_7\text{-}i)_2$	Ti: Org.Verb.1-205/6
$C_{12}H_{22}O_3Ti$	$CH_3C_5H_4Ti(OC_2H_5)_3$	Ti: Org.Verb.1-158, 164
$C_{12}H_{22}O_4STi$	$CH_3Ti(SO_2C_5H_5)(OC_3H_7\text{-}i)_2$	Ti: Org.Verb.1-60
–	$C_5H_5Ti(OC_3H_7\text{-}i)_2(SO_2CH_3)$	Ti: Org.Verb.1-190
$C_{12}H_{22}O_4S_2$	$C_5H_{11}OC(O)SS(O)COC_5H_{11}$	C: MVol.D4-236
$C_{12}H_{22}O_4S_4U$	$UO_2[S_2COC_5H_{11}]_2$	C: MVol.D4-261
$C_{12}H_{22}O_4Sn$	$(C_2H_5)_3SnC(COOCH_3)CHCOOCH_3$	Sn: Org.Verb.2-196
–	$[Sn(C_4H_9)_2OCO(CH_2)_2COO]_x$	Sn: Org.Verb.6-94
$C_{12}H_{22}O_6S_2Ti$	$CH_3SO_2Ti(SO_2C_5H_5)(OC_3H_7\text{-}i)_2$	Ti: Org.Verb.1-60
$C_{12}H_{22}SiSn$	$(CH_3)_3SnC_6H_4Si(CH_3)_3$	Sn: Org.Verb.2-142, 144
$C_{12}H_{22}Sn$	$(CH_2CHCH_2)_3SnC_3H_7$	Sn: Org.Verb.2-348
–	$(CH_2CH)_3SnC_6H_{13}$	Sn: Org.Verb.2-347
–	$(CH_3)(i\text{-}C_3H_7)Sn(CH(CH_2)_4)(CHCCH_2)$	Sn: Org.Verb.3-95, 97
–	$(CH_3)(i\text{-}C_3H_7)Sn(CH(CH_2)_4)(CH_2CCH)$	Sn: Org.Verb.3-95, 97
–	$(CH_3)_3SnCHCH(CH_2)_5CCH$	Sn: Org.Verb.2-115

Formula	Compound	Reference
$C_{12}H_{22}Sn$	$(C_2H_5)_3SnCCC(CH_3)CHCH_3$	Sn: Org.Verb.2-215
–	$(C_2H_5)_3SnCCC_3H_4CH_3$	Sn: Org.Verb.2-225
–	$(C_2H_5)_3SnCHCHCCC_2H_5$	Sn: Org.Verb.2-215
–	$(C_3H_7)_2SnC_6H_8$	Sn: Org.Verb.3-109/10
$C_{12}H_{23}KOSe_2$	$K[Se_2COC_{11}H_{23}]$	C: MVol.D6-221
$C_{12}H_{23}NO_7U$	$UO_2(CH_3COO)_2 \cdot CH_3CON(C_3H_7)_2$	U: SVol.E1-107, 109
$C_{12}H_{23}NSn$	$(C_2H_5)_3SnCCCHCHN(CH_3)_2$	Sn: Org.Verb.2-225
–	$(C_2H_5)_3SnCHCHCCN(CH_3)_2$	Sn: Org.Verb.2-226
$C_{12}H_{23}N_3O_2Sn$	$(C_2H_5)_3SnCH_2CH_2CH(CONH)_2CNH$	Sn: Org.Verb.2-170, 177
$C_{12}H_{23}ORhSn_2$	$[Rh(CO)(C_5H_5)[(CH_3)_3Sn]_2]$	Sn: Org.Verb.6-33
$C_{12}H_{24}I_2MnO_3$	$MnI_2 \cdot 3\ C_4H_8O$	Mn: MVol.D1-144
$C_{12}H_{24}I_4N_8Sn$	$SnI_4 \cdot 2\ (CH_2)_6N_4$	Sn: MVol.C5-230
$C_{12}H_{24}MnN_2O_{12}$	$Mn(NO_3)_2 \cdot (C_2H_4O)_6 \cdot 6\ H_2O$	Mn: MVol.D1-158/9
$C_{12}H_{24}MnN_{10}O_6$	$Mn(NO_3)_2 \cdot 2\ (CH_2)_6N_4 \cdot 10\ H_2O$	Mn: MVol.C3-291
$C_{12}H_{24}MnO_{12}^{3-}$	$[Mn(O_2C_4H_6(OH)_2)_3]^{3-}$	Mn: MVol.D1-38/9
$C_{12}H_{24}N_2Ni$	$(CH_3)_2C_3H_3Ni(N(CH_2CH_2)_3N)CH_3$	Ni: Org.Verb.2-35, 37
$C_{12}H_{24}N_2NiO_6P_2$	$(NCCHCH_2)_2Ni(P(OCH_3)_3)_2$	Ni: Org.Verb.1-413
$C_{12}H_{24}N_2O_2S$	$C_5H_{10}NSO_2NCH_3(C_6H_{11})$	S: S-N-Verb.1-173
–	$(C_6H_{11}NH)_2SO_2$	S: S-N-Verb.1-169
$C_{12}H_{24}N_2O_4S$	$(CH_3CO)(C_4H_9)NSO_2N(C_4H_9)(COCH_3)$	S: S-N-Verb.1-173/4
$C_{12}H_{24}N_2O_{10}U$	$UO_2(NO_3)_2 \cdot 2\ CH_3COC_4H_9$	U: SVol.E1-79/80
$C_{12}H_{24}N_2O_{14}U$	$UO_2(NO_3)_2 \cdot (CH_2OCH_2)_6 \cdot n\ H_2O$	U: SVol.E1-71/2, 77/8
$C_{12}H_{24}N_2Sn$	$(C_3H_7)_3SnC_3H_3N_2$	Sn: Org.Verb.2-269
$C_{12}H_{24}N_4NiO_4Ti$	$(CH_3)_2N[((CH_3)_2N)_3TiO]CNi(CO)_3$	Ni: Org.Verb.1-334/5
$C_{12}H_{24}N_4O_6S_2$	$O(C_2H_4)_2NSO_2N(C_2H_4)_2NSO_2N(C_2H_4)_2O$	S: S-N-Verb.1-174
$C_{12}H_{24}N_4S_8Sn$	$Sn((CH_3)_2NCSS)_4$	Sn: MVol.C6-117
$C_{12}H_{24}N_6NiO_2P_2$	$(CO)_2Ni(P(N(CH_3)_2)_2CN)_2$	Ni: Org.Verb.1-126
$C_{12}H_{24}N_6O_4S_2U$	$UO_2(NCS)_2 \cdot 2\ CO(N(CH_3)_2)_2$	U: SVol.E1-92/4
$C_{12}H_{24}N_6O_6S_3$	$[NSO(N(CH_2CH_2)_2O)]_3$	S: S-N-Verb.1-19
$C_{12}H_{24}N_6S_2Se_2Te$	$Te((CH_3)_2NCSN(CH_3)_2)_2(SeCN)_2$	Te: SVol.B3-148/9
$C_{12}H_{24}N_8O_2P_2S$	$SO_2[NP(NC_2H_4)_3]_2$	S: S-N-Verb.1-192
$C_{12}H_{24}NiO_2P_2$	$(CO)_2Ni(C_2H_5)_2P(CH_2)_2P(C_2H_5)_2$	Ni: Org.Verb.1-149, 152
$C_{12}H_{24}NiO_6P_2$	$(CO)_2Ni(C_2H_5O)_2P(CH_2)_2P(OC_2H_5)_2$	Ni: Org.Verb.1-150, 153
$C_{12}H_{24}NiO_{12}P_4$	$Ni(C_2H_4O_2POC_2H_4OPC_2H_4O_2)_2$	Ni: Org.Verb.1-245
$C_{12}H_{24}Ni_2$	$(CH_3CHCHCHCH_3NiCH_3)_2$	Ni: Org.Verb.2-314/5
–	$(CH_3CHCHCH_2NiC_2H_5)_2$	Ni: Org.Verb.2-299
$C_{12}H_{24}OSn$	$(C_2H_5)_3SnCCC(CH_3)_2OCH_3$	Sn: Org.Verb.2-226
–	$(C_2H_5)_3SnC(CH(OH)CH_3)CCHCH_3$	Sn: Org.Verb.2-196
–	$(C_2H_5)_3SnCHCH(CH_2)_2OCHCH_2$	Sn: Org.Verb.2-196
–	$(C_2H_5)_3SnCH(CH_2)_4CO$	Sn: Org.Verb.2-181/2
–	$(C_2H_5)_3SnCH_2CH_2CCCH(OH)CH_3$	Sn: Org.Verb.2-226
–	$(C_4H_9)_2Sn(CHCH_2)(CH(O)CH_2)$	Sn: Org.Verb.3-88/9
$C_{12}H_{24}O_2Sn$	$(C_2H_5)_3SnCCCH(CH_3)OCH_2OCH_3$	Sn: Org.Verb.2-226
–	$(C_2H_5)_3SnCCCH_2CH_2OCH_2OCH_3$	Sn: Org.Verb.2-226
$C_{12}H_{24}O_4Sn$	$(CH_3)_3SnCH(CH_3)CH(COOC_2H_5)_2$	Sn: Org.Verb.2-44
–	$(C_2H_5)_2Sn(CH_2CH_2COOCH_3)_2$	Sn: Org.Verb.3-27
–	$(C_2H_5)_2Sn(CH_2COOC_2H_5)_2$	Sn: Org.Verb.3-27
–	$(C_3H_7)_2Sn(CH_2COOCH_3)_2$	Sn: Org.Verb.3-37
–	$(C_4H_9)_2Sn(OCOCH_3)_2$	Sn: Org.Verb.6-89, 95
$C_{12}H_{24}O_{12}U_2$	$2\ UO_3 \cdot [CH_2OCH_2]_6$	U: SVol.E1-78

Formula	Compound	Reference
$C_{12}H_{24}SSn$	$(C_2H_5)_3SnCHCHCHCHSC_2H_5$	Sn: Org.Verb.2-197
$C_{12}H_{24}SiSn$	$(CH_3)_3SnC_5H_4CH_2Si(CH_3)_3$	Sn: Org.Verb.2-100
–	$(C_2H_5)_3SnCHCHSi(CH_3)_2CCH$	Sn: Org.Verb.2-226
$C_{12}H_{24}Sn$	$((CH_2)_5CH)_2SnH_2$	Sn: Org.Verb.4-112/3
–	$(CH_3)_2Sn(CH(CH_2)_4)_2$	Sn: Org.Verb.3-9
–	$(C_2H_5)_3SnCCC(CH_3)_3$	Sn: Org.Verb.2-215
–	$(C_2H_5)_3SnCCC_4H_9$	Sn: Org.Verb.2-212, 215
–	$(C_2H_5)_3SnC(CHCH_2)CHC_2H_5$	Sn: Org.Verb.2-188
–	$(C_2H_5)_3SnC(CHCH_3)CHCHCH_3$	Sn: Org.Verb.2-187
–	$(C_2H_5)_3SnC(CH_3)CHCHCHCH_3$	Sn: Org.Verb.2-188
–	$(C_2H_5)_3SnC(C_2H_5)CHCHCH_2$	Sn: Org.Verb.2-188
–	$(C_2H_5)_3SnCH(CH_3)CH_2CCCH_3$	Sn: Org.Verb.2-215
–	$(C_2H_5)_3SnCH_2CH_2CCC_2H_5$	Sn: Org.Verb.2-215
–	$(C_3H_7)_2Sn(CH_2CHCH_2)_2$	Sn: Org.Verb.3-37
–	$(C_4H_9)_2Sn(CHCH_2)_2$	Sn: Org.Verb.3-41/2
$C_{12}H_{25}NSn$	$(i\text{-}C_3H_7)_3SnCH(CH_3)CN$	Sn: Org.Verb.2-271
–	$(C_3H_7)_3SnCH_2CH_2CN$	Sn: Org.Verb.2-251/2
–	$(i\text{-}C_3H_7)_3SnCH_2CH_2CN$	Sn: Org.Verb.2-271
$C_{12}H_{25}N_3O_4S_2$	$HN(SO_2NHC_6H_{11})_2$	S: S-N-Verb.1-161
$C_{12}H_{25}N_3Ti$	$CH_3C_5H_4Ti(N(CH_3)_2)_3$	Ti: Org.Verb.1-171/2, 175/9
$C_{12}H_{25}NaO_2Sn$	$(C_3H_7)_3SnCH_2CH_2COONa$	Sn: Org.Verb.2-252, 257
$C_{12}H_{25}O_3PSn$	$(C_2H_5)_3SnCCPO(OC_2H_5)_2$	Sn: Org.Verb.2-226, 234
$C_{12}H_{26}MnO_{12}$	$Mn(OC_6H_8(OH)_5)_2$	Mn: MVol.D1-38/9
$C_{12}H_{26}NO_2Sn$	$(C_2H_5)_3SnCH(COOCH_3)CHN(CH_3)_2$	Sn: Org.Verb.2-170
$C_{12}H_{26}N_2OSn$	$[CH_2CHC(NH_2)NOSn(C_3H_7)_3]_n$	Sn: Org.Verb.5-114
$C_{12}H_{26}N_2O_6Sn$	$Sn((OC_2H_4)_2NC_2H_4OH)_2$	Sn: MVol.C6-3
$C_{12}H_{26}N_2O_{11}U$	$UO_2(NO_3)_2 \cdot (C_4H_9OC_2H_4)_2O \cdot 3\ H_2O$	U: SVol.E1-71
$C_{12}H_{26}N_2Sn$	$[(C_4H_9)_2SnN(CH_2CH_2)_2N]_x$	Sn: Org.Verb.6-89, 97
$C_{12}H_{26}N_4O_2Sn$	$[Sn(C_4H_9)_2N_2H_2C(O)(CH_2)_2C(O)N_2H_2]_x$	Sn: Org.Verb.6-89, 98
$C_{12}H_{26}N_4O_4S_2$	$C_5H_{10}NSO_2NH(CH_2)_2NHSO_2NC_5H_{10}$	S: S-N-Verb.1-171
$C_{12}H_{26}N_4O_{10}U$	$UO_2(NO_3)_2 \cdot 2\ CH_3CON(C_2H_5)_2$	U: SVol.E1-107/9
–	$UO_2(NO_3)_2 \cdot 2\ C_3H_7CON(CH_3)_2$	U: SVol.E1-110/1
$C_{12}H_{26}OSiSn$	$(C_2H_5)_3SnCCCH_2OSi(CH_3)_3$	Sn: Org.Verb.2-227
$C_{12}H_{26}OSn$	$(C_2H_5)_3SnCHCHC(OH)(CH_3)C_2H_5$	Sn: Org.Verb.2-197
–	$(C_2H_5)_3SnCHCHOC_4H_9$	Sn: Org.Verb.2-197
–	$(C_2H_5)_3Sn(CH_2)_3OCH_2CHCH_2$	Sn: Org.Verb.2-170
–	$(C_2H_5)_3SnOC_6H_{11}$	Sn: Org.Verb.5-105
–	$(C_3H_7)_3SnCHCHCH_2OH$	Sn: Org.Verb.2-260
–	$(C_3H_7)_3SnCH_2COCH_3$	Sn: Org.Verb.2-250/1
$C_{12}H_{26}O_2SSn$	$(C_2H_5)_3SnCH_2CH_2SCH_2COOC_2H_5$	Sn: Org.Verb.2-183
$C_{12}H_{26}O_2Sn$	$(C_2H_5)_3SnCH(CH_3)CH_2COOC_2H_5$	Sn: Org.Verb.2-171, 177
–	$(C_2H_5)_3SnCH_2CH_2CH_2COOC_2H_5$	Sn: Org.Verb.2-170, 177
–	$(C_2H_5)_3SnCH_2COO(i\text{-}C_4H_9)$	Sn: Org.Verb.2-170, 177
–	$(C_2H_5)_3Sn(CH_2)_3OCH_2CH(O)CH_2$	Sn: Org.Verb.2-170
–	$(C_3H_7)_3SnCH_2COOCH_3$	Sn: Org.Verb.2-251
–	$(C_{12}H_{25}SnOOH)_x$	Sn: Org.Verb.6-253
$C_{12}H_{26}SSn$	$(C_2H_5)_3SnCHCHSC_4H_9$	Sn: Org.Verb.2-197, 204
$C_{12}H_{26}SiSn$	$(C_2H_5)_3SnCCCH_2Si(CH_3)_3$	Sn: Org.Verb.2-226, 234
$C_{12}H_{26}Sn$	$(CH_3)(C_2H_5)Sn(i\text{-}C_3H_7)(C_6H_{11})$	Sn: Org.Verb.3-97
–	$(CH_3)_2Sn(s\text{-}C_4H_9)(C_6H_{11})$	Sn: Org.Verb.3-75, 78

Formula	Compound		Reference
$C_{12}H_{26}Sn$	$(CH_3)_3SnC(C_5H_{11})CHC_2H_5$	Sn:	Org.Verb.2-82
–	$(CH_3)_3SnCHCHC_7H_{15}$	Sn:	Org.Verb.2-82
–	$(C_2H_5)_3SnC(C_4H_9)CH_2$	Sn:	Org.Verb.2-188
–	$(C_2H_5)_3SnCHCHCH_2CH_2CH_2CH_3$	Sn:	Org.Verb.2-188, 190
–	$(C_2H_5)_3SnCH_2C(CH_3)C(CH_3)_2$	Sn:	Org.Verb.2-188
–	$(C_2H_5)_3SnCH_2CH(CH_3)C(CH_3)CH_2$	Sn:	Org.Verb.2-188
–	$(C_2H_5)_3SnCH_2CH_2CH_2CH_2CHCH_2$	Sn:	Org.Verb.2-188
–	$(i\text{-}C_3H_7)_3SnCHCH_2CH_2$	Sn:	Org.Verb.2-270/1
–	$(C_3H_7)_3SnCH_2CHCH_2$	Sn:	Org.Verb.2-259/60
–	$(i\text{-}C_3H_7)_3SnCH_2CHCH_2$	Sn:	Org.Verb.2-271
–	$(C_4H_9)_2Sn(CH_2)_4$	Sn:	Org.Verb.3-102/3
$C_{12}H_{27}LiSn$	$(C_4H_9)_3SnLi$	Sn:	Org.Verb.5-129
$C_{12}H_{27}NOSn$	$(C_2H_5)_3SnCH(CHO)CH(CH_3)N(CH_3)_2$	Sn:	Org.Verb.2-171
–	$(C_2H_5)_3SnCH_2CON(C_2H_5)_2$	Sn:	Org.Verb.2-171, 177
–	$(C_3H_7)_3SnCH_2CH_2CONH_2$	Sn:	Org.Verb.2-252
$C_{12}H_{27}NO_2Sn$	$(C_2H_5)_3SnCH(COOCH_3)CH_2N(CH_3)_2$	Sn:	Org.Verb.2-171, 177
$C_{12}H_{27}NO_3Sn$	$(C_4H_9)_3SnNO_3$	Sn:	Org.Verb.5-132
$C_{12}H_{27}NSn$	$(CH_3)_3SnNC(C_4H_9)_2$	Sn:	Org.Verb.5-84
$C_{12}H_{27}N_2O_9PU$	$UO_2(NO_3)_2 \cdot (C_4H_9)_3PO$	U:	SVol.E1-151, 159
$C_{12}H_{27}N_2O_{12}PU$	$UO_2(NO_3)_2 \cdot (C_4H_9O)_3PO$	U:	SVol.E1-182
$C_{12}H_{27}N_3Sn$	$(C_4H_9)_3SnN_3$	Sn:	Org.Verb.5-131
$C_{12}H_{27}NiO_3PPb_3$	$(CO)_3NiP(Pb(CH_3)_3)_3$	Ni:	Org.Verb.1-163/4
$C_{12}H_{27}NiO_3PSi_3$	$(CO)_3NiP(Si(CH_3)_3)_3$	Ni:	Org.Verb.1-163/4
$C_{12}H_{27}NiO_3PSn_3$	$(CO)_3NiP(Sn(CH_3)_3)_3$	Ni:	Org.Verb.1-163/4
$C_{12}H_{27}NiO_3SbSi_3$	$(CO)_3NiSb(Si(CH_3)_3)_3$	Ni:	Org.Verb.1-187, 189
$C_{12}H_{27}NiO_3SbSn_3$	$(CO)_3NiSb(Sn(CH_3)_3)_3$	Ni:	Org.Verb.1-188, 189
$C_{12}H_{27}O_{11}PRe_2U$	$UO_2(ReO_4)_2 \cdot (C_4H_9)_3PO$	U:	SVol.E1-150, 159
$C_{12}H_{28}I_2N_4O_6U$	$UO_2I_2 \cdot 4\ HCON(CH_3)_2$	U:	SVol.E1-100/1
$C_{12}H_{28}I_4N_4O_4U$	$UI_4 \cdot 4\ HCON(CH_3)_2$	U:	SVol.E1-99/100
$C_{12}H_{28}Li_2NiO_2$	$Li_2[Ni(CH_3)_4] \cdot 2\ C_4H_8O$	Ni:	Org.Verb.1-104
$C_{12}H_{28}NO_{11}UV_3$	$C_{12}H_{25}NH_3[UO_2(VO_3)_3] \cdot n\ H_2O$	U:	SVol.C3-302/3
$C_{12}H_{28}N_2O_2S$	$(C_2H_5)(C_4H_9)NSO_2N(C_2H_5)C_4H_9$	S:	S-N-Verb.1-173
–	$(C_2H_5)(i\text{-}C_4H_9)NSO_2N(C_2H_5)(C_4H_9\text{-}i)$	S:	S-N-Verb.1-173
–	$(C_2H_5)_2NSO_2N(C_4H_9)_2$	S:	S-N-Verb.1-173
$C_{12}H_{28}N_2O_2Sn$	$Sn(OC_2H_4N(C_2H_5)_2)_2$	Sn:	MVol.C5-30
$C_{12}H_{28}N_2O_3Si_2Sn$	$O(Si(CH_3)_2CH_2)_2Sn(ONC(CH_3)_2)_2$	Sn:	Org.Verb.6-207
$C_{12}H_{28}N_2O_{10}U$	$UO_2(NO_3)_2 \cdot 2\ (C_3H_7)_2O$	U:	SVol.E1-71, 76
–	$UO_2(NO_3)_2 \cdot 2\ (C_3H_7)_2O \cdot 2\ H_2O$	U:	SVol.E1-71, 76
$C_{12}H_{28}N_2O_{12}U$	$UO_2(NO_3)_2 \cdot 2\ C_2H_4(OC_2H_5)_2 \cdot 2\ H_2O$	U:	SVol.E1-71, 76
$C_{12}H_{28}N_2O_{15}W_4$	$(C_6H_{14}NO)_2W_4O_{13}$	W:	SVol.B3-244
$C_{12}H_{28}N_4O_4S_2$	$(C_2H_5)_2NSO_2N(CH_2CH_2)_2NSO_2N(C_2H_5)_2$	S:	S-N-Verb.1-174
$C_{12}H_{28}OSiSn$	$(C_2H_5)_3SnCHCHSi(CH_3)_2OC_2H_5$	Sn:	Org.Verb.2-197
$C_{12}H_{28}OSn$	$(C_2H_5)_3SnCH_2CH_2OC_4H_9$	Sn:	Org.Verb.2-171
–	$(C_2H_5)_3SnCH_2CH_2(OC_4H_9\text{-}i)$	Sn:	Org.Verb.2-171
$C_{12}H_{28}O_2S_6Sn$	$Sn(SCH_2CH_2S)_2 \cdot 2\ (C_2H_5)_2SO$	Sn:	MVol.C6-102/3
$C_{12}H_{28}O_2Sn$	$(C_4H_9)_2Sn(CH_2OCH_3)_2$	Sn:	Org.Verb.3-43
$C_{12}H_{28}O_4Sn$	$(C_2H_5)_2Sn[OOC(CH_3)_3]_2$	Sn:	Org.Verb.6-58
$C_{12}H_{28}O_5U$	$UO_2(OC_4H_9)_2 \cdot C_4H_9OH$	U:	SVol.E1-59, 64
$C_{12}H_{28}SSn$	$(C_2H_5)_3SnCH_2CH_2SC(CH_3)_3$	Sn:	Org.Verb.2-171, 177
–	$(C_2H_5)_3SnCH_2CH_2SC_4H_9$	Sn:	Org.Verb.2-171, 177

$C_{12}H_{28}Si_2Sn$ .... $(CH_2CH)_2Sn[CH_2Si(CH_3)_3]_2$ .... Sn: Org.Verb.3-71
$C_{12}H_{28}Sn$ .... $(CH_3)_2Sn(C(CH_3)_2C_2H_5)_2$ .... Sn: Org.Verb.3-10
– .... $(CH_3)_2Sn(CH(CH_3)C_3H_7)_2$ .... Sn: Org.Verb.3-10, 17
– .... $(CH_3)_2Sn(C_5H_{11})_2$ .... Sn: Org.Verb.3-10
– .... $(C_2H_5)_2Sn(C_4H_9)_2$ .... Sn: Org.Verb.3-24
– .... $(C_2H_5)_2Sn(i-C_4H_9)_2$ .... Sn: Org.Verb.3-27
– .... $(C_2H_5)_2Sn(t-C_4H_9)_2$ .... Sn: Org.Verb.3-28
– .... $(C_2H_5)_3SnC_6H_{13}$ .... Sn: Org.Verb.2-159
– .... $(C_3H_7)_2Sn(i-C_3H_7)_2$ .... Sn: Org.Verb.3-37
– .... $(C_3H_7)_4Sn$ .... Sn: Org.Verb.1-82/90
– .... $(i-C_3H_7)_4Sn$ .... Sn: Org.Verb.1-90/3
– .... $(C_4H_9)_2Sn(CH_3)(i-C_3H_7)$ .... Sn: Org.Verb.3-88/9
– .... $(t-C_4H_9)_2Sn(C_2H_5)_2$ .... Sn: Org.Verb.6-107
– .... $(C_4H_9)_3SnH$
Chemical reactions .... Sn: Org.Verb.4-46/65
Formation and preparation .... Sn: Org.Verb.4-41/3
Physical properties .... Sn: Org.Verb.4-45
Spectra .... Sn: Org.Verb.4-43/5
Uses .... Sn: Org.Verb.4-65/6
– .... $(C_4H_9)_3SnD$ .... Sn: Org.Verb.4-41, 45, 53, 58, 60/2
– .... $(i-C_4H_9)_3SnH$ .... Sn: Org.Verb.4-66/7
– .... $(i-C_4H_9)_3SnD$ .... Sn: Org.Verb.4-66
– .... $(s-C_4H_9)_3SnH$ .... Sn: Org.Verb.4-67
– .... $(t-C_4H_9)_3SnH$ .... Sn: Org.Verb.4-67
$C_{12}H_{28}Sn^+$ .... $Sn(C_3H_7)_4^+$ .... Sn: Org.Verb.1-86
$C_{12}H_{28}Ti$ .... $(C_3H_7)_4Ti$ .... Ti: Org.Verb.1-90, 98
– .... $(C_4H_9)_2(C_2H_5)_2Ti$ .... Ti: Org.Verb.1-90
$C_{12}H_{29}NO_2Sn$ ...... $(C_2H_5)_3Sn(CH_2)_3OCH_2CH(OH)CH_2NH_2$ ..... Sn: Org.Verb.2-171
$C_{12}H_{29}NSn$ .... $(C_3H_7)_3SnCH_2CH_2CH_2NH_2$ .... Sn: Org.Verb.2-252, 257
– .... $(C_4H_9)_3SnNH_2$ .... Sn: Org.Verb.5-129
$C_{12}H_{29}N_3O_4S_2$ ...... $HN(SO_2N(C_3H_7)_2)_2$ .... S: S-N-Verb.1-161
– .... $HN(SO_2N(i-C_3H_7)_2)_2$ .... S: S-N-Verb.1-161
$C_{12}H_{30}IOPSn$ .... $(C_2H_5)_3SnI \cdot (C_2H_5)_3PO$ .... Sn: Org.Verb.5-107
$C_{12}H_{30}I_4O_4P_2Sn$ ..... $SnI_4 \cdot 2\ C_2H_5(H)P(O)(OC_4H_9-n)$ .... Sn: MVol.C6-151/2
$C_{12}H_{30}I_4O_6P_2Sn$ ..... $SnI_4 \cdot 2\ C_2H_5P(O)(OC_2H_5)_2$ .... Sn: MVol.C6-162/3
– .... $SnI_4 \cdot 2\ P(OC_2H_5)_3$ .... Sn: MVol.C6-141/2
$C_{12}H_{30}IrO_6P_3S_6$ ..... $Ir(S_2P(OC_2H_5)_2)_3$ .... Ir: SVol.2-170
$C_{12}H_{30}IrO_6P_3Se_6$ .... $Ir(Se_2P(OC_2H_5)_2)_3$ .... Ir: SVol.2-175
$C_{12}H_{30}N_2NiP_2$ ...... $Ni(P(C_2H_5)_3)_2N_2$ .... Ni: Org.Verb.1-79
$C_{12}H_{30}N_2O_{10}P_2U$ .... $UO_2(NO_3)_2 \cdot 2\ (C_2H_5)_3PO$ .... U: SVol.E1-149, 156/7
$C_{12}H_{30}N_2O_{11}U$ ...... $UO_2(NO_3)_2 \cdot 3\ C_4H_9OH$ .... U: SVol.E1-59, 63
– .... $UO_2(NO_3)_2 \cdot 3\ C_4H_9OH \cdot 2\ H_2O$ .... U: SVol.E1-59, 63
– .... $UO_2(NO_3)_2 \cdot 3\ C_4H_9OH \cdot 3\ H_2O$ .... U: SVol.E1-59, 63
$C_{12}H_{30}N_2O_{16}P_2U$ .... $UO_2(NO_3)_2 \cdot 2\ (C_2H_5O)_3PO$ .... U: SVol.E1-179/82, 184
$C_{12}H_{30}N_2Sn$ .... $(C_2H_5)_2Sn[N(C_2H_5)_2]_2$ .... Sn: Org.Verb.6-59
$C_{12}H_{30}N_4O_4S_6Te$ .... $Te((CH_3)_2NCSN(CH_3)_2)_2(S_2O_2CH_3)_2$ .... Te: SVol.B3-149
$C_{12}H_{30}N_4O_8U$ .... $UO_2(NO_3)_2 \cdot 2\ (C_2H_5)_3N$ .... U: SVol.E1-20, 22/3
$C_{12}H_{30}N_4O_{14}U$ ...... $UO_2(NO_3)_2 \cdot 2\ [N(C_2H_4OH)_3]$ .... U: SVol.E1-20, 23
$C_{12}H_{30}N_4O_{15}S_3U$ .... $U(NO_3)_4 \cdot 3\ (C_2H_5)_2SO$ .... U: SVol.E1-210/1

Formula	Compound	Reference
$C_{12}H_{30}N_6O_2Sn_3$	$(C_2H_5)_2Sn[OSn(C_2H_5)_2N_3]_2$	Sn: Org.Verb.6-59
$C_{12}H_{30}Ni_2O_4P_2$	$(CH_3Ni(P(CH_3)_3)O_2CCH_3)_2$	Ni: Org.Verb.2-252/3
$C_{12}H_{30}OSi_2Sn$	$(CH_3)_2Sn(CH_2CH_2CH_2Si(CH_3)_2)_2O$	Sn: Org.Verb.3-110
$C_{12}H_{30}OSn_2$	$[(C_2H_5)_3Sn]_2O$	Sn: Org.Verb.5-105
$C_{12}H_{30}O_5Sn_2$	$[(CH_3)_3COOSn(CH_3)_2]_2O$	Sn: Org.Verb.6-27
$C_{12}H_{30}O_9U_2$	$U_2O_5(OC_3H_7)_2 \cdot 2\ C_3H_7OH$	U: SVol.E1-59, 64
$C_{12}H_{30}SSn_2$	$[(C_2H_5)_3Sn]_2S$	Sn: Org.Verb.5-105
$C_{12}H_{30}S_3Sn_3$	$[(C_2H_5)_2SnS]_3$	Sn: Org.Verb.6-56, 59
$C_{12}H_{30}Sn_2Zn$	$[(C_2H_5)_3Sn]_2Zn$	Sn: Org.Verb.5-102
$C_{12}H_{32}I_4N_4Sn$	$SnI_4 \cdot 2\ NH_2(CH_2)_6NH_2$	Sn: MVol.C5-185/6
$C_{12}H_{32}N_4O_4S_2$	$(C_2H_5)_2NH_2[N(SO_2N(C_2H_5)_2)_2]$	S: S-N-Verb.1-162
$C_{12}H_{32}N_4Sn$	$Sn(CH_2CH_2CH_2NH_2)_4$	Sn: Org.Verb.1-125
$C_{12}H_{32}N_8S_4Te^{2+}$	$Te(SC(NH_2)_2)_2((CH_3)_2NCSN(CH_3)_2)_2{}^{2+}$	Te: SVol.B3-149/50
$C_{12}H_{32}N_{10}O_{12}U$	$UO_2(NO_3)_2 \cdot 4\ CO(NHCH_3)_2$	U: SVol.E1-92, 95
$C_{12}H_{32}Si_2Sn$	$(CH_3)_2Sn[(CH_2)_3SiH(CH_3)_2]_2$	Sn: Org.Verb.3-10
$C_{12}H_{33}IrP_2$	$Ir(P(C_2H_5)_3)_2H_3$	Ir: SVol.2-195
$C_{12}H_{34}MnN_2O_{17}$	$[MnNO_3(H_2O)_5][(C_2H_4O)_6]NO_3 \cdot H_2O$	Mn: MVol.D1-158/9
$C_{12}H_{34}Ni_2O_2P_2$	$(CH_3Ni(P(CH_3)_3)OC_2H_5)_2$	Ni: Org.Verb.2-252
$C_{12}H_{34}O_6Si_4Sn$	$[(OSi(CH_3)_2)_4OSn(C_2H_5)_2O]_x$	Sn: Org.Verb.6-53
$C_{12}H_{34}Si_3Sn$	$[(CH_3)_3SiCH_2]_3SnH$	Sn: Org.Verb.4-68
$C_{12}H_{35}IrP_2$	$Ir(P(C_2H_5)_3)_2H_5$	Ir: SVol.2-195
$C_{12}H_{35}O_{13}Sn_4{}^+$	$[(C_2H_5Sn)_4(OCOCH_3)_2(OH)_9]^+$	Sn: Org.Verb.6-230
$C_{12}H_{36}I_4N_6O_2P_2Sn$	$SnI_4 \cdot 2\ ((CH_3)_2N)_3PO$	Sn: MVol.C6-171/2
$C_{12}H_{36}I_4N_6Sn$	$SnI_4 \cdot 3\ NH_2(CH_2)_4NH_2$	Sn: MVol.C5-185
$C_{12}H_{36}MnO_6S_6{}^{2+}$	$[Mn(CH_3SOCH_3)_6]^{2+}$	Mn: MVol.C5-254, 288
$C_{12}H_{36}N_2O_2SSi_4$	$[(CH_3)_3Si]_2NSO_2N[Si(CH_3)_3]_2$	S: S-N-Verb.1-174
$C_{12}H_{36}N_2O_2SSn_4$	$[(CH_3)_3Sn]_2NSO_2N[Sn(CH_3)_3]_2$	S: S-N-Verb.1-174
$C_{12}H_{36}N_3O_{13}P_4U^+$	$[U(NO_3)_3((CH_3)_3PO)_4]^+$	U: SVol.E1-156
$C_{12}H_{36}N_4O_{16}P_4U$	$U(NO_3)_4 \cdot 4\ (CH_3)_3PO$	U: SVol.E1-148, 156
$C_{12}H_{36}N_4SSi_2Sn_2{}^+$	$[(CH_3)_3SiN]_2S[NSn(CH_3)_3]_2{}^+$	S: S-N-Verb.1-210
$C_{12}H_{36}N_6O_8P_2SU$	$UO_2SO_4 \cdot 2\ [(CH_3)_2N]_3PO$	U: SVol.E1-163, 165
$C_{12}H_{36}N_6O_{12}U$	$UO_2(NO_3)_2 \cdot 4\ (CH_3)_3NO$	U: SVol.E1-131/2, 135
$C_{12}H_{36}N_8O_8P_2U$	$UO_2(NO_2)_2 \cdot 2\ [(CH_3)_2N]_3PO$	U: SVol.E1-163, 165
$C_{12}H_{36}N_8O_{10}P_2U$	$UO_2(NO_3)_2 \cdot 2\ [(CH_3)_2N]_3PO$	U: SVol.E1-161, 163, 165
$C_{12}H_{36}N_{10}O_{14}P_2U$	$U(NO_3)_4 \cdot 2\ [(CH_3)_2N]_3PO$	U: SVol.E1-162, 164
$C_{12}H_{36}N_{10}O_{16.5}P_3U$	$U(NO_3)_4 \cdot 1.5\ [((CH_3)_2N)_2PO]_2O$	U: SVol.E1-166/7
$C_{12}H_{36}N_{12}O_4P_2U$	$UO_2(N_3)_2 \cdot 2\ [(CH_3)_2N]_3PO$	U: SVol.E1-163, 165
$C_{12}H_{36}NiO_{12}P_4$	$Ni(P(OCH_3)_3)_4$	Ni: Org.Verb.1-240, 242
		Ni: Org.Verb.2-99
$C_{12}H_{36}NiP_4$	$Ni(P(CH_3)_3)_4$	Ni: Org.Verb.1-11, 72
		Ni: Org.Verb.2-89, 91
$C_{12}H_{36}Ni_2O_4P_4$	$(CH_3Ni(P(CH_3)_3)O_2P(CH_3)_2)_2$	Ni: Org.Verb.2-251
$C_{12}H_{36}SiSn_4$	$[(CH_3)_3Sn]_4Si$	Sn: Org.Verb.5-84
$C_{12}H_{36}Si_4Sn$	$[(CH_3)_3Si]_4Sn$	Sn: MVol.C2-235
$C_{12}H_{42}Ir_2N_{30}O_{12}S_3$	$[Ir((NH_2C(NH))_2NH)_3]_2(SO_4)_3 \cdot 9\ H_2O$	Ir: SVol.2-94
$C_{12}H_{48}Ir_2N_{24}O_{12}S_{15}$	$[Ir(SC(NH_2)_2)_6]_2(SO_4)_3$	Ir: SVol.2-168
$C_{12}H_{48}N_{12}Na_4Sn_9$	$Na_4Sn_9 \cdot 6\ C_2H_8N_2$	Sn: MVol.C5-13/4
$C_{12}Ir_2Mn_3N_{12}$	$Mn_3[Ir(CN)_6]_2 \cdot 12\ H_2O$	Ir: SVol.2-183, 184
$C_{12}Ir_2N_{12}Ni_3$	$Ni_3[Ir(CN)_6]_2 \cdot 12\ H_2O$	Ir: SVol.2-183, 184
–	$Ni_3[Ir(CN)_6]_2 \cdot 18\ H_2O$	Ir: SVol.2-183, 184

$C_{12}Ir_2N_{12}Zn_3$ .... $Zn_3[Ir(CN)_6]_2 \cdot 12\ H_2O$ .... Ir: SVol.2-183, 184
$C_{12}Ir_4O_{12}$ .... $Ir_4(CO)_{12}$ .... Ir: SVol.2-113
$C_{12}K_2N_6S_6Sn$ .... $K_2[Sn((NC)_2CCS_2)_3] \cdot H_2O$ .... Sn: MVol.C6-88
$C_{12}K_2Ni_6O_{12}$ .... $K_2[Ni_6(CO)_{12}]$ .... Ni: Org.Verb.2-392
$C_{12}MnO_{12}^{2-}$ .... $[Mn(C_6O_6)_2]^{2-}$ .... Mn: MVol.D1-133
$C_{12}N_8Ni$ .... $Ni((NC)_2CC(CN)_2)_2$ .... Ni: Org.Verb.1-230
$C_{12}N_9Ti$ .... $(C(CN)_3)_3Ti$ .... Ti: Org.Verb.1-82
$C_{12}Na_2Ni_6O_{12}$ ...... $Na_2[Ni_6(CO)_{12}]$ .... Ni: Org.Verb.2-392
$C_{12}Ni_4O_{18}P_4$ .... $P_4O_6(Ni(CO)_3)_4$ .... Ni: Org.Verb.2-385, 387
$C_{12}Ni_6O_{12}^{2-}$ .... $[Ni_6(CO)_{12}]^{2-}$ .... Ni: Org.Verb.2-384, 392
$C_{12.5}H_{27.5}N_{6.5}O_{14.5}U$ .. $U(NO_3)_4 \cdot 2.5\ C_2H_5CON(CH_3)_2$ .... U: SVol.E1-109/11
$C_{12.8}H_{51.2}N_{12.8}Na_4Sn_9$ $Na_4Sn_9 \cdot 6.4\ C_2H_8N_2$ .... Sn: MVol.C5-13/4

$C_{13}ClF_3H_{30}NiP_2$ ..... $CF_3Ni(P(C_2H_5)_3)_2Cl$ .... Ni: Org.Verb.1-28
$C_{13}ClF_5H_5NO$ ...... $HOC_6H_3ClCHNC_6F_5$ .... F: PerFHalOrg.7-59
$C_{13}ClF_6FeH_4NO_2S_2$ .. $ClC_6H_4NCFe(CO)_2S_2C_2(CF_3)_2$ .... Fe: Org.Verb.B4-2
$C_{13}ClF_8HN_2$ .... $(ClC_6F_4)(NCC_6F_4)NH$ .... F: PerFHalOrg.7-14/6, 42
$C_{13}ClF_8N$ .... $ClC(C_6F_4)_2N$ .... F: PerFHalOrg.6-135/6, 149
$C_{13}ClFeH_7O_5$ .... $C_6H_5COCHCHClFe(CO)_4$ .... Fe: Org.Verb.B4-267/8, 276, 293/4
– .... $HC(O)CHCHC_6H_4ClFe(CO)_4$ .... Fe: Org.Verb.B4-273, 291/2
$C_{13}ClFeH_7O_6$ .... $HOC(O)CHCHC_6H_4ClFe(CO)_4$ .... Fe: Org.Verb.B4-274, 292/3
$C_{13}ClFeH_9O_4$ .... $ClC_6H_4C(CH_3)CH_2Fe(CO)_4$ .... Fe: Org.Verb.B4-258
$C_{13}ClFeH_{11}O$ .... $C_5H_5FeC_5H_4CClCHCHO$ .... Fe: Org.Verb.A2-173/4, 176
– .... $C_5H_5FeC_5H_4CHCHCOCl$ .... Fe: Org.Verb.A3-159
– .... $C_5H_5FeC_5H_4COCHCHCl$ .... Fe: Org.Verb.A2-248/9
$C_{13}ClFeH_{11}O_7$ ...... $[(C_6H_5)(CH_3)C_3H_3Fe(CO)_3]ClO_4$ .... Fe: Org.Verb.B5-85
$C_{13}ClFeH_{13}$ .... $C_5H_5FeC_5H_4CClCHCH_3$ .... Fe: Org.Verb.A1-385, 386
$C_{13}ClFeH_{13}O$ .... $C_5H_5FeC_5H_4CH(CH_3)COCl$ .... Fe: Org.Verb.A3-159
– .... $C_5H_5FeC_5H_4(CH_2)_2COCl$ .... Fe: Org.Verb.A3-159
– .... $C_5H_5FeC_5H_4COCHClCH_3$ .... Fe: Org.Verb.A2-243, 246
– .... $C_5H_5FeC_5H_4COCH_2CH_2Cl$ .... Fe: Org.Verb.A2-242/3, 245/6
$C_{13}ClFeH_{15}$ .... $C_5H_5FeC_5H_4CCl(CH_3)_2$ .... Fe: Org.Verb.A1-381, 383
– .... $C_5H_5FeC_5H_4CH_2CH_2CH_2Cl$ .... Fe: Org.Verb.A1-381, 383
$C_{13}ClFeH_{15}O$ .... $C_5H_5FeC_5H_4CH_2CHOHCH_2Cl$ .... Fe: Org.Verb.A2-74, 79
$C_{13}ClFe_2GeH_{17}N_4O_4$. $(C_3H_5Fe(NO)_2)_2Ge(Cl)CH_2C_6H_5$ .... Fe: Org.Verb.C2-165
– .... $(C_3H_5Fe(NO)_2)_2Ge(Cl)C_6H_4CH_3$ .... Fe: Org.Verb.C2-165
$C_{13}ClFe_2H_{17}N_4O_4Pb$. $(C_3H_5Fe(NO)_2)_2Pb(Cl)CH_2C_6H_5$ .... Fe: Org.Verb.C2-165
– .... $(C_3H_5Fe(NO)_2)_2Pb(Cl)C_6H_4CH_3$ .... Fe: Org.Verb.C2-165
$C_{13}ClFe_2H_{17}N_4O_4Sn$. $(C_3H_5Fe(NO)_2)_2Sn(Cl)CH_2C_6H_5$ .... Fe: Org.Verb.C2-165
– .... $(C_3H_5Fe(NO)_2)_2Sn(Cl)C_6H_4CH_3$ .... Fe: Org.Verb.C2-165
$C_{13}ClH_{12}N_3O_5S_2$ .... $ClSO_2NHCONHSO_2N(C_6H_5)_2$ .... S: S-N-Verb.1-125
$C_{13}ClH_{13}Sn$ .... $(C_6H_5)_2(CH_3)SnCl$ .... Sn: Org.Verb.5-216, 218
$C_{13}ClH_{15}NOSn$ ..... $C_4H_9SnCl(OC_9H_6N)$ .... Sn: Org.Verb.6-246
$C_{13}ClH_{16}NNiO_3$ ..... $[C_6H_5CH_2N(CH_3)_3][(CO)_3NiCl]$ .... Ni: Org.Verb.1-191, 192
$C_{13}ClH_{16}NOSSn$ .... $CH_3ClSn(C_3H_5(COCH_3)C_7H_5NS)$ .... Sn: Org.Verb.6-218/9
$C_{13}ClH_{17}N_2Sn$ ...... $(CH_3)_3SnCl \cdot NC_5H_4C_5H_4N$ .... Sn: Org.Verb.5-87
$C_{13}ClH_{17}O_2Sn$ ...... $(CH_3)_2[C_2H_5OOCCHC(C_6H_5)]SnCl$ .... Sn: Org.Verb.5-195
$C_{13}ClH_{18}NOSn$ ..... $(C_2H_5)_2ClSn(OCH(CH_3)C_6H_4CN)$ .... Sn: Org.Verb.6-54
$C_{13}ClH_{20}NSn$ .... $(CH_3)_3SnCClC(C_6H_5)N(CH_3)_2$ .... Sn: Org.Verb.2-83

Formula	Compound	Reference
$C_{13}ClH_{20}N_6PaS_5$	$[C_8H_{17}NH_3][Pa(SCN)_5Cl]$	Pa: SVol.2-166
$C_{13}ClH_{21}OSn$	$(CH_3)[(CH_3)_3CCH_2](CH_3OC_6H_4)SnCl$	Sn: Org.Verb.5-225, 227
$C_{13}ClH_{23}O_2Ti$	$C_5H_5Ti(OC_4H_9)_2Cl$	Ti: Org.Verb.1-185/6
$C_{13}ClH_{25}OSn$	$(i\text{-}C_4H_9)_2(CH_3OCHCHCHCH)SnCl$	Sn: Org.Verb.5-213/4
$C_{13}ClH_{25}O_2Sn$	$(C_4H_9)_2(CH_3COOCH_2CHCH)SnCl$	Sn: Org.Verb.5-209
$C_{13}ClH_{25}Sn$	$(C_4H_9)_2(C_5H_7)SnCl$	Sn: Org.Verb.5-209
–	$(i\text{-}C_4H_9)_2(C_5H_7)SnCl$	Sn: Org.Verb.5-213/4
–	$(C_6H_{11})_2(CH_3)SnCl$	Sn: Org.Verb.5-221
$C_{13}ClH_{26}NO_4S$	$ClSO_2NHCOOC_{12}H_{25}$	S: S-N-Verb.1-124/5
$C_{13}ClH_{27}OSn$	$(C_4H_9)_2[(CH_3)_2(HO)CCHCH]SnCl$	Sn: Org.Verb.5-209
$C_{13}ClH_{27}O_2Sn$	$(C_4H_9)_2[CH_3COO(CH_2)_3]SnCl$	Sn: Org.Verb.5-209
$C_{13}ClH_{27}Sn$	$(C_4H_9)_2[CH_2C(CH_3)CH_2CH_2]SnCl$	Sn: Org.Verb.5-209
–	$(i\text{-}C_4H_9)_2[CH_2C(CH_3)CH_2CH_2]SnCl$	Sn: Org.Verb.5-213, 215
–	$(C_4H_9)_2[CH_2CHCH(CH_3)CH_2]SnCl$	Sn: Org.Verb.5-209
–	$(i\text{-}C_4H_9)_2[CH_2CHCH(CH_3)CH_2]SnCl$	Sn: Org.Verb.5-213, 215
–	$(C_4H_9)_2[CH_3CHC(CH_3)CH_2]SnCl$	Sn: Org.Verb.5-209
–	$(i\text{-}C_4H_9)_2[CH_3CHC(CH_3)CH_2]SnCl$	Sn: Org.Verb.5-213, 215
–	$(C_4H_9)_2(CH_3CHCHCH_2CH_2)SnCl$	Sn: Org.Verb.5-209
–	$(i\text{-}C_4H_9)_2(CH_3CHCHCH_2CH_2)SnCl$	Sn: Org.Verb.5-213, 215
–	$(C_4H_9)_2(CH_3CH_2CHCHCH_2)SnCl$	Sn: Org.Verb.5-209
–	$(i\text{-}C_4H_9)_2(CH_3CH_2CHCHCH_2)SnCl$	Sn: Org.Verb.5-213, 215
–	$(C_4H_9)_2[(CH_3)_2CCHCH_2]SnCl$	Sn: Org.Verb.5-209
–	$(i\text{-}C_4H_9)_2[(CH_3)_2CCHCH_2]SnCl$	Sn: Org.Verb.5-213, 215
$C_{13}ClH_{29}O_2Sn$	$(C_4H_9)_2[HOCH_2CH_2O(CH_2)_3]SnCl$	Sn: Org.Verb.5-210
$C_{13}ClH_{29}Sn$	$(CH_3)(C_4H_9)(C_8H_{17})SnCl$	Sn: Org.Verb.5-225
–	$(C_4H_9)_2(C_5H_{11})SnCl$	Sn: Org.Verb.5-210
–	$(C_4H_9)_3SnCH_2Cl$	Sn: Org.Verb.2-280
$C_{13}ClH_{30}NOSn$	$(C_3H_7)_3SnCl \cdot NH(CH_2CH_2)_2O$	Sn: Org.Verb.5-114
$C_{13}ClH_{32}OPSn$	$(C_2H_5)_3SnCl \cdot (C_2H_5)_2(C_3H_7O)P$	Sn: Org.Verb.5-107
$C_{13}ClH_{34}NiO_{13}P_3S$	$[Cl(CH_3)C_3H_3Ni(P(OCH_3)_3)_3]HSO_4$	Ni: Org.Verb.2-26
$C_{13}ClH_{35}O_4P_2Sn$	$(CH_3)_3SnCl \cdot 2\ (C_2H_5)(C_3H_7O)HPO$	Sn: Org.Verb.5-87
$C_{13}ClH_{39}NiO_3P_4$	$CH_3Ni(P(CH_3)_3)_3(P(OCH_3)_3)Cl$	Ni: Org.Verb.1-18, 24
$C_{13}ClH_{39}NiP_4$	$CH_3Ni(P(CH_3)_3)_4Cl$	Ni: Org.Verb.1-12, 20
$C_{13}Cl_2F_2H_9NO$	$NC_5F_2Cl_2(OCH_2C_6H_4CH_3)$	F: PerFHalOrg.5-136
		F: PerFHalOrg.6-14
$C_{13}Cl_2F_2H_9NO_2$	$NC_5F_2Cl_2(OCH_2C_6H_4OCH_3)$	F: PerFHalOrg.5-136
		F: PerFHalOrg.6-14
$C_{13}Cl_2F_2H_{18}NO_3PS$	$NC_5F_2Cl_2(OPS(OC_4H_9)_2)$	F: PerFHalOrg.5-212
$C_{13}Cl_2F_2H_{18}NO_4P$	$NC_5F_2Cl_2(OPO(OC_4H_9)_2)$	F: PerFHalOrg.5-212
$C_{13}Cl_2F_5HN_4$	$NC_5F_2Cl_2NHC_6F_3(CN)_2$	F: PerFHalOrg.5-188/9
		F: PerFHalOrg.6-13
$C_{13}Cl_2F_5H_6N$	$Cl_2C_6H_3CH_2NHC_6F_5$	F: PerFHalOrg.7-59
$C_{13}Cl_2F_8H_4Si$	$(C_6F_4Cl)_2SiCH_3H$	F: PerFHalOrg.4-77
$C_{13}Cl_2FeGeH_{10}NiO_3$	$C_5H_5Ni(CO)GeCl_2Fe(CO)_2C_5H_5$	Ni: Org.Verb.2-164/6
$C_{13}Cl_2FeH_{10}NiO_3Sn$	$C_5H_5Ni(CO)SnCl_2Fe(CO)_2C_5H_5$	Ni: Org.Verb.2-164/6
$C_{13}Cl_2FeH_{10}O_2Sn$	$(C_6H_5)Cl_2Sn(CO)_2FeC_5H_5$	Sn: Org.Verb.6-275, 277
$C_{13}Cl_2FeH_{12}$	$C_5H_5FeC_5H_4C(CH_3)CCl_2$	Fe: Org.Verb.A1-385, 386
–	$C_5H_5FeC_5H_4C_3H_3Cl_2$	Fe: Org.Verb.A1-392
$C_{13}Cl_2GeH_{25}NiP$	$C_5H_5Ni(P(C_2H_5)_3)Ge(C_2H_5)Cl_2$	Ni: Org.Verb.2-131
$C_{13}Cl_2H_8O_4Sn$	$SnCl_2(O_2C_6H_2(OH)COC_6H_5)$	Sn: MVol.C5-90/1

Formula	Compound	Reference
$C_{13}Cl_2H_9NOSSn$	$[SnCl_2(SC_6H_4NCHC_6H_4O)]$	Sn: MVol.C6-91/2
$C_{13}Cl_2H_9NO_2Sn$	$[SnCl_2(OC_6H_4CHNC_6H_4O)]$	Sn: MVol.C6-21
$C_{13}Cl_2H_{10}N_2Sn$	$SnCl_2 \cdot C_{13}H_8NNH_2$	Sn: MVol.C5-40
$C_{13}Cl_2H_{12}Sn$	$(C_6H_5)(C_6H_5CH_2)SnCl_2$	Sn: Org.Verb.6-201, 205
–	$(C_6H_5)_2(CH_2Cl)SnCl$	Sn: Org.Verb.5-216, 218
$C_{13}Cl_2H_{13}NO_2U$	$UO_2Cl_2 \cdot (C_6H_5)(C_6H_5CH_2)NH$	U: SVol.E1-19, 22
$C_{13}Cl_2H_{13}N_3O_2P_2S$	$[O_2S(NPClC_6H_5)_2NCH_3]$	S: S-N-Verb.1-34
$C_{13}Cl_2H_{14}N_2O_2U$	$[UO_2Cl_2(CH_2(CH_2C_5H_4N)_2)]$	U: SVol.E1-41, 43
$C_{13}Cl_2H_{14}Sn$	$(i\text{-}C_3H_7)(C_{10}H_7)SnCl_2$	Sn: Org.Verb.6-199
$C_{13}Cl_2H_{15}N_5O_2P_2S$	$[O_2S[NPCl(NHC_6H_5)]_2NCH_3]$	S: S-N-Verb.1-34
$C_{13}Cl_2H_{20}OSn$	$(C_4H_9)(C_6H_5)SnCl_2 \cdot (CH_3)_2CO$	Sn: Org.Verb.6-205
$C_{13}Cl_2H_{23}NSn$	$(C_4H_9)_2SnCl_2 \cdot C_5H_5N$	Sn: Org.Verb.6-99
$C_{13}Cl_2H_{24}O_2S_2Sn$	$(C_3H_7)(C_6H_5)SnCl_2 \cdot 2\ (CH_3)_2SO$	Sn: Org.Verb.6-204
$C_{13}Cl_2H_{28}Sn$	$(CH_3)(C_{12}H_{25})SnCl_2$	Sn: Org.Verb.6-197
–	$(C_4H_9)_3SnCHCl_2$	Sn: Org.Verb.2-280
$C_{13}Cl_2H_{30}OSiSn$	$(C_4H_9)_2[(CH_3)_2ClSiO(CH_2)_3]SnCl$	Sn: Org.Verb.5-210
$C_{13}Cl_2H_{30}Sn_2$	$(C_2H_5)_2ClSn(CH_2)_5SnCl(C_2H_5)_2$	Sn: Org.Verb.6-289
$C_{13}Cl_2H_{31}O_3PSSn$	$(C_2H_5)_2SnCl_2 \cdot (C_4H_9O)_2(CH_3S)PO$	Sn: Org.Verb.6-61
$C_{13}Cl_3F_3H_{11}N_3O_2S$	$((CH_2)_5NSO_2)(CF_3)C_7Cl_3HN_2$	F: PerFHalOrg.6-160, 162
$C_{13}Cl_3F_6HN_2$	$(NCC_6F_4)(C_6F_2Cl_3)NH$	F: PerFHalOrg.7-14/6, 42
$C_{13}Cl_3FeH_{13}O$	$C_5H_5FeC_5H_4C(CCl_3)(CH_3)OH$	Fe: Org.Verb.A2-58
–	$C_5H_5FeC_5H_4CH_2CHOHCCl_3$	Fe: Org.Verb.A2-73
$C_{13}Cl_3H_7O_3Sn$	$SnCl_3(C_{13}H_7O_3) \cdot 0.5\ C_6H_6$	Sn: MVol.C5-163
$C_{13}Cl_3H_7O_4Sn$	$SnCl_3 \cdot C_{13}H_7O_4$	Sn: MVol.C5-163
$C_{13}Cl_3H_9O_2Sn$	$SnCl_3(OC_6H_4COC_6H_5)$	Sn: MVol.C5-90/1
$C_{13}Cl_3H_9O_3Sn$	$SnCl_3(OC_6H_3(OH)COC_6H_5) \cdot 0.25\ C_6H_6$	Sn: MVol.C5-90
–	$SnCl_3(OC_6H_4COC_6H_4OH)$	Sn: MVol.C5-90/1
$C_{13}Cl_3H_{10}MnN$	$(CH(C_6H_4)_2NH)MnCl_3$	Mn: MVol.C5-208/9
$C_{13}Cl_3H_{11}N_2Sn$	$CH_3SnCl_3 \cdot C_{12}H_8N_2$	Sn: Org.Verb.6-222
$C_{13}Cl_3H_{11}N_4SSn$	$SnCl_3(C_6H_5NNCSNNHC_6H_5)$	Sn: MVol.C6-122
$C_{13}Cl_3H_{13}N_2O_2Sn$	$CH_3C_6H_4SnCl_3 \cdot NO_2C_6H_4NH_2$	Sn: Org.Verb.6-282
–	$C_6H_5SnCl_3 \cdot NH_2C_6H_3(CH_3)NO_2$	Sn: Org.Verb.6-279
$C_{13}Cl_3H_{13}N_2Ti$	$CH_2CHCH_2TiCl_3 \cdot NC_5H_4C_5H_4N$	Ti: Org.Verb.1-36/7, 41, 127
$C_{13}Cl_3H_{14}N_3O_2Sn$	$CH_3C_6H_4SnCl_3 \cdot (NH_2)_2C_6H_3NO_2$	Sn: Org.Verb.6-282
–	$C_2H_5SnCl_3 \cdot NC_5H_4NNC_6H_3(OH)_2$	Sn: Org.Verb.6-232
$C_{13}Cl_3H_{15}N_2Ti$	$C_3H_7TiCl_3 \cdot NC_5H_4C_5H_4N$	Ti: Org.Verb.1-34/5, 41
$C_{13}Cl_3H_{16}IrN_2O$	$Ir(C_5H_5N)_2Cl_3 \cdot CH_3COCH_3$	Ir: SVol.2-69, 73, 74
$C_{13}Cl_3H_{17}N_2Sn$	$C_3H_7SnCl_3 \cdot 2\ C_5H_5N$	Sn: Org.Verb.6-236
–	$C_4H_9SnCl_3 \cdot NH_2C_9H_6N$	Sn: Org.Verb.6-247/8
$C_{13}Cl_3H_{25}IrNS_2$	$Ir(S(C_2H_5)_2)_2(C_5H_5N)Cl_3$	Ir: SVol.2-165
$C_{13}Cl_3H_{27}Sn$	$(C_4H_9)_3SnCCl_3$	Sn: Org.Verb.2-278, 280
$C_{13}Cl_3H_{29}N_2Sn$	$C_3H_7SnCl_3 \cdot 2\ NH(CH_2)_5$	Sn: Org.Verb.6-236
$C_{13}Cl_3H_{29}SiSn$	$(C_4H_9)_3SiCH_2SnCl_3$	Sn: Org.Verb.6-260
$C_{13}Cl_3H_{35}N_4Sn$	$CH_3SnCl_3 \cdot 2\ (CH_3)_2NC_2H_4N(CH_3)_2$	Sn: Org.Verb.6-222
$C_{13}Cl_4F_2H_9N_3O_2S$	$NC_5F_2Cl_2NHC_6H_2Cl_2(SO_2N(CH_3)_2)$	F: PerFHalOrg.5-140
$C_{13}Cl_4F_3H_8N_3O$	$NC_5F_2Cl_2NHC_5FCl_2(OC_3H_7)N$	F: PerFHalOrg.5-138/9
$C_{13}Cl_4F_4HN_3O_2$	$(NCC_6F_4)(NO_2C_6Cl_4)NH$	F: PerFHalOrg.7-14/6, 42
–	$(NO_2C_6F_4)(NCC_6Cl_4)NH$	F: PerFHalOrg.7-15/6, 43
$C_{13}Cl_4H_7IOSn$	$SnCl_4 \cdot C_{13}H_7IO$	Sn: MVol.C5-100
$C_{13}Cl_4H_8OSn$	$SnCl_4 \cdot C_{13}H_8O$	Sn: MVol.C5-98

Formula	Compound	Element	Reference
$C_{13}Cl_4H_8O_3Sn$	$SnCl_4 \cdot C_{13}H_8O_3$	Sn:	MVol.C5-163
$C_{13}Cl_4H_{10}IrKN_2$	$K[Ir(CH_3C_{12}H_7N_2)Cl_4]$	Ir:	SVol.2-93
$C_{13}Cl_4H_{10}N_2O_2Te$	$TeCl_4 \cdot C_6H_4(NO_2)CHNC_6H_5$	Te:	SVol.B3-172/3
–	$TeCl_4 \cdot C_6H_5CHNC_6H_4NO_2$	Te:	SVol.B3-172/3
$C_{13}Cl_4H_{10}N_2Sn$	$SnCl_4 \cdot C_{13}H_8NNH_2$	Sn:	MVol.C6-30
$C_{13}Cl_4H_{11}NTe$	$TeCl_4 \cdot C_6H_5CHNC_6H_5$	Te:	SVol.B3-172/3
$C_{13}Cl_4H_{12}N_2OSn$	$SnCl_4 \cdot C_6H_5NNC_6H_4OCH_3$	Sn:	MVol.C6-31/2
$C_{13}Cl_4H_{12}N_2O_2Sn$	$ClC_6H_4SnCl_3 \cdot NH_2C_6H_3(CH_3)NO_2$	Sn:	Org.Verb.6-283
$C_{13}Cl_4H_{12}N_2Sn$	$SnCl_4 \cdot C_6H_5CHNNHC_6H_5$	Sn:	MVol.C6-28/9
$C_{13}Cl_4H_{16}O_2Sn$	$SnCl_4 \cdot CH_2C(CH_3)COOCH_3 \cdot C_6H_5CHCH_2$	Sn:	MVol.C5-138
$C_{13}Cl_4H_{17}NO_2Sn$	$SnCl_4 \cdot C_6H_5CONHCOC_5H_{11}$	Sn:	MVol.C6-50/1
$C_{13}Cl_4H_{20}N_2O_2Sn$	$SnCl_4 \cdot (C_2H_5)_2C(OH)CON_2H_2C_6H_4CH_3$	Sn:	MVol.C6-61
$C_{13}Cl_4H_{24}O_4Sn$	$SnCl_4 \cdot C_7H_{14}(COOC_2H_5)_2$	Sn:	MVol.C5-141/3
$C_{13}Cl_4H_{28}S_2Sn$	$SnCl_4 \cdot C_4H_9S(CH_2)_5SC_4H_9$	Sn:	MVol.C6-97/8
$C_{13}Cl_4H_{28}S_2Te$	$TeCl_4 \cdot C_4H_9S(CH_2)_5SC_4H_9$	Te:	SVol.B3-175/6
$C_{13}Cl_5FeH_5O_5$	$(C_6Cl_5)(C_2H_5O)CFe(CO)_4$	Fe:	Org.Verb.B4-121/3, 125
$C_{13}Cl_5H_{10}NTe$	$TeCl_4 \cdot C_6H_4ClCHNC_6H_5$	Te:	SVol.B3-172/3
–	$TeCl_4 \cdot C_6H_5CHNC_6H_4Cl$	Te:	SVol.B3-172/3
$C_{13}Cl_5H_{19}N_2Sn$	$[C_6H_5NH_3]_2[CH_3SnCl_5]$	Sn:	Org.Verb.6-222
$C_{13}Cl_6H_{16}Sn$	$(C_2H_5)_3SnC_7HCl_6$	Sn:	Org.Verb.2-207
$C_{13}Cl_6H_{27}NiO_4P_3$	$CONi(P(OC_4H_9)Cl_2)_3$	Ni:	Org.Verb.1-116, 117
$C_{13}Cl_7H_{11}N_4SSn_2$	$SnCl_3(C_6H_5NNCSNNHC_6H_5) \cdot SnCl_4$	Sn:	MVol.C6-122
$C_{13}Cl_8H_{12}N_4OSn_2$	$2\ SnCl_4 \cdot C_6H_5NNCONHNHC_6H_5$	Sn:	MVol.C6-35
$C_{13}Cl_8H_{28}S_2Te_2$	$2\ TeCl_4 \cdot C_4H_9S(CH_2)_5SC_4H_9$	Te:	SVol.B3-176
$C_{13}Cl_9Fe_2H_{16}NO_7Si_2Sn$	$[(C_2H_5)_3NH][(CO)_7Fe_2Si_2SnCl_9]$	Fe:	Org.Verb.C1-60
$C_{13}CoFeH_{10}NO_4$	$[Co(C_5H_5)_2][Fe(CO)_3NO]$	Fe:	Org.Verb.B1-190
$C_{13}CoH_{25}N_2S_7$	$Co[S_2CSC_2H_5][S_2CN(C_2H_5)_2]_2$	C:	MVol.D4-270
$C_{13}Co_3FeH_{19}O_8P^+$	$[HFeCo_3(CO)_5P(OC_3H_7)_2OC_2H_4]^+$	Fe:	Org.Verb.B1-204
$C_{13}Co_3FeH_{22}O_7P^+$	$[HFeCo_3(CO)_4P(OC_3H_7)_3]^+$	Fe:	Org.Verb.B1-204
$C_{13}Co_3FeO_{13}P$	$(CO)_4FeP(Co(CO)_3)_3$	Fe:	Org.Verb.B2-92, 112
$C_{13}CrFeH_{11}NO_2^+$	$[C_5H_5FeC_5H_4C(NH_2)Cr(CO)_2]^+$	Fe:	Org.Verb.A1-346
$C_{13}CrFeH_{12}O_2^+$	$[C_5H_5FeC_5H_4C(OCH_3)Cr(CO)]^+$	Fe:	Org.Verb.A1-346
$C_{13}CrFeH_{12}O_9P_2$	$(CO)_4FeP(CH_3)_2P(CH_3)_2Cr(CO)_5$	Fe:	Org.Verb.B2-92, 112
$C_{13}CrFeH_{14}O^+$	$[C_5H_5FeC_5H_4C(OC_2H_5)Cr]^+$	Fe:	Org.Verb.A1-346
$C_{13}CrFeH_{15}N^+$	$[C_5H_5FeC_5H_4C(N(CH_3)_2)Cr]^+$	Fe:	Org.Verb.A1-346
$C_{13}CrH_{16}O_3Sn$	$(CH_3)_3SnCH_2C_6H_5Cr(CO)_3$	Sn:	Org.Verb.2-24
$C_{13}CrH_{18}O_5Sn$	$[(CH_3)_3C]_2SnCr(CO)_5 \cdot C_4H_8O$	Sn:	Org.Verb.6-107
$C_{13}FH_{15}Sn$	$(CH_3)_3SnC_{10}H_6F$	Sn:	Org.Verb.2-147
$C_{13}F_2H_{11}Ni_2^+$	$C_3F_2H(NiC_5H_5)_2^+$	Ni:	Org.Verb.2-359
$C_{13}F_2H_{29}N_3O_5S_2$	$[(C_3H_7)_4N][FSO_2NCONHSO_2F]$	S:	S-N-Verb.1-103
$C_{13}F_3H_6NO_3$	$OC_6F_3N(O)C_6H_3OCH_3$	F:	PerFHalOrg.7-198
$C_{13}F_3H_{10}N_3$	$CF_3NNC_6H_4C_6H_4NH_2$	F:	PerFHalOrg.7-158/9
$C_{13}F_3H_{10}P$	$CF_3P(C_6H_5)_2$	F:	PerFHalOrg.3-101
$C_{13}F_3H_{11}Ni_2$	$(C_5H_5Ni)_2HCCCF_3$	Ni:	Org.Verb.2-356, 359
$C_{13}F_3H_{11}Ni_2^+$	$C_3F_3H(NiC_5H_5)_2^+$	Ni:	Org.Verb.2-359
$C_{13}F_3H_{16}N_2O_3P$	$CF_3P(O)(OH)_2 \cdot 2\ C_6H_5NH_2$	F:	PerFHalOrg.3-50
$C_{13}F_3H_{19}N_2O$	$(NH_2)(C_8H_{17}O)C_5F_3N$	F:	PerFHalOrg.5-214/5
$C_{13}F_3H_{19}Sn$	$(C_2H_5)_3SnC_6H_4CF_3$	Sn:	Org.Verb.2-240
$C_{13}F_3H_{21}N_4O_3S$	$C_2N_2S(CF_3)NHCON(CH_3)C_2H_3(OC_3H_7)_2$	F:	PerFHalOrg.5-107

Formula	Compound		Reference
$C_{13}F_3H_{27}NO_4P$	$CF_3N(O)P(OC_4H_9)_3$	F:	PerFHalOrg.7-169
$C_{13}F_3H_{27}P_2$	$(C_4H_9)_3PPCF_3$	F:	PerFHalOrg.3-7
$C_{13}F_3H_{27}Sn$	$(C_4H_9)_3SnCF_3$	Sn:	Org.Verb.2-280
$C_{13}F_3H_{28}N_2P$	$CF_3P[N(n\text{-}C_3H_7)_2]_2$	F:	PerFHalOrg.3-102, 115
$C_{13}F_3H_{30}NO_7P_2$	$CF_3NO \cdot 2\ (C_2H_5O)_3P$	F:	PerFHalOrg.7-169
$C_{13}F_3H_{36}N_6NiOP_3$	$CONi(P(N(CH_3)_2)_2F)_3$	Ni:	Org.Verb.1-113
$C_{13}F_4FeH_5LiO_2$	$LiC_6F_4[C_5H_5Fe(CO)_2]$	F:	PerFHalOrg.4-8, 12, 39
$C_{13}F_4FeH_8O_4$	$(C_7H_8CF_2CF_2)Fe(CO)_4$	Fe:	Org.Verb.B4-339, 349/50
$C_{13}F_4H_6N_4$	$C_6H_5C_2HN_3C_5F_4N$	F:	PerFHalOrg.5-213
$C_{13}F_4H_9NO_3$	$OC_6H_8NC_6F_4COOH$	F:	PerFHalOrg.7-198/9
$C_{13}F_4H_{10}N_4$	$(CH_3)_2NC_6H_4NNC_5F_4N$	F:	PerFHalOrg.5-214
$C_{13}F_4H_{15}N_2$	$NC_5F_4NH(CCH_3(CH_2)_4CCH_3)$	F:	PerFHalOrg.5-137
$C_{13}F_4H_{16}N_2$	$(CH_2)_8NC_5F_4N$	F:	PerFHalOrg.5-213
–	$C_8H_{15}NHC_5F_4N$	F:	PerFHalOrg.5-137
–	$NC_5F_4NHC_6H_9(CH_3)_2$	F:	PerFHalOrg.5-213
$C_{13}F_5FeH_5O_2$	$C_5H_5Fe(CO)_2C_6F_5$	F:	PerFHalOrg.4-37, 38, 111
$C_{13}F_5FeH_5O_5$	$(C_6F_5)(C_2H_5O)CFe(CO)_4$	Fe:	Org.Verb.B4-121/3, 125
$C_{13}F_5FeH_6NO_4$	$(C_6F_5)(N(CH_3)_2)CFe(CO)_4$	Fe:	Org.Verb.B4-122/3, 125
$C_{13}F_5FeH_9$	$C_5H_5FeC_5H_4CFCFCF_3$	Fe:	Org.Verb.A1-385, 386
$C_{13}F_5FeH_9O$	$C_5H_5FeC_5H_4COC_2F_5$	Fe:	Org.Verb.A2-241/3, 245
$C_{13}F_5FeH_{11}O$	$C_5H_5FeC_5H_4C(CF_3)(CHF_2)OH$	Fe:	Org.Verb.A2-58/9
–	$C_5H_5FeC_5H_4CHOHC_2F_5$	Fe:	Org.Verb.A2-44
$C_{13}F_5H_5N_2O_2$	$NO_2C_6H_4CHNC_6F_5$	F:	PerFHalOrg.7-59
$C_{13}F_5H_5O$	$C_6F_5COC_6H_5$	F:	PerFHalOrg.4-101, 102
$C_{13}F_5H_6NO$	$HOC_6H_4CHNC_6F_5$	F:	PerFHalOrg.7-59
$C_{13}F_5H_6NO_3$	$C_6F_5CH(OH)C_6H_4NO_2$	F:	PerFHalOrg.4-101
$C_{13}F_5H_7O$	$(C_6H_5)(C_6F_5)CHOH$	F:	PerFHalOrg.4-47, 100
$C_{13}F_5H_8N$	$C_6F_5NHC_6H_4CH_3$	F:	PerFHalOrg.7-197
$C_{13}F_5H_8NO$	$C_6F_5NHC_6H_4OCH_3$	F:	PerFHalOrg.7-197
$C_{13}F_5O_5Re$	$(CO)_5ReCCC_6F_5$	F:	PerFHalOrg.4-33
$C_{13}F_6FeH_{10}O$	$C_5H_5FeC_5H_4C(CF_3)_2OH$	Fe:	Org.Verb.A2-56
$C_{13}F_6H_8N_2NiO$	$(CF_3)_2CONi(NC_5H_4C_5H_4N)$	Ni:	Org.Verb.1-61
$C_{13}F_6H_8N_2NiS$	$(CF_3)_2CSNi(NC_5H_4C_5H_4N)$	Ni:	Org.Verb.1-61
$C_{13}F_6H_9N$	$C_4H_9C_9F_6N$	F:	PerFHalOrg.6-159
$C_{13}F_6H_{18}N_2Ni$	$(CF_3CFCF_2)Ni(CNC_4H_9\text{-}t)_2$	Ni:	Org.Verb.1-382, 383
$C_{13}F_6H_{18}N_2NiO$	$(t\text{-}C_4H_9NC)_2Ni(OC(CF_3)_2)$	Ni:	Org.Verb.1-318, 321
$C_{13}F_6H_{18}N_2NiO_2$	$(t\text{-}C_4H_9NC)_2Ni(O_2C(CF_3)_2)$	Ni:	Org.Verb.1-315, 317
$C_{13}F_6H_{18}N_2NiO_3$	$(t\text{-}C_4H_9NC)_2Ni(CO_3(CF_3)_2)$	Ni:	Org.Verb.1-317, 321
$C_{13}F_6H_{18}N_2O$	$NH_2C(CF_3)_2C_6H_8N(CH_2CH_2)_2O$	F:	PerFHalOrg.7-69, 73
$C_{13}F_6H_{19}N_3Ni$	$(t\text{-}C_4H_9NC)_2Ni(NHC(CF_3)_2)$	Ni:	Org.Verb.1-318, 322
$C_{13}F_6H_{22}N_2O_8P_2$	$C_3F_6[C(O)NHPO(OC_2H_5)_2]_2$	F:	PerFHalOrg.7-63
$C_{13}F_6H_{30}N_3NiOP_3$	$CONi(P(N(C_2H_5)_2)F_2)_3$	Ni:	Org.Verb.1-113
$C_{13}F_7FeH_9$	$C_5H_5FeC_5H_4(C_3F_7\text{-}n)$	Fe:	Org.Verb.A1-381, 383
$C_{13}F_7H_3N_2O$	$CH_3OC_6F_3N_2C_6F_4$	F:	PerFHalOrg.6-160
$C_{13}F_7H_6NO_2$	$(CH_3O)(CH_3OC_6F_4)C_5F_3N$	F:	PerFHalOrg.5-178
$C_{13}F_7H_6N_3O$	$NH_2C_6F_4C(O)C_6F_3(NH_2)_2$	F:	PerFHalOrg.7-12/3, 40
$C_{13}F_7H_7N_2O$	$NC_5F_3(OC_3H_7\text{-}n)C_5F_4N$	F:	PerFHalOrg.5-222/3
–	$NC_5F_3(OC_3H_7\text{-}i)C_5F_4N$	F:	PerFHalOrg.5-222/3
$C_{13}F_8FeH_{15}O_6P$	$(CF_2CF_2)_2Fe(CO)_3P(OC_2H_5)_3$	Fe:	Org.Verb.B4-202, 207
$C_{13}F_8HNO$	$CO(C_6F_4)_2NH$	F:	PerFHalOrg.6-135/6, 149, 155/6, 161

Formula	Compound		Reference
$C_{13}F_8HN_3O_2$	$(NO_2C_6F_4)(NCC_6F_4)NH$	F:	PerFHalOrg.7-14/6, 42
$C_{13}F_8H_4N_2O_3$	$(CH_3O)(NO_2)C_6F_3N(H)C_6F_5$	F:	PerFHalOrg.7-59
$C_{13}F_8H_{10}N_2$	$N(CF_2)_4CN(C_2H_5)C_6H_5$	F:	PerFHalOrg.5-144
$C_{13}F_8Li_2O_2$	$LiC_6F_4C_6F_4C(O)OLi$	F:	PerFHalOrg.4-7/8
$C_{13}F_8NNaO$	$CO(C_6F_4)_2NNa$	F:	PerFHalOrg.6-135/6
$C_{13}F_8O$	$C_6F_4COC_6F_4$	F:	PerFHalOrg.4-53
$C_{13}F_9^+$	$[C_6F_4CFC_6F_4]^+$	F:	PerFHalOrg.5-95
$C_{13}F_9FeH_{12}NO_3$	$[N(CH_3)_4][C_6F_9Fe(CO)_3]$	Fe:	Org.Verb.B5-74, 76
$C_{13}F_9HN_2$	$(NCC_6F_4)(C_6F_5)NH$	F:	PerFHalOrg.7-14/5, 42
$C_{13}F_9HO_3$	$C_6F_5OC_6F_4COOH$	F:	PerFHalOrg.4-104
$C_{13}F_9H_2NO$	$C_6F_5C(O)C_6F_4NH_2$	F:	PerFHalOrg.7-12/3, 37, 55
$C_{13}F_9H_2NO_2$	$(HOCOC_6F_4)(C_6F_5)NH$	F:	PerFHalOrg.7-14/5, 43
$C_{13}F_{10}HNO$	$C_6F_5NHC(O)C_6F_5$	F:	PerFHalOrg.7-23/4, 49
$C_{13}F_{10}HNO_2$	$C_6F_5C(O)N(OH)C_6F_5$	F:	PerFHalOrg.7-104
$C_{13}F_{10}H_2N_2O$	$C_6F_5NHC(O)NHC_6F_5$	F:	PerFHalOrg.7-24, 49
$C_{13}F_{10}H_2O$	$(C_6F_5)_2CHOH$	F:	PerFHalOrg.4-99
$C_{13}F_{10}H_3N$	$(C_6F_5)_2NCH_3$	F:	PerFHalOrg.7-60
$C_{13}F_{10}H_3OP$	$(C_6F_5)_2P(OCH_3)$	F:	PerFHalOrg.3-129, 133
$C_{13}F_{10}H_3P$	$(C_6F_5)_2PCH_3$	F:	PerFHalOrg.3-129, 130
$C_{13}F_{10}H_4NP$	$(C_6F_5)_2PNHCH_3$	F:	PerFHalOrg.3-125, 129, 132
$C_{13}F_{10}H_4Si$	$(C_6F_5)_2SiHCH_3$	F:	PerFHalOrg.4-24, 80
$C_{13}F_{10}N^+$	$[C_6F_5CNC_6F_5]^+$	F:	PerFHalOrg.5-94/5
$C_{13}F_{10}NOP$	$(C_6F_5)_2PNCO$	F:	PerFHalOrg.3-60/1, 66, 68
$C_{13}F_{10}NP$	$(C_6F_5)_2PCN$	F:	PerFHalOrg.3-165, 170, 173/4
$C_{13}F_{10}NPS$	$(C_6F_5)_2PNCS$	F:	PerFHalOrg.3-61, 66, 68
$C_{13}F_{10}NTl$	$(C_6F_5)_2TlCN$	F:	PerFHalOrg.4-131/2, 135/6
$C_{13}F_{10}N_2$	$C_6F_3(CF_3)NNC_6F_4$	F:	PerFHalOrg.6-136, 149
$C_{13}F_{10}N_2^+$	$[C_6F_5C(N)NC_6F_5]^+$	F:	PerFHalOrg.5-94/5
$C_{13}F_{10}O$	$(C_6F_5)_2CO$	F:	PerFHalOrg.4-51, 98
$C_{13}F_{11}HN_2O_2$	$(CF_3C_6F_4)(NO_2C_6F_4)NH$	F:	PerFHalOrg.7-14/6, 42
$C_{13}F_{12}$	$CF_3C_6F_4C_6F_5$	F:	PerFHalOrg.4-16
$C_{13}F_{12}FeH_{23}N_7P_2$	$[CH_3N(CNHCH_3)_2Fe(CNCH_3)_4][PF_6]_2$	Fe:	Org.Verb.B4-352/3
–	$[CD_3N(CNHCH_3)_2Fe(CNCH_3)_4][PF_6]_2$	Fe:	Org.Verb.B4-352/3
$C_{13}F_{12}FeH_{24}N_8P_2$	$[CH_3C_2HN_2(NHCH_3)_2Fe(CNCH_3)_4][PF_6]_2$	Fe:	Org.Verb.B4-353/4
$C_{13}F_{12}HN$	$(CF_3C_6F_4)(C_6F_5)NH$	F:	PerFHalOrg.7-14/5, 42
$C_{13}F_{12}H_{11}O_{10}PU$	$UO_2(CH(COCF_3)_2)_2 \cdot (CH_3O)_3PO$	U:	SVol.E1-181, 184
$C_{13}F_{12}H_{13}N_3O_{10}$	$3\ ON(CF_2)_2COOH \cdot (C_2H_5)_2O$	F:	PerFHalOrg.7-192
$C_{13}F_{12}H_{14}N_4$	$C_3HN_2(CF_3)_4(NHCH_2N(CH_2)_5)$	F:	PerFHalOrg.5-106
$C_{13}F_{12}H_{16}N_2NiO_2$	$(t\text{-}C_4H_9NC)Ni(HN(CH_3)_2)(C_2O_2(CF_3)_4)$	Ni:	Org.Verb.1-304, 307
$C_{13}F_{12}Ni_3O_9P_4$	$(PCF_3)_4[Ni(CO)_3]_3$	Ni:	Org.Verb.1-246
$C_{13}F_{14}N_2$	$(NC_5F_4CF_2)_2CF_2$	F:	PerFHalOrg.5-156/7, 169
$C_{13}F_{15}H_6NO_2$	$(CH_3O)_2[(CF_3)_2CF]_2C_5FN$	F:	PerFHalOrg.5-177
$C_{13}F_{17}H_{13}O_3Sn$	$CF_3(CF_2)_7CH_2CH_2Sn(OCH_3)_3$	Sn:	Org.Verb.6-264
$C_{13}F_{17}N$	$NC_5F_4(CCF_3)_3CFCF_3$	F:	PerFHalOrg.5-155/6, 167
$C_{13}F_{18}H_5N_3O_3$	$C_6H_5C[ON(CF_3)_2]_3$	F:	PerFHalOrg.7-144

Formula	Compound		Reference
$C_{13}F_{18}H_8O_6P_2$	$C_5F_6(P(O)(OCH_2CF_3)_2)_2$	F:	PerFHalOrg.3-57, 59
$C_{13}F_{20}H_4N_4O_4$	$(C_3F_7C(O)ONC(NH_2)CF_2)_2CF_2$	F:	PerFHalOrg.7-5/6, 32
$C_{13}F_{20}N_4$	$[i\text{-}C_3F_7]_2[(CF(CF_3)CF_2)CN]C_3N_3$	F:	PerFHalOrg.6-88/9, 108
$C_{13}F_{20}N_4O_2$	$CF_2(CF_2CNOC(C_3F_7)N)_2$	F:	PerFHalOrg.5-69/71, 91
–	$CF_2(CF_2COC(C_3F_7)NN)_2$	F:	PerFHalOrg.5-69/70, 90
–	$CF_2(CF_2CONC(C_3F_7)N)_2$	F:	PerFHalOrg.5-69/71, 91
$C_{13}F_{21}H_4N_5O$	$(C_3F_7)_2(OH)C_3N_3 \cdot NH_2C(NH)C_3F_7$	F:	PerFHalOrg.6-79/80, 101
$C_{13}F_{21}H_9Si$	$(CH_3)_3SiC_{10}F_{21}$	F:	PerFHalOrg.4-86
$C_{13}F_{21}N$	$[(CF_3)_2CF]_2(CF_3)_2C_5FN$	F:	PerFHalOrg.5-153/4, 164
–	$[(CF_3)_2CF]_2(C_2F_5)C_5F_2N$	F:	PerFHalOrg.5-153/4, 165/6, 176
–	$(C_2F_5)_4C_5FN$	F:	PerFHalOrg.5-149/50, 160
–	$(C_4F_9)_2C_5F_3N$	F:	PerFHalOrg.5-154, 166
$C_{13}F_{22}N_2$	$NC(i\text{-}C_3F_7)CFC(i\text{-}C_3F_7)C(i\text{-}C_3F_7)N$	F:	PerFHalOrg.6-16, 36, 55
–	$NC(i\text{-}C_3F_7)NC(i\text{-}C_3F_7)CFC(i\text{-}C_3F_7)$	F:	PerFHalOrg.6-23/4, 45
$C_{13}F_{24}H_6N_2Ni_2O_3P_4$	$Ni_2(CO)_3[(CF_3)_2PN(CH_3)P(CF_3)_2]_2$	Ni:	Org.Verb.2-258
$C_{13}F_{24}H_8NiOP_4$	$CONi(P(CF_3)_2(CH_2)_2P(CF_3)_2)_2$	Ni:	Org.Verb.1-120
$C_{13}F_{27}NO$	$CF_3(CF_2)_{12}NO$	F:	PerFHalOrg.7-179, 185
–	$O(CF_2CF_2)_2NCF_2(CF_2)_7CF_3$	F:	PerFHalOrg.6-2, 7
$C_{13}F_{29}N_5O_4$	$N[CF(NO(CF_3)_2)]_4CF$	F:	PerFHalOrg.5-184, 200
–	$NC_5F_5(ON(CF_3)_2)_4$	F:	PerFHalOrg.7-99/100
$C_{13}FeGe_3H_{27}O_4P$	$(CO)_4FeP(Ge(CH_3)_3)_3$	Fe:	Org.Verb.B2-85, 91, 112
$C_{13}FeH_2O_{13}OsRu_2$	$FeRu_2Os(CO)_{13}H_2$	Fe:	Org.Verb.B2-197/8
$C_{13}FeH_2O_{13}Os_2Ru$	$FeRuOs_2(CO)_{13}H_2$	Fe:	Org.Verb.B2-197/8
$C_{13}FeH_2O_{13}Os_3$	$FeOs_3(CO)_{13}H_2$	Fe:	Org.Verb.B2-198
$C_{13}FeH_2O_{13}Ru_3$	$FeRu_3(CO)_{13}H_2$	Fe:	Org.Verb.B2-195/7
$C_{13}FeH_5LiO_5$	$Li[CC(C_6H_5)C(O)Fe(CO)_4]$	Fe:	Org.Verb.B3-132
$C_{13}FeH_5O_{10}Re$	$CH_3COCHCHRe(CO)_5Fe(CO)_4$	Fe:	Org.Verb.B4-278, 298
$C_{13}FeH_7MnO_6$	$(CO)_3MnC_7H_7Fe(CO)_3$	Fe:	Org.Verb.B5-152/5
$C_{13}FeH_7NO_4$	$(CO)_4Fe(C_9H_7N)$	Fe:	Org.Verb.B2-71, 75
–	$(NCCHCHC(C_6H_5)O)Fe(CO)_3$	Fe:	Org.Verb.B5-133, 146
$C_{13}FeH_7NO_6S_2$	$NO_2C_6H_4C_3H_3(O)Fe(CO)_3(CS_2)$	Fe:	Org.Verb.B4-204
$C_{13}FeH_7NO_7$	$HC(O)CHCHC_6H_4NO_2Fe(CO)_4$	Fe:	Org.Verb.B4-273, 291/2
$C_{13}FeH_7NO_8$	$HOC(O)CHCHC_6H_4NO_2Fe(CO)_4$	Fe:	Org.Verb.B4-274, 292/3
$C_{13}FeH_7O_5^-$	$[C_6H_5C(O)C_3H_2(O)Fe(CO)_3]^-$	Fe:	Org.Verb.B5-72, 75
$C_{13}FeH_7O_6Re$	$(CO)_3ReC_7H_7Fe(CO)_3$	Fe:	Org.Verb.B5-152/5
$C_{13}FeH_8O_5$	$C_6H_5COCHCH_2Fe(CO)_4$	Fe:	Org.Verb.B4-237, 249
–	$HC(O)CHCHC_6H_5Fe(CO)_4$	Fe:	Org.Verb.B4-268, 272/3, 290/1
$C_{13}FeH_8O_6$	$(C_6H_5)(CH_3COO)CFe(CO)_4$	Fe:	Org.Verb.B4-122, 125
–	1) $C_7H_8(CO)_2Fe(CO)_4$	Fe:	Org.Verb.B4-335
	2) $C_7H_8(CO)_2Fe(CO)_4$	Fe:	Org.Verb.B4-339, 349
–	$HOC(O)CHCHC_6H_5Fe(CO)_4$	Fe:	Org.Verb.B4-274, 292/3
$C_{13}FeH_9KO_5$	$K[C_6H_5CH_2CH_2C(O)Fe(CO)_4]$	Fe:	Org.Verb.B4-152, 162
$C_{13}FeH_9LiO_5$	$Li[(CH_3)_2C_6H_3C(O)Fe(CO)_4]$	Fe:	Org.Verb.B4-153, 162
$C_{13}FeH_9NO_4$	$(CH_3)_2C_6H_3NCFe(CO)_4$	Fe:	Org.Verb.B4-6, 8
$C_{13}FeH_9NO_5$	$C_6H_5NHCOCHCH_2Fe(CO)_4$	Fe:	Org.Verb.B4-237, 249
$C_{13}FeH_9O^+$	$[FeC_5H_4C_7H_5CO]^+$	Fe:	Org.Verb.A3-127
$C_{13}FeH_9O_5^-$	$[(CH_3)_2C_6H_3C(O)Fe(CO)_4]^-$	Fe:	Org.Verb.B4-153, 162

Formula	Compound	Reference
$C_{13}FeH_9O_5^-$	$[C_6H_5CH_2CH_2C(O)Fe(CO)_4]^-$	Fe: Org.Verb.B4-152, 162
$C_{13}FeH_{10}N_2O_3$	$(CO)_3Fe(NC_5H_5)_2$	Fe: Org.Verb.B1-148
$C_{13}FeH_{10}N_2O_3Te$	$COFe(NO)_2Te(C_6H_5)_2$	Fe: Org.Verb.B1-42/4, 46
$C_{13}FeH_{10}N_2O_5S_2$	$[Fe(C_6H_5SO_2N)_2CO \cdot H_2O]_n$	Fe: Org.Verb.B3-204
$C_{13}FeH_{10}NiO_3$	$C_5H_5Ni(CO)_2Fe(C_5H_5)CO$	Ni: Org.Verb.2-169/70
$C_{13}FeH_{10}Ni_2O_3^+$	$(C_5H_5)_2Ni_2Fe(CO)_3^+$	Ni: Org.Verb.2-351
$C_{13}FeH_{10}O$	$C_5H_5FeC_5H_4CCCHO$	Fe: Org.Verb.A2-173/4
–	$C_5H_5FeC_5H_4COCCH$	Fe: Org.Verb.A2-248/9
$C_{13}FeH_{10}OS_3$	$(C_5H_5)Fe(CO)[S_2CSC_6H_5]$	C: MVol.D4-269
$C_{13}FeH_{10}O_2$	$C_5H_5FeC_5H_4CCCOOH$	Fe: Org.Verb.A3-77/9
$C_{13}FeH_{10}O_3S_2$	$(C_6H_5CH_2SCCHCH_2S)Fe(CO)_3$	Fe: Org.Verb.B4-213, 220/1
$C_{13}FeH_{10}O_4$	$(CH_3CHCHC(C_6H_5)O)Fe(CO)_3$	Fe: Org.Verb.B5-132, 141
–	$CH_3CHCHC_6H_5Fe(CO)_4$	Fe: Org.Verb.B4-267, 272
–	$(CH_3C_6H_4CHCHCHO)Fe(CO)_3$	Fe: Org.Verb.B5-122/5, 130, 140
–	$(C_6H_5CHCHC(CH_3)O)Fe(CO)_3$	Fe: Org.Verb.B5-122/7, 132/3, 143/6
–	$C_9H_{10}Fe(CO)_4$	Fe: Org.Verb.B4-304, 312
$C_{13}FeH_{10}O_4^{2-}$	$[(C_6H_5CHCHC(CH_3)O)Fe(CO)_3]^{2-}$	Fe: Org.Verb.B5-143
$C_{13}FeH_{10}O_5$	$CH_3CHCHOC_6H_5Fe(CO)_4$	Fe: Org.Verb.B4-272, 289
–	$(CH_3OC_6H_4CHCHCHO)Fe(CO)_3$	Fe: Org.Verb.B5-122/7, 130/1, 140
–	$(C_6H_5)(C_2H_5O)CFe(CO)_4$	Fe: Org.Verb.B4-121/2, 124
$C_{13}FeH_{10}O_6$	1) $C_7H_{10}(CO)_2Fe(CO)_4$	Fe: Org.Verb.B4-335, 343
	2) $C_7H_{10}(CO)_2Fe(CO)_4$	Fe: Org.Verb.B4-339, 349
$C_{13}FeH_{11}^+$	$[FeC_5H_4CHCHC_6H_5]^+$	Fe: Org.Verb.A2-179
$C_{13}FeH_{11}NO$	$[C(C_5H_4FeC_5H_5)NCOHCH]_n$	Fe: Org.Verb.A2-221
–	$CH_3COC_5H_4FeC_5H_4CN$	Fe: Org.Verb.A2-211
–	$C_5H_5FeC_5H_4CHCH(CN)O$	Fe: Org.Verb.A2-168/9
–	$C_5H_5FeC_5H_4CNOCHCH$	Fe: Org.Verb.A2-249
–	$C_5H_5FeC_5H_4COCH_2CN$	Fe: Org.Verb.A3-110
$C_{13}FeH_{11}NO_4$	$(C_6H_5)(N(CH_3)_2)CFe(CO)_4$	Fe: Org.Verb.B4-122, 125
$C_{13}FeH_{11}NO_4S$	$(CO)_4FeSC(C_6H_5)N(CH_3)_2$	Fe: Org.Verb.B2-128/31
$C_{13}FeH_{11}NaO_2$	$C_5H_5FeC_5H_4COCHCHONa$	Fe: Org.Verb.A2-225, 296
$C_{13}FeH_{11}O^+$	$[C_5H_5FeC_5H_4CHCHCO]^+$	Fe: Org.Verb.A3-123
–	$[FeC_5H_4C_6H_4COCH_3]^+$	Fe: Org.Verb.A3-21
–	$[FeHC_5H_4C_6H_4CH_2CO]^+$	Fe: Org.Verb.A3-126/7
$C_{13}FeH_{11}O_2^+$	$[FeC_5H_4COC_6H_4OCH_3]^+$	Fe: Org.Verb.A2-296
–	$[FeC_5H_4C_6H_4COOCH_3]^+$	Fe: Org.Verb.A3-126
$C_{13}FeH_{11}O_3^+$	$[(C_6H_5)(CH_3)C_3H_3Fe(CO)_3]^+$	Fe: Org.Verb.B5-82, 85/6
$C_{13}FeH_{11}O_3^{2-}$	$[C_5H_5FeC_5H_3(O)COCOCH_3]^{2-}$	Fe: Org.Verb.A2-302/3
$C_{13}FeH_{11}O_4^+$	$[(C_6H_5CHCHC(CH_3)OH)Fe(CO)_3]^+$	Fe: Org.Verb.B5-151
$C_{13}FeH_{12}$	$C_5H_5FeC_5H_4CCCH_3$	Fe: Org.Verb.A1-300/1
–	$C_5H_5FeC_5H_4CHCCH_2$	Fe: Org.Verb.A1-296
–	$C_5H_5FeC_5H_4CH_2CCH$	Fe: Org.Verb.A1-300/1
$C_{13}FeH_{12}HgO_3$	$CH_3COOHgC_5H_4FeC_5H_4CHO$	Fe: Org.Verb.A2-150
–	$C_5H_5FeC_5H_3(CHO)(HgOOCCH_3)$	Fe: Org.Verb.A2-150
$C_{13}FeH_{12}LiNO_4$	$[(CO)_4FeC_6H_4CH_2N(CH_3)_2]Li$	Fe: Org.Verb.B2-188
$C_{13}FeH_{12}NO_4^-$	$[(CO)_4FeC_6H_4CH_2N(CH_3)_2]^-$	Fe: Org.Verb.B2-188
$C_{13}FeH_{12}N_2$	$[C(C_5H_5FeC_5H_4)CHNC(NH_2)]_n$	Fe: Org.Verb.A2-221

$C_{13}FeH_{12}N_2$	$[C(C_5H_5FeC_5H_4)(CH_3)NCN]_n$	Fe: Org.Verb.A2-221
–	$C_5H_5FeC_5H_4C_3H_3N_2$	Fe: Org.Verb.A2-296/7
$C_{13}FeH_{12}N_2O_2$	$C_5H_5FeC_5H_4CHNHCONHCO$	Fe: Org.Verb.A2-168/9
$C_{13}FeH_{12}N_3O_4P$	$(CO)_4FeP(CH_2CH_2CN)_3$	Fe: Org.Verb.B2-87, 104
$C_{13}FeH_{12}O$	$C_5H_4FeC_5H_4(CH_2)_2CO$	Fe: Org.Verb.A3-101
–	$C_5H_5FeC_5H_4CCCH_2OH$	Fe: Org.Verb.A2-74, 79
–	$C_5H_5FeC_5H_4CHCHCHO$	Fe: Org.Verb.A2-174, 176/7
–	$C_5H_5FeC_5H_4CHOHCCH$	Fe: Org.Verb.A2-45, 49
–	$C_5H_5FeC_5H_4COCHCH_2$	Fe: Org.Verb.A2-248/50
$C_{13}FeH_{12}O_2$	$CH_3COC_5H_4FeC_5H_4CHO$	Fe: Org.Verb.A2-150
–	$C_5H_5FeC_5H_3(CHO)(COCH_3)$	Fe: Org.Verb.A2-150
–	$C_5H_5FeC_5H_4CHCHCOOH$	Fe: Org.Verb.A3-77/8
–	$C_5H_5FeC_5H_4COCHCHOH$	Fe: Org.Verb.A2-295, 296
–	$C_5H_5FeC_5H_4COCOCH_3$	Fe: Org.Verb.A3-31/2
$C_{13}FeH_{12}O_2^-$	$[C_5H_5FeC_5H_4COCOCH_3]^-$	Fe: Org.Verb.A2-302/3
$C_{13}FeH_{12}O_3$	$CH_3COC_5H_4FeC_5H_4COOH$	Fe: Org.Verb.A3-58/9
–	$CH_3C_9H_9Fe(CO)_3$	Fe: Org.Verb.B5-145
–	$C_5H_5FeC_5H_4COCH_2COOH$	Fe: Org.Verb.A3-87/8
–	$C_5H_5FeC_5H_4COCOOCH_3$	Fe: Org.Verb.A3-134/5
–	$C_{10}H_{12}Fe(CO)_3$	Fe: Org.Verb.B5-144/5
$C_{13}FeH_{12}O_4$	$CH_3OOCC_5H_4FeC_5H_4COOH$	Fe: Org.Verb.A3-58/9
$C_{13}FeH_{12}O_5$	$C_2H_5OC(O)C_7H_7Fe(CO)_3$	Fe: Org.Verb.B5-77, 78/9
–	$C_9H_{12}OFe(CO)_4$	Fe: Org.Verb.B4-318
$C_{13}FeH_{13}IO$	$C_5H_5FeC_5H_4COCH_2CH_2I$	Fe: Org.Verb.A2-243
$C_{13}FeH_{13}NO$	$C_5H_5FeC_5H_4CHOHCH_2CN$	Fe: Org.Verb.A2-154
$C_{13}FeH_{13}NO_2$	$C_5H_5FeC_5H_4C(CH_3)NCOOH$	Fe: Org.Verb.A2-221
–	$C_5H_5FeC_5H_4CHC(CH_3)NO_2$	Fe: Org.Verb.A2-169
$C_{13}FeH_{13}N_3O_2$	$C_5H_5FeC_5H_4COO(CH_2)_2N_3$	Fe: Org.Verb.A3-153
$C_{13}FeH_{13}O^+$	$[C_6H_6FeC_5H_4COCH_3]^+$	Fe: Org.Verb.A2-186, 208
$C_{13}FeH_{13}O_2$	$[C_5H_5FeC_5H_4(CH_2)_2COO]$	Fe: Org.Verb.A3-68
$C_{13}FeH_{13}O_2^-$	$[C_5H_5FeC_5H_4(CH_2)_2COO]^-$	Fe: Org.Verb.A3-68
$C_{13}FeH_{13}O_4P$	$(CO)_4FeP(CH_3)_2CH_2C_6H_5$	Fe: Org.Verb.B2-89, 111
$C_{13}FeH_{14}$	$C_5H_5FeC_5H_4C(CH_3)CH_2$	Fe: Org.Verb.A1-282, 291/2
–	$C_5H_5FeC_5H_4CHCHCH_3$	Fe: Org.Verb.A1-290/1
–	$C_5H_5FeC_5H_4CH(CH_2)_2$	Fe: Org.Verb.A1-337, 340
–	$C_5H_5FeC_5H_4CH_2CHCH_2$	Fe: Org.Verb.A1-291, 293/4
$C_{13}FeH_{14}^+$	$[C_5H_4FeC_5H_4(CH_2)_3]^+$	Fe: Org.Verb.A2-79
–	$[C_5H_5FeC_5H_4C_3H_5]^+$	Fe: Org.Verb.A3-165
$C_{13}FeH_{14}LiO_7P$	$Li[C_6H_5C(O)Fe(CO)_3P(OCH_3)_3]$	Fe: Org.Verb.B4-140/1
$C_{13}FeH_{14}NNaO_3$	$[C_5H_5FeC_5H_4CHOHC(CH_3)NO_2]Na$	Fe: Org.Verb.A2-169
$C_{13}FeH_{14}N_2O$	$C_5H_5FeC_5H_4CHNNHCOCH_3$	Fe: Org.Verb.A2-167
$C_{13}FeH_{14}N_2O_4$	$(CO)_4FeC_7H_8N_2(CH_3)_2$	Fe: Org.Verb.B2-183/4, 187
–	$C_5H_5FeC_5H_4CH(CH_2NO_2)_2$	Fe: Org.Verb.A2-169
$C_{13}FeH_{14}O$	$C_5H_5FeC_5H_4CHCHCH_2OH$	Fe: Org.Verb.A2-74, 79
–	$C_5H_5FeC_5H_4CH(CH_3)CHO$	Fe: Org.Verb.A2-174, 177
–	$C_5H_5FeC_5H_4CHOHCHCH_2$	Fe: Org.Verb.A2-45, 48
–	$C_5H_5FeC_5H_4CH_2CHCHOH$	Fe: Org.Verb.A2-74
–	$C_5H_5FeC_5H_4CH_2COCH_3$	Fe: Org.Verb.A3-1/2
–	$C_5H_5FeC_5H_4CH_2C_2H_3O$	Fe: Org.Verb.A3-173
–	$C_5H_5FeC_5H_4(CH_2)_2CHO$	Fe: Org.Verb.A2-175

$C_{13}FeH_{14}O$	$C_5H_5FeC_5H_4COC_2H_5$	Fe: Org.Verb.A2-185/6, 232/5
–	$C_5D_5FeC_5D_4COC_2H_5$	Fe: Org.Verb.A2-232
–	$C_5H_5FeC_5H_4OCH_2CHCH_2$	Fe: Org.Verb.A2-108/10
$C_{13}FeH_{14}O^+$	$[C_5H_5FeC_5H_4C_3H_5O]^+$	Fe: Org.Verb.A3-165
$C_{13}FeH_{14}OW^+$	$[C_5H_5FeC_5H_4C(OC_2H_5)W]^+$	Fe: Org.Verb.A1-347
$C_{13}FeH_{14}O_2$	$C_2H_5C_5H_4FeC_5H_4COOH$	Fe: Org.Verb.A3-58/9
–	$C_5H_5FeC_5H_4CH(CH_3)COOH$	Fe: Org.Verb.A3-72
–	$C_5H_5FeC_5H_4CH(OCH_2)_2$	Fe: Org.Verb.A3-178
–	$C_5H_5FeC_5H_4CH_2COOCH_3$	Fe: Org.Verb.A3-115/7
–	$C_5H_5FeC_5H_4CH_2OOCCH_3$	Fe: Org.Verb.A2-97/8
–	$C_5H_5FeC_5H_4(CH_2)_2COOH$	Fe: Org.Verb.A3-68/70, 100/1
–	$C_5H_5FeC_5H_4(CH_2)_2OOCH$	Fe: Org.Verb.A2-103/4
–	$C_5H_5FeC_5H_4CH_2CD_2OOCH$	Fe: Org.Verb.A2-103
–	$C_5H_5FeC_5H_4COOC_2H_5$	Fe: Org.Verb.A3-108, 111
–	$C_5H_5FeC_5H_4OOCC_2H_5$	Fe: Org.Verb.A2-94
$C_{13}FeH_{14}O_3$	$C_5H_5FeC_5H_4CH(OCH_3)COOH$	Fe: Org.Verb.A3-84/6
–	$C_5H_5FeC_5H_4CH(OCH_3)COOH \cdot H_2O$	Fe: Org.Verb.A3-84/6
$C_{13}FeH_{14}O_4$	$(CH_3)_2(CH_2)C_7H_6OFe(CO)_3$	Fe: Org.Verb.B5-135, 147/8
$C_{13}FeH_{14}O_7$	$(CH_3)_2C(OCO)_2CCHCH(CH_3)_2Fe(CO)_3$	Fe: Org.Verb.B5-134/5, 147
$C_{13}FeH_{14}O_7P^-$	$[C_6H_5C(O)Fe(CO)_3P(OCH_3)_3]^-$	Fe: Org.Verb.B4-140/1
$C_{13}FeH_{15}{}^+$	$[C_5H_5FeC_5H_4C(CH_3)_2]^+$	Fe: Org.Verb.A1-350/3, 357/9
		Fe: Org.Verb.A2-53
		Fe: Org.Verb.A3-123
		Fe: Org.Verb.A6-70/2
–	$[C_5H_5FeC_5H_4CH(CH_3)CH_2]^+$	Fe: Org.Verb.A1-349
		Fe: Org.Verb.A2-105
$C_{13}FeH_{15}IMgO$	$[C_5H_5FeC_5H_4C(CH_3)_2][OMgI]$	Fe: Org.Verb.A6-70
$C_{13}FeH_{15}Li$	$C_3H_7C_5H_4FeC_5H_4Li$	Fe: Org.Verb.A1-268
$C_{13}FeH_{15}NO$	$C_5H_5FeC_5H_4CON(CH_3)_2$	Fe: Org.Verb.A3-154
$C_{13}FeH_{15}NO_5$	$(CH_2)_5CHNHC(O)CHCH_2Fe(CO)_4$	Fe: Org.Verb.B4-237, 249
$C_{13}FeH_{15}N_3$	$C_5H_5FeC_5H_4C(CH_3)_2N_3$	Fe: Org.Verb.A1-383
$C_{13}FeH_{15}N_3O$	$C_5H_5FeC_5H_4C(CH_3)NNHCONH_2$	Fe: Org.Verb.A2-221
–	$C_5H_5FeC_5H_4CH_2CHNNHCONH_2$	Fe: Org.Verb.A2-173
$C_{13}FeH_{15}N_3S$	$C_5H_5FeC_5H_4CHNNHCSNHCH_3$	Fe: Org.Verb.A2-167
$C_{13}FeH_{15}N_5O_2$	$C_5H_5FeC_5H_4C(CH_3)NNHC(NH)NHNO_2$	Fe: Org.Verb.A2-221
$C_{13}FeH_{15}O^+$	$[C_5H_5FeC_5H_4COHC_2H_5]^+$	Fe: Org.Verb.A2-233
$C_{13}FeH_{15}O_5{}^+$	$[(CH_3)_4(CH_3CO)C_3Fe(CO)_4]^+$	Fe: Org.Verb.B5-102, 106
$C_{13}FeH_{16}$	$C_5H_5FeC_5H_4CH(CH_3)_2$	Fe: Org.Verb.A1-270/2, 282
–	$C_5H_5FeC_5H_4CH_2CH_2CH_3$	Fe: Org.Verb.A1-267/8
$C_{13}FeH_{16}{}^{2+}$	$[HC_5H_5FeC_5H_4C(CH_3)_2]^{2+}$	Fe: Org.Verb.A1-351, 353
$C_{13}FeH_{16}N^+$	$[C_5H_5FeC_5H_4CHN(CH_3)_2]^+$	Fe: Org.Verb.A2-169
$C_{13}FeH_{16}NOS$	$[Fe(C_5H_5)_2SCN] \cdot C_2H_5OH$	Fe: Org.Verb.A1-206
$C_{13}FeH_{16}N_2O_3$	$(CH_3)_3C_3N_2CH_2C(CH_3)CH_2Fe(CO)_3$	Fe: Org.Verb.B4-216, 225
$C_{13}FeH_{16}N_2O_4$	$CH_2CHC(N(CH_2)_5)N(CH_3)COFe(CO)_3$	Fe: Org.Verb.B5-159/60
$C_{13}FeH_{16}O$	$C_5H_5FeC_5H_4C(CH_3)_2OH$	Fe: Org.Verb.A2-4, 52/4
–	$C_5H_5FeC_5H_4CH(CH_3)CH_2OH$	Fe: Org.Verb.A2-74, 79
–	$C_5H_5FeC_5H_4CH(CH_3)OCH_3$	Fe: Org.Verb.A2-117/8
–	$C_5H_5FeC_5H_4CHOHC_2H_5$	Fe: Org.Verb.A2-43, 44

Formula	Compound	Reference
$C_{13}FeH_{16}O$	$C_5H_5FeC_5H_4CH_2CHOHCH_3$	Fe: Org.Verb.A2-74, 79
–	$C_5H_5FeC_5H_4CH_2OC_2H_5$	Fe: Org.Verb.A2-113/4
–	$C_5H_5FeC_5H_4(CH_2)_3OH$	Fe: Org.Verb.A2-3/4, 74, 79
$C_{13}FeH_{16}O^+$	$[C_5H_5FeC_5H_4(CH_2)_3OH]^+$	Fe: Org.Verb.A2-79
$C_{13}FeH_{16}O_2$	$C_5H_5FeC_5H_4CH(OCH_3)_2$	Fe: Org.Verb.A2-119, 123
–	$C_5H_5FeC_5H_4CH_2O(CH_2)_2OH$	Fe: Org.Verb.A2-113/4
$C_{13}FeH_{16}O_2P$	$[C_3H_5Fe(CO)_2(P(CH_3)_2C_6H_5)]$	Fe: Org.Verb.B5-69/70
$C_{13}FeH_{16}O_3Si$	$(CH_3)_3SiC_7H_7Fe(CO)_3$	Fe: Org.Verb.B5-77, 78/9
$C_{13}FeH_{16}O_4$	$(CH_3)(C(CH_3)_2)C_6H_7OFe(CO)_3$	Fe: Org.Verb.B5-135, 147
–	$C_9H_{16}Fe(CO)_4$	Fe: Org.Verb.B4-302, 305, 314
$C_{13}FeH_{17}N$	$C_5H_5FeC_5H_4CH(CH_3)NHCH_3$	Fe: Org.Verb.A2-30
–	$C_5H_5FeC_5H_4CH_2N(CH_3)_2$	Fe: Org.Verb.A1-132, 167, 246
		Fe: Org.Verb.A2-19, 149
$C_{13}FeH_{17}N^+$	$[C_5H_5FeC_5H_4CH_2N(CH_3)_2]^+$	Fe: Org.Verb.A1-346
$C_{13}FeH_{17}NaO_5$	$Na[CH_3(CH_2)_5CH(CH_3)C(O)Fe(CO)_4]$	Fe: Org.Verb.B4-151, 160/1
–	$Na[CH_3(CH_2)_7C(O)Fe(CO)_4]$	Fe: Org.Verb.B4-151, 160
$C_{13}FeH_{17}O_5^-$	$[C_8H_{17}C(O)Fe(CO)_4]^-$	Fe: Org.Verb.B4-151, 156, 160/1, 167
$C_{13}FeH_{18}N_2O_3$	$(t\text{-}C_4H_9NC)_2Fe(CO)_3$	Fe: Org.Verb.B4-18/9
$C_{13}FeH_{18}N_2O_4S$	$(CO)_4FeSC(NHC_6H_{11})N(CH_3)_2$	Fe: Org.Verb.B2-128/30
$C_{13}FeH_{18}N_2O_5$	$[(CH_2)_4NH_2][(CH_2)_4NC(O)Fe(CO)_4]$	Fe: Org.Verb.B2-173/4
		Fe: Org.Verb.B4-155, 165
–	$(CO)_5Fe(NH(CH_2)_4)_2$	Fe: Org.Verb.B3-249
		Fe: Org.Verb.B4-165
$C_{13}FeH_{18}O_4$	$(CH_2C(CH_3)C(C_6H_{13})O)Fe(CO)_3$	Fe: Org.Verb.B5-132, 141
$C_{13}FeH_{18}O_9P_2$	$(CO)_3Fe(P(CH_2O)_3CCH_3)_2$	Fe: Org.Verb.B1-148/51
–	$(CO)_3Fe(P(OCH_2)_3CCH_3)_2$	Fe: Org.Verb.B1-148/50, 154, 163
$C_{13}FeH_{19}LiO_4$	$Li[(CO)_4FeC_9H_{19}]$	Fe: Org.Verb.B2-174
$C_{13}FeH_{19}NaO_4$	$Na[(CO)_4FeC_9H_{19}]$	Fe: Org.Verb.B2-174
$C_{13}FeH_{19}O_4$	$C_3H_5Fe(O_2C_3H(CH_3)_2)_2$	Fe: Org.Verb.B5-4/5
$C_{13}FeH_{19}O_4^-$	$[C_9H_{19}C(O)Fe(CO)_3]^-$	Fe: Org.Verb.B4-109
$C_{13}FeH_{20}N_2O_3$	$(CO)_3Fe(CH_3)_3CNCHCHNC(CH_3)_3$	Fe: Org.Verb.B1-206
$C_{13}FeH_{20}N_4O_3$	$((CH_2NCH_3)_2C)_2Fe(CO)_3$	Fe: Org.Verb.B4-137
$C_{13}FeH_{20}N_4O_3^+$	$[((CH_2NCH_3)_2C)_2Fe(CO)_3]^+$	Fe: Org.Verb.B4-172
$C_{13}FeH_{20}O_5$	$(CO)_3Fe(CH_3C_4H_7O)_2$	Fe: Org.Verb.B1-146/7
$C_{13}FeH_{21}N_3O_6$	$[(N(CH_3)_2)_3C][CH_3OC(O)Fe(CO)_4]$	Fe: Org.Verb.B4-154, 162/3
$C_{13}FeH_{21}O_4P$	$(CO)_4FeP(C_3H_7)_3$	Fe: Org.Verb.B2-87
$C_{13}FeH_{21}O_6P$	$(C_6H_9COCH_3)Fe(CO)_2P(OCH_3)_3$	Fe: Org.Verb.B5-117, 119
$C_{13}FeH_{23}N_7^{2+}$	$[CH_3N(CNHCH_3)_2Fe(CNCH_3)_4]^{2+}$	Fe: Org.Verb.B4-352/3
–	$[(CH_3NH)_2CFe(CNCH_3)_5]^{2+}$	Fe: Org.Verb.B4-172/3
$C_{13}FeH_{24}N_4O_5Sn$	$[(CH_3)_2N][((CH_3)_2N)_3SnO]CFe(CO)_4$	Fe: Org.Verb.B4-122, 124, 126
$C_{13}FeH_{24}N_4O_5Ti$	$[(CH_3)_2N][((CH_3)_2N)_3TiO]CFe(CO)_4$	Fe: Org.Verb.B4-122, 124, 126/7
$C_{13}FeH_{24}N_8^{2+}$	$[CH_3C_2HN_2(NHCH_3)_2Fe(CNCH_3)_4]^{2+}$	Fe: Org.Verb.B4-353/4
$C_{13}FeH_{24}O_{11}P_2S_2$	$OCFe(P(OCH_3)_3)_2S_2C_2(COOCH_3)_2$	Fe: Org.Verb.B4-138
$C_{13}FeH_{26}NO_5P$	$C_3H_5Fe(CO)(NO)P(OCH(CH_3)_2)_3$	Fe: Org.Verb.B5-6/9, 20/1

$C_{13}FeH_{27}N_2O_3P$	$COFe(NO)_2P(C_4H_9)_3$	Fe:	Org.Verb.B1-42/4, 46
$C_{13}FeH_{27}N_2O_6P$	$COFe(NO)_2P(OC_4H_9)_3$	Fe:	Org.Verb.B1-42/3, 45/8
$C_{13}FeH_{27}O_4PSi_3$	$(CO)_4FeP(Si(CH_3)_3)_3$	Fe:	Org.Verb.B2-85, 91, 112
$C_{13}FeH_{27}O_4PSn_3$	$(CO)_4FeP(Sn(CH_3)_3)_3$	Fe:	Org.Verb.B2-85, 92, 112
$C_{13}FeH_{28}N_4O_5$	$(CO)_5Fe(NH(CH_3)_2)_4$	Fe:	Org.Verb.B3-249
$C_{13}FeH_{30}N_5O_3P_2{}^+$	$[(CO)_3FeP_2(N(CH_3)_2)_5]^+$	Fe:	Org.Verb.B1-164
$C_{13}FeH_{30}OP_2{}^+$	$[COFe(P(C_2H_5)_3)_2]^+$	Fe:	Org.Verb.B1-94
$C_{13}FeH_{36}I_2O_{13}P_4$	$[COFeI(P(OCH_3)_3)_4]I$	Fe:	Org.Verb.B1-32
$C_{13}FeH_{36}N_6OP_2{}^+$	$[(CO)Fe(P(N(CH_3)_2)_3)_2]^+$	Fe:	Org.Verb.B1-164
$C_{13}FeH_{36}O_{13}P_4$	$COFe(P(OCH_3)_3)_4$	Fe:	Org.Verb.B1-29
$C_{13}Fe_2GeH_{17}IN_4O_4$	$(C_3H_5Fe(NO)_2)_2Ge(I)CH_2C_6H_5$	Fe:	Org.Verb.C2-165
–	$(C_3H_5Fe(NO)_2)_2Ge(I)C_6H_4CH_3$	Fe:	Org.Verb.C2-165
$C_{13}Fe_2H_6O_6S_2$	$(CO)_3FeSC_6H_3(CH_3)SFe(CO)_3$	Fe:	Org.Verb.C1-115, 117, 122/3
$C_{13}Fe_2H_7NO_6S$	$(CO)_3Fe(CH_3C_6H_4NS)Fe(CO)_3$	Fe:	Org.Verb.C2-46
$C_{13}Fe_2H_8N_2O_7$	$(CO)_7Fe_2(NC(CH_3)C_2H_2C(CH_3)N)$	Fe:	Org.Verb.C1-200/1
$C_{13}Fe_2H_8NaO_{10}$	$Na[CH_2CHCOOC_2H_5(Fe_2(CO)_8)]$	Fe:	Org.Verb.C2-14/5
$C_{13}Fe_2H_8O_6S_2$	$(CO)_3Fe(SC_7H_8S)Fe(CO)_3$	Fe:	Org.Verb.C2-44
$C_{13}Fe_2H_8O_{10}$	$C_2H_5OCHCH_2Fe(CO)_4 \cdot Fe(CO)_5$	Fe:	Org.Verb.B4-241
$C_{13}Fe_2H_9N_{11}$	$[(CH_3NC)_2Fe_2(CN)_9]H_3$	Fe:	Org.Verb.B4-10
$C_{13}Fe_2H_9O_6P$	$Fe_2(CO)_6(P(CH_3)C_6H_5)H$	Fe:	Org.Verb.B2-110
$C_{13}Fe_2H_9O_7P$	$(CO)_3Fe(P(CH_3)C_6H_5)(OH)Fe(CO)_3$	Fe:	Org.Verb.C1-185/7, 189
$C_{13}Fe_2H_{10}N_2O_6$	$CHCHN_2C_5H_8Fe_2(CO)_6$	Fe:	Org.Verb.C2-122, 125
$C_{13}Fe_2H_{10}N_2O_7$	$(CO)_3Fe(NC(CH_2)_5)(OCN)Fe(CO)_3$	Fe:	Org.Verb.C1-133, 141
$C_{13}Fe_2H_{11}O_6P$	$(CO)_4FeP(CH_3)_2Fe(CO)_2C_5H_5$	Fe:	Org.Verb.B2-115
$C_{13}Fe_2H_{12}O_7S$	$(CO)_6Fe_2(S(C(CH_3)_2)_2CO)$	Fe:	Org.Verb.C1-124/5
$C_{13}Fe_2H_{13}{}^+$	$[Fe_2(C_5H_5)_2(C_3H_3)]^+$	Fe:	Org.Verb.A1-93/4
$C_{13}Fe_2H_{13}NO_7$	$(HCCHC(OC_2H_5)N(CH_3)_2)Fe_2(CO)_6$	Fe:	Org.Verb.C2-156, 160
$C_{13}Fe_2H_{14}N_2O_7$	$(CO)_3Fe(C_3H_7NCONC_3H_7)Fe(CO)_3$	Fe:	Org.Verb.C1-143/4
$C_{13}Fe_2H_{14}O_4S_2$	$(CO)_3Fe(SCH_3)_2Fe(CO)(C_7H_8)$	Fe:	Org.Verb.C1-85
$C_{13}Fe_2H_{14}O_8Sb_2$	$(CO)_4Fe(CH_2(Sb(CH_3)_2)_2)Fe(CO)_4$	Fe:	Org.Verb.C1-37, 41
$C_{13}Fe_2H_{14}O_8Si_2$	$((CH_3)_2Si)_2CH_2Fe_2(CO)_8$	Fe:	Org.Verb.B3-138
$C_{13}Fe_2H_{15}N_3O_6$	$(CO)_6Fe_2[(NCH_3)_2CNC_4H_9]$	Fe:	Org.Verb.C1-143/4, 146
$C_{13}Fe_2H_{17}IN_4O_4Pb$	$(C_3H_5Fe(NO)_2)_2Pb(I)CH_2C_6H_5$	Fe:	Org.Verb.C2-165
–	$(C_3H_5Fe(NO)_2)_2Pb(I)C_6H_4CH_3$	Fe:	Org.Verb.C2-165
$C_{13}Fe_2H_{17}IN_4O_4Sn$	$(C_3H_5Fe(NO)_2)_2Sn(I)CH_2C_6H_5$	Fe:	Org.Verb.C2-165
–	$(C_3H_5Fe(NO)_2)_2Sn(I)C_6H_4CH_3$	Fe:	Org.Verb.C2-165
$C_{13}Fe_2H_{21}O_5PS_2$	$(CO)_3Fe(SCH_3)_2Fe(CO)_2P(C_2H_5)_3$	Fe:	Org.Verb.C1-94
$C_{13}GeH_{24}Sn$	$(CH_3)_3SnC_6H_4CH_2Ge(CH_3)_3$	Sn:	Org.Verb.2-129
$C_{13}Ge_2H_{27}NiO_3P$	$(CO)_3NiP(C_4H_9\text{-}t)(Ge(CH_3)_3)_2$	Ni:	Org.Verb.1-166, 168
$C_{13}H_5N_4S_4Ti^-$	$[C_5H_5Ti(SC(CN)C(CN)S)_2]^-$	Ti:	Org.Verb.1-200
$C_{13}H_5N_5O_3S_4U^{2-}$	$[UO_2(C_4N_2S_2)_2(C_5H_5NO)]^{2-}$	U:	SVol.E1-143
$C_{13}H_8N_2NiO_2$	$(NC_5H_4C_5H_4N)Ni(C_3O_2)$	Ni:	Org.Verb.1-369
$C_{13}H_9I_3Ti$	$C_{13}H_9TiI_3$	Ti:	Org.Verb.1-141, 152
$C_{13}H_{10}NO_2Pu^{3+}$	$Pu(C_6H_5CON(O)C_6H_5)^{3+}$	Np:	TrU.D1-168/9
$C_{13}H_{10}NiO_3Ru$	$(C_5H_5)_2RuNi(CO)_3$	Ni:	Org.Verb.2-207
$C_{13}H_{10}OTi$	$(C_6H_5C(O)C_6H_5)Ti$	Ti:	Org.Verb.1-133
$C_{13}H_{11}N_3Ni$	$(CH_2CHCN)Ni(NC_5H_4C_5H_4N)$	Ni:	Org.Verb.1-371
$C_{13}H_{12}N_2O$	$(C_6H_5NH)_2CO$	C:	MVol.D3-101
$C_{13}H_{12}N_2Se$	$(C_6H_5NH)_2CSe$	C:	MVol.D6-201

$C_{13}H_{12}N_4Sn$ .... $(CH_3)_3SnC_6H_3(CN)_4$ .... Sn: Org.Verb.5-79
$C_{13}H_{12}NiS_2$ .... $C_5H_5NiSC(S)CH_2C_6H_5$ .... Ni: Org.Verb.2-114
$C_{13}H_{12}O_3Sn$ .... $(C_4H_3O)_3SnCH_3$ .... Sn: Org.Verb.2-435
$C_{13}H_{12}S_3Sn$ .... $(C_4H_3S)_3SnCH_3$ .... Sn: Org.Verb.2-435
$C_{13}H_{12}Ti$ .... $(C_6H_5)(C_6H_5CH_2)Ti$ .... Ti: Org.Verb.1-69
$C_{13}H_{13}LiSn$ .... $(C_6H_5)_2(CH_3)SnLi$ .... Sn: Org.Verb.5-216
$C_{13}H_{13}N_3O_3S$ .... $(C_6H_5)_2NSO_2NHCONH_2$ .... S: S-N-Verb.1-171
$C_{13}H_{13}Ni_2{}^+$ .... $Ni_2(C_5H_5)_2C_3H_3{}^+$ .... Ni: Org.Verb.2-200
$C_{13}H_{13}Sn^+$ .... $(C_6H_5)_2SnCH_3{}^+$ .... Sn: Org.Verb.2-355
$C_{13}H_{14}N_2O_6SU$ ..... $[UO_2SO_4(CH_2(CH_2C_5H_4N)_2)]$ .... U: SVol.E1-41, 43
$C_{13}H_{14}N_4O_8U$ .... $[UO_2(NO_3)_2(CH_2(CH_2C_5H_4N)_2)]$ .... U: SVol.E1-41, 43
$C_{13}H_{14}Ni$ .... $C_6H_5CH_2C_6H_5NiH_2$ .... Ni: Org.Verb.2-222, 225
$C_{13}H_{14}Ni_2$ .... $(C_5H_5Ni)_2HCCCH_3$ .... Ni: Org.Verb.2-356, 359
$C_{13}H_{14}Sn$ .... $(CH_3)_3SnC_6H_4CCCCH$ .... Sn: Org.Verb.2-129/30
– .... $(C_6H_5)_2(CH_3)SnH$ .... Sn: Org.Verb.4-92
$C_{13}H_{15}I_3S_3V$ .... $(C_5H_5)_2V[I_3][S_2CSC_2H_5]$ .... C: MVol.D4-269
$C_{13}H_{15}Ni^+$ .... $[CH_3C_5H_4NiC_7H_8]^+$ .... Ni: Org.Verb.2-189
$C_{13}H_{15}NiO_3P$ .... $(CO)_3NiP(C_6H_5)(C_2H_5)_2$ .... Ni: Org.Verb.1-172
$C_{13}H_{15}NiO_5P$ .... $(CO)_3NiP(OC_2H_5)_2C_6H_5$ .... Ni: Org.Verb.1-177
$C_{13}H_{15}S_3V$ .... $(C_5H_5)_2V[S_2CSC_2H_5]$ .... C: MVol.D4-269
$C_{13}H_{16}N_2S_2Sn$ .... $(CH_3)_3SnC_6H_4CH(SCN)CH_2SCN$ .... Sn: Org.Verb.2-129
$C_{13}H_{16}Ni$ .... $C_5H_5NiC_8H_{11}$ .... Ni: Org.Verb.2-182/3
$C_{13}H_{16}NiO$ .... $C_5H_5NiC_7H_8OCH_3$ .... Ni: Org.Verb.2-186
$C_{13}H_{16}NiO_2$ .... $C_8H_9Ni(CH(COCH_3)_2)$ .... Ni: Org.Verb.2-8
$C_{13}H_{16}NiS_2$ .... $C_{12}H_{16}NiCS_2$ .... Ni: Org.Verb.2-244, 246
$C_{13}H_{16}O_2SSn$ .... $(CH_2CH)_3SnOS(O)C_6H_4CH_3$ .... Sn: Org.Verb.5-156
$C_{13}H_{16}Sn$ .... $(CH_3)_3SnC_{10}H_7$ .... Sn: Org.Verb.2-146/8
$C_{13}H_{17}IrOSn$ .... $(CH_2CHCH_2)_2Sn[Ir(CO)(CH_3C_5H_4)]$ .... Sn: Org.Verb.5-50
$C_{13}H_{17}NNi$ .... $(CH_3)_2C_3H_3Ni(NCC_6H_5)CH_3$ .... Ni: Org.Verb.2-35
– .... $C_8H_{12}Ni(NC_5H_5)$ .... Ni: Org.Verb.2-231, 233
$C_{13}H_{17}NOSSn$ .... $(CH_3)_2Sn(SC_6H_4NC(CH_3)CHC(CH_3)O)$ .... Sn: Org.Verb.6-31
$C_{13}H_{17}N_2O_5U$ .... $[UO_2(C_5H_8(O)NC_2H_4COO)(C_5H_5N)]$ .... U: SVol.E1-31, 37
$C_{13}H_{17}N_3O_6U$ .... $UO_2(NC_5H_3(COO)_2) \cdot 2\ C_3H_5NH_2$ .... U: SVol.E1-18, 21, 25
$C_{13}H_{17}Ni^+$ .... $[C_5H_5NiC_8H_{12}]^+$ .... Ni: Org.Verb.2-189/90
$C_{13}H_{17}Pa^+$ .... $Pa(C_5H_5)_2C_3H_7{}^+$ .... Pa: SVol.2-99
$C_{13}H_{17}Ti$ .... $C_5H_5Ti(C_5H_5C_3H_7\text{-i})$ .... Ti: Org.Verb.1-212
$C_{13}H_{18}N_6Ni$ .... $(t\text{-}C_4H_9NC)_2NiN_2C(CN)_2$ .... Ni: Org.Verb.1-308, 309
$C_{13}H_{18}Ni$ .... $C_5H_5NiC_8H_{13}$ .... Ni: Org.Verb.2-183/4
$C_{13}H_{18}NiO$ .... $C_{12}H_{18}NiCO$ .... Ni: Org.Verb.2-232, 238
$C_{13}H_{18}Sn$ .... $(CH_3)_3SnC(CH_3)CCHC_6H_5$ .... Sn: Org.Verb.2-82
– .... $(CH_3)_3SnC(C_6H_5)CCHCH_3$ .... Sn: Org.Verb.2-82
– .... $(CH_3)_3SnC_9H_6CH_3$ .... Sn: Org.Verb.2-103, 106/7
$C_{13}H_{19}NO_2PbS_4$ ..... $Pb[S_2COC_3H_7]_2 \cdot C_5H_5N$ .... C: MVol.D4-245, 254
$C_{13}H_{19}N_2Ti$ .... $(CH_3)_3Ti \cdot 2\ C_5H_5N$ .... Ti: Org.Verb.1-83
$C_{13}H_{19}Ni$ .... $C_8H_{12}NiC_4H_4CH_3$ .... Ni: Org.Verb.2-226/9, 231
$C_{13}H_{20}NiO_2$ .... $C_8H_{13}Ni(CH(COCH_3)_2)$ .... Ni: Org.Verb.2-8, 71/2
$C_{13}H_{20}NiO_3$ .... $C_8H_{13}ONi(OCCH_3)_2CH$ .... Ni: Org.Verb.1-122
$C_{13}H_{20}OSn$ .... $(CH_3)_3Sn(CH_2)_3COC_6H_5$ .... Sn: Org.Verb.2-44
$C_{13}H_{20}O_2Sn$ .... $(C_2H_5)_3SnC_6H_4COOH$ .... Sn: Org.Verb.2-241, 243
$C_{13}H_{20}O_7U$ .... $UO_2(CH(COCH_3)_2)_2 \cdot (CH_3)_2CO$ .... U: SVol.E1-79/80

Formula	Compound	Ref.
$C_{13}H_{20}Sn$	$(C_2H_5)_2SnC_9H_{10}$	Sn: Org.Verb.3-112
$C_{13}H_{21}NNi$	$(CH_3)_2C_3H_3Ni(NH(CH_3)C_6H_5)CH_3$	Ni: Org.Verb.2-35
$C_{13}H_{21}NNiO$	$(t\text{-}C_4H_9)_2C_5H_3NiNO$	Ni: Org.Verb.2-349
$C_{13}H_{21}N_3O_2S$	$(C_3H_7)_2NSO_2NHNCHC_6H_5$	S: S-N-Verb.1-222
$C_{13}H_{22}NOSn$	$(CH_3)_3SnC_6H_4N(O)C_4H_9$	Sn: Org.Verb.2-144
$C_{13}H_{22}N_2Sn$	$(CH_3)_3SnC[N(CH_3)_2]NC_6H_4CH_3$	Sn: Org.Verb.2-17
$C_{13}H_{22}OSn$	$(CH_3)((CH_3)_3CCH_2)(CH_3OC_6H_4)SnH$	Sn: Org.Verb.4-92
–	$(CH_3)_3Sn(CH_2)_4OC_6H_5$	Sn: Org.Verb.2-50/1
–	$(C_2H_5)_3SnC_6H_4OCH_3$	Sn: Org.Verb.2-241, 243
$C_{13}H_{22}O_2SSn$	$(C_2H_5)_3SnSO_2C_6H_4CH_3$	Sn: Org.Verb.5-105
$C_{13}H_{22}O_3SSn$	$(CH_3)_3Sn(CH_2)_3OSO_2C_6H_4CH_3$	Sn: Org.Verb.2-45/6
$C_{13}H_{22}O_7U$	$UO_2(CH(COCH_3)_2)_2 \cdot C_3H_7OH$	U: SVol.E1-60, 64/5
$C_{13}H_{22}Sn$	$(CH_3)(C_3H_7)Sn(i\text{-}C_3H_7)(C_6H_5)$	Sn: Org.Verb.3-97
–	$(CH_3)_2Sn(C_3H_7)(CH(CH_3)C_6H_5)$	Sn: Org.Verb.3-75, 76
–	$(CH_3)_2Sn(i\text{-}C_3H_7)(CH(CH_3)C_6H_5)$	Sn: Org.Verb.3-75, 77
–	$(CH_3)_2Sn(s\text{-}C_4H_9)(CH_2C_6H_5)$	Sn: Org.Verb.3-75, 78
–	$(CH_3)_2Sn(t\text{-}C_4H_9)(CH_2C_6H_5)$	Sn: Org.Verb.3-75, 79
–	$(CH_3)_3Sn(CH_2)_4C_6H_5$	Sn: Org.Verb.2-50/1
–	$(CH_3)_3SnC_6H_4C(CH_3)_3$	Sn: Org.Verb.2-129/30
–	$(C_2H_5)_3SnCCC_5H_7$	Sn: Org.Verb.2-227
–	$(C_2H_5)_3SnCH_2C_6H_5$	Sn: Org.Verb.2-171, 177
–	$(C_2H_5)_3SnC_6H_4CH_3$	Sn: Org.Verb.2-240, 243
–	$(C_3H_7)_2Sn(CH_3)(C_6H_5)$	Sn: Org.Verb.3-87
–	$(i\text{-}C_3H_7)_2Sn(CH_3)(C_6H_5)$	Sn: Org.Verb.3-92/3
–	$(C_3H_7)_3SnCCCCH$	Sn: Org.Verb.2-264
$C_{13}H_{23}NOSn$	$(CH_3)_3SnC_6H_4N(OH)C(CH_3)_3$	Sn: Org.Verb.2-142, 144
$C_{13}H_{23}NO_6U$	$UO_2(CH(COCH_3)_2)_2 \cdot N(CH_3)_3$	U: SVol.E1-17, 20, 23
$C_{13}H_{23}NiO_3P$	$(CO)_3Ni(CH_2P(C_3H_7\text{-}i)_3)$	Ni: Org.Verb.1-182, 183
$C_{13}H_{24}N_2OSSn$	$(C_2H_5)_3Sn(CH_2)_2C_4H_2N_2(O)(S)CH_3$	Sn: Org.Verb.2-172, 178
$C_{13}H_{24}N_2O_2Sn$	$(C_2H_5)_3Sn(CH_2)_2C_4H_2N_2(O)_2CH_3$	Sn: Org.Verb.2-172, 178
–	$(C_2H_5)_3SnC_3N_2(CH_3)_2COOCH_3$	Sn: Org.Verb.2-247, 249
$C_{13}H_{24}N_4NiO_4$	$[C(N(CH_3)_2)_3][(CO)_3NiC(O)N(CH_3)_2]$	Ni: Org.Verb.1-193, 194
$C_{13}H_{24}O_2Sn$	$(C_2H_5)_3SnCCC(CH_3)_2OOCCH_3$	Sn: Org.Verb.2-227
–	$(C_2H_5)_3SnCCCH_2OCH_2CH_2OCHCH_2$	Sn: Org.Verb.2-227, 234/5
–	$(C_2H_5)_3SnCH_2COOC(CH_3)_2CCH$	Sn: Org.Verb.2-172, 178
–	$(C_2H_5)_3SnC_3(CH_3)_2COOCH_3$	Sn: Org.Verb.2-207
–	$(C_3H_7)_3SnCCCOOCH_3$	Sn: Org.Verb.2-264
$C_{13}H_{24}O_3Ti$	$C_2H_5C_5H_4Ti(OC_2H_5)_3$	Ti: Org.Verb.1-158, 164
$C_{13}H_{24}O_4RuSiSn$	$[(C_2H_5)_3Si][(CH_3)_3Sn]Ru(CO)_4$	Sn: Org.Verb.5-82
$C_{13}H_{24}O_5Sn$	$(C_2H_5)_3SnC_2(COOCH_3)_2OCH_3$	Sn: Org.Verb.2-197
$C_{13}H_{24}SiSn$	$(CH_3)_3SnCH_2C_6H_4Si(CH_3)_3$	Sn: Org.Verb.2-27
–	$(CH_3)_3SnC_6H_4CH_2Si(CH_3)_3$	Sn: Org.Verb.2-129
$C_{13}H_{24}Sn$	$(CH_2CHCH_2)_3SnC_4H_9$	Sn: Org.Verb.2-346, 348
–	$(CH_2CH_2CH)_3SnC_4H_9$	Sn: Org.Verb.2-340
–	$(CH_3)_3SnC_5(CH_3)_5$	Sn: Org.Verb.2-99, 105
–	$(CH_3)_3SnC_{10}H_{15}$	Sn: Org.Verb.2-64, 68
–	$(C_2H_5)_3SnC(CH_2)C_5H_7$	Sn: Org.Verb.2-197
–	$(C_2H_5)_3SnCHCC(CH_2)_4$	Sn: Org.Verb.2-197
–	$(C_2H_5)_3SnCHCHC_5H_7$	Sn: Org.Verb.2-197
–	$(C_2H_5)_3SnC_5H_7CCH_2$	Sn: Org.Verb.2-181/2

Formula	Compound	Reference
$C_{13}H_{24}Sn$	$(C_3H_7)_3SnCCCHCH_2$	Sn: Org.Verb.2-263/4
–	$(C_4H_9)_2Sn(CHCH)_2CH_2$	Sn: Org.Verb.3-106
$C_{13}H_{25}KOS_2$	$K[S_2COC_{12}H_{25}]$	C: MVol.D4-253
$C_{13}H_{25}KOSe_2$	$K[Se_2COC_{12}H_{25}]$	C: MVol.D6-221
$C_{13}H_{25}NO_2S$	$HNCS \cdot 2\ CH_3COC_4H_9 \cdot H_2O$	C: MVol.D6-30
$C_{13}H_{25}NO_4Sn$	$(CH_3)_3SnC_2(COOC_2H_5)_2N(CH_3)_2$	Sn: Org.Verb.2-83
$C_{13}H_{25}NSn$	$(C_2H_5)_3SnC(CN)(CH_2)_5$	Sn: Org.Verb.2-181/2
$C_{13}H_{25}N_3OSn$	$(C_2H_5)_3Sn(CH_2)_2C_4H_2N_2(O)(NH)CH_3$	Sn: Org.Verb.2-172, 178
$C_{13}H_{25}OSe_2^-$	$[Se_2COC_{12}H_{25}]^-$	C: MVol.D6-220
$C_{13}H_{26}NiOP_2$	$CH_3Ni(P(CH_3)_3)_2OC_6H_5$	Ni: Org.Verb.1-12, 19
$C_{13}H_{26}OSn$	$(C_2H_5)_3SnC(CCHCH_3)CH(OH)C_2H_5$	Sn: Org.Verb.2-198
–	$(C_2H_5)_3SnC(CCHCH_3)C(OH)(CH_3)_2$	Sn: Org.Verb.2-198
–	$(C_2H_5)_3SnC(CHCH_2)C(OH)CH(CH_3)_2$	Sn: Org.Verb.2-198, 204
–	$(C_2H_5)_3SnC(CH_2)C(OH)(CH_2)_4$	Sn: Org.Verb.2-197
–	$(C_2H_5)_3SnCH_2CH_2CCCH(OH)C_2H_5$	Sn: Org.Verb.2-227
–	$(C_2H_5)_3SnCH_2CH_2CCC(OH)(CH_3)_2$	Sn: Org.Verb.2-227
$C_{13}H_{26}O_2Sn$	$(CH_3)_3SnCH(C_6H_{11})COOC_2H_5$	Sn: Org.Verb.2-25
–	$(C_2H_5)_3SnCCC(CH_3)_2OCH_2OCH_3$	Sn: Org.Verb.2-227
–	$(C_3H_7)_3SnC(COOCH_3)CH_2$	Sn: Org.Verb.2-260, 262
–	$(C_3H_7)_3SnCHCHCOOCH_3$	Sn: Org.Verb.2-260, 262
$C_{13}H_{26}O_4Sn$	$(CH_3)_3Sn(CH_2)_3CH(COOC_2H_5)_2$	Sn: Org.Verb.2-50/1
$C_{13}H_{26}SiSn$	$(C_2H_5)_3SnCHCHCCSi(CH_3)_3$	Sn: Org.Verb.2-228
$C_{13}H_{26}Sn$	$(CH_3)_3SnC_7H_8(CH_3)_3$	Sn: Org.Verb.2-63
–	$(C_3H_7)_3SnCHCCHCH_3$	Sn: Org.Verb.2-260
–	$(C_3H_7)_3SnCHCHCHCH_2$	Sn: Org.Verb.2-260
$C_{13}H_{27}NOSn$	$(C_4H_9)_3SnNCO$	Sn: Org.Verb.5-131/2
$C_{13}H_{27}NSSn$	$(C_4H_9)_3SnNCS$	Sn: Org.Verb.5-131/2
$C_{13}H_{27}NSn$	$(C_2H_5)_3SnCCCH_2N(C_2H_5)_2$	Sn: Org.Verb.2-228, 235
–	$(C_3H_7)_3SnCH_2CH_2CH_2CN$	Sn: Org.Verb.2-252, 257
–	$(C_4H_9)_3SnCN$	Sn: Org.Verb.2-308
$C_{13}H_{27}N_2P$	$(C_4H_9)_3PNCN$	F: PerFHalOrg.5-22
$C_{13}H_{27}N_2PS_3$	$(C_4H_9S)_3PNCN$	F: PerFHalOrg.5-23
$C_{13}H_{27}N_3O_3S_4U$	$UO_2(SCSN(C_2H_5)_2)_2 \cdot HCON(CH_3)_2$	U: SVol.E1-100, 102
$C_{13}H_{27}N_3Ti$	$C_2H_5C_5H_4Ti(N(CH_3)_2)_3$	Ti: Org.Verb.1-172, 175/8
$C_{13}H_{27}NiO_3PSi_2$	$(CO)_3NiP(C_4H_9\text{-t})(Si(CH_3)_3)_2$	Ni: Org.Verb.1-166, 168
$C_{13}H_{27}NiO_3PSn_2$	$(CO)_3NiP(C_4H_9\text{-t})(Sn(CH_3)_3)_2$	Ni: Org.Verb.1-166, 168
$C_{13}H_{28}N_4O_2Sn$	$[Sn(C_4H_9)_2N_2H_2C(O)(CH_2)_3C(O)N_2H_2]_x$	Sn: Org.Verb.6-89, 98
$C_{13}H_{28}N_4O_{14}SU$	$UO_2[(C_{10}H_{16}N_3O_6S)NO_3] \cdot 3\ CH_3OH$	U: SVol.E1-59, 63
$C_{13}H_{28}OSiSn$	$(C_2H_5)_3SnCCCH(CH_3)OSi(CH_3)_3$	Sn: Org.Verb.2-228
–	$(C_2H_5)_3SnCCCH_2CH_2OSi(CH_3)_3$	Sn: Org.Verb.2-228
$C_{13}H_{28}OSn$	$(C_2H_5)_3SnC(C_2H_5)CHC(OH)(CH_3)_2$	Sn: Org.Verb.2-198
–	$(C_2H_5)_3SnCHCHC(OH)(CH_3)C_3H_7$	Sn: Org.Verb.2-198
–	$(C_2H_5)_3SnCHCHC(OH)(CH_3)(i\text{-}C_3H_7)$	Sn: Org.Verb.2-198
–	$(C_3H_7)_3SnCHCHOC_2H_5$	Sn: Org.Verb.2-261/2
–	$(C_3H_7)_3SnCH_2CH_2COCH_3$	Sn: Org.Verb.2-252, 257
–	$(C_4H_9)_2((CH_3)_2C(OH)CHCH)SnH$	Sn: Org.Verb.4-91
$C_{13}H_{28}O_2Sn$	$(C_3H_7)_3SnCH_2CH_2COOCH_3$	Sn: Org.Verb.2-252, 257
–	$(C_3H_7)_3SnCH_2COOC_2H_5$	Sn: Org.Verb.2-253, 257
$C_{13}H_{28}Si_2Sn$	$(CH_3)_3SnC_5H_4Si(CH_3)_2Si(CH_3)_3$	Sn: Org.Verb.2-100
$C_{13}H_{28}Sn$	$(CH_3)_3SnC_6H_{10}C(CH_3)_3$	Sn: Org.Verb.2-61/2

Formula	Compound	Reference
$C_{13}H_{28}Sn$	$(C_2H_5)_2Sn(i\text{-}C_3H_7)(C_6H_{11})$	Sn: Org.Verb.3-86/7
–	$(C_3H_7)_3SnCHCHC_2H_5$	Sn: Org.Verb.2-260
–	$(C_4H_9)_2Sn(CH_2)_5$	Sn: Org.Verb.3-105
$C_{13}H_{29}NOSn$	$(C_2H_5)_3SnCONHC_6H_{13}$	Sn: Org.Verb.2-172
–	$(C_3H_7)_3SnCH_2CH_2C(CH_3)NOH$	Sn: Org.Verb.2-253
–	$(C_3H_7)_3SnCH_2CON(CH_3)_2$	Sn: Org.Verb.2-253
–	$(C_4H_9)_3SnONCH_2$	Sn: Org.Verb.5-132
$C_{13}H_{29}NO_2S$	$[(C_3H_7)_3NH][SOCOC_3H_7]$	C: MVol.D4-237
$C_{13}H_{29}NO_2Sn$	$(C_4H_9)_2Sn(OCH_2CH_2)_2NCH_3$	Sn: Org.Verb.6-89, 94
$C_{13}H_{29}N_3O_3S_4U$	$[UO_2(S_2CN(C_2H_5)_2)_2((CH_3)_3NO)]$	U: SVol.E1-5, 131/2, 135, 141/2
$C_{13}H_{29}N_3O_3Se_4U$	$UO_2(Se_2CN(C_2H_5)_2)_2 \cdot (CH_3)_3NO$	U: SVol.E1-131/2, 135
$C_{13}H_{29}NiP$	$(CH_2CH_2)_2NiP(C_3H_7\text{-}i)_3$	Ni: Org.Verb.1-411
–	$C_3H_5Ni(P(C_3H_7)_3)CH_3$	Ni: Org.Verb.2-30
$C_{13}H_{29}NiP^+$	$Ni(CH_2P(C_4H_9)_3)^+$	Ni: Org.Verb.1-182
$C_{13}H_{29}SbSn$	$(CH_3)_3SnCHCHSb(C_4H_9)_2$	Sn: Org.Verb.2-83
$C_{13}H_{30}N_2O_{14}P_2U$	$UO_2(NO_3)_2 \cdot CH_2(PO(OC_3H_7)_2)_2$	U: SVol.E1-188/9
$C_{13}H_{30}N_2Sn_2$	$(C_2H_5)_3SnNCNSn(C_2H_5)_3$	Sn: Org.Verb.5-104, 106
$C_{13}H_{30}OSn$	$(C_2H_5)_3SnCH_2CH_2C(C_2H_5)_2OH$	Sn: Org.Verb.2-172
–	$(C_4H_9)_3SnCH_2OH$	Sn: Org.Verb.2-280
–	$(C_4H_9)_3SnOCH_3$	Sn: Org.Verb.5-131
$C_{13}H_{30}O_3Sn$	$C_4H_9Sn(OC_3H_7)_3$	Sn: Org.Verb.6-246
$C_{13}H_{30}O_3Ti$	$CH_3Ti(OC_4H_9)_3$	Ti: Org.Verb.1-44
–	$CH_3Ti(OC_4H_9\text{-}t)_3$	Ti: Org.Verb.1-44, 48/9
–	$CD_3Ti(OC_4H_9\text{-}t)_3$	Ti: Org.Verb.1-44, 49
$C_{13}H_{30}SiSn$	$(C_2H_5)_3SnCH_2CH_2Si(CH_3)_2CH_2CHCH_2$	Sn: Org.Verb.2-173
$C_{13}H_{30}Sn$	$(CH_3)_3SnC_{10}H_{21}$	Sn: Org.Verb.2-53/4
–	$(C_2H_5)_3SnC_7H_{15}$	Sn: Org.Verb.2-159
–	$(C_3H_7)_3SnC_4H_9$	Sn: Org.Verb.2-250
–	$(C_3H_7)_3Sn(i\text{-}C_4H_9)$	Sn: Org.Verb.2-253
–	$(C_3H_7)_3Sn(t\text{-}C_4H_9)$	Sn: Org.Verb.2-253
–	$(i\text{-}C_3H_7)_3Sn(t\text{-}C_4H_9)$	Sn: Org.Verb.2-271
–	$(C_4H_9)_2Sn(C_2H_5)(i\text{-}C_3H_7)$	Sn: Org.Verb.3-88/9
–	$(C_4H_9)_3SnCH_3$	Sn: Org.Verb.2-272
		Sn: Org.Verb.6-218
–	$(i\text{-}C_4H_9)_3SnCH_3$	Sn: Org.Verb.2-327
–	$(s\text{-}C_4H_9)_3SnCH_3$	Sn: Org.Verb.2-326
–	$(t\text{-}C_4H_9)_3SnCH_3$	Sn: Org.Verb.2-329
$C_{13}H_{31}LiO_4Ti$	$Li[(CH_3)_5Ti] \cdot 2\ C_4H_8O_2$	Ti: Org.Verb.1-120/1
$C_{13}H_{31}NSn$	$(C_3H_7)_3SnN(C_2H_5)_2$	Sn: Org.Verb.5-114
$C_{13}H_{32}N_2S_3$	$[(C_2H_5)_3NH]_2CS_3$	C: MVol.D4-217
$C_{13}H_{32}OSi_2Sn$	$(C_2H_5)_3SnCHCHSi(CH_3)_2OSi(CH_3)_3$	Sn: Org.Verb.2-198
$C_{13}H_{33}N_3Ti$	$CH_3Ti(N(C_2H_5)_2)_3$	Ti: Org.Verb.1-52, 57
–	$CD_3Ti(N(C_2H_5)_2)_3$	Ti: Org.Verb.1-52, 57
$C_{13}H_{34}NiO_9P_3^+$	$[CH_3C_3H_4Ni(P(OCH_3)_3)_3]^+$	Ni: Org.Verb.2-25/6
$C_{13}H_{35}NiO_{13}P_3S$	$[CH_3C_3H_4Ni(P(OCH_3)_3)_3]HSO_4$	Ni: Org.Verb.2-26
$C_{13}H_{36}Si_2Sn_2$	$[(CH_3)_3Si]_2C[Sn(CH_3)_3]_2$	Sn: Org.Verb.5-79
$C_{13}H_{36}Si_3Sn$	$((CH_3)_3SiCH_2)_3SnCH_3$	Sn: Org.Verb.2-336, 338
–	$(CH_3)_3SnC(Si(CH_3)_3)_3$	Sn: Org.Verb.2-17
$C_{13}H_{39}INiP_4$	$CH_3Ni(P(CH_3)_3)_4I$	Ni: Org.Verb.1-12, 20

Formula	Compound	Reference
$C_{13.2}H_{31.2}O_{2.4}Sn_{1.4}$	$CH_3O[Sn(C_4H_9)_2O]_{1.4}CH_3$	Sn: Org.Verb.6-89/91
$C_{13.5}Cl_2H_{27}N_3O_5U$	$UO_2Cl_2 \cdot 1.5\ C_3H_6(CON(CH_3)_2)_2$	U: SVol.E1-113/4
$C_{13.5}Cl_4H_{27}N_3O_3U$	$UCl_4 \cdot 1.5\ CH_2(CH_2CON(CH_3)_2)_2$	U: SVol.E1-112/4
$C_{14}CaCl_6H_{21}N_7Sn$	$[Ca(CH_3CN)_7][SnCl_6]$	Sn: MVol.C3-176
$C_{14}CdFeH_8N_2O_4$	$(NC_5H_4C_5H_4N)CdFe(CO)_4$	Fe: Org.Verb.B3-126
$C_{14}CdFeH_{10}N_2O_4$	$Cd(C_5H_5N)_2Fe(CO)_4$	Fe: Org.Verb.B2-13
$C_{14}CeCl_{10}H_{56}N_7$	$CeCl_3 \cdot 7\ [(CH_3)_2NH_2]Cl$	Sc: MVol.C5-213/5
–	$CeCl_3 \cdot 7\ [(CH_3)_2NH_2]Cl \cdot 2\ H_2O$	Sc: MVol.C5-213/5
$C_{14}CfH_{16}N_2O_{12}^{3-}$	$Cf(N(CH_2CH_2COO)(CH_2COO)_2)_2^{3-}$	Np: TrU.D1-164
$C_{14}CfH_{18}N_3O_{10}^{2-}$	$Cf(CH_2COO)N(C_2H_4N(CH_2COO)_2)_2^{2-}$	Np: TrU.D1-166
$C_{14}ClCo_2H_5O_8Sn$	$(C_6H_5)ClSn[Co(CO)_4]_2$	Sn: Org.Verb.6-275, 278
$C_{14}ClFH_{14}Sn$	$(CH_3)_2Sn(C_6H_4F)(C_6H_4Cl)$	Sn: Org.Verb.3-82, 83
$C_{14}ClF_2H_{19}Sn$	$(C_2H_5)_3SnC_6H_4CFCFCl$	Sn: Org.Verb.2-241
$C_{14}ClF_6H_{10}NP_2$	$(CF_3)_2PNP(C_6H_5)_2Cl$	F: PerFHalOrg.3-69
$C_{14}ClF_8H_7OSi$	$ClC_6F_4OC_6F_4Si(CH_3)_2H$	F: PerFHalOrg.4-85
$C_{14}ClF_{11}$	$C_6F_5CClCFC_6F_5$	F: PerFHalOrg.4-16
–	$ClFCCFC_6F_4C_6F_5$	F: PerFHalOrg.4-16
$C_{14}ClF_{12}H_6N_3O$	$(CF_3)_2CNHC(CF_3)_2NC(NHCOC_6H_4Cl)$	F: PerFHalOrg.5-105
$C_{14}ClF_{29}N_4O_4$	$ClC_6F_5[ON(CF_3)_2]_4$	F: PerFHalOrg.7-98/9, 116
$C_{14}ClFeH_7O_6S$	$C_{10}H_7SO_2Fe(CO)_4Cl$	Fe: Org.Verb.B3-187
$C_{14}ClFeH_9O_5$	$ClCHCHCOC_6H_4CH_3Fe(CO)_4$	Fe: Org.Verb.B4-267/8, 271, 288/9
$C_{14}ClFeH_9O_6$	$ClCHCHCOC_6H_4OCH_3Fe(CO)_4$	Fe: Org.Verb.B4-271, 288/9
$C_{14}ClFeH_{13}O$	$C_5H_5FeC_5H_4CClC(CH_3)CHO$	Fe: Org.Verb.A2-174
–	$C_5H_5FeC_5H_4CClCHCOCH_3$	Fe: Org.Verb.A3-24/5
–	$C_5H_5FeC_5H_4C_3H_4COCl$	Fe: Org.Verb.A3-159
$C_{14}ClFeH_{13}O_3$	$(CH_3)_2(C_6H_5)C_3H_2Fe(CO)_3Cl$	Fe: Org.Verb.B5-46, 54, 64/5
$C_{14}ClFeH_{15}$	$C_5H_5FeC_5H_4CClCHC_2H_5$	Fe: Org.Verb.A1-385, 386
–	$C_5H_5FeC_5H_4CHC(CH_3)CH_2Cl$	Fe: Org.Verb.A1-385, 387
$C_{14}ClFeH_{15}O$	$C_5H_5FeC_5H_4C(CH_3)_2COCl$	Fe: Org.Verb.A3-159
–	$C_5H_5FeC_5H_4(CH_2)_3COCl$	Fe: Org.Verb.A3-159
–	$C_5H_5FeC_5H_4COCH_2CHClCH_3$	Fe: Org.Verb.A2-243
–	$C_5H_5FeC_5H_4CO(CH_2)_3Cl$	Fe: Org.Verb.A2-244, 246
$C_{14}ClFeH_{16}MgO_7P$	$MgCl[C_6H_5CH_2C(O)Fe(CO)_3P(OCH_3)_3]$	Fe: Org.Verb.B4-142
$C_{14}ClFeH_{18}O$	$(C_5H_5)_2Fe(C_4H_8O)Cl$	Fe: Org.Verb.B1-7
$C_{14}ClFe_2GeH_{23}N_4O_4$	$(C_3H_5Fe(NO)_2)_2Ge(Cl)C_8H_{13}$	Fe: Org.Verb.C2-165
$C_{14}ClFe_2H_{23}N_4O_4Pb$	$(C_3H_5Fe(NO)_2)_2Pb(Cl)C_7H_{10}CH_3$	Fe: Org.Verb.C2-165
$C_{14}ClFe_2H_{23}N_4O_4Sn$	$(C_3H_5Fe(NO)_2)_2Sn(Cl)C_7H_{10}CH_3$	Fe: Org.Verb.C2-165
$C_{14}ClH_{11}NORhSn$	$Rh(C_8H_6N)(CO)SnCl(C_5H_5)$	Sn: Org.Verb.6-264
$C_{14}ClH_{12}IrN_3NaO_7$	$Na[Ir(NO)Cl(C_7H_6O_3N)_2]$	Ir: SVol.2-49
$C_{14}ClH_{12}IrN_5NaO_5$	$Na[Ir(NO)Cl(C_7H_6O_2N_2)_2]$	Ir: SVol.2-49
$C_{14}ClH_{12}MnNO_9$	$[MnCl(C_{14}H_5O_3(OH)NO_2)(H_2O)_3]$	Mn: MVol.D1-137
$C_{14}ClH_{12}NOSSn$	$CH_3ClSn(OC_6H_4C_7H_5NS)$	Sn: Org.Verb.6-218/9
$C_{14}ClH_{12}NO_2Sn$	$CH_3ClSn(OC_6H_3(C_6H_4O)CHNH)$	Sn: Org.Verb.6-218/9
$C_{14}ClH_{12}NSn$	$(C_6H_5)_2(NCCH_2)SnCl$	Sn: Org.Verb.5-216, 218
$C_{14}ClH_{13}Sn$	$(C_6H_5)_2(CH_2CH)SnCl$	Sn: Org.Verb.5-216, 218
$C_{14}ClH_{13}Sn^+$	$C_{14}H_{13}SnCl^+$	Sn: Org.Verb.5-144
$C_{14}ClH_{14}Sn^+$	$(C_6H_5CH_2)_2SnCl^+$	Sn: Org.Verb.6-119
$C_{14}ClH_{15}N_2NiS$	$C_3H_5Ni[SC(NH_2)(NHC_{10}H_7)]Cl$	Ni: Org.Verb.2-275
$C_{14}ClH_{15}Sn$	$(C_6H_5)_2(C_2H_5)SnCl$	Sn: Org.Verb.5-216, 218

Formula	Compound	Reference
$C_{14}ClH_{19}Sn$	$(C_2H_5)_3SnCCC_6H_4Cl$	Sn: Org.Verb.2-228
$C_{14}ClH_{21}OSn$	$(C_2H_5)_3SnCH_2COC_6H_4Cl$	Sn: Org.Verb.2-173, 178
$C_{14}ClH_{23}OSn$	$(C_4H_9)_2SnCl(OC_6H_5)$	Sn: Org.Verb.6-88
$C_{14}ClH_{23}Sn$	$(C_4H_9)_2(C_6H_5)SnCl$	Sn: Org.Verb.5-210
–	$(t\text{-}C_4H_9)_2(C_6H_5)SnCl$	Sn: Org.Verb.5-217, 221
$C_{14}ClH_{24}O_2PSn$	$(C_2H_5)_2[C_2H_5O(C_6H_5)P(O)C_2H_4]SnCl$	Sn: Org.Verb.5-203
$C_{14}ClH_{25}O_2Ti$	$(CH_3)_5C_5Ti(OC_2H_5)_2Cl$	Ti: Org.Verb.1-187/8
$C_{14}ClH_{26}NSn$	$(C_3H_7)_3SnCl \cdot C_5H_5N$	Sn: Org.Verb.5-114
$C_{14}ClH_{27}Sn$	$(C_4H_9)_3SnCCCl$	Sn: Org.Verb.5-129
–	$(C_6H_{11})_2(C_2H_5)SnCl$	Sn: Org.Verb.5-221
$C_{14}ClH_{29}OSn$	$(i\text{-}C_4H_9)_2ClSnOCH(CH_2)_5$	Sn: Org.Verb.6-105
$C_{14}ClH_{29}O_2Sn$	$(i\text{-}C_4H_9)_2(C_2H_5OCOCH_2CH(CH_3))SnCl$	Sn: Org.Verb.5-213, 215
$C_{14}ClH_{29}Sn$	$(C_4H_9)_2(CH_2C(CH_3)CH(CH_3)CH_2)SnCl$	Sn: Org.Verb.5-210
–	$(i\text{-}C_4H_9)_2(CH_2C(CH_3)CH(CH_3)CH_2)SnCl$	Sn: Org.Verb.5-213, 215
–	$(C_4H_9)_2((CH_3)_2CC(CH_3)CH_2)SnCl$	Sn: Org.Verb.5-210
–	$(i\text{-}C_4H_9)_2((CH_3)_2CC(CH_3)CH_2)SnCl$	Sn: Org.Verb.5-213, 215
–	$(C_4H_9)_2(C_6H_{11})SnCl$	Sn: Org.Verb.5-210, 212
$C_{14}ClH_{31}Sn$	$(CH_3)_2(C_{12}H_{25})SnCl$	Sn: Org.Verb.5-196
–	$(C_3H_7)_2(C_8H_{17})SnCl$	Sn: Org.Verb.5-205/6
$C_{14}ClH_{32}NSn$	$(C_3H_7)_3SnCl \cdot (CH_2)_5NH$	Sn: Org.Verb.5-114
$C_{14}ClH_{32}O_3PSn$	$(C_4H_9)_2((C_2H_5O)_2P(O)CH_2CH_2)SnCl$	Sn: Org.Verb.5-210
$C_{14}ClH_{33}OSn_2$	$(C_2H_5)_3SnOSnCl(C_4H_9)_2$	Sn: Org.Verb.5-104
$C_{14}ClH_{34}OPSn$	$(C_2H_5)_3SnCl \cdot (C_2H_5)_2(C_4H_9O)P$	Sn: Org.Verb.5-107
$C_{14}ClH_{36}NO_2Si_2Sn$	$[(CH_3)_3SnCH_2CH(NH(Si(CH_3)_3)_2)COOC_2H_5]Cl$	Sn: Org.Verb.2-38
$C_{14}ClH_{39}Si_4Sn$	$[((CH_3)_3Si)_2CH]_2SnHCl$	Sn: Org.Verb.6-290
$C_{14}Cl_2F_2H_9NO_3$	$NC_5F_2Cl_2(OCH_2C_6H_4OCOCH_3)$	F: PerFHalOrg.5-136 F: PerFHalOrg.6-14
$C_{14}Cl_2F_2H_{12}Sn$	$(FC_6H_4CH_2)_2SnCl_2$	Sn: Org.Verb.6-134, 138/9
$C_{14}Cl_2F_2H_{21}N_3$	$N(CFCCl)_2CNH(CH_2)_9NH_2$	F: PerFHalOrg.5-140/1
$C_{14}Cl_2F_4H_{14}Ni_2O_2$	$(CH_3CHCHCH_2NiCl)_2 \cdot C_6F_4O_2$	Ni: Org.Verb.2-300, 302
$C_{14}Cl_2F_{10}$	$Cl_2CCFC_6F_4C_6F_5$	F: PerFHalOrg.4-17
$C_{14}Cl_2F_{12}H_5N_3O$	$(CF_3)_2CNHC(CF_3)_2NC(NHCOC_6H_3Cl_2)$	F: PerFHalOrg.5-105
$C_{14}Cl_2F_{16}GeH_6$	$(CH_3)_2Ge(C_6F_8Cl)_2$	F: PerFHalOrg.4-27
$C_{14}Cl_2FeH_8Hg_2N_2O_2$	$(CO)_2Fe(C_{12}H_8N_2)(HgCl)_2$	Fe: Org.Verb.B1-116/7
$C_{14}Cl_2FeH_8Hg_2N_2O_4$	$(CO)_4Fe(HgCl)_2(NC_5H_4C_5H_4N)$	Fe: Org.Verb.B2-153/5, 158
$C_{14}Cl_2FeH_8N_2O_2$	$(CO)_2Fe(N_2C_{12}H_8)Cl_2$	Fe: Org.Verb.B1-101
$C_{14}Cl_2FeH_8N_4O_2$	$(ClC_6H_4NC)_2Fe(NO)_2$	Fe: Org.Verb.B4-11, 13
$C_{14}Cl_2FeH_{10}Hg_2N_2O_4$	$(CO)_4Fe(HgCl \cdot NC_5H_5)_2$	Fe: Org.Verb.B2-153/5, 158, 161
$C_{14}Cl_2FeH_{10}N_2$	$(C_6H_5NC)_2FeCl_2$	Fe: Org.Verb.B4-11/2
$C_{14}Cl_2FeH_{12}$	$C_5H_5FeC_5H_4CClCHCClCH_2$	Fe: Org.Verb.A1-385, 387
$C_{14}Cl_2FeH_{14}$	$C_5H_5FeC_5H_4C_3H_2Cl_2CH_3$	Fe: Org.Verb.A1-392/3
$C_{14}Cl_2FeH_{14}Hg_2N_2O_2$	$(CO)_2Fe(NC_5H_4CH_3)_2(HgCl)_2$	Fe: Org.Verb.B1-116/7
$C_{14}Cl_2FeH_{14}O_2$	$C_5H_5FeC_5H_4CH(CHCl_2)OOCCH_3$	Fe: Org.Verb.A2-98, 100
$C_{14}Cl_2FeH_{17}N_5O_5$	$COFe(NC_5H_3Cl_2)(C_2HN_2O_2(CH_3)_2)_2$	Fe: Org.Verb.B1-33
$C_{14}Cl_2FeH_{18}Si$	$C_5H_5FeC_5H_4CH(CH_3)CH_2SiCl_2CH_3$	Fe: Org.Verb.A1-292
$C_{14}Cl_2FeH_{26}Hg_2N_2O_2$	$(CO)_2Fe(NH_2CH(CH_2)_5)_2(HgCl)_2$	Fe: Org.Verb.B1-116/7
$C_{14}Cl_2FeH_{30}O_2P_2$	$(CO)_2Fe(P(C_2H_5)_3)_2Cl_2$	Fe: Org.Verb.B1-101/3
$C_{14}Cl_2H_6O_4Sn$	$SnCl_2(C_{14}H_6O_4) \cdot C_6H_6$	Sn: MVol.C5-114

Formula	Compound	Reference
$C_{14}Cl_2H_6O_5Sn$	$SnCl_2(C_{14}H_6O_5) \cdot C_6H_6$	Sn: MVol.C5-114
$C_{14}Cl_2H_8MnO_4$	$Mn(OC_6H_3(Cl)CHO)_2$	Mn: MVol.D1-53/4
$C_{14}Cl_2H_8NiO_6P_2$	$(CO)_2Ni(P(O_2C_6H_4)Cl)_2$	Ni: Org.Verb.1-241
$C_{14}Cl_2H_9MnO_3Sn$	$(C_6H_5)[C_5H_4Mn(CO)_3]SnCl_2$	Sn: Org.Verb.6-202
$C_{14}Cl_2H_{10}N_2S_2Sn$	$SnCl_2 \cdot 2\ C_7H_5NS$	Sn: MVol.C5-53
$C_{14}Cl_2H_{10}N_2S_4Sn$	$SnCl_2 \cdot 2\ C_7H_5NS_2$	Sn: MVol.C5-53
$C_{14}Cl_2H_{10}O_2S_2Sn$	$SnCl_2(C_7H_5OS)_2$	Sn: MVol.C6-115
$C_{14}Cl_2H_{10}O_4Sn$	$SnCl_2(C_7H_5O_2)_2$	Sn: MVol.C5-97
–	$SnCl_2(OC_6H_4CHO)_2$	Sn: MVol.C5-81
$C_{14}Cl_2H_{10}Sn$	$(C_6H_4CH)_2SnCl_2$	Sn: Org.Verb.6-208
$C_{14}Cl_2H_{12}N_2O_2U$	$[UO_2Cl_2((CH_3)_2C_{12}H_6N_2)]$	U: SVol.E1-44/6
$C_{14}Cl_2H_{12}N_2O_4Sn$	$(O_2NC_6H_4CH_2)_2SnCl_2$	Sn: Org.Verb.6-135
$C_{14}Cl_2H_{12}N_4S_2Sn$	$SnCl_2 \cdot 2\ NH_2C_7H_4NS \cdot 2\ HCl$	Sn: MVol.C5-53
$C_{14}Cl_2H_{12}Sn$	$C_2H_4(C_6H_4)_2SnCl_2$	Sn: Org.Verb.6-209
$C_{14}Cl_2H_{13}NSn$	$(C_6H_5)_2(CH_3NCCl)SnCl$	Sn: Org.Verb.5-216, 218
$C_{14}Cl_2H_{14}N_2O_6U$	$UO_2Cl_2 \cdot 2\ NH_2C_6H_4COOH$	U: SVol.E1-18, 21
$C_{14}Cl_2H_{14}N_2Sn$	$(CH_2CH)_2SnCl_2 \cdot NC_5H_4C_5H_4N$	Sn: Org.Verb.6-146
–	$(CH_3)_2SnCl_2 \cdot C_{12}H_8N_2$	Sn: Org.Verb.6-35, 38
$C_{14}Cl_2H_{14}N_4Sn$	$(CH_3)_2SnCl_2 \cdot 2\ NC_5H_4CN$	Sn: Org.Verb.6-36, 38
$C_{14}Cl_2H_{14}N_6O_8U$	$UO_2Cl_2 \cdot 2\ NO_2C_6H_4CON_2H_3 \cdot n\ H_2O$	U: SVol.E1-114/6
$C_{14}Cl_2H_{14}O_2Sn$	$(CH_3OC_6H_4)_2SnCl_2$	Sn: Org.Verb.6-183/5, 187
$C_{14}Cl_2H_{14}S_2Sn$	$SnCl_2(SCH_2C_6H_5)_2$	Sn: MVol.C6-80
$C_{14}Cl_2H_{14}Sn$	$(CH_3C_6H_4)_2SnCl_2$	Sn: Org.Verb.6-180/3
–	$(C_6H_5CH_2)_2SnCl_2$	Sn: Org.Verb.6-118/24
$C_{14}Cl_2H_{14}Ti$	$(C_6H_5CH_2)_2TiCl_2$	Ti: Org.Verb.1-72/3
–	$(C_7H_7)_2TiCl_2$	Ti: Org.Verb.1-72
$C_{14}Cl_2H_{15}NO_2U$	$UO_2Cl_2 \cdot (C_6H_5CH_2)_2NH$	U: SVol.E1-19, 22
$C_{14}Cl_2H_{15}N_2O_5PS$	$(CH_3)_2NSO_2NHPO(OC_6H_4Cl)_2$	S: S-N-Verb.1-184/5
$C_{14}Cl_2H_{16}MnO_4$	$MnCl_2 \cdot 2\ O(CCH_3CH)_2CO$	Mn: MVol.D1-148
–	$MnCl_2 \cdot 2\ O(CCH_3CH)_2CO \cdot 2\ H_2O$	Mn: MVol.D1-149
$C_{14}Cl_2H_{16}MnO_6$	$Mn(C_4H_3O_2(CH_2Cl)COCH_3)_2$	Mn: MVol.D1-145
$C_{14}Cl_2H_{16}N_2O_2Sn$	$(CH_2CH)_2SnCl_2 \cdot 2\ C_5H_5NO$	Sn: Org.Verb.6-146
$C_{14}Cl_2H_{16}N_2Sn$	$(CH_2CH)_2SnCl_2 \cdot 2\ C_5H_5N$	Sn: Org.Verb.6-145
$C_{14}Cl_2H_{16}N_4S_2Te$	$Te(C_6H_5NHCSNH_2)_2Cl_2$	Te: SVol.B3-151
$C_{14}Cl_2H_{16}OSSn$	$(C_6H_5)_2SnCl_2 \cdot (CH_3)_2SO$	Sn: Org.Verb.6-174, 177
$C_{14}Cl_2H_{17}NOSn$	$(C_6H_5)_2SnCl_2 \cdot NH_2CH_2CH_2OH$	Sn: Org.Verb.6-174, 177
$C_{14}Cl_2H_{18}N_2O_2U$	$UO_2Cl_2 \cdot 2\ CH_3C_6H_4NH_2$	U: SVol.E1-18, 21
$C_{14}Cl_2H_{18}N_2O_4U$	$UO_2Cl_2 \cdot 2\ CH_3OC_6H_4NH_2$	U: SVol.E1-18, 21
–	$UO_2Cl_2 \cdot 2\ NC_5H_4C_2H_4OH$	U: SVol.E1-29, 34
$C_{14}Cl_2H_{18}N_2Sn$	$(C_2H_5)_2SnCl_2 \cdot NC_5H_4C_5H_4N$	Sn: Org.Verb.6-61/2
–	$SnCl_2 \cdot 2\ CH_3C_6H_4NH_2$	Sn: MVol.C5-13
$C_{14}Cl_2H_{18}Ni_2O_2$	$(CH_3CHCHCH_2NiCl)_2 \cdot C_6H_4O_2$	Ni: Org.Verb.2-300, 301
$C_{14}Cl_2H_{20}N_2Sn$	$(CH_3)_2SnCl_2 \cdot 2\ CH_3C_5H_4N$	Sn: Org.Verb.6-35, 38
–	$(CH_3)_2SnCl_2 \cdot 2\ C_6H_5NH_2$	Sn: Org.Verb.6-36, 38
–	$(C_2H_5)_2SnCl_2 \cdot (CH_3)_2C_8H_4N_2$	Sn: Org.Verb.6-62
–	$(C_2H_5)_2SnCl_2 \cdot 2\ C_5H_5N$	Sn: Org.Verb.6-62
$C_{14}Cl_2H_{20}O_7U$	$UO_2Cl_2 \cdot C_6H_4(OC_2H_4OC_2H_4)_2O \cdot 2\ H_2O$	U: SVol.E1-73, 78
$C_{14}Cl_2H_{20}O_{15}U$	$UO_2(ClO_4)_2 \cdot C_6H_4(OC_2H_4OC_2H_4)_2O \cdot 7\ H_2O$	U: SVol.E1-73, 78
$C_{14}Cl_2H_{21}NO_2Sn$	$C_4H_9SnCl_2(CH(COCH_3)_2) \cdot C_5H_5N$	Sn: Org.Verb.6-246

Formula	Compound	Element	Reference
$C_{14}Cl_2H_{22}Ni_2$	$(C_7H_{11}NiCl)_2$	Ni:	Org.Verb.2-320
$C_{14}Cl_2H_{23}NS_2Ti$	$C_5H_5Ti(S_2CN(n\text{-}C_4H_9)_2)Cl_2$	Ti:	Org.Verb.1-198/9
$C_{14}Cl_2H_{24}N_2Sn$	$[NC(CH_2)_6]_2SnCl_2$	Sn:	Org.Verb.6-135
$C_{14}Cl_2H_{24}O_6U$	$UCl_2(CH(COCH_3)_2)_2 \cdot C_2H_4(OCH_3)_2$	U:	SVol.E1-70, 74/5
$C_{14}Cl_2H_{25}NSn$	$(C_4H_9)_2SnCl_2 \cdot CH_3C_5H_4N$	Sn:	Org.Verb.6-99
$C_{14}Cl_2H_{26}N_2O_4Sn$	$(CH_3)_2SnCl_2 \cdot C_2H_4(NH(CH_2CO)_2CH_3)_2$	Sn:	Org.Verb.6-36
$C_{14}Cl_2H_{26}O_2S_2Sn$	$(C_4H_9)(C_6H_5)SnCl_2 \cdot 2\ (CH_3)_2SO$	Sn:	Org.Verb.6-205
$C_{14}Cl_2H_{26}O_4Sn$	$[C_3H_7OCOCH(CH_3)CH_2]_2SnCl_2$	Sn:	Org.Verb.6-135
–	$[C_3H_7OCOCH_2CH(CH_3)]_2SnCl_2$	Sn:	Org.Verb.6-135
$C_{14}Cl_2H_{28}O_2SSn$	$(C_4H_9)(C_7H_{15}OCOC_2H_4S)SnCl_2$	Sn:	Org.Verb.6-245
–	$(C_4H_9)(C_8H_{17}OCOCH_2S)SnCl_2$	Sn:	Org.Verb.6-245
$C_{14}Cl_2H_{28}Sn$	$(CH_3)_2C(C_2H_4)_2Sn(Cl)C_2H_4C(CH_3)_2C_2H_4Cl$	Sn:	Org.Verb.5-230
$C_{14}Cl_2H_{30}N_2O_4U$	$UO_2Cl_2 \cdot 2\ (CH_3)_2C_2H_3CON(CH_3)_2$	U:	SVol.E1-110/1
–	$UO_2Cl_2 \cdot 2\ (CH_3)_3CCON(CH_3)_2$	U:	SVol.E1-110/1
$C_{14}Cl_2H_{30}Sn$	$(C_4H_9)(C_{10}H_{21})SnCl_2$	Sn:	Org.Verb.6-200
–	$(C_7H_{15})_2SnCl_2$	Sn:	Org.Verb.6-110
$C_{14}Cl_2H_{32}N_2O_{14}P_2U$	$UO_2(NO_3)_2 \cdot 2\ (C_3H_7O)_2(CH_2Cl)PO$	U:	SVol.E1-186/7
$C_{14}Cl_2H_{32}OSn_2$	$(C_3H_7)_2SnCl_2 \cdot (C_4H_9)_2SnO$	Sn:	Org.Verb.6-68
–	$(C_4H_9)_2SnCl_2 \cdot (C_3H_7)_2SnO$	Sn:	Org.Verb.6-88
$C_{14}Cl_2H_{32}Sn_2$	$(C_2H_5)_2ClSn(CH_2)_6SnCl(C_2H_5)_2$	Sn:	Org.Verb.6-54, 289
$C_{14}Cl_2H_{33}OPSn$	$(CH_3)_2SnCl_2 \cdot (C_4H_9)_3PO$	Sn:	Org.Verb.6-37/8
–	$(CH_3)_2SnCl_2 \cdot (C_6H_{13})_2HPO$	Sn:	Org.Verb.6-37
$C_{14}Cl_2H_{33}O_2PSn$	$(C_2H_5)_2SnCl_2 \cdot (C_2H_5)_2(C_6H_{13}O)PO$	Sn:	Org.Verb.6-61/2
$C_{14}Cl_2H_{33}PSn$	$(CH_3)_2SnCl_2 \cdot (C_4H_9)_3P$	Sn:	Org.Verb.6-37
$C_{14}Cl_2H_{34}O_2S_2Sn$	$(CH_3)_2SnCl_2 \cdot 2\ (C_3H_7)_2SO$	Sn:	Org.Verb.6-37
$C_{14}Cl_2H_{34}O_6P_2Sn$	$SnCl_2 \cdot 2\ CH_3PO(OCH(CH_3)_2)_2$	Sn:	MVol.C5-60/1
$C_{14}Cl_2H_{34}O_6Si_2Sn$	$[(C_2H_5O)_3SiCH_2]_2SnCl_2$	Sn:	Org.Verb.6-135
$C_{14}Cl_2H_{34}Si_2Sn$	$[(CH_3)_2(C_4H_9)SiCH_2]_2SnCl_2$	Sn:	Org.Verb.6-135
$C_{14}Cl_2H_{36}O_6P_2Sn$	$(C_4H_9)_2SnCl_2 \cdot 2\ (CH_3O)_2P(O)CH_3$	Sn:	Org.Verb.6-100
$C_{14}Cl_2H_{37}NSn_2$	$2\ (C_2H_5)_3SnCl \cdot (CH_3)_2NH$	Sn:	Org.Verb.5-107
$C_{14}Cl_2H_{38}N_4Sn$	$(CH_3)_2SnCl_2 \cdot 2\ C_2H_4(N(CH_3)_2)_2$	Sn:	Org.Verb.6-36
$C_{14}Cl_2H_{38}O_4Sn_4$	$H[(C_2H_5)_2SnO]_3OH \cdot (CH_3)_2SnCl_2$	Sn:	Org.Verb.6-21
$C_{14}Cl_2H_{38}Si_4Sn$	$[((CH_3)_3Si)_2CH]_2SnCl_2$	Sn:	Org.Verb.6-136
$C_{14}Cl_3FH_{12}Sn$	$(CH_3)_2Sn(C_6H_4F)(C_6H_2Cl_3)$	Sn:	Org.Verb.3-82
$C_{14}Cl_3F_3H_{15}N_3O_2S$	$((C_3H_7)_2NSO_2)(CF_3)C_7Cl_3HN_2$	F:	PerFHalOrg.6-162
$C_{14}Cl_3F_7HN_3$	$N_2C_4Cl_3HNC_{10}F_7$	F:	PerFHalOrg.7-15/6, 43
$C_{14}Cl_3F_{25}H_4Sn$	$CF_3(CF_2)_{11}CH_2CH_2SnCl_3$	Sn:	Org.Verb.6-260
$C_{14}Cl_3FeH_{13}O_2$	$C_5H_5FeC_5H_4CH(CCl_3)OOCCH_3$	Fe:	Org.Verb.A2-98/9
–	$C_5H_5FeC_5H_4COCH_2CHOHCCl_3$	Fe:	Org.Verb.A2-292
$C_{14}Cl_3GeH_{26}NiP$	$C_5H_5Ni(P(C_3H_7\text{-}n)_3)GeCl_3$	Ni:	Org.Verb.2-130
$C_{14}Cl_3H_7O_3Sn$	$SnCl_3(C_{14}H_7O_3)$	Sn:	MVol.C5-113
$C_{14}Cl_3H_7O_4Sn$	$SnCl_3(C_{14}H_7O_4)$	Sn:	MVol.C5-114
–	$SnCl_3(C_{14}H_7O_4) \cdot 0.25\ C_6H_6$	Sn:	MVol.C5-113
$C_{14}Cl_3H_{11}O_2Sn$	$SnCl_3(OC_6H_3(CH_3)COC_6H_5)$	Sn:	MVol.C5-90/1
$C_{14}Cl_3H_{11}O_3Sn$	$SnCl_3(OC_6H_3(OCH_3)COC_6H_5)$	Sn:	MVol.C5-90/1
$C_{14}Cl_3H_{11}O_6Sn$	$SnCl_2(OC_6H_4COOH)_2 \cdot HCl$	Sn:	MVol.C5-123/4
$C_{14}Cl_3H_{11}Sn$	$C_6H_5CHC(C_6H_5)SnCl_3$	Sn:	Org.Verb.6-264, 266
$C_{14}Cl_3H_{15}N_2O_2Sn$	$CH_3C_6H_4SnCl_3 \cdot NH_2C_6H_3(CH_3)NO_2$	Sn:	Org.Verb.6-282
–	$C_6H_5SnCl_3 \cdot NH_2C_6H_2(CH_3)_2NO_2$	Sn:	Org.Verb.6-280
$C_{14}Cl_3H_{15}N_2Sn$	$(CH_3)_2SnCl_2 \cdot NC_5H_4CHNC_6H_4Cl$	Sn:	Org.Verb.6-36, 38

Formula	Compound	Reference
$C_{14}Cl_3H_{16}NSn$	$[C_6H_5NH_2CH_2CH_2Sn(C_6H_5)Cl_2]Cl$	Sn: Org.Verb.6-202
$C_{14}Cl_3H_{17}N_2Sn$	$C_4H_9SnCl_3 \cdot NC_5H_4C_5H_4N$	Sn: Org.Verb.6-247/8
$C_{14}Cl_3H_{17}Ni_2O_2$	$(CH_3CHCHCH_2NiCl)_2 \cdot C_6H_3ClO_2$	Ni: Org.Verb.2-300, 301
$C_{14}Cl_3H_{19}N_2Sn$	$C_4H_9SnCl_3 \cdot 2\ C_5H_5N$	Sn: Org.Verb.6-247/8
$C_{14}Cl_3H_{20}IrN_2S$	$Ir(S(C_2H_5)_2)(C_5H_5N)_2Cl_3$	Ir: SVol.2-165
$C_{14}Cl_3H_{21}N_2O_2Sn$	$[SnCl_3((C_3H_7)_2C(OH)CONNHC_6H_5)]$	Sn: MVol.C6-60
$C_{14}Cl_3H_{23}SSn_2$	$C_6H_5SnCl_2SSnCl(C_4H_9)_2$	Sn: Org.Verb.6-274
$C_{14}Cl_3H_{26}NiPSn$	$C_5H_5Ni(P(C_3H_7\text{-}n)_3)SnCl_3$	Ni: Org.Verb.2-130
$C_{14}Cl_3H_{27}NiO_2P_2$	$(CO)_2Ni(PCl_3)(P(C_4H_9\text{-}n)_3)$	Ni: Org.Verb.1-146
$C_{14}Cl_3H_{29}O_4Sn$	$C_4H_9SnCl_3 \cdot 2\ HCOOC_4H_9$	Sn: Org.Verb.6-247/8
$C_{14}Cl_3H_{29}Sn$	$C_{14}H_{29}SnCl_3$	Sn: Org.Verb.6-253
$C_{14}Cl_3H_{30}INiP_2$	$CCl_2CClNi(P(C_2H_5)_3)_2I$	Ni: Org.Verb.1-28, 30
$C_{14}Cl_3H_{31}OSSn$	$C_2H_5SnCl_3 \cdot (C_6H_{13})_2SO$	Sn: Org.Verb.6-232
$C_{14}Cl_3H_{32}OPSn$	$C_2H_5SnCl_3 \cdot (C_4H_9)_3PO$	Sn: Org.Verb.6-232
$C_{14}Cl_3H_{32}O_3PSn$	$C_2H_5SnCl_3 \cdot C_4H_9(C_4H_9O)_2PO$	Sn: Org.Verb.6-232
$C_{14}Cl_3H_{32}O_4PSn$	$C_2H_5SnCl_3 \cdot (C_4H_9O)_3PO$	Sn: Org.Verb.6-232
$C_{14}Cl_3H_{35}O_4Sn_3$	$2\ (C_2H_5)_3SnCl \cdot C_2H_5(Cl)SnO \cdot O_3$	Sn: Org.Verb.5-105
$C_{14}Cl_3H_{35}O_6P_2Sn$	$C_2H_5SnCl_3 \cdot 2\ OP(OC_2H_5)_2C_2H_5$	Sn: Org.Verb.6-232
$C_{14}Cl_4F_2H_8N_2Sn$	$SnCl_4 \cdot 2\ FC_6H_4CN$	Sn: MVol.C6-71/2
$C_{14}Cl_4F_3H_{10}N_3O$	$NC_5F_2Cl_2NHC_5FCl_2(OC_4H_9)N$	F: PerFHalOrg.5-138/9
$C_{14}Cl_4F_3H_{10}N_3O_2$	$NC_5F_2Cl_2NHC_5FCl_2(OC_2H_4OC_2H_5)N$	F: PerFHalOrg.5-138/9
$C_{14}Cl_4F_4HN_3$	$(NCC_6F_4)(NCC_6Cl_4)NH$	F: PerFHalOrg.7-15/6, 42
$C_{14}Cl_4H_8N_4O_4Sn$	$SnCl_4 \cdot 2\ NO_2C_6H_4CN$	Sn: MVol.C6-71/2
$C_{14}Cl_4H_8O_2Sn$	$SnCl_4 \cdot C_{14}H_8O_2$	Sn: MVol.C5-113, 115
$C_{14}Cl_4H_{10}N_2O_2Sn$	$SnCl_4 \cdot 2\ C_6H_5OCN$	Sn: MVol.C6-76/7
$C_{14}Cl_4H_{10}N_2O_6Sn$	$SnCl_4 \cdot 2\ C_6H_4(NO_2)CHO$	Sn: MVol.C5-82
$C_{14}Cl_4H_{10}N_2S_2Sn$	$SnCl_4 \cdot 2\ C_6H_4SCHN$	Sn: MVol.C6-111/2
$C_{14}Cl_4H_{10}N_2S_4Sn$	$SnCl_4 \cdot 2\ C_6H_4SC(SH)N$	Sn: MVol.C6-112
$C_{14}Cl_4H_{10}N_2Sn$	$(C_6H_5NCCl)_2SnCl_2$	Sn: Org.Verb.6-134
–	$SnCl_4 \cdot 2\ C_6H_5CN$	Sn: MVol.C6-70/1
$C_{14}Cl_4H_{10}NiO_2P_2$	$(CO)_2Ni(P(C_6H_5)Cl_2)_2$	Ni: Org.Verb.1-132, 133
$C_{14}Cl_4H_{10}O_2Sn$	$SnCl_4 \cdot C_6H_5COCOC_6H_5$	Sn: MVol.C5-109
$C_{14}Cl_4H_{10}O_3Sn$	$SnCl_4 \cdot (C_6H_5CO)_2O$	Sn: MVol.C5-125
$C_{14}Cl_4H_{11}NO_2Sn$	$SnCl_4 \cdot C_6H_5COC(NOH)C_6H_5$	Sn: MVol.C6-36
–	$SnCl_4 \cdot C_6H_5CONHCOC_6H_5$	Sn: MVol.C6-50/1
$C_{14}Cl_4H_{12}IrKN_2$	$K[Ir((CH_3)_2C_{12}H_6N_2)Cl_4]$	Ir: SVol.2-93
$C_{14}Cl_4H_{12}N_2OSn$	$SnCl_4 \cdot HOC_9H_6N \cdot C_5H_5N$	Sn: MVol.C5-212
$C_{14}Cl_4H_{12}N_2O_2Sn$	$SnCl_4 \cdot C_6H_5CONHNHCOC_6H_5$	Sn: MVol.C6-59/60
$C_{14}Cl_4H_{12}N_2Sn$	$SnCl_4 \cdot C_6H_5CHNNCHC_6H_5$	Sn: MVol.C6-30/1
$C_{14}Cl_4H_{12}N_4S_2Sn$	$SnCl_4 \cdot 2\ C_6H_4NHC(SH)N$	Sn: MVol.C6-104
$C_{14}Cl_4H_{12}N_4Sn$	$SnCl_4 \cdot 2\ C_7H_6N_2$	Sn: MVol.C5-223/4
$C_{14}Cl_4H_{12}O_2Sn$	$SnCl_4 \cdot 2\ C_6H_5CHO$	Sn: MVol.C5-79/80
$C_{14}Cl_4H_{12}O_4Sn$	$SnCl_4 \cdot 2\ C_6H_5COOH$	Sn: MVol.C5-121/3
–	$SnCl_4 \cdot 2\ HOC_6H_4CHO$	Sn: MVol.C5-80, 82
–	$SnCl_4 \cdot 2\ HOC_6H_4CHO \cdot 2\ H_2O$	Sn: MVol.C5-80
–	$SnCl_4 \cdot 2\ C_7H_6O_2$	Sn: MVol.C5-94
$C_{14}Cl_4H_{12}Sn$	$(ClC_6H_4CH_2)_2SnCl_2$	Sn: Org.Verb.6-134/5, 139/40
$C_{14}Cl_4H_{13}NOTe$	$TeCl_4 \cdot C_6H_5CHNC_6H_4OCH_3$	Te: SVol.B3-172/3
$C_{14}Cl_4H_{14}I_2Ni_2O_2$	$(CH_3CHCHCH_2NiI)_2 \cdot C_6Cl_4O_2$	Ni: Org.Verb.2-301, 302
$C_{14}Cl_4H_{14}IrN_2S_2$	$Ir(C_6H_5CSNH_2)_2Cl_4$	Ir: SVol.2-171

Formula	Compound	Element	Reference
$C_{14}Cl_4H_{14}N_2O_2Sn$	$SnCl_4 \cdot 2\ CH_2CHNC_4H_4C(O)$	Sn:	MVol.C5-204
–	$SnCl_4 \cdot 2\ C_6H_5CONH_2$	Sn:	MVol.C6-46/7
–	$SnCl_4 \cdot 2\ NC_5H_4OCHCH_2$	Sn:	MVol.C5-203/4
$C_{14}Cl_4H_{14}N_2O_4Sn$	$SnCl_4 \cdot 2\ HOC_6H_4CONH_2$	Sn:	MVol.C6-48
–	$SnCl_4 \cdot 2\ NH_2C_6H_4COOH$	Sn:	MVol.C6-13
$C_{14}Cl_4H_{14}N_2O_6Sn$	$SnCl_4 \cdot 2\ NO_2C_6H_4OCH_3$	Sn:	MVol.C5-152/3
$C_{14}Cl_4H_{14}N_2Sn$	$(CHClCH)_2SnCl_2 \cdot 2\ C_5H_5N$	Sn:	Org.Verb.6-148
$C_{14}Cl_4H_{14}N_4S_2Sn$	$SnCl_2 \cdot 2\ NH_2C_7H_4NS \cdot 2\ HCl$	Sn:	MVol.C5-53
$C_{14}Cl_4H_{14}SSn$	$SnCl_4 \cdot (C_6H_5CH_2)_2S$	Sn:	MVol.C6-94/5
$C_{14}Cl_4H_{14}S_2Sn$	$SnCl_4 \cdot C_6H_5S(CH_2)_2SC_6H_5$	Sn:	MVol.C6-98
$C_{14}Cl_4H_{15}N_3Sn$	$SnCl_4 \cdot C_6H_5NNC_6H_4N(CH_3)_2$	Sn:	MVol.C6-31/2
–	$SnCl_4 \cdot C_6H_5NNC_6H_4N(CH_3)_2 \cdot 2\ H_2O$	Sn:	MVol.C6-31/3
$C_{14}Cl_4H_{16}N_2OTe$	$TeCl_4 \cdot CH_3(NH_2)C_6H_3C_6H_3(NH_2)OCH_3$	Te:	SVol.B3-169
$C_{14}Cl_4H_{16}N_2Te$	$TeCl_4 \cdot CH_3(NH_2)C_6H_3C_6H_3(NH_2)CH_3$	Te:	SVol.B3-169
$C_{14}Cl_4H_{16}N_4O_2Sn$	$SnCl_4 \cdot 2\ C_6H_5NHCONH_2$	Sn:	MVol.C6-58
$C_{14}Cl_4H_{16}N_4Sn$	$SnCl_4 \cdot 2\ C_6H_5CHNNH_2$	Sn:	MVol.C6-28/9
$C_{14}Cl_4H_{16}Ni_2O_2$	$(CH_3CHCHCH_2NiCl)_2 \cdot C_6H_2Cl_2O_2$	Ni:	Org.Verb.2-300, 302
$C_{14}Cl_4H_{16}O_2Sn$	$SnCl_4 \cdot 2\ CH_3C_6H_4OH$	Sn:	MVol.C5-76
–	$SnCl_4 \cdot 2\ C_6H_5OCH_3$	Sn:	MVol.C5-152
$C_{14}Cl_4H_{16}O_4Sn$	$SnCl_4 \cdot 2\ (CH_3)_2C_5H_2O_2$	Sn:	MVol.C5-160
$C_{14}Cl_4H_{16}S_2Sn$	$SnCl_4 \cdot 2\ C_6H_5CH_2SH$	Sn:	MVol.C6-80
–	$SnCl_4 \cdot 2\ C_6H_5SCH_3$	Sn:	MVol.C6-94
$C_{14}Cl_4H_{18}N_2O_2Sn$	$SnCl_4 \cdot 2\ (CH_3)_2C_5H_3NO$	Sn:	MVol.C5-208/9
–	$SnCl_4 \cdot 2\ C_2H_5C_5H_4NO$	Sn:	MVol.C5-208/9
–	$SnCl_4 \cdot 2\ NH_2C_6H_4OCH_3$	Sn:	MVol.C6-7
$C_{14}Cl_4H_{18}N_2O_2Te$	$TeCl_4 \cdot 2\ CH_3OC_6H_4NH_2$	Te:	SVol.B3-166/7
–	$TeCl_4 \cdot 2\ (CH_3)_2C_5H_3NO$	Te:	SVol.B3-171
$C_{14}Cl_4H_{18}N_2O_4Sn$	$SnCl_4 \cdot 2\ C_4H_2O(CHO)N(CH_3)_2$	Sn:	MVol.C6-5
$C_{14}Cl_4H_{18}N_2Sn$	$SnCl_4 \cdot 2\ C_6H_5CH_2NH_2$	Sn:	MVol.C5-181
$C_{14}Cl_4H_{18}N_2Te$	$TeCl_4 \cdot 2\ CH_3C_6H_4NH_2$	Te:	SVol.B3-166/7
–	$TeCl_4 \cdot 2\ (CH_3)(C_6H_5)NH$	Te:	SVol.B3-167
–	$TeCl_4 \cdot 2\ C_6H_5CH_2NH_2$	Te:	SVol.B3-166/7
$C_{14}Cl_4H_{18}N_6O_{10}U_2$	$(C_{10}H_{10}N_2)[(UO_2)_2Cl_4(C_2O_4)((NH_2)_2CO)_2]$	U:	SVol.E1-84, 91
$C_{14}Cl_4H_{18}O_4P_2Sn$	$SnCl_4 \cdot 2\ C_6H_5(H)P(O)OCH_3$	Sn:	MVol.C6-153
$C_{14}Cl_4H_{18}O_5Sn$	$SnCl_4 \cdot CH_3C_6H_4C(COOC_2H_5)_2OH$	Sn:	MVol.C5-146/7
$C_{14}Cl_4H_{19}NO_2Sn$	$SnCl_4 \cdot C_6H_5CONHCOC_6H_{13}$	Sn:	MVol.C6-50/1
$C_{14}Cl_4H_{20}Ni_2$	$(H(CH_3)_3C_4NiCl_2)_2$	Ni:	Org.Verb.2-324
$C_{14}Cl_4H_{20}O_4Sn$	$SnCl_4 \cdot 2\ CH_3(CHCH)_2COOCH_3$	Sn:	MVol.C5-138
$C_{14}Cl_4H_{20}O_8Sn$	$SnCl_4 \cdot C_2(COOC_2H_5)_4$	Sn:	MVol.C5-144
$C_{14}Cl_4H_{21}O_4PSn$	$SnCl_4 \cdot C_6H_5COCH_2P(O)(OC_3H_7)_2$	Sn:	MVol.C6-166
$C_{14}Cl_4H_{22}N_2Ni_2O_2$	$(C_6H_{11}NCl_2NiCO)_2$	Ni:	Org.Verb.2-394
$C_{14}Cl_4H_{22}N_2Sn$	$[C_5H_5NH]_2[(C_2H_5)_2SnCl_4]$	Sn:	Org.Verb.6-61
$C_{14}Cl_4H_{22}O_8Sn$	$SnCl_4 \cdot C_2H_2(COOC_2H_5)_4$	Sn:	MVol.C5-144
$C_{14}Cl_4H_{26}O_4Sn$	$SnCl_4 \cdot C_8H_{16}(COOC_2H_5)_2$	Sn:	MVol.C5-141/3
$C_{14}Cl_4H_{28}O_4Sn$	$SnCl_4 \cdot 2\ CH_3COOC_5H_{11}$	Sn:	MVol.C5-131/2
–	$SnCl_4 \cdot 2\ C_2H_5COOC_4H_9$	Sn:	MVol.C5-134/5
–	$SnCl_4 \cdot 2\ C_3H_7COOC_3H_7$	Sn:	MVol.C5-135/6
–	$SnCl_4 \cdot 2\ C_5H_{11}COOCH_3$	Sn:	MVol.C5-136/7
–	$SnCl_4 \cdot 2\ C_6H_{13}COOH$	Sn:	MVol.C5-120/1
$C_{14}Cl_4H_{30}N_2O_2U$	$UCl_4 \cdot 2\ (CH_3)_2C_2H_3CON(CH_3)_2$	U:	SVol.E1-109/11

Formula	Compound	Reference
$C_{14}Cl_4H_{30}N_2O_2U$	$UCl_4 \cdot 2\ (CH_3)_3CCON(CH_3)_2$	U: SVol.E1-109/11
$C_{14}Cl_4H_{30}N_2O_4Sn$	$SnCl_4 \cdot 2\ CH_3COOH \cdot 2\ (CH_2)_5NH$	Sn: MVol.C5-193
$C_{14}Cl_4H_{30}NiP_2$	$CCl_2CClNi(P(C_2H_5)_3)_2Cl$	Ni: Org.Verb.1-28, 30
$C_{14}Cl_4H_{30}O_8P_2Sn$	$SnCl_4 \cdot 2\ (CH_3)_3CCOP(O)(OCH_3)_2$	Sn: MVol.C6-164
$C_{14}Cl_4H_{30}STe$	$TeCl_4 \cdot (C_7H_{15})_2S$	Te: SVol.B3-175
$C_{14}Cl_4H_{30}S_2Sn$	$SnCl_4 \cdot C_4H_9S(CH_2)_6SC_4H_9$	Sn: MVol.C6-97/8
$C_{14}Cl_4H_{30}S_2Te$	$TeCl_4 \cdot C_4H_9S(CH_2)_6SC_4H_9$	Te: SVol.B3-175
$C_{14}Cl_4H_{32}N_2O_6U$	$[CH_3COOC_2H_4N(CH_3)_3]_2UO_2Cl_4$	U: SVol.C9-154/5
$C_{14}Cl_4H_{32}N_4O_2U$	$UCl_4 \cdot 2\ (CH_3)_2NCON(C_2H_5)_2$	U: SVol.E1-91/2, 94
$C_{14}Cl_4H_{33}NO_3SiSn$	$SnCl_4 \cdot (C_4H_9O)_3SiNHC_2H_5$	Sn: MVol.C6-130/1
$C_{14}Cl_4H_{34}MnN_4$	$[CH_3N(C_2H_4)_2N(CH_3)_2]_2MnCl_4$	Mn: MVol.C5-204/5
$C_{14}Cl_4H_{34}O_2P_2S_2Sn$	$SnCl_4 \cdot 2\ (C_2H_5)_2P(O)SC_3H_7$	Sn: MVol.C6-158/60
–	$SnCl_4 \cdot 2\ (i\text{-}C_3H_7)_2P(O)SCH_3$	Sn: MVol.C6-158/60
$C_{14}Cl_4H_{34}O_2P_2Sn$	$SnCl_4 \cdot 2\ (C_2H_5)_2(C_3H_7)PO$	Sn: MVol.C6-143/4
–	$SnCl_4 \cdot 2\ (C_2H_5)_2POC_3H_7$	Sn: MVol.C6-139/40
$C_{14}Cl_4H_{34}O_6P_2S_2Sn$	$SnCl_4 \cdot 2\ (i\text{-}C_3H_7O)_2(CH_3S)PO$	Sn: MVol.C6-174/5
$C_{14}Cl_4H_{34}O_6P_2Sn$	$SnCl_4 \cdot 2\ CH_3P(O)(OC_3H_7\text{-}i)_2$	Sn: MVol.C6-162/3
$C_{14}Cl_4H_{34}P_2S_2Sn$	$SnCl_4 \cdot 2\ (C_2H_5)_2PSC_3H_7$	Sn: MVol.C6-141
$C_{14}Cl_4H_{34}P_2S_4Sn$	$SnCl_4 \cdot 2\ (CH(CH_3)_2)_2P(S)SCH_3$	Sn: MVol.C6-158/60
$C_{14}Cl_4H_{36}MnN_2$	$(C_7H_{15}NH_3)_2MnCl_4$	Mn: MVol.C5-182
$C_{14}Cl_4H_{36}N_2O_4P_2Sn$	$SnCl_4 \cdot 2\ (CH_3)_2CHOPH(O)N(C_2H_5)_2$	Sn: MVol.C6-167
$C_{14}Cl_4H_{42}O_7S_7U$	$UCl_4 \cdot 7\ (CH_3)_2SO$	U: SVol.E1-202/3, 207
$C_{14}Cl_5F_5O_2$	$C_6F_5C(O)C(O)C_6Cl_5$	F: PerFHalOrg.4-53
$C_{14}Cl_5H_{10}OU$	$UCl_5 \cdot C_{14}H_{10}O$	U: SVol.E1-80/1
$C_{14}Cl_5H_{10}O_2U$	$UCl_5 \cdot C_6H_5COCOC_6H_5$	U: SVol.E1-80/1
$C_{14}Cl_5H_{18}NO_2Sn$	$SnCl_4 \cdot ClC_6H_4CONHCOC_6H_{13}$	Sn: MVol.C6-50/1
$C_{14}Cl_5H_{21}N_7NaSn$	$[Na(CH_3CN)_6][SnCl_5(CH_3CN)]$	Sn: MVol.C6-67/8
$C_{14}Cl_6F_{19}N_3$	$(C_5F_8Cl_3)_2(CF_3)C_3N_3$	F: PerFHalOrg.6-87/8, 107
$C_{14}Cl_6FeH_{11}O_4$	$[Fe(C_5H_5)_2][CCl_3COO \cdot CCl_3COOH]$	Fe: Org.Verb.A1-227
$C_{14}Cl_6FeH_{12}O_2$	$Fe(C_5H_4CHOHCCl_3)_2$	Fe: Org.Verb.A1-163
		Fe: Org.Verb.A2-48
$C_{14}Cl_6FeH_{22}N_6Pt$	$[(C_2H_5NC)_2(CH_3NC)_4Fe]PtCl_6$	Fe: Org.Verb.B4-107
$C_{14}Cl_6H_8N_2Sn$	$SnCl_4 \cdot 2\ ClC_6H_4CN$	Sn: MVol.C6-71/2
$C_{14}Cl_6H_{10}Sn$	$(Cl_2C_6H_3CH_2)_2SnCl_2$	Sn: Org.Verb.6-134
$C_{14}Cl_6H_{12}N_2Sn$	$[C_6H_5CNH]_2[SnCl_6]$	Sn: MVol.C6-71
$C_{14}Cl_6H_{14}Ni_2O_2$	$(CH_3CHCHCH_2NiCl)_2 \cdot C_6Cl_4O_2$	Ni: Org.Verb.2-301, 302
$C_{14}Cl_6H_{14}Sn$	$(C_7H_7)_2[SnCl_6]$	Sn: MVol.C3-109
$C_{14}Cl_6H_{18}Ni_2O_4$	$(CH_3CHCHCHCH_3NiO_2CCCl_3)_2$	Ni: Org.Verb.2-314
$C_{14}Cl_6H_{18}S_2Sn$	$[C_6H_5CH_2SH_2]_2[SnCl_6]$	Sn: MVol.C6-80
$C_{14}Cl_6H_{20}N_2Sn$	$[CH_3C_6H_4NH_3]_2[SnCl_6]$	Sn: MVol.C3-103/4
$C_{14}Cl_6H_{20}N_2Te$	$[(CH_3)_2C_5H_3NH]_2TeCl_6$	Te: SVol.B2-116
$C_{14}Cl_6H_{20}O_8Sn_{1.5}$	$1.5\ SnCl_4 \cdot C_2(COOC_2H_5)_4$	Sn: MVol.C5-144
$C_{14}Cl_6H_{21}MgN_7Sn$	$[Mg(CH_3CN)_7][SnCl_6]$	Sn: MVol.C3-165
$C_{14}Cl_6H_{22}O_8Sn_{1.5}$	$1.5\ SnCl_4 \cdot C_2H_2(COOC_2H_5)_4$	Sn: MVol.C5-144
$C_{14}Cl_6H_{36}N_2Sn$	$[CH_3(C_2H_5)_3N]_2[SnCl_6]$	Sn: MVol.C3-106
–	$[(CH_3)_3(C_4H_9)N]_2[SnCl_6]$	Sn: MVol.C3-106
$C_{14}Cl_8H_8Sn$	$(CH_3)_2Sn(C_6HCl_4)_2$	Sn: Org.Verb.3-10, 18
$C_{14}Cl_8H_{12}N_4O_4Sn$	$[NO_2C_6H_4CClNH_2]_2[SnCl_6]$	Sn: MVol.C6-72
$C_{14}Cl_8H_{14}N_2Sn$	$[C_6H_5CClNH_2]_2[SnCl_6]$	Sn: MVol.C6-71
$C_{14}Cl_8H_{20}N_2Sn_2$	$2\ SnCl_4 \cdot C_2H_4(NC_4H_2(CH_3)_2)_2$	Sn: MVol.C5-190

Formula	Compound		Reference
$C_{14}Cl_8H_{20}O_8Sn_2$	$2\ SnCl_4 \cdot C_2(COOC_2H_5)_4$	Sn:	MVol.C5-144
$C_{14}Cl_8H_{22}O_8Sn_2$	$2\ SnCl_4 \cdot C_2H_2(COOC_2H_5)_4$	Sn:	MVol.C5-144
$C_{14}Cl_8H_{30}S_2Te_2$	$2\ TeCl_4 \cdot C_4H_9S(CH_2)_6SC_4H_9$	Te:	SVol.B3-176
$C_{14}Cl_{10}H_{56}LaN_7$	$LaCl_3 \cdot 7\ [(CH_3)_2NH_2]Cl$	Sc:	MVol.C5-213/5
–	$LaCl_3 \cdot 7\ [(CH_3)_2NH_2]Cl \cdot 2\ H_2O$	Sc:	MVol.C5-213/5
$C_{14}Cl_{10}H_{56}N_7Nd$	$NdCl_3 \cdot 7\ [(CH_3)_2NH_2]Cl$	Sc:	MVol.C5-213/5
–	$NdCl_3 \cdot 7\ [(CH_3)_2NH_2]Cl \cdot 2\ H_2O$	Sc:	MVol.C5-213/5
$C_{14}Cl_{10}H_{56}N_7Pr$	$PrCl_3 \cdot 7\ [(CH_3)_2NH_2]Cl$	Sc:	MVol.C5-213/5
–	$PrCl_3 \cdot 7\ [(CH_3)_2NH_2]Cl \cdot 2\ H_2O$	Sc:	MVol.C5-213/5
$C_{14}Cl_{12}FeH_{16}N_6$	$(CH_3NC)_4Fe(CN)_2 \cdot 4\ CHCl_3$	Fe:	Org.Verb.B4-68
$C_{14}Cl_{18}H_{12}NiO_8P_2$	$(CO)_2Ni(P(OCH_2CCl_3)_3)_2$	Ni:	Org.Verb.1-139, 140
$C_{14}CmH_{16}N_2O_{12}^{3-}$	$Cm(N(CH_2CH_2COO)(CH_2COO)_2)_2^{3-}$	Np:	TrU.D1-164
$C_{14}CmH_{18}N_2O_8^{-}$	$CmC_6H_{10}N_2(CH_2COO)_4^{-}$	Np:	TrU.D1-166
$C_{14}CmH_{18}N_3O_{10}^{2-}$	$Cm(CH_2COO)N(C_2H_4N(CH_2COO)_2)_2^{2-}$	Np:	TrU.D1-166
$C_{14}CmH_{19}N_2O_8$	$CmC_6H_{10}N_2(CH_2COOH)(CH_2COO)_3$	Np:	TrU.D1-166
$C_{14}CmH_{19}N_3O_{10}^{-}$	$CmH(CH_2COO)N(C_2H_4N(CH_2COO)_2)_2^{-}$	Np:	TrU.D1-166
$C_{14}CmH_{20}N_2O_{10}^{-}$	$CmC_2H_4(OC_2H_4N(CH_2COO)_2)_2^{-}$	Np:	TrU.D1-166
$C_{14}CoF_6H_{14}O_4PSn$	$[(OCOC_5H_4CoC_5H_4COOSn(CH_3)_2)PF_6]_x$	Sn:	Org.Verb.6-29
$C_{14}CoFe_2H_5O_9$	$[(CO)_4FeFe(CO)_3](CO)_2CoC_5H_5$	Fe:	Org.Verb.C1-211/2
$C_{14}Co_2H_{18}O_6SiSn$	$(CH_3)_3SnC_2Si(CH_3)_3Co_2(CO)_6$	Sn:	Org.Verb.2-154, 156
$C_{14}Co_3FeH_{10}O_{14}P$	$HFeCo_3(CO)_{11}(P(OCH_3)_3)$	Fe:	Org.Verb.B1-204
$C_{14}Co_3FeH_{19}O_9P^{+}$	$[HFeCo_3(CO)_6P(OC_3H_7)_2OC_2H_4]^{+}$	Fe:	Org.Verb.B1-204
$C_{14}Co_3FeH_{22}O_8P^{+}$	$[HFeCo_3(CO)_5P(OC_3H_7)_3]^{+}$	Fe:	Org.Verb.B1-204
$C_{14}Co_3H_5NiO_9$	$C_5H_5NiCo_3(CO)_9$	Ni:	Org.Verb.2-115
$C_{14}Co_4Ni_2O_{14}^{2-}$	$[Ni_2Co_4(CO)_{14}]^{2-}$	Ni:	Org.Verb.2-266/7
$C_{14}CrFeH_{11}NO_3^{+}$	$[C_5H_5FeC_5H_4C(NH_2)Cr(CO)_3]^{+}$	Fe:	Org.Verb.A1-346
$C_{14}CrFeH_{12}N_2O_{10}Sn$	$(CO)_5CrSn[OCN(CH_3)_2]_2Fe(CO)_3$	Fe:	Org.Verb.B3-143
$C_{14}CrFeH_{12}O_3^{+}$	$[C_5H_5FeC_5H_4C(OCH_3)Cr(CO)_2]^{+}$	Fe:	Org.Verb.A1-346
$C_{14}CrFeH_{14}O_2^{+}$	$[C_5H_5FeC_5H_4C(OC_2H_5)Cr(CO)]^{+}$	Fe:	Org.Verb.A1-346
$C_{14}CrFeH_{15}NO^{+}$	$[C_5H_5FeC_5H_4C(N(CH_3)_2)Cr(CO)]^{+}$	Fe:	Org.Verb.A1-346
$C_{14}CrFeH_{16}N_6S_4$	$[Fe(C_5H_5)_2][Cr(SCN)_4(NH_3)_2]$	Fe:	Org.Verb.A1-225
$C_{14}CrH_{16}N_6NiS_4$	$[Ni(C_5H_5)_2][Cr(SCN)_4(NH_3)_2]$	Ni:	Org.Verb.2-213
$C_{14}CrH_{23}N_3O_3Ti$	$C_5H_5Ti(N(CH_3)_2)_3Cr(CO)_3$	Ti:	Org.Verb.1-203
$C_{14}Cr_2F_{12}N_2O_8P_2$	$[(CO)_4CrP(CF_3)_2CN]_2$	F:	PerFHalOrg.3-174
$C_{14}CuF_{24}H_4N_6$	$[HNC(C_2F_5)NC(C_3F_7)NH]_2Cu$	F:	PerFHalOrg.7-58
$C_{14}CuFeH_{11}O_2$	$C_5H_5FeC_5H_4COOCH_2CCCu$	Fe:	Org.Verb.A3-111
$C_{14}FH_{13}Sn$	$(C_6H_5)_2(CH_2CH)SnF$	Sn:	Org.Verb.5-40
$C_{14}FH_{15}Sn$	$(CH_3)_2Sn(C_6H_5)(C_6H_4F)$	Sn:	Org.Verb.3-80
$C_{14}FH_{29}Sn$	$(C_4H_9)_2(C_6H_{11})SnF$	Sn:	Org.Verb.5-40
$C_{14}F_2H_8NiO_6P_2$	$(CO)_2Ni(C_6H_4O_2PF)_2$	Ni:	Org.Verb.1-142
$C_{14}F_2H_{14}O_2S$	$(CH_3C_6H_4O)_2SF_2$	S:	SVol.2-44, 83
$C_{14}F_2H_{14}Sn$	$(CH_3)_2Sn(C_6H_4F)_2$	Sn:	Org.Verb.3-10, 18
–	$(C_6H_5CH_2)_2SnF_2$	Sn:	Org.Verb.5-49/50
$C_{14}F_3FeH_9O_5$	$C_6H_5C_3H_4Fe(CO)_3OCOCF_3$	Fe:	Org.Verb.B5-43/4, 48, 52, 62
$C_{14}F_3FeH_{11}O_2$	$C_5H_5FeC_5H_4COCH_2COCF_3$	Fe:	Org.Verb.A3-31/2
$C_{14}F_3FeH_{12}O_2^{+}$	$[C_5H_5FeC_5H_4C(CH_3)OOCCF_3]^{+}$	Fe:	Org.Verb.A2-96
$C_{14}F_3FeH_{13}O$	$C_5H_5FeC_5H_4(CH_2)_2COCF_3$	Fe:	Org.Verb.A3-3
$C_{14}F_3FeH_{13}O_2S$	$C_5H_5FeC_5H_4CH(CF_3)SCH_2COOH$	Fe:	Org.Verb.A2-47
$C_{14}F_3FeH_{15}O$	$C_5H_5FeC_5H_4(CH_2)_2OCH_2CF_3$	Fe:	Org.Verb.A2-130/1, 135

$C_{14}F_3H_{27}Sn$	$(n\text{-}C_4H_9)_3SnCFCF_2$	F:	PerFHalOrg.4-73
		Sn:	Org.Verb.2-309/10
$C_{14}F_4FeH_{12}O_4$	$(C_8H_{12}CF_2CF_2)Fe(CO)_4$	Fe:	Org.Verb.B4-337, 346/7
$C_{14}F_4Fe_2O_8$	$C_6F_4Fe_2(CO)_8$	Fe:	Org.Verb.C2-77/8
$C_{14}F_4H_{10}NO_4P$	$CF_3CFNOPO(OC_6H_5)_2$	F:	PerFHalOrg.7-199
$C_{14}F_4H_{10}N_3NaO$	$(CH_3)_2NC_6H_4NNC_6F_4ONa$	F:	PerFHalOrg.7-61/2
$C_{14}F_4H_{10}Sn$	$(C_6H_5)_2(CF_2CF)SnF$	Sn:	Org.Verb.5-40
$C_{14}F_4H_{11}N_3O$	$(CH_3)_2NC_6H_4NNC_6F_4OH$	F:	PerFHalOrg.7-61/2
$C_{14}F_4H_{11}N_3O_2S$	$[C_6H_5CH_2NHCSNH_2][C_5F_4(COOH)N]$	F:	PerFHalOrg.5-217
$C_{14}F_4H_{12}O_2Sn$	$SnF_4 \cdot 2\ C_6H_5CHO$	Sn:	MVol.C5-79/80
$C_{14}F_4H_{14}I_2Ni_2O_2$	$(CH_3CHCHCH_2NiI)_2 \cdot C_6F_4O_2$	Ni:	Org.Verb.2-300
$C_{14}F_4H_{16}O_4Sn$	$SnF_4 \cdot 2\ (CH_3)_2C_5H_2O_2$	Sn:	MVol.C5-160
$C_{14}F_4H_{18}N_2O_2Sn$	$SnF_4 \cdot 2\ (CH_3)_2C_5H_3NO$	Sn:	MVol.C5-208/9
–	$SnF_4 \cdot 2\ C_2H_5C_5H_4NO$	Sn:	MVol.C5-208/9
$C_{14}F_4H_{18}Ni$	$C_{12}H_{18}NiC_2F_4$	Ni:	Org.Verb.2-232
$C_{14}F_5FeH_5O_3$	$C_5H_5Fe(CO)_2C(O)C_6F_5$	F:	PerFHalOrg.4-37
–	$(C_6F_5)C_5H_5Fe(CO)_3$	F:	PerFHalOrg.4-40
$C_{14}F_5FeH_8NO$	$C_5H_5Fe(CO)(CNCH_3)C_6F_5$	F:	PerFHalOrg.4-38
$C_{14}F_5H_{10}N$	$C_6F_5NHC_6H_3(CH_3)_2$	F:	PerFHalOrg.7-197
$C_{14}F_5H_{10}N_3$	$(CH_3)_2NC_6H_4NNC_6F_5$	F:	PerFHalOrg.7-61/2
$C_{14}F_5H_{11}Si$	$C_6H_5(CH_3)_2SiC_6F_5$	F:	PerFHalOrg.4-79
$C_{14}F_5H_{12}NO_4$	$C_6F_5NHC(COOC_2H_5)CH(COOC_2H_5)$	F:	PerFHalOrg.7-59/60
$C_{14}F_5H_{20}N_2P$	$C_6F_5P[N(C_2H_5)_2]_2$	F:	PerFHalOrg.3-125, 131
–	$C_6F_5P[NHC(CH_3)_3]_2$	F:	PerFHalOrg.3-125, 131
$C_{14}F_5H_{27}Sn$	$(C_4H_9)_3SnC_2F_5$	Sn:	Org.Verb.2-280
$C_{14}F_6FeH_{11}O_4PS_2$	$(CO)_2Fe(P(OCH_3)_2C_6H_5)S_2C_2(CF_3)_2$	Fe:	Org.Verb.B1-101/2, 108
$C_{14}F_6FeH_{14}O_2Sn$	$(CH_3)_3SnC(CF_3)C(CF_3)Fe(CO)_2C_5H_5$	Sn:	Org.Verb.2-83, 89
$C_{14}F_6FeH_{15}NO_3P_2$	$[C_4H_4PC_6H_5(CH_3)_2Fe(CO)_2NO][PF_6]$	Fe:	Org.Verb.B5-28/31
$C_{14}F_6FeH_{18}P$	$[C_5H_5FeC_5H_4C_4H_9\text{-}n]PF_6$	Fe:	Org.Verb.A1-278
$C_{14}F_6FeH_{30}NO_3P_3$	$[(CO)_2Fe(P(C_2H_5)_3)_2NO]PF_6$	Fe:	Org.Verb.B1-130/3
$C_{14}F_6Fe_2H_{12}O_6P_2$	$(CO)_6Fe_2(CF_3CCCF_3)(P(CH_3)_2)_2$	Fe:	Org.Verb.C2-67/70
$C_{14}F_6H_8N_4$	$CF_3NNC_6H_4C_6H_4NNCF_3$	F:	PerFHalOrg.7-158/9
$C_{14}F_6H_9NO_3$	$(CH_3O)_2(CH_3OC_6F_4)C_5F_2N$	F:	PerFHalOrg.5-178
$C_{14}F_6H_{10}Ni$	$C_5H_5Ni(C_5H_5CF_3CCCF_3)$	Ni:	Org.Verb.2-357
–	$C_5H_5NiC_7H_5(CF_3)_2$	Ni:	Org.Verb.2-184
$C_{14}F_6H_{10}Ni_2$	$(C_5H_5Ni)_2CF_3CCCF_3$	Ni:	Org.Verb.2-356, 357
$C_{14}F_6H_{11}N_3Ni$	$(CF_3)_2CN(CH_3)Ni(NC_5H_4C_5H_4N)$	Ni:	Org.Verb.1-62, 64
$C_{14}F_6H_{12}N_2P_2$	$(CF_3)_2PNP(C_6H_5)_2NH_2$	F:	PerFHalOrg.3-69
$C_{14}F_6H_{18}N_2Ni$	$C_4F_6Ni(CNC_4H_9\text{-}t)_2$	Ni:	Org.Verb.1-381
$C_{14}F_6H_{18}Ni_2O_4$	$(CH_3CHCHCHCH_3NiO_2CCF_3)_2$	Ni:	Org.Verb.2-314
$C_{14}F_6H_{20}MnO_6S_2$	$[Mn(CF_3COCHCOCH_3)_2(SO(CH_3)_2)_2]$	Mn:	MVol.D1-110
$C_{14}F_6H_{21}N_3Ni$	$(t\text{-}C_4H_9NC)_2Ni(N(CH_3)C(CF_3)_2)$	Ni:	Org.Verb.1-319
$C_{14}F_6H_{22}N_2NiP_2$	$(C_6H_{11}NC)_2Ni(PF_3)_2$	Ni:	Org.Verb.1-311, 314
$C_{14}F_6Li_4O_4$	$LiC_6F_3(COOLi)C_6F_3(COOLi)Li$	F:	PerFHalOrg.4-7, 14
$C_{14}F_7FeH_9O$	$C_5H_5FeC_5H_4COC_3F_7$	Fe:	Org.Verb.A2-241/3
$C_{14}F_7FeH_{11}IO_3P$	$(CO)_3Fe(C_3F_7\text{-}i)(P(CH_3)_2C_6H_5)I$	Fe:	Org.Verb.B1-209
$C_{14}F_7FeH_{11}O$	$C_5H_5FeC_5H_4CHOHC_3F_7$	Fe:	Org.Verb.A2-44
$C_{14}F_7H_4NO_2$	$C_6F_3(OCH_3)C(O)C_6F_4NH$	F:	PerFHalOrg.6-155
$C_{14}F_7H_6N_3$	$C_6F_4N_2C_6F_3N(CH_3)_2$	F:	PerFHalOrg.6-155, 160
$C_{14}F_7H_9N_2O$	$NC_5F_3(OC(CH_3)_3)C_5F_4N$	F:	PerFHalOrg.5-222/3

Formula	Compound		Reference
$C_{14}F_7H_9N_2O$	$NC_5F_3(OC_4H_9)C_5F_4N$	F:	PerFHalOrg.5-222/3
$C_{14}F_8HN_3$	$(NCC_6F_4)_2NH$	F:	PerFHalOrg.7-14/6, 42
$C_{14}F_8H_2N_2O_3$	$NC(C_6F_4OH)OC(C_6F_4OH)N$	F:	PerFHalOrg.5-64/5, 88
$C_{14}F_8H_4N_2O_2$	$[C(O)NH_2]_2C_6F_4C_6F_4$	F:	PerFHalOrg.7-19, 22, 48
$C_{14}F_8H_4O_2$	$HC_6F_4C_6F_4COOCH_3$	F:	PerFHalOrg.4-55
$C_{14}F_8H_6Sn$	$(CH_3)_2Sn(C_{12}F_8)$	F:	PerFHalOrg.4-29
		Sn:	Org.Verb.3-115
$C_{14}F_8H_8OSi$	$HC_6F_4OC_6F_4Si(CH_3)_2H$	F:	PerFHalOrg.4-85
$C_{14}F_8H_8O_2S$	$(CF_3C_6H_4O)_2SF_2$	S:	SVol.2-44, 83
$C_{14}F_8H_{18}N_2Ni$	$C_4F_8Ni(CNC_4H_9\text{-}t)_2$	Ni:	Org.Verb.1-382
$C_{14}F_9NO_2$	$C_6F_4C(COC_6F_5)NO$	F:	PerFHalOrg.6-126, 139
–	$C_6F_4C(O)C(C_6F_5)NO$	F:	PerFHalOrg.6-137, 150
$C_{14}F_{10}$	$C_6F_5CCC_6F_5$	F:	PerFHalOrg.4-111
$C_{14}F_{10}GeH_6$	$(C_2H_5)(C_6F_5)_2GeH$	F:	PerFHalOrg.4-179
–	$(C_6F_5)_2Ge(CH_3)_2$	F:	PerFHalOrg.4-180
$C_{14}F_{10}GeH_6O_2$	$(C_6F_5)_2Ge(OCH_3)_2$	F:	PerFHalOrg.4-180
$C_{14}F_{10}HNO_2$	$C_6F_5C[C(O)C_6F_5]NOH$	F:	PerFHalOrg.7-88
$C_{14}F_{10}HN_3$	$NC(C_6F_5)NHC(C_6F_5)N$	F:	PerFHalOrg.5-71/2, 92, 99
$C_{14}F_{10}H_3O_2Tl$	$(C_6F_5)_2TlOC(O)CH_3$	F:	PerFHalOrg.4-136, 139
$C_{14}F_{10}H_5OP$	$(C_6F_5)_2P(OC_2H_5)$	F:	PerFHalOrg.3-129, 133
$C_{14}F_{10}H_5P$	$(C_6F_5)_2PC_2H_5$	F:	PerFHalOrg.3-129, 130
$C_{14}F_{10}H_6Si$	$(C_6F_5)_2Si(CH_3)_2$	F:	PerFHalOrg.4-21/2, 81, 162
$C_{14}F_{10}H_6Sn$	$(CH_3)_2Sn(C_6F_5)_2$	F:	PerFHalOrg.4-30, 36, 82, 84
		Sn:	Org.Verb.3-4/5
$C_{14}F_{10}H_9N_5O_2S$	$(C_2F_5)_2(CH_3C_6H_4SO_2NHNH)C_3N_3$	F:	PerFHalOrg.6-112
$C_{14}F_{10}N_2O$	$NC(C_6F_5)OC(C_6F_5)N$	F:	PerFHalOrg.5-64/5, 88, 94/5
$C_{14}F_{10}N_2O^+$	$[NCOC(C_6F_5)_2N]^+$	F:	PerFHalOrg.5-95
$C_{14}F_{10}N_2S$	$NC(C_6F_5)SC(C_6F_5)N$	F:	PerFHalOrg.5-67/8, 89, 94/5
$C_{14}F_{10}O_2$	$C_6F_5C(O)C(O)C_6F_5$	F:	PerFHalOrg.4-52, 53
$C_{14}F_{11}HN_2$	$(CF_3C_6F_4)(NCC_6F_4)NH$	F:	PerFHalOrg.7-14/6, 42
$C_{14}F_{11}I$	$C_6F_5C_6F_4CFCFI$	F:	PerFHalOrg.4-17
$C_{14}F_{12}FeH_8O_5$	$(CO)_3FeOC(CF_3)_2C_4H_5CH_2C(CF_3)_2OH$	Fe:	Org.Verb.B5-89, 92/3
–	$(CO)_3FeOC(CF_3)_2OC(CF_3)_2C_4H_5CH_3$	Fe:	Org.Verb.B5-90, 92/3
$C_{14}F_{12}FeH_{24}N_8P_2$	$[CH_3C_3HN_2(NHCH_3)_2Fe(CNCH_3)_4][PF_6]_2$	Fe:	Org.Verb.B4-355
$C_{14}F_{12}FeH_{25}N_7P_2$	$[C(NHCH_3)N(CH_3)C(NHC_2H_5)Fe(CNCH_3)_4][PF_6]_2$	Fe:	Org.Verb.B4-352/3
–	$[C_2H_5N(CNHCH_3)_2Fe(CNCH_3)_4][PF_6]_2$	Fe:	Org.Verb.B4-352/3
$C_{14}F_{12}Fe_2O_6S_4$	$(CO)_3Fe(SC(CF_3)C(CF_3)S)_2Fe(CO)_3$	Fe:	Org.Verb.C1-24/5
$C_{14}F_{12}Ge_2H_{12}$	$(CH_3)_2Ge(C_5F_6)_2Ge(CH_3)_2$	F:	PerFHalOrg.4-26
$C_{14}F_{12}HNO$	$C_6F_5C_6F_4NHC(O)CF_3$	F:	PerFHalOrg.7-23/4, 49
$C_{14}F_{12}H_7N_3$	$(CF_3)_2CNHC(CF_3)_2NC(NCHC_6H_5)$	F:	PerFHalOrg.5-109
$C_{14}F_{12}H_7N_3O$	$(CF_3)_2CNHC(CF_3)_2NC(NHCOC_6H_5)$	F:	PerFHalOrg.5-105
$C_{14}F_{12}H_8N_4O$	$(C_6H_5CONHN)CC(CF_3)_2NHC(CF_3)_2NH$	F:	PerFHalOrg.5-100/1
$C_{14}F_{12}H_9N_3$	$(CF_3)_2CNHC(CF_3)_2NC(NHCH_2C_6H_5)$	F:	PerFHalOrg.5-104
$C_{14}F_{12}H_{10}Ni_2O_2S_8$	$[Ni(S_2COC_2H_5)(S_2C_2(CF_3)_2)]_2$	C:	MVol.D4-260

Formula	Compound		Reference
$C_{14}F_{12}H_{10}Ni_2P_2$	$(C_5H_5NiP(CF_3)_2)_2$	Ni:	Org.Verb.2-339, 340
$C_{14}F_{12}H_{10}O_2P_2Ti$	$[(CF_3)_2PO]_2Ti(C_5H_5)_2$	F:	PerFHalOrg.3-37, 39
$C_{14}F_{12}H_{12}N_2Ni$	$C_8H_{12}Ni(C(CF_3)_2NNC(CF_3)_2)$	Ni:	Org.Verb.2-84
$C_{14}F_{12}H_{14}N_2NiO_2$	$(i\text{-}C_3H_7NC)_2Ni(C_2O_2(CF_3)_4)$	Ni:	Org.Verb.1-315, 316
$C_{14}F_{12}H_{15}N_3O$	$(CF_3)_2CNHC(CF_3)_2NC(NHCOC_6H_{13})$	F:	PerFHalOrg.5-105
$C_{14}F_{12}H_{16}N_2$	$(NH_2C(CF_3)_2CH_2CHCH)_2C_2H_4$	F:	PerFHalOrg.7-69, 72
$C_{14}F_{12}H_{16}N_4O$	$(CF_3)_2CNHC(CF_3)_2NC(NHCON(C_3H_7)_2)$	F:	PerFHalOrg.5-108/9
$C_{14}F_{12}H_{18}NiO_2P_2$	$(CO)_2Ni(P(C_4H_9\text{-}i)(CF_3)_2)_2$	Ni:	Org.Verb.1-129
$C_{14}F_{12}Mo_2N_2O_8P_2$	$[(CO)_4MoP(CF_3)_2CN]_2$	F:	PerFHalOrg.3-174
$C_{14}F_{13}O_2Tl$	$(C_6F_5)_2TlOC(O)CF_3$	F:	PerFHalOrg.4-131/3, 136/8
$C_{14}F_{14}$	$(CF_3)_2C_6F_3C_6F_5$	F:	PerFHalOrg.4-16
$C_{14}F_{14}Fe_2HgO_8$	$C_3F_7(CO)_4FeHgFe(CO)_4C_3F_7$	Fe:	Org.Verb.C1-223
$C_{14}F_{14}HN$	$(CF_3C_6F_4)_2NH$	F:	PerFHalOrg.7-14/6, 42
$C_{14}F_{16}H_8N_{10}$	$(NH_2)_2C_3N_3(CF_2)_8C_3N_3(NH_2)_2$	F:	PerFHalOrg.6-92
$C_{14}F_{17}H_{11}O$	$(CH_2)_5C(C_8F_{17})OH$	F:	PerFHalOrg.4-103
$C_{14}F_{20}Fe_2O_6S_2$	$C_3F_7(CO)_3Fe(SCF_3)_2Fe(CO)_3C_3F_7$	Fe:	Org.Verb.C1-222/3
$C_{14}F_{21}N$	$(C_6F_{11})((CF_3)_2CF)C_5F_3N$	F:	PerFHalOrg.5-155/6, 168
$C_{14}F_{21}N_3$	$NC(i\text{-}C_3F_7)NC(i\text{-}C_3F_7)C(CN)C(i\text{-}C_3F_7)$	F:	PerFHalOrg.6-27, 47
$C_{14}F_{22}H_4N_4O_4$	$(C_3F_7C(O)ONCNH_2)_2(CF_2)_4$	F:	PerFHalOrg.7-5/6, 33
$C_{14}F_{22}Hg$	$[FC(CF_2)(CF_2CF_2)_2C]_2Hg$	F:	PerFHalOrg.4-33
$C_{14}F_{22}N_2$	$(CF_3)_2(i\text{-}C_3F_7)_2C_6F_2N_2$	F:	PerFHalOrg.6-128/9, 140
$C_{14}F_{22}N_2O$	$NC(C_6F_{11})OC(C_6F_{11})N$	F:	PerFHalOrg.5-64/5, 88
$C_{14}F_{22}N_3$	$[(CF_2)_8C_3N_3(CF_2)_3]_x$	F:	PerFHalOrg.6-119/20
$C_{14}F_{22}N_4O_2$	$C_2F_4(CF_2CNOC(C_3F_7)N)_2$	F:	PerFHalOrg.5-69/71, 91
–	$C_2F_4(CF_2COC(C_3F_7)NN)_2$	F:	PerFHalOrg.5-69/70, 90
$C_{14}F_{23}H_9OSi$	$(CH_3)_3Si(CF_2)_8OCF(CF_3)_2$	F:	PerFHalOrg.4-87
$C_{14}F_{23}N$	$((CF_3)_2CF)(C_2F_5)_3C_5FN$	F:	PerFHalOrg.5-153/4, 165
–	$((CF_3)_2CF)_3C_5F_2N$	F:	PerFHalOrg.5-151/3, 163/4, 173/4, 176
$C_{14}F_{23}N_3$	$[(CF_2)_8C_3N_3(C_3F_7)]_x$	F:	PerFHalOrg.6-118/21
$C_{14}F_{23}N_3O_2$	$[i\text{-}C_3F_7C_3N_3CF(CF_3)OC_4F_8OCF(CF_3)]_x$	F:	PerFHalOrg.6-120
$C_{14}F_{24}N_2$	$NC(i\text{-}C_3F_7)(CC_2F_5)_2C(i\text{-}C_3F_7)N$	F:	PerFHalOrg.6-16, 37, 55
–	$NC(i\text{-}C_3F_7)NC(i\text{-}C_3F_7)C(C_2F_5)C(C_2F_5)$	F:	PerFHalOrg.6-22/3, 44, 55
$C_{14}F_{25}H_5S_2Sn$	$CF_3(CF_2)_{11}CH_2CH_2Sn(S)SH$	Sn:	Org.Verb.6-264
$C_{14}F_{26}H_3N_3$	$C_6F_{13}C(NH)NC(NH_2)C_6F_{13}$	F:	PerFHalOrg.7-7, 35
$C_{14}F_{26}N_2O$	$NC(C_6F_{13})OC(C_6F_{13})N$	F:	PerFHalOrg.5-64/5, 88
$C_{14}F_{29}HN_4O_5$	$HOC_6F_5[ON(CF_3)_2]_4$	F:	PerFHalOrg.7-98/9
$C_{14}F_{29}H_2N_5O_4$	$NH_2C_6F_5[ON(CF_3)_2]_4$	F:	PerFHalOrg.7-98/9, 116
$C_{14}F_{29}N$	$C_9F_{19}N(CF_2)_5$	F:	PerFHalOrg.5-157/8, 172, 175
$C_{14}F_{29}NO$	$O(CF_2CF_2)_2NCF_2(CF_2)_8CF_3$	F:	PerFHalOrg.6-2, 7
$C_{14}F_{30}NO$	$(C_7F_{15})_2NO$	F:	PerFHalOrg.7-85/6, 103, 122
$C_{14}F_{30}N_4O_4$	1) $C_6F_6[ON(CF_3)_2]_4$	F:	PerFHalOrg.7-98/9, 116
	2) $C_6F_6(ON(CF_3)_2)_4$	F:	PerFHalOrg.7-99, 117
$C_{14}FeGe_2H_{27}O_4P$	$(CO)_4FeP(Ge(CH_3)_3)_2C_4H_9$	Fe:	Org.Verb.B2-85, 91
$C_{14}FeH_6O_6$	$(C(O)C(C_6H_5)CHCO)Fe(CO)_4$	Fe:	Org.Verb.B4-335
$C_{14}FeH_7MnO_7$	$(CO)_3FeC_7H_7Mn(CO)_4$	Fe:	Org.Verb.B5-79

Formula	Compound		Reference
$C_{14}FeH_7NO_5$	$(CO)_4Fe(NCCHCHCOC_6H_5)$	Fe:	Org.Verb.B2-71, 76, 80
		Fe:	Org.Verb.B4-268, 276
$C_{14}FeH_7NO_6$	$C_6H_5N(C(O)CH)_2Fe(CO)_4$	Fe:	Org.Verb.B4-302, 304, 310
$C_{14}FeH_8I_2N_2O_2$	$(CO)_2Fe(N_2C_{12}H_8)I_2$	Fe:	Org.Verb.B1-101/2, 106
$C_{14}FeH_8N_2O_6$	$C_6H_5NHN(C(O)CH)_2Fe(CO)_4$	Fe:	Org.Verb.B4-304, 310
$C_{14}FeH_8O_5$	$C_6H_4C_4H_4OFe(CO)_4$	Fe:	Org.Verb.B4-303, 308/9
$C_{14}FeH_8O_7$	$C_6H_5COCHCHCOOHFe(CO)_4$	Fe:	Org.Verb.B4-267/8, 276, 294
$C_{14}FeH_9LiO_5S_2$	$Li[(CH_2S)_2C(C_6H_5)C(O)Fe(CO)_4]$	Fe:	Org.Verb.B4-157, 168
$C_{14}FeH_9LiO_7$	$Li[(CH_2O)_2C(C_6H_5)C(O)Fe(CO)_4]$	Fe:	Org.Verb.B4-157, 167/8
$C_{14}FeH_9O_4P$	$(CO)_4Fe(C_6H_5P(CHCH)_2)$	Fe:	Org.Verb.C2-49
$C_{14}FeH_9O_5^-$	$[(CHO)(C_6H_5CO)C_3H_3Fe(CO)_3]^-$	Fe:	Org.Verb.B5-74, 76
$C_{14}FeH_9O_5S_2^-$	$[(CH_2S)_2C(C_6H_5)C(O)Fe(CO)_4]^-$	Fe:	Org.Verb.B4-157, 168
$C_{14}FeH_9O_7^-$	$[(CH_2O)_2C(C_6H_5)C(O)Fe(CO)_4]^-$	Fe:	Org.Verb.B4-157, 167/8
$C_{14}FeH_{10}$	$C_5H_5FeC_5H_4(CC)_2H$	Fe:	Org.Verb.A1-300/1
$C_{14}FeH_{10}N_2$	$C_5H_5FeC_5H_4CHC(CN)_2$	Fe:	Org.Verb.A2-161
$C_{14}FeH_{10}N_2O_4S$	$(C_6H_5NC)_2FeSO_4$	Fe:	Org.Verb.B4-12
$C_{14}FeH_{10}N_2O_5$	$(CH_3)(C_6H_5)C_4H_2N_2O_2Fe(CO)_3$	Fe:	Org.Verb.B5-160/1
$C_{14}FeH_{10}N_2O_5S$	$(CO)_4Fe(C(S)N(C_6H_5)COCH(CH_3)NH)$	Fe:	Org.Verb.B2-128/30
$C_{14}FeH_{10}N_2O_7$	$HOC(O)C_2H_2C(O)N_2H_2C_6H_5Fe(CO)_4$	Fe:	Org.Verb.B4-274, 292/3
$C_{14}FeH_{10}N_4O_2$	$(C_6H_5NC)_2Fe(NO)_2$	Fe:	Org.Verb.B4-11/2, 14
$C_{14}FeH_{10}Ni_2O_4^+$	$(C_5H_5)_2Ni_2Fe(CO)_4^+$	Ni:	Org.Verb.2-351
$C_{14}FeH_{10}O_2S_3$	$(C_5H_5)Fe(CO)_2[S_2CSC_6H_5]$	C:	MVol.D4-269
$C_{14}FeH_{10}O_5$	$CH_3COCHCHC_6H_5Fe\ (CO)_4$	Fe:	Org.Verb.B4-268, 277, 297/8
–	$C_6H_5COCH_2CHCH_2Fe(CO)_4$	Fe:	Org.Verb.B4-238, 250
–	$HC(O)CHCHC_6H_4CH_3Fe(CO)_4$	Fe:	Org.Verb.B4-274, 291/2
$C_{14}FeH_{10}O_6$	$(C(C_3H_5)CO)_2Fe(CO)_4$	Fe:	Org.Verb.B4-335, 342
–	$CH_3OC(O)CHCHC_6H_5Fe(CO)_4$	Fe:	Org.Verb.B4-275, 293
–	$(CH_3OOCCHCHC(C_6H_5)O)Fe(CO)_3$	Fe:	Org.Verb.B5-127, 132, 141/2
–	$HC(O)CHCHC_6H_4OCH_3Fe(CO)_4$	Fe:	Org.Verb.B4-273, 291/2
–	$HOC(O)CHCHC_6H_4CH_3Fe(CO)_4$	Fe:	Org.Verb.B4-275, 292/3
$C_{14}FeH_{10}O_7$	$HOC(O)CHCHC_6H_4OCH_3Fe(CO)_4$	Fe:	Org.Verb.B4-274/5, 292/3
$C_{14}FeH_{10}O_7S$	$CH_3COCHCHSO_2C_6H_5Fe(CO)_4$	Fe:	Org.Verb.B4-267/8, 277, 296/7
$C_{14}FeH_{11}KO_5$	$K[C_6H_5CH_2CH(CH_3)C(O)Fe(CO)_4]$	Fe:	Org.Verb.B4-152, 162
$C_{14}FeH_{11}LiO_5$	$Li[(CH_3)_3C_6H_2C(O)Fe(CO)_4]$	Fe:	Org.Verb.B4-153, 162
$C_{14}FeH_{11}NOS_2$	$C_5H_5FeC_5H_4CHCC(O)N(H)C(S)S$	Fe:	Org.Verb.A2-158
$C_{14}FeH_{11}NO_2$	$C_5H_5FeC_5H_4CHC(CN)COOH$	Fe:	Org.Verb.A2-161
$C_{14}FeH_{11}NO_4$	$((CH_3)_3C_6H_2CN)Fe(CO)_4$	Fe:	Org.Verb.B3-195
$C_{14}FeH_{11}NO_7$	$((CH_3O)_3C_6H_2CN)Fe(CO)_4$	Fe:	Org.Verb.B3-195
$C_{14}FeH_{11}O^+$	$[FeC_5H_4C(CHO)CHC_6H_5]^+$	Fe:	Org.Verb.A2-179
$C_{14}FeH_{11}O_4^+$	$[C_6H_5C(O)C_4H_6Fe(CO)_3]^+$	Fe:	Org.Verb.B5-94/7
$C_{14}FeH_{11}O_5^+$	$[(CH_3COO)(C_6H_5)C_3H_3Fe(CO)_3]^+$	Fe:	Org.Verb.B5-94/7
$C_{14}FeH_{11}O_5^-$	$[(CH_3)_3C_6H_2C(O)Fe(CO)_4]^-$	Fe:	Org.Verb.B4-153, 162
–	$[C_6H_5CH_2CH(CH_3)C(O)Fe(CO)_4]^-$	Fe:	Org.Verb.B4-152
$C_{14}FeH_{11}O_7SbW$	$(CO)_4FeSb(CH_3)_2W(CO)_3C_5H_5$	Fe:	Org.Verb.B2-102, 123

Formula	Compound	Reference
$C_{14}FeH_{12}$	$C_5H_5FeC_5H_4CHCHCCH$	Fe: Org.Verb.A1-300/1
$C_{14}FeH_{12}{}^+$	$[C_5H_5FeC_5H_4CHCHCCH]^+$	Fe: Org.Verb.A1-301
$C_{14}FeH_{12}LiNO_5$	$Li[(CH_3)_2NCH_2C_6H_4C(O)Fe(CO)_4]$	Fe: Org.Verb.B4-171
$C_{14}FeH_{12}MoN_2O_{10}Sn$	$(CO)_5MoSn[OCN(CH_3)_2]_2Fe(CO)_3$	Fe: Org.Verb.B3-143
$C_{14}FeH_{12}NO_5{}^-$	$[(CH_3)_2NCH_2C_6H_4C(O)Fe(CO)_4]^-$	Fe: Org.Verb.B4-171
$C_{14}FeH_{12}N_2O$	$C_5H_5FeC_5H_4CHC(CN)CONH_2$	Fe: Org.Verb.A2-161
$C_{14}FeH_{12}N_2OS$	$C_5H_5FeC_5H_4CONHC_3H_2NS$	Fe: Org.Verb.A3-154
$C_{14}FeH_{12}N_2O_2$	$C_5H_5FeC_5H_4C_3H_2N_2COOH$	Fe: Org.Verb.A3-95
$C_{14}FeH_{12}N_2O_{10}SnW$	$(CO)_5WSn[OCN(CH_3)_2]_2Fe(CO)_3$	Fe: Org.Verb.B3-143
$C_{14}FeH_{12}O_2$	$C_5H_5FeC_5H_4CCCOOCH_3$	Fe: Org.Verb.A3-121
–	$C_5H_5FeC_5H_4COOCH_2CCH$	Fe: Org.Verb.A3-108, 111
$C_{14}FeH_{12}O_3$	$C_5H_5FeC_5H_4CHCHCOCOOH$	Fe: Org.Verb.A3-94/6
–	$C_5H_5FeC_5H_4CHOHCCCOOH$	Fe: Org.Verb.A3-84/6
–	$C_5H_5FeC_5H_4COCHCHCOOH$	Fe: Org.Verb.A3-94/6
$C_{14}FeH_{12}O_4$	$CH_3CHCHCH_2C_6H_5Fe(CO)_4$	Fe: Org.Verb.B4-272, 289
–	$C_5H_5FeC_5H_4CHC(COOH)_2$	Fe: Org.Verb.A3-96/8
–	$C_5H_5FeC_5H_4COCH_2COCOOH$	Fe: Org.Verb.A3-94/6
$C_{14}FeH_{12}O_5$	$CH_3CHCHC_6H_4OCH_3Fe(CO)_4$	Fe: Org.Verb.B4-267, 272, 289/90
–	$(CH_3OC_6H_4CHCHC(CH_3)O)Fe(CO)_3$	Fe: Org.Verb.B5-136, 149
$C_{14}FeH_{12}O_7$	$(CH_3OCO)_2C_7H_6Fe(CO)_3$	Fe: Org.Verb.B5-79
$C_{14}FeH_{12}O_{12}$	$(CH_3OCO)_2CC(COOCH_3)_2Fe(CO)_4$	Fe: Org.Verb.B4-321, 325
$C_{14}FeH_{13}{}^+$	$[FeC_{14}H_{13}]^+$	Fe: Org.Verb.A2-65
$C_{14}FeH_{13}N$	$C_5H_5FeC_5H_4C(CH_3)CHCN$	Fe: Org.Verb.A2-219
–	$C_5H_5FeC_5H_4C_4H_3NH$	Fe: Org.Verb.A1-377
–	$C_5H_5FeC_5H_4NC_4H_4$	Fe: Org.Verb.A1-376
$C_{14}FeH_{13}NO$	$C_5H_5FeC_5H_4CHCHCHOHCN$	Fe: Org.Verb.A2-177
$C_{14}FeH_{13}NO_4$	$((CH_3)_2NC_6H_4CHCHCHO)Fe(CO)_3$	Fe: Org.Verb.B5-122/5, 130, 140
–	$C_6H_5CHCHC(O)N(CH_3)_2Fe(CO)_3$	Fe: Org.Verb.B5-158
$C_{14}FeH_{13}NO_7$	$(CH_3OCO)_2CCH_2Fe(CO)_3NC_5H_5$	Fe: Org.Verb.B4-199, 205
$C_{14}FeH_{13}N_3O$	$C_5H_5FeC_5H_4CHNNHCOCH_2CN$	Fe: Org.Verb.A2-167
$C_{14}FeH_{13}O_2{}^+$	$[FeC_5H_4C_6H_4CH_2COOCH_3]^+$	Fe: Org.Verb.A3-127
–	$[FeC_5H_4C_6H_4COOC_2H_5]^+$	Fe: Org.Verb.A3-126
$C_{14}FeH_{13}O_2{}^-$	$[C_5H_5FeC_5H_4COCHCOCH_3]^-$	Fe: Org.Verb.A3-34
$C_{14}FeH_{13}O_3{}^+$	$[(CH_3)_2(C_6H_5)C_3H_2Fe(CO)_3]^+$	Fe: Org.Verb.B5-65
$C_{14}FeH_{13}O_3{}^{2-}$	$[C_5H_5FeC_5H_3(O)COCOC_2H_5]^{2-}$	Fe: Org.Verb.A2-302
$C_{14}FeH_{14}$	$(CH_3C_6H_4)_2Fe$	Fe: Org.Verb.B1-13
–	$C_5H_5FeC_5H_4CCC_2H_5$	Fe: Org.Verb.A1-300/1
–	$C_5H_5FeC_5H_4C(CH_2)CHCH_2$	Fe: Org.Verb.A1-291, 294
–	$C_5H_5FeC_5H_4C(CH_3)CCH_2$	Fe: Org.Verb.A1-296
–	$C_5H_5FeC_5H_4CHCCHCH_3$	Fe: Org.Verb.A1-296
–	$C_5H_5FeC_5H_4CHCHCHCH_2$	Fe: Org.Verb.A1-291, 294
–	$C_5H_5FeC_5H_4CH(CH_3)CCH$	Fe: Org.Verb.A1-300/1
–	$C_5H_5FeC_5H_4CH_2CH_2CCH$	Fe: Org.Verb.A1-300/1
–	$(C_6H_5CH_2)_2Fe$	Fe: Org.Verb.B1-6/7
$C_{14}FeH_{14}HgO_3$	$CH_3COC_5H_4FeC_5H_4HgOOCCH_3$	Fe: Org.Verb.A2-211
–	$C_5H_5FeC_5H_3(COCH_3)(HgOOCCH_3)$	Fe: Org.Verb.A2-211
$C_{14}FeH_{14}Hg_2O_4$	$Fe(C_5H_4HgOOCCH_3)_2$	Fe: Org.Verb.A1-174/5
$C_{14}FeH_{14}N_2$	$C_5H_5FeC_5H_4CCHC(CH_3)NNH$	Fe: Org.Verb.A3-34

$C_{14}FeH_{14}N_2$	$[Fe(C_7H_7N)_2]_n$	Fe:	Org.Verb.B3-206
$C_{14}FeH_{14}N_2O_4S$	$(CO)_4FeSC(NHC_6H_4CH_3)N(CH_3)_2$	Fe:	Org.Verb.B2-128/30
$C_{14}FeH_{14}N_4O$	$C_5H_5FeC_5H_4C_3H_2N_2CONHNH_2 \cdot H_2O$	Fe:	Org.Verb.A3-134
$C_{14}FeH_{14}O$	$C_5H_4FeC_5H_4CH(CH_3)CH_2CO$	Fe:	Org.Verb.A3-101
–	$C_5H_4FeC_5H_4CH_2CH(CH_3)CO$	Fe:	Org.Verb.A3-101
–	$C_5H_5FeC_5H_3CH(CH_3)CH_2CO$	Fe:	Org.Verb.A3-101
–	$C_5H_5FeC_5H_3(CH_2)_3CO$	Fe:	Org.Verb.A3-101
–	$C_5H_5FeC_5H_4C(CH_3)(CCH)OH$	Fe:	Org.Verb.A2-58/9, 62
–	$C_5H_5FeC_5H_4CHCHCOCH_3$	Fe:	Org.Verb.A3-8/10
–	$C_5H_5FeC_5H_4COC(CH_3)CH_2$	Fe:	Org.Verb.A2-248
–	$C_5H_5FeC_5H_4COCHCHCH_3$	Fe:	Org.Verb.A2-248, 250
–	$C_5H_5FeC_5H_4COCH(CH_2)_2$	Fe:	Org.Verb.A2-289/90
$C_{14}FeH_{14}O^+$	$[C_5H_5FeC_5H_4CHCHCOCH_3]^+$	Fe:	Org.Verb.A3-10
$C_{14}FeH_{14}O_2$	$(CH_3OC_6H_4)_2Fe$	Fe:	Org.Verb.B1-13
–	$C_5H_5FeC_5H_3(COCH_3)_2$	Fe:	Org.Verb.A2-210
–	$C_5H_5FeC_5H_4C(CH_2)OOCCH_3$	Fe:	Org.Verb.A2-98/100
–	$C_5H_5FeC_5H_4C(CH_3)CHCOOH$	Fe:	Org.Verb.A3-77/9
–	$C_5H_5FeC_5H_4CHCHCOOCH_3$	Fe:	Org.Verb.A3-121, 123
–	$C_5H_5FeC_5H_4CHCHCOOCD_3$	Fe:	Org.Verb.A3-123
–	$C_5H_5FeC_5H_4CH(CH_2)_2COO$	Fe:	Org.Verb.A3-165/6
–	$C_5H_5FeC_5H_4CH_2OOCCHCH_2$	Fe:	Org.Verb.A2-97/8
–	$C_5H_5FeC_5H_4COCH_2COCH_3$	Fe:	Org.Verb.A3-31/2, 34
–	$C_5H_5FeC_5H_4C_3H_4COOH$	Fe:	Org.Verb.A3-82/4, 100, 103
–	$C_5H_5FeC_5H_4OOCCHCHCH_3$	Fe:	Org.Verb.A2-94
–	$Fe(C_5H_4COCH_3)_2$	Fe:	Org.Verb.A1-374
		Fe:	Org.Verb.A2-210
$C_{14}FeH_{14}O_2^+$	$[C_5H_5FeC_5H_4COCH_2COCH_3]^+$	Fe:	Org.Verb.A3-34
$C_{14}FeH_{14}O_2^-$	$[C_5H_5FeC_5H_4COCOC_2H_5]^-$	Fe:	Org.Verb.A2-302
$C_{14}FeH_{14}O_2W^+$	$[C_5H_5FeC_5H_4C(OC_2H_5)W(CO)]^+$	Fe:	Org.Verb.A1-347
$C_{14}FeH_{14}O_3$	$C_2H_5COC_5H_4FeC_5H_4COOH$	Fe:	Org.Verb.A3-58/9
–	$C_5H_5FeC_5H_4CO(CH_2)_2COOH$	Fe:	Org.Verb.A3-87/8, 90, 104
$C_{14}FeH_{14}O_3^+$	$[(CH_3)_2(C_6H_5)C_3H_2Fe(CO)_3H]^+$	Fe:	Org.Verb.B5-65
$C_{14}FeH_{14}O_3S$	$C_5H_5FeC_5H_4COCH_2SCH_2COOH$	Fe:	Org.Verb.A2-245
$C_{14}FeH_{14}O_4$	$C_5H_5FeC_5H_4CH_2CH(COOH)_2$	Fe:	Org.Verb.A3-96/8, 105
–	$C_{10}H_{14}Fe(CO)_4$	Fe:	Org.Verb.B4-318
$C_{14}FeH_{14}O_4Sn$	$[(CH_3)_2SnOCOC_5H_4FeC_5H_4COO]_x$	Sn:	Org.Verb.6-29
$C_{14}FeH_{14}O_8$	$(CH_3)_2C(OCO)_2CCH(i\text{-}C_3H_7)Fe(CO)_4$	Fe:	Org.Verb.B4-317
–	$(C_2H_5OCO)_2C_3H_2CH_2Fe(CO)_4$	Fe:	Org.Verb.B4-259, 260/3
–	$(C_2H_5OCO)_2C_3D_2CH_2Fe(CO)_4$	Fe:	Org.Verb.B4-260/3
$C_{14}FeH_{15}^+$	$[C_5H_5FeC_5H_4CHC_3H_5]^+$	Fe:	Org.Verb.A2-50, 120
$C_{14}FeH_{15}NO$	$C_5H_5FeC_5H_4C(CH_3)(OH)CH_2CN$	Fe:	Org.Verb.A2-227
$C_{14}FeH_{15}NO_3P^+$	$[C_4H_4PC_6H_5(CH_3)_2Fe(CO)_2NO]^+$	Fe:	Org.Verb.B5-28/31
$C_{14}FeH_{15}N_3O$	$C_5H_5FeC_5H_4CHCHCHNNHCONH_2$	Fe:	Org.Verb.A2-176
$C_{14}FeH_{15}O_2$	$[C_5H_5FeC_5H_4(CH_2)_3COO]$	Fe:	Org.Verb.A3-68
$C_{14}FeH_{15}O_2^+$	$[C_5H_5FeC_5H_4C(CH_3)CH_2COOH]^+$	Fe:	Org.Verb.A3-49, 123
–	$[C_5H_5FeC_5H_4C(CH_3)CHDCOOD]^+$	Fe:	Org.Verb.A3-79
–	$[C_5H_5FeC_5H_4CHCH_2CH_2COOH]^+$	Fe:	Org.Verb.A3-49
$C_{14}FeH_{15}O_2^-$	$[C_5H_5FeC_5H_4(CH_2)_3COO]^-$	Fe:	Org.Verb.A3-68

Formula	Compound		Reference
$C_{14}FeH_{15}O_4P$	$(CO)_4FeP(C_2H_5)_2C_6H_5$	Fe:	Org.Verb.B2-89
$C_{14}FeH_{16}$	$C_5H_5FeC_5H_4C(CH_3)CHCH_3$	Fe:	Org.Verb.A1-291/2
–	$C_5H_5FeC_5H_4C(C_2H_5)CH_2$	Fe:	Org.Verb.A1-291, 293
–	$C_5H_5FeC_5H_4CHC(CH_3)_2$	Fe:	Org.Verb.A1-291
–	$C_5H_5FeC_5H_4CH_2CH_2CHCH_2$	Fe:	Org.Verb.A1-293
$C_{14}FeH_{16}{}^{2+}$	$[Fe(C_6H_5CH_3)_2]^{2+}$	Fe:	Org.Verb.A2-186, 208
$C_{14}FeH_{16}I_2N_4O_2$	$(CO)_2Fe(NH_2C_6H_4NH_2)_2I_2$	Fe:	Org.Verb.B1-101/3
$C_{14}FeH_{16}N_2$	$C_5H_5FeC_5H_4CHCH_2C(CH_3)NNH$	Fe:	Org.Verb.A3-10
–	$C_5H_5FeC_5H_4CH(CN)N(CH_3)_2$	Fe:	Org.Verb.A2-169
$C_{14}FeH_{16}N_2O_2S_4$	$Fe[S_2COCH_3]_2(C_5H_5N)_2$	C:	MVol.D4-259
$C_{14}FeH_{16}O$	$[C(C_5H_5FeC_5H_4)(OC_2H_5)CH_2]_n$	Fe:	Org.Verb.A2-228
–	$C_5H_5FeC_5H_3C_4H_7OH$	Fe:	Org.Verb.A1-358
–	$C_5H_5FeC_5H_4C(CH_3)(CHCH_2)OH$	Fe:	Org.Verb.A2-59, 62
–	$C_5H_5FeC_5H_4CHCHCHOHCH_3$	Fe:	Org.Verb.A2-75, 80
–	$C_5H_5FeC_5H_4CHCH(CH_2)_2OH$	Fe:	Org.Verb.A2-75, 80
–	$C_5H_5FeC_5H_4CH(CH_2)_3O$	Fe:	Org.Verb.A3-172/3
–	$C_5H_5FeC_5H_4CH(CH_3)C_2H_3O$	Fe:	Org.Verb.A3-173
–	$C_5H_5FeC_5H_4CHOHCHCHCH_3$	Fe:	Org.Verb.A2-45, 48
–	$C_5H_5FeC_5H_4CHOHCH(CH_2)_2$	Fe:	Org.Verb.A2-4, 46, 50
–	$C_5H_5FeC_5H_4CHOHCH_2CHCH_2$	Fe:	Org.Verb.A2-45, 49
–	$C_5H_5FeC_5H_4CH_2CH(CH_3)CHO$	Fe:	Org.Verb.A2-174, 177
–	$C_5H_5FeC_5H_4CH_2OCH_2CHCH_2$	Fe:	Org.Verb.A2-113/4
–	$C_5H_5FeC_5H_4(CH_2)_2COCH_3$	Fe:	Org.Verb.A3-3/4
–	$C_5H_5FeC_5H_4COCH(CH_3)_2$	Fe:	Org.Verb.A2-186, 237/8
–	$C_5H_5FeC_5H_4COC_3H_7$-n	Fe:	Org.Verb.A2-185, 235/6
–	$C_5H_5FeC_5H_4C_3H_4CH_2OH$	Fe:	Org.Verb.A2-75, 80
$C_{14}FeH_{16}O^+$	$[C_5H_5FeC_5H_4COC_3H_7$-n$]^+$	Fe:	Org.Verb.A2-236
$C_{14}FeH_{16}OS$	$C_5H_5FeC_5H_4COCH_2SC_2H_5$	Fe:	Org.Verb.A2-245
$C_{14}FeH_{16}O_2$	$C_3H_7C_5H_4FeC_5H_4COOH$	Fe:	Org.Verb.A3-58/9
–	$C_5H_5FeC_5H_4C(CH_3)(OCH_2)_2$	Fe:	Org.Verb.A3-178
–	$C_5H_5FeC_5H_4C(CH_3)_2COOH$	Fe:	Org.Verb.A3-71/2
–	$C_5H_5FeC_5H_4CH(CH_3)CH_2COOH$	Fe:	Org.Verb.A3-71/2, 100/1
–	$C_5H_5FeC_5H_4CH(CH_3)OOCCH_3$	Fe:	Org.Verb.A1-359, 362
		Fe:	Org.Verb.A2-98, 100
–	$C_5H_5FeC_5H_4CH_2CH(CH_3)COOH$	Fe:	Org.Verb.A3-71/2, 75, 100/1
–	$C_5H_5FeC_5H_4CH_2COOC_2H_5$	Fe:	Org.Verb.A3-115/7
–	$C_5H_5FeC_5H_4CH_2OOCC_2H_5$	Fe:	Org.Verb.A2-98
–	$C_5H_5FeC_5H_4(CH_2)_2COOCH_3$	Fe:	Org.Verb.A3-115/7
–	$C_5H_5FeC_5H_4(CH_2)_2OOCCH_3$	Fe:	Org.Verb.A2-104
–	$C_5H_5FeC_5H_4(CH_2)_3COOH$	Fe:	Org.Verb.A3-68/70, 100/1
–	$C_5H_5FeC_5H_4COC(CH_3)_2OH$	Fe:	Org.Verb.A2-295, 297
$C_{14}FeH_{16}O_2S$	$C_5H_5FeC_5H_4CH(CH_3)SCH_2COOH$	Fe:	Org.Verb.A2-30
–	$C_5H_5FeC_5H_4COCH_2SCH_2CH_2OH$	Fe:	Org.Verb.A2-245
$C_{14}FeH_{16}O_3$	$C_5H_5FeC_5H_3(CH(CH_3)OCH_3)(COOH)$	Fe:	Org.Verb.A2-117
–	$C_5H_5FeC_5H_4C(CH_3)(OH)CH_2COOH$	Fe:	Org.Verb.A3-84
–	$C_5H_5FeC_5H_4CHOH(CH_2)_2COOH$	Fe:	Org.Verb.A3-84/6, 104
–	$FeC_{10}H_8(OCH_3)(COOC_2H_5)$	Fe:	Org.Verb.A2-109
$C_{14}FeH_{16}O_4Si_2$	$(CO)_4Fe(Si(CH_3)_2)_2C_6H_4$	Fe:	Org.Verb.B2-144, 150

Formula	Compound	Reference
$C_{14}FeH_{16}O_7P^-$	$[C_6H_5CH_2C(O)Fe(CO)_3P(OCH_3)_3]^-$	Fe: Org.Verb.B4-142
$C_{14}FeH_{16}S_2$	$C_5H_5FeC_5H_4CH(SCH_2)_2CH_2$	Fe: Org.Verb.A2-169
$C_{14}FeH_{17}^+$	$[CH_3C_5H_4FeC_5H_3(CH_3)CHCH_3]^+$	Fe: Org.Verb.A1-359/60
–	$[CH_3C_5H_4FeC_5H_4C(CH_3)_2]^+$	Fe: Org.Verb.A1-357
–	$[(CH_3)_3C_6H_3FeC_5H_5]^+$	Fe: Org.Verb.A2-150, 186
–	$[C_5H_5FeC_5H_3(CH_3)C(CH_3)_2]^+$	Fe: Org.Verb.A1-357
–	$[C_5H_5FeC_5H_4C(CH_3)C_2H_5]^+$	Fe: Org.Verb.A1-292, 351
		Fe: Org.Verb.A2-58, 62
–	$[C_5H_5FeC_5H_4CHCH(CH_3)_2]^+$	Fe: Org.Verb.A1-291, 350/3
		Fe: Org.Verb.A2-47, 80
–	$[C_6H_6FeC_5H_4C_3H_7\text{-}n]^+$	Fe: Org.Verb.A1-268
$C_{14}FeH_{17}NO_6$	$(CH_3COO)(N(C_2H_5)_2CO)C_4H_4Fe(CO)_3$	Fe: Org.Verb.B5-76
$C_{14}FeH_{17}NS$	$C_5H_5FeC_5H_4CH_2CSN(CH_3)_2$	Fe: Org.Verb.A2-228
$C_{14}FeH_{17}N_3O_3^{2+}$	$[CH_3NCFe(H_2O)_3(C_{12}H_8N_2)]^{2+}$	Fe: Org.Verb.B4-10
$C_{14}FeH_{17}N_3S$	$C_5H_5FeC_5H_4CHNNHCSNHC_2H_5$	Fe: Org.Verb.A2-167
$C_{14}FeH_{17}O^+$	$[C_5H_5FeC_5H_4CH(CH_2)_3OH]^+$	Fe: Org.Verb.A2-80, 90
$C_{14}FeH_{17}O_4^+$	$[(CH_3)_2C_6H_9C_2H_2Fe(CO)_4]^+$	Fe: Org.Verb.B5-100, 105
$C_{14}FeH_{18}$	$C_5H_5FeC_5H_4C(CH_3)_3$	Fe: Org.Verb.A1-272/5
–	$C_5H_5FeC_5H_4CH(CH_3)C_2H_5$	Fe: Org.Verb.A1-275
–	$C_5H_5FeC_5H_4CH_2CH(CH_3)_2$	Fe: Org.Verb.A1-275/6
–	$C_5H_5FeC_5H_4(CH_2)_3CH_3$	Fe: Org.Verb.A1-268/9
$C_{14}FeH_{18}^+$	$[C_5H_5FeC_5H_4C(CH_3)_3]^+$	Fe: Org.Verb.A1-278
–	$[C_5H_5FeC_5H_4C_4H_9\text{-}n]^+$	Fe: Org.Verb.A1-278
$C_{14}FeH_{18}I_3$	$[C_5H_5FeC_5H_4C_4H_9\text{-}n]I_3$	Fe: Org.Verb.A1-278
$C_{14}FeH_{18}N_2$	$C_5H_5FeC_5H_4C(CH_3)NN(CH_3)_2$	Fe: Org.Verb.A2-221
$C_{14}FeH_{18}N_4O_2^{2+}$	$[(CH_3NC)_2Fe(H_2O)_2(NC_5H_4C_5H_4N)]^{2+}$	Fe: Org.Verb.B4-28
$C_{14}FeH_{18}O$	$C_5H_5FeC_5H_4C(CH_3)(C_2H_5)OH$	Fe: Org.Verb.A2-58/9, 62
–	$C_5H_5FeC_5H_4C(CH_3)_2OCH_3$	Fe: Org.Verb.A2-119, 123
–	$C_5H_5FeC_5H_4CH(CH_3)OC_2H_5$	Fe: Org.Verb.A2-117/8
–	$C_5H_5FeC_5H_4CHOHCH(CH_3)_2$	Fe: Org.Verb.A2-43/4, 47
–	$C_5H_5FeC_5H_4CHOHC_3H_7\text{-}n$	Fe: Org.Verb.A2-44
–	$C_5H_5FeC_5H_4CH_2C(CH_3)_2OH$	Fe: Org.Verb.A2-75, 80
–	$C_5H_5FeC_5H_4CH_2OCH(CH_3)_2$	Fe: Org.Verb.A2-113
–	$C_5H_5FeC_5H_4(CH_2)_2CHOHCH_3$	Fe: Org.Verb.A2-75, 80
–	$C_5H_5FeC_5H_4(CH_2)_4OH$	Fe: Org.Verb.A2-3/4, 75, 81
$C_{14}FeH_{18}O_2$	$C_5H_5FeC_5H_4CH(CH_3)O(CH_2)_2OH$	Fe: Org.Verb.A2-117/8
–	$C_5H_5FeC_5H_4CHOH(CH_2)_3OH$	Fe: Org.Verb.A2-87, 90
$C_{14}FeH_{18}O_4$	$CH_3CHCHCH_2CH(CH_2)_5Fe(CO)_4$	Fe: Org.Verb.B4-272, 289
–	$(CH_3)_3CCCC(CH_3)_3Fe(CO)_4$	Fe: Org.Verb.B4-328
		Fe: Org.Verb.C2-38
$C_{14}FeH_{19}N$	$C_5H_5FeC_5H_4CH(CH_3)N(CH_3)_2$	Fe: Org.Verb.A1-353, 359, 361/2, 382
		Fe: Org.Verb.A2-30, 100
$C_{14}FeH_{19}NO$	$FeC_{10}H_8(OCH_3)(CH_2N(CH_3)_2)$	Fe: Org.Verb.A2-109
$C_{14}FeH_{19}N_2O_4^+$	$[C_3H_3(NC_5H_{10})NHC(OC_2H_5)Fe(CO)_3]^+$	Fe: Org.Verb.B5-161
$C_{14}FeH_{19}N_5O_5$	$COFe(NC_5H_5)(C_2HN_2O_2(CH_3)_2)_2$	Fe: Org.Verb.B1-33
$C_{14}FeH_{19}NaO_5$	$Na[CH_3(CH_2)_8C(O)Fe(CO)_4]$	Fe: Org.Verb.B4-151, 161
$C_{14}FeH_{19}O_5^-$	$[C_9H_{19}C(O)Fe(CO)_4]^-$	Fe: Org.Verb.B4-151, 161
$C_{14}FeH_{20}IN$	$[C_5H_5FeC_5H_4CH_2N(CH_3)_3]I$	Fe: Org.Verb.A1-167
$C_{14}FeH_{20}N_2O_2S_4$	$(CO)_2Fe(S_2CNC_5H_{10})_2$	Fe: Org.Verb.B1-114/5

$C_{14}FeH_{20}N_4O_4$ ...... $(CH_2NCH_3)_2CC(NCH_3CH_2)_2Fe(CO)_4$ .... Fe: Org.Verb.B4-321
$C_{14}FeH_{20}N_6$ .... $(C_2H_5NC)_4Fe(CN)_2$ .... Fe: Org.Verb.B4-48/51, 58
$C_{14}FeH_{20}N_6O_2P_2$ .... $Fe(NO)_2(C_2H_4(P(C_2H_4CN)_2)_2)$ .... Fe: Org.Verb.B1-142
$C_{14}FeH_{20}N_6O_5$ ...... $COFe(NC_5H_4NH_2)(C_2HN_2O_2(CH_3)_2)_2$ .... Fe: Org.Verb.B1-33
$C_{14}FeH_{20}Si$ .... $C_5H_5FeC_5H_4CH_2Si(CH_3)_3$ .... Fe: Org.Verb.A2-112
$C_{14}FeH_{21}O_3P$ .... $C_8H_{12}Fe(CO)_3P(CH_3)_3$ .... Fe: Org.Verb.B4-201, 207
$C_{14}FeH_{22}Hg_2I_6N_6$ .... $[(C_2H_5NC)_2(CH_3NC)_4Fe]I_2 \cdot 2\ HgI_2$ .... Fe: Org.Verb.B4-107
$C_{14}FeH_{22}I_2N_6$ .... $[(C_2H_5NC)_2(CH_3NC)_4Fe]I_2$ .... Fe: Org.Verb.B4-107
$C_{14}FeH_{22}N_6^{2+}$ ...... $[(C_2H_5NC)_2(CH_3NC)_4Fe]^{2+}$ .... Fe: Org.Verb.B4-107
$C_{14}FeH_{22}N_8O_6$ ...... $[(C_2H_5NC)_2(CH_3NC)_4Fe][NO_3]_2$ .... Fe: Org.Verb.B4-107
$C_{14}FeH_{23}N_3O_6$ ...... $[(N(CH_3)_2)_3C][C_2H_5OC(O)Fe(CO)_4]$ .... Fe: Org.Verb.B4-154, 163
$C_{14}FeH_{24}N_4O_5$ ...... $[(N(CH_3)_2)_3C][(CH_3)_2NC(O)Fe(CO)_4]$ .... Fe: Org.Verb.B4-155, 164
$C_{14}FeH_{24}N_6O_8S_2$ .... $[(CH_3NC)_6Fe][CH_3OSO_3]_2$ .... Fe: Org.Verb.B4-97
$C_{14}FeH_{24}N_8^{2+}$ ...... $[CH_3C_3HN_2(NHCH_3)_2Fe(CNCH_3)_4]^{2+}$ .... Fe: Org.Verb.B4-355
$C_{14}FeH_{24}O_{10}P_2S_2$ ... $(C_2H_5OCOC_3HS_2)Fe(CO)_2(P(OCH_3)_3)_2$ ..... Fe: Org.Verb.B4-137/8
$C_{14}FeH_{25}N_2O_3P$ ..... $(CH_2NCH_3)_2CFe(CO)_3P(C_2H_5)_3$ .... Fe: Org.Verb.B4-115, 118
$C_{14}FeH_{25}N_2O_3P^+$ .... $[(CH_2NCH_3)_2CFe(CO)_3P(C_2H_5)_3]^+$ .... Fe: Org.Verb.B4-171/2
$C_{14}FeH_{25}N_7^{2+}$ ...... $[CH_3NHCN(CH_3)C(NHC_2H_5)Fe(CNCH_3)_4]^{2+}$. Fe: Org.Verb.B4-352/3
– .... $[C_2H_5N(CNHCH_3)_2Fe(CNCH_3)_4]^{2+}$ .... Fe: Org.Verb.B4-352/3
$C_{14}FeH_{27}O_4PSi_2$ .... $(CO)_4FeP(Si(CH_3)_3)_2C_4H_9$ .... Fe: Org.Verb.B2-85, 91
$C_{14}FeH_{27}O_4PSn_2$ .... $(CO)_4FeP(Sn(CH_3)_3)_2C_4H_9$ .... Fe: Org.Verb.B2-85, 92, 112
$C_{14}FeH_{27}S_7$ .... $[Fe(S_2CSC_4H_9)_2(SC_4H_9)]_2$ .... C: MVol.D4-269
$C_{14}FeH_{30}I_2O_2P_2$ ..... $(CO)_2Fe(P(C_2H_5)_3)_2I_2$ .... Fe: Org.Verb.B1-101/2, 104, 109
$C_{14}FeH_{30}I_2O_8P_2$ ..... $(CO)_2Fe(P(OC_2H_5)_3)_2I_2$ .... Fe: Org.Verb.B1-101/2, 105, 110
$C_{14}FeH_{30}NO_3P_2^+$ .... $[(CO)_2Fe(P(C_2H_5)_3)_2NO]^+$ .... Fe: Org.Verb.B1-130/3
$C_{14}FeH_{30}O_2P_2^+$ ..... $[(CO)_2Fe(P(C_2H_5)_3)_2]^+$ .... Fe: Org.Verb.B1-94
$C_{14}FeH_{30}O_4P_2S$ ..... $(CO)_2Fe(P(C_2H_5)_3)_2SO_2$ .... Fe: Org.Verb.B1-94
$C_{14}FeH_{30}O_4P_2S^+$ .... $[(CO)_2Fe(P(C_2H_5)_3)_2SO_2]^+$ .... Fe: Org.Verb.B1-94
$C_{14}FeH_{33}O_8P_3$ ...... $(CO)_2Fe(P(C_2H_5)_3)(P(OCH_3)_3)_2$ .... Fe: Org.Verb.B1-94
$C_{14}FeH_{36}N_6O_2P_2^+$ ... $[(CO)_2Fe(P(N(CH_3)_2)_3)_2]^+$ .... Fe: Org.Verb.B1-164
$C_{14}FeH_{36}P_4$ .... $CH_2CH_2Fe(P(CH_3)_2CH_2CH_2P(CH_3)_2)_2$ ...... Fe: Org.Verb.B4-180, 183/4
$C_{14}FeH_{38}P_2Si_2^+$ ..... $[Fe(P(CH_3)_2C_2H_4Si(CH_3)_3)_2]^+$ .... Fe: Org.Verb.B1-160
$C_{14}FeMnO_{14}Re$ ..... $(CO)_4FeMnRe(CO)_{10}$ .... Fe: Org.Verb.B2-165
$C_{14}FeMn_2O_{14}$ ...... $(CO)_4Fe(Mn(CO)_5)_2$ .... Fe: Org.Verb.B2-163/4
$C_{14}FeO_{14}Re_2$ .... $(CO)_4Fe(Re(CO)_5)_2$ .... Fe: Org.Verb.B2-165
$C_{14}Fe_2GeH_{23}IN_4O_4$ .. $(C_3H_5Fe(NO)_2)_2Ge(I)C_7H_{10}CH_3$ .... Fe: Org.Verb.C2-165
$C_{14}Fe_2Ge_3H_{18}O_8$ .... $(CO)_8Fe_2(Ge_2(CH_3)_4)(Ge(CH_3)_2)$ .... Fe: Org.Verb.C1-42, 48, 55
$C_{14}Fe_2H_5O_9Rh$ ..... $(CO)_8Fe_2Rh(CO)C_5H_5$ .... Fe: Org.Verb.C1-213
$C_{14}Fe_2H_6K_8N_{12}$ ..... $K_8[(CN)_5FeC_4Fe(CN)_5] \cdot 2\ NH_3$ .... Fe: Org.Verb.C2-66
$C_{14}Fe_2H_6N_2O_6$ ...... $(CO)_3Fe(NCHC_6H_4CHN)Fe(CO)_3$ .... Fe: Org.Verb.C1-132, 140
$C_{14}Fe_2H_6O_6S$ .... $(CHCHC_6H_4S)Fe_2(CO)_6$ .... Fe: Org.Verb.C2-128/9
– .... $C_4H_3(C_4H_3S)Fe_2(CO)_6$ .... Fe: Org.Verb.C2-127
$C_{14}Fe_2H_6O_6S_2$ ...... $(CO)_3FeSCHC(C_6H_5)SFe(CO)_3$ .... Fe: Org.Verb.C1-115/6, 120
– .... $C_4H_3(C_4H_3S)SFe_2(CO)_6$ .... Fe: Org.Verb.C2-126, 127/8
$C_{14}Fe_2H_6O_7S$ .... $(CO)_3Fe(C_6H_3(CH_2O)CHS)Fe(CO)_3$ .... Fe: Org.Verb.C2-108, 114, 118
$C_{14}Fe_2H_8N_2O_6$ ...... $(CO)_3Fe(NHCH_2C_6H_4CHN)Fe(CO)_3$ .... Fe: Org.Verb.C1-134, 142
$C_{14}Fe_2H_8O_7S$ .... $(CO)_6Fe_2(CH_3OC(C_6H_5)S)$ .... Fe: Org.Verb.C1-238/40

Formula	Compound	Reference
$C_{14}Fe_2H_8O_7S$	$(C_6H_4CH(OCH_3)S)Fe_2(CO)_6$	Fe: Org.Verb.C2-112
$C_{14}Fe_2H_9NO_6$	$(CO)_3Fe(C_6H_4CH_2NCH_3)Fe(CO)_3$	Fe: Org.Verb.C2-114, 118/9
$C_{14}Fe_2H_{10}N_2O_8$	$(CO)_4Fe((CH_2)_5CN_2)Fe(CO)_4$	Fe: Org.Verb.C1-34, 37
$C_{14}Fe_2H_{10}O_4$	$(C_5H_5Fe(CO)_2)_2$	Fe: Org.Verb.A2-245
		Fe: Org.Verb.B3-151
$C_{14}Fe_2H_{11}NO_7$	$(C_5H_5N)(CO)_2Fe(CH_3CO)_2Fe(CO)_3$	Fe: Org.Verb.C1-232/4
$C_{14}Fe_2H_{12}N_2NiO_{10}$	$Ni[(CH_3)_2NC(O)Fe(CO)_4]_2$	Fe: Org.Verb.B4-175/6
		Fe: Org.Verb.C1-247
$C_{14}Fe_2H_{12}O_6$	$C_4(CH_3)_4(Fe(CO)_3)_2$	Fe: Org.Verb.B3-150
$C_{14}Fe_2H_{12}O_{10}S_2$	$(CO)_6Fe_2(C_2(COOCH_3)_2)(SCH_3)_2$	Fe: Org.Verb.C2-67/8
$C_{14}Fe_2H_{14}O_6S_2$	$[Fe(SO_3C_6H_4CH_3)_2Fe]_n$	Fe: Org.Verb.B3-187
$C_{14}Fe_2H_{17}NO_7$	$((CH_2)_5NH)(CO)_2Fe(CH_3CO)_2Fe(CO)_3$	Fe: Org.Verb.C1-232/4
$C_{14}Fe_2H_{18}HgO_8Si_2$	$[(CH_3)_3Si(CO)_4Fe]_2Hg$	Fe: Org.Verb.C1-59/60
$C_{14}Fe_2H_{18}I_2O_6S_2$	$I(CO)_3Fe(SC_4H_9)_2Fe(CO)_3I$	Fe: Org.Verb.C1-6/7, 9
$C_{14}Fe_2H_{18}N_2O_6$	$(CO)_3Fe(C_4H_9NNC_4H_9)Fe(CO)_3$	Fe: Org.Verb.C1-149/51, 154
$C_{14}Fe_2H_{18}O_6S_2$	$(CO)_3Fe(SC_4H_9)_2Fe(CO)_3$	Fe: Org.Verb.C1-78/80, 82, 88
$C_{14}Fe_2H_{18}O_8P_2Pt$	$(CO)_8Fe_2Pt(P(CH_3)_3)_2$	Fe: Org.Verb.C1-216/7
$C_{14}Fe_2H_{18}O_8Sn_2$	$(CH_3)_3Sn(CO)_4FeFe(CO)_4Sn(CH_3)_3$	Fe: Org.Verb.C1-72
$C_{14}Fe_2H_{20}N_2O_6$	$(CO)_3Fe(C_4H_9NH)_2Fe(CO)_3$	Fe: Org.Verb.C1-130, 136
$C_{14}Fe_2H_{20}O_6P_2$	$(CO)_3Fe(P(C_2H_5)_2)_2Fe(CO)_3$	Fe: Org.Verb.C1-164/7, 174
$C_{14}Fe_2H_{20}O_6P_4$	$(CO)_3Fe(PC_2H_5)_4Fe(CO)_3$	Fe: Org.Verb.C1-164, 168, 177
$C_{14}Fe_2H_{20}O_8Si_2$	$(HOC)_2Fe_2(CO)_6(Si(CH_3)_3)_2$	Fe: Org.Verb.C2-72
$C_{14}Fe_2H_{23}IN_4O_4Pb$	$(C_3H_5Fe(NO)_2)_2Pb(I)C_7H_{10}CH_3$	Fe: Org.Verb.C2-165
$C_{14}Fe_2H_{23}IN_4O_4Sn$	$(C_3H_5Fe(NO)_2)_2Sn(I)C_7H_{10}CH_3$	Fe: Org.Verb.C2-165
$C_{14}Fe_2H_{33}IO_{12}P_3S_2^+$	$[(CO)_3(P(OCH_3)_3)_3Fe_2I(SCH_3)_2]^+$	Fe: Org.Verb.C1-12, 14, 17
$C_{14}Fe_2H_{33}O_{12}P_3S_2$	$(CO)_3(P(OCH_3)_3)_3Fe_2(SCH_3)_2$	Fe: Org.Verb.C1-111/4
$C_{14}Fe_2K_8N_{10}$	$K_8[(CN)_5FeC_4Fe(CN)_5] \cdot 2\ NH_3$	Fe: Org.Verb.C2-66
$C_{14}Fe_3H_9O_{14}P$	$Fe_3(CO)_{11}P(OCH_3)_3$	Fe: Org.Verb.B2-115
$C_{14}Fe_3H_{24}N_6O_8$	$[Fe(C_2H_8N_2)_3][Fe_2(CO)_8]$	Fe: Org.Verb.C2-8
$C_{14}Fe_3H_{24}O_{16}$	$(Fe_3(CH_3COO)_6OH \cdot OH_2)CH_3CO_2$	Fe: Org.Verb.B3-118
$C_{14}GeH_{27}NiO_3P$	$(CO)_3NiP(C_4H_9\text{-}t)_2Ge(CH_3)_3$	Ni: Org.Verb.1-166, 168/9
$C_{14}GeH_{30}SiSn$	$(CH_3)_3SnC_5H_3(Ge(CH_3)_3)Si(CH_3)_3$	Sn: Org.Verb.2-100
$C_{14}GeH_{30}Sn$	$(C_2H_5)_3SnCCGe(C_2H_5)_3$	Sn: Org.Verb.2-229
$C_{14}GeH_{31}N_3Ti$	$(CH_3)_3GeC_5H_4Ti(N(CH_3)_2)_3$	Ti: Org.Verb.1-172, 175/8
$C_{14}H_6N_2O_8Sn$	$Sn(NC_5H_3(COO)_2)_2 \cdot H_2O$	Sn: MVol.C5-206
$C_{14}H_8MnNO_2^+$	$[Mn(C_6H_4(CO)_2CC_5H_4N)]^+$	Mn: MVol.D1-128
$C_{14}H_8MnN_2O_8$	$Mn(OC_6H_3(NO_2)CHO)_2$	Mn: MVol.D1-53/4
$C_{14}H_8N_2NiO_2$	$(CO)_2Ni(C_{12}H_8N_2)$	Ni: Org.Verb.1-148
$C_{14}H_8N_2O_6U$	$[UO_2(C_2O_4)(C_{12}H_8N_2)] \cdot H_2O$	U: SVol.E1-44/6
$C_{14}H_8O_6Sn$	$SnO(OH)(C_{14}H_7O_4) \cdot 0.5\ C_5H_5N$	Sn: MVol.C5-113
–	$Sn(O_2C_6H_3CHO)_2$	Sn: MVol.C5-81
$C_{14}H_{10}I_2O_4Sn$	$[SnI_2(C_7H_5O_2)_2]$	Sn: MVol.C5-97
–	$SnI_2(OC_6H_4CHO)_2$	Sn: MVol.C5-81
$C_{14}H_{10}IrKN_2O_8$	$K[Ir(C_2O_4)_2(C_5H_5N)_2]$	Ir: SVol.2-35
$C_{14}H_{10}K_2NiO_2P_2$	$K_2[(CO)_2Ni(C_6H_5PPC_6H_5)] \cdot C_4H_8O$	Ni: Org.Verb.1-157/8
$C_{14}H_{10}MnO_4$	$Mn(OC(CH)_5CO)_2$	Mn: MVol.D1-59/60
–	$Mn(OC_6H_4CHO)_2 \cdot n\ H_2O$	Mn: MVol.D1-52/3
$C_{14}H_{10}MnO_{10}S_2$	$Mn(OC_6H_3(SO_3H)CHO)_2$	Mn: MVol.D1-53/4

$C_{14}H_{10}MoO_2S_3$	$(C_5H_5)Mo(CO)_2[S_2CSC_6H_5]$	C: MVol.D4-269
$C_{14}H_{10}N_2Ni_2$	$(C_5H_5Ni)_2NCCCCN$	Ni: Org.Verb.2-356, 358
$C_{14}H_{10}N_2O_3Sn$	$[Sn(OC_6H_4CHNNHCOC_6H_4O)] \cdot NH_3$	Sn: MVol.C5-40
$C_{14}H_{10}NiO_4$	$Ni(O_2CC_6H_5)_2$	Ni: Org.Verb.2-294
$C_{14}H_{10}O_2S_3W$	$(C_5H_5)W(CO)_2[S_2CSC_6H_5]$	C: MVol.D4-269
$C_{14}H_{12}K_2NiO_2P_2$	$K_2[(CO)_2Ni(P(H)C_6H_5)_2] \cdot 2\ C_4H_8O$	Ni: Org.Verb.1-157/8
$C_{14}H_{12}MnN_2O_4$	$Mn(OC_6H_3(CH_3)NO)_2$	Mn: MVol.D1-40
$C_{14}H_{12}MnN_2O_6$	$Mn(OC_6H_3(OCH_3)NO)_2$	Mn: MVol.D1-40
$C_{14}H_{12}MnN_2O_8$	$Mn(OC_6H_3(OCH_3)NO_2)_2$	Mn: MVol.D1-41
$C_{14}H_{12}N_2NpO_6^-$	$NpO_2(NC_5H_4CH_2COO)_2^-$	Np: TrU.D1-156
$C_{14}H_{12}N_2O_2Sn$	$Sn(C_6H_5CHNO)_2$	Sn: MVol.C5-41/2
$C_{14}H_{12}N_2O_4Sn$	$Sn(C_6H_5CONHO)_2$	Sn: MVol.C5-42/3
–	$Sn(NH_2C_6H_4COO)_2$	Sn: MVol.C5-32
$C_{14}H_{12}N_2O_5U$	$[UO_2(OH)(CH_3COO)(C_{12}H_8N_2)]$	U: SVol.E1-44/6
$C_{14}H_{12}N_4O_8U$	$[UO_2(NO_3)_2((CH_3)_2C_{12}H_6N_2)]$	U: SVol.E1-44/6
$C_{14}H_{12}NiO_3$	$Ni(C_5H_5)_2 \cdot C_4H_2O_3$	Ni: Org.Verb.2-208
$C_{14}H_{12}OSn$	$C_2H_4(C_6H_4)_2SnO$	Sn: Org.Verb.6-209
$C_{14}H_{12}O_2Sn$	$[C_6H_5CHC(C_6H_5)SnOOH]_x$	Sn: Org.Verb.6-264
$C_{14}H_{12}S_4Sn$	$[Sn(CH_3C_6H_3S_2)_2]_x$	Sn: MVol.C6-86
$C_{14}H_{13}N_3Ni$	$NCC(CH_3)CH_2Ni(NC_5H_4C_5H_4N)$	Ni: Org.Verb.1-371, 373
$C_{14}H_{13}N_3O_3Sn$	$[Sn(OC_6H_4CHNNHCOC_6H_4O)] \cdot NH_3$	Sn: MVol.C5-40
$C_{14}H_{13}NiO_2$	$C_5H_5Ni(CO)_2C_7H_8$	Ni: Org.Verb.2-174
$C_{14}H_{13}Sn^+$	$(C_6H_5)_2SnCHCH_2^+$	Sn: Org.Verb.2-386
$C_{14}H_{14}I_2Sn$	$(CH_3)_2Sn(C_6H_4I)_2$	Sn: Org.Verb.3-11
–	$(C_6H_5CH_2)_2SnI_2$	Sn: Org.Verb.5-146
$C_{14}H_{14}I_6Ni_2O_2$	$(CH_3CHCHCH_2NiI)_2 \cdot C_6I_4O_2$	Ni: Org.Verb.2-301
$C_{14}H_{14}Li_2Ni$	$C_2H_4Ni(LiC_6H_5)_2 \cdot (C_2H_5)_2O$	Ni: Org.Verb.1-341
$C_{14}H_{14}N_2NiO_2$	$CH_3COOCHCH_2Ni(NC_5H_4C_5H_4N)$	Ni: Org.Verb.1-368, 369
–	$CH_3OOCCHCH_2Ni(NC_5H_4C_5H_4N)$	Ni: Org.Verb.1-368, 369
$C_{14}H_{14}N_2O_2S_2Sn$	$(CH_2CH)_2Sn(SC_5H_4NO)_2$	Sn: Org.Verb.6-145
$C_{14}H_{14}N_2O_6U$	$[UO_2(CH_3COO)_2(C_{10}H_8N_2)]$	U: SVol.E1-39/42
$C_{14}H_{14}N_2O_7SU$	$UO_2(NC_5H_4COO)_2 \cdot (CH_3)_2SO$	U: SVol.E1-205, 208
$C_{14}H_{14}N_2O_9SU$	$UO_2(OCOC_5H_4NO)_2 \cdot (CH_3)_2SO$	U: SVol.E1-205, 208
$C_{14}H_{14}N_2O_9U$	$UO_2(O(CH_2COO)_2) \cdot 2\ C_5H_5NO$	U: SVol.E1-135, 141/3
$C_{14}H_{14}N_4S_3$	$[C_6H_5NHNHC(S)]_2S$	C: MVol.D6-171
$C_{14}H_{14}N_8O_{14}U$	$UO_2(NO_3)_2 \cdot 2\ NO_2C_6H_4CON_2H_3 \cdot n\ H_2O$	U: SVol.E1-114/6
$C_{14}H_{14}Na_2O_8S_2Sn$	$(C_2H_5)_2Sn(O_2C_{10}H_4(SO_3Na)_2)$	Sn: Org.Verb.6-58
$C_{14}H_{14}Ni$	$(C_6H_5CH_2)_2Ni$	Ni: Org.Verb.1-71
$C_{14}H_{14}NiO_2$	$C_5H_5NiC_7H_6(COOCH_3)$	Ni: Org.Verb.2-184/5
$C_{14}H_{14}NiO_2P_2$	$(CO)_2Ni(P(C_6H_5)H_2)_2$	Ni: Org.Verb.1-132
$C_{14}H_{14}NiO_6S_2$	$Ni(O_3SC_6H_4CH_3)_2$	Ni: Org.Verb.2-284, 297
$C_{14}H_{14}NiS_2$	$Ni(SCH_2C_6H_5)_2$	Ni: Org.Verb.2-90
$C_{14}H_{14}Ni_2O_2$	$(CH_3C_5H_4NiCO)_2$	Ni: Org.Verb.2-349
–	$(C_5H_5Ni)_2NCCCOOCH_3$	Ni: Org.Verb.2-356, 359
$C_{14}H_{14}OSn$	$[(CH_3C_6H_4)_2SnO]_x$	Sn: Org.Verb.6-180/2
–	$(CH_3)_2Sn(C_6H_4OC_6H_4)$	Sn: Org.Verb.3-119, 122
–	$[(C_6H_5CH_2)_2SnO]_x$	Sn: Org.Verb.6-123
$C_{14}H_{14}O_2SSn$	$(CH_3)_2Sn(C_6H_4S(O_2)C_6H_4)$	Sn: Org.Verb.3-120, 123
$C_{14}H_{14}O_2Sn$	$[Sn(C_6H_5)_2OCH_2CH_2O]_x$	Sn: Org.Verb.6-162/4
$C_{14}H_{14}O_3Sn$	$[(CH_3OC_6H_4)_2SnO]_x$	Sn: Org.Verb.6-183

Formula	Compound	Reference
$C_{14}H_{14}O_4S_2Sn$	$Sn(CH_3C_6H_4SO_2)_2$	Sn: MVol.C5-45/6
$C_{14}H_{14}SSn$	$(CH_3)_2Sn(C_6H_4SC_6H_4)$	Sn: Org.Verb.3-120, 123
$C_{14}H_{14}Se_2$	$C_6H_5CH_2SeSeCH_2C_6H_5$	Se: SVol.A1-186
$C_{14}H_{14}Sn$	$(CH_3)_2Sn(C_6H_4C_6H_4)$	Sn: Org.Verb.3-113
–	$(C_6H_4CH_2CH_2C_6H_4)SnH_2$	Sn: Org.Verb.4-120
$C_{14}H_{14}Ti$	$(CH_3C_6H_4)_2Ti$	Ti: Org.Verb.1-68
–	$(C_6H_5CH_2)_2Ti$	Ti: Org.Verb.1-67
$C_{14}H_{15}IN_2NiS$	$C_3H_5Ni(SC(NH_2)NHC_{10}H_7)I$	Ni: Org.Verb.2-281
$C_{14}H_{15}NNiO$	$C_6H_5(CH_3)_2CC_5H_4NiNO$	Ni: Org.Verb.2-113
$C_{14}H_{15}N_3O_8U$	$[UO_2NO_3(HOC_6H_4COO)(CH_3)_2NC_5H_4N]$	U: SVol.E1-36
$C_{14}H_{15}N_4O_9PS$	$(CH_3)_2NSO_2NHPO(OC_6H_4NO_2)_2$	S: S-N-Verb.1-184/5
$C_{14}H_{15}NiO_6P$	$(CO)_3NiP(OC_6H_5)(OCH_2)_2C(CH_3)_2$	Ni: Org.Verb.1-181
$C_{14}H_{15}Sn^+$	$(C_6H_5)_2SnC_2H_5{}^+$	Sn: Org.Verb.2-355
		Sn: Org.Verb.3-25
$C_{14}H_{16}N_2NiO_2S_4$	$Ni[S_2COCH_3]_2(C_5H_5N)_2$	C: MVol.D4-260
$C_{14}H_{16}N_2Ni_2$	$(C_5H_5NiCNCH_3)_2$	Ni: Org.Verb.2-352/3
$C_{14}H_{16}N_2O_2S$	$(C_6H_5CH_2NH)_2SO_2$	S: S-N-Verb.1-169
$C_{14}H_{16}N_2O_2S_4Zn$	$Zn[S_2COCH_3]_2(C_5H_5N)_2$	C: MVol.D4-257
$C_{14}H_{16}N_2S_4Sn$	$Sn(SCH_2CH_2S)_2 \cdot NC_5H_4C_5H_4N$	Sn: MVol.C6-82/3
$C_{14}H_{16}N_6O_{10}U$	$UO_2(NO_3)_2 \cdot 2\ C_6H_5CONHNH_2$	U: SVol.E1-114/6
–	$UO_2(NO_3)_2 \cdot 2\ C_6H_5CONHNH_2 \cdot 2\ H_2O$	U: SVol.E1-114/6
$C_{14}H_{16}N_6O_{12}U$	$UO_2(NO_3)_2 \cdot 2\ HOC_6H_4CON_2H_3 \cdot 2\ H_2O$	U: SVol.E1-114/6
$C_{14}H_{16}Ni_2$	$C_3H_5NiC_8H_6NiC_3H_5$	Ni: Org.Verb.2-369/70
–	$(C_5H_5Ni)_2CHCC_2H_5$	Ni: Org.Verb.2-356, 359
–	$(C_5H_5Ni)_2(CH_3)CCCH_3$	Ni: Org.Verb.2-355, 357
$C_{14}H_{16}OSn$	$(C_6H_5)_2(C_2H_5)SnOH$	Sn: Org.Verb.5-216
$C_{14}H_{16}O_2S_4Ti$	$(C_5H_5)_2Ti[S_2COCH_3]_2$	C: MVol.D4-258
$C_{14}H_{16}O_2Sn$	$(CH_3)_2Sn(C_6H_4OH)_2$	Sn: Org.Verb.3-11
$C_{14}H_{16}O_4Sn$	$(C_5H_5)_2Sn(OCOCH_3)_2$	Sn: Org.Verb.6-143
$C_{14}H_{16}Se_2Sn$	$(CH_3)_2Sn(SeC_6H_5)_2$	Sn: Org.Verb.6-31
$C_{14}H_{16}Sn$	$(CH_3C_6H_4)_2SnH_2$	Sn: Org.Verb.4-118
–	$(CH_3)_2Sn(C_6H_5)_2$	Sn: Org.Verb.3-4
–	$(C_6H_5CH_2)_2SnH_2$	Sn: Org.Verb.4-113
		Sn: Org.Verb.6-123
$C_{14}H_{16}Ti$	$(CH_3)_2(C_6H_5)_2Ti$	Ti: Org.Verb.1-93, 106
$C_{14}H_{17}N_2O_5PS$	$(CH_3)_2NSO_2NHPO(OC_6H_5)_2$	S: S-N-Verb.1-184/5
$C_{14}H_{17}N_3S_4Sn$	$Sn(SCH_2CH_2S)_2 \cdot (NC_5H_4)_2NH$	Sn: MVol.C6-82/3
$C_{14}H_{17}S_3V$	$(C_5H_5)_2V[S_2CSC_3H_7]$	C: MVol.D4-269
$C_{14}H_{18}MnN_2O_4$	$[Mn(CH(COCH_3)_2)_2(N(CHCH)_2N)]_n$	Mn: MVol.D1-81/2
$C_{14}H_{18}MnO_4$	$[Mn(CH_2C(CH_3)COCHCOCH_3)_2]$	Mn: MVol.D1-127
$C_{14}H_{18}N_2Ni$	$(C_2H_5)_2Ni(NC_5H_4C_5H_4N)$	Ni: Org.Verb.1-70, 73/6
$C_{14}H_{18}N_2O_8SU$	$UO_2SO_4 \cdot 2\ (CH_3)_2C_5H_3NO$	U: SVol.E1-133, 140, 142/3
$C_{14}H_{18}N_2O_8Sn^{2-}$	$[Sn(C_6H_{10}(N(CH_2COO)_2)_2)]^{2-}$	Sn: MVol.C5-38
$C_{14}H_{18}N_2S_4Sn$	$Sn(SCH_2CH_2S)_2 \cdot 2\ C_5H_5N$	Sn: MVol.C6-82/3
$C_{14}H_{18}N_3NpO_{10}{}^-$	$Np(CH_2COO)N(C_2H_4N(CH_2COO)_2)_2{}^-$	Np: TrU.D1-166
$C_{14}H_{18}N_3O_{10}Sn^{3-}$	$[Sn(N(C_2H_4N(CH_2COO)_2)_2CH_2COO)]^{3-}$	Sn: MVol.C5-38
$C_{14}H_{18}N_4NiO_2P_2$	$(CO)_2Ni(P(C_2H_4CN)_2H)_2$	Ni: Org.Verb.1-129, 130
$C_{14}H_{18}N_4NiS_4$	$Ni[S_2CNH_2]_2(CH_3C_5H_4N)_2$	C: MVol.D6-140
$C_{14}H_{18}N_4O_{10}U$	$UO_2(NO_3)_2 \cdot 2\ (CH_3)_2C_5H_3NO$	U: SVol.E1-133, 139, 142/3
$C_{14}H_{18}N_6O_2Sn$	$(C_4H_9)_2Sn[ONC(CN)_2]_2$	Sn: Org.Verb.6-89, 95

$C_{14}H_{18}Ni$ .... $Ni(C_5H_4C_2H_5)_2$ .... Ni: Org.Verb.2-215, 217
$C_{14}H_{18}NiO$ .... $C_5H_5Ni(CO)C_8H_{13}$ .... Ni: Org.Verb.2-183
$C_{14}H_{18}NiO_8P_2$ ...... $(CO)_2Ni(PO_3C_6H_9)_2$ .... Ni: Org.Verb.1-139, 140
$C_{14}H_{18}Sn$ .... $(CH_3)_3SnCH_2C_{10}H_7$ .... Sn: Org.Verb.2-28
– .... $(CH_3)_3SnC_{11}H_9$ .... Sn: Org.Verb.2-104
– .... $(C_2H_5)_2Sn(CHCH)_2C_6H_4$ .... Sn: Org.Verb.3-111, 113
$C_{14}H_{19}NNi$ .... $(CH_3)_2C_3H_3Ni(C_8H_7N)CH_3$ .... Ni: Org.Verb.2-35
$C_{14}H_{19}N_2O_8Sn^-$ ..... $[Sn(C_6H_{10}(N(CH_2COO)_2)_2H)]^-$ .... Sn: MVol. C5-38
$C_{14}H_{19}N_3O_{10}Sn^{2-}$ .... $[Sn(N(C_2H_4N(CH_2COO)_2)_2CH_2COOH)]^{2-}$ .... Sn: MVol.C5-38
$C_{14}H_{19}Ni^+$ .... $[CH_3C_5H_4NiC_8H_{12}]^+$ .... Ni: Org.Verb.2-189/90
$C_{14}H_{20}N_2Ni$ .... $[Ni(CH(CN)C_2H_4CHCHCH_3)_2]_x$ .... Ni: Org.Verb.1-389
$C_{14}H_{20}N_2NiO_3$ ...... $((CHCO)_2O)Ni(CNC_4H_9\text{-}t)_2$ .... Ni: Org.Verb.1-382, 383
$C_{14}H_{20}N_2O_8Sn$ ...... $[Sn(C_6H_{10}(N(CH_2COO)_2)_2H_2)]$ .... Sn: MVol.C5-38
$C_{14}H_{20}N_2O_{10}Sn^{2-}$ .... $[Sn(C_2H_4(OC_2H_4N(CH_2COO)_2)_2)]^{2-}$ .... Sn: MVol.C5-38
$C_{14}H_{20}N_2O_{13}U$ ...... $UO_2(NO_3)_2 \cdot C_6H_4(OC_2H_4OC_2H_4)_2O \cdot 5\ H_2O$ U: SVol.E1-73, 78
$C_{14}H_{20}N_2Ti$ .... $(CH_3)_4Ti \cdot NC_5H_4C_5H_4N$ .... Ti: Org.Verb.1-114, 118
$C_{14}H_{20}N_3O_{10}Sn^-$ .... $[SnH_2(N(C_2H_4N(CH_2COO)_2)_2CH_2COO)]^-$ .... Sn: MVol.C5-38
$C_{14}H_{20}N_4Ni$ .... $(NCCHCHCN)Ni(CNC_4H_9\text{-}t)_2$ .... Ni: Org.Verb.1-382, 383
$C_{14}H_{20}N_4Ni_2O_2$ ..... $[(CO)NiNCN(CH_2)_5]_2$ .... Ni: Org.Verb.2-256/7
$C_{14}H_{20}N_4S_4Sn$ ...... $Sn(SCH_2CH_2S)_2 \cdot 2\ NC_5H_4NH_2$ .... Sn: MVol.C6-82/3
$C_{14}H_{20}N_6O_8U$ .... $UO_2(NO_3)_2 \cdot 2\ (NH_2)_2C_6H_3CH_3$ .... U: SVol.E1-23/4
$C_{14}H_{20}N_7PaS_6$ ...... $[C_8H_{17}NH_3][Pa(SCN)_6]$ .... Pa: SVol.2-166
$C_{14}H_{20}Na_2O_8S_2Sn$ ... $(C_4H_9)_2Sn(O_2C_6H_2(SO_3Na)_2)$ .... Sn: Org.Verb.6-89, 94
$C_{14}H_{20}Ni$ .... $CH_3C_5H_4NiC_8H_{13}$ .... Ni: Org.Verb.2-183/4
$C_{14}H_{20}NiO$ .... $C_5H_5NiC_8H_{12}OCH_3$ .... Ni: Org.Verb.2-184
$C_{14}H_{20}Ni_2S_2$ .... $(C_5H_5NiSC_2H_5)_2$ .... Ni: Org.Verb.2-338, 339
$C_{14}H_{20}O_4S_8$ .... $S_2(C(S)OC_5H_{10}OCS_2)_2$ .... C: MVol.D4-251
$C_{14}H_{20}O_4Sn$ .... $(C_4H_9)(C_6H_5)Sn(OCOCH_3)_2$ .... Sn: Org.Verb.5-50
$C_{14}H_{20}Sn$ .... $(CH_3)_3SnC_9H_5(CH_3)_2$ .... Sn: Org.Verb.2-103, 106
– .... $(C_2H_5)_2Sn(C_5H_5)_2$ .... Sn: Org.Verb.3-28
– .... $(C_2H_5)_3SnCCC_6H_5$ .... Sn: Org.Verb.2-219/20
– .... $(C_3H_7)_2Sn(CCCHCH_2)_2$ .... Sn: Org.Verb.3-37
$C_{14}H_{21}MnN_{10}O_2$ ..... $[Mn(C_6H_5COCHCOCH_3)(C_2H_6N_5)_2]$ .... Mn: MVol.D1-119
$C_{14}H_{21}NNi$ .... $(CH_3)_2C_3H_3Ni(C_8H_9N)CH_3$ .... Ni: Org.Verb.2-35
$C_{14}H_{21}N_2O_{10}Sn^-$ .... $[Sn(C_2H_4(OC_2H_4N(CH_2COO)_2)_2H)]^-$ .... Sn: MVol.C5-38
$C_{14}H_{21}O_{14}Sn_3^-$ ...... $Sn_3(CH_3COO)_7^-$ .... Sn: MVol.C2-219/20
$C_{14}H_{22}MnO_3$ .... $[MnOH(OC_6H_2(t\text{-}C_4H_9)_2OH)]$ .... Mn: MVol.D1-45/6
$C_{14}H_{22}MnO_4$ .... $[Mn(C_2H_5COCHCOC_2H_5)_2]$ .... Mn: MVol.D1-125
$C_{14}H_{22}MnO_8$ .... $[Mn(CH(COCH_2OCH_3)_2)_2]$ .... Mn: MVol.D1-112
$C_{14}H_{22}N_2NiO_2$ ...... $[Ni(O_2)(CNC_6H_{11})_2]_x$ .... Ni: Org.Verb.2-397
$C_{14}H_{22}N_2O_{10}Sn$ ..... $[Sn(C_2H_4(OC_2H_4N(CH_2COO)_2)_2H_2)]$ .... Sn: MVol.C5-38
$C_{14}H_{22}N_2Ti$ .... $(CH_3)_4Ti \cdot 2\ C_5H_5N$ .... Ti: Org.Verb.1-114, 117
$C_{14}H_{22}Ni$ .... $C_3H_5NiC_4(CH_3)_4C_3H_5$ .... Ni: Org.Verb.2-60
– .... $Ni(C_7H_{11})_2$ .... Ni: Org.Verb.2-67
$C_{14}H_{22}NiO_8P_2$ ...... $(CO)_2Ni(P(OCH_2)_3CC_2H_5)_2$ .... Ni: Org.Verb.1-139/40, 241
$C_{14}H_{22}Ni_2P_2$ .... $(C_5H_5NiP(CH_3)_2)_2$ .... Ni: Org.Verb.2-338, 340
$C_{14}H_{22}OSn$ .... $(C_2H_5)_3SnCH_2COC_6H_5$ .... Sn: Org.Verb.2-173, 178
$C_{14}H_{22}O_2Sn$ .... $(C_4H_9)_2Sn(OC_6H_4O)$ .... Sn: Org.Verb.6-89, 93
$C_{14}H_{22}O_8U$ .... $UO_2(CH(COCH_3)_2)_2 \cdot C_4H_8O_2$ .... U: SVol.E1-71, 77
$C_{14}H_{22}SiSn$ .... $(CH_3)_3SnC_6H_4CCSi(CH_3)_3$ .... Sn: Org.Verb.2-129/30

Formula	Compound	Reference
$C_{14}H_{22}Sn$	$(CH_3)_2Sn(C_6H_5)(C_6H_{11})$	Sn: Org.Verb.3–80
–	$(C_2H_5)_3SnCHCHC_6H_5$	Sn: Org.Verb.2-198, 204
–	$(C_2H_5)_3SnC_6H_4CHCH_2$	Sn: Org.Verb.2-241, 243
–	$(C_3H_7)_2(C_6H_5CHCH)SnH$	Sn: Org.Verb.4-91
–	$C_6H_4(CH_2CH_2)_2Sn(C_2H_5)_2$	Sn: Org.Verb.3-113
$C_{14}H_{23}MoN_3O_3Ti$	$C_5H_5Ti(N(CH_3)_2)_3Mo(CO)_3$	Ti: Org.Verb.1-203
$C_{14}H_{23}NNi$	$(CH_3)_2C_3H_3Ni(NC_5H_2(CH_3)_3)CH_3$	Ni: Org.Verb.2-35, 37
–	$(CH_3)_2C_3H_3Ni(NH_2CH(C_6H_5)CH_3)CH_3$	Ni: Org.Verb.2-35, 37
$C_{14}H_{23}N_3O_3TiW$	$C_5H_5Ti(N(CH_3)_2)_3W(CO)_3$	Ti: Org.Verb.1-203
$C_{14}H_{23}NiP$	$(CH_3)_2C_3H_3Ni(P(CH_3)_2C_6H_5)CH_3$	Ni: Org.Verb.2-36, 38
$C_{14}H_{24}MnN_2O_3Se_2$	$Mn(NCSe)_2 \cdot 3\ C_4H_8O$	Mn: MVol.D1-144
$C_{14}H_{24}N_2NiO_4$	$Ni(CH(COCH_3)_2)_2 \cdot 2\ C_2H_4NH$	Ni: Org.Verb.2-7, 282
$C_{14}H_{24}N_2O_5S_2U$	$UO_2(NCS)_2 \cdot 3\ (CH_2)_4O$	U: SVol.E1-71, 77
$C_{14}H_{24}N_2O_8S_2U$	$UO_2(NCS)_2 \cdot 3\ C_4H_8O_2$	U: SVol.E1-71, 77
$C_{14}H_{24}N_2O_8Se_2U$	$UO_2(NCSe)_2 \cdot 3\ C_4H_8O_2$	U: SVol.E1-71, 77
$C_{14}H_{24}N_2Sn$	$[(C_4H_9)_2SnNHC_6H_4NH]_x$	Sn: Org.Verb.6-89, 97
$C_{14}H_{24}N_4NiO_2$	$Ni(O_2)(CNC_4H_9)_2(CNCH_3)_2$	Ni: Org.Verb.2-397
$C_{14}H_{24}N_8O_5S_2U$	$UO_2(NCS)_2 \cdot 3\ CO(NHCH_2)_2CH_2$	U: SVol.E1-92/3
$C_{14}H_{24}N_8O_6Sn$	$[(CH_3)_4N]_2[Sn(NCO)_6]$	Sn: MVol.C3-126
$C_{14}H_{24}N_{10}O_2S_2U$	$UO_2(NCS)_2 \cdot 2\ (CH_2)_6N_4$	U: SVol.E1-23/6
$C_{14}H_{24}Ni_2O_4$	$(CH_3CHCHCHCH_3NiO_2CCH_3)_2$	Ni: Org.Verb.2-314
–	$(CH_3CO)_2CHNi(C_2H_5)_2NiCH(COCH_3)_2$	Ni: Org.Verb.2-268
$C_{14}H_{24}OSn$	$(C_2H_5)_3SnC_6H_4CH(OH)CH_3$	Sn: Org.Verb.2-241
$C_{14}H_{24}O_2Sn$	$(CH_3)_3SnC_6H_4CH(OC_2H_5)_2$	Sn: Org.Verb.2-129/30
–	$(C_2H_5)_3SnCCCH_2OCH(CH_3)OCH_2CCH$	Sn: Org.Verb.2-228
$C_{14}H_{24}O_3SSn$	$(CH_3)_3Sn(CH_2)_4OSO_2C_6H_4CH_3$	Sn: Org.Verb.2-50/1
$C_{14}H_{24}O_7U$	$UO_2(CH(COCH_3)_2)_2 \cdot C_4H_9OH$	U: SVol.E1-60, 64/5
$C_{14}H_{24}Sn$	$(CH_2CH)_2Sn[CH_2CH(CH_3)CHCH_2]_2$	Sn: Org.Verb.3-72
–	$(CH_3)(i\text{-}C_3H_7)Sn(C_4H_9)(C_6H_5)$	Sn: Org.Verb.3-97
–	$(CH_3)(i\text{-}C_3H_7)Sn(i\text{-}C_4H_9)(C_6H_5)$	Sn: Org.Verb.3-97
–	$(CH_3)(i\text{-}C_3H_7)Sn(s\text{-}C_4H_9)(C_6H_5)$	Sn: Org.Verb.3-97
–	$(CH_3)_2Sn(i\text{-}C_4H_9)(CH(CH_3)C_6H_5)$	Sn: Org.Verb.3-75, 78
–	$(CH_3)_2Sn(s\text{-}C_4H_9)(CH(CH_3)C_6H_5)$	Sn: Org.Verb.3-75, 79
–	$(CH_3)_2Sn(t\text{-}C_4H_9)(CH(CH_3)C_6H_5)$	Sn: Org.Verb.3-75, 79
–	$(CH_3)_2Sn(s\text{-}C_4H_9)(CH_2CH_2C_6H_5)$	Sn: Org.Verb.3-75, 79
–	$(CH_3)_3SnC_6H_4CH_2C(CH_3)_3$	Sn: Org.Verb.2-129
–	$(C_2H_5)_2Sn(CCC_3H_7)_2$	Sn: Org.Verb.3-28, 33
–	$(C_2H_5)_3SnCCC_6H_9$	Sn: Org.Verb.2-228
–	$(C_2H_5)_3SnCH_2CH_2C_6H_5$	Sn: Org.Verb.2-173
–	$(C_4H_9)_2Sn(CCCH_3)_2$	Sn: Org.Verb.3-43
–	$(C_4H_9)_2Sn(CH_2CCH)_2$	Sn: Org.Verb.3-43
$C_{14}H_{25}NSn$	$(C_2H_5)_3SnC_6H_4N(CH_3)_2$	Sn: Org.Verb.2-241, 243
$C_{14}H_{25}N_3OS$	$C_6H_5NSO(NHC(CH_3)_3)_2$	S: S-N-Verb. 1-206
$C_{14}H_{26}MnO_6$	$[Mn(CH(COCH_3)_2)_2(C_2H_5OH)_2]$	Mn: MVol.D1-75
$C_{14}H_{26}N_2S_4Sn$	$(CH_2CH)_2Sn[S_2CN(C_2H_5)_2]_2$	Sn: Org.Verb.6-145
$C_{14}H_{26}N_2SiSn$	$(CH_3)_3SiNC(C_6H_5)N(CH_3)Sn(CH_3)_3$	Sn: Org.Verb.5-84
$C_{14}H_{26}N_2Sn$	$(C_4H_9)_2Sn(CH_2CH_2CN)_2$	Sn: Org.Verb.3-44
–	$(i\text{-}C_4H_9)_2Sn(CH_2CH_2CN)_2$	Sn: Org.Verb.3-69
$C_{14}H_{26}N_4O_2S_4U$	$UO_2(S_2CN(C_2H_5)_2)_2 \cdot CH_3C_3H_3N_2$	U: SVol.E1-38/9
$C_{14}H_{26}NiO_2S_4$	$Ni[S_2COC_6H_{13}]_2$	C: MVol.D4-260

Formula	Compound	Reference
$C_{14}H_{26}NiO_8P_2$	$(CO)_2Ni[P((OCH_2)_2C(CH_3)_2)OCH_3]_2$	Ni: Org.Verb.1-139, 140
$C_{14}H_{26}OSn$	$(C_2H_5)_3SnCCCHCHOC_4H_9$	Sn: Org.Verb.2-229
–	$(C_2H_5)_3SnCHCHC(OH)C(CH_2)_4$	Sn: Org.Verb.2-199
$C_{14}H_{26}O_2PdS_4$	$Pd[S_2COC_6H_{13}]_2$	C: MVol.D4-261
$C_{14}H_{26}O_2S_4$	$C_6H_{13}OC(S)SSC(S)OC_6H_{13}$	C: MVol.D4-251
$C_{14}H_{26}O_3Ti$	$C_5H_5Ti(OC_3H_7\text{-n})_3$	Ti: Org.Verb.1-157, 161
–	$C_5H_5Ti(OC_3H_7\text{-i})_3$	Ti: Org.Verb.1-157, 161
$C_{14}H_{26}O_4S_2$	$C_6H_{13}OC(O)SS(O)COC_6H_{13}$	C: MVol.D4-236
$C_{14}H_{26}O_4S_4U$	$UO_2[S_2COC_6H_{13}]_2$	C: MVol.D4-261
$C_{14}H_{26}O_4Sn$	$(C_2H_5)_3SnC(COOC_2H_5)CHCOOC_2H_5$	Sn: Org.Verb.2-199, 204
–	$[Sn(C_4H_9)_2OCO(CH_2)_4COO]_x$	Sn: Org.Verb.6-89, 94/5
$C_{14}H_{26}O_5Sn$	$(C_2H_5)_3SnC_2(COOCH_3)_2OC_2H_5$	Sn: Org.Verb.2-199
$C_{14}H_{26}SSn$	$(C_2H_5)_3SnCCCHCHSC_4H_9$	Sn: Org.Verb.2-229, 235
$C_{14}H_{26}Sn$	$(CH_2CH)_3SnC_8H_{17}$	Sn: Org.Verb.2-347
–	$(C_2H_5)_3SnC(CH_2)C_6H_9$	Sn: Org.Verb.2-198
–	$(C_2H_5)_3SnCHCC(CH_2)_5$	Sn: Org.Verb.2-198
–	$(C_2H_5)_3SnCHCHC_6H_9$	Sn: Org.Verb.2-199
–	$(C_2H_5)_3SnC_6H_9CCH_2$	Sn: Org.Verb.2-181/2
–	$(C_4H_9)_2SnC_6H_8$	Sn: Org.Verb.3-110
$C_{14}H_{27}Ir_3O_{18}$	$[Ir_3O(CH_3COO)_6(H_2O)_3]CH_3COO$	Ir: SVol.2-32, 123
$C_{14}H_{27}NSn$	$(C_2H_5)_3SnCCCHCHN(C_2H_5)_2$	Sn: Org.Verb.2-229
$C_{14}H_{27}NiO_3P$	$C_8H_{12}NiP(OC_2H_5)_3$	Ni: Org.Verb.2-226/9, 231
$C_{14}H_{27}NiO_3PSi$	$(CO)_3NiP(C_4H_9\text{-t})_2Si(CH_3)_3$	Ni: Org.Verb.1-166, 168/9
$C_{14}H_{27}NiO_3PSn$	$(CO)_3NiP(C_4H_9\text{-t})_2Sn(CH_3)_3$	Ni: Org.Verb.1-166, 169
$C_{14}H_{27}O_7PU$	$UO_2C_2O_4 \cdot (C_4H_9)_3PO$	U: SVol.E1-151, 159
$C_{14}H_{28}N_2O_2S$	$CH_3(C_6H_{11})NSO_2NCH_3(C_6H_{11})$	S: S-N-Verb.1-173
$C_{14}H_{28}N_4O_4S_2$	$(CH_2)_5NSO_2N(CH_2CH_2)_2NSO_2N(CH_2)_5$	S: S-N-Verb.1-174
$C_{14}H_{28}NiOP_2$	$CH_3Ni(P(CH_3)_3)_2OC_6H_4CH_3$	Ni: Org.Verb.1-12, 19
$C_{14}H_{28}Ni_2S_8$	$[Ni(S_2CSC_3H_7)(SC_3H_7)]_2$	C: MVol.D4-270
$C_{14}H_{28}OSn$	$(C_2H_5)_3SnC(CCHCH_3)CH(OH)C_3H_7$	Sn: Org.Verb.2-199
–	$(C_2H_5)_3SnC(CHCH_2)CHC(OH)(CH_3)C_2H_5$	Sn: Org.Verb.2-199
–	$(C_2H_5)_3SnC(C(OCH_3)(CH_3)_2)CCHCH_3$	Sn: Org.Verb.2-199
–	$(C_2H_5)_3SnC(C(OH)(CH_3)C_2H_5)CCHCH_3$	Sn: Org.Verb.2-199
–	$(C_2H_5)_3SnCH_2CH_2CCC(CH_3)_2OCH_3$	Sn: Org.Verb.2-229
–	$(C_2H_5)_3SnCH_2CH_2CCCH(OH)C_3H_7$	Sn: Org.Verb.2-229
–	$(C_2H_5)_3SnCH_2CH_2CCC(OH)(CH_3)C_2H_5$	Sn: Org.Verb.2-229
$C_{14}H_{28}O_2Sn$	$(C_3H_7)_3SnC(COOCH_3)CHCH_3$	Sn: Org.Verb.2-261
$C_{14}H_{28}O_3Sn$	$(C_2H_5)_3Sn(CH_2)_2CH(COCH_3)COOC_2H_5$	Sn: Org.Verb.2-173, 178
$C_{14}H_{28}O_3Ti$	$CHCTi(OC_4H_9)_3$	Ti: Org.Verb.1-45
$C_{14}H_{28}O_4Sn$	$(CH_3)_3Sn(CH_2)_3C(CH_3)(COOC_2H_5)_2$	Sn: Org.Verb.2-55, 57
–	$(C_2H_5)_2Sn(CH_2CH(CH_3)COOCH_3)_2$	Sn: Org.Verb.3-28
–	$(C_2H_5)_2Sn(CH_2COOC_3H_7)_2$	Sn: Org.Verb.3-28
–	$(C_3H_7)_2Sn(CH_2CH_2COOCH_3)_2$	Sn: Org.Verb.3-36/7
–	$(C_3H_7)_2Sn(CH_2COOC_2H_5)_2$	Sn: Org.Verb.3-38
–	$(C_4H_9)_2Sn(CH_2COOCH_3)_2$	Sn: Org.Verb.3-44
$C_{14}H_{28}O_5Ti$	$CH_3COCH_2COCH_2Ti(OC_3H_7\text{-i})_3$	Ti: Org.Verb.1-45
$C_{14}H_{28}Si_2Sn$	$(CH_3)_3SnC_6H_4Si(CH_3)_2Si(CH_3)_3$	Sn: Org.Verb.2-143, 145
$C_{14}H_{28}Sn$	$(CH_3)_2C(CH_2CH_2)_2Sn(CH_2CH_2)_2C(CH_3)_2$	Sn: Org.Verb.3-127
–	$(CH_3)_2Sn(C_6H_{11})_2$	Sn: Org.Verb.3-11, 18
–	$(C_2H_5)_3SnC(C_2H_5)CCHC_3H_7$	Sn: Org.Verb.2-188

Formula	Compound	Reference
$C_{14}H_{28}Sn$	$(C_2H_5)_3SnCH_2CH_2C_6H_9$	Sn: Org.Verb.2-173
–	$(C_4H_9)_2Sn(CHCH_2CH_2)_2$	Sn: Org.Verb.3-44
–	$(C_4H_9)_2Sn(CH_2CHCH_2)_2$	Sn: Org.Verb.3-42, 44
–	$(i\text{-}C_4H_9)_2Sn(CH_2CHCH_2)_2$	Sn: Org.Verb.3-69
–	$(C_4H_9)_3SnCCH$	Sn: Org.Verb.2-315/6
$C_{14}H_{29}NSn$	$(C_2H_5)_3SnCHCHCHCHN(C_2H_5)_2$	Sn: Org.Verb.2-199
–	$(C_4H_9)_3SnCH_2CN$	Sn: Org.Verb.2-281, 299
$C_{14}H_{29}N_3Ti$	$(CH_3)_2CHC_5H_4Ti(N(CH_3)_2)_3$	Ti: Org.Verb.1-172, 175/8
$C_{14}H_{29}NiOP^+$	$(CO)Ni(CH_2P(C_4H_9)_3)^+$	Ni: Org.Verb.1-182
$C_{14}H_{29}NiP$	$CH_3C_3H_4Ni(P(C_2H_5)_3)C_3H_4CH_3$	Ni: Org.Verb.2-31/2
$C_{14}H_{30}I_4O_8P_2Sn$	$SnI_4 \cdot 2\ (CH_3)_3CCOP(O)(OCH_3)_2$	Sn: MVol.C6-164
$C_{14}H_{30}N_2O_2Sn$	$(C_4H_9)_2Sn(CH_2CH_2CONH_2)_2$	Sn: Org.Verb.3-44
$C_{14}H_{30}N_2S_4Sn$	$(C_6H_{13})_2Sn(S_2CNH_2)_2$	C: MVol.D6-140
$C_{14}H_{30}N_2Sn$	$[(C_4H_9)_2SnN(CH_2CHCH_3)_2N]_x$	Sn: Org.Verb.6-89, 98
$C_{14}H_{30}N_2Ti$	$(CH_2CHCH_2)_2Ti(N(C_2H_5)_2)_2$	Ti: Org.Verb.1-78
$C_{14}H_{30}N_3O_{12}PU$	$UO_2(NO_3)_2 \cdot (C_3H_7O)_2POCH_2CON(C_3H_7)_2$	U: SVol.E1-188/9
$C_{14}H_{30}N_4O_4Ti$	$(N(CH_3)_2)_2C_3O(COOCH_3)Ti(N(CH_3)_2)_2OCH_3$	Ti: Org.Verb.1-60
$C_{14}H_{30}N_4O_6S_2$	$(O(CH_2CH_2)_2NSO_2NH)_2C_6H_{12}$	S: S-N-Verb.1-171
$C_{14}H_{30}N_4O_{10}U$	$UO_2(NO_3)_2 \cdot 2\ C_4H_9CON(CH_3)_2$	U: SVol.E1-110/1
$C_{14}H_{30}NiO_2P_2$	$(CO)_2Ni(P(C_2H_5)_3)_2$	Ni: Org.Verb.1-129, 130
$C_{14}H_{30}NiO_2Sb_2$	$(CO)_2Ni(Sb(C_2H_5)_3)_2$	Ni: Org.Verb.1-144
$C_{14}H_{30}NiO_8P_2$	$(CO)_2Ni(P(OC_2H_5)_3)_2$	Ni: Org.Verb.1-139, 140
$C_{14}H_{30}OS_2Sn$	$(C_3H_7)_3Sn[S_2COC_4H_9]$	C: MVol.D4-253
–	$(C_4H_9)_3Sn[S_2COCH_3]$	C: MVol.D4-253
$C_{14}H_{30}OSiSn$	$(C_2H_5)_3SnCCC(CH_3)_2OSi(CH_3)_3$	Sn: Org.Verb.2-230
$C_{14}H_{30}OSi_2Sn$	$(C_2H_5)_3Sn(CH)_2Si(CH_3)_2OSi(CH_3)_2CCH$	Sn: Org.Verb.2-230
$C_{14}H_{30}OSn$	$(C_4H_9)_3SnCHOCH_2$	Sn: Org.Verb.2-322, 324
$C_{14}H_{30}O_2Sn$	$(C_3H_7)_3SnCH_2CH_2OCH_2CHOCH_2$	Sn: Org.Verb.2-253
–	$(C_3H_7)_3SnCH_2COOC_3H_7$	Sn: Org.Verb.2-253, 257
$C_{14}H_{30}O_8U$	$UO_2(CH_3COO)_2 \cdot 2\ C_5H_{11}OH$	U: SVol.E1-59, 63
$C_{14}H_{30}PbSn$	$(C_2H_5)_3SnCCPb(C_2H_5)_3$	Sn: Org.Verb.2-230
$C_{14}H_{30}SiSn$	$(C_2H_5)_3SnC(CCHCH_3)Si(CH_3)_2C_2H_5$	Sn: Org.Verb.2-200
–	$(C_2H_5)_3SnCH(CH_3)CCSi(CH_3)_2C_2H_5$	Sn: Org.Verb.2-230
–	$(C_2H_5)_3SnCH_2CHCCHSi(CH_3)_2C_2H_5$	Sn: Org.Verb.2-200
–	$(C_2H_5)_3SnCH_2CH_2CCSi(CH_3)_2C_2H_5$	Sn: Org.Verb.2-230
$C_{14}H_{30}Si_2Sn$	$(CH_3)_3SnC_5H_3(Si(CH_3)_3)_2$	Sn: Org.Verb.2-100
$C_{14}H_{30}Sn$	$(CH_3)(i\text{-}C_3H_7)Sn(C_4H_9)(C_6H_{11})$	Sn: Org.Verb.3-97
–	$((CH_3)_3CCH_2)_2Sn(CH_2)_4$	Sn: Org.Verb.3-103
–	$(C_2H_5)_3Sn(CH_2)_6CHCH_2$	Sn: Org.Verb.2-188
–	$(C_4H_9)_2(CH_2CHCH_2CH_2CH_2CH_2)SnH$	Sn: Org.Verb.4-91
–	$(C_4H_9)_2Sn(CH_2)_6$	Sn: Org.Verb.3-109
–	$(C_4H_9)_3SnCHCH_2$	Sn: Org.Verb.2-307
–	$(i\text{-}C_4H_9)_3SnCHCH_2$	Sn: Org.Verb.2-328
$C_{14}H_{30}Sn_2$	$[(C_2H_5)_3Sn]_2C_2$	Sn: Org.Verb.5-102
$C_{14}H_{31}N_3OSn$	$(C_3H_7)_3SnCH_2CH_2C(CH_3)NNHCONH_2$	Sn: Org.Verb.2-253
$C_{14}H_{31}N_3O_2S_4U$	$UO_2(S_2CN(C_2H_5)_2)_2 \cdot (C_2H_5)_2NH$	U: SVol.E1-19, 22
$C_{14}H_{31}N_3SiTi$	$(CH_3)_3SiC_5H_4Ti(N(CH_3)_2)_3$	Ti: Org.Verb.1-172, 175/8
$C_{14}H_{31}N_3SnTi$	$(CH_3)_3SnC_5H_4Ti(N(CH_3)_2)_3$	Ti: Org.Verb.1-171
$C_{14}H_{31}NiP$	$C_3H_5Ni(P(C_3H_7)_3)C_2H_5$	Ni: Org.Verb.2-30
$C_{14}H_{32}NO_{11}UV_3$	$C_{14}H_{29}NH_3[UO_2(VO_3)_3] \cdot n\ H_2O$	U: SVol.C3-302/3

Formula	Compound	Reference
$C_{14}H_{32}N_2Sn$	$[Sn(C_4H_9)_2NH(CH_2)_6NH]_x$	Sn: Org.Verb.6-89, 97
$C_{14}H_{32}OSn$	$(C_2H_5)_3SnCH_2CH(CH_3)C(C_2H_5)_2OH$	Sn: Org.Verb.2-174
–	$(C_4H_9)_3SnCH_2CH_2OH$	Sn: Org.Verb.2-281
–	$(C_4H_9)_3SnCH_2OCH_3$	Sn: Org.Verb.2-281
$C_{14}H_{32}O_2S_6Sn$	$Sn(SCH_2CH(S)CH_3)_2 \cdot 2\ (C_2H_5)_2SO$	Sn: MVol.C6-102/3
$C_{14}H_{32}O_2SiSn$	$(C_2H_5)_3SnCH_2COOSi(C_2H_5)_3$	Sn: Org.Verb.2-174, 178
–	$(C_3H_7)_3SnCH_2COOSi(CH_3)_3$	Sn: Org.Verb.2-253, 257
$C_{14}H_{32}O_2Sn$	$(C_4H_9)_2Sn(CH_2CH_2CH_2OH)_2$	Sn: Org.Verb.3-42, 45
$C_{14}H_{32}O_3Si_3Ti$	$C_5H_5Ti(OSi(CH_3)_3)_3$	Ti: Org.Verb.1-170
$C_{14}H_{32}O_3Sn$	$C_2H_5Sn(OC_4H_9)_3$	Sn: Org.Verb.6-230/1
$C_{14}H_{32}O_3Ti$	$C_2H_5Ti(OC_4H_9)_3$	Ti: Org.Verb.1-45
$C_{14}H_{32}O_4Sn$	$(C_2H_5)_2Sn(CH_2CH_2CH_2OCH_2CH_2OH)_2$	Sn: Org.Verb.3-29
$C_{14}H_{32}SSn$	$(C_4H_9)_3SnCH_2CH_2SH$	Sn: Org.Verb.2-281
–	$(C_4H_9)_3SnCH_2SCH_3$	Sn: Org.Verb.2-281, 299
$C_{14}H_{32}Sn$	$(CH_3)_2Sn(C_4H_9)(C_8H_{17})$	Sn: Org.Verb.3-77
–	$(C_2H_5)_2Sn(CH_2C(CH_3)_3)_2$	Sn: Org.Verb.3-29
–	$(C_2H_5)_2Sn(C_5H_{11})_2$	Sn: Org.Verb.3-28
–	$(C_2H_5)_2Sn(i\text{-}C_5H_{11})_2$	Sn: Org.Verb.3-28
–	$(C_2H_5)_3SnCH_2CH(CH_3)CH_2C(CH_3)_3$	Sn: Org.Verb.2-160
–	$(C_2H_5)_3SnC_8H_{17}$	Sn: Org.Verb.2-160
–	$(C_3H_7)_2Sn(C_4H_9)_2$	Sn: Org.Verb.3-36, 38
–	$(i\text{-}C_3H_7)_2Sn(C_4H_9)_2$	Sn: Org.Verb.3-40
–	$(C_4H_9)_3SnC_2H_5$	Sn: Org.Verb.2-272/3
–	$(i\text{-}C_4H_9)_3SnC_2H_5$	Sn: Org.Verb.2-327
$C_{14}H_{32}Ti$	$(C_4H_9)_3(C_2H_5)Ti$	Ti: Org.Verb.1-90
$C_{14}H_{33}NSn$	$(C_4H_9)_3SnN(CH_3)_2$	Sn: Org.Verb.5-129
$C_{14}H_{33}N_3Ti$	$CH_2CHTi(N(C_2H_5)_2)_3$	Ti: Org.Verb.1-53
$C_{14}H_{34}I_4O_6P_2Sn$	$SnI_4 \cdot 2\ CH_3P(O)(OCH(CH_3)_2)_2$	Sn: MVol.C6-162/3
$C_{14}H_{34}N_2O_{14}P_2U$	$UO_2(NO_3)_2 \cdot 2\ (C_3H_7O)_2(CH_3)PO$	U: SVol.E1-186/7
$C_{14}H_{34}N_2O_{16}U_2$	$[UO_2(CH_3COO)_2((CH_3)_3NO)(H_2O)]_2$	U: SVol.E1-131, 135
$C_{14}H_{34}N_4O_4S_2$	$(C_2H_5)_2NSO_2NH(CH_2)_6NHSO_2N(C_2H_5)_2$	S: S-N-Verb.1-171
$C_{14}H_{34}NiP_2$	$C_2H_4Ni(P(C_2H_5)_3)_2$	Ni: Org.Verb.1-349, 350
$C_{14}H_{34}OSi_2Sn$	$(C_4H_9)_2Sn(CH_2Si(CH_3)_2)_2O$	Sn: Org.Verb.3-107
		Sn: Org.Verb.6-207
$C_{14}H_{35}NO_2Si_2Sn$	$(CH_3)_3SnCH_2CH(N[Si(CH_3)_3]_2)COOC_2H_5$	Sn: Org.Verb.2-38/9
$C_{14}H_{35}N_3Ti$	$C_2H_5Ti(N(C_2H_5)_2)_3$	Ti: Org.Verb.1-52, 57
$C_{14}H_{35}NiO_{13}P_3S$	$[C_5H_7Ni(P(OCH_3)_3)_3]HSO_4$	Ni: Org.Verb.2-27
$C_{14}H_{36}N_6NiO_2P_2$	$(CO)_2Ni(P(N(CH_3)_2)_3)_2$	Ni: Org.Verb.1-125, 126
$C_{14}H_{36}N_8O_4P_2S_2U$	$UO_2(NCS)_2 \cdot 2\ [(CH_3)_2N]_3PO$	U: SVol.E1-163/4
$C_{14}H_{36}Si_2Sn$	$(CH_3)_2Sn[CH_2CH(CH_3)CH_2SiH(CH_3)_2]_2$	Sn: Org.Verb.3-11
–	$(C_4H_9)_2Sn[CH_2SiH(CH_3)_2]_2$	Sn: Org.Verb.3-45, 54
$C_{14}H_{37}NiO_{13}P_3S$	$[(CH_3)_2C_3H_3Ni(P(OCH_3)_3)_3]HSO_4$	Ni: Org.Verb.2-26
$C_{14}H_{38}O_3Si_2Sn_2$	$[(CH_3)_3SiOSn(C_2H_5)_2]_2O$	Sn: Org.Verb.6-53
$C_{14}H_{38}Si_3Sn$	$[(CH_3)_3SiCH_2]_3SnC_2H_5$	Sn: Org.Verb.2-336
$C_{14}H_{39}NNiP_4S$	$CH_3Ni(P(CH_3)_3)_4NCS$	Ni: Org.Verb.1-18, 24
$C_{14}H_{40}O_7Si_5Sn$	$[(OSi(CH_3)_2)_5OSn(C_2H_5)_2O]_x$	Sn: Org.Verb.6-53
$C_{14}H_{42}N_4SSi_2Sn_2$	$[(CH_3)_3SiN]_2S[N(CH_3)Sn(CH_3)_3]_2$	S: S-N-Verb.1-210/1
$C_{14}H_{42}N_4SSi_2Sn_2^+$	$[(CH_3)_3SiN]_2S[N(CH_3)Sn(CH_3)_3]_2^+$	S: S-N-Verb.1-210
$C_{14}H_{42}Ni_2O_2P_2Si_2$	$(CH_3Ni(P(CH_3)_3)OSi(CH_3)_3)_2$	Ni: Org.Verb.2-251
$C_{14}H_{45}IrN_6O_2$	$[Ir(NCO)H(C_3H_{10}NH)_4]NCO \cdot H_2O$	Ir: SVol.2-64

Formula	Compound	Reference
$C_{14}K_6O_{28}Sn_2$	$K_6Sn_2(C_2O_4)_7 \cdot 4\ H_2O$	Sn: MVol.C3-78/9
–	$K_6Sn_2(C_2O_4)_7 \cdot 5\ H_2O$	Sn: MVol.C3-79
$C_{14}MnNiO_{14}Re$	$Ni(CO)_4 \cdot MnRe(CO)_{10}$	Ni: Org.Verb.1-228
$C_{14}Mn_2NiO_{14}$	$Ni(CO)_4 \cdot Mn_2(CO)_{10}$	Ni: Org.Verb.1-228
$C_{14}Mo_2Ni_4O_{14}^{2-}$	$[Mo_2Ni_4(CO)_{14}]^{2-}$	Ni: Org.Verb.1-228
$C_{14}NiO_{14}Re_2$	$Ni(CO)_4 \cdot Re_2(CO)_{10}$	Ni: Org.Verb.1-228
$C_{14.5}Cl_{1.5}H_{16.5}N_3O_6S_{1.5}U$	$UO_2(CH_3COO)_2 \cdot 1.5\ NH_2CSNHC_6H_4Cl \cdot 0.5\ H_2O$	U: SVol.E1-220/1
$C_{14.5}Cl_3H_{10.5}O_3Sn$	$SnCl_3(OC_6H_3(OH)COC_6H_5) \cdot 0.25\ C_6H_6$	Sn: MVol.C5-90/1
$C_{14.5}H_{16.8}MnN_{1.7}O_2S_4$	$Mn[S_2COC_2H_5]_2(NC_5H_4C_5H_4N)_{0.85}$	C: MVol.D4-258
$C_{15}CeH_{15}$	$Ce(C_5H_5)_3$	Sc: MVol.C3-183
$C_{15}ClF_2H_{20}N_3O$	$((CH_2)_5N)((CH_2)_5NO)C_5F_2ClN$	F: PerFHalOrg.5-138
$C_{15}ClF_6H_{12}NO_7U$	$UO_2(CF_3COCHCOCH_3)_2 \cdot ClC_5H_4NO$	U: SVol.E1-133, 139, 142/3
$C_{15}ClF_{12}H_6NO_7U$	$UO_2(CH(COCF_3)_2)_2 \cdot ClC_5H_4NO$	U: SVol.E1-133, 139, 142/3
$C_{15}ClF_{12}H_7O_4Ti$	$C_5H_5Ti(CF_3COCHCOCF_3)_2Cl$	Ti: Org.Verb.1-192, 194
$C_{15}ClFeH_{10}NO_4Sn$	$(CO)_3Fe(NO)Sn(C_6H_5)_2Cl$	Fe: Org.Verb.B1-197/8
$C_{15}ClFeH_{13}HgO_3$	$CH_3COOHgC_5H_4FeC_5H_4CClCHCHO$	Fe: Org.Verb.A2-176
$C_{15}ClFeH_{15}O_2$	$CH_3COC_5H_4FeC_5H_4COCH_2CH_2Cl$	Fe: Org.Verb.A2-210
–	$C_5H_5FeC_5H_4CO(CH_2)_3COCl$	Fe: Org.Verb.A3-159
$C_{15}ClFeH_{17}$	$C_5H_5FeC_5H_4CClC(CH_3)C_2H_5$	Fe: Org.Verb.A1-385, 387
$C_{15}ClFeH_{17}O$	$C_5H_5FeC_5H_4CH(CClCH_2)OC_2H_5$	Fe: Org.Verb.A2-118, 120
–	$C_5H_5FeC_5H_4CHCClCH_2OC_2H_5$	Fe: Org.Verb.A2-131
–	$C_5H_5FeC_5H_4CH_2CH(C_2H_5)COCl$	Fe: Org.Verb.A3-159
–	$C_5H_5FeC_5H_4(CH_2)_4COCl$	Fe: Org.Verb.A3-159
–	$C_5H_5FeC_5H_4COC(CH_3)_2CH_2Cl$	Fe: Org.Verb.A2-244, 246
–	$C_5H_5FeC_5H_4COCH_2C(CH_3)_2Cl$	Fe: Org.Verb.A2-244, 246
$C_{15}ClFeH_{17}O_2$	$C_5H_5FeC_5H_4C(CH_3)OCH(CH_2Cl)CH_2O$	Fe: Org.Verb.A3-179
$C_{15}ClFeH_{18}NO_2$	$C_5H_5FeC_5H_4CHClCH_2NHCOOC_2H_5$	Fe: Org.Verb.A1-287
$C_{15}ClFeH_{20}NO$	$C_5H_5FeC_5H_4CO(CH_2)_2N(CH_3)_2 \cdot HCl$	Fe: Org.Verb.A2-228
$C_{15}ClFe_2GeH_{27}N_4O_4$	$((CH_3)_2C_3H_3Fe(NO)_2)_2Ge(Cl)C_5H_9$	Fe: Org.Verb.C2-165
$C_{15}ClFe_2H_9O_6$	$CH_2C(CH_3)C_6H_4Cl[Fe(CO)_3]_2$	Fe: Org.Verb.B3-146
$C_{15}ClFe_2H_{27}N_4O_4Pb$	$((CH_3)_2C_3H_3Fe(NO)_2)_2Pb(Cl)C_5H_9$	Fe: Org.Verb.C2-165
$C_{15}ClFe_2H_{27}N_4O_4Sn$	$((CH_3)_2C_3H_3Fe(NO)_2)_2Sn(Cl)C_5H_9$	Fe: Org.Verb.C2-165
$C_{15}ClH_{10}MnO_2^+$	$[Mn(OC_6H_3(Cl)COCHCHC_6H_5)]^+$	Mn: MVol.D1-58/9
$C_{15}ClH_{10}NiO_3P$	$(CO)_3NiP(C_6H_5)_2Cl$	Ni: Org.Verb.1-171, 172
$C_{15}ClH_{14}NSn$	$(C_6H_5CH_2)_2ClSnCN$	Sn: Org.Verb.6-123
$C_{15}ClH_{15}Sn$	$(C_5H_5)_3SnCl$	Sn: Org.Verb.5-157
$C_{15}ClH_{18}NO_6U$	$[UO_2(CH(COCH_3)_2)_2(ClC_5H_4N)]$	U: SVol.E1-31, 36/7
$C_{15}ClH_{19}MnNO_5$	$[MnCl(CH(COCH_3)_2)_2(C_5H_5NO)]$	Mn: MVol.D1-106
$C_{15}ClH_{19}O_4Ti$	$C_5H_5Ti((CH_3CO)_2CH)_2Cl$	Ti: Org.Verb.1-191/2, 194
$C_{15}ClH_{19}O_8Ti$	$C_5H_5Ti(CH(COCH_3)_2)_2ClO_4$	Ti: Org.Verb.1-192, 194
$C_{15}ClH_{23}OSSn$	$[(CH_3)_3SnCH(S(CH_2)_4)COC_6H_5]Cl$	Sn: Org.Verb.2-25, 29
$C_{15}ClH_{23}O_2Sn$	$(C_4H_9)_2ClSnOCOC_6H_5$	Sn: Org.Verb.5-132
$C_{15}ClH_{24}OSn$	$(C_3H_7)_2ClSnOC(CH_3)C_6H_4CH_3$	Sn: Org.Verb.5-114
$C_{15}ClH_{25}O_2Ti$	$C_5H_5Ti(OC_5H_{10})_2Cl$	Ti: Org.Verb.1-185/6
$C_{15}ClH_{25}SSn$	$(C_3H_7)_3SnSC_6H_4Cl$	Sn: Org.Verb.5-114
$C_{15}ClH_{25}Sn$	$(C_3H_7)_3SnC_6H_4Cl$	Sn: Org.Verb.2-267
–	$(C_4H_9)_2(C_6H_5CH_2)SnCl$	Sn: Org.Verb.5-210
–	$(t\text{-}C_4H_9)_2(CH_3C_6H_4)SnCl$	Sn: Org.Verb.5-217, 221

$C_{15}ClH_{26}O_2PSn$ ..... $(C_2H_5)_2[C_2H_5O(C_6H_5)P(O)(CH_2)_3]SnCl$ ..... Sn: Org.Verb.5-203
$C_{15}ClH_{28}NSn$ .... $(C_3H_7)_3SnCl \cdot CH_3C_5H_4N$ .... Sn: Org.Verb.5-114
– .... $(C_3H_7)_3SnCl \cdot C_6H_5NH_2$ .... Sn: Org.Verb.5-114
$C_{15}ClH_{29}O_2Sn$ ...... $(C_4H_9)_2[CH_3COOC(CH_3)_2CHCH]SnCl$ ...... Sn: Org.Verb.5-211
$C_{15}ClH_{29}Sn$ .... $(C_6H_{11})_2(C_3H_7)SnCl$ .... Sn: Org.Verb.5-222
$C_{15}ClH_{32}NiP$ .... $C_3H_5Ni(P(C_4H_9)_3)Cl$ .... Ni: Org.Verb.2-12
$C_{15}ClH_{33}Sn$ .... $[(CH_3)_3CCH_2]_3SnCl$ .... Sn: Org.Verb.5-139/40
– .... $(C_5H_{11})_3SnCl$ .... Sn: Org.Verb.5-139
– .... $(i\text{-}C_5H_{11})_3SnCl$ .... Sn: Org.Verb.5-139
$C_{15}ClH_{36}NSn$ .... $[(C_4H_9)_3SnCH_2NH(CH_3)_2]Cl$ .... Sn: Org.Verb.2-283
$C_{15}ClH_{42}O_9P_3Sn$ .... $(CH_3)_3SnCl \cdot 3\ (C_2H_5O)_2HPO$ .... Sn: Org.Verb.5-87
$C_{15}ClH_{45}IrO_{15}P_5$ .... $[Ir(P(OCH_3)_3)_5]Cl$ .... Ir: SVol.2-235
$C_{15}Cl_2FH_{20}N_3$ ...... $[(CH_2)_5N]_2C_5FCl_2N$ .... F: PerFHalOrg.5-138
$C_{15}Cl_2F_2H_6N_4O_4$ .... $NC_5F_2Cl_2NHC_{10}H_5(NO_2)_2$ .... F: PerFHalOrg.5-215
$C_{15}Cl_2F_9HN_2$ .... $NC_5F_2Cl_2NHC_{10}F_7$ .... F: PerFHalOrg.5-188/9
F: PerFHalOrg.6-13
$C_{15}Cl_2F_{12}H_7N_3O$ .... $(CF_3)_2CNHC(CF_3)_2NC(NHCOCH_2C_6H_3Cl_2)$ .. F: PerFHalOrg.5-105
$C_{15}Cl_2FeH_{16}$ .... $C_5H_5FeC_5H_4C_3HCl_2(CH_3)_2$ .... Fe: Org.Verb.A1-392, 394
$C_{15}Cl_2Fe_2H_{18}N_2O_7P_2$. $(CO)_7Fe_2(N_2P_2Cl_2(C_4H_9)_2)$ .... Fe: Org.Verb.C1-200/2, 205
$C_{15}Cl_2GeH_{32}OSn$ .... $(C_4H_9)_2[(CH_3)_2ClGeOC(CH_3)_2CHCH]SnCl$ .. Sn: Org.Verb.5-211
– .... $(C_4H_9)_2[(C_2H_5)_2ClGeOCH_2CHCH]SnCl$ ..... Sn: Org.Verb.5-211
$C_{15}Cl_2H_{10}O_5Sn$ ..... $SnCl_2 \cdot (HO)_2C_9H_4OC_6H_3O(OH)$ .... Sn: MVol.C5-11
$C_{15}Cl_2H_{10}O_6Sn$ ..... $SnCl_2 \cdot (HO)_2C_9H_4OC_6H_2O(OH)_2$ .... Sn: MVol.C5-11
$C_{15}Cl_2H_{11}N_3Sn$ ..... $SnCl_2 \cdot NC_5H_3(C_5H_4N)_2$ .... Sn: MVol.C5-22
– .... $SnCl_2 \cdot NC_5H_4CHNC_9H_6N$ .... Sn: MVol.C5-39
$C_{15}Cl_2H_{12}N_3O_5U$ .... $UO_2Cl_2 \cdot 1.5\ C_{10}H_8N_2O_2$ .... U: SVol.E1-134, 141, 144
$C_{15}Cl_2H_{12}STi$ .... $C_5H_5Ti(SC_{10}H_7)Cl_2$ .... Ti: Org.Verb.1-189
$C_{15}Cl_2H_{15}N_3O_5U$ .... $UO_2Cl_2 \cdot 3\ C_5H_5NO$ .... U: SVol.E1-132, 135, 142/3
$C_{15}Cl_2H_{15}N_3O_5U_2$ .... $U_2O_5Cl_2 \cdot 3\ C_5H_5N$ .... U: SVol.E1-29, 34
$C_{15}Cl_2H_{16}IrN_3O$ ..... $Ir(C_5H_5N)_3Cl_2(OH)$ .... Ir: SVol.2-72
$C_{15}Cl_2H_{16}OSn$ ...... $(C_6H_5)_2SnCl_2 \cdot (CH_3)_2CO$ .... Sn: Org.Verb.6-174, 177
$C_{15}Cl_2H_{16}Sn$ .... $(C_6H_5)_2[Cl(CH_2)_3]SnCl$ .... Sn: Org.Verb.5-216, 218
$C_{15}Cl_2H_{17}IIrN_3O$ .... $[Ir(C_5H_5N)_3Cl_2(H_2O)]I \cdot H_2O$ .... Ir: SVol.2-72
$C_{15}Cl_2H_{17}IrN_3O^+$ .... $[Ir(C_5H_5N)_3Cl_2(H_2O)]^+$ .... Ir: SVol.2-72
$C_{15}Cl_2H_{17}MnN_2O_4$ ... $[MnCl_2(CH(COCH_3)_2)(C_5H_5NO)_2]$ .... Mn: MVol.D1-106
$C_{15}Cl_2H_{17}NOSn$ ..... $(C_6H_5)_2SnCl_2 \cdot HC(O)NHC_2H_5$ .... Sn: Org.Verb.6-174/5
$C_{15}Cl_2H_{18}IrN_3O_2$ .... $[Ir(C_5H_5N)_3Cl(OH)(H_2O)]Cl \cdot 2\ H_2O$ .... Ir: SVol.2-73
– .... $[Ir(C_5H_5N)_3Cl_2(H_2O)](OH)$ .... Ir: SVol.2-72
$C_{15}Cl_2H_{18}N_2OSn$ .... $(CH_3)_2SnCl_2 \cdot NC_5H_4CHNC_6H_4OCH_3$ ...... Sn: Org.Verb.6-36, 38
$C_{15}Cl_2H_{18}N_2Sn$ ..... $(CH_3)_2SnCl_2 \cdot NC_5H_4CHNC_6H_4CH_3$ .... Sn: Org.Verb.6-36, 38
$C_{15}Cl_2H_{26}N_2O_2Sn$ ... $(CH_3)(C_6H_5)SnCl_2 \cdot 2\ NH(CH_2CH_2)_2O$ ..... Sn: Org.Verb.6-195
$C_{15}Cl_2H_{27}N_3Ni$ ...... $(t\text{-}C_4H_9NC)_3NiCl_2$ .... Ni: Org.Verb.1-323
$C_{15}Cl_2H_{28}N_2O_4Sn$ ... $(CH_3)_2SnCl_2 \cdot CH_3C_2H_3(NH(CH_2CO)_2CH_3)_2$. Sn: Org.Verb.6-36
$C_{15}Cl_2H_{35}N_5O_{15}U$ .... $UO_2(ClO_4)_2 \cdot 5\ HCON(CH_3)_2$ .... U: SVol.E1-100/1
$C_{15}Cl_2H_{35}O_{30}P_5U$ .... $UO_2(ClO_4)_2 \cdot 5\ (CH_2O)_2PO(OCH_3) \cdot H_2O$ .. U: SVol.E1-185
$C_{15}Cl_2H_{36}MnO_3$ ..... $MnCl_2 \cdot 3\ C_2H_5CH(CH_3)CH_2OH$ .... Mn: MVol.D1-33
– .... $MnCl_2 \cdot 3\ C_5H_{11}OH$ .... Mn: MVol.D1-33
$C_{15}Cl_2H_{36}N_6O_5U$ .... $UO_2Cl_2 \cdot 3\ CO(NHC_2H_5)_2$ .... U: SVol.E1-92/3
$C_{15}Cl_2H_{39}NSn_2$ ..... $2\ (C_2H_5)_3SnCl \cdot N(CH_3)_3$ .... Sn: Org.Verb.5-107
$C_{15}Cl_2H_{40}N_{10}O_7U$ .... $UO_2Cl_2 \cdot 5\ CO(NHCH_3)_2$ .... U: SVol.E1-92/3

$C_{15}Cl_2H_{40}N_{10}O_{15}U$ ... $UO_2(ClO_4)_2 \cdot 5\ CO(NHCH_3)_2$ ... U: SVol.E1-93, 95
$C_{15}Cl_2H_{45}O_{30}P_5U$ ... $UO_2(ClO_4)_2 \cdot 5\ (CH_3O)_3PO$ ... U: SVol.E1-181, 184
$C_{15}Cl_3Fe_3H_3O_{15}$ ... $[(CO)_5FeH]_3Cl_3$ ... Fe: Org.Verb.B3-252
$C_{15}Cl_3GeH_{20}NiP$ ... $C_5H_5Ni(P(C_2H_5)_2C_6H_5)GeCl_3$ ... Ni: Org.Verb.2-130
$C_{15}Cl_3H_9O_4Sn$ ... $SnCl_3(CH_3OC_{14}H_6O_3)$ ... Sn: MVol.C5-114
$C_{15}Cl_3H_{10}NaO_7SSn$ .. $CH_3SnCl_3 \cdot C_6H_4(CO)_2C_6H(OH)_2SO_3Na$ ... Sn: Org.Verb.6-222/3
$C_{15}Cl_3H_{15}IrN_3$ ... $Ir(C_5H_5N)_3Cl_3$ ... Ir: SVol.2-69
– ... $[Ir(C_5H_5N)_3Cl_3]_n$ ... Ir: SVol.2-69
$C_{15}Cl_3H_{15}IrN_3O$ ... $Ir(C_{12}H_8N_2)Cl_3((CH_3)_2NCHO)$ ... Ir: SVol.2-92
$C_{15}Cl_3H_{15}N_2Ti$ ... $C_5H_5TiCl_3 \cdot 2\ C_5H_5N$ ... Ti: Org.Verb.1-141, 152
$C_{15}Cl_3H_{17}IrN_3O$ ... $[Ir(C_5H_5N)_3Cl_2(H_2O)]Cl$ ... Ir: SVol.2-69
– ... $[Ir(C_5H_5N)_3Cl_2(H_2O)]Cl \cdot 2\ H_2O$ ... Ir: SVol.2-72
$C_{15}Cl_3H_{18}MnO_6$ ... $[Mn(CCl(COCH_3)_2)_3]$ ... Mn: MVol.D1-112
$C_{15}Cl_3H_{19}N_2Ti$ ... $C_5H_{11}TiCl_3 \cdot NC_5H_4C_5H_4N$ ... Ti: Org.Verb.1-34/5, 41
$C_{15}Cl_3H_{20}NiPSn$ ... $C_5H_5Ni(P(C_2H_5)_2C_6H_5)SnCl_3$ ... Ni: Org.Verb.2-130
$C_{15}Cl_3H_{21}N_2Sn$ ... $C_3H_7SnCl_3 \cdot 2\ CH_3C_5H_4N$ ... Sn: Org.Verb.6-236
– ... $C_3H_7SnCl_3 \cdot 2\ C_6H_5NH_2$ ... Sn: Org.Verb.6-236
$C_{15}Cl_3H_{23}O_4Ti$ ... $C_5H_5Ti(OC_4H_9)_2OCOCCl_3$ ... Ti: Org.Verb.1-190
$C_{15}Cl_3H_{30}IrS_3$ ... $IrCl_3(C_5H_{10}S)_3$ ... Ir: SVol.2-166
$C_{15}Cl_3H_{36}O_3Sn^+$ ... $[SnCl_3 \cdot 3\ C_5H_{11}OH]^+$ ... Sn: MVol.C5-72
$C_{15}Cl_4Fe_2H_{25}N_5O$ ... $(C_2H_5NC)_5Fe_2OCl_4$ ... Fe: Org.Verb.B4-96
$C_{15}Cl_4H_{10}OSn$ ... $SnCl_4 \cdot C_{15}H_{10}O$ ... Sn: MVol.C5-99
$C_{15}Cl_4H_{11}N_3Te$ ... $TeCl_4 \cdot NC_5H_3(C_5H_4N)_2$ ... Te: SVol.B3-171
$C_{15}Cl_4H_{12}O_2Sn$ ... $SnCl_4 \cdot HOC_6H_4CHCHCOC_6H_5$ ... Sn: MVol.C5-91/2
$C_{15}Cl_4H_{13}NO_2Sn$ ... $SnCl_4 \cdot C_6H_5COCH_2CONHC_6H_5$ ... Sn: MVol.C6-49/50
$C_{15}Cl_4H_{13}N_3O_2Sn$ ... $SnCl_4 \cdot CH_3OC_6H_4C_3HN_2(OH)C_5H_4N$ ... Sn: MVol.C5-222
$C_{15}Cl_4H_{15}N_3O_2Te$ ... $TeCl_4 \cdot (CH_3)_2NC_6H_4CHNC_6H_4NO_2$ ... Te: SVol.B3-172/3
$C_{15}Cl_4H_{16}IrN_3$ ... $(C_5H_5NH)[Ir(C_5H_5N)_2Cl_4]$ ... Ir: SVol.2-70
$C_{15}Cl_4H_{16}N_2Sn$ ... $SnCl_4 \cdot C_6H_5CHNC_6H_4N(CH_3)_2$ ... Sn: MVol.C6-18
$C_{15}Cl_4H_{16}N_2Te$ ... $TeCl_4 \cdot (CH_3)_2NC_6H_4CHNC_6H_5$ ... Te: SVol.B3-172/3
$C_{15}Cl_4H_{20}N_5U$ ... $UCl_4 \cdot 2.5\ C_6H_4(NH_2)_2$ ... U: SVol.E1-23/5
$C_{15}Cl_4H_{27}N_3OSn$ ... $SnCl_4 \cdot HOC_6H_2(CH_2N(CH_3)_2)_3$ ... Sn: MVol.C6-4
$C_{15}Cl_4H_{27}N_3U$ ... $[UCl_4(C_4H_9CN)_3]$ ... U: SVol.E1-48/50
$C_{15}Cl_4H_{30}N_3PSSn$ ... $SnCl_4 \cdot (C_5H_{10}N)_3PS$ ... Sn: MVol.C6-149
$C_{15}Cl_4H_{40}Mn_2O_5$ ... $2\ MnCl_2 \cdot 5\ C_3H_7OH$ ... Mn: MVol.D1-33
$C_{15}Cl_5H_{10}OU$ ... $UCl_5 \cdot C_{15}H_{10}O$ ... U: SVol.E1-80/1
$C_{15}Cl_5H_{15}N_3U$ ... $UCl_5 \cdot 3\ C_5H_5N$ ... U: SVol.E1-28, 33/4
$C_{15}Cl_5H_{17}N_3OU$ ... $(C_5H_5NH)_2UOCl_5 \cdot C_5H_5N$ ... U: SVol.E1-34
$C_{15}Cl_5H_{21}O_3Ti$ ... $C_6Cl_5Ti(OC_3H_7\text{-}i)_3$ ... Ti: Org.Verb.1-46
$C_{15}Cl_6DyH_{18}N_3$ ... $[C_5H_5NH]_3DyCl_6$ ... Sc: MVol.C5-217
$C_{15}Cl_6ErH_{18}N_3$ ... $[C_5H_5NH]_3ErCl_6$ ... Sc: MVol.C5-217
$C_{15}Cl_6EuH_{18}N_3$ ... $EuCl_3 \cdot 3\ [C_5H_5NH]Cl$ ... Sc: MVol.C5-213/5, 217
$C_{15}Cl_6GdH_{18}N_3$ ... $[C_5H_5NH]_3GdCl_6$ ... Sc: MVol.C5-217
$C_{15}Cl_6H_{10}N_4O_4Sn$ ... $SnCl_4 \cdot 2\ NO_2C_6H_4CN \cdot CH_2Cl_2$ ... Sn: MVol.C6-72
$C_{15}Cl_6H_{10}O_5Sn_3$ ... $3\ SnCl_2 \cdot (HO)_2C_9H_4OC_6H_3O(OH)$ ... Sn: MVol.C5-11
$C_{15}Cl_6H_{10}O_6Sn_3$ ... $3\ SnCl_2 \cdot (HO)_2C_9H_4OC_6H_2O(OH)_2$ ... Sn: MVol.C5-11
$C_{15}Cl_6H_{18}N_3Nd$ ... $[C_5H_5NH]_3NdCl_6$ ... Sc: MVol.C5-217
– ... $NdCl_3 \cdot 3\ [C_5H_5NH]Cl$ ... Sc: MVol.C5-213/5, 217
$C_{15}Cl_6H_{18}N_3Pr$ ... $PrCl_3 \cdot 3\ [C_5H_5NH]Cl \cdot x\ C_2H_5OH$ ... Sc: MVol.C5-213/5, 217
$C_{15}Cl_6H_{18}N_3Yb$ ... $[C_5H_5NH]_3YbCl_6$ ... Sc: MVol.C5-217

Formula	Compound		Reference
$C_{15}Cl_6H_{18}N_3Yb$	$YbCl_3 \cdot 3\,[C_5H_5NH]Cl$	Sc:	MVol.C5-213/5, 217
$C_{15}Cl_6H_{18}Sn$	$(C_2H_5)_3SnCCC_7H_3Cl_6$	Sn:	Org.Verb.2-230
$C_{15}Cl_6H_{39}IrN_3$	$(C_5H_{10}NH_3)_3[IrCl_6] \cdot H_2O$	Ir:	SVol.2-127
$C_{15}Cl_7H_{13}N_4Sn$	$SnCl_4 \cdot 2\,C_7H_6N_2 \cdot CHCl_3$	Sn:	MVol.C5-224
$C_{15}Cl_8H_{11}N_3Te_2$	$2\,TeCl_4 \cdot NC_5H_3(C_5H_4N)_2$	Te:	SVol.B3-171
$C_{15}Cl_8H_{15}N_3U_2$	$2\,UCl_4 \cdot 3\,C_5H_5N = UCl_4 \cdot 1.5\,C_5H_5N$	U:	SVol.E1-28, 33
$C_{15}Cl_8H_{50}LaN_5$	$LaCl_3 \cdot 5\,[(CH_3)_3NH]Cl$	Sc:	MVol.C5-214, 221
$C_{15}Cl_9H_9Sn$	$(CH_3)_3SnC_6Cl_4C_6Cl_5$	Sn:	Org.Verb.2-148
$C_{15}Cl_9H_{18}La_2N_3$	$2\,LaCl_3 \cdot 3\,[C_5H_5NH]Cl \cdot 2\,C_2H_5OH$	Sc:	MVol.C5-213/5, 217
$C_{15}CoH_{27}S_9$	$Co[S_2CSC_4H_9]_3$	C:	MVol.D4-267, 270
$C_{15}CoH_{36}N_3S_9$	$[(CH_3)_4N]_3[Co(CS_3)_3]$	C:	MVol.D4-225
$C_{15}Co_2FeH_5O_9P$	$(CO)_3FeP(C_6H_5)(Co(CO)_3)_2$	Fe:	Org.Verb.C2-55/6
$C_{15}Co_2H_{18}O_6Sn$	$(CH_3)_3SnC_2C(CH_3)_3Co_2(CO)_6$	Sn:	Org.Verb.2-154, 156
$C_{15}Co_2H_{48}N_{12}S_9$	$Co_2(C_2H_8N_2)_6(CS_3)_3$	C:	MVol.D4-225
$C_{15}Co_3FeH_{19}O_{10}P^+$	$[HFeCo_3(CO)_7P(OC_3H_7)_2OC_2H_4]^+$	Fe:	Org.Verb.B1-204
$C_{15}Co_3FeH_{22}O_9P^+$	$[HFeCo_3(CO)_6P(OC_3H_7)_3]^+$	Fe:	Org.Verb.B1-204
$C_{15}CrFeH_{12}O_4{}^+$	$[C_5H_5FeC_5H_4C(OCH_3)Cr(CO)_3]^+$	Fe:	Org.Verb.A1-346
$C_{15}CrFeH_{14}O_3{}^+$	$[C_5H_5FeC_5H_4C(OC_2H_5)Cr(CO)_2]^+$	Fe:	Org.Verb.A1-346
$C_{15}CrFeH_{15}NO_2{}^+$	$[C_5H_5FeC_5H_4C(N(CH_3)_2)Cr(CO)_2]^+$	Fe:	Org.Verb.A1-346
$C_{15}CrFeH_{17}N^+$	$[C_5H_5FeC_5H_4C(NC_4H_8)Cr]^+$	Fe:	Org.Verb.A1-346
$C_{15}CrH_{20}O_3Sn$	$(CH_3)_2Sn(s\text{-}C_4H_9)(C_6H_5Cr(CO)_3)$	Sn:	Org.Verb.3-75, 79
$C_{15}CrH_{27}O_3S_6$	$Cr[S_2COC_4H_9]_3$	C:	MVol.D4-258
$C_{15}Cr_2H_{48}N_{24}O_{12}S_9$	$[Cr(OC(NH_2)_2)_6]_2(CS_3)_3$	C:	MVol.D4-225
$C_{15}CsF_9H_{12}MnO_6$	$Cs[Mn(CF_3COCHCOCH_3)_3]$	Mn:	MVol.D1-109
$C_{15}CsF_{18}H_3MnO_6$	$Cs[Mn(CH(COCF_3)_2)_3]$	Mn:	MVol.D1-111
$C_{15}CsH_{21}MnO_6$	$Cs[Mn(CH(COCH_3)_2)_3]$	Mn:	MVol.D1-75
$C_{15}CuH_{35}O_7P_2S_2$	$[(C_2H_5O)_3P]_2Cu[S_2COC_2H_5]$	C:	MVol.D4-257
$C_{15}ErH_{18}I_6N_3$	$(C_5H_5NH)_3ErI_6$	Sc:	MVol.C6-273
$C_{15}EuH_{15}$	$Eu(C_5H_5)_3$	Sc:	MVol.B1-42
$C_{15}EuH_{18}I_6N_3$	$(C_5H_5NH)_3EuI_6$	Sc:	MVol.C6-273
$C_{15}FH_{11}O_2$	$(C_6H_5)_2CCFCOOH$	F:	PerFHalOrg.4-89
$C_{15}FH_{17}OSn$	$(CH_3)_2Sn(C_6H_4F)(C_6H_4OCH_3)$	Sn:	Org.Verb.3-83
$C_{15}FH_{17}Sn$	$(CH_3)_3SnC_6H_4C_6H_4F$	Sn:	Org.Verb.2-146, 148
$C_{15}FH_{25}Sn$	$(C_3H_7)_3SnC_6H_4F$	Sn:	Org.Verb.2-267
$C_{15}FH_{33}Sn$	$(C_5H_{11})_3SnF$	Sn:	Org.Verb.5-35/6
$C_{15}F_2H_{36}N_3P$	$[(C_2H_5)_2NCH_2]_3PF_2$	F:	PerFHalOrg.5-22
$C_{15}F_3FeH_{14}O_2{}^+$	$[C_5H_5FeC_5H_4CH(CH_2)_2OOCCF_3]^+$	Fe:	Org.Verb.A2-80
$C_{15}F_3Fe_2H_7O_7S$	$(CF_3C_6H_3CH(OCH_3)S)Fe_2(CO)_6$	Fe:	Org.Verb.C2-113
–	$(CO)_6Fe_2(CH_3OC(C_6H_4CF_3)S)$	Fe:	Org.Verb.C1-238/40
$C_{15}F_3H_{10}NO_2$	$CF_3NC(O)C(C_6H_5)_2O$	F:	PerFHalOrg.7-168
$C_{15}F_3H_{17}NiO_2$	$C_6H_6Ni(C_3H_5)(C_4H_6)O_2CCF_3$	Ni:	Org.Verb.2-222, 224
–	$C_6H_6Ni(C_7H_{11})O_2CCF_3$	Ni:	Org.Verb.2-221, 224
$C_{15}F_3H_{20}N_3$	$[(CH_2)_5N]_2C_5F_3N$	F:	PerFHalOrg.5-136
$C_{15}F_3H_{27}Si_3$	$[(CH_3)_3Si]_3C_6F_3$	F:	PerFHalOrg.4-25
$C_{15}F_3H_{30}INiP_2$	$CF_3CCNi(P(C_2H_5)_3)_2I$	Ni:	Org.Verb.1-28, 30
$C_{15}F_4FeH_8O_5$	$(CH(CHO)CH(C_6H_5)C_2F_4)Fe(CO)_4$	Fe:	Org.Verb.B4-337, 345/6
$C_{15}F_4H_{12}N_2$	$C_{10}H_{12}NC_5F_4N$	F:	PerFHalOrg.5-213
$C_{15}F_4H_{13}N_3O_2S$	$NH_2C_6F_4COOH \cdot C_6H_5CH_2NHCSNH_2$	F:	PerFHalOrg.7-61
$C_{15}F_5FeH_5O_2$	$C_5H_5Fe(CO)_2CCC_6F_5$	F:	PerFHalOrg.4-33
$C_{15}F_5FeH_8NO_2$	$C_5H_5Fe(CO)_2C(C_6F_5)NCH_3$	F:	PerFHalOrg.4-38

$C_{15}F_5FeH_8NO_2$	$C_6F_5C_5H_5Fe(CO)_2CNCH_3$	F:	PerFHalOrg.4-38
$C_{15}F_5FeH_{11}N_2$	$C_5H_5Fe(CNCH_3)_2C_6F_5$	F:	PerFHalOrg.4-38
$C_{15}F_5H_5MoO_4$	$C_5H_5Mo(CO)_3C(O)C_6F_5$	F:	PerFHalOrg.4-35
$C_{15}F_5H_5O_4W$	$C_5H_5W(CO)_3C(O)C_6F_5$	F:	PerFHalOrg.4-36
$C_{15}F_5H_7MoO_2$	$C_7H_7Mo(CO)_2C_6F_5$	F:	PerFHalOrg.4-111
$C_{15}F_5H_{12}N$	$C_6F_5NHC_6H_2(CH_3)_3$	F:	PerFHalOrg.7-197
$C_{15}F_5H_{21}O_3Ti$	$[(CH_3)_2CHO]_3TiC_6F_5$	F:	PerFHalOrg.4-35
		Ti:	Org.Verb.1-46, 50
$C_{15}F_6FeH_{12}O_2$	$CH_3COC_5H_4FeC_5H_4C(CF_3)_2OH$	Fe:	Org.Verb.A2-210
$C_{15}F_6FeH_{12}O_4$	$(C_8H_{12}CF_2CFCF_3)Fe(CO)_4$	Fe:	Org.Verb.B4-337, 347
$C_{15}F_6FeH_{18}N_2OS_2$	$(t\text{-}C_4H_9NC)_2Fe(CO)S_2C_2(CF_3)_2$	Fe:	Org.Verb.B4-15
$C_{15}F_6FeH_{24}NO_2P_3$	$[(CH_3)_3PC_8H_6Fe(CO)(NO)P(CH_3)_3][PF_6]$	Fe:	Org.Verb.B5-11/3
$C_{15}F_6FeH_{25}NO_3P_2$	$[C_4H_4P(n\text{-}C_3H_7)_3Fe(CO)_2NO][PF_6]$	Fe:	Org.Verb.B5-28/31
$C_{15}F_6H_6N_3$	$C_6H_5NNHC_9F_6N$	F:	PerFHalOrg.6-159
$C_{15}F_6H_7NO_3$	$CO(C_6F_3(OCH_3))_2NH$	F:	PerFHalOrg.6-155/6
$C_{15}F_6H_{11}N$	$C_6H_5C_6H_4C(CF_3)_2NH_2$	F:	PerFHalOrg.7-69/70
$C_{15}F_6H_{11}NO$	$C_6H_5OC_6H_4C(CF_3)_2NH_2$	F:	PerFHalOrg.7-69, 71
$C_{15}F_6H_{12}N_2O_9U$	$UO_2(CF_3COCHCOCH_3)_2 \cdot NO_2C_5H_4NO$	U:	SVol.E1-133, 139, 142/3
$C_{15}F_6H_{13}NO_6U$	$[UO_2(CF_3COCHCOCH_3)_2(C_5H_5N)]$	U:	SVol.E1-31, 36/7
$C_{15}F_6H_{13}NO_7U$	$UO_2(CF_3COCHCOCH_3)_2 \cdot C_5H_5NO$	U:	SVol.E1-132, 136, 142/3
$C_{15}F_6H_{15}Ni_2P$	$[(C_5H_5)_3Ni_2]PF_6$	Ni:	Org.Verb.2-364/5
$C_{15}F_6H_{22}NiO_6P_2S$	$(CF_3)_2CSNi(P(OCH_2)_3CC_2H_5)_2$	Ni:	Org.Verb.1-61
$C_{15}F_6H_{22}NiO_7P_2$	$(CF_3)_2CONi(P(OCH_2)_3CC_2H_5)_2$	Ni:	Org.Verb.1-60
$C_{15}F_6H_{28}N_3NiPS_2$	$[(t\text{-}C_4H_9NC)_2NiS_2CN(C_2H_5)_2]PF_6$	Ni:	Org.Verb.1-319
$C_{15}F_6H_{30}NiOP_2$	$(CF_3)_2CONi(P(C_2H_5)_3)_2$	Ni:	Org.Verb.1-60
$C_{15}F_6H_{30}NiP_2$	$(CF_3CFCF_2)Ni(P(C_2H_5)_3)_2$	Ni:	Org.Verb.1-364, 365
$C_{15}F_7FeH_8IN_2O_2$	$(CO)_2Fe(C_3F_7)(NC_5H_4C_5H_4N)I$	Fe:	Org.Verb.B1-121/2, 126
$C_{15}F_7FeH_{10}IN_2O_2$	$(CO)_2Fe(C_3F_7)(NC_5H_5)_2I$	Fe:	Org.Verb.B1-121/2, 126
$C_{15}F_7HN_4$	$((NC)_2C_6F_3)(NCC_6F_4)NH$	F:	PerFHalOrg.7-14/6, 42
$C_{15}F_7H_{16}N_2O_3P$	$C_3F_7P(O)(OH)_2 \cdot 2\ C_6H_5NH_2$	F:	PerFHalOrg.3-50
$C_{15}F_7H_{27}Sn$	$(C_4H_9)_3SnC_3F_7$	Sn:	Org.Verb.2-281
$C_{15}F_8FeH_{10}O$	$C_5H_5FeC_5H_4CO(CF_2)_4H$	Fe:	Org.Verb.A2-244
$C_{15}F_8FeH_{12}O$	$C_5H_5FeC_5H_4CHOH(CF_2)_4H$	Fe:	Org.Verb.A2-44
$C_{15}F_8H_6O_2$	$HC_6F_4C_6F_4COOC_2H_5$	F:	PerFHalOrg.4-54
$C_{15}F_9FeH_9O$	$C_5H_5FeC_5H_4COC_4F_9$	Fe:	Org.Verb.A2-241/2
$C_{15}F_9H_{12}IrO_6$	$Ir(CF_3COCHCOCH_3)_3$	Ir:	SVol.2-37
$C_{15}F_9H_{12}KMnO_6$	$K[Mn(CF_3COCHCOCH_3)_3]$	Mn:	MVol.D1-109
$C_{15}F_9H_{12}MnO_6$	$[Mn(CF_3COCHCOCH_3)_3]$	Mn:	MVol.D1-109/10
$C_{15}F_9H_{17}Sn$	$(CF_3CH_2CH_2)_3SnC_6H_5$	Sn:	Org.Verb.2-336, 338
$C_{15}F_9H_{26}N_3O_4Te$	$[C_5H_5NH][TeF_2(OH)_3O] \cdot 2\ C_5H_5N \cdot 7\ HF$	Te:	SVol.B2-47/8
$C_{15}F_{10}H_3P$	$(C_6F_5)_2PCCCH_3$	F:	PerFHalOrg.3-129, 130
$C_{15}F_{10}H_4O_3$	$(C_6F_5)_2C(OH)C(O)OCH_3$	F:	PerFHalOrg.4-50, 52
$C_{15}F_{10}H_6NP$	$(C_6F_5)_2PN(CH_2)(CHCH_3)$	F:	PerFHalOrg.3-125, 129, 132
$C_{15}F_{10}N_2S^+$	$[NC(C_6F_5)CSC(C_6F_5)N]^+$	F:	PerFHalOrg.5-95
$C_{15}F_{11}HN_2$	$NHC(C_6F_5)CFC(C_6F_5)N$	F:	PerFHalOrg.5-61, 83
$C_{15}F_{12}H_2N_2O$	$NHC(OH)(C_6F_5)CF_2C(C_6F_5)N$	F:	PerFHalOrg.5-60, 83
$C_{15}F_{12}H_4N_2O_2$	$F_2C(C(OH)(C_6F_5)NH)_2 \cdot N_2H_4$	F:	PerFHalOrg.5-61, 83
$C_{15}F_{12}H_6N_2O_9U$	$UO_2(CH(COCF_3)_2)_2 \cdot NO_2C_5H_4NO$	U:	SVol.E1-133, 139, 142/3
$C_{15}F_{12}H_7NO_7U$	$UO_2(CH(COCF_3)_2)_2 \cdot C_5H_5NO$	U:	SVol.E1-132, 136, 142/3

$C_{15}F_{12}H_8N_4O_2$ ...... $F_2C(C(OH)(C_6F_5)NH)_2 \cdot N_2H_4$ .... F: PerFHalOrg.5-61, 83
$C_{15}F_{12}H_9N_3O_2$ ...... $(CF_3)_2CNHC(CF_3)_2NC(NHCOC_6H_4OCH_3)$ ... F: PerFHalOrg.5-105
$C_{15}F_{12}H_{10}N_4O_2$ ..... $(CF_3)_2CNHC(CF_3)_2NHC(NCONHC_6H_4OCH_3)$. F: PerFHalOrg.5-106
$C_{15}F_{14}$ .... $C_6F_5C_6F_4CF_2CFCF_2$ .... F: PerFHalOrg.4-16
$C_{15}F_{15}H_{20}N_5O_5$ ..... $(C_2H_4 \cdot CF_3NO)_5$ .... F: PerFHalOrg.7-163
$C_{15}F_{18}FeH_3O_6$ ...... $Fe(CH(COCF_3)_2)_3$ .... Fe: Org.Verb.B3-181/2
$C_{15}F_{18}H_3MnO_6$ ..... $[Mn(CH(COCF_3)_2)_3]$ .... Mn: MVol.D1-111
$C_{15}F_{18}H_3MnO_6^-$ ..... $[Mn(CF_3COCHCOCF_3)_3]^-$ .... Mn: MVol.D1-111
$C_{15}F_{18}H_6N_6O_3$ ...... $(NH_2C(O)C_3F_6)_3C_3N_3$ .... F: PerFHalOrg.6-87/8, 107
$C_{15}F_{18}H_9N_5O_5U$ ..... $UO_2(OC(CF_3)_2C_3H_3N_2)_2 \cdot (CF_3)_2CO \cdot NH_3$. U: SVol.E1-13
$C_{15}F_{18}Ni_4O_3$ .... $(CF_3CCCF_3)_3Ni_4(CO)_3$ .... Ni: Org.Verb.2-386/7
$C_{15}F_{19}N$ .... $(i\text{-}C_3F_7)_2C_9F_5N$ .... F: PerFHalOrg.6-131/2, 143/4, 151, 158
$C_{15}F_{21}H_{13}O_3Sn$ ..... $CF_3(CF_2)_9CH_2CH_2Sn(OCH_3)_3$ .... Sn: Org.Verb.6-264
$C_{15}F_{22}N_2$ .... $[(CF_3)_2CF]_3(CN)C_5FN$ .... F: PerFHalOrg.5-197, 210
$C_{15}F_{24}H_4N_6O$ .... $[(CF_3)_2CNHC(CF_3)_2NCNH]_2CO$ .... F: PerFHalOrg.5-108
$C_{15}F_{24}N_2$ .... $(CF_3)_3(i\text{-}C_3F_7)_2C_6FN_2$ .... F: PerFHalOrg.6-128/9
$C_{15}F_{25}N$ .... $[(CF_3)_2CF]_2(CF_3)_2(C_2F_5)C_5N$ .... F: PerFHalOrg.5-153/4, 156/7, 164, 176
– .... $[(CF_3)_2CF]_3(CF_3)C_5FN$ .... F: PerFHalOrg.5-153/4, 164
– .... $(C_2F_5)_5C_5N$ .... F: PerFHalOrg.5-149/50, 156/7, 160, 169/70, 176
$C_{15}F_{27}N_3$ .... $(t\text{-}C_4F_9)_3C_3N_3$ .... F: PerFHalOrg.6-83/4, 103, 114/5
$C_{15}F_{27}N_3O_3$ .... $(C_2F_5OC_2F_4)_3C_3N_3$ .... F: PerFHalOrg.6-86/8, 106
$C_{15}F_{27}P_3$ .... $[(CF_3)_2P]_3C_6(CF_3)_3$ .... F: PerFHalOrg.3-164
$C_{15}F_{29}HN_4O_5$ .... $HC(O)C_6F_5[ON(CF_3)_2]_4$ .... F: PerFHalOrg.7-136
$C_{15}F_{29}HN_4O_6$ .... $HOC(O)C_6F_5[ON(CF_3)_2]_4$ .... F: PerFHalOrg.7-98/9, 116
$C_{15}F_{29}H_2NO_2$ .... $(CF_3)_2CFO(CF_2)_{11}CO(NH_2)$ .... F: PerFHalOrg.7-18/20, 63
$C_{15}F_{30}H_2N_2$ .... $(C_7F_{15})_2C(NH)_2$ .... F: PerFHalOrg.5-6
$C_{15}F_{32}N_4O_4$ .... $CF_3C_6F_5[ON(CF_3)_2]_4$ .... F: PerFHalOrg.7-98/9, 116
$C_{15}F_{35}P_5$ .... $(C_3F_7P)_5$ .... F: PerFHalOrg.3-1/3, 6
$C_{15}FeGeH_{27}O_4P$ .... $(CO)_4FeP(Ge(CH_3)_3)(C_4H_9)_2$ .... Fe: Org.Verb.B2-85, 91
$C_{15}FeH_8N_2O_3$ ...... $(CO)_3Fe(N_2C_{12}H_8)$ .... Fe: Org.Verb.B1-168/70
$C_{15}FeH_9NO_6$ .... $((CHCO)_2O)Fe(CO)_3(CNCH_2C_6H_5)$ .... Fe: Org.Verb.B4-351
$C_{15}FeH_{10}I_2O_3Te$ .... $(CO)_3Fe(Te(C_6H_5)_2)I_2$ .... Fe: Org.Verb.B1-180/3
$C_{15}FeH_{10}N_2O_6$ ...... $(C_6H_5NCO(CH)_2CONCH_3)Fe(CO)_4$ .... Fe: Org.Verb.B4-305, 312
$C_{15}FeH_{10}Ni_2O_5$ ..... $(C_5H_5)_2Ni_2Fe(CO)_5$ .... Ni: Org.Verb.2-350/1
$C_{15}FeH_{10}Ni_2O_5^+$ .... $(C_5H_5)_2Ni_2Fe(CO)_5^+$ .... Ni: Org.Verb.2-351
$C_{15}FeH_{10}O$ .... $C_5H_5FeC_5H_4(CC)_2CHO$ .... Fe: Org.Verb.A2-174
$C_{15}FeH_{10}O_2$ .... $C_5H_5FeC_5H_4(CC)_2COOH$ .... Fe: Org.Verb.A3-77/9
$C_{15}FeH_{10}O_4$ .... $C_6H_4C_5H_6Fe(CO)_4$ .... Fe: Org.Verb.B4-302
$C_{15}FeH_{10}O_7$ .... $C_6H_5COCHCHCOOCH_3Fe(CO)_4$ .... Fe: Org.Verb.B4-267/8, 276, 294
$C_{15}FeH_{11}NbO_5$ ..... $(C_5H_5)_2(CO)NbHFe(CO)_4$ .... Fe: Org.Verb.B3-142
$C_{15}FeH_{12}I_2O$ .... $(C_6H_5CHCHC(C_6H_5)O)FeI_2$ .... Fe: Org.Verb.B5-146
$C_{15}FeH_{12}N_2$ .... $C_5H_5FeC_5H_4C(CH_3)C(CN)_2$ .... Fe: Org.Verb.A2-220
$C_{15}FeH_{12}N_2O_2S$ ..... $C_5H_5FeC_5H_4CHC(CONH)_2CS$ .... Fe: Org.Verb.A2-158
$C_{15}FeH_{12}N_2O_3$ ...... $C_5H_5FeC_5H_4CHC(CONH)_2CO$ .... Fe: Org.Verb.A2-158
$C_{15}FeH_{12}O$ .... $C_5H_5FeC_5H_4(CC)_2CH_2OH$ .... Fe: Org.Verb.A2-75, 81

Formula	Compound	Reference
$C_{15}FeH_{12}O$	$C_5H_5FeC_5H_4CHOHCCCCH$	Fe: Org.Verb.A2-45
$C_{15}FeH_{12}O_2$	$C_5H_5FeC_5H_4COC_4H_3O$	Fe: Org.Verb.A3-162
$C_{15}FeH_{12}O_6$	$C_2H_5OC(O)CHCHC_6H_5Fe(CO)_4$	Fe: Org.Verb.B4-278, 298
$C_{15}FeH_{13}LiO_5$	$Li[(CH_3)_3CC_6H_4C(O)Fe(CO)_4]$	Fe: Org.Verb.B4-153, 162
$C_{15}FeH_{13}NO_5$	$(CH_3)_2NCOCHCHC_6H_5Fe(CO)_4$	Fe: Org.Verb.B4-278, 298
–	$HC(O)CHCHC_6H_4N(CH_3)_2Fe(CO)_4$	Fe: Org.Verb.B4-273
$C_{15}FeH_{13}N_2OS$	$COFe(C_{12}H_8N_2)SC_2H_5$	Fe: Org.Verb.B1-34
$C_{15}FeH_{13}O_4{}^+$	$[(C_6H_5)(CH_3)_2C_3H_2Fe(CO)_4]^+$	Fe: Org.Verb.B5-101, 105
$C_{15}FeH_{13}O_5{}^-$	$[(CH_3)_3CC_6H_4C(O)Fe(CO)_4]^-$	Fe: Org.Verb.B4-153, 162
$C_{15}FeH_{14}$	$C_5H_5FeC_5H_4CCC(CH_3)CH_2$	Fe: Org.Verb.A2-81
–	$C_5H_5FeC_5H_4C_5H_5$	Fe: Org.Verb.A1-338
$C_{15}FeH_{14}N_2O_5S_2$	$[Fe(CH_3C_6H_4SO_2N)_2CO \cdot H_2O]_n$	Fe: Org.Verb.C2-47
$C_{15}FeH_{14}N_2O_8$	$(CO)_4FeC_7H_8N_2(COOCH_3)_2$	Fe: Org.Verb.B2-183/4, 187
$C_{15}FeH_{14}O$	$C_5H_5FeC_5H_4(CHCH)_2CHO$	Fe: Org.Verb.A2-174, 177/8
–	$C_5H_5FeC_5H_4CH_2C_4H_3O$	Fe: Org.Verb.A3-162
–	$C_5H_5FeC_5H_4COCHCHCHCH_2$	Fe: Org.Verb.A2-248, 250
–	$C_5H_5FeC_5H_4C_5H_5O$	Fe: Org.Verb.A3-43, 46
$C_{15}FeH_{14}OS$	$C_5H_5FeC_5H_4C_4H_2SCH_2OH$	Fe: Org.Verb.A2-81
$C_{15}FeH_{14}O_2$	$C_5H_5FeC_5H_4CCCOOC_2H_5$	Fe: Org.Verb.A3-121, 123
–	$C_5H_5FeC_5H_4(CHCH)_2COOH$	Fe: Org.Verb.A3-78
$C_{15}FeH_{14}O_3$	$C_5H_5FeC_5H_4COCHCHCOOCH_3$	Fe: Org.Verb.A3-135
–	$C_5H_5FeC_5H_4COCH_2OOCCHCH_2$	Fe: Org.Verb.A2-295, 297
–	$C_{12}H_{14}Fe(CO)_3$	Fe: Org.Verb.B5-144
$C_{15}FeH_{14}O_3W^+$	$[C_5H_5FeC_5H_4C(OC_2H_5)W(CO)_2]^+$	Fe: Org.Verb.A1-347
$C_{15}FeH_{14}O_4$	$CH_3CHCHCH_2CH_2C_6H_5Fe(CO)_4$	Fe: Org.Verb.B4-272, 289
$C_{15}FeH_{14}O_4Si$	$C_6H_5CCSi(CH_3)_3Fe(CO)_4$	Fe: Org.Verb.B4-328
$C_{15}FeH_{14}S$	$C_5H_5FeC_5H_4CH_2C_4H_3S$	Fe: Org.Verb.A2-47
$C_{15}FeH_{15}$	$(C_5H_5)_3Fe$	Fe: Org.Verb.B1-9
$C_{15}FeH_{15}KO_3$	$C_5H_5FeC_5H_4COCH_2CH(CH_3)COOK$	Fe: Org.Verb.A2-227
$C_{15}FeH_{15}N$	$C_5H_5FeC_5H_3C_4H_7CN$	Fe: Org.Verb.A1-358
$C_{15}FeH_{15}NO$	$C_5H_5FeC_5H_4COCH_2CH(CN)CH_3$	Fe: Org.Verb.A2-250
$C_{15}FeH_{15}Ni^+$	$[FeNi(C_5H_5)_3]^+$	Fe: Org.Verb.A1-93/4
		Ni: Org.Verb.2-135, 201
$C_{15}FeH_{15}O_2{}^+$	$[FeC_5H_4C_6H_4COOC_3H_7]^+$	Fe: Org.Verb.A3-126
$C_{15}FeH_{15}O_3{}^{2-}$	$[C_5H_5FeC_5H_3(O)COCOCH(CH_3)_2]^{2-}$	Fe: Org.Verb.A2-302
$C_{15}FeH_{15}O_6P$	$HOC(O)CHCH_2Fe(CO)_4 \cdot P(CH_3)_2C_6H_5$	Fe: Org.Verb.B4-245/6
$C_{15}FeH_{16}$	$C_5H_5FeC_5H_4C(CH_2)C_3H_5$	Fe: Org.Verb.A1-333/4
–	$C_5H_5FeC_5H_4C(CH_3)CHCHCH_2$	Fe: Org.Verb.A1-291, 294
–	$C_5H_5FeC_5H_4C_5H_7$	Fe: Org.Verb.A1-338, 340/1
$C_{15}FeH_{16}N_2O$	$C_5H_5FeC_5H_4CHNNHCOC_3H_5$	Fe: Org.Verb.A2-167
$C_{15}FeH_{16}O$	$C_5H_5FeC_5H_3CH_2C(CH_3)_2CO$	Fe: Org.Verb.A3-101
–	$C_5H_5FeC_5H_3CH_2CH(CH_3)CH_2CO$	Fe: Org.Verb.A3-102
–	$C_5H_5FeC_5H_3CH_2CH(C_2H_5)CO$	Fe: Org.Verb.A3-101
–	$C_5H_5FeC_5H_3(CH_2)_2CH(CH_3)CO$	Fe: Org.Verb.A3-102
–	$C_5H_5FeC_5H_3(CH_2)_4CO$	Fe: Org.Verb.A3-103
–	$C_5H_5FeC_5H_4CCC(CH_3)_2OH$	Fe: Org.Verb.A2-75, 81
–	$C_5H_5FeC_5H_4CCCH_2OC_2H_5$	Fe: Org.Verb.A2-131, 135
–	$C_5H_5FeC_5H_4C(CH_3)CHCOCH_3$	Fe: Org.Verb.A3-24/5
–	$C_5H_5FeC_5H_4C(CH_3)(CH_2CCH)OH$	Fe: Org.Verb.A2-59, 62
–	$C_5H_4FeC_5H_4C(CH_3)_2CH_2CO$	Fe: Org.Verb.A3-101

Formula	Compound	Reference
$C_{15}FeH_{16}O$	$C_5H_5FeC_5H_4CH(CCH)OC_2H_5$	Fe: Org.Verb.A2-118, 120
–	$C_5H_5FeC_5H_4CHC(CH_3)COCH_3$	Fe: Org.Verb.A3-24/5
–	$C_5H_5FeC_5H_4CHCHCOC_2H_5$	Fe: Org.Verb.A3-8/10
–	$C_5H_4FeC_5H_4CH(CH_3)CH(CH_3)CO$	Fe: Org.Verb.A3-101
–	$C_5H_4FeC_5H_4CH_2C(CH_3)_2CO$	Fe: Org.Verb.A3-101
–	$C_5H_4FeC_5H_4CH_2CH(C_2H_5)CO$	Fe: Org.Verb.A3-101
–	$C_5H_5FeC_5H_4COCHC(CH_3)_2$	Fe: Org.Verb.A2-248, 250
–	$C_5H_5FeC_5H_4COC_4H_7$	Fe: Org.Verb.A2-290
–	$C_5H_5FeC_5H_4C_5H_6OH$	Fe: Org.Verb.A2-75, 81
$C_{15}FeH_{16}O_2$	$CH_3COC_5H_4FeC_5H_4COC_2H_5$	Fe: Org.Verb.A2-210
–	$C_5H_5FeC_5H_4C(CHCH_3)COOCH_3$	Fe: Org.Verb.A3-121, 123
–	$C_5H_5FeC_5H_4C(CH_3)C(CH_3)COOH$	Fe: Org.Verb.A3-77/8
–	$C_5H_5FeC_5H_4C(CH_3)CHCOOCH_3$	Fe: Org.Verb.A3-121, 123/4
–	$C_5H_5FeC_5H_4C(CH_3)(CH_2)_2COO$	Fe: Org.Verb.A3-166
–	$C_5H_5FeC_5H_4CHC(CH_3)COOCH_3$	Fe: Org.Verb.A3-122, 124
–	$C_5H_5FeC_5H_4CHCHCH_2COOCH_3$	Fe: Org.Verb.A3-122, 124/5
–	$C_5H_5FeC_5H_4CHCHCOOC_2H_5$	Fe: Org.Verb.A3-121, 123
–	$C_5H_5FeC_5H_4CH(CH_3)OOCCHCH_2$	Fe: Org.Verb.A2-98, 100/1
–	$C_5H_5FeC_5H_4CHOHCCCHOHCH_3$	Fe: Org.Verb.A2-88
–	$C_5H_5FeC_5H_4CHOHCCOC_2H_5$	Fe: Org.Verb.A2-133
–	$C_5H_5FeC_5H_4CH_2OOCC(CH_3)CH_2$	Fe: Org.Verb.A2-97/8
–	$C_5H_5FeC_5H_4(CH_2)_2OOCCHCH_2$	Fe: Org.Verb.A2-103/4
–	$C_5H_5FeC_5H_4CO(CH_2)_2COCH_3$	Fe: Org.Verb.A3-32, 34
–	$C_5H_5FeC_5H_4C(OC_2H_5)CHCHO$	Fe: Org.Verb.A2-174, 178
–	$C_5H_5FeC_5H_4C_3H_3(CH_3)COOH$	Fe: Org.Verb.A3-82/4
$C_{15}FeH_{16}O_2^-$	$[C_5H_5FeC_5H_4COCOCH(CH_3)_2]^-$	Fe: Org.Verb.A2-302
$C_{15}FeH_{16}O_3$	$(CH_3)_4C_8H_4Fe(CO)_3$	Fe: Org.Verb.B5-145/6
–	$C_3H_7COC_5H_4FeC_5H_4COOH$	Fe: Org.Verb.A3-58/9
–	$C_5H_5FeC_5H_4CH_2CH(COCH_3)COOH$	Fe: Org.Verb.A3-94
–	$C_5H_5FeC_5H_4(CH_2)_2COCH_2COOH$	Fe: Org.Verb.A3-94
–	$C_5H_5FeC_5H_4COCH(CH_3)CH_2COOH$	Fe: Org.Verb.A3-87/8, 90, 104
–	$C_5H_5FeC_5H_4COCH_2CH(CH_3)COOH$	Fe: Org.Verb.A3-87/8, 90, 104
–	$C_5H_5FeC_5H_4COCH_2COOC_2H_5$	Fe: Org.Verb.A3-134/5
–	$C_5H_5FeC_5H_4CO(CH_2)_2COOCH_3$	Fe: Org.Verb.A3-135, 140
–	$C_5H_5FeC_5H_4CO(CH_2)_3COOH$	Fe: Org.Verb.A3-87, 89/90, 104
$C_{15}FeH_{16}O_4$	$C_5H_5FeC_5H_4CH(CH_3)CH(COOH)_2$	Fe: Org.Verb.A3-96/8
–	$C_5H_5FeC_5H_4CH_2C(COOH)_2CH_3$	Fe: Org.Verb.A3-96/8
–	$C_5H_5FeC_5H_4COOCH(CH_2O)_2CH_2$	Fe: Org.Verb.A3-179
–	$C_5H_5FeC_5H_4COOCH_2C_3H_5O_2$	Fe: Org.Verb.A3-178
$C_{15}FeH_{16}O_6$	$((CH_3)_3C_8H_7O_2)Fe(CO)_4$	Fe: Org.Verb.B4-128, 132, 135
$C_{15}FeH_{16}O_8$	$(CO)_3FeC_6H_7(COOCH_3)_2(C(O)CH_3)$	Fe: Org.Verb.B4-211
$C_{15}FeH_{17}^+$	$[C_5H_5FeC_5H_4C(CH_3)C_3H_5]^+$	Fe: Org.Verb.A2-62
$C_{15}FeH_{17}NO_2$	$C_5H_5FeC_5H_4CON(CH_2CH_2)_2O$	Fe: Org.Verb.A3-154
$C_{15}FeH_{17}NO_3$	$C_5H_5FeC_5H_4CONHCH_2COOC_2H_5$	Fe: Org.Verb.A3-154
$C_{15}FeH_{17}NO_5$	$[(CH_3)_4N][C_6H_5C(O)Fe(CO)_4]$	Fe: Org.Verb.B4-152, 161
$C_{15}FeH_{17}NW^+$	$[C_5H_5FeC_5H_4C(NC_4H_8)W]^+$	Fe: Org.Verb.A1-347

Formula	Compound	Reference
$C_{15}FeH_{17}N_3O$	$C_5H_5FeC_5H_4CHCHC(CH_3)NNHCONH_2$	Fe: Org.Verb.A3-10
$C_{15}FeH_{17}N_3S$	$C_5H_5FeC_5H_4CHNNHCSNHCH_2CHCH_2$	Fe: Org.Verb.A2-167
$C_{15}FeH_{17}O_2$	$[C_5H_5FeC_5H_4(CH_2)_4COO]$	Fe: Org.Verb.A3-68
$C_{15}FeH_{17}O_2^+$	$[C_5H_5FeC_5H_4C(CH_3)CH_2COOCH_3]^+$	Fe: Org.Verb.A3-123
–	$[C_5H_5FeC_5H_4C(CH_3)CD_2COOCH_3]^+$	Fe: Org.Verb.A3-124
$C_{15}FeH_{17}O_2^-$	$[C_5H_5FeC_5H_4(CH_2)_4COO]^-$	Fe: Org.Verb.A3-68
$C_{15}FeH_{17}O_4P$	$(C(O)CH_2CH_2CH_2)Fe(CO)_3P(CH_3)_2C_6H_5$	Fe: Org.Verb.B4-203, 207
$C_{15}FeH_{17}O_7P$	$C_6H_5C_2H_2CH(O)Fe(CO)_3P(OCH_3)_3$	Fe: Org.Verb.B4-199, 205/6
$C_{15}FeH_{18}$	$C_5H_5FeC_5H_4C(CH_3)C(CH_3)_2$	Fe: Org.Verb.A1-291/3
–	$C_5H_5FeC_5H_4C(CH_3)CHC_2H_5$	Fe: Org.Verb.A2-62
–	$C_5H_5FeC_5H_4C(C_3H_7\text{-}n)CH_2$	Fe: Org.Verb.A1-291, 293
–	$C_5H_5FeC_5H_4C(C_3H_7\text{-}i)CH_2$	Fe: Org.Verb.A1-291, 293
–	$C_5H_5FeC_5H_4CH(CH_2)_4$	Fe: Org.Verb.A1-338, 341
–	$C_5H_5FeC_5H_4CH_2CHC(CH_3)_2$	Fe: Org.Verb.A1-291, 294
–	$C_5H_5FeC_5H_4C_3H_3(CH_3)_2$	Fe: Org.Verb.A1-337
$C_{15}FeH_{18}N_6O_5$	$COFe(NC_5H_4CN)(C_2HN_2O_2(CH_3)_2)_2$	Fe: Org.Verb.B1-33
$C_{15}FeH_{18}O$	$C_5H_5FeC_5H_4C(CH_2)(CH_2)_3OH$	Fe: Org.Verb.A2-76
–	$C_5H_5FeC_5H_4C(CH_3)CH(CH_2)_2OH$	Fe: Org.Verb.A2-75, 81/2
–	$C_5H_5FeC_5H_4C(CH_3)(C_3H_5)OH$	Fe: Org.Verb.A2-59, 62
–	$C_5H_5FeC_5H_4CH(CH_2)_4O$	Fe: Org.Verb.A3-176
–	$C_5H_5FeC_5H_4CH(C_3H_5)OCH_3$	Fe: Org.Verb.A2-118
–	$C_5H_5FeC_5H_4CHOHC_4H_7$	Fe: Org.Verb.A2-4, 46, 50
–	$C_5H_5FeC_5H_4(CH_2)_2COC_2H_5$	Fe: Org.Verb.A3-3/4
–	$C_5H_5FeC_5H_4(CH_2)_3COCH_3$	Fe: Org.Verb.A3-24/5
–	$C_5H_5FeC_5H_4COC(CH_3)_3$	Fe: Org.Verb.A2-185, 238/40
–	$C_5H_5FeC_5H_4COCH(CH_3)C_2H_5$	Fe: Org.Verb.A2-238/9
–	$C_5H_5FeC_5H_4COCH_2CH(CH_3)_2$	Fe: Org.Verb.A2-185, 238/9
–	$C_5H_5FeC_5H_4(COC_4H_9\text{-}n)$	Fe: Org.Verb.A2-235/6
–	$C_5H_5FeC_5H_4C_3H_3(CH_3)CH_2OH$	Fe: Org.Verb.A2-75, 81
–	$C_5H_5FeC_5H_4C_5H_8OH$	Fe: Org.Verb.A2-68, 75, 81
$C_{15}FeH_{18}O^+$	$[C_5H_5FeC_5H_4COC_4H_9\text{-}t]^+$	Fe: Org.Verb.A2-239
$C_{15}FeH_{18}O^-$	$[C_5H_5FeC_5H_4COC(CH_3)_3]^-$	Fe: Org.Verb.A2-301
$C_{15}FeH_{18}OS$	$C_5H_5FeC_5H_4COCH_2SC_3H_7$	Fe: Org.Verb.A2-245
$C_{15}FeH_{18}O_2$	$C_4H_9C_5H_4FeC_5H_4COOH$	Fe: Org.Verb.A3-58/9
–	$C_5H_5FeC_5H_3(CH_3)CH(CH_3)OCOCH_3$	Fe: Org.Verb.A1-357/8
–	$C_5H_5FeC_5H_4C(CH_3)_2CH_2COOH$	Fe: Org.Verb.A3-72, 100/1
–	$C_5H_5FeC_5H_4C(CH_3)_2OOCCH_3$	Fe: Org.Verb.A2-98
–	$C_5H_5FeC_5H_4CH(CH_3)CH(CH_3)COOH$	Fe: Org.Verb.A3-72, 100/1
–	$C_5H_5FeC_5H_4CH(CH_3)CH_2OOCCH_3$	Fe: Org.Verb.A2-104/5
–	$C_5H_5FeC_5H_4CH(CH_3)(CH_2)_2COOH$	Fe: Org.Verb.A3-73, 75
–	$C_5H_5FeC_5H_4CHOHCHCHOC_2H_5$	Fe: Org.Verb.A2-133, 137
–	$C_5H_5FeC_5H_4CH_2C(CH_3)_2COOH$	Fe: Org.Verb.A3-72, 100/1
–	$C_5H_5FeC_5H_4CH_2CH(CH_3)CH_2COOH$	Fe: Org.Verb.A3-71, 73, 100, 102
–	$C_5H_5FeC_5H_4CH_2CH(CH_3)OOCCH_3$	Fe: Org.Verb.A2-104/5
–	$C_5H_5FeC_5H_4CH_2CH(C_2H_5)COOH$	Fe: Org.Verb.A3-72, 100/1
–	$C_5H_5FeC_5H_4CH_2OOC(C_3H_7\text{-}n)$	Fe: Org.Verb.A2-98
–	$C_5H_5FeC_5H_4(CH_2)_2CH(CH_3)COOH$	Fe: Org.Verb.A3-71, 73, 100, 102
–	$C_5H_5FeC_5H_4(CH_2)_2COOC_2H_5$	Fe: Org.Verb.A3-115/7

$C_{15}FeH_{18}O_2$	$C_5H_5FeC_5H_4(CH_2)_3COOCH_3$	Fe: Org.Verb.A3-115, 118
–	$C_5H_5FeC_5H_4(CH_2)_4COOH$	Fe: Org.Verb.A3-68/70, 101, 103
–	$C_5H_5FeC_5H_4COC(CH_3)_2CH_2OH$	Fe: Org.Verb.A2-292
–	$C_5H_5FeC_5H_4OCH(CH_2)_4O$	Fe: Org.Verb.A3-176
$C_{15}FeH_{18}O_2S$	$C_5H_5FeC_5H_4CH(C_2H_5)SCH_2COOH$	Fe: Org.Verb.A2-43
$C_{15}FeH_{18}O_3$	$C_5H_5FeC_5H_4CHOHCH_2COOC_2H_5$	Fe: Org.Verb.A3-143
–	$C_5H_5FeC_5H_4CH_2OOCCH_2CHOHCH_3$	Fe: Org.Verb.A2-98
–	$C_5H_5FeC_5H_4COCH_2CH(OCH_3)_2$	Fe: Org.Verb.A2-295, 297
$C_{15}FeH_{18}O_3S$	$C_5H_5FeC_5H_4COCH_2SCH_2CHOHCH_2OH$	Fe: Org.Verb.A2-245
$C_{15}FeH_{18}O_9P_2$	$(CO)_3Fe(PO_3C_6H_9)_2$	Fe: Org.Verb.B1-148/50, 155, 163
$C_{15}FeH_{18}S$	$C_5H_5FeC_5H_4CSC(CH_3)_3$	Fe: Org.Verb.A2-240
$C_{15}FeH_{19}^+$	$[(CH_3)_3C_6H_3FeC_5H_4CH_3]^+$	Fe: Org.Verb.A2-150
–	$[(CH_3)_4C_6H_2FeC_5H_5]^+$	Fe: Org.Verb.A2-186, 208
–	$[C_5H_5FeC_5H_4C(CH_3)CH(CH_3)_2]^+$	Fe: Org.Verb.A1-351
–	$[C_5H_5FeC_5H_4CHC(CH_3)_3]^+$	Fe: Org.Verb.A1-350/1
$C_{15}FeH_{19}N$	$C_5H_5FeC_5H_4CHCHCH_2N(CH_3)_2$	Fe: Org.Verb.A1-287
$C_{15}FeH_{19}N^+$	$[C_5H_5FeC_5H_4CH_2NC_4H_8]^+$	Fe: Org.Verb.A1-346
$C_{15}FeH_{19}N_3O$	$C_5H_5FeC_5H_4(CH_2)_2C(CH_3)NNHCONH_2$	Fe: Org.Verb.A3-4
$C_{15}FeH_{19}NaO_7$	$Na[C_5H_{11}OCO(CH_2)_4C(O)Fe(CO)_4]$	Fe: Org.Verb.B4-155, 163
$C_{15}FeH_{19}NiO_2P$	$C_5H_5Ni(CO)_2Fe(C_5H_5)P(CH_3)_3$	Ni: Org.Verb.2-169
$C_{15}FeH_{19}O^+$	$[C_5H_5FeC_5H_4C(CH_3)(CH_2)_3OH]^+$	Fe: Org.Verb.A2-81/2, 90
–	$[C_5H_5FeC_5H_4CH(CH_2)_4OH]^+$	Fe: Org.Verb.A2-90
$C_{15}FeH_{19}O_6P$	$(C_6H_5CHCHCOCH_3)Fe(CO)_2P(OCH_3)_3$	Fe: Org.Verb.B5-115
$C_{15}FeH_{19}O_7^-$	$[C_5H_{11}OCO(CH_2)_4C(O)Fe(CO)_4]^-$	Fe: Org.Verb.B4-155, 163
$C_{15}FeH_{20}$	$C_5H_5FeC_5H_4C(CH_3)_2C_2H_5$	Fe: Org.Verb.A1-275/6
–	$C_5H_5FeC_5H_4CH(CH_3)(C_3H_7\text{-n})$	Fe: Org.Verb.A1-275
–	$C_5H_5FeC_5H_4CH_2C(CH_3)_3$	Fe: Org.Verb.A1-275/6
–	$C_5H_5FeC_5H_4(CH_2)_2CH(CH_3)_2$	Fe: Org.Verb.A1-275/6
–	$C_5H_5FeC_5H_4(CH_2)_4CH_3$	Fe: Org.Verb.A1-268/9
$C_{15}FeH_{20}O$	$C_5H_5FeC_5H_4C(CH_3)(C_3H_7\text{-n})OH$	Fe: Org.Verb.A2-59, 62
–	$C_5H_5FeC_5H_4C(CH_3)(C_3H_7\text{-i})OH$	Fe: Org.Verb.A2-58/9, 63
–	$C_5H_5FeC_5H_4C(CH_3)_2OC_2H_5$	Fe: Org.Verb.A2-119, 123
–	$C_5H_5FeC_5H_4CH(C_3H_7)OCH_3$	Fe: Org.Verb.A2-118
–	$C_5H_5FeC_5H_4CHOHC(CH_3)_3$	Fe: Org.Verb.A2-4, 44, 47
–	$C_5H_5FeC_5H_4CHOHC_4H_9\text{-n}$	Fe: Org.Verb.A2-44, 47
–	$C_5H_5FeC_5H_4CH_2OC(CH_3)_3$	Fe: Org.Verb.A2-113/4
–	$C_5H_5FeC_5H_4CH_2OC_4H_9\text{-n}$	Fe: Org.Verb.A2-113
$C_{15}FeH_{20}OSi$	$C_5H_5FeC_5H_4COCH_2Si(CH_3)_3$	Fe: Org.Verb.A3-110
$C_{15}FeH_{20}O_2$	$C_5H_5FeC_5H_4C(CH_3)(OH)(CH_2)_3OH$	Fe: Org.Verb.A2-87, 90
–	$C_5H_5FeC_5H_4CH(CH_2CH_2CH_2OH)OCH_3$	Fe: Org.Verb.A2-118, 120
–	$C_5H_5FeC_5H_4CHOHC(CH_3)_2CH_2OH$	Fe: Org.Verb.A2-87, 90
–	$C_5H_5FeC_5H_4CHOH(CH_2)_4OH$	Fe: Org.Verb.A2-87, 90
–	$C_5H_5FeC_5H_4CH_2CHOHCH_2OC_2H_5$	Fe: Org.Verb.A2-133, 137
–	$C_5H_5FeC_5H_4CH_2O(CH_2)_2OC_2H_5$	Fe: Org.Verb.A2-113
$C_{15}FeH_{21}N$	$C_5H_5FeC_5H_4CH(C_2H_5)N(CH_3)_2$	Fe: Org.Verb.A2-43
–	$C_5H_5FeC_5H_4CH_2N(C_2H_5)_2$	Fe: Org.Verb.A1-167
$C_{15}FeH_{21}NO$	$C_5H_5FeC_5H_4CHOHCH_2CH_2N(CH_3)_2$	Fe: Org.Verb.A1-287
		Fe: Org.Verb.A2-30

Formula	Compound	Reference
$C_{15}FeH_{21}N_2O_4^+$	$[CH_2CHC(N(CH_2)_5)N(CH_3)C(OC_2H_5)Fe(CO)_3]^+$	Fe: Org.Verb.B5-161
$C_{15}FeH_{21}N_5O_5$	$COFe(NC_5H_4CH_3)(C_2HN_2O_2(CH_3)_2)_2$	Fe: Org.Verb.B1-33
$C_{15}FeH_{21}NaO_5$	$Na[C_{10}H_{21}C(O)Fe(CO)_4]$	Fe: Org.Verb.B4-151
$C_{15}FeH_{21}O_5^-$	$[C_{10}H_{21}C(O)Fe(CO)_4]^-$	Fe: Org.Verb.B4-151
$C_{15}FeH_{21}O_6$	$Fe(CH(COCH_3)_2)_3$	Fe: Org.Verb.B3-181
$C_{15}FeH_{22}IN$	$C_5H_5FeC_5H_4CH(CH_3)N(CH_3)_3I$	Fe: Org.Verb.A1-359, 362
$C_{15}FeH_{22}LiNO_7$	$Li[(CH_3)_2NC(O)Fe(CO)_4] \cdot 2\ C_4H_8O$	Fe: Org.Verb.B4-155, 163
$C_{15}FeH_{22}O_9P_2$	$(CO)_3Fe(P(OCH_2)_3CC_2H_5)_2$	Fe: Org.Verb.B1-148/50, 154, 163
$C_{15}FeH_{22}Si$	$C_5H_5FeC_5H_4CH(CH_3)Si(CH_3)_3$	Fe: Org.Verb.A2-117
$C_{15}FeH_{23}O_2P$	$(C_7H_8)Fe(CO)_2P(C_2H_5)_3$	Fe: Org.Verb.B1-158
$C_{15}FeH_{24}NO_2P_2^+$	$[(CH_3)_3PC_8H_6Fe(CO)(NO)P(CH_3)_3]^+$	Fe: Org.Verb.B5-11/3
$C_{15}FeH_{24}O_{12}P_2S_2$	$(CH_3OCOCS)_2CFe(CO)_2(P(OCH_3)_3)_2$	Fe: Org.Verb.B4-137/8
$C_{15}FeH_{25}NO_3P^+$	$[C_4H_4P(C_3H_7\text{-}n)_3Fe(CO)_2NO]^+$	Fe: Org.Verb.B5-28/31
$C_{15}FeH_{27}$	$((CH_3)_2C_3H_3)_3Fe$	Fe: Org.Verb.B5-108
$C_{15}FeH_{27}NO_4Sn$	$(CO)_3Fe(NO)Sn(C_4H_9\text{-}n)_3$	Fe: Org.Verb.B1-197/8
$C_{15}FeH_{27}O_4PSi$	$(CO)_4FeP(Si(CH_3)_3)(C_4H_9)_2$	Fe: Org.Verb.B2-85, 91
$C_{15}FeH_{27}O_4PSn$	$(CO)_4FeP(Sn(CH_3)_3)(C_4H_9\text{-}t)_2$	Fe: Org.Verb.B2-85, 92, 112
$C_{15}FeH_{27}S_9$	$Fe[S_2CSC_4H_9]_3$	C: MVol.D4-266/7, 269
$C_{15}FeH_{30}O_3P_2$	$(CO)_3Fe(P(C_2H_5)_3)_2$	Fe: Org.Verb.B1-148/51, 158
$C_{15}FeH_{30}O_8P_2S_2$	$CS_2Fe(CO)_2(P(OC_2H_5)_3)_2$	Fe: Org.Verb.B5-143
$C_{15}FeH_{30}O_9P_2$	$(CO)_3Fe(P(OC_2H_5)_3)_2$	Fe: Org.Verb.B1-148/50, 154, 163
$C_{15}FeH_{34}NP_2^+$	$[(CH_3)_3CNCFe(C_2H_4(P(C_2H_5)_2)_2)H]^+$	Fe: Org.Verb.B4-10
$C_{15}FeH_{35}NP_4$	$NCCHCH_2Fe(P(CH_3)_2C_2H_4P(CH_3)_2)_2$	Fe: Org.Verb.B4-181, 183/4
$C_{15}FeH_{36}N_6O_3P_2$	$(CO)_3Fe(P(N(CH_3)_2)_3)_2$	Fe: Org.Verb.B1-148/50, 155, 164
$C_{15}FeH_{36}N_6O_3P_2^+$	$[(CO)_3Fe(P(N(CH_3)_2)_3)_2]^+$	Fe: Org.Verb.B1-164
$C_{15}FeH_{36}O_9P_3^+$	$[C_6H_9Fe(P(OCH_3)_3)_3]^+$	Fe: Org.Verb.B5-3
$C_{15}FeH_{38}OP_2Si_2^+$	$[(CO)Fe(P(CH_3)_2C_2H_4Si(CH_3)_3)_2]^+$	Fe: Org.Verb.B1-160
$C_{15}Fe_2GeH_{27}IN_4O_4$	$((CH_3)_2C_3H_3Fe(NO)_2)_2Ge(I)C_5H_9$	Fe: Org.Verb.C2-165
$C_{15}Fe_2H_6N_2O_7$	$(CO)_3Fe(CO)(N_2C_8H_6)Fe(CO)_3$	Fe: Org.Verb.C1-200/1
$C_{15}Fe_2H_6O_8$	$(CO)_4Fe(CHC_6H_5)Fe(CO)_4$	Fe: Org.Verb.C1-225/6
$C_{15}Fe_2H_{10}O_7S$	$(CH_3C_6H_3CH(OCH_3)S)Fe_2(CO)_6$	Fe: Org.Verb.C2-113
–	$(CO)_3Fe(C_6H_4CH(OC_2H_5)S)Fe(CO)_3$	Fe: Org.Verb.C2-107, 112, 118
–	$(CO)_6Fe_2(CH_3OC(C_6H_4CH_3)S)$	Fe: Org.Verb.C1-238/40
–	$(CO)_6Fe_2(C_2H_5OC(C_6H_5)S)$	Fe: Org.Verb.C1-238/9, 241, 243
$C_{15}Fe_2H_{10}O_8S$	$(CH_3OC_6H_3CH(OCH_3)S)Fe_2(CO)_6$	Fe: Org.Verb.C2-113
–	$(CO)_6Fe_2(CH_3OC(C_6H_4OCH_3)S)$	Fe: Org.Verb.C1-238/9, 241
$C_{15}Fe_2H_{11}NO_6S$	$(CO)_6Fe_2((CH_3)_2NC(C_6H_5)S)$	Fe: Org.Verb.C1-238/40
$C_{15}Fe_2H_{12}^+$	$[FeC_5H_4C_5H_4FeC_5H_4]^+$	Fe: Org.Verb.A6-15, 17
$C_{15}Fe_2H_{12}N_4O_6$	$(CO)_6Fe_2(N_2C_3(CH_3)_4(CN)_2)$	Fe: Org.Verb.C1-149/50, 153
$C_{15}Fe_2H_{13}O_4P$	$C_5H_5Fe(P(CHCCH_3)_2)Fe(CO)_4$	Fe: Org.Verb.C2-54
$C_{15}Fe_2H_{14}N_2O_6$	$C(CH_3)C(CH_3)N_2C_5H_8Fe_2(CO)_6$	Fe: Org.Verb.C2-122, 125
$C_{15}Fe_2H_{14}N_2O_7$	$(C_6H_5N_2H_3)(CO)_2Fe(CH_3CO)_2Fe(CO)_3$	Fe: Org.Verb.C1-232/4
$C_{15}Fe_2H_{14}O_6S_2$	$(CO)_6Fe_2S_2C_7H_8(CH_3)_2$	Fe: Org.Verb.C1-115/6, 118, 208/9

Formula	Compound		Reference
$C_{15}Fe_2H_{15}^+$	$[Fe_2(C_5H_5)_3]^+$	Fe:	Org.Verb.A1-93/4
$C_{15}Fe_2H_{15}N_7O_4$	$Fe_2(NO)_4(NC_5H_5)_3$	Fe:	Org.Verb.B1-141
$C_{15}Fe_2H_{16}N_2O_6$	$(CO)_6Fe_2(NC(CH_3)_2)(NC(CH_2)_5)$	Fe:	Org.Verb.C1-133, 141
$C_{15}Fe_2H_{17}N_3O_6$	$(CO)_6Fe_2[(NCH_3)_2CNC_6H_{11}]$	Fe:	Org.Verb.C1-143/4, 146
$C_{15}Fe_2H_{17}O_5PS_2$	$(CO)_3Fe(SCH_3)_2Fe(CO)_2(P(CH_3)_2C_6H_5)$	Fe:	Org.Verb.C1-95, 99
$C_{15}Fe_2H_{18}N_2O_7$	$(CO)_3Fe(C_4H_9NCONC_4H_9)Fe(CO)_3$	Fe:	Org.Verb.C1-143/5
$C_{15}Fe_2H_{20}O_{12}P_2$	$(CO)_7Fe_2((C_2H_5O)_2POP(OC_2H_5)_2)$	Fe:	Org.Verb.C1-200, 202
$C_{15}Fe_2H_{27}IN_4O_4Pb$	$((CH_3)_2C_3H_3Fe(NO)_2)_2Pb(I)C_5H_9$	Fe:	Org.Verb.C2-165
$C_{15}Fe_2H_{27}IN_4O_4Sn$	$((CH_3)_2C_3H_3Fe(NO)_2)_2Sn(I)C_5H_9$	Fe:	Org.Verb.C2-165
$C_{15}Fe_2H_{27}O_5P_3$	$(CO)_3Fe(P(CH_3)_2)_2Fe(CO)_2P(C_2H_5)_3$	Fe:	Org.Verb.C1-182/3
$C_{15}Fe_3H_5N_3O_{12}Sn$	$C_6H_5Sn[Fe(CO)_3NO]_3$	Sn:	Org.Verb.6-275, 277
$C_{15}Fe_5O_{14}^{2-}$	$[Fe_5(CO)_{14}C]^{2-}$	Fe:	Org.Verb.B3-143
$C_{15}GdH_{18}I_6N_3$	$(C_5H_5NH)_3GdI_6$	Sc:	MVol.C6-273
$C_{15}GeH_{32}Sn$	$(C_2H_5)_3SnCCCH_2Ge(C_2H_5)_3$	Sn:	Org.Verb.2-231, 235
$C_{15}H_9N_{11}O_8U$	$[UO_2(NO_3)_2(C_3N_3(C_4H_3N_2)_3)] \cdot 7\ H_2O$	U:	SVol.E1-46/8
$C_{15}H_{10}KNiO_3P$	$K[(CO)_3NiP(C_6H_5)_2]$	Ni:	Org.Verb.1-191, 192
$C_{15}H_{10}MnNO_2^+$	$[Mn(C_6H_4(CO)_2CC_5H_3(CH_3)N)]^+$	Mn:	MVol.D1-128
$C_{15}H_{10}MnNi_2O_5^-$	$[(C_5H_5)_2Ni_2Mn(CO)_5]^-$	Ni:	Org.Verb.2-350
$C_{15}H_{10}MnO_7^{2+}$	$[Mn(C_9H_2O_2(OH)_3C_6H_3(OH)_2)]^{2+}$	Mn:	MVol.D1-152
$C_{15}H_{11}I_4N_3Sn$	$SnI_4 \cdot NC_5H_3(C_5H_4N)_2$	Sn:	MVol.C5-220
$C_{15}H_{11}MnO_2^+$	$[Mn(C_6H_5COCHCOC_6H_5)]^+$	Mn:	MVol.D1-112
$C_{15}H_{11}NiO_3P$	$(CO)_3NiP(C_6H_5)_2H$	Ni:	Org.Verb.1-171, 172
$C_{15}H_{12}N_5O_8U$	$[UO_2(NO_3)_2(C_{10}H_8N_2)_{1.5}]$	U:	SVol.E1-41, 43
$C_{15}H_{12}N_8O_{17}U$	$UO_2(NO_3)_2 \cdot 3\ NO_2C_5H_4NO$	U:	SVol.E1-133, 142/3
$C_{15}H_{13}MnO_3^+$	$[Mn(OC_6H_3(COCH_3)OCH_2C_6H_5)]^+$	Mn:	MVol.D1-57/8
$C_{15}H_{13}Ni_3^+$	$C_5H_5(C_5H_4)_2Ni_3^+$	Ni:	Org.Verb.2-381
$C_{15}H_{13}Pa^+$	$Pa(C_5H_5)_2C_5H_3^+$	Pa:	SVol.2-99
$C_{15}H_{14}NiS_4$	$[Ni(S_2CSCH_2C_6H_5)(SCH_2C_6H_5)]_2$	C:	MVol.D4-266
$C_{15}H_{14}O_2Sn$	$(C_6H_5)_2Sn(CH_2CH_2COO)$	Sn:	Org.Verb.2-359, 380
$C_{15}H_{14}O_7U$	$UO_2(C_7H_5O_2)_2 \cdot CH_3OH$	U:	SVol.E1-60, 64/5
$C_{15}H_{15}I_3IrN_3$	$Ir(C_5H_5N)_3I_3$	Ir:	SVol.2-70
$C_{15}H_{15}NNi_3^+$	$(C_5H_5)_3Ni_3N^+$	Ni:	Org.Verb.2-378
$C_{15}H_{15}NOSSn$	$(CH_3)_2Sn(SC_6H_4NCHC_6H_4O)$	Sn:	Org.Verb.6-31
$C_{15}H_{15}NO_2Sn$	$(CH_3)_2Sn(OC_6H_4NCHC_6H_4O)$	Sn:	Org.Verb.6-31
$C_{15}H_{15}NO_5U$	$UO_2(OC_6H_4CHNC_6H_4O) \cdot C_2H_5OH$	U:	SVol.E1-61, 68/9
$C_{15}H_{15}N_3O_6SU$	$UO_2SO_4 \cdot 3\ C_5H_5N$	U:	SVol.E1-29, 35
$C_{15}H_{15}N_5O_{11}U$	$UO_2(NO_3)_2 \cdot 3\ C_5H_5NO$	U:	SVol.E1-132, 135, 142/3
$C_{15}H_{15}Ni_2^+$	$[Ni_2(C_5H_5)_3]^+$	Ni:	Org.Verb.2-188, 200/1, 364/5
$C_{15}H_{15}Ni_3^+$	$(C_5H_5)_3Ni_3^+$	Ni:	Org.Verb.2-381
$C_{15}H_{15}Ni_3S_2$	$(C_5H_5)_3Ni_3S_2$	Ni:	Org.Verb.2-380
$C_{15}H_{15}OPa^+$	$[PaO(C_5H_5)_3]^+$	Pa:	SVol.2-99
$C_{15}H_{15}O_6Pa$	$Pa(OH)_4((C_6H_5CO)_2CH)$	Pa:	SVol.2-180/1
$C_{15}H_{15}Pa$	$Pa(C_5H_5)_3$	Pa:	SVol.2-99
$C_{15}H_{15}Pa^+$	$Pa(C_5H_5)_3^+$	Pa:	SVol.2-99
$C_{15}H_{15}Yb$	$Yb(C_5H_5)_3$	Sc:	MVol.B1-42
$C_{15}H_{16}N_2NiO_2$	$(CH_2C(CH_3)COOCH_3)Ni(NC_5H_4C_5H_4N)$	Ni:	Org.Verb.1-368
$C_{15}H_{16}NiO_2$	$C_5H_5NiC_7H_6COOC_2H_5$	Ni:	Org.Verb.2-184/5
$C_{15}H_{16}Ni_2$	$(C_5H_5Ni)_2C_5H_6$	Ni:	Org.Verb.2-364
$C_{15}H_{16}Ni_2O_2$	$(C_5H_5Ni)_2HCCCOOC_2H_5$	Ni:	Org.Verb.2-356, 359

Formula	Compound	Reference
$C_{15}H_{16}O_2Sn$	$[Sn(C_6H_5)_2(O(CH_2)_3O)]_x$	Sn: Org.Verb.6-162/3, 165
$C_{15}H_{16}Pa^+$	$Pa(C_5H_5)_3H^+$	Pa: SVol.2-99
$C_{15}H_{16}Sn$	$(CH_3)_2Sn(C_6H_4CH_2C_6H_4)$	Sn: Org.Verb.3-111, 115
		Sn: Org.Verb.6-16
$C_{15}H_{17}ISn$	$(CH_3)_3SnC_6H_4C_6H_4I$	Sn: Org.Verb.2-146, 148
$C_{15}H_{17}NSn$	$(CH_3)_2Sn(C_6H_4N(CH_3)C_6H_4)$	Sn: Org.Verb.3-117, 122
$C_{15}H_{17}N_3Ni$	$NCCHCH_2Ni(C_{10}H_8N_2)(CH_3)_2$	Ni: Org.Verb.1-380/1
$C_{15}H_{17}Pa^+$	$Pa(C_5H_5)_3H_2{}^+$	Pa: SVol.2-99
$C_{15}H_{17.5}I_4N_{2.5}Sn$	$SnI_4 \cdot 2.5\ C_6H_5NH_2$	Sn: MVol.C5-174
$C_{15}H_{18}HoI_6N_3$	$(C_5H_5NH)_3HoI_6$	Sc: MVol.C6-273
$C_{15}H_{18}I_6LaN_3$	$(C_5H_5NH)_3LaI_6$	Sc: MVol.C6-273
$C_{15}H_{18}I_6N_3Nd$	$(C_5H_5NH)_3NdI_6$	Sc: MVol.C6-273
$C_{15}H_{18}I_6N_3Pr$	$(C_5H_5NH)_3PrI_6$	Sc: MVol.C6-273
$C_{15}H_{18}I_6N_3Sm$	$(C_5H_5NH)_3SmI_6$	Sc: MVol.C6-273
$C_{15}H_{18}I_6N_3Tm$	$(C_5H_5NH)_3TmI_6$	Sc: MVol.C6-273
$C_{15}H_{18}I_6N_3Yb$	$(C_5H_5NH)_3YbI_6$	Sc: MVol.C6-273
$C_{15}H_{18}N_2O_2S$	$C_5H_{10}NSO_2NHC_{10}H_7$	S: S-N-Verb.1-171
$C_{15}H_{18}N_2Sn$	$[(CH_3)_2SnNHC_6H_4CH_2C_6H_4NH]_x$	Sn: Org.Verb.6-25
$C_{15}H_{18}N_6Ni_3O_9P_4$	$P_4(NCH_3)_6(Ni(CO)_3)_3$	Ni: Org.Verb.2-373/4
$C_{15}H_{18}Ni_2$	$(C_5H_5Ni)_2HCCC_3H_7\text{-n}$	Ni: Org.Verb.2-356, 359
$C_{15}H_{18}Pa^+$	$Pa(C_5H_5)_3H_3{}^+$	Pa: SVol.2-99
$C_{15}H_{18}Sn$	$(CH_3)_3SnC_6H_4C_6H_5$	Sn: Org.Verb.2-146, 148/9, 152
–	$(C_6H_5)_2Sn(CH_3)(C_2H_5)$	Sn: Org.Verb.3-90
$C_{15}H_{19}MnNO_4$	$[Mn(CH(COCH_3)_2)_2(C_5H_5N)]$	Mn: MVol.D1-78/9
$C_{15}H_{19}NNi$	$(CH_3)_2C_3H_3Ni(C_9H_7N)CH_3$	Ni: Org.Verb.2-35, 37
$C_{15}H_{19}NO_6U$	$[UO_2(CH(COCH_3)_2)_2(C_5H_5N)]$	U: SVol.E1-30, 36/7
$C_{15}H_{19}NO_7U$	$UO_2(CH(COCH_3)_2)_2 \cdot C_5H_5NO$	U: SVol.E1-132, 136, 142/3
–	$[UO_2(CH(COCH_3)_2)_2(HOC_5H_4N)]$	U: SVol.E1-31, 36/7
$C_{15}H_{19}NiO_5P$	$(CO)_3NiP(OC_3H_7\text{-i})_2C_6H_5$	Ni: Org.Verb.1-177
$C_{15}H_{19}PSn$	$(CH_3)_3SnP(C_6H_5)_2$	Sn: Org.Verb.5-81
$C_{15}H_{20}N_2O_6U$	$[UO_2(CH(COCH_3)_2)_2(NH_2C_5H_4N)]$	U: SVol.E1-31, 36/7
$C_{15}H_{20}N_5Ti^{2+}$	$[C_5H_5Ti(CH_3CN)_5]^{2+}$	Ti: Org.Verb.1-133
$C_{15}H_{20}NiO_4$	$CONi(C(O)C_8H_{13})((OCCH_3)_2CH)$	Ni: Org.Verb.1-122
$C_{15}H_{20}Sn$	$[CH_2C(CH_3)]_3SnC_6H_5$	Sn: Org.Verb.2-348
–	$(CH_2CHCH_2)_3SnC_6H_5$	Sn: Org.Verb.2-348
–	$(CH_2CH_2CH)_3SnC_6H_5$	Sn: Org.Verb.2-340
$C_{15}H_{21}IrO_6$	$Ir(CH_3COCHCOCH_3)_3$	Ir: SVol.2-37
$C_{15}H_{21}IrS_6$	$Ir(CH_3CSCHCSCH_3)_3$	Ir: SVol.2-168
$C_{15}H_{21}KMnO_6$	$K[Mn(CH(COCH_3)_2)_3]$	Mn: MVol.D1-75
$C_{15}H_{21}MnNaO_6$	$Na[Mn(CH(COCH_3)_2)_3]$	Mn: MVol.D1-74/5
$C_{15}H_{21}MnO_6$	$[Mn(CH(COCH_3)_2)_3]$	Mn: MVol.D1-85/98
$C_{15}H_{21}MnO_6{}^+$	$[Mn(CH(COCH_3)_2)_3]^+$	Mn: MVol.D1-92/3
$C_{15}H_{21}MnO_6{}^-$	$[Mn(CH(COCH_3)_2)_3]^-$	Mn: MVol.D1-66
$C_{15}H_{21}Mn_2O_6{}^+$	$[Mn_2(CH(COCH_3)_2)_3]^+$	Mn: MVol.D1-69/70, 93
$C_{15}H_{21}N_3O_6U$	$[UO_2(CH(COCH_3)_2)_2((NH_2)_2C_5H_3N)]$	U: SVol.E1-31, 36/7
$C_{15}H_{21}NpO_6{}^+$	$Np(CH_3COCHCOCH_3)_3{}^+$	Np: TrU.D2-242
$C_{15}H_{21}O_6Pu^+$	$Pu(CH_3COCHCOCH_3)_3{}^+$	Np: TrU.D1-167/8
$C_{15}H_{21}O_7Pa$	$PaO((CH_3CO)_2CH)_3$	Pa: SVol.2-177/8
$C_{15}H_{22}Ni$	$C_{15}H_{22}Ni$	Ni: Org.Verb.2-244, 247

$C_{15}H_{22}OSSn$	$(CH_3)_3SnC[S(CH_2)_4]COC_6H_5$	Sn: Org.Verb.2-25, 29
$C_{15}H_{22}OSn$	$(C_2H_5)_2(C_2H_5OCH_2)SnCCC_6H_5$	Sn: Org.Verb.5-204
–	$(C_2H_5)_3SnCCC_6H_4OCH_3$	Sn: Org.Verb.2-230
$C_{15}H_{22}O_7Pa$	$Pa(OH)((CH_3CO)_2CH)_3$	Pa: SVol.2-176
$C_{15}H_{22}Sn$	$(C_2H_5)_3SnCCC_6H_4CH_3$	Sn: Org.Verb.2-230
–	$(C_2H_5)_3SnC_9H_7$	Sn: Org.Verb.2-245
$C_{15}H_{23}IN_2Ni$	$[C_5H_5Ni(CNC(CH_3)_3)_2]I$	Ni: Org.Verb.2-172
$C_{15}H_{23}NiO_3PSn_2$	$(CO)_3NiP(C_6H_5)(Sn(CH_3)_3)_2$	Ni: Org.Verb.1-171, 172
$C_{15}H_{24}IrN_3O_3S_6$	$Ir((OC_4H_8N)CS_2)_3$	Ir: SVol.2-169
$C_{15}H_{24}IrN_3S_6$	$Ir(C_4H_8NCS_2)_3$	Ir: SVol.2-169
$C_{15}H_{24}IrN_3S_9$	$Ir((SC_4H_8N)CS_2)_3$	Ir: SVol.2-169
$C_{15}H_{24}OSn$	$(C_2H_5)_3SnCH_2COC_6H_4CH_3$	Sn: Org.Verb.2-174, 178
$C_{15}H_{24}O_2Sn$	$(C_2H_5)_3SnCH_2COC_6H_4OCH_3$	Sn: Org.Verb.2-174, 178
$C_{15}H_{24}O_{8.5}U$	$UO_2(CH(COCH_3)_2)_2 \cdot 1.25\ C_4H_8O_2$	U: SVol.E1-71, 77
$C_{15}H_{24}SiSn$	$(CH_3)_3SnCH_2C_6H_4CCSi(CH_3)_3$	Sn: Org.Verb.2-28, 30
$C_{15}H_{24}Sn$	$(C_2H_5)_3SnCH_2CHCHC_6H_5$	Sn: Org.Verb.2-200, 205
–	$(C_2H_5)_3SnC_6H_4C(CH_3)CH_2$	Sn: Org.Verb.2-242
–	$(C_2H_5)_3SnC_6H_4CH_2CHCH_2$	Sn: Org.Verb.2-242
$C_{15}H_{25}ISn$	$(C_3H_7)_3SnC_6H_4I$	Sn: Org.Verb.2-267/8
$C_{15}H_{25}NO_8U$	$HUO_2(CH(COCH_3)_2)_3 \cdot NH_3$	U: SVol.E1-13, 16
$C_{15}H_{25}N_2S_4Ti$	$C_5H_5Ti(S_2CN(C_2H_5)_2)_2$	Ti: Org.Verb.1-197
$C_{15}H_{25}N_3NiOS_4$	$[(C_2H_5)_4N][Ni(S_2COC_2H_5)(S_2C_2(CN)_2)]$	C: MVol.D4-260
$C_{15}H_{25}N_3O_2S_4U$	$[UO_2(S_2CN(C_2H_5)_2)_2(C_5H_5N)]$	U: SVol.E1-33, 37/8
$C_{15}H_{25}N_3Ti$	$C_9H_7Ti(N(CH_3)_2)_3$	Ti: Org.Verb.1-172, 179
$C_{15}H_{26}O_3SSn$	$(CH_3)_3Sn(CH_2)_5OSO_2C_6H_4CH_3$	Sn: Org.Verb.2-55, 57
$C_{15}H_{26}O_3Ti$	$C_6H_5Ti(OC_3H_7\text{-}i)_3$	Ti: Org.Verb.1-45/6, 49
$C_{15}H_{26}O_4Ti$	$C_6H_5OTi(OC_3H_7)_3$	Ti: Org.Verb.1-49
$C_{15}H_{26}SSn$	$(C_2H_5)_3SnCH_2CH_2CH_2SC_6H_5$	Sn: Org.Verb.2-174
$C_{15}H_{26}Sn$	$(CH_3)_2Sn(i\text{-}C_5H_{11})(CH(CH_3)C_6H_5)$	Sn: Org.Verb.3-75, 79
–	$(C_2H_5)_3SnCH_2CH(CH_3)C_6H_5$	Sn: Org.Verb.2-174
–	$(C_3H_7)_3SnC_6H_5$	Sn: Org.Verb.2-266, 267
–	$(i\text{-}C_3H_7)_3SnC_6H_5$	Sn: Org.Verb.2-270/1
–	$(C_4H_9)_2Sn(CH_3)(C_6H_5)$	Sn: Org.Verb.3-88/9
$C_{15}H_{27}IrN_6S_6$	$Ir((NHC_4H_8N)CS_2)_3$	Ir: SVol.2-169
$C_{15}H_{27}N_3NiO$	$(t\text{-}C_4H_9NC)_2Ni(OCNC_4H_9\text{-}t)$	Ni: Org.Verb.1-310, 313
$C_{15}H_{27}N_3NiO_2S$	$(t\text{-}C_4H_9NC)_3NiSO_2$	Ni: Org.Verb.1-325, 326
$C_{15}H_{27}N_3OSn$	$(C_4H_9)_3SnONC(CN)_2$	Sn: Org.Verb.5-133
$C_{15}H_{27}N_4NiO^+$	$[(t\text{-}C_4H_9NC)_3NiNO]^+$	Ni: Org.Verb.1-326
$C_{15}H_{27}N_5NiO_4$	$[(t\text{-}C_4H_9NC)_3Ni(NO)]NO_3$	Ni: Org.Verb.1-325, 326
$C_{15}H_{27}NiO_3P$	$(CO)_3NiP(C_4H_9\text{-}n)_3$	Ni: Org.Verb.1-166, 169
–	$(CO)_3NiP(C_4H_9\text{-}s)_3$	Ni: Org.Verb.1-166, 169
–	$(CO)_3NiP(C_4H_9\text{-}t)_3$	Ni: Org.Verb.1-166, 169
$C_{15}H_{27}NiO_3Sb$	$(CO)_3NiSb(C(CH_3)_3)_3$	Ni: Org.Verb.1-188, 190
$C_{15}H_{27}NiO_6P$	$(CO)_3NiP(OC_4H_9\text{-}n)_3$	Ni: Org.Verb.1-178, 179
$C_{15}H_{27}O_3PS_6$	$P[S_2COC_4H_9]_3$	C: MVol.D4-254
$C_{15}H_{27}O_3PS_7$	$P(S)(S_2COC_4H_9)_3$	C: MVol.D4-254
$C_{15}H_{27}O_3PSe_6$	$P[Se_2COC_4H_9]_3$	C: MVol.D6-221
$C_{15}H_{27}O_3S_6Sb$	$Sb[S_2COC_4H_9]_3$	C: MVol.D4-254
$C_{15}H_{27}O_3S_6Tl$	$Tl[S_2COC_4H_9]_3$	C: MVol.D4-253
$C_{15}H_{27}O_4PS_6$	$P(O)(S_2COC_4H_9)_3$	C: MVol.D4-254

Formula	Compound		Reference
$C_{15}H_{27}Ti$	$(CH_3CHCHCHCH_3)_3Ti$	Ti:	Org.Verb.1-128
$C_{15}H_{28}OSn$	$(C_2H_5)_3SnCHCHC(OH)C(CH_2)_5$	Sn:	Org.Verb.2-200
$C_{15}H_{28}O_3Sn$	$(C_2H_5)_3SnCCCH_2O[(CH_2)_2O]_2CHCH_2$	Sn:	Org.Verb.2-230, 235
$C_{15}H_{28}O_3Ti$	$CH_3C_5H_4Ti(OC_3H_7\text{-}i)_3$	Ti:	Org.Verb.1-159, 166
$C_{15}H_{28}O_4Sn$	$(CH_3)_3SnC_2(COOC_2H_5)_2(t\text{-}C_4H_9)$	Sn:	Org.Verb.2-83
$C_{15}H_{28}O_5Sn$	$(C_2H_5)_3SnC_2(COOCH_3)_2OC_3H_7$	Sn:	Org.Verb.2-200
–	$(C_2H_5)_3SnC_2(COOC_2H_5)_2OCH_3$	Sn:	Org.Verb.2-200
$C_{15}H_{28}O_6Sn$	$(CH_3OOCCH_2CH_2)_3SnC_3H_7$	Sn:	Org.Verb.2-336
$C_{15}H_{29}NSn$	$(C_4H_9)_3SnC(CN)CH_2$	Sn:	Org.Verb.2-309/10
$C_{15}H_{29}N_3O_3S_4U$	$UO_2(S_2CN(C_2H_5)_2)_2 \cdot CH_3N(CH_2)_3CO$	U:	SVol.E1-119
$C_{15}H_{29}N_3S_2Sn$	$[(C_2H_5)_4N][(CH_3)_3SnSC(CN)C(CN)S]$	Sn:	Org.Verb.5-84
$C_{15}H_{29}NiO_2P^+$	$(CO)_2Ni(CH_2P(C_4H_9)_3)^+$	Ni:	Org.Verb.1-182
$C_{15}H_{30}IrN_3S_6$	$Ir(S_2CN(C_2H_5)_2)_3$	Ir:	SVol.2-169
$C_{15}H_{30}NiO_6P_2$	$C_5H_5Ni(P(OC_2H_5)_3)PO(OC_2H_5)_2$	Ni:	Org.Verb.2-134
$C_{15}H_{30}OSn$	$(C_2H_5)_3SnC(CCHCH_3)C(OCH_3)(CH_3)C_2H_5$	Sn:	Org.Verb.2-200
–	$(C_2H_5)_3Sn(CH_2)_2CCC(OCH_3)(CH_3)C_2H_5$	Sn:	Org.Verb.2-231
$C_{15}H_{30}O_2Sn$	$(C_2H_5)_3SnCCCH_2OCH(CH_3)OC_4H_9$	Sn:	Org.Verb.2-231
$C_{15}H_{30}O_4Sn$	$(C_2H_5)_3SnCH(CH_3)CH(COOC_2H_5)_2$	Sn:	Org.Verb.2-174, 178
–	$(C_2H_5)_3SnCH_2CH_2CH(COOC_2H_5)_2$	Sn:	Org.Verb.2-174, 178
$C_{15}H_{30}Pd_3S_{12}$	$[Pd(S_2CSC_2H_5)(SC_2H_5)]_3$	C:	MVol.D4-266, 270
$C_{15}H_{30}Sn$	$(C_2H_5)_3SnC(C_2H_5)CCHC_4H_9$	Sn:	Org.Verb.2-188
–	$(C_2H_5)_3SnC(C_3H_7)CCHC_3H_7$	Sn:	Org.Verb.2-188
$C_{15}H_{31}NSn$	$(C_4H_9)_3SnCH(CH_3)CN$	Sn:	Org.Verb.2-281
–	$(C_4H_9)_3SnCH_2CH_2CN$	Sn:	Org.Verb.2-282, 299
$C_{15}H_{31}N_3Ti$	$(CH_3)_3CC_5H_4Ti(N(CH_3)_2)_3$	Ti:	Org.Verb.1-172, 175/8
$C_{15}H_{32}N_2OSn$	$[CH_2CHC(NH_2)NOSn(C_4H_9)_3]_n$	Sn:	Org.Verb.5-132
$C_{15}H_{32}N_4O_4S_2$	$(CH_2)_5NH_2[N(SO_2N(CH_2)_5)_2]$	S:	S-N-Verb.1-162
$C_{15}H_{32}OSn$	$(CH_3)(i\text{-}C_3H_7)Sn(C_2H_4C(CH_3)_2OH)(C_6H_{11})$	Sn:	Org.Verb.3-98
–	$(C_3H_7)_3SnCHCHOC_4H_9$	Sn:	Org.Verb.2-261/2
–	$(C_4H_9)_3SnCH_2COCH_3$	Sn:	Org.Verb.2-276/7
$C_{15}H_{32}O_2SSn$	$(C_4H_9)_3SnSCH_2COOCH_3$	Sn:	Org.Verb.5-132
$C_{15}H_{32}O_2Sn$	$(C_2H_5)_3SnCHCHCH_2OCH(CH_3)OC_4H_9$	Sn:	Org.Verb.2-200
–	$(C_3H_7)_3SnCH(CH_3)CH_2COOC_2H_5$	Sn:	Org.Verb.2-254, 257
–	$(C_3H_7)_3Sn(CH_2)_3COOC_2H_5$	Sn:	Org.Verb.2-254
–	$(C_3H_7)_3Sn(CH_2)_3OCH_2CHOCH_2$	Sn:	Org.Verb.2-254
–	$(C_4H_9)_3SnCH_2CH_2COOH$	Sn:	Org.Verb.2-282, 299
–	$(C_4H_9)_3SnCH_2COOCH_3$	Sn:	Org.Verb.2-278, 279
$C_{15}H_{32}SiSn$	$(C_2H_5)_3SnCCCH_2Si(C_2H_5)_3$	Sn:	Org.Verb.2-231, 235
$C_{15}H_{32}Sn$	$((CH_3)_3CCH_2)_2Sn(CH_2)_5$	Sn:	Org.Verb.3-105
–	$(C_2H_5)(i\text{-}C_3H_7)Sn(C_4H_9)(C_6H_{11})$	Sn:	Org.Verb.3-100
–	$(C_3H_7)_3SnCHCHC_4H_9$	Sn:	Org.Verb.2-261
–	$(C_3H_7)_3SnCH(CH_2)_5$	Sn:	Org.Verb.2-254, 257
–	$(C_4H_9)_2Sn(CH_2CH_2)_2C(CH_3)_2$	Sn:	Org.Verb.3-106
–	$(C_4H_9)_2Sn(CH_3)(C_6H_{11})$	Sn:	Org.Verb.3-88/9
–	$(C_4H_9)_3SnCHCH_2CH_2$	Sn:	Org.Verb.2-306
–	$(C_4H_9)_3SnCH_2CHCH_2$	Sn:	Org.Verb.2-307/8
–	$(i\text{-}C_4H_9)_3SnCH_2CHCH_2$	Sn:	Org.Verb.2-328
$C_{15}H_{32.5}N_{6.5}O_{14.5}U$	$U(NO_3)_4 \cdot 2.5\ CH_3CON(C_2H_5)_2$	U:	SVol.E1-107/8
$C_{15}H_{33}NOSn$	$(C_4H_9)_3SnCH_2CH_2CONH_2$	Sn:	Org.Verb.2-282
$C_{15}H_{33}N_3SiTi$	$(CH_3)(Si(CH_3)_3)C_5H_3Ti(N(CH_3)_2)_3$	Ti:	Org.Verb.1-172, 176

Formula	Compound	Reference
$C_{15}H_{33}NiO_3PSi_3$	$(CO)_3NiP(CH_2Si(CH_3)_3)_3$	Ni: Org.Verb.1-165, 167
$C_{15}H_{33}NiP$	$(CH_3)_2C_3H_3Ni(P(C_3H_7\text{-}i)_3)CH_3$	Ni: Org.Verb.2-36, 38
$C_{15}H_{33}Ti^+$	$[((CH_3)_3CCH_2)_3Ti]^+$	Ti: Org.Verb.1-99
$C_{15}H_{34}OSn$	$(C_4H_9)_3SnC(CH_3)_2OH$	Sn: Org.Verb.2-283
–	$(C_4H_9)_3SnCH_2CH(CH_3)OH$	Sn: Org.Verb.2-282
–	$(C_4H_9)_3SnCH_2CH_2CH_2OH$	Sn: Org.Verb.2-282
–	$(C_4H_9)_3SnCH_2OC_2H_5$	Sn: Org.Verb.2-283
–	$(i\text{-}C_4H_9)_3Sn(OC_3H_7\text{-}i)$	Sn: Org.Verb.5-138
$C_{15}H_{34}O_2SiSn$	$(C_2H_5)_3SnCH(COOCH_3)Si(C_2H_5)_3$	Sn: Org.Verb.2-174
$C_{15}H_{34}O_5U$	$UO_2(OC_5H_{11})_2 \cdot C_5H_{11}OH$	U: SVol.E1-59, 64
$C_{15}H_{34}SSn$	$(C_4H_9)_3SnCH_2CH_2CH_2SH$	Sn: Org.Verb.2-283
$C_{15}H_{34}Sn$	$((CH_3)_3CCH_2)_3SnH$	Sn: Org.Verb.4-67
–	$(CH_3)_3SnC_{12}H_{25}$	Sn: Org.Verb.2-53/4
–	$(C_2H_5)_3SnC_9H_{19}$	Sn: Org.Verb.2-160
–	$(C_4H_9)_2Sn(i\text{-}C_3H_7)(s\text{-}C_4H_9)$	Sn: Org.Verb.3-88/9
–	$(C_4H_9)_3SnC_3H_7$	Sn: Org.Verb.2-273
–	$(C_4H_9)_3Sn(i\text{-}C_3H_7)$	Sn: Org.Verb.2-274
–	$(i\text{-}C_4H_9)_3SnC_3H_7$	Sn: Org.Verb.2-327
–	$(i\text{-}C_4H_9)_3Sn(i\text{-}C_3H_7)$	Sn: Org.Verb.2-327
–	$(i\text{-}C_5H_{11})_2Sn(C_2H_5)(C_3H_7)$	Sn: Org.Verb.3-92/3
–	$(C_5H_{11})_3SnH$	Sn: Org.Verb.4-67
$C_{15}H_{35}ISSn$	$[(C_4H_9)_3SnCH_2S(CH_3)_2]I$	Sn: Org.Verb.2-283, 299
$C_{15}H_{35}NO_2SSn$	$(C_4H_9)_3SnCH_2SO_2N(CH_3)_2$	Sn: Org.Verb.2-283, 299
$C_{15}H_{35}NSn$	$(C_4H_9)_3SnCH_2CH_2CH_2NH_2$	Sn: Org.Verb.2-283, 299
–	$(C_4H_9)_3SnCH_2N(CH_3)_2$	Sn: Org.Verb.2-283, 299
$C_{15}H_{35}N_3Ti$	$CH_2C(CH_3)Ti(N(C_2H_5)_2)_3$	Ti: Org.Verb.1-53, 58
–	$CH_2CHCH_2Ti(N(C_2H_5)_2)_3$	Ti: Org.Verb.1-53, 57
–	$CH_3CHCHTi(N(C_2H_5)_2)_3$	Ti: Org.Verb.1-54
$C_{15}H_{36}INSn$	$[(C_3H_7)_3Sn(CH_2)_3N(CH_3)_3]I$	Sn: Org.Verb.2-254
$C_{15}H_{36}N_6O_9SU$	$UO_2SO_4 \cdot 3\ CO(NHC_2H_5)_2$	U: SVol.E1-93, 95
$C_{15}H_{36}OSiSn$	$(i\text{-}C_4H_9)_3SnOSi(CH_3)_3$	Sn: Org.Verb.5-138
$C_{15}H_{36}O_6U$	$UO(OC_3H_7)_4 \cdot C_3H_7OH$	U: SVol.E1-59, 64
$C_{15}H_{36}SiSn$	$(C_4H_9)_3SnCH_2SiH(CH_3)_2$	Sn: Org.Verb.2-283, 299
$C_{15}H_{37}N_3Ti$	$(n\text{-}C_3H_7)Ti(N(C_2H_5)_2)_3$	Ti: Org.Verb.1-52
–	$(i\text{-}C_3H_7)Ti(N(C_2H_5)_2)_3$	Ti: Org.Verb.1-52
$C_{15}H_{38}O_2Si_3Sn$	$[(CH_3)_3SiCH_2]_3SnCH_2COOCH_3$	Sn: Org.Verb.2-336
$C_{15}H_{39}NiO_{13}P_3S$	$[(CH_3)_3C_3H_2Ni(P(OCH_3)_3)_3]HSO_4$	Ni: Org.Verb.2-26, 27
–	$[C_2H_5(CH_3)C_3H_3Ni(P(OCH_3)_3)_3]HSO_4$	Ni: Org.Verb.2-26
$C_{15}H_{40}N_{12}O_{13}U$	$UO_2(NO_3)_2 \cdot 5\ CO(NHCH_3)_2$	U: SVol.E1-92, 95
–	$UO_2(NO_3)_2 \cdot 5\ NH_2CONHC_2H_5$	U: SVol.E1-92/3, 95
$C_{15}H_{45}N_4O_{17}P_5U$	$U(NO_3)_4 \cdot 5\ (CH_3)_3PO$	U: SVol.E1-148, 156
$C_{15}H_{45}NiO_2P_5$	$CH_3Ni(P(CH_3)_3)_4O_2P(CH_3)_2$	Ni: Org.Verb.1-18, 24
$C_{15.5}Cl_3H_{8.5}O_4Sn$	$SnCl_3(C_{14}H_7O_4) \cdot 0.25\ C_6H_6$	Sn: MVol.C5-113
$C_{16}CdFeH_8N_2O_4$	$(C_{12}H_8N_2)CdFe(CO)_4$	Fe: Org.Verb.B3-126
$C_{16}CdFeH_{14}N_2O_4$	$(CH_3C_5H_4N)_2CdFe(CO)_4$	Fe: Org.Verb.B3-126
$C_{16}CdH_{20}N_2O_2S_4$	$Cd[S_2COC_2H_5]_2(C_5H_5N)_2$	C: MVol.D4-258
$C_{16}CeCl_3H_{32}O_{20}$	$Ce(ClO_4)_3 \cdot 4\ C_4H_8O_2 \cdot 9\ H_2O$	Sc: MVol.C5-96/7
$C_{16}CeCl_6H_{34}O_8$	$H_2CeCl_6 \cdot 4\ C_4H_8O_2$	Sc: MVol.C5-1/2
$C_{16}CeCl_6H_{40}N_2$	$[(C_2H_5)_4N]_2CeCl_6$	Sc: MVol.C5-223
$C_{16}ClCoH_{10}O_4Sn$	$(C_6H_5)_2ClSnCo(CO)_4$	Sn: Org.Verb.6-162/3, 173

$C_{16}ClF_3H_8N_2O_2$	$ClC_6F_3(OH)NNC_{10}H_6OH$	F:	PerFHalOrg.7-61/2
$C_{16}ClF_5H_{10}Ti$	$(C_5H_5)_2Ti(C_6F_5)Cl$	F:	PerFHalOrg.4-35, 112
$C_{16}ClF_{20}H_{15}O_3Sn$	$CF_2Cl(CF_2)_9(CH_2)_3Sn(OCH_3)_3$	Sn:	Org.Verb.6-265
$C_{16}ClFeH_{10}O_4P$	$(CO)_4FeP(C_6H_5)_2Cl$	Fe:	Org.Verb.B2-84, 95
$C_{16}ClFeH_{12}N_3O_6$	$Fe(C_5H_5)_2 \cdot ClC_6H_2(NO_2)_3$	Fe:	Org.Verb.A1-238
$C_{16}ClFeH_{13}$	$C_5H_5FeC_5H_4C_6H_4Cl$	Fe:	Org.Verb.A1-390
$C_{16}ClFeH_{13}N_2O_4$	$Fe(C_5H_5)_2 \cdot ClC_6H_3(NO_2)_2$	Fe:	Org.Verb.A1-238
$C_{16}ClFeH_{14}N_2O_2$	$Fe(OC_6H_4CHNCH_2)_2Cl$	Fe:	Org.Verb.B1-5
$C_{16}ClFeH_{14}O_2Zr$	$C_5H_5FeC_5H_4COOZr(C_5H_5)Cl$	Fe:	Org.Verb.A3-110
$C_{16}ClFeH_{15}NO_2P$	$C_3H_5Fe(CO)(NO)P(C_6H_5)_2Cl$	Fe:	Org.Verb.B5-6/9, 20
$C_{16}ClFeH_{15}O$	$C_5H_5FeC_5H_4CClCHCHCHCOCH_3$	Fe:	Org.Verb.A3-25
–	$C_5H_5FeC_5H_4COC_5H_6Cl$	Fe:	Org.Verb.A2-289/90
$C_{16}ClFeH_{16}Zr$	$C_5H_5FeC_5H_4CH_2Zr(C_5H_5)Cl$	Fe:	Org.Verb.A3-110
$C_{16}ClFeH_{17}O_2$	$C_5H_5FeC_5H_4CO(CH_2)_4COCl$	Fe:	Org.Verb.A3-159
$C_{16}ClFeH_{18}O$	$Fe(C_6H_5)_2Cl \cdot C_4H_8O$	Fe:	Org.Verb.A1-215
$C_{16}ClFeH_{19}O$	$C_5H_5FeC_5H_4CH_2CH(C_3H_7)COCl$	Fe:	Org.Verb.A3-160
–	$C_5H_5FeC_5H_4COCH(CH_2Cl)CH(CH_3)_2$	Fe:	Org.Verb.A2-244, 246
–	$C_5H_5FeC_5H_4COCH_2CHClCH(CH_3)_2$	Fe:	Org.Verb.A2-244, 246
$C_{16}ClFeH_{20}NO_5$	$[C_5H_5FeC_5H_4C(CH_3)NC_4H_8O]ClO_4$	Fe:	Org.Verb.A2-221
$C_{16}ClFeH_{21}S_2$	$C_5H_5FeC_5H_4C(SC_2H_5)_2CH_2Cl$	Fe:	Org.Verb.A2-245
$C_{16}ClFeH_{23}O_8P_2S_2$	$OCFe(P(OCH_3)_3)_2S_2C_2(C_6H_4Cl)CHO$	Fe:	Org.Verb.B4-138
$C_{16}ClFeH_{25}N_6O_4$	$[(C_2H_5NC)_5FeCN]ClO_4$	Fe:	Org.Verb.B4-78, 82
$C_{16}ClFeH_{31}NO_2P$	$ClC_3H_4Fe(CO)(NO)P(C_4H_9)_3$	Fe:	Org.Verb.B5-108/11
$C_{16}ClFe_2GeH_{17}N_4O_4$	$(C_3H_5Fe(NO)_2)_2Ge(Cl)C_{10}H_7$	Fe:	Org.Verb.C2-165
$C_{16}ClFe_2GeH_{23}N_4O_4$	$((CH_3)_2C_3H_3Fe(NO)_2)_2Ge(Cl)C_6H_5$	Fe:	Org.Verb.C2-165
$C_{16}ClFe_2H_{17}N_4O_4Pb$	$(C_3H_5Fe(NO)_2)_2Pb(Cl)C_{10}H_7$	Fe:	Org.Verb.C2-165
$C_{16}ClFe_2H_{17}N_4O_4Sn$	$(C_3H_5Fe(NO)_2)_2Sn(Cl)C_{10}H_7$	Fe:	Org.Verb.C2-165
$C_{16}ClFe_2H_{23}N_4O_4Pb$	$((CH_3)_2C_3H_3Fe(NO)_2)_2Pb(Cl)C_6H_5$	Fe:	Org.Verb.C2-165
$C_{16}ClFe_2H_{23}N_4O_4Sn$	$((CH_3)_2C_3H_3Fe(NO)_2)_2Sn(Cl)C_6H_5$	Fe:	Org.Verb.C2-165
$C_{16}ClH_{13}N_2Ni$	$C_6H_5Ni(NC_5H_4C_5H_4N)Cl \cdot 0.5\ C_6H_5Cl$	Ni:	Org.Verb.1-33, 40
$C_{16}ClH_{14}IrN_5$	$Ir(C_{11}H_9N_4)(C_5H_5N)Cl$	Ir:	SVol.2-96
$C_{16}ClH_{14}N_2O_2Sn^+$	$[SnCl(C_2H_4(NCHC_6H_4O)_2)]^+$	Sn:	MVol.C6-23
$C_{16}ClH_{15}Sn$	$(C_5H_5)_2(C_6H_5)SnCl$	Sn:	Org.Verb.5-223
$C_{16}ClH_{16}NO_5U$	$UO_2(OC_6H_3ClCHNCH_2C_6H_4O) \cdot C_2H_5OH$	U:	SVol.E1-62, 68/9
$C_{16}ClH_{19}Sn$	$(CH_3)(C_6H_5)(C_6H_5CH(CH_3)CH_2)SnCl$	Sn:	Org.Verb.5-225, 227
–	$(C_6H_5)_2(C_4H_9)SnCl$	Sn:	Org.Verb.5-216, 218
–	$(C_6H_5)_2(t\text{-}C_4H_9)SnCl$	Sn:	Org.Verb.5-216, 219
$C_{16}ClH_{22}IrN_2P_2$	$Ir(N_2)(P(CH_3)_2C_6H_5)_2Cl$	Ir:	SVol.2-39
$C_{16}ClH_{22}N_3NiO$	$(t\text{-}C_4H_9NC)_2Ni(ONC_6H_4Cl)$	Ni:	Org.Verb.1-310, 313
$C_{16}ClH_{22}N_3SSn$	$(C_6H_5)_2ClSnN(CH_3)S(NCH_3)_2CH_3$	Sn:	Org.Verb.6-162/3, 172
$C_{16}ClH_{25}Sn$	$(i\text{-}C_4H_9)_2(C_6H_5CHCH)SnCl$	Sn:	Org.Verb.5-213, 215
$C_{16}ClH_{26}OSn$	$(C_4H_9)_2ClSnOC(CH_3)C_6H_5$	Sn:	Org.Verb.5-132
$C_{16}ClH_{28}NSn$	$(C_4H_9)_2ClSnN(CH_3)CH_2C_6H_5$	Sn:	Org.Verb.6-88
$C_{16}ClH_{30}NSn$	$(C_3H_7)_3SnCl \cdot C_6H_5CH_2NH_2$	Sn:	Org.Verb.5-114
$C_{16}ClH_{30}N_3Ni$	$(t\text{-}C_4H_9NC)_3Ni(CH_3)Cl$	Ni:	Org.Verb.1-307
$C_{16}ClH_{31}Sn$	$(C_6H_{11})_2(C_4H_9)SnCl$	Sn:	Org.Verb.5-217, 222
$C_{16}ClH_{32}NiOP$	$C_3H_5Ni(CO)(P(C_4H_9\text{-}n)_3)Cl$	Ni:	Org.Verb.1-169
$C_{16}ClH_{35}OSiSn$	$(C_4H_9)_2[(CH_3)_3SiOC(CH_3)_2CHCH]SnCl$	Sn:	Org.Verb.5-211
$C_{16}ClH_{35}OSn$	$(i\text{-}C_4H_9)_2ClSnOCH_2C_7H_{15}$	Sn:	Org.Verb.6-105
$C_{16}ClH_{35}Sn$	$(C_4H_9)_2(C_8H_{17})SnCl$	Sn:	Org.Verb.5-211

Formula	Compound	Reference
$C_{16}ClH_{35}Sn$	$(i\text{-}C_4H_9)_2(C_8H_{17})SnCl$	Sn: Org.Verb.5-213, 215
–	$(C_8H_{17})_2SnHCl$	Sn: Org.Verb.6-293
$C_{16}ClH_{37}O_4Sn_2$	$C_8H_{17}(OH)_2SnOSn(Cl)(OH)C_8H_{17}$	Sn: Org.Verb.6-252
$C_{16}ClH_{38}OPSn$	$(C_2H_5)_3SnCl \cdot (C_2H_5)_2(C_6H_{13}O)P$	Sn: Org.Verb.5-107
$C_{16}Cl_2F_{21}H_{13}Sn$	$(C_4H_9)(C_{10}F_{21}CH_2CH_2)SnCl_2$	Sn: Org.Verb.6-200
$C_{16}Cl_2FeH_8Hg_2N_2O_4$	$(CO)_4Fe(HgCl)_2(C_{12}H_8N_2)$	Fe: Org.Verb.B2-153/5, 158
$C_{16}Cl_2FeH_{12}O_2$	$C_5H_5FeC_5H_4CClC(CHO)CClCHCHO$	Fe: Org.Verb.A2-174, 178
$C_{16}Cl_2FeH_{14}N_2$	$(CH_3C_6H_4NC)_2FeCl_2$	Fe: Org.Verb.B4-11
$C_{16}Cl_2FeH_{16}O_2$	$C_5H_5FeC_5H_4C_3H_2Cl_2COOC_2H_5$	Fe: Org.Verb.A3-131
$C_{16}Cl_2FeH_{19}N_5O_9$	$[(CH_3NC)_3Fe(C_{10}H_8N_2)H_2O][ClO_4]_2$	Fe: Org.Verb.B4-45
$C_{16}Cl_2FeH_{20}N_4$	$(CH_2CHCH_2NC)_4FeCl_2$	Fe: Org.Verb.B4-48/51
$C_{16}Cl_2FeH_{26}Hg_2N_2O_4$	$(CO)_4Fe(HgCl)_2 \cdot (C_6H_{11}NH_2)_2$	Fe: Org.Verb.B2-153/5, 158
$C_{16}Cl_2FeH_{28}O_4Sn_2$	$[(C_3H_7)_2SnCl]_2Fe(CO)_4$	Fe: Org.Verb.B3-139/40
$C_{16}Cl_2Fe_2H_{18}O_8Sn_2$	$[C_4H_9ClSnFe(CO)_4]_2$	Fe: Org.Verb.B3-139/40
		Fe: Org.Verb.C1-43
$C_{16}Cl_2H_8Mn_2O_6Sn$	$[(CO)_3MnC_5H_4]_2SnCl_2$	Sn: Org.Verb.6-184/5, 191
$C_{16}Cl_2H_{12}N_2Ni$	$ClC_6H_4Ni(NC_5H_4C_5H_4N)Cl$	Ni: Org.Verb.1-34, 40
$C_{16}Cl_2H_{12}O_8Sn_2$	$Sn_2Cl_2 \cdot C_{16}H_{12}O_8$	Sn: MVol.C5-166/7
$C_{16}Cl_2H_{13}NO_2Sn$	$[SnCl_2(C_6H_5COCHC(CH_3)NC_6H_4O)]$	Sn: MVol.C6-27/8
$C_{16}Cl_2H_{14}N_2O_2Sn$	$[SnCl_2(C_2H_4(NCHC_6H_4O)_2)]$	Sn: MVol.C6-22/3
$C_{16}Cl_2H_{14}N_2O_2Sn^+$	$[SnCl_2(C_2H_4(NCHC_6H_4O)_2)]^+$	Sn: MVol.C6-23
$C_{16}Cl_2H_{14}N_2Sn$	$(CH_2CH)_2SnCl_2 \cdot C_{12}H_8N_2$	Sn: Org.Verb.6-146
–	$(C_6H_5)_2SnCl_2 \cdot N(CHCH)_2N$	Sn: Org.Verb.6-174, 176
$C_{16}Cl_2H_{14}O_2Sn$	$(C_6H_5)_2[ClCH_2COOC(CH_2)]SnCl$	Sn: Org.Verb.5-216, 218
$C_{16}Cl_2H_{14}Ti$	$(C_8H_7)_2TiCl_2$	Ti: Org.Verb. 1-72, 75
$C_{16}Cl_2H_{16}IIrN_8$	$[Ir(C_4H_4N_2)_4Cl_2]I$	Ir: SVol.2-77
$C_{16}Cl_2H_{16}IrN_8^+$	$[Ir(C_4H_4N_2)_4Cl_2]^+$	Ir: SVol.2-77, 138
$C_{16}Cl_2H_{16}IrN_9O_3$	$[Ir(C_4H_4N_2)_4Cl_2]NO_3$	Ir: SVol.2-77
$C_{16}Cl_2H_{16}N_2Sn$	$(CH_3)_2Sn(CClNC_6H_5)_2$	Sn: Org.Verb.3-11
$C_{16}Cl_2H_{17}IrN_{10}O_6$	$[Ir(C_4H_4N_2)_3(C_4H_4N_2H)Cl_2](NO_3)_2 \cdot 0.5\ H_2O$	Ir: SVol.2-77
$C_{16}Cl_2H_{17}NOSn$	$(C_6H_5)(C_6H_5N(COCH_3)CH_2CH_2)SnCl_2$	Sn: Org.Verb.6-202
$C_{16}Cl_2H_{18}N_2Sn$	$(C_2H_5)_2SnCl_2 \cdot C_{12}H_8N_2$	Sn: Org.Verb.6-62
$C_{16}Cl_2H_{18}N_3Ti$	$CH_3TiCl_2 \cdot 3\ C_5H_5N$	Ti: Org.Verb.1-9
$C_{16}Cl_2H_{18}Ni_2$	$(C_8H_9NiCl)_2$	Ni: Org.Verb.2-320
$C_{16}Cl_2H_{18}O_2S_2Sn$	$(C_6H_5)_2SnCl_2 \cdot OS(CH_2CH_2)_2SO$	Sn: Org.Verb.6-174, 178
$C_{16}Cl_2H_{18}O_2Sn$	$(CH_3OC_6H_4CH_2)_2SnCl_2$	Sn: Org.Verb.6-136
–	$(C_2H_5OC_6H_4)_2SnCl_2$	Sn: Org.Verb.6-184/5, 188
$C_{16}Cl_2H_{18}Sn$	$(CH_3C_6H_4CH_2)_2SnCl_2$	Sn: Org.Verb.6-127
–	$((CH_3)_2C_6H_3)_2SnCl_2$	Sn: Org.Verb.6-184/5, 188
–	$(CH_3)_2Sn(CH_2C_6H_4Cl)_2$	Sn: Org.Verb.3-11, 18/9
–	$(C_2H_5)_2Sn(C_6H_4Cl)_2$	Sn: Org.Verb.3-29
$C_{16}Cl_2H_{19}NOSn$	$(C_6H_5)_2SnCl_2 \cdot CH_3C(O)NHC_2H_5$	Sn: Org.Verb.6-174/5
$C_{16}Cl_2H_{19}N_2O_5PS$	$(C_2H_5)_2NSO_2NHPO(OC_6H_4Cl)_2$	S: S-N-Verb.1-185
$C_{16}Cl_2H_{20}N_2OSn$	$(CH_3)_2SnCl_2 \cdot C_6H_5CH_2(CH_3)NCOC_5H_4N$	Sn: Org.Verb.6-36
$C_{16}Cl_2H_{20}N_2Sn$	$[(CH_3)_2NC_6H_4]_2SnCl_2$	Sn: Org.Verb.6-184/5, 188
$C_{16}Cl_2H_{20}N_4O_4U$	$UO_2Cl_2 \cdot 2\ CH_3CONHC_6H_4NH_2$	U: SVol.E1-96/7, 99
–	$UO_2Cl_2 \cdot 2\ (CH_3)_2NC_6H_4NO$	U: SVol.E1-144/5
$C_{16}Cl_2H_{20}N_4O_6U$	$UO_2Cl_2 \cdot 2\ CH_3OC_6H_4CON_2H_3 \cdot n\ H_2O$	U: SVol.E1-114/6
$C_{16}Cl_2H_{20}OSSn$	$(C_6H_5)_2SnCl_2 \cdot (C_2H_5)_2SO$	Sn: Org.Verb.6-174, 177
$C_{16}Cl_2H_{21}NOSn$	$(CH_3C_6H_4)_2SnCl_2 \cdot NH_2C_2H_4OH$	Sn: Org.Verb.6-180, 181, 183

Formula	Compound	Reference
$C_{16}Cl_2H_{21}NO_2Sn$	$(C_6H_5)_2SnCl_2 \cdot HN(CH_2CH_2OH)_2$	Sn: Org.Verb.6-174, 177
$C_{16}Cl_2H_{21}NSn$	$[(C_6H_5)_2((CH_3)_2NHC_2H_4)SnCl]Cl$	Sn: Org.Verb.5-216, 219
$C_{16}Cl_2H_{21}NSn_2$	$(C_6H_5)_2ClSnN(C_2H_5)SnCl(CH_3)_2$	Sn: Org.Verb.6-161
$C_{16}Cl_2H_{22}N_2O_2U$	$UO_2Cl_2 \cdot 2\ NH_2C_6H_3(CH_3)_2$	U: SVol.E1-18, 21
$C_{16}Cl_2H_{22}N_2O_4U$	$UO_2Cl_2 \cdot 2\ (CH_3)_3C_5H_2NO$	U: SVol.E1-133, 140, 142/3
–	$UO_2Cl_2 \cdot 2\ C_2H_5OC_6H_4NH_2$	U: SVol.E1-18, 21
$C_{16}Cl_2H_{22}N_2Sn$	$(C_3H_7)_2SnCl_2 \cdot NC_5H_4C_5H_4N$	Sn: Org.Verb.6-69
$C_{16}Cl_2H_{22}Ni_2O_2$	$(CH_3CHCHCHCH_3NiCl)_2 \cdot C_6H_4O_2$	Ni: Org.Verb.2-316
$C_{16}Cl_2H_{22}O_2S_2Sn$	$(C_6H_5)_2SnCl_2 \cdot 2\ (CH_3)_2SO$	Sn: Org.Verb.6-174, 177
$C_{16}Cl_2H_{24}N_2Sn$	$(C_2H_5)_2SnCl_2 \cdot 2\ C_6H_5NH_2$	Sn: Org.Verb.6-62
–	$(C_3H_7)_2SnCl_2 \cdot (CH_3)_2C_8H_4N_2$	Sn: Org.Verb.6-69
–	$(C_3H_7)_2SnCl_2 \cdot 2\ C_5H_5N$	Sn: Org.Verb.6-69
–	$(C_6H_5)_2SnCl_2 \cdot 2\ (CH_3)_2NH$	Sn: Org.Verb.6-174/5
$C_{16}Cl_2H_{26}Ni_2$	$(CH_3CHCH(CH_2)_2CHCHCH_2NiCl)_2$	Ni: Org.Verb.2-297
–	1) $(C_8H_{13}NiCl)_2$	Ni: Org.Verb.2-320
	2) $(C_8H_{13}NiCl)_2$	Ni: Org.Verb.2-321
$C_{16}Cl_2H_{28}N_2O_2Sn$	$(C_2H_5)(C_6H_5)SnCl_2 \cdot 2\ NH(CH_2CH_2)_2O$	Sn: Org.Verb.6-204
$C_{16}Cl_2H_{28}N_2O_6Sn$	$[C_2H_5OC(O)CH_2NHC(O)CH(CH_3)CH_2]_2SnCl_2$	Sn: Org.Verb.6-136
$C_{16}Cl_2H_{28}N_4Ni$	$[(i\text{-}C_3H_7NC)_4Ni]Cl_2$	Ni: Org.Verb.1-333
$C_{16}Cl_2H_{30}Ni_2$	$((CH_3)_3CCH_2CHCHCH_2NiCl)_2$	Ni: Org.Verb.2-297
$C_{16}Cl_2H_{32}MnO_{12}$	$Mn(ClO_4)_2 \cdot 4\ C_4H_8O \cdot 2\ H_2O$	Mn: MVol.D1-142/3
$C_{16}Cl_2H_{32}MnO_{16}$	$Mn(ClO_4)_2 \cdot 4\ C_4H_8O_2 \cdot 2\ H_2O$	Mn: MVol.D1-154
$C_{16}Cl_2H_{32}N_8O_8S_4Te$	$Te(SC(NHCH_2)_2CH_2)_4(ClO_4)_2$	Te: SVol.B3-161
$C_{16}Cl_2H_{32}N_8S_4Te$	$Te(SC(NHCH_2)_2CH_2)_4Cl_2 \cdot 2\ H_2O$	Te: SVol.B3-161
$C_{16}Cl_2H_{32}O_8Sn^{2+}$	$[SnCl_2(CH_3COOC_2H_5)_4]^{2+}$	Sn: MVol.C5-130
$C_{16}Cl_2H_{32}Sn$	$(C_4H_9)_3SnCH_2CHCCl_2CH_2$	Sn: Org.Verb.2-283
$C_{16}Cl_2H_{34}N_2O_4U$	$UO_2Cl_2 \cdot 2\ CH_3CON(C_3H_7)_2$	U: SVol.E1-107/8
$C_{16}Cl_2H_{34}N_2Sn$	$SnCl_2 \cdot 2\ C_6H_{11}N(CH_3)_2$	Sn: MVol.C5-12
$C_{16}Cl_2H_{34}Sn$	$(C_4H_9CH(C_2H_5)CH_2)_2SnCl_2$	Sn: Org.Verb.6-126
–	$(C_4H_9)(C_{12}H_{25})SnCl_2$	Sn: Org.Verb.6-200
–	$(C_8H_{17})_2SnCl_2$	Sn: Org.Verb.6-111/8
–	$(i\text{-}C_8H_{17})_2SnCl_2$	Sn: Org.Verb.6-125/6
$C_{16}Cl_2H_{36}MnO_{14}$	$[Mn(C_4H_8O)_4 \cdot (H_2O)_2](ClO_4)_2$	Mn: MVol.D1-142/3
$C_{16}Cl_2H_{36}N_2Sn$	$(C_3H_7)_2SnCl_2 \cdot 2\ (CH_2)_5NH$	Sn: Org.Verb.6-69
$C_{16}Cl_2H_{36}OSSn$	$(C_2H_5)_2SnCl_2 \cdot (C_6H_{13})_2SO$	Sn: Org.Verb.6-61
$C_{16}Cl_2H_{36}OSn_2$	$[(C_4H_9)_2ClSn]_2O$	Sn: Org.Verb.6-85, 88, 91, 95, 290
–	$[(s\text{-}C_4H_9)_2ClSn]_2O$	Sn: Org.Verb.6-106
–	$(C_4H_9)_2SnO \cdot (C_4H_9)_2SnCl_2$	Sn: Org.Verb.6-21
$C_{16}Cl_2H_{36}O_4P_2Sn$	$[SnCl_2((C_4H_9)_2PO_2)_2]_n$	Sn: MVol.C6-151
$C_{16}Cl_2H_{36}O_{12}U$	$UO_2(ClO_4)_2 \cdot 2\ (C_4H_9)_2O$	U: SVol.E1-71, 76/7
$C_{16}Cl_2H_{36}SSn_2$	$[(C_4H_9)_2ClSn]_2S$	Sn: Org.Verb.5-132
		Sn: Org.Verb.6-88, 89, 96
$C_{16}Cl_2H_{36}S_{2.5}Sn_2$	$[(C_4H_9)_2ClSn]_2S_{2.5}$	Sn: Org.Verb.6-89, 96
$C_{16}Cl_2H_{36}S_4Sn_2$	$[(C_4H_9)_2ClSn]_2S_4$	Sn: Org.Verb.6-89, 96
$C_{16}Cl_2H_{36}Sn_2$	$[(C_4H_9)_2ClSn]_2$	Sn: Org.Verb.6-84, 88, 97, 290
–	$[(i\text{-}C_4H_9)_2ClSn]_2$	Sn: Org.Verb.6-104
$C_{16}Cl_2H_{37}OPSn$	$(C_2H_5)_2SnCl_2 \cdot (C_4H_9)_3PO$	Sn: Org.Verb.6-61

Formula	Compound	Reference
$C_{16}Cl_2H_{37}O_3PSn$	$(C_2H_5)_2SnCl_2 \cdot (C_4H_9O)_2(C_4H_9)PO$	Sn: Org.Verb.6-61
$C_{16}Cl_2H_{40}I_4N_2Sn$	$[(C_2H_5)_4N]_2[SnCl_2I_4]$	Sn: MVol.C3-121
$C_{16}Cl_2H_{40}IrNO_3S_4$	$[Ir(S(C_2H_5)_2)_4Cl_2]NO_3$	Ir: SVol.2-165
$C_{16}Cl_2H_{40}N_2Sn$	$(C_8H_{17})_2SnCl_2 \cdot 2\ NH_3$	Sn: Org.Verb.6-115
$C_{16}Cl_2H_{40}O_2P_2S_2Sn$	$(CH_3)_2SnCl_2 \cdot 2\ (C_3H_7)_2(CH_3S)PO$	Sn: Org.Verb.6-37
$C_{16}Cl_2H_{40}O_{14}U$	$UO_2(ClO_4)_2 \cdot 4\ (C_2H_5)_2O \cdot H_2O$	U: SVol.E1-71, 76/7
$C_{16}Cl_2H_{40}Sn_4$	$[(C_2H_5)_2ClSnSn(C_2H_5)_2]_2$	Sn: Org.Verb.6-52
$C_{16}Cl_2H_{41}NSn_2$	$2\ (C_2H_5)_3SnCl \cdot (C_2H_5)_2NH$	Sn: Org.Verb.5-107
$C_{16}Cl_2H_{42}IrO_6P_3$	$Ir(P(t\text{-}C_4H_9)_2C_2H_5)(P(OCH_3)_3)_2HCl_2$	Ir: SVol.2-197/8
$C_{16}Cl_2H_{42}N_6O_2P_2Sn$	$(CH_2CH)_2SnCl_2 \cdot 2\ [(CH_3)_2N]_3PO$	Sn: Org.Verb.6-146
$C_{16}Cl_3ErH_{32}O_{20}$	$Er(ClO_4)_3 \cdot 4\ C_4H_8O_2 \cdot 9\ H_2O$	Sc: MVol.C5-96/7
$C_{16}Cl_3EuH_{32}O_{20}$	$Eu(ClO_4)_3 \cdot 4\ C_4H_8O_2 \cdot 9\ H_2O$	Sc: MVol.C5-96/7
$C_{16}Cl_3F_3H_{19}N_3O_2S$	$((C_4H_9)_2NSO_2)(CF_3)C_7Cl_3HN_2$	F: PerFHalOrg.6-162
$C_{16}Cl_3F_5H_{10}N_2Ti$	$C_6F_5TiCl_3 \cdot 2\ C_5H_5N$	Ti: Org.Verb.1-36/7, 41
$C_{16}Cl_3F_{21}H_{12}OSn$	$CF_3(CF_2)_9(CH_2)_3O(CH_2)_3SnCl_3$	Sn: Org.Verb.6-260
$C_{16}Cl_3F_{21}H_{12}Sn$	$CF_3(CF_2)_9(CH_2)_6SnCl_3$	Sn: Org.Verb.6-260
$C_{16}Cl_3GdH_{32}O_{20}$	$Gd(ClO_4)_3 \cdot 4\ C_4H_8O_2 \cdot 9\ H_2O$	Sc: MVol.C5-96/7
$C_{16}Cl_3H_8NO_2SSn$	$SnCl_3(COC_6H_4NCCSC_6H_4CO)$	Sn: MVol.C6-111
$C_{16}Cl_3H_9N_2O_2Sn$	$SnCl_3(C_{16}H_9N_2O_2)$	Sn: MVol.C5-192
$C_{16}Cl_3H_{10}O_3Sn$	$SnCl_3(C_{13}H_7O_3) \cdot 0.5\ C_6H_6$	Sn: MVol.C5-163
$C_{16}Cl_3H_{11}N_2OSn$	$SnCl_3(C_{16}H_{11}N_2O)$	Sn: MVol.C5-192
$C_{16}Cl_3H_{13}N_2Sn$	$C_6H_5SnCl_3 \cdot NC_5H_4C_5H_4N$	Sn: Org.Verb.6-279
$C_{16}Cl_3H_{13}N_2Ti$	$C_6H_5TiCl_3 \cdot NC_5H_4C_5H_4N$	Ti: Org.Verb.1-36/7
$C_{16}Cl_3H_{13}O_3Sn$	$SnCl_3(OC_6H_3(OCH_3)COCHCHC_6H_5)$	Sn: MVol.C5-91/2
$C_{16}Cl_3H_{15}IrN_5$	$Ir(C_{11}H_{10}N_4)(C_5H_5N)Cl_3 \cdot H_2O$	Ir: SVol.2-96
$C_{16}Cl_3H_{15}N_2Sn$	$C_6H_5SnCl_3 \cdot 2\ C_5H_5N$	Sn: Org.Verb.6-279
$C_{16}Cl_3H_{15}N_2Ti$	$C_6H_5TiCl_3 \cdot 2\ C_5H_5N$	Ti: Org.Verb.1-34/5, 41
$C_{16}Cl_3H_{16}IrN_8$	$[Ir(C_4H_4N_2)_4Cl_2]Cl \cdot 2\ H_2O$	Ir: SVol.2-77
–	$[Ir(C_4H_4N_2)_4Cl_2]Cl \cdot 4\ H_2O$	Ir: SVol.2-77
–	$[Ir(C_4H_4N_2)_4Cl_2]Cl \cdot 5\ H_2O$	Ir: SVol.2-77
$C_{16}Cl_3H_{17}N_2Sn$	$C_4H_9SnCl_3 \cdot C_{12}H_8N_2$	Sn: Org.Verb.6-247/8
$C_{16}Cl_3H_{18}IrS_3$	$IrCl_3(S(CH_2CH_2SC_6H_5)_2)$	Ir: SVol.2-167
$C_{16}Cl_3H_{19}OSn_2$	$Cl_2(C_4H_9)SnOSn(C_6H_5)_2Cl$	Sn: Org.Verb.6-245
$C_{16}Cl_3H_{22}NOsP_2S$	$[OsCl_3(NS)(C_6H_5P(CH_3)_2)_2]$	S: SVol.2-261
$C_{16}Cl_3H_{22}NP_2ReS$	$[ReCl_3(NS)(C_6H_5P(CH_3)_2)_2]$	S: SVol.2-261
$C_{16}Cl_3H_{22}NSn$	$[(CH_3)_4N][(C_6H_5)_2SnCl_3]$	Sn: Org.Verb.6-174/5
$C_{16}Cl_3H_{23}N_2Sn$	$C_4H_9SnCl_3 \cdot 2\ CH_3C_5H_4N$	Sn: Org.Verb.6-247/8
$C_{16}Cl_3H_{25}N_2O_2Sn$	$[SnCl_3((C_4H_9)_2C(OH)CONNHC_6H_5)]$	Sn: MVol.C6-60
$C_{16}Cl_3H_{25}N_2O_4PSn^+$	$[SnCl_3(C_5H_5N)_2((C_2H_5O)_3PO)]^+$	Sn: MVol.C6-170
$C_{16}Cl_3H_{27}N_2Sn$	$C_6H_5SnCl_3 \cdot 2\ (CH_2)_5NH$	Sn: Org.Verb.6-279
$C_{16}Cl_3H_{32}LaO_{20}$	$La(ClO_4)_3 \cdot 4\ C_4H_8O_2 \cdot 9\ H_2O$	Sc: MVol.C5-96/7
$C_{16}Cl_3H_{32}MnO_4$	$MnCl_3 \cdot 4\ C_4H_8O$	Mn: MVol.D1-142
$C_{16}Cl_3H_{32}NdO_{20}$	$Nd(ClO_4)_3 \cdot 4\ C_4H_8O_2 \cdot 9\ H_2O$	Sc: MVol.C5-96/7
$C_{16}Cl_3H_{32}O_{20}Pr$	$Pr(ClO_4)_3 \cdot 4\ C_4H_8O_2 \cdot 9\ H_2O$	Sc: MVol.C5-96/7
$C_{16}Cl_3H_{32}O_{20}Sm$	$Sm(ClO_4)_3 \cdot 4\ C_4H_8O_2 \cdot 9\ H_2O$	Sc: MVol.C5-96/7, 115
$C_{16}Cl_3H_{32}O_{20}Y$	$Y(ClO_4)_3 \cdot 4\ C_4H_8O_2 \cdot 9\ H_2O$	Sc: MVol.C5-96/7
$C_{16}Cl_3H_{33}O_4Sn$	$C_4H_9SnCl_3 \cdot 2\ CH_3COOC_4H_9$	Sn: Org.Verb.6-247/8
$C_{16}Cl_3H_{36}NSn$	$[(n\text{-}C_4H_9)_4N]SnCl_3$	Sn: MVol.C3-95/6
$C_{16}Cl_3H_{36}PSn$	$C_4H_9SnCl_3 \cdot (C_4H_9)_3P$	Sn: Org.Verb.6-247/8

Formula	Compound		Reference
$C_{16}Cl_3H_{40}IrO_4S_4$	$[Ir(S(C_2H_5)_2)_4Cl_2]ClO_4$	Ir:	SVol.2-165
$C_{16}Cl_4Co_2FeGe_2H_{10}O_6$	$((C_5H_5Co)_2(CO)_2)(GeCl_2)_2Fe(CO)_4$	Fe:	Org.Verb.B3-149
$C_{16}Cl_4F_2H_{40}N_2Te$	$[(C_2H_5)_4N]_2[TeCl_4F_2]$	Te:	SVol.B3-87
$C_{16}Cl_4F_3H_6N_3O$	$NC_5F_2Cl_2NHC_5FCl_2(OC_6H_5)N$	F:	PerFHalOrg.5-138/9
$C_{16}Cl_4F_3H_6N_3S$	$NC_5F_2Cl_2NHC_5FCl_2(SC_6H_5)N$	F:	PerFHalOrg.5-138/9
$C_{16}Cl_4F_4H_{14}N_4$	$NC_5F_2Cl_2NH(CH_2)_6NHC_5F_2Cl_2N$	F:	PerFHalOrg.5-141
$C_{16}Cl_4FeH_{10}O_2$	$[Fe(C_5H_5)_2][C_6Cl_4O_2]$	Fe:	Org.Verb.A1-229
$C_{16}Cl_4H_8S_4Sn$	$Sn(C_4H_2ClS)_4$	Sn:	Org.Verb.1-177
$C_{16}Cl_4H_{10}N_2O_2Sn$	$SnCl_4 \cdot C_{16}H_{10}N_2O_2$	Sn:	MVol.C5-192
$C_{16}Cl_4H_{10}N_2O_4Sn$	$SnCl_4 \cdot 2\ C_6H_4(CO)_2NH$	Sn:	MVol.C6-54
$C_{16}Cl_4H_{12}N_4O_4Sn$	$SnCl_4 \cdot 2\ NO_2C_6H_4CH_2CN$	Sn:	MVol.C6-72/3
$C_{16}Cl_4H_{12}O_4Sn$	$SnCl_4 \cdot (CH_3O)_2C_{14}H_6O_2$	Sn:	MVol.C5-114
–	$SnCl_4 \cdot 2\ C_8H_6O_2$	Sn:	MVol.C5-165/6
$C_{16}Cl_4H_{12}O_6Sn$	$SnCl_4 \cdot 2\ C_7H_5O_2CHO$	Sn:	MVol.C5-82
$C_{16}Cl_4H_{14}N_2Ni_2O_2$	$(CH_3CHCHCH_2NiCl)_2 \cdot C_6Cl_2(CN)_2O_2$	Ni:	Org.Verb.2-301
$C_{16}Cl_4H_{14}N_2O_2Sn$	$SnCl_4 \cdot 2\ CH_3C_6H_4OCN$	Sn:	MVol.C6-76/7
–	$SnCl_4 \cdot 2\ CH_3OC_6H_4CN$	Sn:	MVol.C6-71/2
$C_{16}Cl_4H_{14}N_2Sn$	$SnCl_4 \cdot 2\ CH_3C_6H_4CN$	Sn:	MVol.C6-71/2
–	$SnCl_4 \cdot 2\ C_6H_5CH_2CN$	Sn:	MVol.C6-72/3
–	$SnCl_4 \cdot (C_6H_5)_2C_4H_4N_2$	Sn:	MVol.C5-226
–	$SnCl_4 \cdot 2\ C_8H_6NH$	Sn:	MVol.C5-191
$C_{16}Cl_4H_{14}O_2Sn$	$SnCl_4 \cdot CH_3C_6H_4COCOC_6H_4CH_3$	Sn:	MVol.C5-109
$C_{16}Cl_4H_{14}O_4Sn$	$SnCl_4 \cdot CH_3OC_6H_4COCOC_6H_4OCH_3$	Sn:	MVol.C5-109
–	$SnCl_4 \cdot (C_6H_5COO)_2C_2H_4$	Sn:	MVol.C5-145
$C_{16}Cl_4H_{14}O_6Sn$	$SnCl_4 \cdot C_{16}H_{14}O_6$	Sn:	MVol.C5-166/7
$C_{16}Cl_4H_{15}NO_2Sn$	$SnCl_4 \cdot C_6H_5COCH_2CONHC_6H_4CH_3$	Sn:	MVol.C6-49/50
$C_{16}Cl_4H_{15}N_3OSn$	$SnCl_4 \cdot C_6H_5C_3HN_2(OH)C_2H_4C_5H_4N$	Sn:	MVol.C5-222
$C_{16}Cl_4H_{16}N_2O_2Sn$	$SnCl_4 \cdot C_2H_4(NCHC_6H_4OH)_2$	Sn:	MVol.C6-22/3
$C_{16}Cl_4H_{16}O_2Sn$	$SnCl_4 \cdot 2\ CH_3COC_6H_5$	Sn:	MVol.C5-88
–	$SnCl_4 \cdot 2\ C_6H_5OCHCH_2$	Sn:	MVol.C5-153
–	$SnCl_4 \cdot 2\ C_8H_8O$	Sn:	MVol.C5-161
$C_{16}Cl_4H_{16}O_4Sn$	$SnCl_4 \cdot 2\ CH_3COC_6H_4OH$	Sn:	MVol.C5-88/9
–	$SnCl_4 \cdot 2\ CH_3COOC_6H_5$	Sn:	MVol.C5-132
–	$SnCl_4 \cdot 2\ CH_3C_6H_4COOH$	Sn:	MVol.C5-124
–	$SnCl_4 \cdot 2\ CH_3OC_6H_4CHO$	Sn:	MVol.C5-81, 82
–	$SnCl_4 \cdot 2\ C_6H_5COOCH_3$	Sn:	MVol.C5-144/5
$C_{16}Cl_4H_{16}O_6Sn$	$SnCl_4 \cdot 2\ CH_3COC_6H_3(OH)_2$	Sn:	MVol.C5-89
$C_{16}Cl_4H_{16}S_4Sn$	$SnCl_4 \cdot 4\ S(CH)_4$	Sn:	MVol.C6-109
$C_{16}Cl_4H_{17}IrN_8O_8$	$[Ir(C_4H_4N_2)_3(C_4H_4N_2H)Cl_2](ClO_4)_2$	Ir:	SVol.2-77
$C_{16}Cl_4H_{18}N_2O_2Sn$	$SnCl_4 \cdot 2\ CH_3CONHC_6H_5$	Sn:	MVol.C6-44/5
–	$[SnCl_4 \cdot 2\ NH_2C_6H_4OCHCH_2]_n$	Sn:	MVol.C6-9
$C_{16}Cl_4H_{18}N_2O_4Sn$	$SnCl_4 \cdot 2\ (CH_3CO)_2C_4H_2NH$	Sn:	MVol.C5-190
–	$SnCl_4 \cdot 2\ CH_3NHC_6H_4COOH$	Sn:	MVol.C6-13
$C_{16}Cl_4H_{18}N_2O_6Sn$	$[NO_2C_6H_4C_6H_4NH_3][HSnCl_4(CH_3COO)_2]$	Sn:	MVol.C5-118
$C_{16}Cl_4H_{20}MnN_6$	$[(CH_3)_2NC_6H_4N_2]_2MnCl_4$	Mn:	MVol.C5-205/6
$C_{16}Cl_4H_{20}N_4Sn$	$(C_2H_5)_2SnCl_2 \cdot 2\ C_6H_5N_2Cl$	Sn:	Org.Verb.6-61
$C_{16}Cl_4H_{20}O_2Sn$	$SnCl_4 \cdot 2\ C_6H_5CH_2OCH_3$	Sn:	MVol.C5-151/2
$C_{16}Cl_4H_{20}Si_2Sn$	$Cl(CH_2(Si(CH_3)C_6H_5)_2CH_2)SnCl_3$	Sn:	Org.Verb.6-254, 260
$C_{16}Cl_4H_{22}IrP_2$	$IrCl_4(P(CH_3)_2C_6H_5)_2$	Ir:	SVol.2-215

Formula	Compound	Reference
$C_{16}Cl_4H_{22}IrP_2^-$	$[Ir(P(CH_3)_2C_6H_5)_2Cl_4]^-$	Ir: SVol.2-215
$C_{16}Cl_4H_{22}N_2O_2Sn$	$SnCl_4 \cdot 2\ (CH_3)_3C_5H_2NO$	Sn: MVol.C5-208/9
$C_{16}Cl_4H_{22}N_2O_2Te$	$TeCl_4 \cdot 2\ C_2H_5OC_6H_4NH_2$	Te: SVol.B3-166/7
$C_{16}Cl_4H_{22}N_2Sn$	$SnCl_4 \cdot 2\ C_6H_5N(CH_3)_2 \cdot CH_3COCl$	Sn: MVol.C5-180
$C_{16}Cl_4H_{22}N_2Te$	$TeCl_4 \cdot 2\ (C_2H_5)(C_6H_5)NH$	Te: SVol.B3-167
–	$TeCl_4 \cdot 2\ C_6H_5N(CH_3)_2$	Te: SVol.B3-168
$C_{16}Cl_4H_{22}O_4P_2Sn$	$SnCl_4 \cdot 2\ (C_6H_5)(CH_3)P(O)OCH_3$	Sn: MVol.C6-153/4
–	$SnCl_4 \cdot 2\ C_6H_5(H)P(O)OC_2H_5$	Sn: MVol.C6-153
–	$SnCl_4 \cdot 2\ C_6H_5P(OCH_3)_2$	Sn: MVol.C6-141
$C_{16}Cl_4H_{22}O_4Sn$	$SnCl_4 \cdot C_6H_4(COOC_4H_9)_2$	Sn: MVol.C5-147
$C_{16}Cl_4H_{22}P_2S_2Sn$	$SnCl_4 \cdot 2\ (C_6H_5)(CH_3)_2PS$	Sn: MVol.C6-148/9
$C_{16}Cl_4H_{22}P_2Sn$	$SnCl_4 \cdot 2\ C_6H_5P(CH_3)_2$	Sn: MVol.C6-136/7
$C_{16}Cl_4H_{24}Ni_2$	$((CH_3)_4C_4NiCl_2)_2$	Ni: Org.Verb.2-325/8
$C_{16}Cl_4H_{24}Ni_2O_{16}$	$((CH_3)_4C_4Ni(ClO_4)_2)_2$	Ni: Org.Verb.2-329
$C_{16}Cl_4H_{25}N_2O_4PSn$	$[SnCl_3(C_5H_5N)_2((C_2H_5O)_3PO)]Cl$	Sn: MVol.C6-170
$C_{16}Cl_4H_{26}N_2Sn$	$[C_5H_5NH]_2[(C_3H_7)_2SnCl_4]$	Sn: Org.Verb.6-68
–	$[C_5H_5NH]_2[(i\text{-}C_3H_7)_2SnCl_4]$	Sn: Org.Verb.6-70
$C_{16}Cl_4H_{27}NOSn$	$SnCl_4 \cdot HOC_6H_3(C(CH_3)_2C_2H_5)CH_2N(C_2H_5)_2$	Sn: MVol.C6-4
$C_{16}Cl_4H_{28}N_4U$	$[UCl_4(C_3H_7CN)_4]$	U: SVol.E1-48/50
$C_{16}Cl_4H_{28}O_2Sn$	$SnCl_4 \cdot 2\ C_8H_{14}O$	Sn: MVol.C5-161
$C_{16}Cl_4H_{28}O_4Sn$	$SnCl_4 \cdot 2\ CH_3COOC_6H_{11}$	Sn: MVol.C5-132
$C_{16}Cl_4H_{30}P_2Sn$	$SnCl_4 \cdot C_6H_5P(C_2H_5)_2 \cdot P(C_2H_5)_3$	Sn: MVol.C6-136/7
$C_{16}Cl_4H_{32}O_4Sn$	$SnCl_4 \cdot 2\ CH_3COOC_6H_{13}$	Sn: MVol.C5-131/2
–	$SnCl_4 \cdot 2\ C_3H_7COOC_4H_9$	Sn: MVol.C5-135/6
$C_{16}Cl_4H_{34}N_2O_2U$	$UCl_4 \cdot 2\ CH_3CON(C_3H_7)_2$	U: SVol.E1-106/8
$C_{16}Cl_4H_{34}N_2O_8Sn$	$SnCl_4 \cdot 2\ (CH_3)_2CHCH_2CH(NH_2)COOH \cdot 2\ CH_3COOH$	Sn: MVol.C6-12
$C_{16}Cl_4H_{34}SSn$	$SnCl_4 \cdot (C_8H_{17})_2S$	Sn: MVol.C6-93/4
$C_{16}Cl_4H_{36}N_4O_4U$	$UCl_4 \cdot 4\ CH_3CONHC_2H_5$	U: SVol.E1-96/8
$C_{16}Cl_4H_{36}O_2S_2Sn$	$SnCl_4 \cdot 2\ (n\text{-}C_4H_9)_2SO$	Sn: MVol.C6-102
$C_{16}Cl_4H_{36}O_2S_2U$	$UCl_4 \cdot 2\ (C_4H_9)_2SO$	U: SVol.E1-209/11
$C_{16}Cl_4H_{36}O_2Sn$	$SnCl_4 \cdot 2\ (C_4H_9)_2O$	Sn: MVol.C5-151/2
–	$SnCl_4 \cdot 2\ C_8H_{17}OH$	Sn: MVol.C5-73
$C_{16}Cl_4H_{36}O_4SiSn$	$SnCl_4 \cdot Si(OC_4H_9)_4$	Sn: MVol.C6-126/7
$C_{16}Cl_4H_{36}O_6Si_2Sn$	$SnCl_4 \cdot 2\ CH_2CHSi(OC_2H_5)_3$	Sn: MVol.C6-127/8
$C_{16}Cl_4H_{36}S_2Sn$	$SnCl_4 \cdot 2\ (C_4H_9)_2S$	Sn: MVol.C6-93/4
$C_{16}Cl_4H_{38}N_2Sn$	$SnCl_4 \cdot 2\ (C_4H_9)_2NH$	Sn: MVol.C5-171
$C_{16}Cl_4H_{38}O_2P_2S_2Sn$	$SnCl_4 \cdot 2\ (C_2H_5)_2P(O)SC_4H_9$	Sn: MVol.C6-158/60
–	$SnCl_4 \cdot 2\ (C_4H_9)_2POSH$	Sn: MVol.C6-156/7
$C_{16}Cl_4H_{38}O_2P_2Sn$	$SnCl_4 \cdot 2\ (C_4H_9)_2HPO$	Sn: MVol.C6-146/7
$C_{16}Cl_4H_{38}O_4P_2Sn$	$SnCl_4 \cdot 2\ (C_4H_9)_2P(O)OH$	Sn: MVol.C6-150/1
$C_{16}Cl_4H_{38}O_6P_2S_2Sn$	$SnCl_4 \cdot 2\ (C_4H_9O)_2(HS)PO$	Sn: MVol.C6-174/5
$C_{16}Cl_4H_{38}O_6P_2Sn$	$SnCl_4 \cdot 2\ HP(O)(OC_4H_9)_2$	Sn: MVol.C6-160/1
$C_{16}Cl_4H_{40}I_2N_2Sn$	$[(C_2H_5)_4N]_2[SnCl_4I_2]$	Sn: MVol.C3-121
$C_{16}Cl_4H_{40}I_2N_2Te$	$[(C_2H_5)_4N]_2[TeCl_4I_2]$	Te: SVol.B3-87
$C_{16}Cl_4H_{40}I_2N_2U$	$[N(C_2H_5)_4]_2UCl_4I_2$	U: SVol.C9-184
$C_{16}Cl_4H_{40}MnN_2$	$[(C_2H_5)_4N]_2MnCl_4$	Mn: MVol.C5-84/7, 194
–	$(C_8H_{17}NH_3)_2MnCl_4$	Mn: MVol.C5-182
$C_{16}Cl_4H_{40}MoO_4$	$MoCl_4 \cdot 4\ (C_2H_5)_2O$	Mo: SVol.B1-142
$C_{16}Cl_4H_{40}N_2O_2U$	$[N(C_2H_5)_4]_2UO_2Cl_4$	U: SVol.C9-93/8, 101

Formula	Compound		Reference
$C_{16}Cl_4H_{40}N_8O_4U$	$UCl_4 \cdot 4\ CH_3NHCON(CH_3)_2$	U:	SVol.E1-91/2, 94
$C_{16}Cl_4H_{40}O_2P_2U$	$[P(C_2H_5)_4]_2UO_2Cl_4$	U:	SVol.C9-59, 93, 101
$C_{16}Cl_4H_{40}O_6Si_2Sn$	$SnCl_4 \cdot 2\ C_2H_5Si(OC_2H_5)_3$	Sn:	MVol.C6-127/8
$C_{16}Cl_4H_{44}N_4U$	$[N(C_2H_5)_4]_2UCl_4(NH_2)_2$	U:	SVol.C9-59, 109
–	$UCl_4 \cdot 4\ C_4H_9NH_2$	U:	SVol.E1-18, 20
$C_{16}Cl_4H_{48}N_8Sn$	$SnCl_4 \cdot 4\ NH_2(CH_2)_4NH_2$	Sn:	MVol.C5-185
$C_{16}Cl_5H_{12}NOSn$	$SnCl_4 \cdot C_9H_7N \cdot C_6H_5COCl$	Sn:	MVol.C5-209/10
$C_{16}Cl_5H_{12}N_4U$	$[UCl_5(C_8H_6N_2)_2]$	U:	SVol.E1-46/8
$C_{16}Cl_5H_{15}O_6Sn$	$SnCl_4 \cdot 2\ C_6H_5COOH \cdot CH_2ClCOOH$	Sn:	MVol.C5-122/3
$C_{16}Cl_5H_{18}N_3OSU$	$[(CH_3)_2C_{14}H_{12}N_3S]UOCl_5$	U:	SVol.C9-92
$C_{16}Cl_5H_{24}MnN_2$	$[(CH_3)_3C_5H_2NH]_2MnCl_5$	Mn:	MVol.C5-212/3
$C_{16}Cl_5H_{36}NSn$	$[(C_4H_9)_4N]SnCl_5$	Sn:	MVol.C3-101
$C_{16}Cl_5H_{40}InN_2$	$[N(C_2H_5)_4]_2InCl_5$ solid solutions $[N(C_2H_5)_4]_2(In,Mn)Cl_5$	Mn:	MVol.C5-228
$C_{16}Cl_5H_{40}MnN_2$	$[(C_2H_5)_4N]_2MnCl_5$	Mn:	MVol.C5-211
–	$[(C_2H_5)_4N]_2MnCl_5$ solid solutions $[N(C_2H_5)_4]_2(Mn,In)Cl_5$	Mn:	MVol.C5-228
$C_{16}Cl_5H_{40}N_2OPa$	$[N(C_2H_5)_4]_2PaOCl_5$	Pa:	SVol.2-61/2
$C_{16}Cl_5H_{40}N_2OU$	$[N(C_2H_5)_4]_2UOCl_5$	U:	SVol.C9-90/2
$C_{16}Cl_6FeH_{16}O_6$	$[Fe(C_5H_5)_2][CHCl_2COOH]_3$	Fe:	Org.Verb.A1-227
$C_{16}Cl_6FeH_{40}N_4O_2Sn_2$	$[N(C_2H_5)_4]_2[Fe(NO)_2(SnCl_3)_2]$	Fe:	Org.Verb.B1-143
$C_{16}Cl_6H_{10}O_2Sn$	$SnCl_4 \cdot C_{10}H_4Cl_2O_2 \cdot C_6H_6$	Sn:	MVol.C5-112
$C_{16}Cl_6H_{12}N_2Sn$	$SnCl_4 \cdot 2\ ClC_6H_4CH_2CN$	Sn:	MVol.C6-72/3
$C_{16}Cl_6H_{14}Sn$	$(CH_3)_3SnC_7Cl_6C_6H_5$	Sn:	Org.Verb.2-102, 106
$C_{16}Cl_6H_{16}N_2O_6Sn$	$[NO_2(C_6H_4)_2NH_3][HSnCl_4(CH_2ClCOO)_2]$	Sn:	MVol.C5-119
$C_{16}Cl_6H_{18}Ni_2O_2$	$(CH_3CHCHCHCH_3NiCl)_2 \cdot C_6Cl_4O_2$	Ni:	Org.Verb.2-317
$C_{16}Cl_6H_{24}Li_2N_8Sn$	$[Li(CH_3CN)_4]_2[SnCl_6]$	Sn:	MVol.C3-5
$C_{16}Cl_6H_{24}N_8SnSr$	$[Sr(CH_3CN)_8][SnCl_6]$	Sn:	MVol.C3-187/8
$C_{16}Cl_6H_{36}NU$	$[N(C_4H_9)_4]UCl_6$	U:	SVol.C9-60
$C_{16}Cl_6H_{40}IrN_2$	$((C_2H_5)_4N)_2[IrCl_6]$	Ir:	SVol.2-127
$C_{16}Cl_6H_{40}N_2Pa$	$[N(C_2H_5)_4]_2PaCl_6$	Pa:	SVol.2-42, 44, 56/7
$C_{16}Cl_6H_{40}N_2Pr$	$[(C_2H_5)_4N]_2PrCl_6$	Sc:	MVol.C5-223
$C_{16}Cl_6H_{40}N_2Sb$	$[(C_2H_5)_4N]_2[SbCl_6]$ solid solutions $[(C_2H_5)_4N]_2[(Sb,Sn)Cl_6]$	Sn:	MVol.C3-106
$C_{16}Cl_6H_{40}N_2Sn$	$[(CH_3)_2(C_3H_7)_2N]_2[SnCl_6]$	Sn:	MVol.C3-106
–	$[(C_2H_5)_4N]_2[SnCl_6]$	Sn:	MVol.C3-105/6
–	$[(C_2H_5)_4N]_2[SnCl_6]$ solid solutions $[(C_2H_5)_4N]_2[(Sn,Sb)Cl_6]$	Sn:	MVol.C3-106
–	$[(i\text{-}C_4H_9)_2NH_2]_2[SnCl_6]$	Sn:	MVol.C3-104/5
$C_{16}Cl_6H_{40}N_2Te$	$[N(C_2H_5)_4]_2TeCl_6$	Te:	SVol.B2-119/20, 124
$C_{16}Cl_6H_{40}N_2U$	$[N(C_2H_5)_4]_2UCl_6$	U:	SVol.C9-52/9, 137
$C_{16}Cl_6H_{40}O_5Sn_2$	$(SnCl_3)_2O \cdot 4\ C_4H_9OH$	Sn:	MVol.C5-74
$C_{16}Cl_6H_{40}P_2U$	$(P(C_2H_5)_4)_2UCl_6$	U:	SVol.C9-52, 59
$C_{16}Cl_7FeH_{40}N_2OS$	$[((C_2H_5)_4N)_2(SOCl)][FeCl_6]$	S:	SVol.1-54
$C_{16}Cl_7H_{32}O_4Sn_2^+$	$[Sn_2Cl_7(C_5H_{11}COOC_2H_5)_2]^+$	Sn:	MVol.C5-136/7
$C_{16}Cl_8F_{16}H_2N_6O_2$	$[OCNCNC(CF_2Cl)_2NHC(CF_2Cl)_2]_2$	F:	PerFHalOrg.5-63/4, 87
$C_{16}Cl_8H_{28}O_8Sn_2$	$[SnCl_4 \cdot C_2H_4(COOC_2H_5)_2]_2$	Sn:	MVol.C5-143
$C_{16}Cl_8H_{32}O_4Sn_2$	$[SnCl_4 \cdot C_5H_{11}COOC_2H_5]_2$	Sn:	MVol.C5-136
$C_{16}Cl_8H_{36}NiO_4P_4$	$Ni(P(OC_4H_9\text{-}n)Cl_2)_4$	Ni:	Org.Verb.1-241, 242

$C_{16}Cl_9FeH_{12}O_6$	$[Fe(C_5H_5)_2][CCl_3COO \cdot 2\ CCl_3COOH]$	Fe:	Org.Verb.A1-227
$C_{16}Cl_9FeH_{13}O_6$	$[Fe(C_5H_5)_2][CCl_3COOH]_3$	Fe:	Org.Verb.A1-227/8
$C_{16}Cl_{10}H_{28}N_2O_8Sn$	$SnCl_4 \cdot 2\ (CH_3)_2CHCH_2CH(NH_2)COOH \cdot 2\ CCl_3COOH$	Sn:	MVol.C6-12
$C_{16}Cl_{18}H_{48}Ir_2N_4Sn_4$	$((CH_3)_4N)_4[Ir_2Cl_6(SnCl_3)_4]$	Ir:	SVol.2-265
$C_{16}CoF_6H_{14}O_4PSn$	$[(CH_2CH)_2Sn(OCOC_5H_4CoC_5H_4COO(PF_6))]_x$	Sn:	Org.Verb.6-145
$C_{16}CoF_6H_{18}O_4PSn$	$[((C_2H_5)_2SnOCOC_5H_4CoC_5H_4COO)PF_6]_x$	Sn:	Org.Verb.6-60
$C_{16}CoH_{18}N_2O_2S_4$	$Co[S_2COC_2H_5]_2(NC_5H_4C_5H_4N)$	C:	MVol.D4-259
$C_{16}Co_2H_{30}S_{14}$	$[Co(S_2CSC_2H_5)_2(SC_2H_5)]_2$	C:	MVol.D4-266, 270
$C_{16}Co_3FeH_{19}O_{11}P^+$	$[HFeCo_3(CO)_8P(OC_3H_7)_2OC_2H_4]^+$	Fe:	Org.Verb.B1-204
$C_{16}Co_3FeH_{19}O_{16}P_2$	$HFeCo_3(CO)_{10}(P(OCH_3)_3)_2$	Fe:	Org.Verb.B1-204
$C_{16}Co_3FeH_{22}O_{10}P^+$	$[HFeCo_3(CO)_7P(OC_3H_7)_3]^+$	Fe:	Org.Verb.B1-204
$C_{16}Co_4F_4O_{10}$	$Co_4(CO)_{10}C_6F_4$	F:	PerFHalOrg.4-70
$C_{16}CrFeH_{11}NO_5$	$C_5H_5FeC_5H_4C(NH_2)Cr(CO)_5$	Fe:	Org.Verb.A1-345/6
$C_{16}CrFeH_{11}NO_5^+$	$[C_5H_5FeC_5H_4C(NH_2)Cr(CO)_5]^+$	Fe:	Org.Verb.A1-346
$C_{16}CrFeH_{12}O_5^+$	$[C_5H_5FeC_5H_4C(OCH_3)Cr(CO)_4]^+$	Fe:	Org.Verb.A1-346
$C_{16}CrFeH_{14}O_4^+$	$[C_5H_5FeC_5H_4C(OC_2H_5)Cr(CO)_3]^+$	Fe:	Org.Verb.A1-346
$C_{16}CrFeH_{15}NO_3^+$	$[C_5H_5FeC_5H_4C(N(CH_3)_2)Cr(CO)_3]^+$	Fe:	Org.Verb.A1-346
$C_{16}CrFeH_{17}NO^+$	$[C_5H_5FeC_5H_4C(NC_4H_8)Cr(CO)]^+$	Fe:	Org.Verb.A1-346
$C_{16}CrFeH_{18}N_6OS_4$	$[C_5H_5FeC_5H_4COCH_3][Cr(NH_3)_2(SCN)_4]$	Fe:	Org.Verb.A2-300
$C_{16}Cr_2Ni_3O_{16}^{2-}$	$[Ni_3Cr_2(CO)_{16}]^{2-}$	Ni:	Org.Verb.2-374
$C_{16}CuF_{28}H_4N_6$	$[N(C(C_3F_7)NH)_2]_2Cu$	F:	PerFHalOrg.7-58
$C_{16}CuFeH_{15}O_4^+$	$[(C_5H_5FeC_5H_4COCHCOCOOC_2H_5)Cu]^+$	Fe:	Org.Verb.A3-140
$C_{16}FFeH_{13}$	$C_5H_5FeC_5H_4C_6H_4F$	Fe:	Org.Verb.A1-390
$C_{16}FH_{16}NSn$	$(C_6H_5)_2(NCCH_2CH_2CH_2)SnF$	Sn:	Org.Verb.5-40
$C_{16}FH_{27}N_2$	$NCFC(C_4H_9)NC(C_4H_9)C(C_4H_9)$	F:	PerFHalOrg.6-60, 69
$C_{16}FH_{31}Sn$	$(C_6H_{11})_2(C_4H_9)SnF$	Sn:	Org.Verb.5-40
$C_{16}FH_{36}N_5O_3S_3$	$[NS(O)F][NS(O)NHC_8H_{17}]_2$	S:	S-N-Verb.1-14
$C_{16}F_2H_{10}N_2S_2$	$NCFC(SC_6H_5)C(SC_6H_5)CFN$	F:	PerFHalOrg.6-60, 67
$C_{16}F_2H_{12}N_4$	$FC(NC(NHC_6H_5))_2CF$	F:	PerFHalOrg.6-59, 65/6
$C_{16}F_2H_{18}Sn$	$(C_2H_5)_2Sn(C_6H_4F)_2$	Sn:	Org.Verb.3-29, 33
$C_{16}F_2H_{34}Sn$	$(C_8H_{17})_2SnF_2$	Sn:	Org.Verb.5-49/50
$C_{16}F_2H_{36}OSn_2$	$[(C_4H_9)_2FSn]_2O$	Sn:	Org.Verb.6-293
$C_{16}F_2H_{36}SSn_2$	$[(C_4H_9)_2FSn]_2S$	Sn:	Org.Verb.5-47
$C_{16}F_2H_{36}Sn_2$	$(C_4H_9)_2SnFSnF(C_4H_9)_2$	Sn:	Org.Verb.5-53
$C_{16}F_3FeH_{13}O_5$	$(CH_3)_2(C_6H_5)C_3H_2Fe(CO)_3OCOCF_3$	Fe:	Org.Verb.B5-54, 64/5
$C_{16}F_3FeH_{15}O_2$	$C_5H_5FeC_5H_4CHCH(CH_2)_2OOCCF_3$	Fe:	Org.Verb.A2-104/5
–	$C_5H_5FeC_5H_4CH(C_3H_5)OOCCF_3$	Fe:	Org.Verb.A2-98, 101
$C_{16}F_3FeH_{16}O_2^+$	$[C_5H_5FeC_5H_4CH(CH_2)_3OOCCF_3]^+$	Fe:	Org.Verb.A2-105, 120
$C_{16}F_4FeH_9O^+$	$[C_5H_5FeC_5H_4C_6F_4O]^+$	Fe:	Org.Verb.A2-135
$C_{16}F_4FeH_{30}O_8P_2$	$F_2CCF_2Fe(CO)_2(P(OC_2H_5)_3)_2$	Fe:	Org.Verb.B4-187, 189
$C_{16}F_4H_{10}Ni$	$C_5H_5NiC_5H_5C_6F_4$	Ni:	Org.Verb.2-181
–	$C_5H_5NiC_7H_5(CF)_4$	Ni:	Org.Verb.2-184/6
$C_{16}F_4H_{16}N_2O_5$	$O(CF_2COONH_3C_6H_5)_2$	F:	PerFHalOrg.6-12
$C_{16}F_4H_{22}N_2O_2Sn$	$SnF_4 \cdot 2\ (CH_3)_3C_5H_2NO$	Sn:	MVol.C5-208/9
$C_{16}F_4H_{24}Ni_2$	$((CH_3)_4C_4NiF_2)_2$	Ni:	Org.Verb.2-329
$C_{16}F_4H_{34}O_4P_2Sn$	$(C_8H_{17})_2Sn(PO_2F_2)_2$	Sn:	Org.Verb.6-114
$C_{16}F_5FeH_9$	$C_5H_5FeC_5H_4C_6F_5$	Fe:	Org.Verb.A1-389/90
$C_{16}F_5FeH_9^+$	$[C_5H_5FeC_5H_4C_6F_5]^+$	Fe:	Org.Verb.A2-287
$C_{16}F_5FeH_{11}N_2O$	$C_5H_5Fe(CO)(CNCH_3)C(C_6F_5)NCH_3$	F:	PerFHalOrg.4-38

Formula	Compound	Element	Reference
$C_{16}F_5FeH_{30}O_6P_2S_2^+$	$[Fe(P(OC_2H_5)_3)_2S_2C_2(CF_3)CF_2]^+$	Fe:	Org.Verb.B1-34/5
$C_{16}F_5H_{13}Si$	$(CH_3)_2Si(CHCHC_6H_5)C_6F_5$	F:	PerFHalOrg.4-20
$C_{16}F_5H_{17}N_2O_2$	$O(CH_2CH_2)_2NC_6H_8N(OH)C_6F_5$	F:	PerFHalOrg.7-198
$C_{16}F_6FeH_{30}N_{10}O_3$	$[COFe(C_6H_8N_8(CH_3)_4)CH_3N_2H_3](OCH_2CF_3)_2$	Fe:	Org.Verb.B1-87
$C_{16}F_6FeH_{30}O_6P_2S_2^+$	$[Fe(P(OC_2H_5)_3)_2S_2C_2(CF_3)_2]^+$	Fe:	Org.Verb.B1-34/5
$C_{16}F_6Fe_2H_{24}O_4P_2S_2$	$[P(CH_3)_3(CO)_2Fe]_2(CF_3CCCF_3)(SCH_3)_2$	Fe:	Org.Verb.C2-67/9
$C_{16}F_6H_8MnO_4S_2$	$[Mn(SC_4H_3COCHCOCF_3)_2]_n$	Mn:	MVol.D1-122
$C_{16}F_6H_8MnO_6$	$[Mn(OC_4H_3COCHCOCF_3)_2]$	Mn:	MVol.D1-122
$C_{16}F_6H_8N_5O_6P$	$NP(OC_6H_4NO_2)_2NC(CF_3)NC(CF_3)$	F:	PerFHalOrg.6-112
$C_{16}F_6H_9NpO_6S_2$	$HNpO_2(SC_4H_3COCHCOCF_3)_2$	Np:	TrU.D2-240, 364
$C_{16}F_6H_9O_5PaS_2$	$PaO(C_4H_3SCOCHCOCF_3) \cdot C_4H_3SCOCH_2COCF_3$	Pa:	SVol.2-179
$C_{16}F_6H_{10}Sn$	$(C_6H_5)_2Sn(CFCF_2)_2$	Sn:	Org.Verb.3-58/9
$C_{16}F_6H_{10}Ti$	$(C_5H_5)_2Ti(CCCF_3)_2$	F:	PerFHalOrg.4-110
$C_{16}F_6H_{11}O_7PaS_2$	$Pa(OH)_3(C_4H_3SCOCHCOCF_3)_2$	Pa:	SVol.2-178/9
$C_{16}F_6H_{12}MnO_6S_2$	$[Mn(SC_4H_3COCHCOCF_3)_2(H_2O)_2]$	Mn:	MVol.D1-123
$C_{16}F_6H_{12}N_5P$	$NP(NHC_6H_5)_2NC(CF_3)NC(CF_3)$	F:	PerFHalOrg.6-112
$C_{16}F_6H_{14}MnN_2O_4S_2$	$[Mn(SC_4H_3COCHCOCF_3)_2(NH_3)_2]$	Mn:	MVol.D1-123/4
$C_{16}F_6H_{14}P_2$	$(C_6H_5)_2PCH_2CH_2P(CF_3)_2$	F:	PerFHalOrg.3-24
$C_{16}F_6H_{15}NO_7U$	$UO_2(CF_3COCHCOCH_3)_2 \cdot CH_3C_5H_4NO$	U:	SVol.E1-133, 138, 142/3
$C_{16}F_6H_{15}NO_8U$	$UO_2(CF_3COCHCOCH_3)_2 \cdot CH_3OC_5H_4NO$	U:	SVol.E1-133, 138, 142/3
$C_{16}F_6H_{18}N_4Ni$	$(CF_3)_2CC(CN)_2Ni(CNC_4H_9\text{-}t)_2$	Ni:	Org.Verb.1-382, 383
$C_{16}F_6H_{28}Sn$	$(C_4H_9)_3SnC(CF_3)CHCF_3$	Sn:	Org.Verb.2-310
$C_{16}F_6H_{30}N_3NiOP_3$	$CONi(P(NC_5H_{10})F_2)_3$	Ni:	Org.Verb.1-113
$C_{16}F_6H_{30}NiP_2$	$(CF_2CF)_2Ni(P(C_2H_5)_3)_2$	Ni:	Org.Verb.1-70, 76
$C_{16}F_6H_{37}NiO_6P_3$	$[CH_3C_3H_4Ni(P(OC_2H_5)_3)_2]PF_6$	Ni:	Org.Verb.2-24
$C_{16}F_6H_{40}N_2Pa$	$[N(C_2H_5)_4]_2PaF_6$	Pa:	SVol.2-40/4
$C_{16}F_7H_{11}O$	$C_3F_7C(C_6H_5)_2OH$	F:	PerFHalOrg.4-92
$C_{16}F_8FeH_8N_2O_2$	$(CF_2CF_2)_2Fe(CO)_2(NC_5H_4C_5H_4N)$	Fe:	Org.Verb.B4-187, 191
$C_{16}F_8FeH_{10}N_2O_2$	$(CF_2CF_2)_2Fe(CO)_2(C_5H_5N)_2$	Fe:	Org.Verb.B4-187, 191
$C_{16}F_8H_6O_3$	$(C_6F_4)_2C(OH)COOC_2H_5$	F:	PerFHalOrg.4-54
$C_{16}F_8H_{10}Ni$	$C_5H_5NiC_5H_5C_6F_8$	Ni:	Org.Verb.2-181/2
$C_{16}F_8H_{12}Si_2$	$(CH_3)_2Si(C_6F_4)_2Si(CH_3)_2$	F:	PerFHalOrg.4-25
$C_{16}F_8H_{14}N_2Ni$	$C_4F_8Ni(CH_3C_5H_4N)_2$	Ni:	Org.Verb.1-344
$C_{16}F_8H_{14}OSi_2$	$O[C_6F_4Si(CH_3)_2H]_2$	F:	PerFHalOrg.4-85
$C_{16}F_8H_{30}NiP_2$	$C_4F_8Ni(P(C_2H_5)_3)_2$	Ni:	Org.Verb.1-345, 346
$C_{16}F_9FeH_{13}O_6$	$[Fe(C_5H_5)_2][CF_3COOH]_3$	Fe:	Org.Verb.A1-227
$C_{16}F_9H_{15}O_6Ti$	$(CH_3)_5C_5Ti(OCOCF_3)_3$	Ti:	Org.Verb.1-168, 170
$C_{16}F_{10}FeO_4$	$(CO)_4Fe(C_6F_5)_2$	Fe:	Org.Verb.B2-192
$C_{16}F_{10}GeH_{10}$	$(C_2H_5)_2Ge(C_6F_5)_2$	F:	PerFHalOrg.4-180
$C_{16}F_{10}H_4N_2O_2$	$C_6F_5N(COCH_2)(CH_2CO)NC_6F_5$	F:	PerFHalOrg.7-60
$C_{16}F_{10}H_6Sn$	$(CH_2CH)_2Sn(C_6F_5)_2$	Sn:	Org.Verb.3-72
$C_{16}F_{10}H_{10}NP$	$(C_6F_5)_2PN(C_2H_5)_2$	F:	PerFHalOrg.3-125, 132
–	$(C_6F_5)_2PNHC(CH_3)_3$	F:	PerFHalOrg.3-125, 129, 132
$C_{16}F_{10}H_{10}O_2Si$	$(C_6F_5)_2Si(OC_2H_5)_2$	F:	PerFHalOrg.4-81
$C_{16}F_{10}H_{10}Si$	$(C_6F_5)_2Si(C_2H_5)_2$	F:	PerFHalOrg.4-163
$C_{16}F_{10}H_{10}Sn$	$(C_2H_5)_2Sn(C_6F_5)_2$	Sn:	Org.Verb.3-29
$C_{16}F_{10}H_{12}OSi_2$	$[C_6F_5(CH_3)_2Si]_2O$	F:	PerFHalOrg.4-23
$C_{16}F_{10}H_{12}Si_2$	$[C_6F_5(CH_3)_2Si]_2$	F:	PerFHalOrg.4-21

Formula	Compound	Element	Reference
$C_{16}F_{11}FeH_9^+$	$[C_5H_5FeC_5H_4C_6F_{11}]^+$	Fe:	Org.Verb.A1-394
$C_{16}F_{11}FeH_9O$	$C_5H_5FeC_5H_4COC_5F_{11}$	Fe:	Org.Verb.A2-241/3
$C_{16}F_{12}FeH_8O_4$	$C_6H_7(C_3HF_6)(C_2F_3CF_3)Fe(CO)_4$	Fe:	Org.Verb.B4-337, 346
$C_{16}F_{12}FeH_{27}N_7P_2$	$[(CH_2)_4N(CH_3NH)CFe(CNCH_3)_5][PF_6]_2$	Fe:	Org.Verb.B4-174
$C_{16}F_{12}Fe_3O_{10}P_2$	$(CF_3)_4C_2P_2Fe_3(CO)_{10}$	F:	PerFHalOrg.3-13
$C_{16}F_{12}HN_3O$	$(C_5F_4N)_3COH$	F:	PerFHalOrg.5-195, 209
$C_{16}F_{12}H_9MnO_5Sn$	$(CH_3)_3SnC_4(CF_3)_4Mn(CO)_5$	Sn:	Org.Verb.2-98, 105
$C_{16}F_{12}H_9NO_7U$	$UO_2(CH(COCF_3)_2)_2 \cdot CH_3C_5H_4NO$	U:	SVol.E1-133, 138, 142/3
$C_{16}F_{12}H_9NO_8U$	$UO_2(CH(COCF_3)_2)_2 \cdot CH_3OC_5H_4NO$	U:	SVol.E1-133, 138, 142/3
$C_{16}F_{12}H_{10}N_2Ni$	$C_5H_5NiC(CF_3)_2NNC(CF_3)_2C_5H_5$	Ni:	Org.Verb.2-186/7
$C_{16}F_{12}H_{11}N_3O$	$(CF_3)_2CNHC(CF_3)_2NC(NC(C_6H_5)OC_2H_5)$	F:	PerFHalOrg.5-109
$C_{16}F_{12}H_{12}N_4$	$C_3HN_2(CF_3)_4NCHC_6H_4N(CH_3)_2$	F:	PerFHalOrg.5-109
$C_{16}F_{12}H_{13}N_5O$	$C_3HN_2(CF_3)_4NHCONHC_6H_4N(CH_3)_2$	F:	PerFHalOrg.5-108/9
$C_{16}F_{12}H_{18}N_2NiO_2$	$(t\text{-}C_4H_9NC)_2Ni(C_2O_2(CF_3)_4)$	Ni:	Org.Verb.1-318, 321
$C_{16}F_{12}H_{18}N_2NiO_4$	$(t\text{-}C_4H_9NC)_2Ni(OC(CF_3)_2)(O_3C(CF_3)_2)$	Ni:	Org.Verb.1-317, 321
$C_{16}F_{12}H_{18}N_4Ni$	$(t\text{-}C_4H_9NC)_2NiN_2(C(CF_3)_2)_2$	Ni:	Org.Verb.1-319, 322
$C_{16}F_{12}H_{19}N_3NiO$	$(t\text{-}C_4H_9NC)_2Ni(NHC(CF_3)_2OC(CF_3)_2)$	Ni:	Org.Verb.1-318, 321/2
$C_{16}F_{12}H_{20}N_4Ni$	$(t\text{-}C_4H_9NC)_2Ni(C(CF_3)_2NHC(CF_3)_2NH)$	Ni:	Org.Verb.1-319, 322
$C_{16}F_{12}H_{20}N_4Pd$	$[(CH_3)_3CNC]_2Pd(C(CF_3)_2NH)_2$	F:	PerFHalOrg.7-74/5
$C_{16}F_{12}H_{20}Sn$	$(C_4H_9)_2Sn[C(CF_3)CHCF_3]_2$	Sn:	Org.Verb.3-45
$C_{16}F_{12}N_2$	$NCFC(C_6F_5)C(C_6F_5)CFN$	F:	PerFHalOrg.6-18/9, 39
$C_{16}F_{14}$	$C_6F_5C_{10}F_9$	F:	PerFHalOrg.4-17
$C_{16}F_{14}H_9N_5O_2S$	$(C_3F_7)_2(CH_3C_6H_4SO_2NHNH)C_3N_3$	F:	PerFHalOrg.6-112
$C_{16}F_{15}H_{25}N_3O_5U$	$U(OCH_2CF_3)_5 \cdot 3\ (CH_2)_2NH$	U:	SVol.E1-38/9
$C_{16}F_{15}H_{28}N_2O_5U$	$U(OCH_2CF_3)_5 \cdot 2\ (CH_3)_3N$	U:	SVol.E1-20, 23
–	$U(OCH_2CF_3)_5 \cdot 2\ C_3H_7NH_2$	U:	SVol.E1-18, 21
$C_{16}F_{15}H_{31}N_3O_5U$	$U(OCH_2CF_3)_5 \cdot 3\ (CH_3)_2NH$	U:	SVol.E1-19, 22
$C_{16}F_{18}H_{13}N_3O_{10}$	$3\ ON(CF_2)_3COOH \cdot (C_2H_5)_2O$	F:	PerFHalOrg.7-192
$C_{16}F_{21}H_8O_{10}PU$	$UO_2(CH(COCF_3)_2)_2 \cdot (CF_3CH_2O)_3PO$	U:	SVol.E1-185, 187
$C_{16}F_{24}H_2N_6O_2$	$[OCNCNC(CF_3)_2NHC(CF_3)_2]_2$	F:	PerFHalOrg.5-63/4, 86/7, 108/10
$C_{16}F_{24}H_8Ni_2O_4P_4$	$(CO)_2Ni[(CF_3)_2PC_2H_4P(CF_3)_2]_2Ni(CO)_2$	Ni:	Org.Verb.2-260
$C_{16}F_{24}N_2$	$NCFC(C_6F_{11})CFC(C_6F_{11})N$	F:	PerFHalOrg.6-18/9, 39
$C_{16}F_{26}N_2$	$(CF_3)_4(i\text{-}C_3F_7)_2C_6N_2$	F:	PerFHalOrg.6-128/9
$C_{16}F_{27}H_4N_5$	$(C_{13}F_{27})(NH_2)_2C_3N_3$	F:	PerFHalOrg.6-79/80
$C_{16}F_{27}N$	$[(CF_3)_2CF]_3(CF_3)_2C_5N$	F:	PerFHalOrg.5-153/4, 164
$C_{16}F_{28}H_4HgN_6$	$[N(C(C_3F_7)NH)_2]_2Hg$	F:	PerFHalOrg.7-58
$C_{16}F_{28}H_4N_6Ni$	$[N(C(C_3F_7)NH)_2]_2Ni$	F:	PerFHalOrg.7-58
$C_{16}F_{28}H_4N_6Zn$	$[N(C(C_3F_7)NH)_2]_2Zn$	F:	PerFHalOrg.7-58
$C_{16}F_{28}N_2$	$NC(i\text{-}C_3F_7)N(CC_3F_7\text{-}i)_3$	F:	PerFHalOrg.6-23/4, 45, 56/7
–	$NC(s\text{-}C_4F_9)NC(s\text{-}C_4F_9)CFC(s\text{-}C_4F_9)$	F:	PerFHalOrg.6-24, 46
$C_{16}F_{29}H_4NO_3$	$(CF_3)_2CFO(CF_2)_{11}C(O)NHCH_2OH$	F:	PerFHalOrg.7-63
$C_{16}F_{29}N_3$	$(C_3F_7)_2(C_7F_{15})C_3N_3$	F:	PerFHalOrg.6-83/4, 104
$C_{16}F_{30}H_2N_2O_2$	$C_7F_{15}C(NH_2)NOC(O)C_7F_{15}$	F:	PerFHalOrg.7-5/6, 32
$C_{16}F_{30}N_2O$	$C_7F_{15}CNC(C_7F_{15})NO$	F:	PerFHalOrg.5-66, 88
–	$NC(C_7F_{15})OC(C_7F_{15})N$	F:	PerFHalOrg.5-64/5, 88
$C_{16}F_{32}N_2$	$(CF_2)_3CF(CF_3)N(CF_2)_6NCF(CF_3)(CF_2)_3$	F:	PerFHalOrg.5-53/4, 76
$C_{16}F_{33}NO$	$O(CF_2CF_2)_2NCF_2(CF_2)_{10}CF_3$	F:	PerFHalOrg.6-2, 7
$C_{16}F_{34}HO_2P$	$(C_8F_{17})_2P(O)OH$	F:	PerFHalOrg.3-43, 46, 50

Formula	Compound		Reference
$C_{16}F_{34}IP$	$(C_8F_{17})_2PI$	F:	PerFHalOrg.3-76, 95
$C_{16}F_{34}NaO_2P$	$(C_8F_{17})_2P(O)ONa$	F:	PerFHalOrg.3-50/1
$C_{16}F_{48}N_{12}O_8P_4$	$[((CF_3)_2NO)_2PN]_4$	F:	PerFHalOrg.7-101, 119
$C_{16}FeGeH_{22}O_3$	$(C_2H_5)_3GeC_7H_7Fe(CO)_3$	Fe:	Org.Verb.B5-77, 79
$C_{16}FeGe_2H_{30}O_4$	$(CO)_4Fe(Ge(C_2H_5)_3)_2$	Fe:	Org.Verb.B2-136, 140
		Fe:	Org.Verb.B3-138
$C_{16}FeH_6NaO_7^-$	$[NaOC(O)C_{10}H_6C(O)Fe(CO)_4]^-$	Fe:	Org.Verb.B4-154, 162
$C_{16}FeH_6Na_2O_7$	$Na[NaOC(O)C_{10}H_6C(O)Fe(CO)_4]$	Fe:	Org.Verb.B4-154, 162
$C_{16}FeH_8LiNO_5$	$Li[C_6H_5C_5H_3NC(O)Fe(CO)_4]$	Fe:	Org.Verb.B4-157, 168
$C_{16}FeH_8NO_5^-$	$[C_6H_5C_5H_3NC(O)Fe(CO)_4]^-$	Fe:	Org.Verb.B4-157, 168
$C_{16}FeH_8N_2O_4$	$(CO)_4Fe(C_6H_4NNC_6H_4)$	Fe:	Org.Verb.B2-71, 76
$C_{16}FeH_8O_4$	$(CHC_{10}H_6CH)Fe(CO)_4$	Fe:	Org.Verb.B4-302, 304, 311
$C_{16}FeH_8O_5$	$C_{10}H_6(CO)(CH_2)Fe(CO)_4$	Fe:	Org.Verb.B4-335, 343
$C_{16}FeH_{10}$	$C_5H_5FeC_5H_4(CC)_3H$	Fe:	Org.Verb.A1-300/1
$C_{16}FeH_{10}Hg_2O_4$	$(CO)_4Fe(HgC_6H_5)_2$	Fe:	Org.Verb.B2-153, 157
$C_{16}FeH_{10}LiNO_5$	$Li[C_6H_5C_5H_5NC(O)Fe(CO)_4]$	Fe:	Org.Verb.B4-157, 168
$C_{16}FeH_{10}LiO_4P$	$(CO)_4FeP(C_6H_5)_2Li$	Fe:	Org.Verb.B2-50, 81/2
$C_{16}FeH_{10}NO_5^-$	$[C_6H_5C_5H_5NC(O)Fe(CO)_4]^-$	Fe:	Org.Verb.B4-157, 168
$C_{16}FeH_{10}N_4$	$[Fe(C_5H_5)_2][(NC)_2CC(CN)_2]$	Fe:	Org.Verb.A1-229, 234/6
$C_{16}FeH_{10}O_4P^-$	$[(CO)_4FeP(C_6H_5)_2]^-$	Fe:	Org.Verb.B2-52, 81
$C_{16}FeH_{10}O_4Te$	$(CO)_4FeTe(C_6H_5)_2$	Fe:	Org.Verb.B2-131, 133
$C_{16}FeH_{10}O_5$	$C_{12}H_{10}OFe(CO)_4$	Fe:	Org.Verb.B4-339, 349
$C_{16}FeH_{10}O_6Rh_2$	$Fe(RhC_5H_5)_2(CO)_6$	Fe:	Org.Verb.B2-198
$C_{16}FeH_{11}MnO_5$	$C_5H_5FeC_5H_3(COCH_3)Mn(CO)_4$	Fe:	Org.Verb.A2-211
$C_{16}FeH_{11}O_4P$	$(CO)_4FeP(C_6H_5)_2H$	Fe:	Org.Verb.B2-83, 89, 109/10
$C_{16}FeH_{11}O_5Re$	$C_5H_5FeC_5H_3(COCH_3)Re(CO)_4$	Fe:	Org.Verb.A2-211
$C_{16}FeH_{12}$	$[C_5H_4FeC_5H_4C_6H_4]_n$	Fe:	Org.Verb.A2-127
$C_{16}FeH_{12}^+$	$[C_5H_4FeC_5H_4C_6H_4]^+$	Fe:	Org.Verb.A3-126
$C_{16}FeH_{12}N_3O_7$	$[Fe(C_5H_5)_2][OC_6H_2(NO_2)_3]$	Fe:	Org.Verb.A1-206, 228
$C_{16}FeH_{12}O_2$	$C_5H_5FeC_5H_4(CC)_2COOCH_3$	Fe:	Org.Verb.A3-122
$C_{16}FeH_{12}O_4$	$(C_6H_5(CHCH)_2CHCHCHO)Fe(CO)_3$	Fe:	Org.Verb.B5-131, 140
–	$C_{12}H_{12}Fe(CO)_4$	Fe:	Org.Verb.B4-303, 306
$C_{16}FeH_{12}O_6$	$(CH_3COO)(C_6H_5CO)C_4H_4Fe(CO)_3$	Fe:	Org.Verb.B5-76
$C_{16}FeH_{12}S$	$C_5H_5FeC_5H_4CCC_4H_3S$	Fe:	Org.Verb.A1-302
$C_{16}FeH_{13}^+$	$[C_5H_5FeC_5H_4C_6H_4]^+$	Fe:	Org.Verb.A3-21, 126
$C_{16}FeH_{13}I$	$C_5H_5FeC_5H_4C_6H_4I$	Fe:	Org.Verb.A1-390, 391
$C_{16}FeH_{13}NO_2$	$C_5H_5FeC_5H_4C_6H_4NO_2$	Fe:	Org.Verb.A1-326
$C_{16}FeH_{13}N_2O_4P$	$(CO)_4FeP(CH_2CH_2CN)_2C_6H_5$	Fe:	Org.Verb.B2-89
$C_{16}FeH_{13}N_3O_6$	$Fe(C_5H_5)_2 \cdot C_6H_3(NO_2)_3$	Fe:	Org.Verb.A1-238
$C_{16}FeH_{13}N_3O_7$	$Fe(C_5H_5)_2 \cdot HOC_6H_2(NO_2)_3$	Fe:	Org.Verb.A1-238
–	$[Fe(C_5H_5)_2][HOC_6H_2(NO_2)_3]$	Fe:	Org.Verb.A1-228/9
$C_{16}FeH_{13}O_4P$	$(CO)_4Fe(C_6H_5P(CHCCH_3)_2)$	Fe:	Org.Verb.C2-49
$C_{16}FeH_{14}$	$C_5H_5FeC_5H_4CHC_5H_4$	Fe:	Org.Verb.A1-333, 335
–	$C_5H_5FeC_5H_4C_6H_5$	Fe:	Org.Verb.A1-320/7
$C_{16}FeH_{14}^+$	$[C_5H_5FeC_5H_4C_6H_5]^+$	Fe:	Org.Verb.A1-324/6, 331/2
$C_{16}FeH_{14}IN_2O_2$	$Fe(OC_6H_4CHNCH_2)_2I$	Fe:	Org.Verb.B1-5
$C_{16}FeH_{14}I_4$	$[C_5H_5FeC_5H_4C_6H_5]I_4$	Fe:	Org.Verb.A1-332
$C_{16}FeH_{14}NNaO_4S_3$	$C_5H_5FeC_5H_4CHCCON(C_2H_4SO_3Na)CS_2$	Fe:	Org.Verb.A2-158

Formula	Compound	Reference
$C_{16}FeH_{14}N_2O_2$	$Fe(OC_6H_4CHNCH_2)_2$	Fe: Org.Verb.B3-201/2
$C_{16}FeH_{14}N_2O_4$	$Fe(C_5H_5)_2 \cdot C_6H_4(NO_2)_2$	Fe: Org.Verb.A1-237
$C_{16}FeH_{14}N_4O_2$	$(CH_3C_6H_4NC)_2Fe(NO)_2$	Fe: Org.Verb.B4-11, 13
$C_{16}FeH_{14}N_4O_4$	$(CH_3OC_6H_4NC)_2Fe(NO)_2$	Fe: Org.Verb.B4-11, 13
$C_{16}FeH_{14}O$	$C_5H_5FeC_5H_4C(CH_3)(CCCCH)OH$	Fe: Org.Verb.A2-59
–	$C_5H_5FeC_5H_4CHCHC_4H_3O$	Fe: Org.Verb.A3-162
–	$C_5H_5FeC_5H_4C_6H_4OH$	Fe: Org.Verb.A2-4, 91/2
–	$C_5H_5FeC_5H_4OC_6H_5$	Fe: Org.Verb.A2-108, 110
$C_{16}FeH_{14}O^+$	$[C_5H_5FeC_5H_4C_6H_4OH]^+$	Fe: Org.Verb.A2-92
$C_{16}FeH_{14}OS_2$	$C_5H_5FeC_5H_4COCH_2SC_4H_3S$	Fe: Org.Verb.A2-245
$C_{16}FeH_{14}O_2$	$C_5H_5FeC_5H_2C_6H_7O_2$	Fe: Org.Verb.A3-105
–	$Fe(C_5H_4COCHCH_2)_2$	Fe: Org.Verb.A1-151, 153
		Fe: Org.Verb.A2-249
–	$Fe(C_5H_5)_2 \cdot C_6H_4O_2$	Fe: Org.Verb.A1-236
–	$[Fe(C_5H_5)_2][OC_6H_4O]$	Fe: Org.Verb.A1-228
$C_{16}FeH_{14}O_2S$	$C_5H_5FeC_5H_4SO_2C_6H_5$	Fe: Org.Verb.A1-168
$C_{16}FeH_{14}O_3$	$C_5H_5FeC_5H_3C_4H_2O(COOH)CH_3$	Fe: Org.Verb.A3-171, 174
–	$C_5H_5FeC_5H_4C(CH_3)CCH_2C(O)OCO$	Fe: Org.Verb.A3-174
$C_{16}FeH_{14}O_3S$	$C_6H_5C_5H_4FeC_5H_4SO_3H$	Fe: Org.Verb.A1-326
$C_{16}FeH_{14}O_4$	$C_5H_5FeC_5H_4CHCHCHC(COOH)_2$	Fe: Org.Verb.A3-96/8
$C_{16}FeH_{14}O_4W^+$	$[C_5H_5FeC_5H_4C(OC_2H_5)W(CO)_3]^+$	Fe: Org.Verb.A1-347
$C_{16}FeH_{14}O_5$	$CH_3OC_6H_4C(C_3H_5)CH_2Fe(CO)_4$	Fe: Org.Verb.B4-258, 260
$C_{16}FeH_{14}S$	$C_5H_5FeC_5H_4SC_6H_5$	Fe: Org.Verb.A1-376
$C_{16}FeH_{15}NO$	$C_5H_5FeC_5H_4CHCCHC(CH_3)C(O)NH$	Fe: Org.Verb.A2-162
$C_{16}FeH_{15}NOS_2$	$C_5H_5FeC_5H_4CHCC(O)N(C_2H_5)C(S)S$	Fe: Org.Verb.A2-158
$C_{16}FeH_{15}NO_2$	$C_5H_5FeC_5H_4CHC(CN)COOC_2H_5$	Fe: Org.Verb.A2-161, 220
$C_{16}FeH_{15}NO_3$	$C_5H_5FeC_5H_4CCHC(COOC_2H_5)NO$	Fe: Org.Verb.A3-140
$C_{16}FeH_{15}NO_6$	$(CO)_4FeCH(COOC_2H_5)NCH(CH_3)C_6H_5$	Fe: Org.Verb.B2-185
$C_{16}FeH_{15}NO_9$	$(CH_3OCO)_3C_2HFe(CO)_3(NC_5H_5)$	Fe: Org.Verb.B4-202
$C_{16}FeH_{15}Ni^+$	$(C_5H_5)_2NiFeC_6H_5{}^+$	Ni: Org.Verb.2-168
$C_{16}FeH_{15}NiO_4{}^+$	$[(C_5H_5FeC_5H_4COCHCOCOOC_2H_5)Ni]^+$	Fe: Org.Verb.A3-140
$C_{16}FeH_{15}OP$	$C_5H_5FeC_5H_4PH(O)C_6H_5$	Fe: Org.Verb.A1-171
$C_{16}FeH_{15}O_4P$	$(CO)_4Fe(C_6H_5P(CH_2CCH_3)_2)$	Fe: Org.Verb.C2-49
$C_{16}FeH_{15}O_4Zn^+$	$[(C_5H_5FeC_5H_4COCHCOCOOC_2H_5)Zn]^+$	Fe: Org.Verb.A3-140
$C_{16}FeH_{16}$	$C_5H_5FeC_5H_4CH_2C_5H_5$	Fe: Org.Verb.A1-333, 335
$C_{16}FeH_{16}{}^+$	$[C_5H_5FeC_5H_4C_5H_4CH_3]^+$	Fe: Org.Verb.A1-347
$C_{16}FeH_{16}Mn^+$	$[C_5H_5FeC_5H_4MnC_5H_4CH_3]^+$	Fe: Org.Verb.A1-347
$C_{16}FeH_{16}N^+$	$[C_5H_5FeC_5H_4CH_2NC_5H_5]^+$	Fe: Org.Verb.A2-20
$C_{16}FeH_{16}N_2O_2$	$C_5H_5FeC_5H_4CCHC(COOC_2H_5)NNH$	Fe: Org.Verb.A3-134
–	$C_5H_5FeC_5H_4C(OH)CHC_4H_5N_2O$	Fe: Org.Verb.A3-140
$C_{16}FeH_{16}O$	$C_5H_5FeC_5H_3C_5H_8CO$	Fe: Org.Verb.A3-103
–	$C_5H_5FeC_5H_4CHC_5H_6O$	Fe: Org.Verb.A3-43
–	$C_5H_5FeC_5H_4COC_5H_7$	Fe: Org.Verb.A2-289/90
$C_{16}FeH_{16}O_2$	$C_5H_5FeC_5H_3CH_2CH(COCH_3)CH_2CO$	Fe: Org.Verb.A3-104
–	$C_5H_5FeC_5H_4CC(CH_3)(CH_2)_2COO$	Fe: Org.Verb.A3-166
–	$C_5H_5FeC_5H_4(CHCH)_2COOCH_3$	Fe: Org.Verb.A3-122
–	$C_5H_5FeC_5H_4CHOHCCCHCHOCH_3$	Fe: Org.Verb.A2-133
–	$C_5H_5FeC_5H_4CH_2OCH_2C_4H_3O$	Fe: Org.Verb.A3-162
–	$[Fe(C_6H_4CH_2OC_2H_4OCH_2C_6H_4)]_n$	Fe: Org.Verb.B1-14/5
$C_{16}FeH_{16}O_4$	$C_5H_5FeC_5H_4C(CHCOOH)(CH_2)_2COOH$	Fe: Org.Verb.A3-96/8

Formula	Compound	Reference
$C_{16}FeH_{16}O_4$	$C_5H_5FeC_5H_4C(CH_3)C(COOH)CH_2COOH$	Fe: Org.Verb.A3-96/8, 105
–	$C_5H_5FeC_5H_4C(COOCH_3)CHCOOCH_3$	Fe: Org.Verb.A3-148
–	$C_5H_5FeC_5H_4CHC(COOCH_3)_2$	Fe: Org.Verb.A3-148
–	$C_5H_5FeC_5H_4COCH_2COCOOC_2H_5$	Fe: Org.Verb.A3-134/5, 140
$C_{16}FeH_{16}O_6$	$C_5H_5FeC_5H_4CH_2C(COOH)_2CH_2COOH$	Fe: Org.Verb.A3-96/9, 105
$C_{16}FeH_{17}NO$	$C_5H_5FeC_5H_4CHCHC(CH_3)(OH)CH_2CN$	Fe: Org.Verb.A3-10
$C_{16}FeH_{17}NOW^+$	$[C_5H_5FeC_5H_4C(NC_4H_8)W(CO)]^+$	Fe: Org.Verb.A1-347
$C_{16}FeH_{17}NO_3$	$C_5H_5FeC_5H_4CHCH_2C(COOC_2H_5)NO$	Fe: Org.Verb.A3-95
–	$C_5H_5FeC_5H_4COCHC(COOC_2H_5)NH_2$	Fe: Org.Verb.A3-134
$C_{16}FeH_{17}NO_4$	$C_5H_5FeC_5H_4COCH_2C(COOC_2H_5)NOH$	Fe: Org.Verb.A3-134
$C_{16}FeH_{17}N_3O$	$C_5H_5FeC_5H_4(CHCH)_2CHNNHCONH_2$	Fe: Org.Verb.A2-177
$C_{16}FeH_{17}O_3{}^{2-}$	$[C_5H_5FeC_5H_3(O)COCOCH_2CH(CH_3)_2]^{2-}$	Fe: Org.Verb.A2-302
$C_{16}FeH_{18}$	$C_5H_5FeC_5H_4C(CHCH_3)C_3H_5$	Fe: Org.Verb.A1-333/5
–	$C_5H_5FeC_5H_4CHCHCH_2C(CH_3)CH_2$	Fe: Org.Verb.A2-50, 82
–	$C_5H_5FeC_5H_4CH_2C_5H_7$	Fe: Org.Verb.A1-333, 335
–	$C_5H_5FeC_5H_4C_6H_9$	Fe: Org.Verb.A1-338, 341
$C_{16}FeH_{18}N_2O$	$C_5H_5FeC_5H_4CHCH_2C(CH_3)NNCOCH_3$	Fe: Org.Verb.A3-10
$C_{16}FeH_{18}N_2O_2$	$C_5H_5FeC_5H_4C_4H_3N_2O \cdot C_2H_5OH$	Fe: Org.Verb.A2-296/7
$C_{16}FeH_{18}N_2O_2S_4$	$Fe[S_2COC_2H_5]_2(NC_5H_4C_5H_4N)$	C: MVol.D4-259
$C_{16}FeH_{18}N_2O_8$	$(CO)_4FeC_5H_5N_2(COOCH_3)_2C_3H_7$	Fe: Org.Verb.B2-183/4, 186
$C_{16}FeH_{18}N_4O_2$	$Fe(NC_5H_4C_5H_4N)(CH_2CHCONH_2)_2$	Fe: Org.Verb.B1-8
$C_{16}FeH_{18}N_4O_2{}^{2+}$	$[(CH_3NC)_2Fe(H_2O)_2(C_{12}H_8N_2)]^{2+}$	Fe: Org.Verb.B4-28
$C_{16}FeH_{18}O$	$C_5H_4FeC_5H_4CH_2CH(C_3H_7)CO$	Fe: Org.Verb.A3-101
–	$C_5H_5FeC_5H_3CH_2CH(CH_3)(CH_2)_2CO$	Fe: Org.Verb.A3-103
–	$C_5H_5FeC_5H_3CH_2CH(C_3H_7)CO$	Fe: Org.Verb.A3-101
–	$C_5H_5FeC_5H_3(CH_2)_2C(CH_3)_2CO$	Fe: Org.Verb.A3-102
–	$C_5H_5FeC_5H_3(CH_2)_2CH(CH_3)CH_2CO$	Fe: Org.Verb.A3-103
–	$C_5H_5FeC_5H_4C(CH_3)(CCH)OC_2H_5$	Fe: Org.Verb.A2-119, 123
–	$C_5H_5FeC_5H_4CHCHCOC_3H_7$-n	Fe: Org.Verb.A3-9/10
–	$C_5H_5FeC_5H_4CH_2C_5H_7O$	Fe: Org.Verb.A3-43, 46
–	$C_5H_5FeC_5H_4COC(CH_2)CH(CH_3)_2$	Fe: Org.Verb.A2-248, 251
–	$C_5H_5FeC_5H_4COCHCHCH(CH_3)_2$	Fe: Org.Verb.A2-248, 250/1
–	$C_5H_5FeC_5H_4COCH_2CHC(CH_3)_2$	Fe: Org.Verb.A2-248, 251
–	$C_5H_5FeC_5H_4COC_3H_3(CH_3)_2$	Fe: Org.Verb.A2-289/90
–	$C_5H_5FeC_5H_4COC_5H_9$	Fe: Org.Verb.A2-290
–	$C_5H_5FeC_5H_4C_6H_9O$	Fe: Org.Verb.A3-43, 46
$C_{16}FeH_{18}O_2$	$CH_3COC_5H_4FeC_5H_4COC_3H_7$	Fe: Org.Verb.A2-210
–	$C_5H_5FeC_5H_4C(CH_3)CHCOOC_2H_5$	Fe: Org.Verb.A3-121, 123/4
–	$C_5H_5FeC_5H_4C(CH_3)(OH)CCCHOHCH_3$	Fe: Org.Verb.A2-88
–	$C_5H_5FeC_5H_4CHCH(CH_2)_2OOCCH_3$	Fe: Org.Verb.A2-104/5
–	$C_5H_5FeC_5H_4CHCHCOOC_3H_7$-n	Fe: Org.Verb.A3-121
–	$C_5H_5FeC_5H_4CHOH(CHCH)_2OCH_3$	Fe: Org.Verb.A2-133
–	$C_5H_5FeC_5H_4CH_2CH(COCH_3)_2$	Fe: Org.Verb.A3-32, 34
–	$C_5H_5FeC_5H_4(CH_2)_2OOCC(CH_3)CH_2$	Fe: Org.Verb.A2-103/4
–	$C_5H_5FeC_5H_4C_3H_3(CH_3)COOCH_3$	Fe: Org.Verb.A3-131, 133
–	$C_5H_5FeC_5H_4C_3H_4CH_2OOCCH_3$	Fe: Org.Verb.A2-104/5
–	$C_5H_5FeC_5H_4C_3H_4COOC_2H_5$	Fe: Org.Verb.A3-131, 133
–	$C_5H_5FeC_5H_4C_5H_8COOH$	Fe: Org.Verb.A3-82/4, 100, 103
–	$[Fe(C_6H_4CH_2OCH_3)_2]_n$	Fe: Org.Verb.B1-15

$C_{16}FeH_{18}O_2^-$ .... $[C_5H_5FeC_5H_4COCOCH_2CH(CH_3)_2]^-$ .... Fe: Org.Verb.A2-302
$C_{16}FeH_{18}O_3$ .... $C_5H_5FeC_5H_4CH_2CH(COCH_3)CH_2COOH$ .... Fe: Org.Verb.A3-94, 100, 104
– .... $C_5H_5FeC_5H_4COC(CH_3)_2CH_2COOH$ .... Fe: Org.Verb.A3-88
– .... $C_5H_5FeC_5H_4COCH(CH_3)CH_2COOCH_3$ ...... Fe: Org.Verb.A3-136
– .... $C_5H_5FeC_5H_4COCH(CH_3)(CH_2)_2COOH$ ...... Fe: Org.Verb.A3-87, 89, 104
– .... $C_5H_5FeC_5H_4COCH_2C(CH_3)_2COOH$ .... Fe: Org.Verb.A3-87/8
– .... $C_5H_5FeC_5H_4COCH_2CH(CH_3)CH_2COOH$ .... Fe: Org.Verb.A3-87, 89
– .... $C_5H_5FeC_5H_4COCH_2CH(CH_3)COOCH_3$ ...... Fe: Org.Verb.A3-136
– .... $C_5H_5FeC_5H_4CO(CH_2)_2COOC_2H_5$ .... Fe: Org.Verb.A3-136, 140
– .... $C_5H_5FeC_5H_4CO(CH_2)_3COOCH_3$ .... Fe: Org.Verb.A3-137
– .... $C_5H_5FeC_5H_4CO(CH_2)_4COOH$ .... Fe: Org.Verb.A3-87, 89/90
– .... $C_5H_5FeC_5H_4C_5H_7(OH)COOH$ .... Fe: Org.Verb.A3-84/6
$C_{16}FeH_{18}O_3S$ .... $C_5H_5FeC_5H_4COCH_2SCH_2COOC_2H_5$ .... Fe: Org.Verb.A2-245
$C_{16}FeH_{18}O_4$ .... $C_5H_5FeC_5H_4C(CH_3)_2CH(COOH)_2$ .... Fe: Org.Verb.A3-96/8
– .... $C_5H_5FeC_5H_4CH(CH_2COOH)(CH_2)_2COOH$ ... Fe: Org.Verb.A3-96/8, 105
– .... $C_5H_5FeC_5H_4CH_2C(COOH)_2C_2H_5$ .... Fe: Org.Verb.A3-96/8
– .... $C_5H_5FeC_5H_4COOCOOCH_2CH(CH_3)_2$ .... Fe: Org.Verb.A3-108, 112
$C_{16}FeH_{18}O_4Sn$ ...... $(C_2H_5)_2Sn(OCOC_5H_4FeC_5H_4COO)$ .... Sn: Org.Verb.6-60
– .... $[(C_2H_5)_2Sn(OCOC_5H_4FeC_5H_4COO)]_x$ ...... Sn: Org.Verb.6-60
$C_{16}FeH_{19}NO$ .... $C_5H_5FeC_5H_4CHNC_4H_7CH_2O$ .... Fe: Org.Verb.A2-170
$C_{16}FeH_{19}NOS$ ...... $C_5H_5FeC_5H_4CH_2CSN(CH_2CH_2)_2O$ .... Fe: Org.Verb.A2-227/8
$C_{16}FeH_{19}N_5O^{2+}$ ..... $[(CH_3NC)_3Fe(C_{10}H_8N_2)H_2O]^{2+}$ .... Fe: Org.Verb.B4-45
$C_{16}FeH_{19}O^+$ .... $[(CH_3)_3C_6H_3FeC_5H_4COCH_3]^+$ .... Fe: Org.Verb.A2-186, 208
$C_{16}FeH_{19}O_2^+$ .... $[C_5H_5FeC_5H_4C(CH_3)CH_2COOC_2H_5]^+$ ...... Fe: Org.Verb.A3-123
– .... $[C_5H_5FeC_5H_4C(CH_3)CD_2COOC_2H_5]^+$ ...... Fe: Org.Verb.A3-124
– .... $[C_5H_5FeC_5H_4C(CH_3)(CH_2)_2COOCH_3]^+$ ..... Fe: Org.Verb.A3-120, 130
– .... $[C_5H_5FeC_5H_4CH(CH_2)_2COOC_2H_5]^+$ .... Fe: Org.Verb.A3-133
$C_{16}FeH_{19}O_7P$ .... $C_6H_5C_2H_2C(O)CH_3Fe(CO)_3P(OCH_3)_3$ ...... Fe: Org.Verb.B4-200, 205/6
$C_{16}FeH_{20}$ .... $C_5H_5FeC_5H_4C(C_4H_9\text{-}t)CH_2$ .... Fe: Org.Verb.A1-291, 293
– .... $C_5H_5FeC_5H_4CHCHC(CH_3)_3$ .... Fe: Org.Verb.A2-47
– .... $C_5H_5FeC_5H_4CH(CH_2)_5$ .... Fe: Org.Verb.A1-338, 341
$C_{16}FeH_{20}NO^+$ ...... $[C_5H_5FeC_5H_4C(CH_3)NC_4H_8O]^+$ .... Fe: Org.Verb.A2-189, 221
$C_{16}FeH_{20}N_2$ .... $C_5H_5FeC_5H_4CH(CN)N(C_2H_5)_2$ .... Fe: Org.Verb.A2-169
$C_{16}FeH_{20}N_2O_2$ ...... $C_5H_5FeC_5H_4CH(CN)N(OC_2H_5)_2$ .... Fe: Org.Verb.A2-169
$C_{16}FeH_{20}N_2O_2S_4$ .... $Fe[S_2COC_2H_5]_2(C_5H_5N)_2$ .... C: MVol.D4-259
$C_{16}FeH_{20}O$ .... $C_5H_5FeC_5H_4C(CH_3)(CH_2)_2CH(CH_3)O$ ...... Fe: Org.Verb.A3-174
– .... $C_5H_5FeC_5H_4C(CH_3)_2CH_2COCH_3$ .... Fe: Org.Verb.A3-24/5
– .... $C_5H_5FeC_5H_4CHCHCH_2C(CH_3)_2OH$ .... Fe: Org.Verb.A2-76, 82
– .... $C_5H_5FeC_5H_4CHCHOC_4H_9\text{-}n$ .... Fe: Org.Verb.A2-130/1
– .... $C_5H_5FeC_5H_4CH(CH_2)_2C(CH_3)_2O$ .... Fe: Org.Verb.A3-174
– .... $C_5H_5FeC_5H_4CH(C_3H_5)OC_2H_5$ .... Fe: Org.Verb.A2-118, 120
– .... $C_5H_5FeC_5H_4CHOHC_3H_3(CH_3)_2$ .... Fe: Org.Verb.A2-46, 50
– .... $C_5H_5FeC_5H_4CHOHC_5H_9$ .... Fe: Org.Verb.A2-4, 46, 50
– .... $C_5H_5FeC_5H_4(CH_2)_2COC_3H_7\text{-}n$ .... Fe: Org.Verb.A3-3/4
– .... $C_5H_5FeC_5H_4COCH(C_2H_5)_2$ .... Fe: Org.Verb.A2-238
– .... $C_5H_5FeC_5H_4COCH_2C(CH_3)_3$ .... Fe: Org.Verb.A2-238, 240
– .... $C_5H_5FeC_5H_4CO(CH_2)_2CH(CH_3)_2$ .... Fe: Org.Verb.A2-185, 238, 240
– .... $C_5H_5FeC_5H_4COC_5H_{11}\text{-}n$ .... Fe: Org.Verb.A2-235
– .... $C_5H_5FeC_5H_4C_6H_{10}OH$ .... Fe: Org.Verb.A2-76, 82

Formula	Compound	Reference
$C_{16}FeH_{20}O_2$	$CH_3C_5H_4FeC_5H_3(CH_3)CH(CH_3)OCOCH_3$	Fe: Org.Verb.A1-357/8
–	$C_5H_5FeC_5H_4CH(CH_3)CH_2COOC_2H_5$	Fe: Org.Verb.A3-117
–	$C_5H_5FeC_5H_4CH(CH_3)(CH_2)_2COOCH_3$	Fe: Org.Verb.A3-116, 118
–	$C_5H_5FeC_5H_4CH(C_3H_7)OOCCH_3$	Fe: Org.Verb.A2-98
–	$C_5H_5FeC_5H_4CH_2CH(CH_3)CH_2COOCH_3$	Fe: Org.Verb.A3-118
–	$C_5H_5FeC_5H_4CH_2CH(CH_3)(CH_2)_2COOH$	Fe: Org.Verb.A3-74, 100, 103
–	$C_5H_5FeC_5H_4CH_2CH(C_3H_7)COOH$	Fe: Org.Verb.A3-72, 75, 100/1
–	$C_5H_5FeC_5H_4CH_2OOCC_4H_9$-n	Fe: Org.Verb.A2-98
–	$C_5H_5FeC_5H_4(CH_2)_2C(CH_3)_2COOH$	Fe: Org.Verb.A3-71, 73, 100, 102
–	$C_5H_5FeC_5H_4(CH_2)_2CH(CH_3)CH_2COOH$	Fe: Org.Verb.A3-74, 100, 103
–	$C_5H_5FeC_5H_4(CH_2)_2CH(CH_3)COOCH_3$	Fe: Org.Verb.A3-118
–	$C_5H_5FeC_5H_4(CH_2)_3COOC_2H_5$	Fe: Org.Verb.A3-115, 118
–	$C_5H_5FeC_5H_4(CH_2)_4COOCH_3$	Fe: Org.Verb.A3-115, 119
–	$C_5H_5FeC_5H_4(CH_2)_5COOH$	Fe: Org.Verb.A3-68/70, 103
–	$C_5H_5FeC_5H_4CO(CH_2)_2C(CH_3)_2OH$	Fe: Org.Verb.A2-295, 297
$C_{16}FeH_{20}O_3$	$C_5H_5FeC_5H_4C(CH_3)(OH)CH_2COOC_2H_5$	Fe: Org.Verb.A3-144
–	$C_5H_5FeC_5H_4CH(OCH_3)(CH_2)_2COOCH_3$	Fe: Org.Verb.A3-145
–	$C_5H_5FeC_5H_4CHOH(CH_2)_2COOC_2H_5$	Fe: Org.Verb.A3-144
$C_{16}FeH_{21}{}^+$	$[(CH_3)_3C_6H_3FeC_5H_4C_2H_5]^+$	Fe: Org.Verb.A2-186, 208
–	$[C_5H_5FeC_5H_4C(CH_3)C(CH_3)_3]^+$	Fe: Org.Verb.A2-58, 63
–	$[C_5H_5FeC_5H_4CHCH_2C(CH_3)_3]^+$	Fe: Org.Verb.A1-350/1 Fe: Org.Verb.A2-47
$C_{16}FeH_{21}N$	$C_5H_5FeC_5H_4CHC(CH_3)CH_2N(CH_3)_2$	Fe: Org.Verb.A1-290 Fe: Org.Verb.A2-43
–	$C_5H_5FeC_5H_4CH_2NC_5H_{10}$	Fe: Org.Verb.A1-167
$C_{16}FeH_{21}NO$	$C_5H_5FeC_5H_4COCH(CH_3)CH_2N(CH_3)_2$	Fe: Org.Verb.A2-233
$C_{16}FeH_{21}N_5O_6$	$COFe(NC_5H_4COCH_3)(C_2HN_2O_2(CH_3)_2)_2$	Fe: Org.Verb.B1-33
$C_{16}FeH_{21}O^+$	$[C_5H_5FeC_5H_4CH(CH_2)_2C(CH_3)_2OH]^+$	Fe: Org.Verb.A2-82, 120
$C_{16}FeH_{22}$	$C_5H_5FeC_5H_4CH(CH_3)C(CH_3)_3$	Fe: Org.Verb.A1-276
–	$C_5H_5FeC_5H_4CH_2CH(C_2H_5)_2$	Fe: Org.Verb.A1-275
–	$C_5H_5FeC_5H_4(CH_2)_5CH_3$	Fe: Org.Verb.A1-268
–	$Fe(C_5H_4C_3H_7$-n$)_2$	Fe: Org.Verb.A1-268
$C_{16}FeH_{22}N_2$	$C_5H_5FeC_5H_4C(CH_3)NN(C_2H_5)_2$	Fe: Org.Verb.A2-221
$C_{16}FeH_{22}O$	$C_5H_5FeC_5H_4C(CH_3)(C_4H_9$-t$)OH$	Fe: Org.Verb.A2-4, 58, 60, 63
–	$C_5H_5FeC_5H_4C(CH_3)_2OCH(CH_3)_2$	Fe: Org.Verb.A2-119, 123
–	$C_5H_5FeC_5H_4CH(CH_3)C(CH_3)_2CH_2OH$	Fe: Org.Verb.A2-76, 82
–	$C_5H_5FeC_5H_4CH(C_4H_9)OCH_3$	Fe: Org.Verb.A2-118
–	$C_5H_5FeC_5H_4CHOHCH_2C(CH_3)_3$	Fe: Org.Verb.A2-44, 47
–	$C_5H_5FeC_5H_4(CH_2)_2C(CH_3)_2OCH_3$	Fe: Org.Verb.A2-131, 135
$C_{16}FeH_{22}O_2$	$C_5H_5FeC_5H_4C(CH_3)(OH)(CH_2)_2CHOHCH_3$	Fe: Org.Verb.A2-87
–	$C_5H_5FeC_5H_4CH_2CH(OC_2H_5)_2$	Fe: Org.Verb.A2-134, 138
–	$Fe(C_5H_4CH_2OC_2H_5)_2$	Fe: Org.Verb.A2-114
$C_{16}FeH_{22}O_2{}^+$	$[C_5H_5FeC_5H_4CH_2CH(OC_2H_5)_2]^+$	Fe: Org.Verb.A1-128/9 Fe: Org.Verb.A2-138
$C_{16}FeH_{22}O_3Si$	$(C_2H_5)_3SiC_7H_7Fe(CO)_3$	Fe: Org.Verb.B5-77, 79
$C_{16}FeH_{23}N_5O_5$	$COFe(NC_5H_3(CH_3)_2)(C_2HN_2O_2(CH_3)_2)_2$	Fe: Org.Verb.B1-33
$C_{16}FeH_{24}Li_2N_2$	$Fe(C_5H_4Li)_2 \cdot (CH_3)_2NCH_2CH_2N(CH_3)_2$	Fe: Org.Verb.A1-173

Formula	Compound	Reference
$C_{16}FeH_{24}Ni^+$	$C_8(CH_3)_8FeNi^+$	Ni: Org.Verb.2-106
$C_{16}FeH_{24}Sn$	$(C_2H_5)_3SnC_5H_4FeC_5H_5$	Sn: Org.Verb.2-248
$C_{16}FeH_{25}I_3N_6$	$[(C_2H_5NC)_5FeCN]I_3$	Fe: Org.Verb.B4-78, 82
$C_{16}FeH_{25}N_6^+$	$[(C_2H_5NC)_5FeCN]^+$	Fe: Org.Verb.B4-78, 82
$C_{16}FeH_{27}N_3Ti$	$C_5H_5FeC_5H_4Ti(N(CH_3)_2)_3$	Ti: Org.Verb.1-55, 58
$C_{16}FeH_{27}N_7^{2+}$	$[(CH_2)_4N(CH_3NH)CFe(CNCH_3)_5]^{2+}$	Fe: Org.Verb.B4-174
$C_{16}FeH_{27}O_4P$	$(CO)_4FeP(C(CH_3)_3)_3$	Fe: Org.Verb.B2-83, 85, 87, 104/5
–	$(CO)_4FeP(C_4H_9)_3$	Fe: Org.Verb.B2-84/5, 87, 104
$C_{16}FeH_{27}O_7P$	$(CO)_4FeP(OC_4H_9)_3$	Fe: Org.Verb.B2-82
$C_{16}FeH_{30}N_2O_2P_2S_2$	$(CO)_2Fe(P(C_2H_5)_3)_2(NCS)_2$	Fe: Org.Verb.B1-101/2, 104
$C_{16}FeH_{30}O_4Pb_2$	$(CO)_4Fe(Pb(C_2H_5)_3)_2$	Fe: Org.Verb.B2-136, 142
$C_{16}FeH_{30}O_4Si_2$	$(CO)_4Fe(Si(C_2H_5)_3)_2$	Fe: Org.Verb.B2-136, 140
$C_{16}FeH_{31}N_3O_6P_2$	$(CO)_3Fe(P(N(CH_3)_2)_3)P(OCH_2)_3CC_3H_7$	Fe: Org.Verb.B1-148/50, 156
$C_{16}FeH_{32}NO_2P$	$C_3H_5Fe(CO)(NO)P(C_4H_9\text{-}n)_3$	Fe: Org.Verb.B5-6/8, 20
$C_{16}FeH_{32}NO_5P$	$C_3H_5Fe(CO)(NO)P(OC_4H_9)_3$	Fe: Org.Verb.B5-6/9, 20/1
$C_{16}FeH_{32}N_6O_{16}S_4$	$(CH_3NC)_4Fe(CN)_2 \cdot 2\ (CH_3O)_2SO_2 \cdot 2\ CH_3OSO_3H$	Fe: Org.Verb.B4-68
$C_{16}FeH_{33}O_4PSi_3$	$(CO)_4FeP(CH_2Si(CH_3)_3)_3$	Fe: Org.Verb.B2-86
$C_{16}FeH_{38}O_2P_4$	$CH_3OC(O)CHCH_2Fe(C_2H_4(P(CH_3)_2)_2)_2$	Fe: Org.Verb.B4-181, 183/4
$C_{16}FeH_{38}O_9P_3$	$C_7H_{11}Fe(P(OCH_3)_3)_3$	Fe: Org.Verb.B5-5
$C_{16}FeH_{38}O_9P_3^+$	$[C_7H_{11}Fe(P(OCH_3)_3)_3]^+$	Fe: Org.Verb.B5-3
$C_{16}FeH_{38}P_4$	$CH_2CHCHCH_2Fe(C_2H_4(P(CH_3)_2)_2)_2$	Fe: Org.Verb.B4-180, 183/4
$C_{16}Fe_2GeH_{17}IN_4O_4$	$(C_3H_5Fe(NO)_2)_2Ge(I)C_{10}H_7$	Fe: Org.Verb.C2-165
$C_{16}Fe_2GeH_{23}IN_4O_4$	$((CH_3)_2C_3H_3Fe(NO)_2)_2Ge(I)C_6H_5$	Fe: Org.Verb.C2-165
$C_{16}Fe_2Ge_2H_{20}O_8$	$[(CO)_4FeGe(C_2H_5)_2]_2$	Fe: Org.Verb.C1-42/5, 49/50
$C_{16}Fe_2Ge_4H_{24}O_8$	$(CO)_4Fe((CH_3)_2GeGe(CH_3)_2)_2Fe(CO)_4$	Fe: Org.Verb.C1-42, 48, 54
$C_{16}Fe_2H_4N_2O_8$	$(CO)_4Fe(CNC_6H_4NC)Fe(CO)_4$	Fe: Org.Verb.C2-71
$C_{16}Fe_2H_8N_2O_6$	$(CO)_3Fe(NHC_{10}H_6NH)Fe(CO)_3$	Fe: Org.Verb.C1-131, 138
$C_{16}Fe_2H_{11}NO_6$	$(C(C_6H_5)CN(CH_3)_2)Fe_2(CO)_6$	Fe: Org.Verb.C2-95/6, 100, 103
–	$(C_6H_4CH_2NCHCHCH_3)Fe_2(CO)_6$	Fe: Org.Verb.C2-114, 119
–	$(C_6H_4CHDNCHCHCH_3)Fe_2(CO)_6$	Fe: Org.Verb.C2-114, 119
–	$(C_6H_4CH_2NCH_2CHCH_2)Fe_2(CO)_6$	Fe: Org.Verb.C2-114, 119
$C_{16}Fe_2H_{12}O_7S$	$(CO)_3Fe(C_6H_4CH(OC_3H_7)S)Fe(CO)_3$	Fe: Org.Verb.C2-105, 113
–	$(CO)_6Fe_2(C_3H_7OC(C_6H_5)S)$	Fe: Org.Verb.C1-238/9, 241
$C_{16}Fe_2H_{12}O_8Sn_2$	$[(CO)_4FeSn(CHCH_2)_2]_2$	Fe: Org.Verb.C1-42, 45, 51
$C_{16}Fe_2H_{14}N_4O_6$	$(CO)_3Fe(C_3H(CH_3)_2N_2)_2Fe(CO)_3$	Fe: Org.Verb.C1-134, 142
$C_{16}Fe_2H_{14}O_4Si$	$(CH_3)_2Si(C_5H_4Fe(CO)_2)_2$	Fe: Org.Verb.B3-138
$C_{16}Fe_2H_{14}O_6S_2$	$(CO)_6Fe_2(S_2C_{10}H_{14})$	Fe: Org.Verb.C1-115/6, 118, 208/9, 238, 241, 243
$C_{16}Fe_2H_{15}^+$	$[CH_3C_5H_4FeC_5H_4C_5H_4Fe]^+$	Fe: Org.Verb.A6-15
$C_{16}Fe_2H_{15}N_3O_{10}$	$[Fe(NCCOOC_2H_5)_3][Fe(CO)_4]$	Fe: Org.Verb.B2-50
$C_{16}Fe_2H_{16}N_2O_{10}$	$(CO)_6Fe_2(N_2C_3H_3(COOCH_3)_2CH(CH_3)_2)$	Fe: Org.Verb.C1-149/50, 152
$C_{16}Fe_2H_{17}IN_4O_4Pb$	$(C_3H_5Fe(NO)_2)_2Pb(I)C_{10}H_7$	Fe: Org.Verb.C2-165
$C_{16}Fe_2H_{17}IN_4O_4Sn$	$(C_3H_5Fe(NO)_2)_2Sn(I)C_{10}H_7$	Fe: Org.Verb.C2-165
$C_{16}Fe_2H_{18}N_2O_8$	$[CH_3N(CH_2CH_2)_3NCH_3][Fe_2(CO)_8]$	Fe: Org.Verb.C2-7
$C_{16}Fe_2H_{18}O_6$	$(C_4H_9CCC_4H_9)Fe_2(CO)_6$	Fe: Org.Verb.C2-72/4
$C_{16}Fe_2H_{18}O_6S_2$	$(CO)_3Fe(SC_5H_9)_2Fe(CO)_3$	Fe: Org.Verb.C1-82

Formula	Compound		Reference
$C_{16}Fe_2H_{18}O_{10}P_2$	$(CO)_6Fe_2(C_2(COOCH_3)_2)(P(CH_3)_2)_2$	Fe:	Org.Verb.C2-67/9
$C_{16}Fe_2H_{20}N_2O_6$	$(CO)_3Fe(CN(C_2H_5)_2)_2Fe(CO)_3$	Fe:	Org.Verb.C1-227/8
–	$(CO)_6Fe_2(N(C_4H_9)CHCHNC_4H_9)$	Fe:	Org.Verb.C1-238/9, 242
$C_{16}Fe_2H_{20}N_2O_8$	$(CO)_3Fe(N(C_2H_5)_2CO)_2Fe(CO)_3$	Fe:	Org.Verb.C1-232/3, 236
$C_{16}Fe_2H_{20}O_4S_2$	$((CH_3)(C_2H_5)C_3H_2SFe(CO)_2)_2$	Fe:	Org.Verb.C2-156, 168/9
$C_{16}Fe_2H_{20}O_5S_2$	$(CH_3)(C_2H_5)C_3H_2SOFe_2(CO)_4SC_3H_2(C_2H_5)(CH_3)$	Fe:	Org.Verb.C2-169
$C_{16}Fe_2H_{20}O_8P_2$	$(CO)_4Fe((C_2H_5)_2PP(C_2H_5)_2)Fe(CO)_4$	Fe:	Org.Verb.C1-34, 38
$C_{16}Fe_2H_{20}O_8Pb_2$	$[(CO)_4FePb(C_2H_5)_2]_2$	Fe:	Org.Verb.C1-42/4, 47, 54
$C_{16}Fe_2H_{20}O_8Sn_2$	$[(CO)_4FeSn(C_2H_5)_2]_2$	Fe:	Org.Verb.C1-42/5, 51/2
$C_{16}Fe_2H_{20}O_{13}P_2$	$(CO)_4Fe((C_2H_5O)_2POP(OC_2H_5)_2)Fe(CO)_4$	Fe:	Org.Verb.C1-35, 39
$C_{16}Fe_2H_{21}NO_8$	$[N(C_2H_5)_4][HFe_2(CO)_8]$	Fe:	Org.Verb.C2-11
$C_{16}Fe_2H_{21}N_3O_6$	$(CO)_6Fe_2(N(CH_3)C(NC_4H_9)NC_4H_9)$	Fe:	Org.Verb.C1-143/4, 146
–	$(CO)_6Fe_2[(NC_3H_7)_2CNC_3H_7]$	Fe:	Org.Verb.C1-143/4, 146, 148
$C_{16}Fe_2H_{22}O_6S_2$	$(CO)_3Fe(SC_5H_{11})_2Fe(CO)_3$	Fe:	Org.Verb.C1-80, 82, 88
$C_{16}Fe_2H_{23}IN_4O_4Pb$	$((CH_3)_2C_3H_3Fe(NO)_2)_2Pb(I)C_6H_5$	Fe:	Org.Verb.C2-165
$C_{16}Fe_2H_{23}IN_4O_4Sn$	$((CH_3)_2C_3H_3Fe(NO)_2)_2Sn(I)C_6H_5$	Fe:	Org.Verb.C2-165
$C_{16}Fe_2H_{26}N_2O_4S_6$	$((C_2H_5)_2NCS_2)_2(CO)_4Fe_2(SCH_3)_2$	Fe:	Org.Verb.C1-4/5
$C_{16}Fe_2H_{27}O_8PS_2$	$(CO)_3Fe(SC_4H_9)_2Fe(CO)_2P(OCH_3)_3$	Fe:	Org.Verb.C1-93, 97, 100
$C_{16}Fe_2H_{28}O_4P_2$	$(C_3H_5Fe(CO)_2P(CH_3)_3)_2$	Fe:	Org.Verb.C2-168
$C_{16}Fe_2H_{28}O_{10}P_2$	$(C_3H_5Fe(CO)_2P(OCH_3)_3)_2$	Fe:	Org.Verb.C2-168
$C_{16}Fe_2H_{30}HgN_2O_6P_2$	$[NO(CO)_2Fe(P(C_2H_5)_3)]_2Hg$	Fe:	Org.Verb.C1-64/6
$C_{16}Fe_2H_{30}HgN_2O_6Sb_2$	$[NO(CO)_2FeSb(C_2H_5)_3]_2Hg$	Fe:	Org.Verb.C1-64/5
$C_{16}Fe_2H_{30}HgN_2O_{12}P_2$	$[NO(CO)_2FeP(OC_2H_5)_3]_2Hg$	Fe:	Org.Verb.C1-64/5
$C_{16}Fe_2H_{30}S_{14}$	$[Fe(S_2CSC_2H_5)_2(SC_2H_5)]_2$	C:	MVol.D4-266, 269
$C_{16}Fe_2H_{32}O_{10}P_2S_2$	$(CO)_4(P(OCH_3)_3)_2Fe_2(SC_3H_7)_2$	Fe:	Org.Verb.C1-104, 109
$C_{16}Fe_2H_{36}HgN_8O_6P_2$	$[NO(CO)_2FeP(N(CH_3)_2)_3]_2Hg$	Fe:	Org.Verb.C1-64/6
$C_{16}Fe_2H_{42}IO_2P_4S_2^+$	$[(P(CH_3)_3)_4(CO)_2Fe_2(SCH_3)_2I]^+$	Fe:	Org.Verb.C1-14
$C_{16}Fe_3H_5O_{11}Rh$	$C_5H_5RhFe_3(CO)_{11}$	Fe:	Org.Verb.C1-213
$C_{16}Fe_3H_{18}O_{16}P_2$	$Fe_3(CO)_{10}(P(OCH_3)_3)_2$	Fe:	Org.Verb.B2-115
$C_{16}Fe_3H_{26}N_6O_8$	$[Fe(NH(C_2H_4NH_2)_2)_2][Fe_2(CO)_8]$	Fe:	Org.Verb.C2-8
$C_{16}Fe_4GeO_{16}$	$Ge(Fe(CO)_4)_4$	Fe:	Org.Verb.B3-128
$C_{16}Fe_4H_2NaO_{16}^-$	$[Na(HFe_2(CO)_8)_2]^-$	Fe:	Org.Verb.C2-10
$C_{16}Fe_4O_{16}Sn$	$Sn(Fe(CO)_4)_4$	Fe:	Org.Verb.B3-128, 139
$C_{16}Fe_4O_{22}P_4$	$[(CO)_4Fe]_4P_4O_6$	Fe:	Org.Verb.B2-117
		Fe:	Org.Verb.C2-28
$C_{16}GeH_{32}Sn$	$(C_2H_5)_3SnCHCHCCGe(C_2H_5)_3$	Sn:	Org.Verb.2-232
$C_{16}Ge_2H_{40}Sn$	$[(C_2H_5)_3Ge]_2Sn(C_2H_5)_2$	Sn:	Org.Verb.6-54
$C_{16}H_8I_2Mn_2O_6Sn$	$[(CO)_3MnC_5H_4]_2SnI_2$	Sn:	Org.Verb.6-191
$C_{16}H_8Mn_2O_7Sn$	$[((CO)_3MnC_5H_4)_2SnO]_x$	Sn:	Org.Verb.6-191
$C_{16}H_8N_2Ni$	$Ni(CN)_2CCC_{12}H_8$	Ni:	Org.Verb.1-337/8
$C_{16}H_8N_6Ni$	$(NC)_2CC(CN)_2Ni(NC_5H_4C_5H_4N)$	Ni:	Org.Verb.1-371, 373
$C_{16}H_8O_8S_4$	$[SC(O)CHCHC(O)]_4$	C:	MVol.D4-215
$C_{16}H_{10}N_2Ni$	$Ni(CN)_2CC(C_6H_5)_2$	Ni:	Org.Verb.1-337/8
$C_{16}H_{10}N_2NiO_2$	$(C_6H_5NC)_2Ni(CO)_2$	Ni:	Org.Verb.1-311, 313
$C_{16}H_{10}N_2S_2Sn$	$(C_6H_5)_2Sn(SC(CN)C(CN)S)$	Sn:	Org.Verb.6-162/3, 170
$C_{16}H_{10}Ni$	$(C_6H_5CC)_2Ni$	Ni:	Org.Verb.1-81, 83
$C_{16}H_{11}MnO_2^+$	$[Mn(C_9H_6(O)COC_6H_5)]^+$	Mn:	MVol.D1-128
$C_{16}H_{12}MnN_2O_6$	$Mn(C_6H_5COCHNO_2)_2$	Mn:	MVol.D1-57

Formula	Compound	Reference
$C_{16}H_{12}N_3NiO_7$	$[Ni(C_5H_5)_2][OC_6H_2(NO_2)_3]$	Ni: Org.Verb.2-213
$C_{16}H_{12}N_4O_{10}U$	$UO_2(NO_3)_2 \cdot C_4(C_5H_3(CH_3)NO)_2$	U: SVol.E1-134, 142
$C_{16}H_{12}N_8S_6Sn$	$[C_5H_6N]_2[Sn(SCN)_6]$	Sn: MVol.C3-127
$C_{16}H_{12}Ni_2$	$(C_8H_6Ni)_2$	Ni: Org.Verb.2-370
$C_{16}H_{12}O_4Sn$	$Sn(C_4H_3O)_4$	Sn: Org.Verb.1-177
$C_{16}H_{12}O_{24}Sn_2^{8-}$	$Sn_2(OCOCH(O)CH(OH)COO)_4^{8-}$	Sn: MVol.C2-225/6
$C_{16}H_{12}S_4Sn$	$Sn(C_4H_3S)_4$	Sn: Org.Verb.1-177
$C_{16}H_{13}IN_2Ni$	$C_6H_5Ni(NC_5H_4C_5H_4N)I$	Ni: Org.Verb.1-38, 44
$C_{16}H_{13}MnO_3^+$	$[Mn(OC_6H_4COCHCHC_6H_4OCH_3)]^+$	Mn: MVol.D1-58/9
$C_{16}H_{13}NiO_3P$	$(CO)_3NiP(C_6H_5)_2CH_3$	Ni: Org.Verb.1-172
$C_{16}H_{13}NiO_4P$	$(CO)_3NiP(OCH_3)(C_6H_5)_2$	Ni: Org.Verb.1-177
$C_{16}H_{14}MnO_4$	$Mn(OC_6H_3(CH_3)CHO)_2$	Mn: MVol.D1-53/4
$C_{16}H_{14}MnO_6$	$Mn(OC_6H_3(OCH_3)CHO)_2$	Mn: MVol.D1-53/4
$C_{16}H_{14}N_2O_2Sn$	$Sn(C_2H_4(NCHC_6H_4O)_2)$	Sn: MVol.C5-39
$C_{16}H_{14}N_2O_2Sn^+$	$[Sn(C_2H_4(NCHC_6H_4O)_2)]^+$	Sn: MVol.C6-23
$C_{16}H_{14}N_2O_6U$	$[UO_2(CH_3COO)_2(C_{12}H_8N_2)]$	U: SVol.E1-44/6
–	$[UO_2(CH_3COO)_2(C_{12}H_8N_2)] \cdot 2.5\ H_2O$	U: SVol.E1-44/6
$C_{16}H_{14}N_4Ni$	$(NCCHCH_2)_2Ni(NC_5H_4C_5H_4N)$	Ni: Org.Verb.1-414, 415
$C_{16}H_{14}NiO_2$	$Ni(C_5H_5)_2 \cdot C_6H_4O_2$	Ni: Org.Verb.2-208
$C_{16}H_{14}O_2S_4Te$	$Te(SC(S)OCH_2C_6H_5)_2$	Te: SVol.B3-136
$C_{16}H_{15}N_3NiO$	$CONi(NC_5H_5)_3$	Ni: Org.Verb.1-112
$C_{16}H_{15}N_3PtS_3$	$Pt(C_5H_5N)_3CS_3$	C: MVol.D4-226
$C_{16}H_{15}Ni_2O$	$[(C_5H_5)_2Ni][C_5H_5NiCO]$	Ni: Org.Verb.2-346
$C_{16}H_{15}Ni_3O^+$	$(C_5H_5)_3Ni_3(CO)^+$	Ni: Org.Verb.2-381
$C_{16}H_{16}N_2NiO_2$	$(CH_2CHCHO)_2Ni(NC_5H_4C_5H_4N)$	Ni: Org.Verb.1-75
–	$Ni(OC_6H_4CHNCH_3)_2$	Ni: Org.Verb.1-65
$C_{16}H_{16}N_2O_5SU$	$UO_2(C_2H_2(NC_6H_4O)_2) \cdot (CH_3)_2SO$	U: SVol.E1-205, 209
$C_{16}H_{16}N_2S_4Sn$	$Sn(SCH_2CH_2S)_2 \cdot C_{12}H_8N_2$	Sn: MVol.C6-82/3
$C_{16}H_{16}Ni$	$Ni(C_8H_8)_2$	Ni: Org.Verb.2-98
$C_{16}H_{16}NiO_4$	$C_5H_5NiC_7H_5(COOCH_3)_2$	Ni: Org.Verb.2-184/5
–	$Ni((CH_3)_2C_6H_2O_2)_2$	Ni: Org.Verb.1-230
$C_{16}H_{16}Ni_2O_4$	$(C_5H_5Ni)_2CH_3O_2CCCCO_2CH_3$	Ni: Org.Verb.2-356, 358
$C_{16}H_{16}O_2Sn$	$[Sn(C_6H_5)_2(OCH_2CHCHCH_2O)]_x$	Sn: Org.Verb.6-162/5
$C_{16}H_{16}O_4S_2Sn$	$(CH_2CH)_2Sn(O_2SC_6H_5)_2$	Sn: Org.Verb.6-144
$C_{16}H_{16}O_4Sn$	$(CH_3)_2Sn(C_6H_4COOH)_2$	Sn: Org.Verb.3-11
–	$(CH_3)_2Sn(OCOC_6H_5)_2$	Sn: Org.Verb.6-22
$C_{16}H_{16}O_6Sn$	$(CH_3)_2Sn(OC(O)C_6H_4OH)_2$	Sn: Org.Verb.6-27
$C_{16}H_{16}O_7SU$	$UO_2(C_7H_5O_2)_2 \cdot (CH_3)_2SO$	U: SVol.E1-205, 208
$C_{16}H_{16}O_7U$	$UO_2(C_7H_5O_2)_2 \cdot C_2H_5OH$	U: SVol.E1-60, 65/6
–	$UO_2(OC_6H_4CHO)_2 \cdot C_2H_5OH$	U: SVol.E1-60, 64/5
$C_{16}H_{16}Pa$	$Pa(C_8H_8)_2$	Pa: SVol.2-101/2
$C_{16}H_{16}S_4Sn$	$[Sn((CH_3)_2C_6H_2S_2)_2]_x$	Sn: MVol.C6-87/8
$C_{16}H_{16}Sn$	$(CH_3)_2SnC_{14}H_{10}$	Sn: Org.Verb.3-116
–	$(C_6H_5)_2Sn(CHCH_2)_2$	Sn: Org.Verb.3-57
$C_{16}H_{17}NO_5U$	$UO_2(OC_6H_4CHNCH_2C_6H_4O) \cdot C_2H_5OH$	U: SVol.E1-62, 68/9
$C_{16}H_{17}NO_6U$	$UO_2(OC_6H_3(OCH_3)CHNCH_2C_6H_4O) \cdot CH_3OH$	U: SVol.E1-61, 65, 68
$C_{16}H_{17}N_2O_6SU$	$[UO_2(C_5H_8(O)NC_6H_4SO_3)(C_5H_5N)]$	U: SVol.E1-31, 37
$C_{16}H_{17}NiO$	$C_3H_5Ni(OC_6H_4C_6H_5)CH_3$	Ni: Org.Verb.2-285
$C_{16}H_{18}Li_2Ni_2O_8$	$[Li((CO)_3NiCOC_4H_9)]_2$	Ni: Org.Verb.2-265/6
$C_{16}H_{18}MnO_4$	$[Mn(C_8H_9O_2)_2] \cdot 2\ H_2O$	Mn: MVol.D1-129

Formula	Compound	Reference
$C_{16}H_{18}N_2NiO_2$	$(CH_2CH_2CHO)_2Ni(NC_5H_4C_5H_4N)$	Ni: Org.Verb.1-414
–	$HOC_6H_4CH_2N(CH_3)NiOC_6H_4CHNCH_3 \cdot C_6H_5CH_3$	Ni: Org.Verb.1-63, 65
$C_{16}H_{18}N_2O_2S_4Zn$	$Zn[S_2COC_2H_5]_2(NC_5H_4C_5H_4N)$	C: MVol.D4-257
$C_{16}H_{18}N_2O_6U$	$[UO_2(CH(COCH_3)_2)_2(NCC_5H_4N)]$	U: SVol.E1-31, 36/7
–	$[UO_2(CH_3COO)_2(C_2H_4(C_5H_4N)_2)]$	U: SVol.E1-41, 43
$C_{16}H_{18}N_2Sn$	$[(C_6H_5)_2Sn(N(CH_2CH_2)_2N)]_x$	Sn: Org.Verb.6-162/3, 171
$C_{16}H_{18}N_4NiO_2$	$(CH_2CHCONH_2)_2Ni(NC_5H_4C_5H_4N)$	Ni: Org.Verb.1-75, 414
$C_{16}H_{18}N_6Ni$	$(NC)_2CC(CN)_2Ni(CNC_4H_9\text{-t})_2$	Ni: Org.Verb.1-383, 384
$C_{16}H_{18}Ni$	$((CH_3)_2C_6H_3)_2Ni$	Ni: Org.Verb.1-88, 90
–	$Ni(C_8H_9)_2$	Ni: Org.Verb.2-67
$C_{16}H_{18}NiO_4$	$C_5H_5NiC_7H_7(COOCH_3)_2$	Ni: Org.Verb.2-186
$C_{16}H_{18}NiPd$	$C_5H_5NiC_3H_4C_3H_4PdC_5H_5$	Ni: Org.Verb.2-176
$C_{16}H_{18}Ni_2$	$C_5H_5NiC_3H_4C_3H_4NiC_5H_5$	Ni: Org.Verb.2-366/7
$C_{16}H_{18}Ni_2O_2$	$(C_2H_5C_5H_4NiCO)_2$	Ni: Org.Verb.2-349
$C_{16}H_{18}OSn$	$(CH_3)_3SnC_6H_4COC_6H_5$	Sn: Org.Verb.2-143, 145
–	$(C_2H_5)_2Sn(C_6H_4OC_6H_4)$	Sn: Org.Verb.3-120, 123
$C_{16}H_{18}OTi$	$(C_6H_5)_2Ti \cdot C_4H_8O$	Ti: Org.Verb.1-69
$C_{16}H_{18}O_2Sn$	$(C_6H_5)_2(C_2H_5)SnOCOCH_3$	Sn: Org.Verb.5-216
–	$[Sn(C_6H_5)_2O(CH_2)_4O]_x$	Sn: Org.Verb.6-162/4
$C_{16}H_{18}O_3Sn$	$(CH_3)_3SnC_{13}H_9O_3$	Sn: Org.Verb.2-65
–	$[(C_2H_5OC_6H_4)_2SnO]_x$	Sn: Org.Verb.6-188
–	$[Sn(C_6H_5)_2OCH_2CH_2OCH_2CH_2O]_x$	Sn: Org.Verb.6-162/4
$C_{16}H_{18}S_3Sn$	$(C_4H_3S)_3SnC_4H_9$	Sn: Org.Verb.2-435
$C_{16}H_{18}Sn$	$(CH_3)_2Sn(C_6H_4CH_2CH_2C_6H_4)$	Sn: Org.Verb.3-116, 122
–	$(CH_3)_3SnCH(C_6H_4)_2$	Sn: Org.Verb.2-65, 68
–	$(C_2H_5)_2Sn(C_6H_4C_6H_4)$	Sn: Org.Verb.3-114
–	$(C_5H_5)_3SnCH_3$	Sn: Org.Verb.2-346, 350
$C_{16}H_{19}MnN_2O_4S$	$[MnSCN(CH(COCH_3)_2)_2(C_5H_5N)]$	Mn: MVol.D1-105
$C_{16}H_{19}NO_4U$	$UO_2(OC_6H_4C_6H_4O) \cdot (C_2H_5)_2NH$	U: SVol.E1-19, 22
$C_{16}H_{19}NSn$	$(CH_3)_3SnC(C_6H_5)NC_6H_5$	Sn: Org.Verb.2-26
$C_{16}H_{19}N_3O_4U$	$UO_2(OC_6H_4CHNCH_2C_6H_4O) \cdot C_2H_8N_2$	U: SVol.E1-23/5
$C_{16}H_{19}N_4O_9PS$	$(C_2H_5)_2NSO_2NHPO(OC_6H_4NO_2)_2$	S: S-N-Verb.1-185
$C_{16}H_{20}I_2O_4Sn$	$[SnI_2(CH(COCH_3)_2)_2] \cdot C_6H_6$	Sn: MVol.C5-106
$C_{16}H_{20}N_2Ni$	$(N(CH_3)_2C_6H_4)_2Ni$	Ni: Org.Verb.1-79
$C_{16}H_{20}N_2NiO_2S_4$	$Ni[S_2COC_2H_5]_2(C_5H_5N)_2$	C: MVol.D4-260
$C_{16}H_{20}N_2NiO_2Se_4$	$Ni[Se_2COC_2H_5]_2(C_5H_5N)_2$	C: MVol.D6-221
$C_{16}H_{20}N_2NiO_4$	$C_5H_5NiC_5H_5N_2(COOC_2H_5)_2$	Ni: Org.Verb.2-187/8
$C_{16}H_{20}N_2NiO_4S_2$	$Ni[SOCOC_2H_5]_2 \cdot 2\ C_5H_5N$	C: MVol.D4-237
$C_{16}H_{20}N_2NiS_6$	$Ni[S_2CSC_2H_5]_2(C_5H_5N)_2$	C: MVol.D4-270
$C_{16}H_{20}N_2O_2S_4Zn$	$Zn[S_2COC_2H_5]_2(C_5H_5N)_2$	C: MVol.D4-257
$C_{16}H_{20}N_2O_7SU$	$UO_2SO_4 \cdot NH_2(C_6H_4)_2NH_2 \cdot C_4H_8O$	U: SVol.E1-23/6
$C_{16}H_{20}N_2O_8S_2U$	$UO_2(NC_5H_4COO)_2 \cdot 2\ (CH_3)_2SO$	U: SVol.E1-205, 208
$C_{16}H_{20}N_2S_4Sn$	$Sn(SCH_2CH(S)CH_3)_2 \cdot NC_5H_4C_5H_4N$	Sn: MVol.C6-83/4
–	$Sn(SCH_2CH_2CH_2S)_2 \cdot NC_5H_4C_5H_4N$	Sn: MVol.C6-84/5
$C_{16}H_{20}N_2Ti$	$(CH_3)_4Ti \cdot C_{12}H_8N_2$	Ti: Org.Verb.1-114, 118
$C_{16}H_{20}N_4NiO_2P_2$	$(CO)_2Ni[(NCC_2H_4)_2P(CH_2)_2P(C_2H_4CN)_2]$	Ni: Org.Verb.1-149, 152
$C_{16}H_{20}N_4O_3S$	$[(CH_3)_2N(CH_2C_6H_5)_2][SO_3N_3]$	S: S-N-Verb.1-213
$C_{16}H_{20}N_4O_8SU$	$UO_2SO_4 \cdot 2\ (CH_3)_2NC_6H_4NO$	U: SVol.E1-144/5
$C_{16}H_{20}N_4O_{10}U$	$UO_2(NO_3)_2 \cdot C_2H_4(CH_2C_5H_3(CH_3)NO)_2$	U: SVol.E1-134, 142

Formula	Compound	Ref.
$C_{16}H_{20}N_6O_{10}U$	$UO_2(NO_3)_2 \cdot 2\ CH_3CONHC_6H_4NH_2$	U: SVol.E1-97
–	$UO_2(NO_3)_2 \cdot 2\ (CH_3)_2NC_6H_4NO$	U: SVol.E1-144/5
$C_{16}H_{20}N_6O_{12}U$	$UO_2(NO_3)_2 \cdot 2\ CH_3OC_6H_4CON_2H_3 \cdot n\ H_2O$	U: SVol.E1-114/6
$C_{16}H_{20}N_{14}O_{20}U$	$UO_2(OC_6H_2(NO_2)_3)_2 \cdot 4\ (NH_2)_2CO$	U: SVol.E1-84/5, 89
$C_{16}H_{20}NiO_2$	$C_8H_{12}NiC_6H_2(CH_3)_2O_2$	Ni: Org.Verb.2-101, 103, 106
$C_{16}H_{20}Ni_2$	$(C_5H_5Ni)_2C_2H_5CCC_2H_5$	Ni: Org.Verb.2-356, 357
–	$(C_5H_5Ni)_2HCCC_4H_9\text{-n}$	Ni: Org.Verb.2-356, 359
$C_{16}H_{20}OTi$	$(C_6H_5)_2Ti \cdot (C_2H_5)_2O$	Ti: Org.Verb.1-69
$C_{16}H_{20}O_2S_4Ti$	$(C_5H_5)_2Ti[S_2COC_2H_5]_2$	C: MVol.D4-258
$C_{16}H_{20}O_2Sn$	$(C_2H_5OC_6H_4)_2SnH_2$	Sn: Org.Verb.4-118
$C_{16}H_{20}O_6S_2Sn$	$(C_6H_5)_2Sn(OSO_2C_2H_5)_2$	Sn: Org.Verb.6-161
$C_{16}H_{20}Sn$	$(CH_3)_2Sn(CH_2C_6H_5)_2$	Sn: Org.Verb.3-11, 19
–	$(CH_3)_2Sn(C_6H_4CH_3)_2$	Sn: Org.Verb.3-11, 19
–	$(CH_3)_2Sn(C_6H_5)(CH(CH_3)C_6H_5)$	Sn: Org.Verb.3-75, 80
–	$(C_2H_5)_2Sn(C_6H_5)_2$	Sn: Org.Verb.3-24/5
–	$(C_6H_5)_2Sn(CH_3)(C_3H_7)$	Sn: Org.Verb.3-90
–	$(C_6H_5)_2Sn(CH_3)(i\text{-}C_3H_7)$	Sn: Org.Verb.3-89/90
$C_{16}H_{20}Sn^+$	$Sn(C_6H_5)_2(C_2H_5)_2{}^+$	Sn: Org.Verb.3-25
$C_{16}H_{21}MnNO_4$	$[Mn(CH(COCH_3)_2)_2(CH_3C_5H_4N)]$	Mn: MVol.D1-79
$C_{16}H_{21}NO_6U$	$[UO_2(CH(COCH_3)_2)_2(CH_3C_5H_4N)]$	U: SVol.E1-31, 36/7
$C_{16}H_{21}NO_7U$	$UO_2(CH(COCH_3)_2)_2 \cdot CH_3C_5H_4NO$	U: SVol.E1-133, 137/8, 142/3
$C_{16}H_{21}NSn$	$(CH_3)_2Sn(C_6H_5)(CH_2CH_2NHC_6H_5)$	Sn: Org.Verb.3-75, 80
$C_{16}H_{21}N_2O_5PS$	$(C_2H_5)_2NSO_2NHPO(OC_6H_5)_2$	S: S-N-Verb.1-185
$C_{16}H_{21}Ni^+$	$[t\text{-}C_4H_9C_5H_4NiC_7H_8]^+$	Ni: Org.Verb.2-189
$C_{16}H_{22}MnN_4O_4$	$[Mn(CH(COCH_3)_2)_2(C_3H_4N_2)_2]$	Mn: MVol.D1-83
$C_{16}H_{22}N_2Ni$	$(C_2H_5)_2Ni(CH_3(C_5H_3N)_2CH_3)$	Ni: Org.Verb.1-70, 76
–	$(C_3H_7)_2Ni(NC_5H_4C_5H_4N)$	Ni: Org.Verb.1-70, 76
$C_{16}H_{22}N_2NiO_2$	$(C_2H_5)_2Ni(CH_3O(C_5H_3N)_2OCH_3)$	Ni: Org.Verb.1-70, 76
$C_{16}H_{22}N_2S_2Sn$	$(C_2H_5)_3SnC_6H_4CH(SCN)CH_2SCN$	Sn: Org.Verb.2-242
$C_{16}H_{22}N_2S_4Sn$	$Sn(SCH_2CH(S)CH_3)_2 \cdot 2\ C_5H_5N$	Sn: MVol.C6-83/4
–	$Sn(SCH_2CH_2CH_2S)_2 \cdot 2\ C_5H_5N$	Sn: MVol.C6-84/5
–	$Sn(SCH_2CH_2S)_2 \cdot 2\ NC_5H_4CH_3$	Sn: MVol.C6-82/3
$C_{16}H_{22}N_4Ni$	$(C_6H_5CC)_2Ni(NH_3)_4$	Ni: Org.Verb.1-81, 83
$C_{16}H_{22}N_4O_{10}U$	$UO_2(NO_3)_2 \cdot 2\ (CH_3)_3C_5H_2NO$	U: SVol.E1-133, 140, 142/3
$C_{16}H_{22}N_4P_2S_2$	$[NS(CH_3)_2NP(C_6H_5)]_2$	S: S-N-Verb.1-6
$C_{16}H_{22}Ni$	$Ni(C_5H_4C_3H_7\text{-n})_2$	Ni: Org.Verb.2-215
–	$Ni(C_5H_4C_3H_7\text{-i})_2$	Ni: Org.Verb.2-215, 217
–	$Ni(C_8H_{11})_2$	Ni: Org.Verb.2-61
$C_{16}H_{22}NiO_2$	$(C_3H_5)_2NiC_6(CH_3)_4O_2$	Ni: Org.Verb.2-71
$C_{16}H_{22}O_2Rh_2Sn_2$	$[Rh(CO)(C_5H_5)[(CH_3)_2Sn]]_2$	Sn: Org.Verb.6-33
$C_{16}H_{22}O_2S_2Sn$	$[(C_4H_9)_2SnSC(O)C_6H_4C(O)S]_x$	Sn: Org.Verb.6-89, 96
$C_{16}H_{22}O_4PbTi$	$[Pb(C_2H_5)_3][CH_3C_5H_4Ti(CO)_4]$	Ti: Org.Verb.1-211
$C_{16}H_{22}O_4Sn$	$[Sn(C_4H_9)_2OCOC_6H_4COO]_x$	Sn: Org.Verb.6-89, 95
$C_{16}H_{22}Sn$	$(CH_3)_2Sn(s\text{-}C_4H_9)(C_{10}H_7)$	Sn: Org.Verb.3-75, 79
$C_{16}H_{23}N_3NiO$	$(t\text{-}C_4H_9NC)_2Ni(ONC_6H_5)$	Ni: Org.Verb.1-310, 313
$C_{16}H_{23}N_3NiO_2$	$[Ni(CNC_4H_9\text{-t})_2C_6H_5NO_2]_x$	Ni: Org.Verb.2-397
$C_{16}H_{24}HfN_4Ni_2O_8$	$[Hf(N(CH_3)_2)_4 \cdot 2\ Ni(CO)_4]_x$	Ni: Org.Verb.2-396
$C_{16}H_{24}I_4Ni_2$	$((CH_3)_4C_4NiI_2)_2$	Ni: Org.Verb.2-329
$C_{16}H_{24}N_2NiO_4$	$(C_2(COOCH_3)_2)Ni(CNC_4H_9\text{-t})_2$	Ni: Org.Verb.1-385

Formula	Compound	Element	Reference
$C_{16}H_{24}N_2NiO_4$	$(C_2H_5NC)_2Ni((OCCH_3)_2CH)_2$	Ni:	Org.Verb.1-315, 316
$C_{16}H_{24}N_4Ni_2O_2$	$[(CO)NiNCN(CH_2)_4C(CH_3)H]_2$	Ni:	Org.Verb.2-256/7
–	$[(CO)NiNCN(CH_2)_6]_2$	Ni:	Org.Verb.2-256/7
$C_{16}H_{24}N_4Ni_2O_8Ti$	$[Ti(N(CH_3)_2)_4 \cdot 2\ Ni(CO)_4]_x$	Ni:	Org.Verb.2-396
$C_{16}H_{24}N_4Ni_2O_8Zr$	$[Zr(N(CH_3)_2)_4 \cdot 2\ Ni(CO)_4]_x$	Ni:	Org.Verb.2-396
$C_{16}H_{24}N_4Ni_2O_{12}$	$((CH_3)_4C_4Ni(NO_3)_2)_2$	Ni:	Org.Verb.2-329
$C_{16}H_{24}N_4O_4P_2S_2$	$[NS(CH_3)_2NP(C_6H_5)O]_2 \cdot H_2O_2$	S:	S-N-Verb.1-6
$C_{16}H_{24}N_4S_4Sn$	$Sn(SCH_2CH(S)CH_3)_2 \cdot 2\ NC_5H_4NH_2$	Sn:	MVol.C6-83/4
$C_{16}H_{24}N_4Sn$	$Sn[CH_2CH(CN)CH_3]_4$	Sn:	Org.Verb.1-124
$C_{16}H_{24}N_{12}Ni_2$	$((CH_3)_4C_4Ni(N_3)_2)_2$	Ni:	Org.Verb.2-329
$C_{16}H_{24}Ni$	$(C_8H_{12})_2Ni$	Ni:	Org.Verb.2-87/98
–	$C_{16}H_{24}Ni$	Ni:	Org.Verb.2-229, 232
–	$[Ni(CH_3CCCH_3)_4]_x$	Ni:	Org.Verb.2-398
$C_{16}H_{24}NiO_2P_2$	$(CO)_2NiC_6H_4(P(C_2H_5)_2)_2$	Ni:	Org.Verb.1-151
$C_{16}H_{24}Ni_2O_6P_2$	$(CO)_3Ni(C_2H_5)_2PC_2H_4P(C_2H_5)_2Ni(CO)_3$	Ni:	Org.Verb.2-263
$C_{16}H_{24}Ni_2O_8S_2$	$((CH_3)_4C_4NiSO_4)_2$	Ni:	Org.Verb.2-329
$C_{16}H_{24}Ni_2S_2$	$(C_5H_5NiSC_3H_7\text{-}n)_2$	Ni:	Org.Verb.2-338, 339
–	$(C_5H_5NiSC_3H_7\text{-}i)_2$	Ni:	Org.Verb.2-338, 339
$C_{16}H_{24}O_2Sn$	$[C_5H_5SnCH(COC(CH_3)_3)_2]$	Sn:	MVol.C5-9
$C_{16}H_{24}O_4Sn$	$(C_6H_{11})_2Sn(OC(O)CH)_2$	Sn:	Org.Verb.6-109
$C_{16}H_{24}O_6Ti$	$(CH_3)_5C_5Ti(OCOCH_3)_3$	Ti:	Org.Verb.1-168, 170
$C_{16}H_{24}O_8S_8$	$S_2(C(S)OC_2H_4OC_2H_4OC_2H_4OCS_2)_2$	C:	MVol.D4-251
$C_{16}H_{24}Sn$	$(CH_3)_2Sn(C_7H_9)_2$	Sn:	Org.Verb.3-12
–	$(C_4H_9)_2Sn(CCCHCH_2)_2$	Sn:	Org.Verb.3-45
$C_{16}H_{25}NSn$	$(CH_3)_3SnC(C_6H_5)NC_6H_{11}$	Sn:	Org.Verb.2-26
$C_{16}H_{25}O_9PU$	$UO_2(CH_3COO)_2 \cdot (C_3H_7O)_2(C_6H_5)PO$	U:	SVol.E1-186/8
$C_{16}H_{26}I_2Ni_2$	$(CH_3CHCH(CH_2)_2CHCHCH_2NiI)_2$	Ni:	Org.Verb.2-297
–	$(CH_3CHCHCH_2CD_2CDCDCD_2NiI)_2$	Ni:	Org.Verb.2-297
–	$(CD_3CDCDCD_2CHCHCHCH_3NiI)_2$	Ni:	Org.Verb.2-315
$C_{16}H_{26}MnO_4$	$[Mn(CH_3COCHCOCH_2CH(CH_3)_2)_2]$	Mn:	MVol.D1-126
–	$[Mn((CH_3)_3CCOCHCOCH_3)_2]$	Mn:	MVol.D1-124
$C_{16}H_{26}N_2NiO_4$	$(C_2H_2(COOCH_3)_2)Ni(CNC_4H_9\text{-}t)_2$	Ni:	Org.Verb.1-383
$C_{16}H_{26}N_2O_4Sn$	$(C_4H_9)_2Sn[CH(CN)COOCH_3]_2$	Sn:	Org.Verb.3-45
$C_{16}H_{26}N_3O_{12}PU$	$UO_2(NO_3)_2 \cdot (C_3H_7O)_2POCH_2CON(C_6H_5)C_2H_5$	U:	SVol.E1-188/9
$C_{16}H_{26}Ni$	$Ni(C_8H_{13})_2$	Ni:	Org.Verb.2-68
$C_{16}H_{26}O_2Sn$	$(C_3H_7)_3SnOCOC_6H_5$	Sn:	Org.Verb.5-116
$C_{16}H_{26}O_4Sn$	$[Sn((CH_3)_3CCOCHCOCH_3)_2]$	Sn:	MVol.C5-9
$C_{16}H_{26}Sn$	$(C_2H_5)_3SnCH_2CH_2CHCHC_6H_5$	Sn:	Org.Verb.2-200, 205
–	$(C_2H_5)_3SnCH_2CH_2C_6H_4CHCH_2$	Sn:	Org.Verb.2-175, 178
$C_{16}H_{27}NO_8U$	$HUO_2(CH(COCH_3)_2)_3 \cdot CH_3NH_2$	U:	SVol.E1-19, 22
$C_{16}H_{27}N_3NiO$	$(t\text{-}C_4H_9NC)_3NiCO$	Ni:	Org.Verb.1-325, 326
$C_{16}H_{27}NiO_{10}P_3$	$CONi(P(OCH_2)_3CCH_3)_3$	Ni:	Org.Verb.1-117, 119
$C_{16}H_{28}MnN_2O_4$	$[Mn(CH(COCH_3)_2)_2(C_3H_5NH_2)_2]$	Mn:	MVol.D1-76
$C_{16}H_{28}Mo_2O_3S_{12}$	$Mo_2O_3[S_2CSC_3H_7]_4$	C:	MVol.D4-266, 269
$C_{16}H_{28}Mo_2O_7S_8$	$Mo_2O_3[S_2COC_3H_7]_4$	C:	MVol.D4-258
$C_{16}H_{28}N_2O_{14}P_2U$	$UO_2(NO_3)_2 \cdot C_6H_5CH_2CH(PO(OC_2H_5)_2)_2$	U:	SVol.E1-188/9
$C_{16}H_{28}N_4Ni$	$(i\text{-}C_3H_7NC)_4Ni$	Ni:	Org.Verb.1-328, 329
$C_{16}H_{28}N_4NiO_6P_2$	$(CO)_2Ni(C_2H_4(P(C_2H_4CONH_2)_2)_2)$	Ni:	Org.Verb.1-149, 152
$C_{16}H_{28}N_4O_4Sn$	$(C_4H_9)_2Sn[C(N_2)COOC_2H_5]_2$	Sn:	Org.Verb.3-45

Formula	Compound	Reference
$C_{16}H_{28}N_4Ti$	$((CH_3)_2C_4H_2N)_2Ti(N(CH_3)_2)_2$	Ti: Org.Verb.1-130/1
–	$(C_4H_4N)_2Ti(N(C_2H_5)_2)_2$	Ti: Org.Verb.1-130
$C_{16}H_{28}O_2Sn$	$(C_2H_5)_3SnCCCH_2OCH(C_3H_7)OCH_2CCH$	Sn: Org.Verb.2-231
$C_{16}H_{28}O_3SSn$	$(CH_3)_3Sn(CH_2)_6OSO_2C_6H_4CH_3$	Sn: Org.Verb.2-56/7
$C_{16}H_{28}O_4Sn$	$(C_2H_5)_2Sn(CH_2CH_2COOCH_2CHCH_2)_2$	Sn: Org.Verb.3-30
$C_{16}H_{28}O_7U$	$UO_2(CH(COC_2H_5)_2)_2 \cdot C_2H_5OH$	U: SVol.E1-60, 64/5
$C_{16}H_{28}O_8Sn$	$Sn(CH_2CH_2COOCH_3)_4$	Sn: Org.Verb.1-124
$C_{16}H_{28}Sn$	$(CH_3)_2Sn(s\text{-}C_4H_9)(CH_2C(CH_3)_2C_6H_5)$	Sn: Org.Verb.3-75, 79
–	$(C_2H_5)_2Sn(CCC_4H_9)_2$	Sn: Org.Verb.3-29, 33
–	$(C_2H_5)_3Sn(CH_2)_4C_6H_5$	Sn: Org.Verb.2-175, 179
–	$(C_2H_5)_3SnC_{10}H_{13}$	Sn: Org.Verb.2-245
–	$(C_3H_7)_2Sn(CCC_3H_7)_2$	Sn: Org.Verb.3-36, 38
–	$(i\text{-}C_3H_7)_3SnCH_2C_6H_5$	Sn: Org.Verb.2-271
–	$(i\text{-}C_3H_7)_3SnC_6H_4CH_3$	Sn: Org.Verb.2-270/1
–	$(t\text{-}C_4H_9)_2Sn(CH_3)(CH_2C_6H_5)$	Sn: Org.Verb.3-92/3
–	$Sn(C(CH_3)CHCH_3)_4$	Sn: Org.Verb.1-137
–	$Sn(CH_2CHCHCH_3)_4$	Sn: Org.Verb.1-137
$C_{16}H_{28}Ti$	$(CH_2C(CH_3)CH_2)_4Ti$	Ti: Org.Verb.1-128
–	$(CH_3C_3H_4)_4Ti$	Ti: Org.Verb.1-128
$C_{16}H_{29}INNiP$	$[C_5H_5Ni(P(C_2H_5)_3)CNC(CH_3)_3]I$	Ni: Org.Verb.2-172
$C_{16}H_{29}Ir_3O_{19}$	$[Ir_3O(CH_3COO)_6(H_2O)_2(CH_3COOH)]CH_3COO$	Ir: SVol.2-32
$C_{16}H_{29}NO_6U$	$UO_2(CH(COCH_3)_2)_2 \cdot N(C_2H_5)_3$	U: SVol.E1-17, 20, 23
$C_{16}H_{29}NSn$	$(C_2H_5)_3SnCH_2C_5H_3NC_4H_9$	Sn: Org.Verb.2-175
–	$(C_3H_7)_3SnCH_2CH_2C_5H_4N$	Sn: Org.Verb.2-254, 257
$C_{16}H_{29}NiO_3P$	$(CO)_3Ni(CH_2P(C_4H_9\text{-}n)_3)$	Ni: Org.Verb.1-182, 183
$C_{16}H_{30}N_2Sn$	$(C_2H_5)(C_6H_5)Sn[N(C_2H_5)_2]_2$	Sn: Org.Verb.6-204
$C_{16}H_{30}OSn$	$(C_2H_5)_3SnC(CCHCH_3)C(OH)(CH_2)_5$	Sn: Org.Verb.2-201
–	$(C_2H_5)_3SnCH_2CH_2CCC_6H_{10}OH$	Sn: Org.Verb.2-231
–	$(C_4H_9)_3SnCCCOCH_3$	Sn: Org.Verb.2-317
$C_{16}H_{30}O_2S_4$	$C_7H_{15}OC(S)SSC(S)OC_7H_{15}$	C: MVol.D4-251
$C_{16}H_{30}O_3Ti$	$(CH_3)_5C_5Ti(OC_2H_5)_3$	Ti: Org.Verb.1-158, 164/5
$C_{16}H_{30}O_4RuSn_2$	$[(C_2H_5)_3Sn]_2Ru(CO)_4$	Sn: Org.Verb.5-106
$C_{16}H_{30}O_5Sn$	$(C_2H_5)_3SnC_2(COOC_2H_5)_2OC_2H_5$	Sn: Org.Verb.2-201
$C_{16}H_{30}O_6Sn$	$(CH_3OOCCH_2CH_2)_3SnC_4H_9$	Sn: Org.Verb.2-336
–	$(CH_3OOCCH_2CH_2)_3Sn(i\text{-}C_4H_9)$	Sn: Org.Verb.2-336
$C_{16}H_{30}O_{12}U$	$UO_2(CH_3COO)_2 \cdot (CH_2OCH_2)_6 \cdot 3\ H_2O$	U: SVol.E1-71/2, 77/8
–	$UO_2(CH_3COO)_2 \cdot (CH_2OCH_2)_6 \cdot 4\ H_2O$	U: SVol.E1-71/2, 77/8
$C_{16}H_{30}O_{23}S_5U_3$	$[UO_2(C_2O_4)]_3 \cdot 5\ (CH_3)_2SO$	U: SVol.E1-204, 208
$C_{16}H_{30}Sn$	$(CH_2CH)_3SnC_{10}H_{21}$	Sn: Org.Verb.2-347
–	$(C_4H_9)_3SnCCCHCH_2$	Sn: Org.Verb.2-317
–	$(C_5H_9)_3SnCH_3$	Sn: Org.Verb.2-340
$C_{16}H_{32}INiOP$	$C_3H_5Ni(CO)(P(C_4H_9\text{-}n)_3)I$	Ni: Org.Verb.2-44
$C_{16}H_{32}MnN_2O_4$	$[Mn(CH(COCH_3)_2)_2(C_3H_7NH_2)_2]$	Mn: MVol.D1-76
$C_{16}H_{32}N_2O_2Sn$	$(C_4H_9)_3SnC(N_2)COOC_2H_5$	Sn: Org.Verb.2-284
$C_{16}H_{32}NiP_2$	$(CHC)_2Ni(P(C_2H_5)_3)_2$	Ni: Org.Verb.1-81/2
$C_{16}H_{32}Ni_2$	$(CH_3CHCHCH_2NiC_4H_9\text{-}n)_2$	Ni: Org.Verb.2-299
$C_{16}H_{32}OSn$	$(C_2H_5)_3SnC(CCHCH_3)C(OH)(CH_3)C_4H_9$	Sn: Org.Verb.2-201
–	$(C_2H_5)_3SnC(CCHCH_3)C(OH)(C_2H_5)C_3H_7$	Sn: Org.Verb.2-201
–	$(C_2H_5)_3SnCH_2CH_2CCC(OH)(CH_3)C_4H_9$	Sn: Org.Verb.2-232
–	$(C_2H_5)_3Sn(CH_2)_2CCC(OH)(C_2H_5)C_3H_7$	Sn: Org.Verb.2-232

Formula	Compound	Reference
$C_{16}H_{32}OSn$	$(C_4H_9)_3SnCCCH_2OCH_3$	Sn: Org.Verb.2-317
–	$(C_6H_{11})_2(C_4H_9)SnOH$	Sn: Org.Verb.5-217
$C_{16}H_{32}O_2SSn$	$(C_4H_9)_3SnC_4H_5SO_2$	Sn: Org.Verb.2-322, 324
$C_{16}H_{32}O_2S_4Sn$	$(C_4H_9)_2Sn[S_2COC_3H_7]_2$	C: MVol.D4-254
$C_{16}H_{32}O_2Sn$	$(C_2H_5)_3SnCCCH(CH_3)OCH(CH_3)OC_4H_9$	Sn: Org.Verb.2-232
–	$(C_2H_5)_3SnCCCH_2CH_2OCH(CH_3)OC_4H_9$	Sn: Org.Verb.2-232
–	$(C_4H_9)_3SnC(CH_2)COOCH_3$	Sn: Org.Verb.2-309/10
–	$(C_4H_9)_3SnCHCHCOOCH_3$	Sn: Org.Verb.2-309/10
$C_{16}H_{32}O_4Sn$	$(C_2H_5)_2Sn((CH_2)_3OCH_2CH(O)CH_2)_2$	Sn: Org.Verb.3-30
–	$(C_2H_5)_3SnCH(C_2H_5)CH(COOC_2H_5)_2$	Sn: Org.Verb.2-175
–	$(C_3H_7)_2Sn(CH_2CH(CH_3)COOCH_3)_2$	Sn: Org.Verb.3-38
–	$(C_4H_9)_2Sn(CH_2CH(CH_3)COOH)_2$	Sn: Org.Verb.3-46
–	$(C_4H_9)_2Sn(CH_2CH_2COOCH_3)_2$	Sn: Org.Verb.3-46
–	$(i\text{-}C_4H_9)_2Sn(CH_2CH_2COOCH_3)_2$	Sn: Org.Verb.3-69
$C_{16}H_{32}O_8Sn^{4+}$	$[Sn(CH_3COOC_2H_5)_4]^{4+}$	Sn: MVol.C5-130
$C_{16}H_{32}Sn$	$(C_2H_5)_2Sn(C_6H_{11})_2$	Sn: Org.Verb.3-30
–	$(C_4H_9)_2Sn(CHCHC_2H_5)_2$	Sn: Org.Verb.3-46
–	$(C_4H_9)_2Sn(CH_2C(CH_3)CH_2)_2$	Sn: Org.Verb.3-46, 54
–	$(C_4H_9)_2Sn(CH_2CH_2CHCH_2)_2$	Sn: Org.Verb.3-46
–	$(C_4H_9)_3SnC(CH_2)CHCH_2$	Sn: Org.Verb.2-310
$C_{16}H_{33}NOSn$	$(C_4H_9)_3SnCH_2CH_2CH_2NCO$	Sn: Org.Verb.2-284
$C_{16}H_{33}NSn$	$(C_4H_9)_3SnCH(CH_3)CH_2CN$	Sn: Org.Verb.2-284, 299
–	$(C_4H_9)_3SnCH(CN)C_2H_5$	Sn: Org.Verb.2-278
$C_{16}H_{33}N_3Ti$	$(CH_3)(C(CH_3)_3)C_5H_3Ti(N(CH_3)_2)_3$	Ti: Org.Verb.1-172, 176
–	$(C_2H_5)_2CHC_5H_4Ti(N(CH_3)_2)_3$	Ti: Org.Verb.1-172, 175, 177/8
$C_{16}H_{33}O_7PU$	$UO_2(CH_3COO)_2 \cdot ((C_2H_5)_2CHCH_2)_2HPO$	U: SVol.E1-168
–	$UO_2(CH_3COO)_2 \cdot (C_4H_9)_3PO$	U: SVol.E1-151, 159
$C_{16}H_{33}O_{10}PU$	$UO_2(CH_3COO)_2 \cdot (C_4H_9O)_3PO$	U: SVol.E1-182
$C_{16}H_{34}N_2O_{13}Sn$	$SnO(C_4H_4O_6) \cdot 2\ N(C_2H_4OH)_3$	Sn: MVol.C6-2
$C_{16}H_{34}N_2Ti$	$(CH_2C(CH_3)CH_2)_2Ti(N(C_2H_5)_2)_2$	Ti: Org.Verb.1-78
$C_{16}H_{34}N_4O_4S_2$	$C_5H_{10}NSO_2NH(CH_2)_6NHSO_2NC_5H_{10}$	S: S-N-Verb.1-171
$C_{16}H_{34}N_4O_{10}U$	$UO_2(NO_3)_2 \cdot 2\ CH_3CON(C_3H_7)_2$	U: SVol.E1-107/9
$C_{16}H_{34}NpO_6P$	$NpO_2(C_4H_9CH(C_2H_5)CH_2)_2PO_4$	Np: TrU.D2-283
$C_{16}H_{34}OSSn$	$(C_4H_9)_3SnCH_2CH_2SCOCH_3$	Sn: Org.Verb.2-284
$C_{16}H_{34}OS_2Sn$	$(C_4H_9)_3Sn[S_2COC_3H_7]$	C: MVol.D4-253
$C_{16}H_{34}OSn$	$(C_2H_5)_3SnOC_{10}H_{19}$	Sn: Org.Verb.5-105
–	$(C_4H_9)_3SnCHCHOC_2H_5$	Sn: Org.Verb.2-309/10
–	$(C_4H_9)_3SnCH(CH_3)COCH_3$	Sn: Org.Verb.2-284, 299
–	$(C_4H_9)_3SnCH_2CH_2CH(O)CH_2$	Sn: Org.Verb.2-284, 300
–	$(C_4H_9)_3SnCH_2COC_2H_5$	Sn: Org.Verb.2-284
–	$[(C_8H_{17})_2SnO]_x$	Sn: Org.Verb.6-113
–	$[(i\text{-}C_8H_{17})_2SnO]_x$	Sn: Org.Verb.6-125
$C_{16}H_{34}O_2Sn$	$(C_2H_5)_3SnCHCHCH(CH_3)OCH(CH_3)OC_4H_9$	Sn: Org.Verb.2-201
–	$(C_2H_5)_3SnCHCHC_2H_4OCH(CH_3)OC_4H_9$	Sn: Org.Verb.2-201
–	$(C_4H_9)_3SnCH_2CH_2COOCH_3$	Sn: Org.Verb.2-285, 300
–	$(C_4H_9)_3SnCH_2CH_2OOCCH_3$	Sn: Org.Verb.2-285
–	$(C_4H_9)_3SnCH_2COOC_2H_5$	Sn: Org.Verb.2-284, 300
–	$(C_4H_9)_3SnOCOC_3H_7$	Sn: Org.Verb.5-133
$C_{16}H_{34}Si_3Sn$	$(CH_3)_3SnC_6H_4Si(CH_3)[Si(CH_3)_3]_2$	Sn: Org.Verb.2-143, 145

Formula	Compound	Reference
$C_{16}H_{34}Sn$	$[(CH_3)_3CCH_2]_2Sn(CH_2)_6$	Sn: Org.Verb.3-109
–	$(C_4H_9)_3SnCH_2C(CH_3)CH_2$	Sn: Org.Verb.2-309/10
–	$(C_4H_9)_3SnCH_2CHCHCH_3$	Sn: Org.Verb.2-308/9
–	$(C_4H_9)_3SnCH_2CH_2CHCH_2$	Sn: Org.Verb.2-308
$C_{16}H_{35}NNiP_2$	$C_3H_5Ni(P(C_2H_5)_3)_2CN$	Ni: Org.Verb.2-22/3
$C_{16}H_{35}NOSn$	$(C_4H_9)_3SnCH_2CON(CH_3)_2$	Sn: Org.Verb.2-285
$C_{16}H_{35}NSn$	$C_2H_5(i\text{-}C_3H_7)Sn(C_6H_{11})((CH_2)_3N(CH_3)_2)$	Sn: Org.Verb.3-99/100
$C_{16}H_{35}NiP$	$(CH_3)_2C_3H_3Ni(CH_3)CH_2P(C_3H_7\text{-}i)_3$	Ni: Org.Verb.2-41
–	$(CH_3)_2C_3H_3Ni[P(C_3H_7\text{-}i)_2C_4H_9\text{-}t]CH_3$	Ni: Org.Verb.2-36
$C_{16}H_{36}I_4N_4O_4U$	$UI_4 \cdot 4\ CH_3CON(CH_3)_2$	U: SVol.E1-103/5
$C_{16}H_{36}N_2O_{10}S_2U$	$UO_2(NO_3)_2 \cdot 2\ (C_4H_9)_2SO$	U: SVol.E1-211, 215
$C_{16}H_{36}N_2O_{10}U$	$UO_2(NO_3)_2 \cdot 2\ (C_4H_9)_2O$	U: SVol.E1-71, 76
$C_{16}H_{36}N_4O_4P_4S_4U$	$U(NCS)_4 \cdot 4\ (CH_3)_3PO$	U: SVol.E1-148, 156
$C_{16}H_{36}N_4O_{11}Pu$	$(C_4H_9)_4NPuO_2(NO_3)_3$	Np: TrU.D2-229
$C_{16}H_{36}N_{10}O_2P_2S_4Sn$	$Sn(SCN)_4 \cdot 2\ ((CH_3)_2N)_3PO$	Sn: MVol.C6-171/2
$C_{16}H_{36}OSn$	$(C_4H_9)_3SnCH(CH_3)CH_2CH_2OH$	Sn: Org.Verb.2-285
–	$(C_4H_9)_3SnCH_2CH(CH_3)CH_2OH$	Sn: Org.Verb.2-285
–	$(C_4H_9)_3SnCH_2CH(OH)C_2H_5$	Sn: Org.Verb.2-285
–	$(C_4H_9)_3SnCH_2CH_2CH(OH)CH_3$	Sn: Org.Verb.2-285
–	$(C_4H_9)_3SnCH_2CH_2CH_2CH_2OH$	Sn: Org.Verb.2-285
–	$(C_4H_9)_3SnCH_2CH_2OC_2H_5$	Sn: Org.Verb.2-286
–	$(C_4H_9)_3SnOC(CH_3)_3$	Sn: Org.Verb.5-131
$C_{16}H_{36}O_2Sn$	$(C_4H_9)_2Sn(OC_4H_9)_2$	Sn: Org.Verb.6-91
–	$(C_4H_9)_3SnOOC_4H_9$	Sn: Org.Verb.5-131
–	$(C_5H_{11})_2Sn(OC_3H_7)_2$	Sn: Org.Verb.6-108
$C_{16}H_{36}O_3Sn$	$C_4H_9Sn(OC_4H_9)_3$	Sn: Org.Verb.6-246
$C_{16}H_{36}O_3Ti$	$C_4H_9Ti(OC_4H_9)_3$	Ti: Org.Verb.1-45
–	$(i\text{-}C_4H_9)Ti(OC_4H_9)_3$	Ti: Org.Verb.1-47, 50
$C_{16}H_{36}O_4Sn$	$(C_4H_9)_2Sn(OOC_4H_9)_2$	Sn: Org.Verb.6-89, 94
$C_{16}H_{36}O_6Sn$	$C_4H_9Sn(OOC_4H_9)_3$	Sn: Org.Verb.6-246
$C_{16}H_{36}O_6Sn_3$	$(CH_3COOSn(C_2H_5)_2O)_2Sn(C_2H_5)_2$	Sn: Org.Verb.6-58
$C_{16}H_{36}Sn$	$(CH_3)_2Sn(C_4H_9)(C_{10}H_{21})$	Sn: Org.Verb.3-77
–	$[(CH_3)_3CCH_2]_3SnCH_3$	Sn: Org.Verb.2-331
–	$(C_2H_5)_3SnC_{10}H_{21}$	Sn: Org.Verb.2-160
–	$[C_3H_7CH(CH_3)]_3SnCH_3$	Sn: Org.Verb.2-329, 331
–	$(C_4H_9)_2Sn(i\text{-}C_4H_9)_2$	Sn: Org.Verb.3-46
–	$(C_4H_9)_2Sn(s\text{-}C_4H_9)_2$	Sn: Org.Verb.3-46
–	$(C_4H_9)_2Sn(t\text{-}C_4H_9)_2$	Sn: Org.Verb.3-47, 54
–	$(C_4H_9)_3SnCH(CH_3)C_2H_5$	Sn: Org.Verb.2-273/4
–	$(C_4H_9)_3Sn(i\text{-}C_4H_9)$	Sn: Org.Verb.2-274
–	$(i\text{-}C_4H_9)_3SnC_4H_9$	Sn: Org.Verb.2-327
–	$(s\text{-}C_4H_9)_3SnC_4H_9$	Sn: Org.Verb.2-326
–	$(t\text{-}C_4H_9)_3SnC_4H_9$	Sn: Org.Verb.2-329
–	$(i\text{-}C_5H_{11})_3SnCH_3$	Sn: Org.Verb.2-330
–	$(C_8H_{17})_2SnH_2$	Sn: Org.Verb.4-112
–	$(i\text{-}C_8H_{17})_2SnH_2$	Sn: Org.Verb.4-112
–	$Sn(n\text{-}C_4H_9)_4$	
	Chemical reactions	Sn: Org.Verb.1-103/8
	Formation	Sn: Org.Verb.1-94/8
	Molecule	Sn: Org.Verb.1-98/100

$C_{16}H_{36}Sn$	$Sn(n\text{-}C_4H_9)_4$	
	Physical properties	Sn: Org.Verb.1-100/3
	Physiology	Sn: Org.Verb.1-108/9
	Preparation	Sn: Org.Verb.1-94/8
	Spectra	Sn: Org.Verb.1-98/100
	Uses	Sn: Org.Verb.1-109/10
–	$^{113}Sn(C_4H_9)_4$	Sn: Org.Verb.1-94
–	$Sn(i\text{-}C_4H_9)_4$	Sn: Org.Verb.1-110/3
–	$Sn(s\text{-}C_4H_9)_4$	Sn: Org.Verb.1-113
–	$Sn(C_4H_9)_4$ solutions	
	$Sn(C_4H_9)_4$-organic solvents	Sn: Org.Verb.1-108
$C_{16}H_{36}Ti$	$(C_4H_9)_4Ti$	Ti: Org.Verb.1-90, 98
$C_{16}H_{37}NO_3S$	$HNCS \cdot 3\ C_5H_{11}OH$	C: MVol.D6-35
$C_{16}H_{37}NSn$	$(C_4H_9)_3SnN(C_2H_5)_2$	Sn: Org.Verb.5-129, 131
–	$(i\text{-}C_4H_9)_3SnN(C_2H_5)_2$	Sn: Org.Verb.5-138
$C_{16}H_{37}N_3O_4S_2$	$HN(SO_2N(CH_2CH(CH_3)_2)_2)_2$	S: S-N-Verb.1-161
–	$HN(SO_2N(C_4H_9)_2)_2$	S: S-N-Verb.1-161
$C_{16}H_{37}N_3O_{17}P_2Pu$	$Pu(NO_3)_3((C_4H_9)_2PO_4)_2H$	Np: TrU.D2-280
$C_{16}H_{37}N_3Ti$	$CH(CH_3)C(CH_3)Ti(N(C_2H_5)_2)_3$	Ti: Org.Verb.1-54, 58
–	$CH_2C(CH_3)CH_2Ti(N(C_2H_5)_2)_3$	Ti: Org.Verb.1-53, 57
–	$CH_2CHCH(CH_3)Ti(N(C_2H_5)_2)_3$	Ti: Org.Verb.1-53, 57/8
–	$CH_3CHCHCH_2Ti(N(C_2H_5)_2)_3$	Ti: Org.Verb.1-53, 57/8
–	$(CH_3)_2CCHTi(N(C_2H_5)_2)_3$	Ti: Org.Verb.1-54, 58
$C_{16}H_{37}NiO_6P_2^+$	$[CH_3C_3H_4Ni(P(OC_2H_5)_3)_2]^+$	Ni: Org.Verb.2-24
$C_{16}H_{38}INSn$	$[(C_4H_9)_3SnCH_2N(CH_3)_3]I$	Sn: Org.Verb.2-286
$C_{16}H_{38}N_2S_4Sn$	$Sn(SCH_2CH_2S)_2 \cdot 2\ (C_2H_5)_3N$	Sn: MVol.C6-82/3
$C_{16}H_{38}N_2Sn$	$(i\text{-}C_4H_9)_2Sn[N(C_2H_5)_2]_2$	Sn: Org.Verb.6-105
$C_{16}H_{38}N_4SSi_2$	$[(CH_3)_3SiN]_2S(NC_5H_{10})_2$	S: S-N-Verb.1-211
$C_{16}H_{38}N_4SSi_2^+$	$[(CH_3)_3SiN]_2S(NC_5H_{10})_2^+$	S: S-N-Verb.1-211
$C_{16}H_{38}N_5O_3PS_4U$	$UO_2(S_2CN(C_2H_5)_2)_2 \cdot [(CH_3)_2N]_3PO$	U: SVol.E1-163, 165, 167
$C_{16}H_{38}N_5O_3PSe_4U$	$UO_2(Se_2CN(C_2H_5)_2)_2 \cdot [(CH_3)_2N]_3PO$	U: SVol.E1-163, 165, 167
$C_{16}H_{38}NiP_2$	$C_3H_5(CH_3)Ni(P(C_2H_5)_3)_2$	Ni: Org.Verb.1-108
$C_{16}H_{38}O_4S_2Sn$	$[(C_4H_9)_3SnCH_2S(CH_3)_2]CH_3SO_4$	Sn: Org.Verb.2-286
$C_{16}H_{38}O_9U_2$	$U_2O_5(OC_4H_9)_2 \cdot 2\ C_4H_9OH$	U: SVol.E1-64
$C_{16}H_{38}Sn_2$	$[(C_2H_5)_2(C_4H_9)Sn]_2$	Sn: Org.Verb.5-41
$C_{16}H_{39}N_3Ti$	$(n\text{-}C_4H_9)Ti(N(C_2H_5)_2)_3$	Ti: Org.Verb.1-52
–	$(t\text{-}C_4H_9)Ti(N(C_2H_5)_2)_3$	Ti: Org.Verb.1-52
$C_{16}H_{40}I_4MnN_2$	$[(C_2H_5)_4N]_2MnI_4$	Mn: MVol.C5-322/3, 324
$C_{16}H_{40}I_6N_2Pa$	$[N(C_2H_5)_4]_2PaI_6$	Pa: SVol.2-42, 44, 72/3
$C_{16}H_{40}I_6N_2Sn$	$[(C_2H_5)_4N]_2[SnI_6]$	Sn: MVol.C3-120
$C_{16}H_{40}I_6N_2Te$	$[(C_2H_5)_4N]_2TeI_6$	Te: SVol.B3-67, 72
$C_{16}H_{40}I_6N_2U$	$[N(C_2H_5)_4]_2UI_6$	U: SVol.C9-177
$C_{16}H_{40}MnN_{14}$	$[(C_2H_5)_4N]_2[Mn(N_3)_4]$	Mn: MVol.C3-263
$C_{16}H_{40}N_2O_{12}U$	$UO_2(NO_3)_2 \cdot 4\ (C_2H_5)_2O \cdot 2\ H_2O$	U: SVol.E1-71, 76
–	$UO_2(NO_3)_2 \cdot 4\ C_4H_9OH \cdot 2\ H_2O$	U: SVol.E1-59, 63
$C_{16}H_{40}N_2O_{19}W_6$	$[N(C_2H_5)_4]_2W_6O_{19}$	W: SVol.B3-241/2
$C_{16}H_{40}N_8NpO_{18}$	$[(C_2H_5)_4N]_2Np(NO_3)_6$	Np: TrU.D2-421
$C_{16}H_{40}N_8O_{18}Pu$	$[(C_2H_5)_4N]_2Pu(NO_3)_6$	Np: TrU.D1-140
		Np: TrU.D2-421

Formula	Compound	Reference
$C_{16}H_{40}N_{20}Sn$	$[(C_2H_5)_4N]_2[Sn(N_3)_6]$	Sn: MVol.C3-83
$C_{16}H_{40}Ni_2P_4$	$Ni(CH_2P(CH_3)_2CH_2)_4Ni$	Ni: Org.Verb.2-269
$C_{16}H_{40}Si_2Sn$	$(CH_3)_2Sn[CH_2Si(CH_3)_2C_4H_9]_2$	Sn: Org.Verb.3-12, 19
–	$(C_4H_9)_2Sn[CH_2Si(CH_3)_3]_2$	Sn: Org.Verb.3-47, 54
$C_{16}H_{41}NiO_{13}P_3S$	$[(CH_3)_4C_3HNi(P(OCH_3)_3)_3]HSO_4$	Ni: Org.Verb.2-27
$C_{16}H_{44}I_4N_4Sn$	$SnI_4 \cdot 4\ C_4H_9NH_2$	Sn: MVol.C5-169
$C_{16}H_{44}Ni_2P_4$	$[CH_3((CH_3)_3P)Ni(CH_2P(CH_3)_2CH_2)]_2$	Ni: Org.Verb.2-269
$C_{16}H_{44}Si_4Sn$	$Sn[CH_2Si(CH_3)_3]_4$	Sn: Org.Verb.1-123
$C_{16}H_{44}Si_4Ti$	$[(CH_3)_3SiCH_2]_4Ti$	Ti: Org.Verb.1-93, 105/6
$C_{16}H_{44}Sn_4Ti$	$[(CH_3)_3SnCH_2]_4Ti$	Ti: Org.Verb.1-95, 108
$C_{16}H_{46}O_8Si_6Sn$	$[(OSi(CH_3)_2)_6OSn(C_2H_5)_2O]_x$	Sn: Org.Verb.6-53
$C_{16}H_{48}I_4N_8Sn$	$SnI_4 \cdot 4\ NH_2(CH_2)_4NH_2$	Sn: MVol.C5-185
$C_{16}H_{48}N_4O_{12}S_4Sn$	$[(CH_3)_4N]_4Sn(SO_3)_4$	Sn: MVol.C3-122
$C_{16}H_{48}N_8O_8S_4$	$(C_4H_9NH_3)_4[NSO_2]_4$	S: S-N-Verb.1-4
$C_{16}H_{58}O_{22}Sn_7$	$[(C_2H_5Sn)_7(OCOCH_3)(OH)_{20}]$	Sn: Org.Verb.6-230
$C_{16}H_{63}O_{23}Sn_8{}^+$	$[(C_2H_5Sn)_8(OH)_{23}]^+$	Sn: Org.Verb.6-230
$C_{16}H_{64}N_{16}Na_4Sn_9$	$Na_4Sn_9 \cdot 8\ C_2H_8N_2$	Sn: MVol.C5-13/4
$C_{16}Mo_2Ni_3O_{16}{}^{2-}$	$[Ni_3Mo_2(CO)_{16}]^{2-}$	Ni: Org.Verb.2-374/6
$C_{16}Ni_3O_{16}W_2{}^{2-}$	$[Ni_3W_2(CO)_{16}]^{2-}$	Ni: Org.Verb.2-374, 376
$C_{16.25}Cl_{2.18}H_{16.25}IrN_{3.25}$	$Ir(C_5H_5N)_{3.25}Cl_{2.18}$	Ir: SVol.2-74
$C_{16.5}Cl_4H_{33}N_3O_3U$	$UCl_4 \cdot 1.5\ (CH_3)_2C(CH_2CON(CH_3)_2)_2$	U: SVol.E1-112/4
$C_{16.5}H_{10.5}N_{0.5}O_6Sn$	$SnO(OH)(C_{14}H_7O_4) \cdot 0.5\ C_5H_5N$	Sn: MVol.C5-113
$C_{17}CdH_{35}NO_3S_6$	$[(C_2H_5)_4N][Cd(S_2COC_2H_5)_3]$	C: MVol.D4-245, 258
$C_{17}ClFeH_{13}O$	$C_5H_5FeC_5H_4COC_6H_4Cl$	Fe: Org.Verb.A2-285/8
–	$C_5H_5FeC_5H_4C_6H_4COCl$	Fe: Org.Verb.A3-160
$C_{17}ClFeH_{13}O_2$	$C_5H_5FeC_5H_4OOCC_6H_4Cl$	Fe: Org.Verb.A2-94
$C_{17}ClFeH_{14}O_5$	$[C_5H_5FeC_5H_4COC_6H_5]ClO_4$	Fe: Org.Verb.A2-301
$C_{17}ClFeH_{15}$	$C_5H_5FeC_5H_4CHClC_6H_5$	Fe: Org.Verb.A1-394
–	$C_5H_5FeC_5H_4CH_2C_6H_4Cl$	Fe: Org.Verb.A1-388/9
$C_{17}ClFeH_{15}O_4$	$[C_5H_5FeC_5H_4CHC_6H_5]ClO_4$	Fe: Org.Verb.A1-350/1, 353
$C_{17}ClFeH_{16}N$	$C_5H_5FeC_5H_4CH_2NHC_6H_4Cl$	Fe: Org.Verb.A2-20
$C_{17}ClFeH_{17}O$	$C_5H_5FeC_5H_4COC_6H_8Cl$	Fe: Org.Verb.A2-290
$C_{17}ClFeH_{17}O_5$	$[C_5H_5FeC_5H_4C(CHCCH_3)_2O]ClO_4$	Fe: Org.Verb.A3-177
$C_{17}ClFeH_{21}O$	$C_5H_5FeC_5H_4CH_2CH(C_4H_9)COCl$	Fe: Org.Verb.A3-160
$C_{17}ClFeH_{22}MgO_7P$	$MgCl[C_6H_5CH_2C(O)Fe(CO)_3P(OC_2H_5)_3]$	Fe: Org.Verb.B4-142
$C_{17}ClFeH_{22}NO_2$	$[C_5H_5FeC_5H_4COC_2H_4NH(C_2H_4)_2O]Cl$	Fe: Org.Verb.A2-227/8
$C_{17}ClFeH_{24}NO$	$C_5H_5FeC_5H_4CO(CH_2)_2N(C_2H_5)_2 \cdot HCl$	Fe: Org.Verb.A2-228
$C_{17}ClFeH_{34}NO_9P_2$	$COFe(NO)(P(OC_2H_5)_3)_2COCH_2CClCH_2$	Fe: Org.Verb.B1-40/2
$C_{17}ClFe_2GeH_{25}N_4O_4$	$((CH_3)_2C_3H_3Fe(NO)_2)_2Ge(Cl)CH_2C_6H_5$	Fe: Org.Verb.C2-165
–	$((CH_3)_2C_3H_3Fe(NO)_2)_2Ge(Cl)C_6H_4CH_3$	Fe: Org.Verb.C2-165
$C_{17}ClFe_2H_{25}N_4O_4Pb$	$((CH_3)_2C_3H_3Fe(NO)_2)_2Pb(Cl)CH_2C_6H_5$	Fe: Org.Verb.C2-165
–	$((CH_3)_2C_3H_3Fe(NO)_2)_2Pb(Cl)C_6H_4CH_3$	Fe: Org.Verb.C2-165
$C_{17}ClFe_2H_{25}N_4O_4Sn$	$((CH_3)_2C_3H_3Fe(NO)_2)_2Sn(Cl)CH_2C_6H_5$	Fe: Org.Verb.C2-165
–	$((CH_3)_2C_3H_3Fe(NO)_2)_2Sn(Cl)C_6H_4CH_3$	Fe: Org.Verb.C2-165
$C_{17}ClH_{13}N_2Ni$	$C_5H_5NiC_6H_4NNC_6H_4Cl$	Ni: Org.Verb.2-159
$C_{17}ClH_{13}N_2O_6Ti$	$C_5H_5Ti(OC_6H_4NO_2)_2Cl$	Ti: Org.Verb.1-185/6
$C_{17}ClH_{15}IrN_5S_2$	$Ir(NCS)_2Cl \cdot 3\ C_5H_5N$	Ir: SVol.2-162
$C_{17}ClH_{15}N_2Ni$	$CH_3C_6H_4Ni(NC_5H_4C_5H_4N)Cl$	Ni: Org.Verb.1-38, 45/6

Formula	Compound	Reference
$C_{17}ClH_{15}N_2Ni$	$C_6H_5CH_2Ni(NC_5H_4C_5H_4N)Cl$	Ni: Org.Verb.1-18, 24
$C_{17}ClH_{15}N_2NiO_2S$	$(CH_3C_6H_4SO_2)Ni(NC_5H_4C_5H_4N)Cl$	Ni: Org.Verb.1-148
$C_{17}ClH_{15}O_2Ti$	$C_5H_5Ti(OC_6H_5)_2Cl$	Ti: Org.Verb.1-185/6
$C_{17}ClH_{15}Sn$	$(C_6H_5)_2(C_5H_5)SnCl$	Sn: Org.Verb.5-216, 219
$C_{17}ClH_{16}NO_2Sn$	$CH_3ClSn(OC_6H_4NCH_2C_2H_2COC_6H_5)$	Sn: Org.Verb.6-218/9
$C_{17}ClH_{18}N_3Ti^+$	$[C_5H_5Ti(Cl)(CH_3CN)(C_5H_5N)_2]^+$	Ti: Org.Verb.1-133
$C_{17}ClH_{19}OSn$	$(C_6H_5)_2(CH_3CO(CH_2)_3)SnCl$	Sn: Org.Verb.5-216, 219
$C_{17}ClH_{20}NOSn$	$(C_6H_5)_2[CH_3CONH(CH_2)_3]SnCl$	Sn: Org.Verb.5-216, 219
–	$(C_6H_5)_2[HONC(CH_3)(CH_2)_3]SnCl$	Sn: Org.Verb.5-216, 219
$C_{17}ClH_{21}N_2NiS$	$C_3H_5Ni[SC(NHCH(CH_3)_2)(NHC_{10}H_7)]Cl$	Ni: Org.Verb.2-275
$C_{17}ClH_{21}Sn$	$(CH_3)(C_6H_5)(C_6H_5C(CH_3)_2CH_2)SnCl$	Sn: Org.Verb.5-225, 227
–	$(C_4H_9)(C_6H_5)(C_6H_5CH_2)SnCl$	Sn: Org.Verb.5-226
$C_{17}ClH_{22}OP_2RhS_2$	$[(CH_3)_2P(C_6H_5)]_2RhCl(S_2CO)$	C: MVol.D4-208
$C_{17}ClH_{23}O_{10}Ti$	$C_5H_5Ti(CH_3COCHCOOC_2H_5)_2ClO_4$	Ti: Org.Verb.1-192, 194
$C_{17}ClH_{27}OSn$	$(i\text{-}C_4H_9)_2(CH_3OC_6H_4CHCH)SnCl$	Sn: Org.Verb.5-213, 215
$C_{17}ClH_{30}N_3NiO$	$(t\text{-}C_4H_9NC)_3Ni(COCH_3)Cl$	Ni: Org.Verb.1-323
$C_{17}ClH_{30}N_3NiO_2$	$(t\text{-}C_4H_9NC)_3Ni(COOCH_3)Cl$	Ni: Org.Verb.1-323
$C_{17}ClH_{32}NSn$	$(C_4H_9)_3SnCl \cdot C_5H_5N$	Sn: Org.Verb.5-134
$C_{17}ClH_{32}NiP$	$C_5H_5Ni(P(C_4H_9\text{-}n)_3)Cl$	Ni: Org.Verb.2-117/8
$C_{17}ClH_{35}NiP_2$	$[C_5H_5Ni(P(C_2H_5)_3)_2]Cl$	Ni: Org.Verb.2-136, 138
$C_{17}ClH_{37}N_4NiS_2$	$C_3H_5Ni(SC(NHC_3H_7)_2)_2Cl$	Ni: Org.Verb.2-21
$C_{17}ClH_{37}Sn$	$(C_8H_{17})_2(CH_3)SnCl$	Sn: Org.Verb.5-221
$C_{17}ClH_{39}P_2Sn$	$(CH_3)ClSn[P(t\text{-}C_4H_9)_2]_2$	Sn: Org.Verb.6-218
$C_{17}ClH_{45}Si_4Sn$	$[((CH_3)_3Si)_2CH]_2(i\text{-}C_3H_7)SnCl$	Sn: Org.Verb.5-217, 222
$C_{17}Cl_2FH_8N_3O_6$	$(NO_2C_6H_4O)_2C_5FCl_2N$	F: PerFHalOrg.5-143
$C_{17}Cl_2FH_{10}NS_2$	$(C_6H_5S)_2C_5FCl_2N$	F: PerFHalOrg.5-141
$C_{17}Cl_2F_2H_9N_5S$	$NC_5F_2Cl_2NHC_4HN_2(CN)SCH_2C_6H_5$	F: PerFHalOrg.5-216
$C_{17}Cl_2F_2H_{27}N_3$	$N(CFCCl)_2CNH(CH_2)_{12}NH_2$	F: PerFHalOrg.5-141
$C_{17}Cl_2FeH_{14}$	$C_5H_5FeC_5H_4CCl_2C_6H_5$	Fe: Org.Verb.A1-394
$C_{17}Cl_2GeH_{36}OSn$	$(C_4H_9)_2[(C_2H_5)_2ClGeOC(CH_3)_2C_2H_2]SnCl$	Sn: Org.Verb.5-211
$C_{17}Cl_2H_{12}N_2Ni$	$C_5H_5Ni(ClC_6H_3NNC_6H_4Cl)$	Ni: Org.Verb.2-159
$C_{17}Cl_2H_{13}NO_4U$	$UO_2(OC_6H_4Cl)_2 \cdot C_5H_5N$	U: SVol.E1-29, 35
$C_{17}Cl_2H_{15}NOSn$	$(C_6H_5)_2SnCl_2 \cdot C_5H_5NO$	Sn: Org.Verb.6-174, 177
$C_{17}Cl_2H_{15}NSn$	$(C_6H_5)_2SnCl_2 \cdot C_5H_5N$	Sn: Org.Verb.6-174, 176
$C_{17}Cl_2H_{16}IrN_9S$	$[Ir(C_4H_4N_2)_4Cl_2]CNS$	Ir: SVol.2-77
$C_{17}Cl_2H_{16}N_2Sn$	$(CH_3)(C_6H_5)SnCl_2 \cdot NC_5H_4C_5H_4N$	Sn: Org.Verb.6-195
$C_{17}Cl_2H_{18}N_2Sn$	$(CH_3)(C_6H_5)SnCl_2 \cdot 2\ C_5H_5N$	Sn: Org.Verb.6-195
$C_{17}Cl_2H_{20}OSSn$	$(C_6H_5)_2SnCl_2 \cdot (CH_2)_5SO$	Sn: Org.Verb.6-174, 178
$C_{17}Cl_2H_{21}NOSn$	$(C_6H_5)_2SnCl_2 \cdot C_2H_5C(O)N(CH_3)_2$	Sn: Org.Verb.6-174/5
–	$(C_6H_5)_2SnCl_2 \cdot HC(O)N(C_2H_5)_2$	Sn: Org.Verb.6-174/5
$C_{17}Cl_2H_{22}KOP_2RhS_2$	$K[(C_6H_5P(CH_3)_2)_2RhCl_2(S_2CO)]$	C: MVol.D4-208
$C_{17}Cl_2H_{24}O_2S_2Sn$	$(C_6H_5)(C_6H_5CH_2)SnCl_2 \cdot 2\ (CH_3)_2SO$	Sn: Org.Verb.6-205
$C_{17}Cl_2H_{30}N_2Sn$	$(CH_3)(C_6H_5)SnCl_2 \cdot 2\ NH(CH_2)_5$	Sn: Org.Verb.6-195
$C_{17}Cl_2H_{34}Sn$	$(C_4H_9)_3SnCH_2CHCHCH_2CHCl_2$	Sn: Org.Verb.2-311
$C_{17}Cl_2H_{44}IrO_6P_3$	$Ir(P(t\text{-}C_4H_9)_2(n\text{-}C_3H_7))(P(OCH_3)_3)_2HCl_2$	Ir: SVol.2-197/8
$C_{17}Cl_3F_3H_{12}N_2Sn$	$CF_3C_6H_4SnCl_3 \cdot NC_5H_4C_5H_4N$	Sn: Org.Verb.6-283
$C_{17}Cl_3GeH_{32}NiP$	$C_5H_5Ni(P(C_4H_9\text{-}n)_3)GeCl_3$	Ni: Org.Verb.2-130
$C_{17}Cl_3GeH_{35}NiP_2$	$[C_5H_5Ni(P(C_2H_5)_3)_2][GeCl_3]$	Ni: Org.Verb.2-137
$C_{17}Cl_3H_{13}O_5Sn$	$SnCl_3(C_7H_5O_2C_2H_2COC_6H_3(OCH_3)O) \cdot 0.5\ C_6H_6$	Sn: MVol.C5-91/2

Formula	Compound		Reference
$C_{17}Cl_3H_{14}IrNP$	$Ir(C_5H_4N(P(C_6H_5)_2))Cl_3 \cdot H_2O$	Ir:	SVol.2-234
$C_{17}Cl_3H_{15}O_4Sn$	$SnCl_3(OC_6H_3(OCH_3)COC_2H_2C_6H_4OCH_3) \cdot 0.25\ C_6H_6$	Sn:	MVol.C5-91/2
$C_{17}Cl_3H_{17}N_2NiO_2Sn$	$CH_3SnCl_3 \cdot Ni(OC_6H_4CHNCH_2)_2$	Sn:	Org.Verb.6-222/3
$C_{17}Cl_3H_{17}N_2Sn$	$CH_3C_6H_4SnCl_3 \cdot 2\ C_5H_5N$	Sn:	Org.Verb.6-282
$C_{17}Cl_3H_{17}O_2Sn$	$SnCl_3(OC_6H_3(t\text{-}C_4H_9)COC_6H_5) \cdot 0.5\ C_6H_6$	Sn:	MVol.C5-90
$C_{17}Cl_3H_{19}N_2O_2Sn$	$CH_3SnCl_3 \cdot C_2H_4(NCHC_6H_4OH)_2$	Sn:	Org.Verb.6-222/3
$C_{17}Cl_3H_{25}N_2Sn$	$C_3H_7SnCl_3 \cdot 2\ C_6H_5CH_2NH_2$	Sn:	Org.Verb.6-236
$C_{17}Cl_3H_{27}N_2OSSn$	$[(C_2H_5)_3NH][SnCl_3(SC_6H_4NC(CH_3)CHCOCH_3)]$	Sn:	MVol.C6-90/1
$C_{17}Cl_3H_{32}NiPSn$	$C_5H_5Ni(P(C_4H_9\text{-}n)_3)SnCl_3$	Ni:	Org.Verb.2-130
$C_{17}Cl_3H_{35}NiP_2Sn$	$[C_5H_5Ni(P(C_2H_5)_3)_2]SnCl_3$	Ni:	Org.Verb.2-137, 139
$C_{17}Cl_4F_3H_8N_3O$	$NC_5F_2Cl_2NHC_5FCl_2(OCH_2C_6H_5)N$	F:	PerFHalOrg.5-138/9
$C_{17}Cl_4H_{10}OSn$	$SnCl_4 \cdot C_{17}H_{10}O$	Sn:	MVol.C5-99
$C_{17}Cl_4H_{10}OTe$	$TeCl_4 \cdot C_{17}H_{10}O$	Te:	SVol.B3-166
$C_{17}Cl_4H_{13}NOTe$	$TeCl_4 \cdot HOC_{10}H_6CHNC_6H_5$	Te:	SVol.B3-172, 174
$C_{17}Cl_4H_{14}OSn$	$SnCl_4 \cdot CO(CHCHC_6H_5)_2$	Sn:	MVol.C5-92/3
–	$SnCl_4 \cdot C_6H_5(CHCH)_2COC_6H_5$	Sn:	MVol.C5-91/2
$C_{17}Cl_4H_{17}NOSn$	$SnCl_4 \cdot (CH_3)_2NC_6H_4COCHCHC_6H_5$	Sn:	MVol.C6-5
$C_{17}Cl_4H_{28}N_2O_2Sn$	$SnCl_4 \cdot (C_4H_9)_2C(OH)CON_2H_2C_6H_4CH_3$	Sn:	MVol.C6-61
$C_{17}Cl_5H_{10}OU$	$UCl_5 \cdot C_{17}H_{10}O$	U:	SVol.E1-80/1
$C_{17}Cl_5H_{43}N_2Sn$	$[(C_2H_5)_4N]_2[CH_3SnCl_5]$	Sn:	Org.Verb.6-222
$C_{17}Cl_5H_{43}N_2Ti$	$[(C_2H_5)_4N]_2[CH_3TiCl_5]$	Ti:	Org.Verb.1-42/3
$C_{17}Cl_6H_{23}MnO_6$	$[Mn(CH(COCH_3)_2)_3] \cdot 2\ CHCl_3$	Mn:	MVol.D1-95
$C_{17}Cl_7H_{17}IrN_3$	$[Ir(C_5H_5N)_3Cl_3] \cdot C_2H_2Cl_4$	Ir:	SVol.2-69
$C_{17}Cl_7H_{19}IrN_3$	$[Ir(C_5H_5N)_3Cl_3] \cdot 2\ CH_2Cl_2$	Ir:	SVol.2-69
$C_{17}Cl_7H_{25}Ni_2$	$((CH_3)_4C_4NiCl_2)_2 \cdot CHCl_3$	Ni:	Org.Verb.2-328
$C_{17}Cl_8H_5S_4Ti^-$	$[C_5H_5Ti(S_2C_6Cl_4)_2]^-$	Ti:	Org.Verb.1-200
$C_{17}Cl_9H_{17}IrN_3$	$Ir(C_5H_5N)_3Cl_3 \cdot 2\ CHCl_3$	Ir:	SVol.2-69
$C_{17}CoF_5H_{15}IO$	$(CH_3)_5C_5Co(CO)(C_6F_5)I$	F:	PerFHalOrg.4-37
$C_{17}CoH_{15}Ni_2O_2$	$(C_5H_5Ni)_2(CO)_2CoC_5H_5$	Ni:	Org.Verb.2-361/2
$C_{17}Co_2FeH_{18}N_2O_7S$	$(CO)_3FeCo_2(CO)_4(CNC(CH_3)_3)_2S$	Fe:	Org.Verb.B1-199
$C_{17}Co_3FeH_{16}O_{11}P$	$HFeCo_3(CO)_{11}(P(C_2H_5)_3)$	Fe:	Org.Verb.B1-201/2
$C_{17}Co_3FeH_{19}O_{12}P^+$	$[HFeCo_3(CO)_9P(OC_3H_7)_2OC_2H_4]^+$	Fe:	Org.Verb.B1-204
$C_{17}Co_3FeH_{22}O_{11}P^+$	$[HFeCo_3(CO)_8P(OC_3H_7)_3]^+$	Fe:	Org.Verb.B1-204
$C_{17}CrF_{10}HO_5P$	$[(C_6F_5)_2PH]Cr(CO)_5$	F:	PerFHalOrg.3-27, 181
$C_{17}CrFeH_{12}O_6$	$C_5H_5FeC_5H_4C(OCH_3)Cr(CO)_5$	Fe:	Org.Verb.A1-345/6
$C_{17}CrFeH_{12}O_6^+$	$[C_5H_5FeC_5H_4C(OCH_3)Cr(CO)_5]^+$	Fe:	Org.Verb.A1-346
$C_{17}CrFeH_{17}NO_2^+$	$[C_5H_5FeC_5H_4C(NC_4H_8)Cr(CO)_2]^+$	Fe:	Org.Verb.A1-346
$C_{17}CrH_{16}O_3Sn$	$(CH_3)_2Sn(C_6H_5)(C_6H_5Cr(CO)_3)$	Sn:	Org.Verb.3-75, 80, 84
$C_{17}CrH_{26}O_6Sn$	$[(CH_3)_3C]_2SnCr(CO)_5 \cdot C_4H_8O$	Sn:	Org.Verb.6-107
$C_{17}FFeH_{13}O$	$C_5H_5FeC_5H_4COC_6H_4F$	Fe:	Org.Verb.A2-285/8
$C_{17}FFeH_{13}O^-$	$[C_5F_5FeC_5H_4COC_6H_4F]^-$	Fe:	Org.Verb.A2-302
$C_{17}FFeH_{15}$	$C_5H_5FeC_5H_4CH_2C_6H_4F$	Fe:	Org.Verb.A1-388/9
$C_{17}FH_{13}N_2Ni$	$C_5H_5NiC_6H_4NNC_6H_4F$	Ni:	Org.Verb.2-159
$C_{17}FH_{21}Sn$	$(CH_3)(C_6H_5)(C_6H_5C(CH_3)_2CH_2)SnF$	Sn:	Org.Verb.5-41
–	$(C_4H_9)(C_6H_5CH_2)(C_6H_5)SnF$	Sn:	Org.Verb.5-41
$C_{17}F_3FeH_{13}$	$C_5H_5FeC_5H_4C_6H_4CF_3$	Fe:	Org.Verb.A1-395
$C_{17}F_3FeH_{15}O_2^+$	$[C_5H_5FeC_5H_4C(OOCCF_3)CC(CH_3)_2]^+$	Fe:	Org.Verb.A2-96
$C_{17}F_3FeH_{15}O_2Pd$	$Pd(C_5H_5FeC_5H_4COCHCOCF_3)C_3H_5$	Fe:	Org.Verb.A3-37, 40

$C_{17}F_3FeH_{17}O_2$	$C_5H_5FeC_5H_4C(CH_2)(CH_2)_3OOCCF_3$	Fe:	Org.Verb.A2-104, 105
–	$C_5H_5FeC_5H_4C(CH_3)CH(CH_2)_2OOCCF_3$	Fe:	Org.Verb.A2-104/5
$C_{17}F_3FeH_{18}O_2^+$	$[C_5H_5FeC_5H_4C(CH_3)(CH_2)_3OOCCF_3]^+$	Fe:	Org.Verb.A2-82, 105
$C_{17}F_3H_{24}O_4PaS$	$PaO(C_4H_3SCOCHCOCF_3) \cdot (C_4H_9)_2CHOH$	Pa:	SVol.2-179
$C_{17}F_3H_{36}N_2P$	$CF_3P[N(n\text{-}C_4H_9)_2]_2$	F:	PerFHalOrg.3-102, 116
$C_{17}F_4FeH_{11}^+$	$[C_5H_5FeC_5H_4CH_2C_6F_4]^+$	Fe:	Org.Verb.A1-389
$C_{17}F_4FeH_{12}O$	$C_5H_5FeC_5H_4C_6F_4OCH_3$	Fe:	Org.Verb.A2-131, 135
$C_{17}F_5FeH_9O$	$C_5H_5FeC_5H_4COC_6F_5$	Fe:	Org.Verb.A2-285/8
$C_{17}F_5FeH_9O^+$	$[C_5H_5FeC_5H_4COC_6F_5]^+$	Fe:	Org.Verb.A2-287
$C_{17}F_5FeH_{11}$	$C_5H_5FeC_5H_4CH_2C_6F_5$	Fe:	Org.Verb.A1-388/9
$C_{17}F_5FeH_{11}O$	$C_5H_5FeC_5H_4CHOHC_6F_5$	Fe:	Org.Verb.A2-46, 50
$C_{17}F_5H_{13}Ti$	$(C_5H_5)_2Ti(CH_3)C_6F_5$	F:	PerFHalOrg.4-35
$C_{17}F_6FeH_{18}$	$C_5H_5FeC_5H_4CH(CH_2CH_2CF_3)_2$	Fe:	Org.Verb.A1-381, 383
$C_{17}F_6FeH_{18}O$	$C_5H_5FeC_5H_4C(CH_2CH_2CF_3)_2OH$	Fe:	Org.Verb.A2-56
$C_{17}F_6FeH_{30}O_7P_2S_2$	$COFe(P(OC_2H_5)_3)_2S_2C_2(CF_3)_2$	Fe:	Org.Verb.B1-34/6
$C_{17}F_6FeH_{30}O_7P_2S_2^+$	$[COFe(P(OC_2H_5)_3)_2S_2C_2(CF_3)_2]^+$	Fe:	Org.Verb.B1-34/5
$C_{17}F_6H_{10}N_2NiO$	$(C_6H_5NC)_2Ni(OC(CF_3)_2)$	Ni:	Org.Verb.1-320, 322
$C_{17}F_6H_{11}N$	$C_{14}H_9C(CF_3)_2NH_2$	F:	PerFHalOrg.7-69, 71
$C_{17}F_6H_{18}N_2O_6P_2$	$C_5F_6(P(O)(OH)_2(C_6H_5NH_2))_2$	F:	PerFHalOrg.3-50/1
$C_{17}F_7FeH_9^+$	$[C_5H_5FeC_5H_4C_6F_4CF_3]^+$	Fe:	Org.Verb.A1-394
$C_{17}F_8H_4N_6O_4$	$(NC_5F_4)_2CNNHC_6H_3(NO_2)_2$	F:	PerFHalOrg.5-217
$C_{17}F_{10}HMoO_5P$	$[(C_6F_5)_2PH]Mo(CO)_5$	F:	PerFHalOrg.3-27, 181
$C_{17}F_{10}HO_5PW$	$[(C_6F_5)_2PH]W(CO)_5$	F:	PerFHalOrg.3-27
$C_{17}F_{10}H_7O_2Tl$	$(C_6F_5)_2TlOC(CH_3)CHC(O)CH_3$	F:	PerFHalOrg.4-137, 144
$C_{17}F_{12}FeH_{10}O$	$C_5H_5FeC_5H_4CO(CF_2)_6H$	Fe:	Org.Verb.A2-244
$C_{17}F_{12}FeH_{12}O$	$C_5H_5FeC_5H_4CHOH(CF_2)_6H$	Fe:	Org.Verb.A2-44
$C_{17}F_{12}FeH_{15}O_6P$	$(CF_3)_4C_4COFe(CO)_2(P(OC_2H_5)_3)$	Fe:	Org.Verb.B1-163
$C_{17}F_{12}FeH_{27}N_5OP_2$	$[COFe(C_{10}H_{12}N_4(CH_3)_4)NCCH_3][PF_6]_2$	Fe:	Org.Verb.B1-87/8
$C_{17}F_{12}H_{19}N_3O_{13}$	$3\ ON(CF_2)_2COOH \cdot 2\ CH_3COOC_2H_5$	F:	PerFHalOrg.7-192
$C_{17}F_{13}FeH_9O$	$C_5H_5FeC_5H_4COC_6F_{13}$	Fe:	Org.Verb.A2-241/2
$C_{17}F_{13}H_4O_2Tl$	$(C_6F_5)_2TlOC(CF_3)CHC(O)CH_3$	F:	PerFHalOrg.4-137, 144
$C_{17}F_{13}Si^+$	$(C_6F_5)_2SiC_5F_3^+$	F:	PerFHalOrg.4-153
$C_{17}F_{16}HO_2Tl$	$(C_6F_5)_2TlOC(CF_3)CHC(O)CF_3$	F:	PerFHalOrg.4-137, 144
$C_{17}F_{23}N$	$(s\text{-}C_4F_9)_2C_9F_5N$	F:	PerFHalOrg.6-131/2, 145
–	$NC_5F_4(CCF_3)_5CFCF_3$	F:	PerFHalOrg.5-155/6, 167
$C_{17}F_{25}N$	$(C_6F_{11})_2C_5F_3N$	F:	PerFHalOrg.5-155/6, 168, 178
$C_{17}F_{29}N$	$((CF_3)_2CF)_4C_5FN$	F:	PerFHalOrg.5-151/3, 164
–	$(C_4F_9)_3C_5F_2N$	F:	PerFHalOrg.5-154, 166/7
$C_{17}F_{29}N_3O_2$	$(CF_3CF_2OCF_2CF_2)_2(C_6F_{11})C_3N_3$	F:	PerFHalOrg.6-86/8, 106
$C_{17}F_{30}N_4O_3$	$(C_2F_5OC_2F_4)_2(O(CF_2CF_2)_2NC_2F_4)C_3N_3$	F:	PerFHalOrg.6-86/8, 106
$C_{17}FeH_8N_2O_5$	$(CO)_5Fe(C_{12}H_8N_2)$	Fe:	Org.Verb.B3-250
$C_{17}FeH_{10}N_2O_3$	$(C_6H_5NC)_2Fe(CO)_3$	Fe:	Org.Verb.B4-18/20
$C_{17}FeH_{10}O_2$	$C_5H_5FeC_5H_4(CC)_3COOH$	Fe:	Org.Verb.A3-77/8
$C_{17}FeH_{10}O_6$	$C_6H_4C_5H_6(C(O))_2Fe(CO)_4$	Fe:	Org.Verb.B4-339, 349
$C_{17}FeH_{11}NO_4S$	$(CO)_4FeSC(NHC_6H_5)C_6H_5$	Fe:	Org.Verb.B2-128/30
$C_{17}FeH_{11}N_3O_2$	$(CO)_2Fe(NC_5H_3(C_5H_4N)_2)$	Fe:	Org.Verb.B1-96, 99
$C_{17}FeH_{12}MoO_5$	$C_5H_5(CO)_2Mo(C_7H_7)Fe(CO)_3$	Fe:	Org.Verb.B3-143
$C_{17}FeH_{12}N_2O_4S$	$(CO)_4FeSC(NH_2)N(C_6H_5)_2$	Fe:	Org.Verb.B2-128/30
$C_{17}FeH_{12}N_4$	$C_5H_5FeC_5H_4CH_3 \cdot C_2(CN)_4$	Fe:	Org.Verb.A1-278

$C_{17}FeH_{12}O$ .... $C_5H_5FeC_5H_3COC_6H_4$ .... Fe: Org.Verb.A3-127
$C_{17}FeH_{12}O^+$ .... $[C_5H_4FeC_5H_4C_6H_4CO]^+$ .... Fe: Org.Verb.A3-126
$C_{17}FeH_{13}^+$ .... $[FeC_5H_4C_6H_4C_6H_5]^+$ .... Fe: Org.Verb.A1-330
$C_{17}FeH_{13}O^+$ .... $[C_5H_5FeC_5H_4C_6H_4CO]^+$ .... Fe: Org.Verb.A3-21, 126
$C_{17}FeH_{13}O^{2+}$ .... $[C_5H_5FeC_5H_4C_6H_4CO]^{2+}$ .... Fe: Org.Verb.A3-21
$C_{17}FeH_{13}O_4P$ .... $(CO)_4FeP(C_6H_5)_2CH_3$ .... Fe: Org.Verb.B2-89, 111
$C_{17}FeH_{13}O_5P$ .... $(CO)_4FeP(C_6H_5)_2OCH_3$ .... Fe: Org.Verb.B2-84, 96, 117
$C_{17}FeH_{14}$ .... $C_5H_5FeC_5H_4CC_6H_5$ .... Fe: Org.Verb.A1-344
$C_{17}FeH_{14}N_2O$ ...... $C_5H_5FeC_5H_4CCHC(C_4H_3O)NNH$ .... Fe: Org.Verb.A3-164
$C_{17}FeH_{14}N_2O_2S$ ..... $C_5H_5FeC_5H_4(CH)_3C(CONH)_2CS$ .... Fe: Org.Verb.A2-177
$C_{17}FeH_{14}N_2O_3$ ...... $C_5H_5FeC_5H_4COCHCHC_4H_3(NO_2)N$ .... Fe: Org.Verb.A2-223
$C_{17}FeH_{14}N_2O_8$ ...... $(CO)_4Fe(C_3H_3N_2(COOCH_3)_2C_6H_5)$ .... Fe: Org.Verb.B2-71, 75, 78/9
$C_{17}FeH_{14}N_4O_4$ ...... $C_5H_5FeC_5H_4CHNNHC_6H_3(NO_2)_2$ .... Fe: Org.Verb.A2-166
$C_{17}FeH_{14}O$ .... $C_5H_5FeC_5H_4COC_6H_5$
Chemical reactions .... Fe: Org.Verb.A2-184/9, 275/84
Chromatography .... Fe: Org.Verb.A2-284/5
Electrochemical behavior .... Fe: Org.Verb.A2-274/5
Formation and preparation .... Fe: Org.Verb.A2-181/2, 268/71
Physical properties .... Fe: Org.Verb.A2-183/4, 271/4
Solubility .... Fe: Org.Verb.A2-273
Uses .... Fe: Org.Verb.A2-285

– .... $C_5H_5FeC_5H_4COC_6D_5$ .... Fe: Org.Verb.A2-270, 271
– .... $C_5D_5FeC_5D_4COC_6H_5$ .... Fe: Org.Verb.A2-270, 271
– .... $C_5H_5FeC_5H_4C_6H_4CHO$ .... Fe: Org.Verb.A2-174/5, 178
– .... $C_5H_5FeC_5H_4C_7H_5O$ .... Fe: Org.Verb.A3-43, 46
$C_{17}FeH_{14}O^+$ .... $[C_5H_5FeC_5H_4COC_6H_5]^+$ .... Fe: Org.Verb.A2-269, 274/5, 278, 301
$C_{17}FeH_{14}O^-$ .... $[C_5H_5FeC_5H_4COC_6H_5]^-$ .... Fe: Org.Verb.A2-274, 276, 302
$C_{17}FeH_{14}OS$ .... $C_5H_5FeC_5H_4C(CHO)CHC_4H_3S$ .... Fe: Org.Verb.A2-48, 173
– .... $C_5H_5FeC_5H_4COCHCHC_4H_3S$ .... Fe: Org.Verb.A2-223
$C_{17}FeH_{14}O_2$ .... $C_5H_5FeC_5H_4CHCHCOC_4H_3O$ .... Fe: Org.Verb.A3-162, 164
– .... $C_5H_5FeC_5H_4COCHCHC_4H_3O$ .... Fe: Org.Verb.A3-161/2, 164
– .... $C_5H_5FeC_5H_4COC_6H_4OH$ .... Fe: Org.Verb.A2-293/4
– .... $C_5H_5FeC_5H_4COOC_6H_5$ .... Fe: Org.Verb.A3-108, 112
– .... $C_5H_5FeC_5H_4C_6H_4COOH$ .... Fe: Org.Verb.A3-80/1, 103
– .... $C_5H_5FeC_5H_4OOCC_6H_5$ .... Fe: Org.Verb.A2-94/5
$C_{17}FeH_{14}O_2^+$ .... $[C_5H_5FeC_5H_4C_6H_4COOH]^+$ .... Fe: Org.Verb.A3-126
$C_{17}FeH_{14}O_2S$ .... $C_5H_5FeC_5H_4COCH_2COC_4H_3S$ .... Fe: Org.Verb.A2-225
$C_{17}FeH_{14}O_3$ .... $CH_3COC_5H_4FeC_5H_4COC_4H_3O$ .... Fe: Org.Verb.A3-162
– .... $C_5H_5FeC_5H_4COCH_2COC_4H_3O$ .... Fe: Org.Verb.A3-163, 164
– .... $C_5H_5FeC_5H_4COC_6H_3(OH)_2$ .... Fe: Org.Verb.A2-292/4, 296
– .... $C_5H_5FeC_5H_4C_6H_3(OH)COOH$ .... Fe: Org.Verb.A3-84/6
$C_{17}FeH_{15}^+$ .... $[C_5H_5FeC_5H_4C(CH)_6]^+$ .... Fe: Org.Verb.A1-339, 341/2
– .... $[C_5H_5FeC_5H_4CHC_6H_5]^+$ .... Fe: Org.Verb.A1-350/3, 357/8, 363
– .... $[C_5H_5FeC_5H_4C_6H_4CH_2]^+$ .... Fe: Org.Verb.A3-127
– .... $[C_5H_5Fe(C_6H_5)_2]^+$ .... Fe: Org.Verb.A1-118

Formula	Compound	Reference
$C_{17}FeH_{15}^{+}$	$[C_6H_6FeC_5H_4C_6H_5]^{+}$	Fe: Org.Verb.A1-325
$C_{17}FeH_{15}I_2O_3P$	$(CO)_3Fe(P(C_6H_5)_2C_2H_5)I_2$	Fe: Org.Verb.B1-180/1
$C_{17}FeH_{15}N$	$C_5H_5FeC_5H_4CHCHC_5H_4N$	Fe: Org.Verb.A2-159
$C_{17}FeH_{15}NO$	$C_5H_5FeC_5H_4C(C_6H_5)NOH$	Fe: Org.Verb.A2-283
–	$C_5H_5FeC_5H_4COCHCHC_4H_4N$	Fe: Org.Verb.A2-223
–	$C_5H_5FeC_5H_4CONHC_6H_5$	Fe: Org.Verb.A2-283
		Fe: Org.Verb.A3-154
–	$C_5H_5FeC_5H_4C_6H_4CONH_2$	Fe: Org.Verb.A3-160
–	$C_5H_5FeC_5H_4NHCOC_6H_5$	Fe: Org.Verb.A2-283
$C_{17}FeH_{15}NO_3$	$C_5H_5FeC_5H_4CHOHC_6H_4NO_2$	Fe: Org.Verb.A2-153
$C_{17}FeH_{15}N_3$	$C_5H_5FeC_5H_4CH(C_6H_5)N_3$	Fe: Org.Verb.A2-39
$C_{17}FeH_{15}N_3O$	$C_5H_5FeC_5H_4CHNNHCOC_5H_4N$	Fe: Org.Verb.A2-167
$C_{17}FeH_{15}NiP^{+}$	$C_5H_5NiFeP(C_6H_5)_2^{+}$	Ni: Org.Verb.2-171
$C_{17}FeH_{15}O^{+}$	$[C_5H_5FeC_5H_4COHC_6H_5]^{+}$	Fe: Org.Verb.A2-279/80
$C_{17}FeH_{16}$	$C_5H_5FeC_5H_4C(CH_3)C_5H_4$	Fe: Org.Verb.A1-333, 335
–	$C_5H_5FeC_5H_4CH_2C_6H_5$	Fe: Org.Verb.A1-302/7
–	$C_5H_5FeC_5H_4C_6H_4CH_3$	Fe: Org.Verb.A1-328/9
–	$C_5H_5FeC_5H_4C_7H_7$	Fe: Org.Verb.A1-339
$C_{17}FeH_{16}^{+}$	$[C_5H_5FeC_5H_4CH_2C_6H_5]^{+}$	Fe: Org.Verb.A1-306
–	$[C_5H_5FeC_5H_4C_6H_4CH_3]^{+}$	Fe: Org.Verb.A1-329
$C_{17}FeH_{16}N^{+}$	$[C_5H_5FeC_5H_4C(C_6H_5)NH_2]^{+}$	Fe: Org.Verb.A2-283
$C_{17}FeH_{16}N_2O$	$C_5H_5FeC_5H_4C_3H_4N_2C_4H_3O$	Fe: Org.Verb.A3-164
$C_{17}FeH_{16}N_2O_2$	$C_5H_5FeC_5H_4CH_2NHC_6H_4NO_2$	Fe: Org.Verb.A2-20
$C_{17}FeH_{16}O$	$C_5H_5FeC_5H_4(CHCH)_3CHO$	Fe: Org.Verb.A2-175, 178
–	$C_5H_5FeC_5H_4CHOHC_6H_5$ (Optically active forms)	Fe: Org.Verb.A2-41
–	$C_5H_5FeC_5H_4CHOHC_6H_5$ (Racemic form)	Fe: Org.Verb.A2-4, 35/41
–	$C_5H_5FeC_5H_4CDOHC_6H_5$	Fe: Org.Verb.A2-36
–	$C_5H_5FeC_5H_4CH_2C_6H_4OH$	Fe: Org.Verb.A2-4, 92
–	$C_5H_5FeC_5H_4CH_2OC_6H_5$	Fe: Org.Verb.A2-113/4
–	$C_5H_5FeC_5H_4C_6H_4CH_2OH$	Fe: Org.Verb.A2-76, 82
–	$C_5H_5FeC_5H_4C_6H_4OCH_3$	Fe: Org.Verb.A2-125/7, 131, 135/6
$C_{17}FeH_{16}O^{+}$	$[C_5H_5FeC_5H_4CHOHC_6H_5]^{+}$	Fe: Org.Verb.A2-38
–	$[C_5H_5FeC_5H_4C_6H_4OCH_3]^{+}$	Fe: Org.Verb.A2-135
$C_{17}FeH_{16}O^{2+}$	$[HC_5H_5FeC_5H_4COHC_6H_5]^{2+}$	Fe: Org.Verb.A2-280
$C_{17}FeH_{16}O_2$	$C_5H_4FeC_5H_3C_7H_9O_2$	Fe: Org.Verb.A3-105
–	$C_5H_5FeC_5H_4(CHCH)_3COOH$	Fe: Org.Verb.A3-77/8
–	$C_5H_5FeC_5H_4CHOHC_6H_4OH$	Fe: Org.Verb.A2-92
–	$C_5H_5FeC_5H_4CH_2C_4H_2OCOCH_3$	Fe: Org.Verb.A3-163
–	$C_5H_5FeC_5H_4CH_2C_6H_3(OH)_2$	Fe: Org.Verb.A2-92
–	$C_5H_5FeC_5H_4COC_6H_7O$	Fe: Org.Verb.A3-174
$C_{17}FeH_{16}O_3$	$C_5H_5FeC_5H_3C_4H_2(CH_3)(COOCH_3)O$	Fe: Org.Verb.A3-145
$C_{17}FeH_{16}O_4$	$C_5H_5FeC_5H_4CHC(COO)_2C(CH_3)_2$	Fe: Org.Verb.A3-179
$C_{17}FeH_{16}S$	$C_5H_5FeC_5H_4SC_6H_4CH_3$	Fe: Org.Verb.A1-376
$C_{17}FeH_{17}LiO_2$	$C_5H_5FeC_5H_4C(LiO)(CHCCH_3)_2O$	Fe: Org.Verb.A3-177
$C_{17}FeH_{17}N$	$C_5H_5FeC_5H_4CH(C_6H_5)NH_2$	Fe: Org.Verb.A2-39
–	$C_5H_5FeC_5H_4CH_2NHC_6H_5$	Fe: Org.Verb.A2-20, 114
$C_{17}FeH_{17}NO$	$C_5H_5FeC_5H_4CHCC(CH_3)C(CH_3)C(O)NH$	Fe: Org.Verb.A2-162
–	$C_5H_5FeC_5H_4CHOHCH_2C_5H_4N$	Fe: Org.Verb.A2-159

Formula	Compound	Reference
$C_{17}FeH_{17}NO_2W^+$	$[C_5H_5FeC_5H_4C(NC_4H_8)W(CO)_2]^+$	Fe: Org.Verb.A1-347
$C_{17}FeH_{17}NO_4$	$C_6H_5CHCHC(O)N(CH_2)_5Fe(CO)_3$	Fe: Org.Verb.B5-158
$C_{17}FeH_{17}N_2O_2$	$CH_3Fe(OC_6H_4CHNCH_2)_2$	Fe: Org.Verb.B1-3
$C_{17}FeH_{17}O_4{}^+$	$[(CH_3)_2CC(C_6H_5)C(CH_3)_2Fe(CO)_4]^+$	Fe: Org.Verb.B4-325
$C_{17}FeH_{17}P$	$C_5H_5FeC_5H_4CH_2PHC_6H_5$	Fe: Org.Verb.A2-20
$C_{17}FeH_{18}$	$C_5H_5FeC_5H_3C_6H_7CH_3$	Fe: Org.Verb.A1-359
–	$C_5H_5FeC_5H_4C_7H_9$	Fe: Org.Verb.A1-339, 342
$C_{17}FeH_{18}N^+$	$[C_5H_5FeC_5H_4CHC_4H(CH_3)_2NH]^+$	Fe: Org.Verb.A2-160
$C_{17}FeH_{18}O$	$C_5H_5FeC_5H_4CHC_6H_8O$	Fe: Org.Verb.A3-43, 46/7
$C_{17}FeH_{18}O_2$	$C_5H_5FeC_5H_4CCHCH_2C(CH_3)_2COO$	Fe: Org.Verb.A3-166
–	$C_5H_5FeC_5H_4C(CH_3)CHCHCHCOOCH_3$	Fe: Org.Verb.A3-122, 125
–	$C_5H_5FeC_5H_4(CHCH)_2COOC_2H_5$	Fe: Org.Verb.A3-123
–	$C_5H_5FeC_5H_4C_5H_6COOCH_3$	Fe: Org.Verb.A3-132
–	$C_5H_5FeC_5H_4C_5H_6OOCCH_3$	Fe: Org.Verb.A2-104/5
$C_{17}FeH_{18}O_3$	$(CH_3)_2C_{12}H_{12}Fe(CO)_3$	Fe: Org.Verb.B5-144
–	$C_5H_4FeC_5H_4CH(C_3H_6COOH)CH_2C(O)$	Fe: Org.Verb.A3-105
$C_{17}FeH_{18}O_4$	$CH_3COC_5H_4FeC_5H_4COC_2H_4COOCH_3$	Fe: Org.Verb.A2-210
–	$C_5H_5FeC_5H_4C(CHCOOH)(CH_2)_3COOH$	Fe: Org.Verb.A3-96/8
–	$C_5H_5FeC_5H_4C(CH_3)C(COOCH_3)CH_2COOH$	Fe: Org.Verb.A3-145
$C_{17}FeH_{18}O_5$	$(CH_3O)_2C_{12}H_{12}Fe(CO)_3$	Fe: Org.Verb.B4-213, 219/20
$C_{17}FeH_{19}{}^+$	$[C_5H_5FeC_5H_4C_7H_{10}]^+$	Fe: Org.Verb.A1-339, 342
		Fe: Org.Verb.A2-69
$C_{17}FeH_{19}NO$	$C_5H_5FeC_5H_4COCHCHCHCHN(CH_3)_2$	Fe: Org.Verb.A2-222, 227
–	$C_5H_5FeC_5H_4COCH(CH_2CN)CH(CH_3)_2$	Fe: Org.Verb.A2-246
–	$C_5H_5FeC_5H_4COCH_2CH(CN)CH(CH_3)_2$	Fe: Org.Verb.A2-251
$C_{17}FeH_{19}NO_3$	$C_5H_5FeC_5H_4COCHC(COOC_2H_5)NHCH_3$	Fe: Org.Verb.A3-134
$C_{17}FeH_{19}O_2{}^-$	$[C_5H_5FeC_5H_4C_6H_{10}COO]^-$	Fe: Org.Verb.A3-84
$C_{17}FeH_{20}$	$C_5H_5FeC_5H_4C(C_4H_9\text{-t})CCH_2$	Fe: Org.Verb.A1-296
–	$C_5H_5FeC_5H_4C_7H_{11}$	Fe: Org.Verb.A1-339/40, 343
$C_{17}FeH_{20}{}^{2+}$	$[HC_5H_5FeC_5H_4C_7H_{10}]^{2+}$	Fe: Org.Verb.A1-339, 343
$C_{17}FeH_{20}N_2O$	$C_5H_5FeC_5H_4CH_2CH(C_3H_7\text{-i})COCHN_2$	Fe: Org.Verb.A3-48
$C_{17}FeH_{20}N_4O_3P_2$	$(CO)_3Fe(C_2H_4(P(CH_2CH_2CN)_2)_2)$	Fe: Org.Verb.B1-170/1, 172
$C_{17}FeH_{20}O$	$C_5H_5FeC_5H_3CH_2C(CH_3)_2(CH_2)_2CO$	Fe: Org.Verb.A3-103
–	$C_5H_5FeC_5H_3CH_2CH(C_3H_7)CH_2CO$	Fe: Org.Verb.A3-102
–	$C_5H_5FeC_5H_3CH_2CH(C_4H_9)CO$	Fe: Org.Verb.A3-101
–	$C_5H_5FeC_5H_3(CH_2)_2CH(C_3H_7)CO$	Fe: Org.Verb.A3-102
–	$C_5H_5FeC_5H_3(CH_2)_3C(CH_3)_2CO$	Fe: Org.Verb.A3-103
–	$C_5H_5FeC_5H_4CHCHCOC(CH_3)_3$	Fe: Org.Verb.A3-8/9, 11
–	$C_5H_5FeC_5H_4CHCHCOCH_2CH(CH_3)_2$	Fe: Org.Verb.A3-8/10
–	$C_5H_5FeC_5H_4CHCHCOC_4H_9\text{-n}$	Fe: Org.Verb.A3-9/10
–	$C_5H_5FeC_5H_4CH_2CH(C_3H_7\text{-i})CHCO$	Fe: Org.Verb.A3-48
–	$C_5H_4FeC_5H_4CH_2CH(C_4H_9)CO$	Fe: Org.Verb.A3-101
–	$C_5H_5FeC_5H_4CH_2C_6H_9O$	Fe: Org.Verb.A3-43, 47
–	$C_5H_5FeC_5H_4COCHCHC(CH_3)_3$	Fe: Org.Verb.A2-248, 251
–	$C_5H_5FeC_5H_4COC_6H_{11}$	Fe: Org.Verb.A2-289, 291
–	$C_5H_5FeC_5H_4C_5H_6OC_2H_5$	Fe: Org.Verb.A2-131, 135
–	$C_5H_5FeC_5H_4C_7H_{10}OH$	Fe: Org.Verb.A2-69
$C_{17}FeH_{20}OS$	$C_5H_5FeC_5H_4COCH_2SC_5H_9$	Fe: Org.Verb.A2-245
$C_{17}FeH_{20}O_2$	$C_2H_5COC_5H_4FeC_5H_4COC_3H_7$	Fe: Org.Verb.A2-236
–	$C_5H_5FeC_5H_4C(CH_3)C(CH_3)COOC_2H_5$	Fe: Org.Verb.A3-122

Formula	Compound		Reference
$C_{17}FeH_{20}O_2$	$C_5H_5FeC_5H_4C(CH_3)CHCOOC_3H_7$-n	Fe:	Org.Verb.A3-122
–	$C_5H_5FeC_5H_4C(CH_3)(OH)CCC(CH_3)_2OH$	Fe:	Org.Verb.A2-88
–	$C_5H_5FeC_5H_4CHCHCOOC_4H_9$-n	Fe:	Org.Verb.A3-121
–	$C_5H_5FeC_5H_4COC_6H_{10}OH$	Fe:	Org.Verb.A2-295, 297
–	$C_5H_5FeC_5H_4C_3H_3(CH_3)COOC_2H_5$	Fe:	Org.Verb.A3-131, 133
–	$C_5H_5FeC_5H_4C_5H_8COOCH_3$	Fe:	Org.Verb.A3-132
–	$C_5H_5FeC_5H_4C_6H_{10}COOH$	Fe:	Org.Verb.A3-82/4
$C_{17}FeH_{20}O_3$	$C_5H_5FeC_5H_4CH_2CH(COCH_3)COOC_2H_5$	Fe:	Org.Verb.A3-138, 140/1
–	$C_5H_5FeC_5H_4C_2H_4COCH_2COOC_2H_5$	Fe:	Org.Verb.A3-138, 140/1
–	$C_5H_5FeC_5H_4COC(CH_3)_2(CH_2)_2COOH$	Fe:	Org.Verb.A3-87, 89, 104
–	$C_5H_5FeC_5H_4COCH(CH_3)C_2H_4COOCH_3$	Fe:	Org.Verb.A3-137
–	$C_5H_5FeC_5H_4COCH(C_3H_7)CH_2COOH$	Fe:	Org.Verb.A3-87/8
–	$C_5H_5FeC_5H_4COCH_2C(CH_3)_2COOCH_3$	Fe:	Org.Verb.A3-136
–	$C_5H_5FeC_5H_4COCH_2CH(CH_3)CH_2COOCH_3$	Fe:	Org.Verb.A3-137
–	$C_5H_5FeC_5H_4COCH_2CH(C_3H_7)COOH$	Fe:	Org.Verb.A3-87/8
–	$C_5H_5FeC_5H_4CO(CH_2)_2C(CH_3)_2COOH$	Fe:	Org.Verb.A3-87, 89, 104
–	$C_5H_5FeC_5H_4CO(CH_2)_3COOC_2H_5$	Fe:	Org.Verb.A3-137
$C_{17}FeH_{20}O_4$	$C_5H_5FeC_5H_4CH(CH_2COOH)(CH_2)_3COOH$	Fe:	Org.Verb.A3-96/9, 105
–	$C_5H_5FeC_5H_4CH_2C(COOH)_2CH(CH_3)_2$	Fe:	Org.Verb.A3-96/9
$C_{17}FeH_{20}O_6$	$[((CH_2)_5CO)_2C]Fe(CO)_4$	Fe:	Org.Verb.B4-128, 131, 135
$C_{17}FeH_{21}{}^+$	$[C_5H_5FeC_5H_4C(CH_3)C_3H_3(CH_3)_2]^+$	Fe:	Org.Verb.A1-332
$C_{17}FeH_{21}NOS$	$C_5H_5FeC_5H_4CH_2CH_2C(S)N(CH_2CH_2)_2O$	Fe:	Org.Verb.A2-232/3
$C_{17}FeH_{21}N_3O$	$C_5H_5FeC_5H_4CH_2C_5H_7NNHCONH_2$	Fe:	Org.Verb.A3-46
$C_{17}FeH_{22}$	$C_5H_5FeC_5H_4C(C_4H_9$-t$)CHCH_3$	Fe:	Org.Verb.A1-291, 293
–	$C_5H_5FeC_5H_4CH_2CH(CH_2)_5$	Fe:	Org.Verb.A1-334
$C_{17}FeH_{22}N_2O_2$	$C_5H_5FeC_5H_4CH(CH(CH_3)NO_2)NC_4H_8$	Fe:	Org.Verb.A2-169
–	$C_5H_5FeC_5H_4CH(CH_2NO_2)NC_5H_{10}$	Fe:	Org.Verb.A2-169
$C_{17}FeH_{22}N_2O_3$	$C_5H_5FeC_5H_4C(CH_3)(CH_2NO_2)NC_4H_8O$	Fe:	Org.Verb.A2-221
$C_{17}FeH_{22}O$	$C_5H_5FeC_5H_4C(CH_3)CHCH_2C(CH_3)_2OH$	Fe:	Org.Verb.A2-76, 82
–	$C_5H_5FeC_5H_4C(CH_3)(CH_2)_2C(CH_3)_2O$	Fe:	Org.Verb.A3-174
–	$C_5H_5FeC_5H_4C(CH_3)(CH_2)_2CH(C_2H_5)O$	Fe:	Org.Verb.A3-174
–	$C_5H_5FeC_5H_4C(CH_3)(C_3H_3(CH_3)_2)OH$	Fe:	Org.Verb.A2-60, 63/4
–	$C_5H_5FeC_5H_4CHOHC_6H_{11}$	Fe:	Org.Verb.A2-4, 46, 50
–	$C_5H_5FeC_5H_4CH_2C_6H_{10}OH$	Fe:	Org.Verb.A2-77, 82
–	$C_5H_5FeC_5H_4(CH_2)_2COCH_2CH(CH_3)_2$	Fe:	Org.Verb.A3-3/4
–	$C_5H_5FeC_5H_4C_6H_9(CH_3)OH$	Fe:	Org.Verb.A2-68
$C_{17}FeH_{22}O_2$	$C_5H_5FeC_5H_4C(CH(CH_2)_2OH)(CH_2)_3OH$	Fe:	Org.Verb.A2-88, 90
–	$C_5H_5FeC_5H_4C(C_3H_6OH)(CH_2)_3O$	Fe:	Org.Verb.A3-172, 175
–	$C_5H_5FeC_5H_4CH(C_4H_9$-t$)OOCCH_3$	Fe:	Org.Verb.A2-98, 101
–	$C_5H_5FeC_5H_4CH_2C(CH_3)_2(CH_2)_2COOH$	Fe:	Org.Verb.A3-74, 100, 103
–	$C_5H_5FeC_5H_4CH_2C(CH_3)_2COOC_2H_5$	Fe:	Org.Verb.A3-118
–	$C_5H_5FeC_5H_4CH_2CH(C_3H_7)CH_2COOH$	Fe:	Org.Verb.A3-73, 100, 102
–	$C_5H_5FeC_5H_4CH_2CH(C_4H_9)COOH$	Fe:	Org.Verb.A3-72, 100/1
–	$C_5H_5FeC_5H_4CH_2OOCC_5H_{11}$-n	Fe:	Org.Verb.A2-98
–	$C_5H_5FeC_5H_4(CH_2)_2C(CH_3)_2COOCH_3$	Fe:	Org.Verb.A3-118
–	$C_5H_5FeC_5H_4(CH_2)_2CH(C_3H_7)COOH$	Fe:	Org.Verb.A3-73, 100, 102
–	$C_5H_5FeC_5H_4(CH_2)_3C(CH_3)_2COOH$	Fe:	Org.Verb.A3-71, 74, 100, 103
–	$C_5H_5FeC_5H_4(CH_2)_4COOC_2H_5$	Fe:	Org.Verb.A3-115, 119

Formula	Compound		Reference
$C_{17}FeH_{22}O_2$	$C_5H_5FeC_5H_4(CH_2)_5COOCH_3$	Fe:	Org.Verb.A3-115, 119
$C_{17}FeH_{22}O_3$	$C_5H_5FeC_5H_4CHOHC(CH_3)_2COOC_2H_5$	Fe:	Org.Verb.A3-144
–	$C_5H_5FeC_5H_4CHOHCH_2COOC(CH_3)_3$	Fe:	Org.Verb.A3-144
–	$C_5H_5FeC_5H_4COCH_2CH(OC_2H_5)_2$	Fe:	Org.Verb.A2-295
–	$C_5H_5FeC_5H_4C_2H(CH_3)_2(OH)COOC_2H_5$	Fe:	Org.Verb.A3-144
$C_{17}FeH_{22}O_6$	$C_5H_5FeC_5H_4CHO_2C_6H_8(OH)_4$	Fe:	Org.Verb.A3-169
–	$C_5H_5FeC_5H_4CH_2OCH(CHOH)_3CH(CH_2OH)O$	Fe:	Org.Verb.A3-170
$C_{17}FeH_{22}O_7P^-$	$[C_6H_5CH_2C(O)Fe(CO)_3P(OC_2H_5)_3]^-$	Fe:	Org.Verb.B4-142
$C_{17}FeH_{23}^+$	$[CH_3C_5H_4FeC_5H_4C(CH_3)C(CH_3)_3]^+$	Fe:	Org.Verb.A1-358/9
–	$[(CH_3)_3CC_5H_4FeC_5H_4C(CH_3)_2]^+$	Fe:	Org.Verb.A1-358/9
–	$[(CH_3)_4C_6H_2FeC_5H_4C_2H_5]^+$	Fe:	Org.Verb.A2-186, 208
–	$[C_5H_5FeC_5H_4C(CH_3)C(CH_3)_2C_2H_5]^+$	Fe:	Org.Verb.A2-64
–	$[C_5H_5FeC_5H_4C(CH_3)CH_2C_4H_9\text{-t}]^+$	Fe:	Org.Verb.A2-64
–	$[C_5H_5FeC_5H_4C(C_2H_5)C(CH_3)_3]^+$	Fe:	Org.Verb.A2-64
–	$[C_{12}H_{18}FeC_5H_5]^+$	Fe:	Org.Verb.A6-86
$C_{17}FeH_{23}N$	$C_5H_5FeC_5H_4CH_2NHC_6H_{11}$	Fe:	Org.Verb.A2-20
$C_{17}FeH_{23}NO_2$	$C_5H_5FeC_5H_4CH(OOCCH_3)C_2H_4N(CH_3)_2$	Fe:	Org.Verb.A1-287
$C_{17}FeH_{23}NO_4Sn$	$(CO)_4FeSn(C_4H_9)_2 \cdot C_5H_5N$	Fe:	Org.Verb.C1-52
$C_{17}FeH_{24}$	$C_5H_5FeC_5H_4CH(CH_3)C_5H_{11}\text{-n}$	Fe:	Org.Verb.A1-276
$C_{17}FeH_{24}N_2O_3$	$(CO)_3Fe(C_2H_2(NC_6H_{11})_2)$	Fe:	Org.Verb.B1-168
$C_{17}FeH_{24}O$	$C_5H_5FeC_5H_4C(CH_3)(CH_2C_4H_9\text{-t})OH$	Fe:	Org.Verb.A2-60, 64
–	$C_5H_5FeC_5H_4C(C_2H_5)(C_4H_9\text{-t})OH$	Fe:	Org.Verb.A2-4, 60, 64
$C_{17}FeH_{24}O_2$	$C_5H_5FeC_5H_4C(CH_3)(OH)(CH_2)_2C(CH_3)_2OH$	Fe:	Org.Verb.A2-87
–	$C_5H_5FeC_5H_4CH(CH_2CH_2C(CH_3)_2OH)OCH_3$	Fe:	Org.Verb.A2-118, 120
$C_{17}FeH_{24}O_3$	$C_5H_5FeC_5H_4CH_2O(CH_2CH_2O)_2C_2H_5$	Fe:	Org.Verb.A2-113
$C_{17}FeH_{25}NaO_5$	$Na[C_{12}H_{25}C(O)Fe(CO)_4]$	Fe:	Org.Verb.B4-151, 161
$C_{17}FeH_{25}O_5^-$	$[C_{12}H_{25}C(O)Fe(CO)_4]^-$	Fe:	Org.Verb.B4-151, 161
$C_{17}FeH_{26}NO_5^-$	$[CH_3(CH_2)_{11}NHC(O)Fe(CO)_4]^-$	Fe:	Org.Verb.B4-156, 166
$C_{17}FeH_{26}OSi$	$C_5H_5FeC_5H_4CHOH(CH_2)_3Si(CH_3)_3$	Fe:	Org.Verb.A2-155
$C_{17}FeH_{26}O_9P_2$	$(CO)_3Fe(P(OCH_2)_3CC_3H_7\text{-n})_2$	Fe:	Org.Verb.B1-148/50, 154
$C_{17}FeH_{31}N_4$	$CH_3FeC_{10}H_{10}N_4(CH_3)_6$	Fe:	Org.Verb.B1-3
$C_{17}FeH_{32}O_2P$	$[C_3H_5Fe(CO)_2P(C_4H_9)_3]$	Fe:	Org.Verb.B5-69/71
$C_{17}FeH_{35}NO_3S_6$	$[(C_2H_5)_4N][Fe(S_2COC_2H_5)_3]$	C:	MVol.D4-259
$C_{17}FeH_{38}O_3P_2Si_2$	$(CO)_3Fe(P(CH_3)_2CH_2CH_2Si(CH_3)_3)_2$	Fe:	Org.Verb.B1-148/50, 152, 160
$C_{17}FeH_{38}O_3P_2Si_2^+$	$[(CO)_3Fe(P(CH_3)_2C_2H_4Si(CH_3)_3)_2]^+$	Fe:	Org.Verb.B1-160
$C_{17}FeH_{39}O_5P_3$	$(CO)_2Fe(P(C_2H_5)_3)_2(P(OCH_3)_3)$	Fe:	Org.Verb.B1-94
$C_{17}FeH_{40}O_9P_3$	$C_8H_{13}Fe(P(OCH_3)_3)_3$	Fe:	Org.Verb.B5-5/6
$C_{17}FeH_{40}O_9P_3^+$	$[C_8H_{13}Fe(P(OCH_3)_3)_3]^+$	Fe:	Org.Verb.B5-3/4
$C_{17}Fe_2GeH_{25}IN_4O_4$	$((CH_3)_2C_3H_3Fe(NO)_2)_2Ge(I)CH_2C_6H_5$	Fe:	Org.Verb.C2-165
–	$((CH_3)_2C_3H_3Fe(NO)_2)_2Ge(I)C_6H_4CH_3$	Fe:	Org.Verb.C2-165
$C_{17}Fe_2H_8N_2O_7$	$(CO)_7Fe_2(NC_5H_4C_5H_4N)$	Fe:	Org.Verb.C1-199/200
$C_{17}Fe_2H_9NO_6S$	$(CO)_3Fe(SC_4H_2CH_2NC_6H_5)Fe(CO)_3$	Fe:	Org.Verb.C2-116
$C_{17}Fe_2H_{11}O_9PPt$	$(CO)_8Fe_2Pt(CO)(P(CH_3)_2C_6H_5)$	Fe:	Org.Verb.C1-216/8
$C_{17}Fe_2H_{12}O_5$	$(CO)_3FeC_7H_7C_5H_5Fe(CO)_2$	Fe:	Org.Verb.B5-79
$C_{17}Fe_2H_{13}NO_6$	$(C_2H_5CCHCH_2NC_6H_5)Fe_2(CO)_6$	Fe:	Org.Verb.C2-109, 117/8
–	$(C_6H_5CCHCHN(CH_3)_2)Fe_2(CO)_6$	Fe:	Org.Verb.C2-156, 159/60
$C_{17}Fe_2H_{13}O_3$	$(C_5H_5FeC_5H_4CHCHCHCH)Fe(CO)_3$	Fe:	Org.Verb.A3-4
$C_{17}Fe_2H_{14}INO_6$	$(CO)_3Fe(C_4H_9(C_6H_5)CN)IFe(CO)_3$	Fe:	Org.Verb.C1-185/7
$C_{17}Fe_2H_{15}^+$	$[CH_2CHC_5H_4FeC_5H_4C_5H_4Fe]^+$	Fe:	Org.Verb.A6-14

Formula	Compound	Element	Reference
$C_{17}Fe_2H_{15}O^+$	$[CH_3COC_5H_4FeC_5H_4C_5H_4Fe]^+$	Fe:	Org.Verb.A6-15
$C_{17}Fe_2H_{18}O_6S$	$((CH_3)_4C_5H_6CCS)Fe_2(CO)_6$	Fe:	Org.Verb.C2-95/9, 100, 102
$C_{17}Fe_2H_{18}O_7$	$(C_4H_9CCC_4H_9)Fe_2(CO)_7$	Fe:	Org.Verb.C2-73
$C_{17}Fe_2H_{24}O_{10}P_2S_2$	$(CO)_4(P(OCH_3)_3)_2Fe_2(SC_6H_3(CH_3)S)$	Fe:	Org.Verb.C1-123/4
$C_{17}Fe_2H_{25}IN_4O_4Pb$	$((CH_3)_2C_3H_3Fe(NO)_2)_2Pb(I)CH_2C_6H_5$	Fe:	Org.Verb.C2-165
–	$((CH_3)_2C_3H_3Fe(NO)_2)_2Pb(I)C_6H_4CH_3$	Fe:	Org.Verb.C2-165
$C_{17}Fe_2H_{25}IN_4O_4Sn$	$((CH_3)_2C_3H_3Fe(NO)_2)_2Sn(I)CH_2C_6H_5$	Fe:	Org.Verb.C2-165
–	$((CH_3)_2C_3H_3Fe(NO)_2)_2Sn(I)C_6H_4CH_3$	Fe:	Org.Verb.C2-165
$C_{17}Fe_2H_{36}N_8O_7P_2$	$[(CO)_2Fe(P(N(CH_3)_2)_3)_2NO][Fe(CO)_3NO]$	Fe:	Org.Verb.B1-188/9, 191
$C_{17}Fe_3H_{18}O_9P_2$	$Fe_3(CO)_9(PC_4H_9)_2$	Fe:	Org.Verb.B2-28
$C_{17}Fe_3H_{18}O_9S_2$	$Fe_3(CO)_9(SC_4H_9)_2$	Fe:	Org.Verb.C1-88
$C_{17}Fe_3H_{20}N_2O_9P_2$	$Fe_3(CO)_9(PN(C_2H_5)_2)_2$	Fe:	Org.Verb.B2-28
$C_{17}Fe_6O_{16}{}^{2-}$	$[Fe_6(CO)_{16}C]^{2-}$	Fe:	Org.Verb.B3-70, 143
$C_{17}H_{12}N_9NaS_7Sn$	$[C_5H_6N]_2[Sn(SCN)_6] \cdot NaSCN$	Sn:	MVol.C3-127
$C_{17}H_{13}N_3Ni$	$C_6H_5Ni(NC_5H_4C_5H_4N)CN$	Ni:	Org.Verb.1-76
$C_{17}H_{13}N_3O_7U$	$UO_2(NC_5H_4COO)_2 \cdot C_5H_5NO$	U:	SVol.E1-132, 142/3
$C_{17}H_{13}N_3O_8U$	$UO_2(NC_5H_3(COO)_2) \cdot 2\ C_5H_5NO$	U:	SVol.E1-132, 136, 142/3
$C_{17}H_{13}N_3O_9U$	$UO_2(OCOC_5H_4NO)_2 \cdot C_5H_5NO$	U:	SVol.E1-132, 142/3
$C_{17}H_{13}NiO_3P$	$(CO)_3NiP(C_6H_5)_2CHCH_2$	Ni:	Org.Verb.1-172
$C_{17}H_{14}N_2Ni$	$C_5H_5NiC_6H_4NNC_6H_5$	Ni:	Org.Verb.2-159
$C_{17}H_{14}N_2Ti$	$C_5H_5TiC_6H_4NNC_6H_5$	Ti:	Org.Verb.1-204
$C_{17}H_{15}NO_4U$	$UO_2(OC_6H_5)_2 \cdot C_5H_5N$	U:	SVol.E1-29, 35
$C_{17}H_{15}N_3Ni$	$C_5H_4NCHCH_2Ni(NC_5H_4C_5H_4N)$	Ni:	Org.Verb.1-371
$C_{17}H_{15}N_5O_5S_2U$	$UO_2(NCS)_2 \cdot 3\ C_5H_5NO$	U:	SVol.E1-132, 135, 142/3
$C_{17}H_{15}NiO_3P$	$(CO)_3NiP(C_6H_5)_2C_2H_5$	Ni:	Org.Verb.1-237
$C_{17}H_{15}NiO_4P$	$(CO)_3NiP(OC_2H_5)(C_6H_5)_2$	Ni:	Org.Verb.1-177
$C_{17}H_{15}Ni_2O_2{}^+$	$(C_5H_5)_3Ni_2(CO)_2{}^+$	Ni:	Org.Verb.2-345
$C_{17}H_{15}Ni_3O_2$	$(C_5H_5)_3Ni_3(CO)_2$	Ni:	Org.Verb.2-381/3
$C_{17}H_{15}Ni_3O_2{}^+$	$(C_5H_5)_3Ni_3(CO)_2{}^+$	Ni:	Org.Verb.2-345, 381
$C_{17}H_{15}Ti$	$C_5H_5Ti(C_6H_5)_2$	Ti:	Org.Verb.1-208
$C_{17}H_{16}O_7U$	$UO_2(C_7H_5O_2)_2 \cdot (CH_3)_2CO$	U:	SVol.E1-79/80
$C_{17}H_{17}NO_8U$	$UO_2(C_7H_5O_2)_2 \cdot NH_2COOC_2H_5$	U:	SVol.E1-120/3
$C_{17}H_{17}N_2O_5U$	$[UO_2(C_5H_8(O)NC_6H_4COO)(C_5H_5N)]$	U:	SVol.E1-31, 37
$C_{17}H_{18}N_2O_5U$	$UO_2(C_2H_4(NCHC_6H_4O)_2) \cdot CH_3OH$	U:	SVol.E1-61, 65/6, 68
$C_{17}H_{18}N_6O_9U$	$[UO_2(NO_3)_2(C_5H_5N)_2(C_6H_5CONHNH_2)] \cdot H_2O$	U:	SVol.E1-29, 35
$C_{17}H_{18}O_2Sn$	$(C_5H_5)_3Sn(OCOCH_3)$	Sn:	Org.Verb.6-143
$C_{17}H_{18}Sn$	$(CH_2CH)_3SnC_{10}H_6CH_3$	Sn:	Org.Verb.2-348
–	$(CH_3)_2Sn(C_9H_7)(C_6H_5)$	Sn:	Org.Verb.3-81, 84/5
–	$(CH_3)_3SnC_{14}H_9$	Sn:	Org.Verb.2-149, 152
–	$(C_5H_5)_3SnCHCH_2$	Sn:	Org.Verb.2-346, 350
$C_{17}H_{19}NO_5U$	$UO_2(OC_6H_4CHNC_6H_4O) \cdot C_4H_9OH$	U:	SVol.E1-62, 69
$C_{17}H_{19}PSSn$	$(CH_3)_3SnCCPS(C_6H_5)_2$	Sn:	Org.Verb.2-115
$C_{17}H_{19}PSn$	$(CH_3)_3SnCCP(C_6H_5)_2$	Sn:	Org.Verb.2-115
$C_{17}H_{20}MoO_3Sn$	$(CH_2CHCH_2)_3SnMo(CO)_3C_5H_5$	Sn:	Org.Verb.5-157
$C_{17}H_{20}N_2O$	$[(CH_3)_2NC_6H_4]_2CO$	C:	MVol.D3-101
$C_{17}H_{20}N_4O_{11}U$	$(C_6H_5NH_3)_2[UO_2(C_2O_4)_2((NH_2)_2CO)] \cdot 2\ H_2O$	U:	SVol.E1-84, 90
$C_{17}H_{20}NiO_2$	$C_7H_8NiC_6(CH_3)_4O_2$	Ni:	Org.Verb.2-102, 104
$C_{17}H_{20}O_3SnW$	$(CH_2CHCH_2)_3SnW(CO)_3C_5H_5$	Sn:	Org.Verb.5-157

$C_{17}H_{20}Sn$ .... $(C_6H_5)_2Sn(CH_2)_5$ .... Sn: Org.Verb.3-105
$C_{17}H_{21}IN_2NiS$ ...... $C_3H_5Ni(SC(NHCH(CH_3)_2)(NHC_{10}H_7))I$ ...... Ni: Org.Verb.2-281
$C_{17}H_{21}ISn$ .... $(CH_3)(C_6H_5)(C_6H_5C(CH_3)_2CH_2)SnI$ .... Sn: Org.Verb.5-41
$C_{17}H_{21}NO_7U$ .... $[UO_2(CH(COCH_3)_2)_2(CH_3COC_5H_4N)]$ ...... U: SVol.E1-31, 36/7
$C_{17}H_{21}NSn$ .... $(CH_3)_2Sn(CH_3C_6H_3N(CH_3)C_6H_3CH_3)$ .... Sn: Org.Verb.3-118, 122
$C_{17}H_{21}N_3Ni$ .... $NCCHCH_2Ni(C_{10}H_8N_2)(C_2H_5)_2$ .... Ni: Org.Verb.1-380/1
$C_{17}H_{22}N_2NiO$ .... $(OHCCHCH_2)Ni(C_{10}H_8N_2)(C_2H_5)_2$ .... Ni: Org.Verb.1-380
$C_{17}H_{22}N_4NiO_2P_2$ .... $(CO)_2Ni(NCC_2H_4)_2P(CH_2)_3P(C_2H_4CN)_2$ ..... Ni: Org.Verb.1-150, 153
$C_{17}H_{22}NiO_2$ .... $C_8H_{12}NiC_6H(CH_3)_3O_2$ .... Ni: Org.Verb.2-103, 106
$C_{17}H_{22}OSn$ .... $(CH_3)(i-C_3H_7)Sn(C_6H_5)(C_6H_4OCH_3)$ .... Sn: Org.Verb.3-98
$C_{17}H_{22}Sn$ .... $(CH_3)(i-C_3H_7)Sn(C_6H_5)(CH_2C_6H_5)$ .... Sn: Org.Verb.3-95, 98
– .... $(CH_3)(C_6H_5(CH_3)_2CCH_2)(C_6H_5)SnH$ .... Sn: Org.Verb.4-92
– .... $(CH_3)_2Sn(C_6H_5)(CH_2CH(CH_3)C_6H_5)$ .... Sn: Org.Verb.3-81
– .... $(C_6H_5)_2Sn(CH_3)(C_4H_9)$ .... Sn: Org.Verb.3-90
– .... $(C_6H_5)_2Sn(CH_3)(i-C_4H_9)$ .... Sn: Org.Verb.3-89/90
$C_{17}H_{23}NO_6U$ .... $[UO_2(CH(COCH_3)_2)_2((CH_3)_2C_5H_3N)]$ .... U: SVol.E1-31, 36/7
– .... $[UO_2(CH(COCH_3)_2)_2(C_2H_5C_5H_4N)]$ .... U: SVol.E1-31, 36/7
– .... $[UO_2(CH_3C(COCH_3)_2)_2(C_5H_5N)]$ .... U: SVol.E1-30, 36/7
$C_{17}H_{23}NO_7U$ .... $UO_2(CH(COCH_3)_2)_2 \cdot (CH_3)_2C_5H_3NO$ ...... U: SVol.E1-133, 140, 142/3
$C_{17}H_{23}NO_8U$ .... $[UO_2(CH_3COCHCOOC_2H_5)_2(C_5H_5N)]$ ...... U: SVol.E1-30, 36/7
$C_{17}H_{23}NiO_3P$ .... $(CO)_3NiP(C_6H_5)(C_4H_9-n)_2$ .... Ni: Org.Verb.1-172
$C_{17}H_{23}NiO_5P$ .... $(CO)_3NiP(OC_4H_9-n)_2C_6H_5$ .... Ni: Org.Verb.1-241
$C_{17}H_{24}N_2NiO_7U$ ..... $[UO_2(C_{16}H_{20}N_2O_4)Ni(CH_3OH)]$ .... U: SVol.E1-63, 69
$C_{17}H_{25}N_2S_4Ti$ ...... $C_5H_5Ti(S_2CN(CH_2)_5)_2$ .... Ti: Org.Verb.1-197
$C_{17}H_{25}N_3NiO$ .... $(t-C_4H_9NC)_2Ni(ONC_6H_4CH_3)$ .... Ni: Org.Verb.1-311, 313
$C_{17}H_{25}Ni^+$ .... $[t-C_4H_9C_5H_4NiC_8H_{12}]^+$ .... Ni: Org.Verb.2-189/90
$C_{17}H_{26}N_2O_7U$ .... $UO_2(C_2H_4(NHCCH_3CHCOCHC(CH_3)O)_2) \cdot CH_3OH$
U: SVol.E1-61, 65/6, 68
$C_{17}H_{26}Sn$ .... $(C_3H_7)_3SnCCC_6H_5$ .... Sn: Org.Verb.2-263/4
$C_{17}H_{28}OSn$ .... $(CH_3)(i-C_3H_7)Sn(C_6H_{11})(C_6H_4OCH_3)$ .... Sn: Org.Verb.3-98
– .... $(C_3H_7)_3SnCH_2COC_6H_5$ .... Sn: Org.Verb.2-254, 257
$C_{17}H_{28}Sn$ .... $(C_3H_7)_3SnCHCHC_6H_5$ .... Sn: Org.Verb.2-261/2
$C_{17}H_{29}MnN_4O_6S_2$ ... $[Mn(CH(COCH_3)_2)_3] \cdot 2\ CS(NH_2)_2$ .... Mn: MVol.D1-108
$C_{17}H_{29}NOSn$ .... $(C_2H_5)_3SnCH(CHO)CH(C_6H_5)N(CH_3)_2$ ...... Sn: Org.Verb.2-175
$C_{17}H_{30}IN_3NiO_2$ ...... $(t-C_4H_9NC)_3Ni(COOCH_3)I$ .... Ni: Org.Verb.1-323
$C_{17}H_{30}OSn$ .... $(C_3H_7)_3SnCH(C_6H_5)OCH_3$ .... Sn: Org.Verb.2-254, 257/8
$C_{17}H_{30}SiSn$ .... $(C_2H_5)_3SnC(C_6H_5)CHSi(CH_3)_3$ .... Sn: Org.Verb.2-201
$C_{17}H_{30}Sn$ .... $(C_3H_7)_3SnCH_2CH_2C_6H_5$ .... Sn: Org.Verb.2-254, 257
– .... $(C_3H_7)_3SnC_6H_3(CH_3)_2$ .... Sn: Org.Verb.2-266, 268
– .... $(C_4H_9)_2Sn(C_2H_5)(CH_2C_6H_5)$ .... Sn: Org.Verb.3-88/9
$C_{17}H_{31}NSn$ .... $(C_4H_9)_3SnC_5H_4N$ .... Sn: Org.Verb.2-322, 324
$C_{17}H_{32}OSn$ .... $(C_2H_5)_3SnC(CCHCH_3)C(OCH_3)(CH_2)_5$ ...... Sn: Org.Verb.2-201
– .... $(C_2H_5)_3SnCH_2CH_2CCC_6H_{10}OCH_3$ .... Sn: Org.Verb.2-232
– .... $(C_3H_7)_3SnCCCHCHOC_4H_9$ .... Sn: Org.Verb.2-264
$C_{17}H_{32}O_3Ti$ .... $C_5H_5Ti(OC_4H_9-n)_3$ .... Ti: Org.Verb.1-157, 161
– .... $C_5H_5Ti(OC_4H_9-t)_3$ .... Ti: Org.Verb.1-157, 161/2
$C_{17}H_{32}O_4Sn$ .... $(C_3H_7)_3SnC(COOC_2H_5)CHCOOC_2H_5$ .... Sn: Org.Verb.2-261
$C_{17}H_{32}SSn$ .... $(C_3H_7)_3SnCCCHCHSC_4H_9$ .... Sn: Org.Verb.2-265
$C_{17}H_{32}Sn$ .... $(C_4H_9)_3SnC_5H_5$ .... Sn: Org.Verb.2-313
$C_{17}H_{33}NSn$ .... $(C_3H_7)_3SnCCCHCHN(C_2H_5)_2$ .... Sn: Org.Verb.2-265

Formula	Compound	Reference
$C_{17}H_{33}NiPS$	$C_5H_5Ni(P(C_4H_9\text{-}n)_3)SH$	Ni: Org.Verb.2-123/5
$C_{17}H_{34}OSn$	$(C_2H_5)_3SnC(CCHCH_3)C(OCH_3)(CH_3)C_4H_9$	Sn: Org.Verb.2-202
–	$(C_2H_5)_3SnC(CCHCH_3)C(OCH_3)(C_2H_5)C_3H_7$	Sn: Org.Verb.2-202
–	$(C_2H_5)_3SnC_2H_4CCC(OCH_3)(CH_3)C_4H_9$	Sn: Org.Verb.2-233
–	$(C_2H_5)_3SnC_2H_4CCC(OCH_3)(C_2H_5)C_3H_7$	Sn: Org.Verb.2-233
–	$(C_4H_9)_3SnCCCH_2CH_2OCH_3$	Sn: Org.Verb.2-317
–	$(C_4H_9)_3SnC_5H_7O$	Sn: Org.Verb.2-305/6
$C_{17}H_{34}O_2SSn$	$(C_4H_9)_3SnC_4H_4(CH_3)SO_2$	Sn: Org.Verb.2-322, 324
$C_{17}H_{34}O_2Sn$	$(CH_3)_3SnCH(CH_3)CH_2COOC_{10}H_{19}$	Sn: Org.Verb.2-44, 46
–	$(C_2H_5)_3SnCCC(CH_3)_2OCH(CH_3)OC_4H_9$	Sn: Org.Verb.2-233
–	$(C_4H_9)_3SnCH(COCH_3)_2$	Sn: Org.Verb.2-286
–	$(C_4H_9)_3SnCH_2COOCH_2CHCH_2$	Sn: Org.Verb.2-286
$C_{17}H_{34}O_3SiTi$	$(CH_3)_3SiC_5H_4Ti(OC_3H_7\text{-}i)_3$	Ti: Org.Verb.1-159, 166
$C_{17}H_{34}O_5Ti$	$CH_3COCH_2COCH_2Ti(OC_4H_9)_3$	Ti: Org.Verb.1-45
$C_{17}H_{35}MnNO_3S_6$	$[(C_2H_5)_4N][Mn(S_2COC_2H_5)_3]$	C: MVol.D4-259
$C_{17}H_{35}NNiO_3S_6$	$[(C_2H_5)_4N][Ni(S_2COC_2H_5)_3]$	C: MVol.D4-260
$C_{17}H_{35}NO_2Sn$	$(C_4H_9)_3SnCH_2CON(CH_3)COCH_3$	Sn: Org.Verb.2-286
$C_{17}H_{35}NO_3PbS_6$	$[(C_2H_5)_4N][Pb(S_2COC_2H_5)_3]$	C: MVol.D4-245, 254
$C_{17}H_{35}NO_3S_6Te$	$[(C_2H_5)_4N][Te(S_2COC_2H_5)_3]$	C: MVol.D4-245, 255 Te: SVol.B3-136
$C_{17}H_{35}NSn$	$(C_4H_9)_3SnCH(CN)CH(CH_3)_2$	Sn: Org.Verb.2-286, 300
$C_{17}H_{35}N_3Ti$	$C_5H_5Ti(N(C_2H_5)_2)_3$	Ti: Org.Verb.1-173, 175/6, 179
$C_{17}H_{36}N_3O_{10}PU$	$UO_2(NO_3)_2 \cdot (C_6H_{13})_2POCON(C_2H_5)_2$	U: SVol.E1-179, 181
$C_{17}H_{36}N_4O_2Sn$	$[Sn(C_4H_9)_2N_2H_2C(O)(CH_2)_7C(O)N_2H_2]_x$	Sn: Org.Verb.6-89, 98
$C_{17}H_{36}N_6O_9U$	$UO_2C_2O_4 \cdot 3\ CO(NHC_2H_5)_2$	U: SVol.E1-93, 95
$C_{17}H_{36}N_8O_2S_5U$	$UO_2(NCS)_2 \cdot 3\ CS(N(CH_3)_2)_2$	U: SVol.E1-220/1
$C_{17}H_{36}OSSn$	$(C_4H_9)_3SnCH_2CH_2CH_2SCOCH_3$	Sn: Org.Verb.2-287
$C_{17}H_{36}OSn$	$(C_4H_9)_3SnCH(CH_3)CH_2COCH_3$	Sn: Org.Verb.2-287, 300
–	$(C_4H_9)_3SnCH(CH_3)COC_2H_5$	Sn: Org.Verb.2-287, 300
–	$(C_4H_9)_3SnCH(C_2H_5)COCH_3$	Sn: Org.Verb.2-287, 300
–	$(C_4H_9)_3SnCH_2COCH(CH_3)_2$	Sn: Org.Verb.2-287, 300
$C_{17}H_{36}O_2Sn$	$[(CH_3)_3CCH_2]_3SnOCOCH_3$	Sn: Org.Verb.5-140
–	$(C_2H_5)_3SnCHCHC(CH_3)_2OCH(CH_3)OC_4H_9$	Sn: Org.Verb.2-202
–	$(C_4H_9)_3SnCH_2CH(CH_3)COOCH_3$	Sn: Org.Verb.2-288
–	$(i\text{-}C_4H_9)_3SnCH_2CH(CH_3)COOCH_3$	Sn: Org.Verb.2-327
–	$(C_4H_9)_3SnCH_2CH_2CH_2OOCCH_3$	Sn: Org.Verb.2-288, 301
–	$(i\text{-}C_4H_9)_3SnCH_2CH_2CH_2OOCCH_3$	Sn: Org.Verb.2-327
–	$(C_4H_9)_3SnCH_2CH_2COOC_2H_5$	Sn: Org.Verb.2-287, 300
–	$(C_4H_9)_3SnCH_2CH_2OCH_2CH(O)CH_2$	Sn: Org.Verb.2-287
$C_{17}H_{36}Sn$	$(C_4H_9)_2Sn(i\text{-}C_3H_7)(C_6H_{11})$	Sn: Org.Verb.3-88/9
$C_{17}H_{37}IN_4NiS_2$	$C_3H_5Ni(SC(NHC_3H_7)_2)_2I$	Ni: Org.Verb.2-21
$C_{17}H_{37}NOSn$	$(C_4H_9)_3SnCH_2N(CH_2CH_2)_2O$	Sn: Org.Verb.2-288
–	$(C_4H_9)_3SnCON(C_2H_5)_2$	Sn: Org.Verb.2-288
$C_{17}H_{37}N_2O_5PS_4U$	$UO_2(S_2CN(C_2H_5)_2)_2 \cdot (C_3H_7O)_2(CH_3)PO$	U: SVol.E1-186/7
$C_{17}H_{38}N_2O_{14}P_2U$	$UO_2(NO_3)_2 \cdot CH_2(PO(OC_4H_9)_2)_2$	U: SVol.E1-179, 181, 188/9
$C_{17}H_{38}OSn$	$(C_4H_9)_3SnC(CH_3)(C_3H_7)OH$	Sn: Org.Verb.2-288
–	$(C_4H_9)_3SnC(C_2H_5)_2OH$	Sn: Org.Verb.2-288
–	$(C_4H_9)_3SnCH_2C(CH_3)_2CH_2OH$	Sn: Org.Verb.2-288

$C_{17}H_{38}OSn$ .... $(C_4H_9)_3SnCH_2CH(C_2H_5)CH_2OH$ .... Sn: Org.Verb.2-288
– .... $(C_4H_9)_3SnCH_2CH_2C(CH_3)_2OH$ .... Sn: Org.Verb.2-288
$C_{17}H_{38}O_2SSn$ ...... $(C_4H_9)_3SnCH_2SO_2C_4H_9$ .... Sn: Org.Verb.2-288, 301
$C_{17}H_{38}O_2Sn$ .... $(C_4H_9)_3Sn(CH_2)_3OCH_2CH_2OH$ .... Sn: Org.Verb.2-288
$C_{17}H_{38}SSn$ .... $(C_4H_9)_3SnCH_2CH_2S(i\text{-}C_3H_7)$ .... Sn: Org.Verb.2-289
$C_{17}H_{38}SiSn$ .... $(C_3H_7)_3SnCHCHSi(C_2H_5)_3$ .... Sn: Org.Verb.2-261/2
$C_{17}H_{38}Si_3Ti$ .... $C_5H_5Ti(CH_2Si(CH_3)_3)_3$ .... Ti: Org.Verb.1-210
$C_{17}H_{38}Sn$ .... $((CH_3)_3CCH_2)_3SnC_2H_5$ .... Sn: Org.Verb.2-329, 331
– .... $(C_3H_7)_3SnC_8H_{17}$ .... Sn: Org.Verb.2-255, 258
– .... $(C_4H_9)_3SnC_5H_{11}$ .... Sn: Org.Verb.2-274
– .... $(C_4H_9)_3Sn(i\text{-}C_5H_{11})$ .... Sn: Org.Verb.2-274
– .... $(i\text{-}C_4H_9)_3Sn(i\text{-}C_5H_{11})$ .... Sn: Org.Verb.2-327
$C_{17}H_{39}N_3Ti$ .... $C(CH_3)_2C(CH_3)Ti(N(C_2H_5)_2)_3$ .... Ti: Org.Verb.1-54, 58
$C_{17}H_{40}OSiSn$ .... $(C_4H_9)_3SnCH_2Si(CH_3)_2OC_2H_5$ .... Sn: Org.Verb.2-289
$C_{17}H_{43}NiO_{13}P_3S$ .... $[(CH_3)_2(C_3H_7)C_3H_2Ni(P(OCH_3)_3)_3]HSO_4$ .... Ni: Org.Verb.2-26
$C_{17.5}Cl_2H_{20.5}IrN_{3.5}O_2$. $[Ir(C_5H_5N)_3Cl(OH)H_2O]Cl \cdot 0.5\ C_5H_5N \cdot 3\ H_2O$
Ir: SVol.2-72
$C_{17.5}Cl_3H_{17.5}IrN_{3.5}$ ... $[Ir(C_5H_5N)_3Cl_3] \cdot 0.5\ C_5H_5N$ .... Ir: SVol.2-69
$C_{17.5}Cl_4H_{32.5}N_{2.5}O_{2.5}U$ $UCl_4 \cdot 2.5\ CH_3N(CH_2)_5CO$ .... U: SVol.E1-118/20
$C_{17.8}Cl_6H_{11.8}O_2Sn$ ... $SnCl_4 \cdot C_{10}H_4Cl_2O_2 \cdot 1.3\ C_6H_6$ .... Sn: MVol.C5-112